Quantitative Analysis of Mineral and Energy Resources

NATO ASI Series

Advanced Science Institutes Series

A Series presenting the results of activities sponsored by the NATO Science Committee, which aims at the dissemination of advanced scientific and technological knowledge, with a view to strengthening links between scientific communities.

The series is published by an international board of publishers in conjunction with the NATO Scientific Affairs Division

A Life Sciences B Physics	Plenum Publishing Corporation London and New York
C Mathematical and Physical Sciences	D. Reidel Publishing Company Dordrecht, Boston, Lancaster and Tokyo
D Behavioural and Social Sciences E Applied Sciences	Martinus Nijhoff Publishers Dordrecht, Boston and Lancaster
F Computer and Systems Sciences G Ecological Sciences H Cell Biology	Springer-Verlag Berlin, Heidelberg, New York, London, Paris, and Tokyo

Series C: Mathematical and Physical Sciences Vol. 223

Quantitative Analysis of Mineral and Energy Resources

edited by

C. F. Chung
Geological Survey of Canada, Ottawa, Canada

A. G. Fabbri
Canada Centre for Remote Sensing, Ottawa, Canada

and

R. Sinding-Larsen
Norwegian Institute of Technology, Trondheim, Norway

D. Reidel Publishing Company

Dordrecht / Boston / Lancaster / Tokyo

Published in cooperation with NATO Scientific Affairs Division

Proceedings of the NATO Advanced Study Institute on
Statistical Treatments for Estimation of Mineral and Energy Resources
Il Ciocco (Lucca), Italy
June 22-July 4, 1986

Library of Congress Cataloging in Publication Data

NATO Advanced Study Institute on Statistical Treatments for Estimation of Mineral and Energy Resources (1986: Lucca, Italy)
Quantitative analysis of mineral and energy resources.

(NATO ASI series. Series C, Mathematical and physical sciences; vol. 223)
"Proceedings of the NATO Advanced Study Institute on Statistical Treatments for Estimation of Mineral and Energy Resources, Il Ciocco, (Lucca) Italy, June 22–July 4, 1986"—T.p. verso.
"Published in cooperation with NATO Scientific Affairs Division."
Includes index.
1. Mines and mineral resources—Statistical methods—Congresses. 2. Power Resources—Statistical methods—Congresses. 3. Geology, Economic—Statistical methods—Congresses. I. Chung, C. F. (Chang-Jo F.), 1945– . II. Fabbri, Andrea G. III. Sinding-Larsen, Richard. IV. Title. V. Series: NATO ASI series. Series C, Mathematical and physical sciences; no. 223.
TN153.N374 1986 622′.1072 87–27519
ISBN 90–277–2635–3

Published by D. Reidel Publishing Company
P.O. Box 17, 3300 AA Dordrecht, Holland

Sold and distributed in the U.S.A. and Canada
by Kluwer Academic Publishers,
101 Philip Drive, Assinippi Park, Norwell, MA 02061, U.S.A.

In all other countries, sold and distributed
by Kluwer Academic Publishers Group,
P.O. Box 322, 3300 AH Dordrecht, Holland

D. Reidel Publishing Company is a member of the Kluwer Academic Publishers Group

Printed in The Netherlands

PREFACE

This volume contains the edited papers prepared by lecturers and participants of the NATO Advanced Study Institute on "Statistical Treatments for Estimation of Mineral and Energy Resources" held at Il Ciocco (Lucca), Italy, June 22 - July 4, 1986.

During the past twenty years, tremendous efforts have been made to acquire quantitative geoscience information from ore deposits, geochemical, geophysical and remotely-sensed measurements. In October 1981, a two-day symposium on "Quantitative Resource Evaluation" and a three-day workshop on "Interactive Systems for Multivariate Analysis and Image Processing for Resource Evaluation" were held in Ottawa, jointly sponsored by the Geological Survey of Canada, the International Association for Mathematical Geology, and the International Geological Correlation Programme. Thirty scientists from different countries in Europe and North America were invited to form a forum for the discussion of quantitative methods for mineral and energy resource assessment.

Since then, not only a multitude of research projects directed toward quantitative analysis in the Earth Sciences, but also recent advances in hardware and software technology, such as high-resolution graphics, data-base management systems and statistical packages on mini- and micro-computers, made it possible to study large geoscience data sets. In addition, methods of image analysis have been utilized to capture data in digital form and to supply a variety of tools for charaterizing natural phenomena.

This upsurge of activities and advancements provided the impetus to organize this ASI under the auspices of the NATO Science Council. The objective of the Institute is to bring together many scientists from various field of research in mathematical geology, creating a forum for disciplinary interaction. Various quantitative methodologies are considered: integration of geochemical, geophysical, lithological and ancillary data; estimation of resources before exploration and after discovery; image analysis including the study of remotely sensed data; and related quantitative methods such as artificial intelligence, spatial statistics, crustal abundance models, and geostatistics.

During the ASI, five parallel one-day workshops were also held on "Current Problems and Future Developments in Mathematical Geology" related to the above topics. The workshop reports offer state-of-the-art and directions for future developments in quantitative studies for resource evaluation.

At the Institute, several new procedures in mathematical geology were presented: methods of integrating spatial characteristics into the traditional multivariate analysis, superpopulation models for biassedly sampled data, analysis of truncated and censored data, Bayesian approaches to geostatistics, spatial point processes, uses of artificial intelligence, and image analysis for feature extration.

In these Proceedings, we as editors have tried to include all the contributions presented at the Institute and to respect any scientific statements made by the authors. Forty papers and five workshop reports in this volume contribute to the current state of knowledge in mathematical geology and its use in resource evaluation. Undoubtly, substantial improvements are still needed in the methodologies presented. As editors we strongly believe that a close cooperation between geologists and mathematical statisticians is needed to instigate such improvements and to generate a new breed of scientists who can use both geological and statistical sciences in an effective manner. We are confident that this ASI represents an important step in this direction.

These ASI Proceedings will provide statisticians with challenging real-life problems, because the application of statistics in geosciences is still very much in its infancy. The geoscientists will find a comprehensive and up-to-date review of the quantitative methodologies which are currently available to deal with geoscience models. We hope that this volume will encourage scientists to participate and contribute in futher developments of this exciting field.

C. F. Chung
A. G. Fabbri
R. Sinding-Larsen

AKNOWLEDGEMENTS

The Scientific Affairs Division of NATO is gratefully acknowledged for having sponsored this Advanced Study Institute. Additional support was provided by Canada Centre for Remote Sensing, Geological Survey of Canada, Istituto di Geologia Marina del CNR di Bologna and the Norwegian Institute of Technology.

The following Institutions also provided financial support:

A.G.I.P. SPA, San Donato Milanese (Milano), Italy
Centro Studi per Problemi Minerari del C.N.R., Torino, Italy
Dip. di Georisorse e Territorio, Politecnico, Torino, Italy
E.N.E.A.-DISP, Roma, Italy
Geomineraria Italiana SRL, Borgo San Dalmazzo (Cuneo), Italy
Mining Italiana SPA, Roma, Italy
National Science Foundation, Washington D.C., USA
S.A.M.I.M., Roma, Italy
Societa' di Geofisica Applicata SRL, Bologna, Italy
STATOIL, Stavanger, Norway
Studio di Informatica Applicata SRL, Bologna, Italy

CONTENTS

PART 3. STATISTICAL MEHTODS FOR OIL AND GAS RESOURCE EVALUATION

PART 4. IMAGE PROCESSING IN EARTH SCIENCES

PART 5. STATISTICAL ANALYSIS OF SPATIAL DATA

PART 6. GEOSTATISTICS

PART 7. OTHER STATISITCAL METHODOLOGIES FOR GEOSCIENCE DATA

PART 8. ARTIFICIAL INTELLIGENCE IN EARTH SCIENCES

PART 9. OTHER CONTRIBUTED PAPERS

WORKSHOP REPORTS

LIST OF PARTICIPANTS AT THE
NATO ADVANCED STUDY INSTITUTE ON
STATISTICAL TREATMENTS FOR ESTIMATION OF MINERAL AND ENERGY RESOURCES

Frits P. Agterberg
Geological Survey of Canada,
601 Booth Street,
Ottawa, Ontario, Canada. K1A 0E8

Unal Artan
MTA Genel Müdurlugü
Maden
Etut ve Arama Dairesi,
ANKARA, Turkey

Andrea Bernasconi
A.G.I.P., S.p.A.,
P.O.Box 12069,
20097 S. Donato Milanese
(Milan), Italy.

Giuseppe Bonifazi
Dip. Ing, Chimica, Materie Prime
e Metallurgia,
Universita' di Roma La Sapienza
Via Eudossiana, 18,
00185 Roma, Italy

Giovanni Bortoluzzi
Istituto di Geologia Marina, CNR
Via Zamboni, 65,
40127 Bologna, Italy

Maria Franca Brigatti
Istituto di Minerologia
Universita'di Modena,
Via Santa Eufemia, 19,
Modena, Italy

R. Bruno
Istituto di Arte Mineraria
Facolta' di Ingegneria,
Universita' di Roma La Sapienza,
Via Eudossiana 18,
00159 ROMA, Italy

Fatma Gül Cakmak
Dept. of Statistics,
Hacettepe Univ. of Ankara,
Cagri sk 15/2, Ayoinlikevler,
Ankara, Turkey

Alan Campbell
Tyee Lake,
Telkwa, B.C., Canada V0J 2X0

Laura Carbognin
Istituto Studio Dinamica
delle Grandi Masse, CNR,
Palazzo Papadopoli,
1364 San Polo,
30125 Venezia, Italy

Giovanni Carosso
Centro Studio Problemi Minerari,
CNR,
c/o Dip. Georisorse e Territorio
del Politecnico di Torino,
Corso Duca degli Abruzzi, 24,
0129 Torino, Italy

Alberto Carrara
Ist. di Ingegneria Civile,
Universita'di Firenze,
Via Santa Marta, 3,
FIRENZE, Italy

Simeon Chow
E53-360, Herman Building,
Sloan School, M.I.T.,
Cambridge, Mass. 02139, U.S.A.

Chang-Jo F. Chung
Geological Survey of Canada,
601 Booth Street,
Ottawa, Ontario, Canada. K1A 0E8

Robert A. Crovelli
U.S. Geological Survey,
P.O. Box 25046,
MS-971 DENVER, Colorado 80225,
U.S.A.

John C. Davis
Kansas Geological Survey
1930 Constant Ave.,
Univ. of Kansas Lawrence,
Kansas, 66046 U.S.A.

Jo Anne DeGraffenreid
Kansas Geological Survey
1930 Constant Ave.,
Univ. of Kansas,
Lawrence, Kansas 66046, U.S.A.

Rao S. Divi
Structural Geology
and Rem. Sensing Dept.,
King Abdulaziz University,
P.O. Box 1744,
JEDDAH, Saudi Arabia

Andrea G. Fabbri
Canada Centre for Remote Sensing,
1790 Woodward Drive,
Ottawa, Ontario, Canada K1A 0Y7

John D. Grace
ARCO Exploration Research,
2300 W. Plano Parkway,
PLANO, Texas 75075, U.S.A.

A. Grassia
CSIRO Institute of Energy
and Earth Resources,
Division of Groundwater Research,
Private Bag, P.O. Wembley,
6014 Western Australia.

John C. Griffiths
Pennsylvania State University,
Deike 310, College of Earth
and Mineral Sciences,
University Park, Pennsylvania
1602,
U.S.A.

De Verle P. Harris
Department of Mineral Economics
Univ. of Arizona,
Tucson, Arizona 85721, U.S.A.

Michael T. Holroyd
DATAPLOTTING Services Inc.,
225 Duncan Road,
Don Mills, Ontario, CANADA, M3B
3K9

Gordon M. Kaufman
E53-380, Herman Building,
Sloan School, M.I.T.,
Cambridge, Mass. 02139, U.S.A.

George S. Koch Jr
Department of Geology,
Univ. of Georgia,
Athens, Georgia 30602, U.S.A.

W. Lanyon
Geothermal Project,
Rosemanowes Quarry,
Herniss, Penryn,
Cornwall TR10 9DU, U.K.

Marco Ligi
Istituto di Geologia Marina, CNR,
Via Zamboni 65,
40127 Bologna, Italy.

Francesco Mastrangelo
Dipart. di Georisorse e
Territorio,
Politecnico di Torino,
Corso Duca degli Abruzzi 24,
10129 TORINO, Italy

D.F. Merriam
Wichita State University,
Dept. of Geology,
Wichita, Kansas, 67208, U.S.A.

Sonjay K. Mitter
Lab. for Information
and Descision Systems,
M.I.T.,
Cambridge, Mass. 02139, U.S.A.

Fernando Muge
Centro de Valorizaco de Recursos Minerais,
I.S.T. - Av. Rovisco Pais no.1,
LISBOA/codex 1096, Portugal

Donald E. Myers
Dept. of Mathematics,
Univ. of Arizona,
TUCSON, Arizona 85721, U.S.A.

Henning Omre
Norwegian Computing Center,
Postbox 335-Blindern,
0314 Oslo 3, Norway

Mustafa Ozcelik
M.T.A. Genel Mdurlug,
Maden Etut ve Arama Dairesi,
Ankara, TURKEY

Demetrios Papacharalampos
Dept. of Geology,
University of Georgia,
ATHENS, Georgia 30602, U.S.A.

Henrique G. Pereira
CVRM - Technical University of Lisbon,
I.S.T. - Av. Rovisco Pais no. 1,
LISBOA/codex 1096, Portugal.

Gianpiero Pinna
Direzione Attivita' Minerarie,
S.A.M.I.M.,
Monte Agruxan,
09016 Iglesias (Sardinia), Italy.

Luciano Poppi
Istituto di Mineralogia,
Universtia' di Bologna,
Piazza di Porta San Donato, 1,
40127 Bologna, Italy

Roberto Potenza
Centro Alpi, CNR,
Via Botticelli, 23,
20133 MILANO, Italy

Francesca F. Quercia
c/o ENEA - DISP,
Com. Naz. Ric. Svil. En. Nuc., Energie Alternative,
Via Vitaliano Brancati, 48,
00144 Roma, Italy

Pino Raspa
Istituto di Arte Mineraria,
Universita' di Roma La Sapienza,
Via Eudossiana 18,
00159 ROMA, Italy

Sonke Rehder
2204, B.G.R.,
Stilleweg 2,
3000 Hannover 51, F.R. Germany

Brian D. Ripley
Dept. of Mathematics,
University of Strathclyde,
26 Richmond Street,
Glasgow, G11XM Scotland, U.K.

Umberto Rossi
AGIP Spa, Servizio STIG,
20097 San Donato Milanese,
(Milano), Italy

Jean Jacques Royer
C.R.P.G.,
15 Rue Notre-Dame des Pauvres,
B.P. 20,
Vandeuvre les Nancy cedex 54501,
France

Stefano Salvi
Ist. Astrofisica Spaziale-CNR,
CP 47,
00044 Frascati, ROMA, Italy

Jean Serra
Centre de Morphologie Mathematique,
35 Rue Saint-Honore,
Fontainbleau 77305, France

Richard Sinding-Larsen
Norwegian Institute of Technology,
Department of Geology,
7034-NTH,Trondheim, Norway

Wolfdietrich Skala
Frei University of Berlin,
Altensteinerstrasse 34A,
1000 Berlin 33, F.R. Germany

Bjorn H. Sollie
Norwegian Computing Centre,
Forskn. v. 1B P.O. 335,
Blindern, 0314 Oslo 3, Norway

Andrew R. Solow
Woods Hole Ocean. Institute,
Woods Hole, Mass. 02545, U.S.A.

Federico Sosso
AGIP Spa.,
20097 San Donato Milanese
(Milano), Italy

Clifford R. Stanley
Dept. of Geological Sciences,
Univ. of British Columbia,
Vancouver, B.C. V6T 2B4, CANADA

Per R. Stokke
Exploration and Production
Division,
STATOIL,
P.O. Box 300, Forus,
4001 STAVANGER, Norway

Michael J. Shulman
Freie Universitaat Berlin,
Malteserstrasse 74-100,
D-100 BERLIN46, West Germany

Peter G. Sutterlin
Dept. of Geology,
Wichita State University,
Wichita, Kansas 67208, U.S.A.

Paul Switzer
Stanford University,
Department of Statistics,
Sequoia Hall,
Stanford, California 9, U.S.A.

Bruno Testa
Centro Alpi, C.N.R.,
Via Botticelli 23,
MILANO 20133, Italy

Amir Toister
Dept. of Geophysics
and Planetary Sciences,
Univ. of Tel Aviv ,
Tel Aviv 69 978, Israel

Hans Wackernagel
Centre de Geostatistique ENSMP,
35 rue Saint Honore,
77305 FONTAINEBLEAU, France

Paul C.C. Wang
Dept. of Mathematics and
Statistics
University of Calgary,
CALGARY, Alberta, Canada T2N 1N4

W. Lynn Watney
Kansas Geological Survey,
1930 Constant Ave.,
Univ. of Kansas,
Lawrence, Kansas, 66044-3896,
U.S.A.

Fritz Wurzer
2402, Joanneur Research Society,
Mineral Resources Research
Division,
Roseggerstrasse 15,
A-8700 Leoben, AUSTRIA

Ahmet Yalniz
Dept. of Statistics,
Hacettepe University,
Cemal Nadir Sok. 23/5, Cankaya,
ANKARA, Turkey

Pengda Zhao
Wuhan College of Geology,
Wuhan, Hubei,
People's Republic of China

APPLICATION OF RECENT DEVELOPMENTS OF REGRESSION ANALYSIS IN REGIONAL MINERAL RESOURCE EVALUATION*

F.P. Agterberg
Geological Survey of Canada
601 Booth Street
Ottawa, Canada K1A 0E8

ABSTRACT. During the past 8 years, several methods for regression diagnostics have become widely used. In this paper these methods are applied to a mineral resource evaluation problem consisting of estimating the frequency of occurrence of mineral deposits of a given type per unit area from measurements on explanatory variables in a region. New results were obtained by constructing partial regression residual plots to check the linearity assumption.

Diagonal elements of the hat matrix help to identify high-leverage cells and clustering of the off-diagonal elements of a hat matrix provides a sequence of cells within a region in order of degree of similarity. Use was made of newly developed computer programs to obtain the hat matrix and modified hat matrix in linear and logistic regression.

1. INTRODUCTION

During the past 20 years, a number of statistical studies have been performed using regression and other multivariate analysis techniques for regional mineral resource evaluation (cf. Harris, 1984). A recent review (Agterberg, 1986) contains the following summary of multivariate mineral resource appraisal work in Canada:

1. Multiple regression of total dollar value on cell data (1967-1971);
2. Two-stage least squares model (1967-1971);
3. Probability index maps and occurrence model for undiscovered deposits (1970-1979);
4. Logistic model for estimating probability of occurrence (1971-1980);
5. Quantification and analysis of geological patterns (1971-1980);
6. Automated coding and development of integrated computer systems for resource evaluation (1978-present).

The purpose of the present study is to re-examine earlier work by applying recent developments of regression analysis to occurrence of

* Geological Survey of Canada Contribution 36586.

C. F. Chung et al. (eds.), *Quantitative Analysis of Mineral and Energy Resources*, 1–28.

massive sulphide deposits and rock types in the Abitibi area of the Canadian Shield which is one of the data sets used extensively in the past. New techniques used in this paper are: (1) partial regression residual plots for the general linear model of least squares in order to examine reliability of estimated regression coefficients; (2) leverage and influence of individual observations on estimated values obtained by the linear and logistic models; and (3) relationship of metal content and concentration within mineral deposits with the probability of occurrence of these deposits. Recent methods of regression analysis by means of the linear model have been reviewed by many authors including Hoaglin and Welsch (1978), Velleman and Welsch (1981), Hocking (1983), and Wrigley (1983). These papers introduce various regression diagnostics. Several of these are derived from the hat matrix H according to which estimated values $\hat{Y}$ of a dependent variable y are related to its observed values Y. If there are n observations, and p explanatory variables, $\hat{Y} = HY$ where $\hat{Y}$ and Y are (nx1) column vectors and H is a (nxn) symmetrical matrix satisfying

$$H = X\,(X'X)^{-1}\,X' \tag{1}$$

The (nxp) matrix X contains the observed values of the explanatory variables x_i for $i = 1, \ldots, p$. The first of these variables has $X_{ij} = 1$ for $j = 1, \ldots, n$ to allow for a constant term in the regression equation. The elements h_{ij} of the hat matrix provide a measure of the amount of leverage exerted by the observation Y_j on the estimated value $\hat{Y}_i$. Gray and Ling (1984) defined

$$h^*_{ij} = h_{ij} + e_i e_j / \sum_{i=1}^{n} e_i^2 \tag{2}$$

as the amount of influence exerted by Y_j on the $\hat{Y}_i$. The elements h^*_{ij} form the so-called modified hat matrix H* in which the effects of the residuals $e_i = Y_i - \hat{Y}_i$ are also considered. Gray and Ling (1984) proposed clustering of the modified hat matrix as a means of finding influential subsets of observations. Their approach was adopted in a study of estimating the probability of occurrence of hydrothermal vents as a function of rock types and tectonic features on the East Pacific Rise near 21° North (Agterberg and Franklin, 1986). The hat matrix was clustered as well and this produced results close to those obtained by clustering of the modified hat matrix. It was concluded in that paper that there exist three types of observations for cells: (1) cells with relatively small leverage and influence on many estimated values $\hat{Y}_i$; (2) cells belonging to relatively small subsets of similar cells which exert strong leverage upon one another; and (3) unique cells with values of the explanatory variables that explain only the observed value Y_i for the cell itself without exerting significant leverage on any other cell in the study area.

It will be seen later in this paper that the preceding types of cells also exist when regression is applied to predict probability of occurrence of massive sulphide deposits in the Abitibi area of the Canadian Shield. A systematic approach to leverage is important because in regional studies of resource evaluation use is made of the principle of geological analogy, viz.: it is assumed that similar mineral deposits occur in similar geological environments. A unique cell would be without counterparts in the same region whereas cells with known deposits that exert leverage on many other cells are useful for extrapolation. Pregibon (1981), Wrigley (1984) and Landwehr, Pregibon and Shoemaker (1984) have developed and applied diagnostics and graphical methods for assessing logistic regression models. Several of these methods will also be applied here. Recently, Wrigley and Dunn (1986) have used graphical diagnostics in logistic oil exploration models.

For the computing required for this paper, use was made of Version 5 of Statistical Analysis System (SAS). SAS regression procedures and clustering procedures (SAS, 1985a) were employed, as

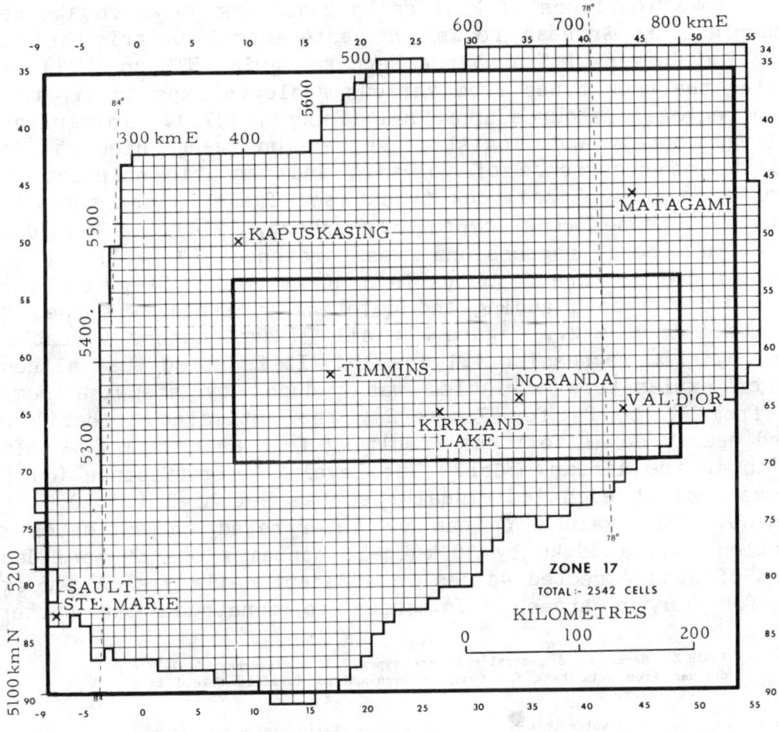

Figure 1. Universal Transverse Mercator (UTM) grid cells measuring 10 km on a side for area on the Canadian Shield underlain by Archean rocks. Outside frame outlines area of Figures 2 and 3. Inside heavy frame outlines area used for Table II and subsequent statistical analysis based on 40-km cells.

well as the SAS matrix procedure (SAS, 1985b). Chung's (1978) computer programme for logistic regression was amplified by this author to allow multiple binomial response and for obtaining logistic regression diagnostics (see Agterberg, in press, in Computers and Geosciences, Special Issue for this NATO ASI).

Regression analysis for estimating probabilities of occurrence and expected frequencies for mineral deposits can be performed without modelling of the explanatory variables (e.g. rock types) as spatial random variables. Nevertheless, regression results are determined by the variance-covariance matrix of the explanatory variables which, in turn, depends strongly on the sizes and shapes of the cells used for coding. This spatial aspect of the estimation of variances and covariances of rock types is considered for several examples in a separate paper by this author elsewhere in this volume.

2. APPLICATION OF REGRESSION TO VOLCANIC AND MAGMATIC SULPHIDE DEPOSITS IN 10-KM CELLS, ABITIBI VOLCANIC BELT ON THE CANADIAN SHIELD

Figure 1 shows locations of 2592 cells measuring 10 km on a side which are underlain by Archean rocks in east-central Ontario and western Quebec. The Universal Transverse Mercator grid (UTM zone 17) was used for coding the rock types from various geologic maps to create a data bank for regional resource appraisal (Fabbri, 1975). In earlier work, the linear model and logistic regression were used to express probability of occurrence of volcanic and magmatic sulphide deposits in terms of 8 generalized rock types (see Table I) and the 28 cross-products of these rock types (Agterberg, 1974). The dependent variable in these regressions was presence ($Y_i = 1$) or absence ($Y_i = 0$) of one or more deposits in a 10-km cell. Stepwise regression with $\alpha = 0.5$ for both adding and deleting variables was used for the linear model (see e.g., Figure 2); all 37 explanatory variables were used in logistic regression (Figure 3). It is noted that although the number of explanatory variables was reduced by stepwise regression (e.g., from 37 to 24 for Figure 2), the estimated probabilities $\hat{Y}_i$ remained nearly equal to those resulting from using all 37 explanatory variables in the linear model. The level of significance (α) for the F-test applied at each individual step was set high ($\alpha = 0.5$) in this experiment. The values $\hat{Y}_i$ can be interpreted as estimates of the expectation that a 10 km by 10 km cell contains one or more deposits. Addition of such expected values in adjacent cells yields new expected values for larger areas. In order to construct the patterns of

Table I. Names of 8 generalized rock types, as in Agterberg (1974); derived from data bank for Superior Province of Canadian Shield by Fabbri (1975).

x_1	Mafic volcanic rocks (and amphibolites)	x_5	Mafic intrusive rocks (and anorthosites)
x_2	Felsic volcanic rocks	x_6	Ultramafics
x_3	Sedimentary rocks	x_7	Granitic gneisses (and granulite complexes)
x_4	Metamorphic rocks	x_8	Felsic intrusive rocks

Figures 2 and 3, use was made of a unit area measuring 40 km on a side. Values of $c\hat{Y}_i$ for overlapping square blocks of sixteen 10-km cells were added and contoured. The constant c in $c\hat{Y}_i$ was set equal to the ratio of the number of 10-km cells with known deposits and expected number of such cells in a "control area". For the example of Figure 2, c = 2.06. This linear transformation was based on the assumption that the probability that a cell contains known deposits is the product of the probability that it contains deposits (including undiscovered deposits) and probability of discovery. The latter was set equal to one in the control area. This assumption can be easily modified. For example, if it were assumed that only half of the cells with deposits have been discovered in the control area, this would be equivalent to multiplying all contour values in Figure 2 by two. Other suggestions for further analysis of this type of contour map have been given by Tukey (1984).

Figure 2. Linear model (stepwise regression, see text) applied to relate occurrence of volcanic polymetallic massive sulphide deposits (crosses) to 8 rock types of Table I. Contours are for expected number of 10-km cells with one or more deposits per larger 40-km cell.

The probability distribution of the values m represented by contours in Figures 2 and 3 is approximately binomial with parameters p = m/N and N = 16. This is because presence or absence of deposits is assumed to be the realization of a Bernoulli random variable of which the mean was estimated by a qualitative response model. The estimated probability for a 10-km cell has its own uncertainty but its variance generally is small in comparison with that of the Bernoulli variable and may be neglected. The value of m = 16p is the sum of 16 probabilities p_i for individual 10-km cells. Strictly speaking, the probability distribution of m would be subnormal binomial with variance Np (1-p) — N var (p_i). Replacement of this distribution by a binomial distribution with variance Np (1-p) is equivalent to slightly underestimating the precision of the contour values of Figures 2 and 3.

In general, the logistic model is more flexible than the linear model for an equal number of explanatory variables. However, this rule does not apply in the situation of Figure 3 where the difference between the patterns of Figures 2 and 3 is relatively small.

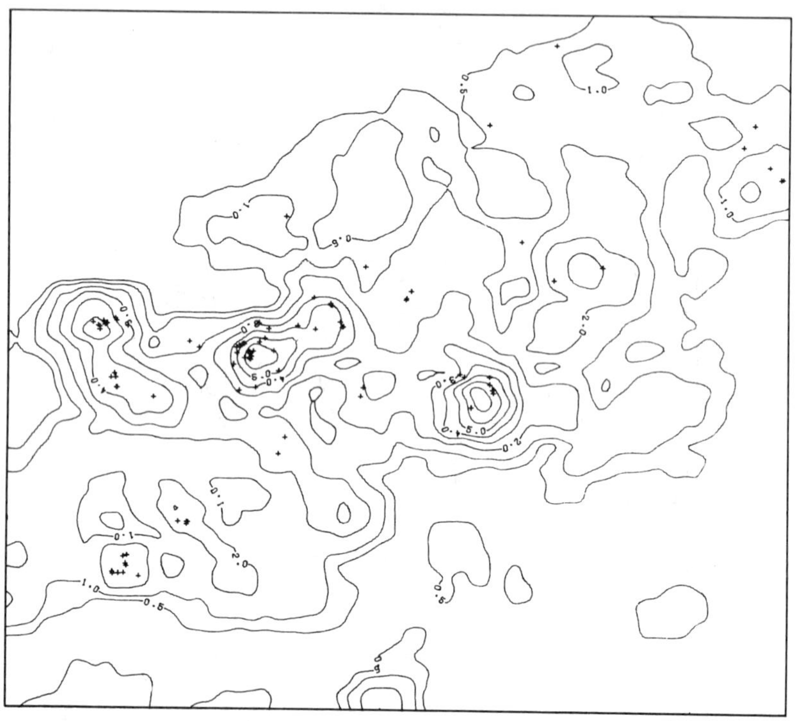

Figure 3. Same as Figure 2; using logistic model with 37 explanatory variables (from Agterberg, 1974).

3. LINEAR AND LOGISTIC MODELS APPLIED TO MASSIVE SULPHIDE DEPOSITS IN 40-km CELLS

The input for this example is given in Table II. There are 40 cells with observations on sizes and grades for copper derived from 69 volcanic polymetallic massive sulphide deposits. The second and third columns of Table II show, for each 40-km cell, the number of 10-km cells with one or more deposits and the number of individual deposits, respectively. Locally there is relatively strong spatial clustering of deposits, especially in the vicinity of Noranda (cell #27) where 19 deposits occur in seven 10-km cells within the same 40-km cell; one of these seven 10-km cells contains as many as 8 separate deposits. Number of deposits per cell can, in principle, be modelled as a Poisson random variable of which the mean depends on the explanatory variables. Strong local clustering of deposits as observed in the vicinity of Noranda, presents a problem because the probability of having so many separate deposits close to one another

Table II. Data for 40 cells (numbered 1 to 40 in column 1) in Abitibi area of Canadian Shield outlined in Figure 1. The relative geographical locations of these 40-km cells are given in Table IV-A. The dependent variable (Y) is number of 10-km cells with volcanic massive sulphide deposits per 40-km cell; D - number of separate volcanic massive sulphide deposits; S - logarithm (base 10) of metric tons of copper; G log (base 10) of copper grade in units of 1000 ppm; U, V - geographic co-ordinates for north-south and west-east directions as in Table IV; X_1 to X_8 - scores for 8 rock types listed in Table I; multiplication of these scores by 1000 gives number of 500 m x 500 m subcells per 40-km cell underlain by a particular rock type (each 40-km cell contains 4800 subcells). The statistics in this table were derived from the data bank described by Fabbri (1975). The data on mineral deposits were revised in 1978 to account for 8 new discoveries and ore reserve estimation at the Kidd Creek deposit in cell #10.

#	Y	D	S	G	U	V	X_1	X_2	X_3	X_4	X_5	X_6	X_7	X_8
1	0	0	0.00	0.00	1	1	0.056	0.000	0.000	0.000	0.000	0.000	0.006	0.000
2	0	0	0.00	0.00	2	1	1.556	0.508	0.013	0.000	0.493	0.000	0.266	1.427
3	0	0	0.00	0.00	3	1	1.556	0.257	0.157	0.000	0.120	0.026	0.291	3.346
4	1	1	1.00	1.00	4	1	2.312	0.263	0.006	0.000	0.045	0.000	0.176	2.623
5	0	0	0.00	0.00	1	2	1.904	0.270	0.158	0.000	0.005	0.099	0.000	0.673
6	2	4	5.07	2.20	2	2	2.354	1.074	0.230	0.000	0.614	0.055	0.001	1.524
7	0	0	0.00	0.00	3	2	0.930	0.443	0.601	0.000	0.369	0.030	0.000	3.737
8	0	0	0.00	0.00	4	2	0.051	0.000	0.012	0.000	0.000	0.000	0.000	5,833
9	0	0	0.00	0.00	1	3	2.060	0.105	0.694	0.000	0.030	0.297	0.000	0.766
10	1	1	6.56	2.40	2	3	2.010	1.565	1.949	0.000	0.079	0.335	0.000	0.071
11	2	2	4.73	1.85	3	3	2.482	0.918	0.176	0.000	0.106	0.343	0.000	1.867
12	0	0	0.00	0.00	4	3	2.147	1.280	0.190	0.000	0.122	0.065	0.000	0.498
13	0	0	0.00	0.00	1	4	1.474	0.000	0.375	0.028	0.035	0.113	0.000	0.763
14	0	0	0.00	0.00	2	4	4.072	0.340	1.195	0.000	0.114	0.141	0.000	0.290
15	0	0	0.00	0.00	3	4	2.589	0.079	0.000	0.000	0.073	0.015	0.000	3.004
16	2	4	3.80	2.03	4	4	1.191	0.000	0.114	0.000	0.032	0.003	0.000	1.947
17	0	0	0.00	0.00	1	5	0.635	0.215	1.184	0.000	0.004	0.030	0.000	0.434
18	2	3	4.19	2.31	2	5	3.283	0.127	0.249	0.000	0.290	0.207	0.000	0.849
19	1	1	1.00	1.00	3	5	4.499	0.469	0.250	0.000	0.459	0.006	0.000	0.306
20	3	5	2.95	2.76	4	5	1.527	0.258	0.103	0.000	0.174	0.025	0.000	3.417
21	2	2	5.34	2.33	1	6	1.350	0.283	0.971	0.027	0.206	0.006	0.000	3.527
22	0	0	0.00	0.00	2	6	3.021	0.145	0.158	0.000	0.047	0.016	0.000	1.314
23	5	7	4.87	2.16	3	6	3.514	0.902	0.079	0.257	0.561	0.000	0.041	0.169
24	1	1	1.00	1.00	4	6	0.456	0.274	0.324	0.522	0.112	0.035	1.716	0.130
25	1	1	1.00	1.00	1	7	2.345	0.066	0.977	0.000	0.103	0.009	0.000	2.798
26	2	2	3.66	2.20	2	7	3.268	0.287	0.486	0.000	0.025	0.058	0.000	2.005
27	7	19	6.32	2.33	3	7	1.969	1.109	0.726	1.147	0.343	0.007	0.000	0.603
28	0	0	0.00	0.00	4	7	0.000	0.000	0.000	1.986	0.000	0.000	3.846	0.058
29	0	0	0.00	0.00	1	8	4.074	0.022	0.431	0.000	0.008	0.005	0.000	1.753
30	0	0	0.00	0.00	2	8	3.416	0.348	0.842	0.000	0.000	0.060	0.018	1.649
31	1	2	3.65	1.45	3	8	1.444	0.098	1.129	1.517	0.022	0.159	0.206	1.104
32	0	0	0.00	0.00	4	8	0.000	0.000	0.000	1.374	0.000	0.000	4.547	0.164
33	0	0	0.00	0.00	1	9	1.344	0.177	1.587	0.000	0.051	0.000	0.000	3.182
34	3	4	4.54	2.09	2	9	3.514	1.286	0.590	0.000	0.128	0.073	0.000	0.620
35	2	4	5.16	1.95	3	9	1.486	0.356	0.544	0.282	0.192	0.422	0.000	2.355
36	1	1	5.14	1.96	4	9	0.071	0.057	0.006	2.041	0.000	0.000	3.355	0.076
37	0	0	0.00	0.00	1	10	2.728	0.038	1.081	0.000	0.054	0.000	0.000	2.384
38	0	0	0.00	0.00	2	10	1.909	0.259	1.880	0.000	0.024	0.017	1.704	0.358
39	1	4	4.86	1.74	3	10	2.660	0.906	0.205	0.000	0.026	0.117	0.008	1.782
40	1	1	4.04	2.10	4	10	0.866	0.154	0.255	2.080	0.054	0.011	1.134	0.197

would be low in this approach. It can be argued that some of the deposits in a local cluster can be counted as a single event, e.g. if they occur within the same favourable stratigraphic horizon. However, such decisions remain to some extent subjective. This type of problem is avoided in the presence-absence approach of the qualitative response model but then number of cells with one or more deposits is predicted instead of actual number of separate mineral deposits. These problems have also been discussed by Chung and Agterberg (1980), and an application of the Poisson regression model to the data of Table II is given by Chung and Agterberg elsewhere in this volume.

Tables III to V show results of four regression experiments relating the vector Y of Table II to the 8 vectors, X_1 to X_8 for rock types. The main difference between these experiments and those described in the previous section is that the dependent variable now represents expected number of events for a larger (40 km x 40 km) cell instead of presence or absence of these events in smaller (10 km x 10 km) cells. The estimated values for stepwise regression (α = 0.3) on 37 explanatory variables (including crossproducts) and those for the corresponding logistic model in Table IV display patterns which are similar to those of Figures 2 and 3. The three peaks in the vicinities of Timmins, Noranda and north of Val d'Or, respectively, are well developed in the patterns of Figures 2 and 3 as in those of Table IV. Individual regression coefficients with standard deviations are given in Table III together with the multiple correlation coefficient squared for the linear model and the deviance of the logistic model. The deviance D (Prebigon, 1981) is approximately χ^2 – distributed with (n-p) degrees of freedom if the fit of the model is good. In the example of Table III, it is well below significance limits of χ^2 corresponding to α = 0.05 or 0.01. This indicates that the fit of the logistic model is satisfactory.

Table III. Regression coefficients with their standard deviations (in brackets) in which Y of Table II is related to the rock types X_1 to X_8; R^2 - multiple correlation coefficient squared; D - deviance for logistic model; df - degrees of freedom; Int. - Intercept; α = 0.30 indicates that a level of significance of 0.30 was applied to F-statistic during addition and deletion of variables. The first stepwise regression experiment had 28 crossproduct variables $x_{ij} = x_i x_j$ for i, j = 1, ..., 8; i ≠ j added to 8 original variables; only two terms (uv^2 and uv^7) were retained in second stepwise regression where octic polynomial was added to individual rock types.

	R^2	Int.	x_1	x_2	x_3	x_4	x_5	x_6	x_7	x_8
8 Variables	0.455	-.225 (0.799)	0.126 (0.215)	1.236 (0.607)	-.164 (0.405)	1.321 (0.488)	2.672 (1.461)	-.455 (2.056)	-.467 (0.302)	0.090 (0.185)

	R^2	Int.	x_1	x_2	x_{12}	x_{15}	x_{16}	x_{18}	x_{45}
With Crossproducts (α = 0.30)	0.797	0.443 (0.276)	-.623 (0.197)	-2.51 (1.018)	1.379 (0.419)	0.874 (0.330)	1.314 (0.611)	0.251 (0.071)	16.83 (2.011)

	D	Int.	x_1	x_2	x_{12}	x_{15}	x_{16}	x_{18}	x_{45}
Logistic Model	16.04 (df:33)	-3.45 (0.529)	-.813 (0.368)	-2.83 (1.661)	1.454 (0.624)	1.025 (0.425)	1.700 (0.804)	0.329 (0.126)	9.372 (2.220)

	R^2	Int.	x_2	x_5	x_7	uv^2	uv^7
With Polynomial (α = 0.30)	0.599	-.630 (0.311)	0.939 (0.449)	2.884 (1.143)	-.677 (0.211)	20.39×10^{-3} (4.250×10^{-3})	-1.62×10^{-8} (0.421×10^{-8})

Table IV. A - Relative geographical locations of the 40 cells of Table II; B - values of dependent variable Y as two-dimensional array; C - Predicted values $\hat{Y}$ using stepwise regression on 8 rock types with crossproducts; D - Predicted values using logistic regression on same variables as used for Table IV-C; coefficients used to compute values of Tables IV-C and IV-D are given in Table III; values for cells with known deposits are printed bold in Tables V-B to IV-D.

A	(v) 1	2	3	4	5	6	7	8	9	10
(u)										
1	1	5	9	13	17	21	25	29	33	37
2	2	6	10	14	18	22	26	30	34	38
3	3	7	11	15	19	23	27	31	35	39
4	4	8	12	16	20	24	28	32	36	40

B	1	2	3	4	5	6	7	8	9	10
1	0	0	0	0	0	**2**	**1**	0	0	0
2	0	**2**	1	0	**2**	0	**2**	0	**3**	0
3	0	0	**2**	0	**1**	**5**	**7**	**1**	**2**	**1**
4	**1**	0	0	**2**	**3**	1	0	0	**1**	**1**

C	1	2	3	4	5	6	7	8	9	10
1	0.4	-.1	0.4	0.1	-.2	**1.0**	**0.9**	-.2	0.6	0.6
2	0.5	**2.1**	**0.7**	0.4	**1.1**	-.0	**0.9**	0.8	**2.5**	-.5
3	0.9	0.5	**2.2**	1.1	**1.6**	**4.7**	**7.0**	**0.8**	**2.2**	**1.5**
4	**0.8**	0.5	0.4	**0.3**	**1.0**	**0.7**	0.4	0.4	**0.3**	**1.7**

D	1	2	3	4	5	6	7	8	9	10
1	0.5	0.2	0.5	0.3	0.2	**0.9**	**0.9**	0.2	0.6	0.6
2	0.5	**2.2**	**0.4**	0.3	**0.9**	0.2	**0.8**	0.6	**2.3**	0.1
3	0.8	0.6	**2.9**	1.0	**1.0**	**5.4**	**7.1**	**0.5**	**2.1**	**1.2**
4	**0.7**	0.5	0.3	**0.4**	**0.9**	**0.4**	0.5	0.5	**0.4**	**0.6**

Table V. Predicted values for second stepwise regression experiment of Table IV. Use of all coefficients gives Table IV-A; B - polynomial terms only; C - rock type variables only; intercept was divided into two parts by setting average of values in Table V-B equal to zero.

A	(v) 1	2	3	4	5	6	7	8	9	10
(u)										
1	-.6	-.3	-.3	-.2	0.1	**0.9**	**0.6**	0.4	0.6	-.0
2	1.1	**2.3**	**1.4**	0.7	**1.3**	1.0	**1.4**	1.6	**2.6**	-.7
3	-.2	1.1	**1.1**	0.6	**2.6**	**3.9**	**4.0**	**2.3**	**2.9**	**1.5**
4	**-.3**	-.3	1.7	**0.8**	**2.1**	**1.5**	0.2	0.1	**0.7**	**0.6**

B	1	2	3	4	5	6	7	8	9	10
1	-1.2	-1.1	-1.0	-.90	-.73	**-.54**	**-.36**	-.26	-.35	-.81
2	-1.2	**-1.1**	**-.86**	-.58	**-.23**	0.15	**0.50**	0.70	**0.52**	-.40
3	-1.2	-1.0	**-.68**	-.26	**0.26**	**0.84**	**1.37**	**1.66**	**1.40**	**0.02**
4	**-1.1**	-.90	-.50	**0.07**	**0.76**	**1.53**	2.23	2.63	**2.27**	**0.43**

C	1	2	3	4	5	6	7	8	9	10
1	0.59	0.87	0.78	0.70	0.81	**1.46**	**0.96**	0.64	0.91	0.79
2	2.32	**3.38**	**2.30**	1.25	**1.55**	0.87	**0.94**	0.91	**2.17**	0.24
3	0.99	2.08	**1.77**	0.88	**2.36**	**3.03**	**2.63**	**0.61**	**1.49**	**1.52**
4	**0.86**	0.60	2.15	**0.69**	**1.34**	**0.02**	-2.0	-2.5	**-1.6**	**0.13**

The first experiment listed in Table III is for multiple regression of y on x_1 to x_8. Figure 4 shows several partial regression residual plots for this run. A plot of this type represents the relationship between two sets of residuals. The first set p_k (plotted in the vertical direction) is obtained when y is predicted from all except the kth explanatory variable. The second set q_k (plotted in the horizontal direction) is obtained when x_k is taken as the dependent variable and is predicted from all the remaining explanatory variables. A least squares regression line fitted to the resulting scattergram should have zero intercept and slope equal to $\hat{\beta}_k$ representing the coefficient of the k-th explanatory variable in the multiple regression solution on all p variables (for proof, see e.g., Velleman and Welsch, 1981). The advantage of this approach is that it is easy to check for model failures such as nonlinearity and heteroscedasticity. Model failure is indicated in Figure 4 where the plots exhibit a parabolic pattern rather than linearity.

It is well known that regression of mineral deposit data on separate rock types usually gives poor results. Two methods (two-stage least squares model and use of crossproduct variables) have been proposed to remedy the situation. For vein-type gold deposits in the Abitibi area, the first of these two methods (two-stage least squares model with addition of a two-dimensional polynomial expression for the geographical co-ordinates) worked best, and for massive sulphide deposits the second approach (use of crossproducts) led to significant improvements. Both methods were applied to the example of Table II

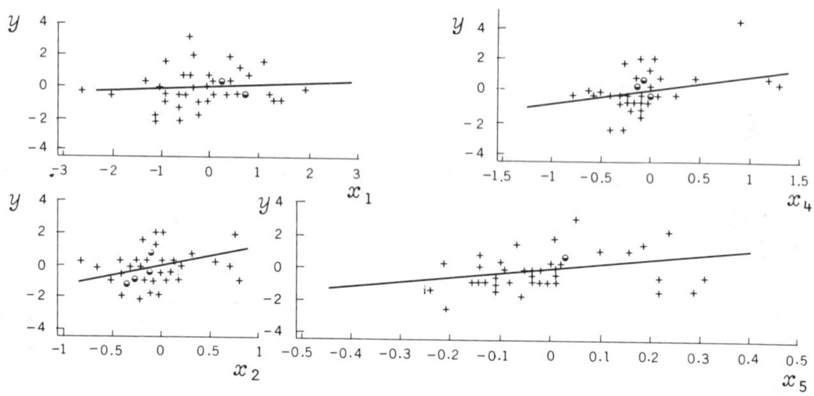

Figure 4. Four partial regression residual plots for y (with values Y in Table II) regressed on 8 rock type variables. The coefficients for this run are given in first row of Table III. Circular symbol represents two or more approximately coinciding data points. Note the nonlinear, curved (parabolic), pattern of data points in these plots indicating that the best-fitting straight lines do not provide a good fit.

with the resulting coefficients shown in Table III and the corresponding partial residual plots in Figures 5 and 6. In Table III, a crossproduct variable $x_i x_j$ is written as x_{ij}. Obvious violations of the linearity assumption are absent from these new scattergrams indicating that the improvements in fit achieved by adding polynomial terms or crossproduct variables are accompanied by elimination of the initial nonlinearity of the regression model. The partial residual plots provide valuable confirmation of basic assumptions which could not be readily tested before. It is noted that the initial nonlinearity also could not be detected by inspection of studentized residuals and Cook's (1979) D measures provided in SAS (1985a) regression procedure outputs.

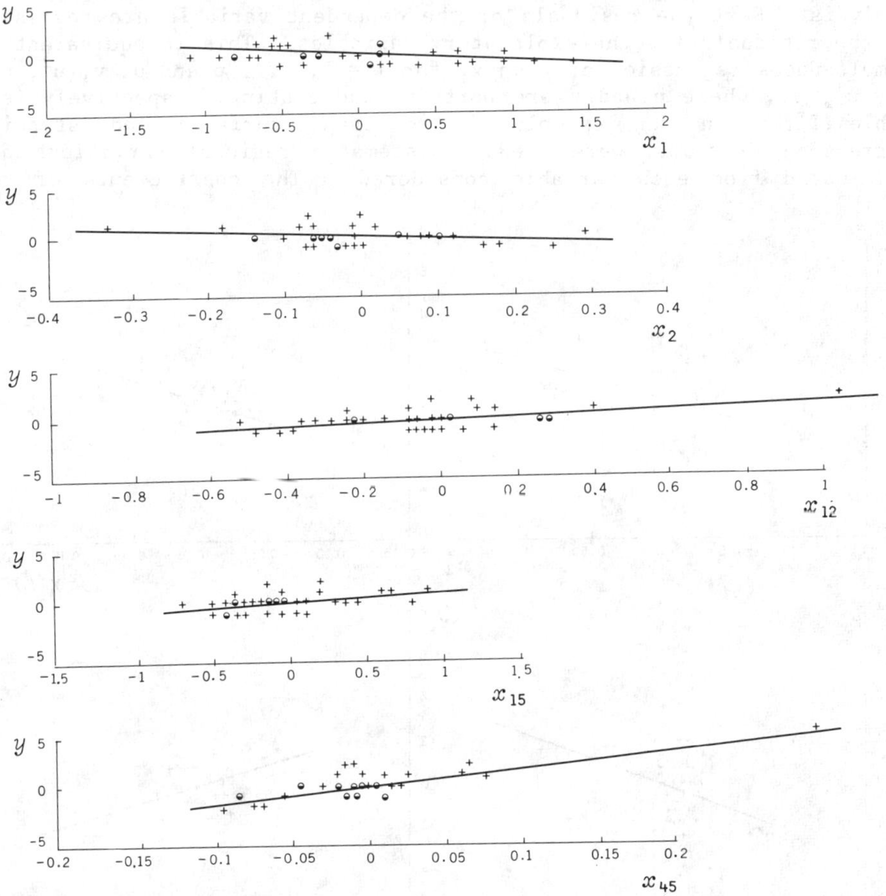

Figure 5. Same as Figure 4; for second experiment listed in Table III. Cross-product variables were added and stepwise regression was used. These plots are linear. The best-fitting straight lines pass through the origin and with coordinates (0,0), have slopes equal to the regression coefficients shown in Table III.

A brief explanation of the two-stage least squares model is as follows: The idea of adding a polynomial of the two geographic co-ordinates (northing and easting) to the explanatory variables in a region was developed because mineral deposits can form broad regional patterns which are more or less independent of the explanatory variables quantified for a region: (1) they may occur in "belts" (e.g., gold deposits in Abitibi area of Canadian Shield) related to deep-seated regional processes that cut across similar lithological environments; or (2) they may occur preferentially in regions with a relatively long exploration and mining history. Suppose that the dependent variable as well as all explanatory variables are first regressed on a two-dimensional polynomial of the geographic co-ordinates, viz., all variables are subjected to "trend-surface" analysis. Next the residuals of the dependent variable are regressed on the residuals for the explanatory variables. This is equivalent to simultaneous regression of y on x_i for i = 1, ..., p and u, v, u^2, uv, v^2, u^3, ... where u and v are northing and easting, respectively (see Table II). In the example, octic trend surfaces and stepwise regression ($\alpha = 0.3$) were used. Systematic regional variations are eliminated from each variable considered. The coefficients of the

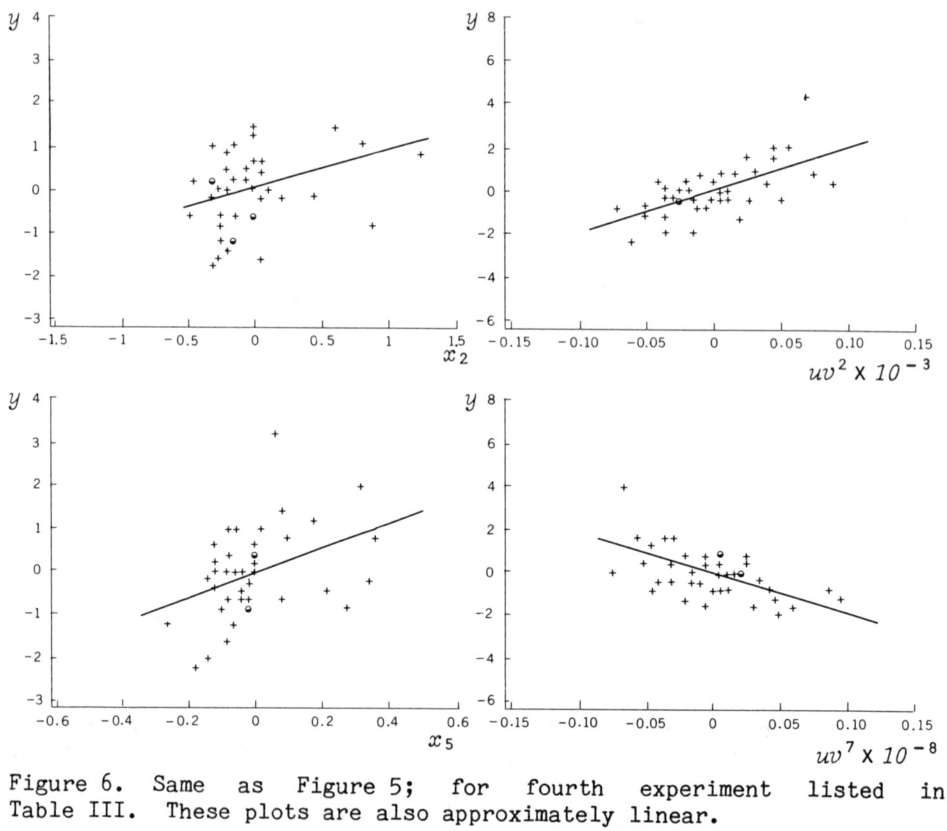

Figure 6. Same as Figure 5; for fourth experiment listed in Table III. These plots are also approximately linear.

geological variables in the combined expression are identical to those in the regression equation based on the use of residuals only. It is noted that the technique of partial residual plots can be regarded as a special case of this two-stage least-squares model.

Table VA shows calculated values for the combined expression for our example. This pattern was divided into two parts as follows: Table VB represents the polynomial part of the combined equation with constant term chosen such that the mean value for Table VB equals zero; Table VC shows the remainder. The values of Tables VB and VC add up to those of Table VA. The coefficients of the variables (x_2, x_5 and x_7) resulting in Table VC are very close to those in the first row of Table III showing results of the regression of Y on the 8 individual rock types. The pattern of Table VA resembles a smoothed version of the patterns of Tables IVC and IVD. The values of Table VB form a trend surface with zero mean. A relatively high peak occurs south of Noranda and minima are developed in the western and northeastern parts of the study region. These patterns suggest that in the vicinity of Noranda, there exist more deposits than can be predicted by means of the initial set of 8 lithological variables. This may reflect the fact that volcanic polymetallic massive sulphide deposits have been mined for a relatively long period of time in the Noranda mining district which may have led to relatively many discoveries. When the pattern of Table VC is compared to the patterns of Tables IVC, IVD and VA it can be seen that its values in the Noranda area are less but those in the Timmins area greater than those in the other patterns. These conclusions are in agreement with those drawn from earlier results derived for the Abitibi area (cf. Agterberg, 1986).

4. HAT MATRIX AND MODIFIED HAT MATRIX FOR LINEAR AND LOGISTIC MODELS APPLIED TO 40-KM CELLS

Formulae for the hat and modified hat matrix were given in the introduction (Equations 1 and 2). The latter can also be derived as follows. Suppose that the vector Y is appended to the matrix X. This gives an (nx (p+1)) matrix Z. Then

$$H^* = Z(Z'Z)^{-1}Z' \tag{3}$$

Both H and H* are symmetric and idempotent with

$$H^2 = H; \quad H^{*2} = H^* \tag{4}$$

It follows immediately that, for the diagonal elements,

$$0 \leq h_{ii}^2 \leq 1; \quad 0 \leq h_{ii}^{*2} \leq 1 \tag{5}$$

Table VI. Hat matrix (above main diagonal consisting of x-es) and modified hat matrix (lower triangle); diagonal elements of the two matrices are shown in separate columns at end of table; all values were multiplied by 100 and rounded off; values greater than 15.00 are printed bold.

#	1	2	3	4	5	6	7	8	9	10	11	12	13	14	15	16	17	18	19	20	21
1	x	0	0	0	2	-2	1	13	3	-5	-1	-3	6	-2	-1	7	7	-2	-1	2	2
2	2	x	5	2	3	**17**	13	0	-3	9	-5	11	0	-4	1	1	7	4	12	5	7
3	4	7	x	7	-1	7	7	3	-2	-0	4	0	-0	-7	9	3	1	-1	-2	8	8
4	-0	1	6	x	0	3	4	0	-3	-3	1	2	-0	-4	10	3	0	-4	-2	7	6
5	2	2	-2	0	x	-4	1	2	6	8	-3	8	6	10	-2	3	7	3	2	-2	-1
6	-2	**17**	8	3	-5	x	14	-1	-7	11	7	11	-6	-10	4	-2	0	5	**18**	9	10
7	2	14	10	4	1	15	x	1	-3	9	1	9	-0	-9	4	1	6	1	0	8	9
8	14	2	5	-0	1	-1	3	x	3	-3	0	-3	6	-3	-0	7	6	-2	-2	3	3
9	4	-2	1	-3	5	-7	-1	4	x	6	14	-6	8	**16**	-1	2	2	**19**	-1	-2	-3
10	-5	8	-3	-3	9	11	8	-6	5	x	**16**	**30**	-1	7	-10	-5	7	3	-7	-2	-1
11	-0	-4	5	1	-3	7	2	1	14	**15**	x	-3	-1	5	5	-2	-6	15	-10	4	3
12	-2	12	2	1	8	11	10	-2	-5	**29**	-2	x	-1	1	-6	-1	9	-11	1	-0	1
13	6	1	0	-0	6	-6	-0	6	8	-1	-1	-1	x	8	-1	5	5	6	3	-1	-1
14	-1	-3	-5	-4	9	-10	-8	-1	**17**	6	5	2	9	x	-5	1	1	**18**	14	-8	-8
15	1	4	15	9	-3	4	8	3	1	-12	6	-4	-0	-2	x	3	-3	1	-1	9	8
16	3	-4	-6	5	4	-3	-4	3	-2	-2	-5	-5	4	-3	-7	x	4	-1	7	3	3
17	6	6	0	0	7	0	5	6	1	8	-6	8	5	1	-4	6	x	-2	0	1	2
18	-4	1	-5	-3	3	5	-2	-5	**17**	4	14	-13	5	**16**	-4	8	-1	x	**17**	0	-1
19	0	14	0	-2	1	**19**	2	-0	1	-8	-9	2	3	**15**	2	-5	-0	14	x	-1	-1
20	-3	-1	-2	9	-0	8	2	-3	-1	2	2	-5	-2	-12	-3	**22**	3	11	-8	x	9
21	-0	4	3	8	-0	9	6	0	-5	1	1	-1	-1	-11	2	13	3	5	-4	**20**	x
22	-1	2	1	5	7	-5	-0	-2	2	-0	-8	5	4	10	4	3	4	1	6	0	0
23	2	3	-3	-1	-4	**16**	-4	2	-5	-9	-2	-4	-1	3	-3	4	-3	8	**31**	4	2
24	5	5	0	-0	6	-0	5	5	1	7	-6	6	5	-1	-5	7	9	0	-3	4	4
25	0	2	7	9	-1	4	4	1	-1	-8	2	-5	0	-4	12	4	-1	2	1	9	8
26	-5	-6	-1	9	2	-3	-4	-5	-1	-1	4	-3	0	2	4	12	-1	6	-5	**16**	9
27	-4	-2	-0	1	-1	-4	-0	-4	-3	3	-1	2	-3	-3	0	-2	-2	-5	-9	-0	1
28	15	2	5	-0	2	-1	3	15	4	-6	0	-2	6	-1	2	3	6	-5	0	-3	-0
29	-8	1	2	10	7	-7	-0	-8	0	-3	-9	4	3	10	9	4	1	3	4	4	3
30	0	-3	6	6	3	-6	-1	0	5	-2	5	4	3	10	12	-5	-2	-5	1	-7	-4
31	5	-2	-1	0	5	-6	-1	4	8	1	3	-3	6	6	-1	6	4	7	-2	3	1
32	15	2	5	-0	2	-1	3	15	4	-6	0	-2	6	-1	2	3	6	-5	0	-3	-0
33	5	5	10	5	0	4	8	6	0	-3	2	2	2	-4	11	-2	2	-6	-1	-1	3
34	5	-14	-4	4	-1	-0	-14	5	-6	5	12	9	-1	6	-3	7	-4	-13	2	3	-1
35	2	-0	5	-1	-2	4	4	3	**15**	5	**24**	-10	2	3	6	-2	-3	**18**	-7	2	3
36	10	0	-2	1	4	-1	1	10	1	-0	-3	-1	6	-4	-6	14	8	2	-3	11	7
37	0	3	9	8	1	0	5	1	1	9	-1	-1	2	2	15	-2	-1	-2	3	-0	2
38	2	3	-4	1	10	-5	0	5	1	9	-11	11	6	9	-7	8	9	1	4	3	1
39	1	-2	7	5	0	4	2	2	1	10	14	10	-1	2	8	-4	-2	-8	-6	-2	-0
40	8	2	3	-1	4	-7	2	8	1	-3	-4	1	6	5	2	-2	5	-4	-1	-9	-4

It can also be shown that

$$\sum_{i=1}^{n} h_{ii} = p; \quad \sum_{i=1}^{n} h^{*}_{ii} = p + 1 \tag{6}$$

For this reason, various authors have suggested that observations with $h_{ii} > 2p/n$ or $h_{ii} > 3p/n$ have high leverage. The complete hat and modified hat matrix for y related to x_1, x_2, x_{12}, x_{15}, x_{16}, x_{18} and x_{45} (cf. Tables II and III) are shown in Table VI. It is noted that there is relatively little difference between H and H*. The average value of h_{ii} is 8/40 = 0.250 and $\bar{h}^{*}_{ii}$ = 9/40 = 0.225.

In Table VI, any values greater than 0.150 are shown in bold print. The largest diagonal elements are for cells #27 (h_{ii} = 0.83; h^{*}_{ii} = 0.83); and #34 (h_{ii} = 0.60; h^{*}_{ii} = 0.61). The estimated values for these cells are 7.0 and 2.5, respectively. These values are close to the corresponding observed values which amount to 7 and 3, respectively. The small residual for cell #27 is largely due to the fact that this cell is "unique". Although off-diagonal elements can be negative, each row (or column) of H and H* is subject to the same type of constraint,

$$\sum_{j=1}^{n} h_{ij} = 1; \quad \sum_{j=1}^{n} h^{*}_{ij} = 1, \quad i = 1, \ldots, n \tag{7}$$

Table VI. (Continued)

22	23	24	25	26	27	28	29	30	31	32	33	34	35	36	37	38	39	40	h_{ii}	h^*_{ii}	#
-1	3	6	0	-2	-4	14	-7	-2	5	14	4	6	2	12	-1	3	-0	7	13	14	1
2	4	6	2	-3	-2	0	1	-5	-1	0	3	-13	-0	2	2	5	-3	-0	**16**	**17**	2
1	-1	1	8	5	-0	2	3	2	0	2	6	0	4	2	6	-2	4	-0	8	13	3
5	-2	-0	9	8	1	0	9	7	0	0	6	4	-0	0	9	1	6	-0	9	9	4
7	-4	6	-1	1	-1	2	7	4	5	2	1	-1	-2	3	1	10	0	5	8	8	5
-5	**16**	-0	4	-3	-4	-2	-7	-6	-6	-2	4	-0	4	-1	-0	-5	3	-8	**32**	**32**	6
-0	-3	6	4	-1	-0	1	0	-4	-0	1	6	-12	3	3	3	2	0	-0	14	**16**	7
-2	3	5	1	-2	-4	13	-8	-2	5	13	4	7	2	12	-1	2	0	6	13	15	8
2	-4	2	-0	2	-3	3	1	3	8	3	-1	-4	15	3	-1	2	-0	4	**18**	**20**	9
-0	-10	7	-9	-3	2	-5	-3	-1	0	-5	-2	5	5	-1	-7	8	11	-2	**45**	**46**	10
-8	-1	-5	2	6	-1	-0	-9	4	3	-0	1	12	**24**	-2	-2	-11	13	-5	**35**	**35**	11
5	-3	7	-5	-1	2	-3	4	2	-2	-3	1	10	-10	0	-2	12	9	-0	**30**	**31**	12
4	-1	5	0	1	-3	6	3	2	6	6	1	-1	2	6	1	6	-2	6	7	7	13
10	4	-0	-3	5	-3	-2	11	8	7	-2	-5	7	3	-2	0	10	0	3	**28**	**29**	14
4	-1	-3	12	10	0	-1	10	8	0	-1	7	-0	5	-2	11	-4	5	-2	**15**	**22**	15
3	0	4	3	2	-2	7	2	2	4	7	4	2	-0	7	3	4	0	5	6	**22**	16
3	-4	9	-1	-3	-2	7	1	-2	4	7	3	-4	-2	8	-0	9	-2	6	9	10	17
1	6	-1	2	1	-5	-2	2	-1	6	-2	-2	-15	**19**	-2	1	-1	-6	-1	**33**	**38**	18
6	**32**	-2	1	-1	-9	-1	5	-1	-2	-1	-3	4	-8	-1	2	5	-7	-3	**47**	**49**	19
0	-0	1	8	4	-1	2	2	1	0	2	7	-3	5	2	6	-3	3	-1	9	**32**	20
0	-0	2	7	3	0	3	2	0	0	3	6	-4	4	3	5	-2	3	3	9	15	21
x	-2	2	4	7	0	-2	**16**	8	3	-2	2	-0	-6	-1	7	10	1	4	11	11	22
-2	x	-2	0	-2	**21**	3	-7	-2	-1	3	-2	**18**	-1	1	-2	-3	-1	9	**41**	**42**	23
2	-1	x	-1	-3	11	6	-0	-3	5	6	2	-8	1	7	-1	8	-3	-1	10	11	24
4	0	-1	x	8	-0	0	9	6	1	0	6	-2	3	-0	10	-2	3	-1	10	10	25
7	0	-2	9	x	1	-2	13	11	2	-2	4	9	2	-3	9	1	8	-1	11	**17**	26
0	**21**	11	-0	1	x	-4	3	1	5	-4	-1	-3	9	-4	0	-1	1	**21**	**83**	**83**	27
-2	2	5	0	-5	-5	x	-8	-2	5	14	4	7	2	12	-1	2	-0	6	14	15	28
16	-6	0	9	14	3	-9	x	14	2	-8	3	-6	-7	-6	14	10	2	2	**28**	**28**	29
8	-3	-4	5	7	1	-0	13	x	2	-2	3	13	-1	-2	8	4	9	1	12	**16**	30
3	-0	5	1	3	5	5	2	1	x	5	2	-1	6	5	1	4	-0	7	6	6	31
-2	2	5	0	-5	-5	15	-9	-0	5	x	4	7	2	12	-1	2	-0	6	14	15	32
2	-3	1	6	0	-1	6	3	5	1	6	x	-0	2	4	6	1	3	2	6	8	33
-0	**19**	-7	-1	11	-3	5	-5	11	-1	5	-2	x	-8	3	-2	-1	**22**	-3	**60**	**61**	34
-6	-1	0	3	0	9	3	-7	-0	6	3	3	-9	x	1	-0	-8	2	2	**25**	**26**	35
-1	2	8	0	2	-4	10	-5	-5	6	10	1	5	0	x	-1	4	-1	6	11	14	36
7	-3	-2	9	6	0	0	13	10	1	0	7	-3	0	-4	x	2	3	0	11	12	37
10	-2	8	-2	3	-1	1	10	2	4	1	-1	0	-9	6	0	x	-1	6	14	**15**	38
1	-2	-4	3	5	1	1	1	11	-1	1	5	**21**	3	-3	4	-2	x	-2	15	**16**	39
3	2	8	-1	-5	**21**	8	1	3	6	8	4	-4	3	3	3	4	-1	x	13	**15**	40

Various clustering experiments (single linkage cluster analysis and Ward's (1963) minimum variance cluster analysis) have been undertaken on H (see Table VII; Figures 7 and 8). These experiments were restricted to cells i with at least one leverage value h_{ij} greater than 0.15. Off-diagonal elements were used only. Distance was defined as $m_{ij} = 1 - h_{ij}$ (see Table VII and Figure 7) or $m'_{ij} = 1 - |h_{ij}|$ (Figure 8). The dendrograms of Figures 7 and 8 are very similar in shape. The order of the cells was not changed when m'_{ij} was used to replace m_{ij} in single linkage cluster analysis. In this method, the distance between two clusters is the minimum distance between an observation in one cluster and an observation in the other cluster (Hartigan, 1981). This type of clustering was recommended by Gray and Ling (1984, p. 310) because it seeks subsets that exhibit nearly block-diagonal structures containing several large positive (or negative) values. Ward's minimum variance clustering analysis, which is the other method for which results are shown in Figures 7 and 8, joins clusters to maximize the likelihood at each level of the hierarchy under the assumptions of (1) multivariate normal mixture; (2) equal spherical covariance matrices; and (3) equal sampling probabilistics. Because of these rather strong assumptions, it is very sensitive to outliers (Milligan, 1980). However, in our applications Ward's method gave nearly the same results as single linkage cluster analysis indicating that the final results are more or less independent of type of clustering technique used.

Table VII. Cells with elements greater than 15.00 in the hat matrix (upper triangle and column for h_{ii} in Table VI) were used in clustering experiments; distances $m_{ij} = 1 - h_{ij}$ for off-diagonal elements only were multiplied by 1000 and rounded off; input to SAS clustering procedures.

	2	6	9	10	11	12	14	15	18	19	22	23	27	29	34	35	39	40
2	0																	
6	830	0																
9	1030	1069	0															
10	909	887	937	0														
11	1046	927	863	845	0													
12	892	894	1058	703	1027	0												
14	1041	1098	838	927	954	988	0											
15	992	963	1013	1098	952	1057	1046	0										
18	962	947	807	974	851	1107	819	988	0									
19	880	817	1009	1074	1103	992	863	1014	834	0								
22	981	1049	982	1002	1075	947	904	961	988	937	0							
23	961	840	1044	1101	1015	1034	962	1008	938	680	1020	0						
27	1022	1038	1025	975	1008	981	1031	996	1051	1094	999	791	0					
29	987	1074	993	1035	1092	958	893	897	983	952	836	1067	972	0				
34	1125	1002	1044	954	875	904	934	1004	1154	965	1002	818	1034	1056	0			
35	1008	960	851	947	760	1102	972	953	808	1075	1061	1006	907	1072	1081	0		
39	1030	968	1004	894	868	911	995	948	1058	1071	990	1006	990	981	779	978	0	
40	1003	1077	962	1019	1052	1003	968	1017	1008	1029	968	863	786	984	1026	981	1025	0

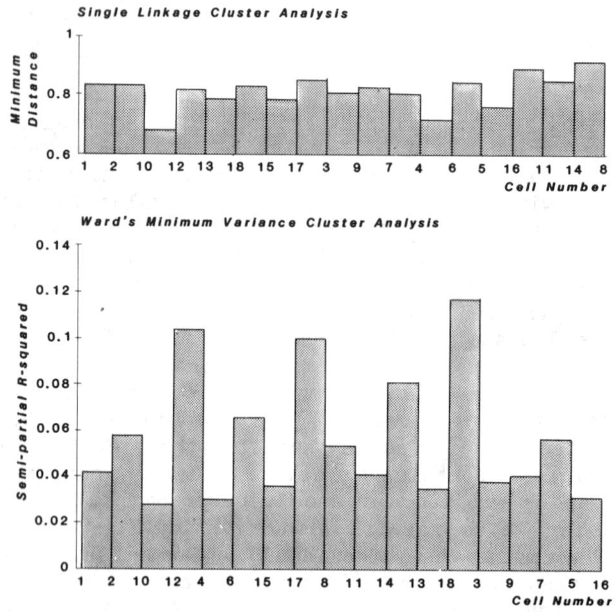

Figure 7. Dendrograms obtained by SAS Tree Procedure applied to results of single linkage cluster analysis, and Ward's minimum variance cluster analysis of reduced hat matrix shown in Table VII.

As pointed out by Pregibon (1981), logistic regression diagnostics can be readily constructed from standard outputs of computing packages using the usual maximum likelihood for the logistic model. Consider a sample of n independent binomial responses $y_i \sim B(N_i, p_i)$. As before, we will set $N_i = N = 16$ for i = 1, ..., n. The probability p_i is assumed to be constant within each 40-km cell labelled i. The vector of scores $S = Y-\hat{Y}$ is made to converge until the likelihood equations

$$X'S = X'(Y-\hat{Y}) = 0 \tag{8}$$

are satisfied. At convergence, we have coefficients β with

$$\beta = (X'VX)^{-1} X'VA \tag{9}$$

where V is a diagonal matrix with nonzero elements $V_{ii} = N\hat{p}_i(1-\hat{p}_i)$ and $A = X\hat{\beta} + V^{-1}S$. Let θ_i (I = 1,..., n) represent the logit of p_i with

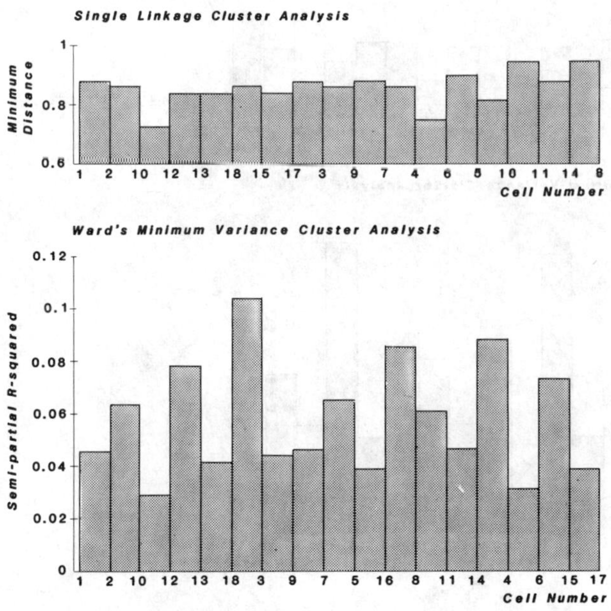

Figure 8. Same as Figure 7; after replacing values in Table VII equal to 1000 + A (A>0) by values equal to 1000-A.

Table VIII. Same as Table VII for hat matrix derived for logistic model using Eq. (13).

	2	6	7	10	11	12	14	15	18	19	23	25	27	34	35	39
2	0															
6	837	0														
7	915	831	0													
10	893	834	920	0												
11	1047	904	1011	900	0											
12	909	870	928	774	824	0										
14	993	1117	1036	906	980	948	0									
15	1011	985	979	1145	977	1068	1036	0								
18	986	1039	1024	964	853	1029	834	1000	0							
19	940	904	999	968	1112	957	869	1000	813	0						
23	1000	850	1045	1148	1034	1089	995	1004	913	709	0					
25	996	985	975	1103	1002	1044	1019	833	991	984	994	0				
27	1004	1047	990	952	1011	965	1010	998	1052	1097	1107	1004	0			
34	1020	1068	1078	945	856	932	965	1016	1162	1039	836	1022	1044	0		
35	1022	991	991	1028	659	1101	992	964	815	1065	985	972	959	1121	0	
39	989	940	975	956	879	912	986	945	1049	1039	1066	964	974	783	1017	0

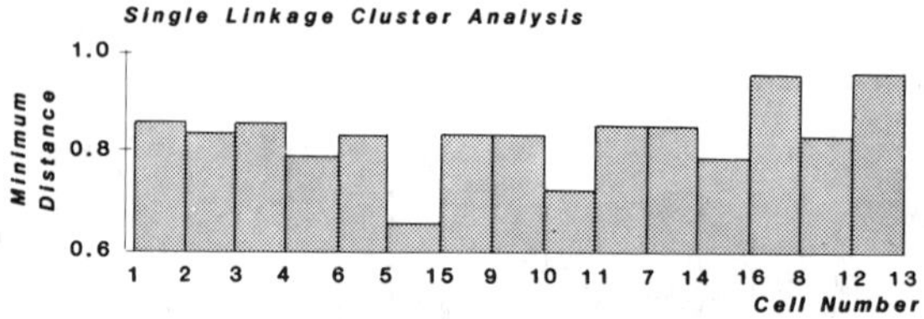

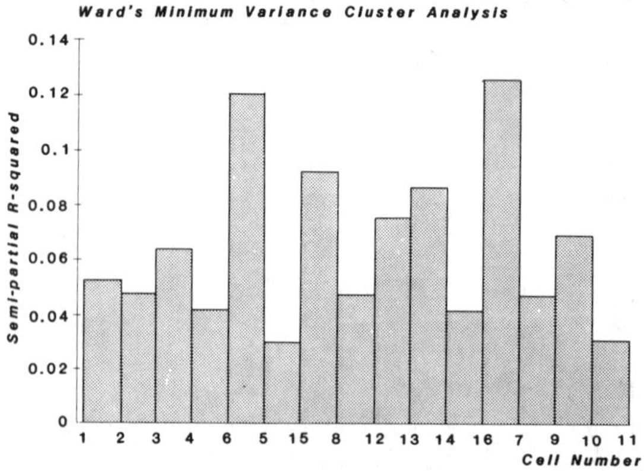

Figure 9. Same as Figure 7; using as input the reduced hat matrix for logistic regression shown in Table VIII.

$$\theta_i = \log_e \frac{p_i}{1-p_i} \tag{10}$$

Let Θ and P represent vectors of logits and probabilities, respectively. The estimated logits satisfy $\hat{\theta} = X\hat{\beta}$. The values $\hat{Y} = N\hat{P}$ can therefore be obtained from $\hat{\beta}$. Other useful statistics that can be obtained are $(X'VX)^{-1}$, which represents the variance-covariance matrix of $\hat{\beta}$, and the following two measures for goodness of fit of the logistic model:

1. Individual components of χ^2 with

$$\chi_i^2 = (y_i - N\hat{p}_i)^2/\{N\hat{p}_i(1-\hat{p}_i)\} \tag{11}$$

2. Individual components of deviance with

$$d_i = -2\{y_i \log_e \hat{p}_i + (N-y_i)\log_e(1-\hat{p}_i)\} \tag{12}$$

These individual components which are approximately distributed as $\chi^2(1)$ with a single degree of freedom and can be added to obtain χ^2 and the deviance

$$D = \sum_{i=1}^{n} d_i.$$

Theoretically these statistics are both distributed as χ^2 with (n-p) degrees of freedom.

The hat matrix of the logistic model satisfies

$$H = V^{\frac{1}{2}}X(X'VX)^{-1}X'V^{\frac{1}{2}} \tag{13}$$

Appending the vector A to X gives a new matrix Z from which the modified hat matrix can be derived with

$$H^* = V^{\frac{1}{2}}Z(Z'VZ)^{-1}Z'V^{\frac{1}{2}} \tag{14}$$

The matrices H and H* for the logistic model are also symmetric and idempotent and have the same other properties as H and H* for the linear model (see before). Both matrices were constructed for the example of this section using Equations 13 and 14 respectively. The new results (not shown here) do not differ much from the results previously shown in Table VI. Table VIII shows $m_{ij} = 1-h_{ij}$ for cells with one or more elements greater than 0.150. Clustering of this submatrix yielded the dendrograms of Figure 9.

In order to obtain a two-dimensional representation of the dendrograms derived from the hat matrix, the following procedure was followed for the sequences of cells shown in the single linkage cluster analysis of Figures 7 and 9, respectively. The cells with leverages greater than 0.150 were renumbered in the dendrograms according to their column numbers in Tables VI and VIII. They are first shown separately in Tables IXA and XA. Next the corresponding sequence numbers representing their order in the dendrograms are shown in Tables IXB and XB. This illustrates that neighboring cells in the dendrogram are to some extent neighbors in the plane (e.g., cells 1 and 2 in both tables). However, there is also a distinct tendency of neighboring cells in the dendrogram to be separated from one another. For example, cells 1 and 18, which occur in the first and last column of Table IXA, respectively, are only 5 spaces apart in the dendrogram where their sequence positions are 1 and 6, respectively.

Table IX. A - Cells with leverages greater than 0.1500 in the hat matrix were numbered from 1 to 18; B - Cells of Table IX-A were renumbered from 1 to 18 using their sequence positions in the dendrogram (Figure &).

A (u) \ (v)	1	2	3	4	5	6	7	8	9	10
1			3					14		
2	1	2	4	7	9	11			15	
3			5	8	10	12	13		16	17
4			6							18

B	1	2	3	4	5	6	7	8	9	10
1			9					17		
2	1	2	12	11	10	16			7	
3			14	18	3	4	5		15	8
4			13							6

Table X. Same as Table IX; using logistic hat matrix.

A (u) \ (v)	1	2	3	4	5	6	7	8	9	10
1							12			
2	1	2	4	7	9				14	
3		3	5	8	10	11	13		15	16
4			6							

B	1	2	3	4	5	6	7	8	9	10
1							15			
2	1	2	4	11	8				12	
3		3	6	14	9	10	16		7	13
4			5							

Finally, Figure 10 is a diagnostic plot of logistic regression results which is similar to a plot originally proposed by Pregibon (1981, Fig. 5, p. 715). The relative contributions to χ^2 of the individual cells (i) are plotted against h_{ii}. The lines labelled $2\bar{h}^*_{ii}$ and $3\bar{h}^*_{ii}$ are based on the average diagonal value (= 0.225, see before) of the modified hat matrix. These lines reflect the straight-line relationship

$$h^*_{ii} = h_{ii} + \chi_i^2/\chi^2 \tag{15}$$

Three cells have $h^*_{ii} > 3\bar{h}^*_{ii}$. In the linear model only cell #27 with $h^*_{ii} = 0.83$ satisfied this unequality (see Table VI). The 95% and 99% fractiles of χ^2 with a single degree of freedom are shown as horizontal lines in Figure 10. These values incorporate the total $\chi^2 = 30.05$. For 32 degrees of freedom this value indicates a good fit of the model (Probability = 0.43 < 0.95 for $\alpha = 0.05$). However, $\chi^2 = 30.05$ exceeds the deviance D = 16.04 (see Table III).

The pattern of Figure 10 identifies residuals which are too large in absolute value for cells #16 and #20. It may be concluded that the statistical testing of the magnitudes of individual residuals (χ_i^2/χ^2) and diagonal leverage values h_{ii} (and h^*_{ii}) are complementary to one another because the cells with large values of h_{ii} and h^*_{ii} have small residuals. A separate study of the hat matrix, therefore, adds valuable information in that "unique" cells such as cell #27 in the present example at Noranda in western Quebec are clearly identified. It is of interest to determine why unique cells are different from other cells in a region. From inspection of the original data for the

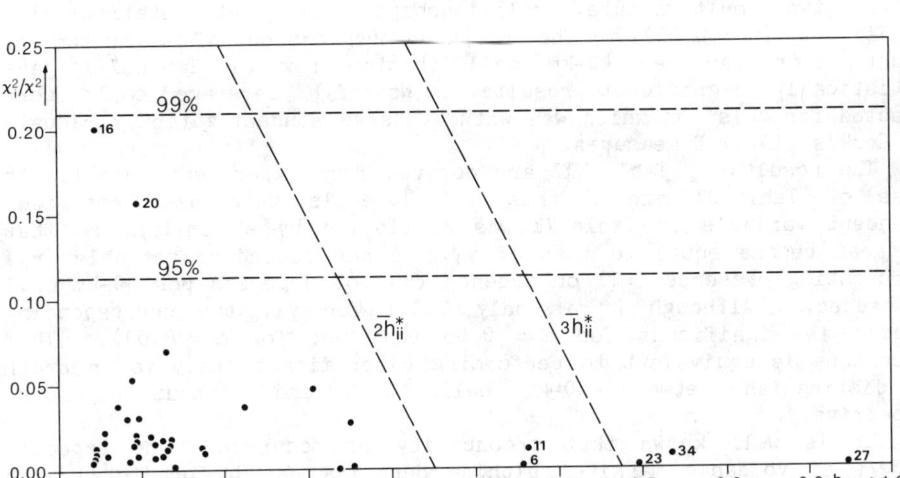

Figure 10. Scatter plot of χ_i^2/χ^2 versus h_{ii} for logistic regression model. The dashed lines with slope -1 correspond to 2x ave (h^*_{ii}) and 3x ave (h^*_{ii}).

explanatory variables in Table II, and the coefficients in Table III, it follows that the values of variables 4 and 5 are relatively large in cells #23 and #27. The values of the coefficient for the product variable x_{45} also are relatively large in these cells thus creating a situation that is "unique" in the region.

5. LINEAR MODEL APPLIED TO COPPER CONTENT AND CONCENTRATION OF VOLCANIC POLYMETALLIC MASSIVE SULPHIDE DEPOSITS IN 40-km CELLS

The statistics of Table II include values for S = $\log_{10}$ (metric tons of copper) and G = $\log_{10}$ (copper concentration in units of 1000 ppm). Because concentration (or "grade") represents amount of copper divided by amount of ore, it follows that $\log_{10}$ (metric tons of copper ore) = S-G + 3. Table XI shows results of regressing the variables y_1 (Y in Table II), Cu (S), Grade (G), Ore (= S-G) and y_2 (D in Table II) on the eight variables retained by stepwise regression ($\alpha = 0.3$) of y_1 on the 8 rock types and their crossproducts. Separate stepwise regression results ($\alpha = 0.30$) for Cu (S) on these 37 explanatory variables are shown at the bottom of Table XI. All results of Table XI are for the 20 cells in Table II that contain at least one deposit. The results for y_1 using 20 cells are comparable to those using 40 cells. In both experiments, the regression coefficient of x_{45} is large. The leverage of the Noranda cells (#27) amounts to 0.972 when only 20 cells are used. The regression coefficients for "ore" are equal to the coefficients of S after subtracting the coefficients of G. The R^2-values of Table XI suggest that grade is less related to rock type than amounts of copper and ore. However, when the R^2-values are converted into F-ratios, it turns out that the latter two multivariate relationships are not statistically significant ($\alpha = 0.05$). The stepwise regression of logarithm of amount of copper per 40-km cell (bottom row of Table XI) gave statistically significant results. No high leverage cells were detected for this run which was without large studentization residuals and Cook's (1979) D measures.

The results of Table XII are derived from experiments similar to those of Table XI except that all 40 cells were used and each dependent variable in Table XI was multiplied by a constant so that its mean became equal to that of y_1. A new dependent variable (y_3) representing presence (=1) or absence (=0) of deposits per 40-km cell was added. Although R^2 is only 0.341 for y_3, the corresponding F-ratio is significant for $\alpha = 0.05$ (but not for $\alpha = 0.01$). This experiment is equivalent to performing discriminant analysis in order to distinguish between 40-km cells with and without deposits, respectively.

It is well known that probability of occurrence and expected numbers of volcanic massive sulphide deposits can be predicted with considerable success. A more difficult problem consists of predicting the sizes and amounts of the metals contained in these deposits. In most earlier quantitative resource appraisal studies it was assumed

Table XI. Regression experiments using several dependent variables on explanatory variables (rock types and their crossproducts) for twenty 40-km cells with massive sulphide deposits; Y_1 - same values as Y in Table II; Cu - log of amount of copper (= S); Grade - $\log_{10}$ of copper concentration value (= G); Ore - log of amount of copper ore (Ore = S-G); Y_2 - number of deposits per cell (= D). Stepwise regression of Y_1 (α = 0.30) was used to select 8 variables including variable for Int. (= Intercept); these same variables were used in the next 4 experiments. Cu* - Stepwise regression of Cu (α = 0.30) was used and this resulted in 7 new variables with larger R^2-value.

	R^2	Int.	x_3	x_{15}	x_{17}	x_{24}	x_{25}	x_{45}	x_{68}
Y_1	0.912	2.481 (0.326)	-.721 (0.299)	-.616 (0.374)	-2.98 (0.585)	-6.77 (2.686)	1.411 (1.236)	34.48 (8.558)	-1.02 (0.595)
Cu (S)	0.480	3.581 (0.892)	0.602 (0.817)	-1.53 (1.023)	-1.76 (1.601)	0.187 (7.352)	5.982 (3.384)	3.368 (23.43)	1.323 (1.628)
Grade (G)	0.286	2.132 (0.319)	-.089 (0.293)	-.506 (0.367)	-.892 (0.573)	-1.69 (2.633)	1.242 (1.212)	6.141 (8.389)	-.218 (0.583)
Ore (S-G)	0.508	1.449 (0.684)	0.690 (0.627)	-1.02 (0.786)	-.871 (1.229)	1.879 (5.643)	4.740 (2.598)	-2.77 (17.98)	1.540 (1.250)
Y_2 (D)	0.964	4.108 (0.549)	-1.66 (0.503)	-1.65 (0.630)	-6.96 (0.986)	-.754 (4.525)	3.943 (2.083)	42.04 (14.42)	-1.32 (1.002)

	R^2	Int.	x_1	x_2	x_4	x_{56}	x_{57}	x_{78}
Cu* (S)	0.795	4.231 (0.774)	-.683 (0.266)	2.174 (0.518)	0.959 (0.413)	15.03 (10.24)	-14.9 (5.584)	-4.71 (2.144)

Table XII. Same as Table XI; using all 40 cells of Table II. Each dependent variable was multiplied by a constant so that its mean value became equal to ave (Y_1). The eight variables selected by stepwise regression of Y_1 were used in all experiments listed; Y_3 - presence or absence of deposits per 40-km cell. Note that the signs of the coefficients are the same; the coefficient of x_{45} is mainly determined by relatively high values of dependent variable in cells #23 and #27.

	R^2	Int.	x_1	x_2	x_{12}	x_{15}	x_{16}	x_{18}	x_{45}
Y_1	0.797	0.443 (0.276)	-.623 (0.197)	-2.51 (1.018)	1.379 (0.419)	0.874 (0.330)	1.314 (0.611)	0.251 (0.071)	16.83 (2.011)
Cu (x 0.5198)	0.534	0.661 (0.340)	-.656 (0.292)	-.869 (1.253)	0.829 (0.515)	0.584 (0.406)	2.308 (0.752)	0.98 (0.087)	7.232 (2.475)
Grade (x 1.0829)	0.408	0.698 (0.351)	-.587 (0.250)	-1.10 (1.295)	0.769 (0.532)	0.742 (0.420)	1.807 (0.777)	0.207 (0.090)	5.671 (2.557)
Ore (x 0.9995)	0.580	0.626 (0.375)	-.720 (0.267)	-.659 (1.382)	0.884 (0.568)	0.437 (0.448)	2.771 (0.830)	0.189 (0.096)	8.672 (2.729)
Y_2 (x 0.5942)	0.846	0.152 (0.317)	-.534 (0.226)	-1.54 (1.168)	1.021 (0.480)	0.569 (0.378)	1.164 (0.701)	0.258 (0.081)	25.42 (2.306)
Y_3 (x 2.0500)	0.341	0.749 (0.346)	-.504 (0.246)	-1.27 (1.274)	0.753 (0.524)	0.748 (0.413)	1.465 (0.765)	0.185 (0.088)	4.694 (2.516)

that these statistics are realizations of random variables which are independent of probability of occurrence. If this assumption is correct for S, G and "ore", the regression coefficients of these dependent variables (see Table XII) should be approximately equal to one another. This assumption will be tested later in this section.

The size statistics display a positively skew frequency distribution with a few giants containing most of the ore, e.g. in the Abitibi area studied at present, using 1978 statistics more than half of the copper (3.652 out of 6.655 million tons in total) occurred in the Kidd Creek deposit near Timmins (cell #10). For comparison, the variation of the copper concentration value (grade) is restricted to a much narrower range.

The logarithmic size and grade statistics of Table II are based on the 69 known deposits used. For the majority (51) of the deposits in this group relatively good size-grade statistics could be based on past production and estimated ore reserves. For the remaining 18 deposits, the amount of copper was either less than 1000 tons or it was not recorded (e.g., for massive sulphide deposits for which zinc and lead were recorded and not copper). In some earlier studies these small deposits were not considered in regression analysis. However, some potentially valuable information is then ignored and they were used in the present study. The statistics of these small deposits have little effect on the cell statistics S and G of Table II if they occur together with one or more larger deposits in a cell. Four cells in Table II contain a single small deposit without recorded statistics for copper; S and G were both set equal to 1.00 in these cells. Thus the small deposits without recorded statistics for copper were assumed to contain 10 tons of copper occurring within 1000 tons of "ore" with a copper grade of 0.1 per cent. On the logarithmic scale, these statistics only slightly exceed the value S = G = 0.00 assigned to the 20 cells without deposits.

In earlier work (cf. Agterberg, 1986), logarithmic measures of size and grade for 10-km cells were used instead of S and G. Let these measures be called S_o and G_o, respectively. Suppose that (1) the amount of copper (S_o) per 10-km cell is independent of probability of occurrence (p_i) for that cell; and (2) the expected number of cells with copper ore per 40-km cell is m = 16 x p ($= \sum_{i=1}^{16} p_i$) with p replacing p_i in each 10-km cell within a 40-km cell.

Implications of the second assumption (2) were discussed earlier in this paper. Under these conditions, the amount of copper expected per 40-km cell ($\hat{S}$) would be 16 times as large as $\hat{S}_o$ per 10-km cell. Because S and S_o are logarithmic measures, their estimated values would differ by $\log_{10}$ 16 = 1.204, and the relation (a) $\hat{S} = \hat{S}_o + 1.204$ would be valid. Suppose instead that the amount of copper (S) per 40-km cell is independent of expected number (m) of cells with copper. This second assumption can be translated into the relation (b) $\hat{S} = \hat{S}_o$. If neither relation (a) for 10-km cells nor (b) for 40-km cells is satisfied, we could still have (c) $\hat{S} = \hat{S}_o + c$ where c is a constant for cells of different size. In terms of the experiments shown in Table XII, the solution for y_1 would be proportional to $\hat{S}_0$. The difference in intercepts between the solutions for y_1 and Cu (S) is $\hat{c} = 0.218$ which is of the order of the standard errors of the intercepts which are compared to one another. It may be concluded that relation (b) could be true, and that relation (a) is probably not true. Similar arguments can be developed for "ore" and "grade". Regardless of the precise value of c, it is of interest to compare the regression solutions of Table XII to one another. The SAS regression procedure allows the performance of multivariate tests. It can test hypotheses involving several dependent variables in the form

$$(L\,\beta - C\,J)\,M = 0 \tag{16}$$

where L is a linear function on the side of the explanatory variables, β is a matrix of parameters, C is a column vector of constants, J is a row vector of ones and M is a linear function on the dependent variable side. These multivariate tests are covered in detail in Kshirsagar (1972), Morrison (1976) and Mardia, Kent and Bibby (1979). To test the hypothesis, two matrices H and E are constructed that correspond to the numerator and denominator of a univariate F-test (SAS, 1985a, p. 681):

$$\begin{aligned} H &= M'\,(LB - CJ)'\,(L\,X'X\,L')^{-1}\,(LB - CJ)\,M \\ E &= M'\,(Y'Y - B'\,X'X\,B)\,M \end{aligned} \tag{17}$$

where B is the matrix of estimated parameters. Four test statistics based on the eigenvalues of $E^{-1}H$ or $(E + H)^{-1}H$ are formed. These are Wilks' lambda, Pillai's trace, the Hotelling-Lawley trace and Roy's maximum root, respectively. Table XIII shows results of nine tests

Table XIII. Pairwise comparison of dependent variables to test the hypothesis that all regression coefficients except intercept are equal. F-ratio (for 7 and 32 degrees of freedom) was converted into probability Prob. (F).

	Wilk's Lambda	Pillai's Trace	Roy's Max. Root	Exact F (df: 7,32)	Prob(F)
Y_1 vs Cu	0.396	0.604	1.527	6.978	0.000
Y_1 vs Grade	0.284	0.716	2.517	11.51	0.000
Y_1 vs Ore	0.519	0.481	0.925	4.228	0.002
Y_1 vs Y_2	0.424	0.576	1.361	6.220	0.000
Y_1 vs Y_3	0.363	0.637	1.755	8.022	0.000
Y_3 vs Cu	0.637	0.363	0.570	2.605	0.030
Y_3 vs Grade	0.877	0.123	0.140	0.640	0.719
Y_3 vs Ore	0.600	0.400	0.667	3.051	0.014
Y_3 vs Y_2	0.294	0.706	2.400	10.97	0.000

performed on the example of this paper. The hypothesis for each test was that all regression coefficients except the intercept are equal for the two regressions of Table XII being compared to one another.

Under these conditions, the Hotelling-Lawley trace and Roy's maximum root are identical, and there is an exact F-ratio (with 7 and 32 degrees of freedom in Table XIII). It turns out that the sets of coefficients for Cu (S), "grade" (G), "ore" and number of deposits are all significantly different from the coefficients of y_1. All other dependent variables were also compared with the coefficients of y_3 (bottom row of Table XII). This is equivalent to estimating, by means of the linear model, the probability that a 40-km cell contains one or more deposits. The F-ratios are considerably less for the comparisons involving y_3. The hypothesis, that the sets of coefficients are equal, is accepted for y_3 versus "grade" (Prob = $0.719 > \alpha = 0.05$). Table XIV shows predicted values of S, "ore", and G, respectively, using the sets of coefficients of Table XII. According to the multivariate tests of Table XIII, these patterns are significantly different from the one for y_1 shown in Table IVC. The main difference is that, in the vicinity of Noranda (cell #27), the values of the patterns of Table XIV are less than those in Table IVC, whereas the reverse holds true in the vicinity of Timmins (cell #11). Thus, for equal probability of occurrence of deposits in a 10-km cell, there would be more and higher-grade copper ore in the latter vicinity. From the results for y_3 shown in Tables XII and XIII, it can be

Table XIV. Predicted values for: A - amount of copper (S); B - copper grade (G); and C - copper ore tonnage. These 3 variables were multiplied by constants as in Table XII so that their means are equal to ave (Y_1).

A (u) \ (v)	1	2	3	4	5	6	7	8	9	10
1	0.6	0.3	1.2	0.3	0.3	**1.0**	**0.7**	-.5	0.7	0.3
2	0.7	**2.1**	**2.3**	0.7	**1.4**	-.1	**0.8**	0.7	**2.3**	-.2
3	1.0	1.0	**3.2**	0.8	**0.6**	**2.5**	**3.7**	**0.9**	**2.5**	**1.8**
4	**0.7**	0.7	1.1	**0.4**	**1.0**	**0.7**	0.6	0.7	**0.6**	**1.0**

B	1	2	3	4	5	6	7	8	9	10
1	0.6	0.3	1.3	0.3	0.2	**0.9**	0.4	-.8	0.6	0.0
2	0.6	**2.1**	**2.9**	0.7	**1.4**	-.3	**0.7**	0.6	**2.6**	-.2
3	0.9	0.9	**3.6**	0.6	0.2	**2.5**	**4.4**	**0.9**	**2.8**	**2.0**
4	0.5	0.6	1.4	**0.2**	**0.9**	0.8	0.6	0.6	**0.5**	**1.1**

C	1	2	3	4	5	6	7	8	9	10
1	0.7	0.3	1.0	0.4	0.3	**1.1**	**0.9**	-.1	0.8	0.6
2	0.9	**2.1**	**1.6**	0.6	**1.5**	0.1	**0.9**	0.8	**2.0**	-.1
3	1.1	1.0	**2.7**	1.1	**1.0**	**2.5**	**3.0**	**0.8**	**2.2**	**1.6**
4	**0.9**	0.7	0.8	**0.5**	**1.2**	**0.6**	0.7	0.7	**0.6**	**0.8**

derived that, for equal probability of occurrence of deposits in a 40-km cell, there would be more copper ore with the same average grade in this part of the region. It is suggested that the nonlinearity of the relationship between the size-grade statistics and probability of occurrence decreases with increasing size of the cells used for coding.

6. CONCLUDING REMARKS

In the approach for this paper, extensive use was made of various regression diagnostics for the linear and logistic model in regional mineral resource evaluation. Use was made of dependent and explanatory variables coded for relatively large cells (40 km x 40 km in size). The use of large cells facilitates the construction and analysis of the (nxn) hat matrix where n is number of cells. In earlier work, smaller cells (10 km x 10 km in size) were generally used to provide better approximation to geological discontinuities such as contacts between rock types on compilation maps.

Three useful new results were obtained in this paper:

1. Partial regression residual plots help to check the linearity assumption;

2. Diagonal elements of the hat matrix identify high-leverage cells, and clustering of the off-diagonal elements of the hat matrix provides a sequence of cells within a region which are related to one another according to degree of similarity; and

3. Multivariate tests applied to dependent variables pairing frequency of occurrence of one or more deposits with amounts of metal or ore and average concentration of metal in the cells, indicate that the random variables defined for size and grade are not necessarily independent of probability of occurrence.

7. REFERENCES

Agterberg, F.P., 1974, Automatic contouring of geological maps to detect target areas for mineral exploration; Jour. Math. Geol., v. 6, pp. 373-395.

Agterberg, F.P., 1986, Canadian experience in application of multivariate analysis techniques; Proceedings, Joint USGS/GSC Workshop on 'Mineral Resource Appraisal' held in Leesburg, Virginia, September 1985, U.S. Geological Survey Circular 980, pp. 173-187.

Agterberg, F.P. and Franklin, J.M., 1986, Estimation of the probability of occurrence of polymetallic massive sulfide deposits on the ocean floor; In: Teleki, P. (ed.), Marine minerals, resource assessment strategies; Reidel, Dordrecht, Holland.

Chung, C.F., 1978, Computer program for the logistic model to estimate the probability of occurrence of discrete events; Geol. Survey of Canada, Paper 78-11, 23 p.

Chung, C.F. and Agterberg, F.P., 1980, Regression models for estimating mineral resources from geological map data; Jour. Math. Geol., v. 12, pp. 473-488.

Cook, R.D., 1979, Influential observations in linear regression; Jour. American Stat. Ass., v. **74**, pp. 169-174.

Fabbri, A.G., 1975, Design and structure of geological data banks for regional mineral potential evaluation; Bulletin Canadian Inst. Mining and Metall., v. **67**, no. 8, pp. 91-98.

Gray, J.B. and Ling, R.F., 1984, K-clustering as a detection tool for influential subsets in regression; Technometrics, v. **26**, pp. 305-318.

Harris, D.P., 1984, Mineral resources appraisal; Oxford University Press, New York, 445 p.

Hartigan, J.A., 1981, Consistency of single linkage for high-density clusters; Jour. American Stat. Ass., v. **76**, pp. 388-394.

Hoaglin, D.C. and Welsch, R.E., 1978, The hat matrix in regression and ANOVA; American Statistician, v. **32**, no. 1, pp. 17-22.

Hocking, R.R., 1983, Developments in linear regression methodology: 1959-1982; Technometrics, v. **25**, pp. 219-229.

Kshirsagar, A.M., 1972, Multivariate analysis; Dekker, New York, 534 p.

Landwehr, J.M., Pregibon, D. and Shoemaker, A.G., 1984, Graphical methods for assessing logistic regression models; Jour. American Stat. Ass., v. **79**, pp. 61-71.

Mardia, K.V., Kent, J.T. and Bibby, J.M., 1979, Multivariate analysis; Academic Press, London.

Milligan, G.W., 1980, An examination of the effect of six types of error perturbation on fifteen clustering algorithms; Psychometrika, v. **45**, pp. 325-342.

Morrison, D.F., 1976, Multivariate statistical methods, 2nd edition; McGraw-Hill, New York, 415 p.

Pregibon, D., 1981, Logistic regression diagnostics; Annals of Statistics, v. **9**, pp. 705-724.

SAS, 1985a, Statistical Analysis System User's Guide: Statistics, Version 5 Edition; SAS Institute, Cary, N.C., 956 p.

SAS, 1985b, The matrix procedure: Language and applications; Techn. Report P-135, SAS Institute, Cary, N.C., 150 p.

Tukey, J.W., 1984, Comments on "Use of spatial analysis in mineral resource evaluation"; Jour. Math. Geol., v. **16**, pp. 591-594.

Velleman, P.F. and Welsch, R.E., 1981, Efficient computing of regression diagnostics; American Statistician, v. **35**, no. 4, pp. 234-242.

Ward, J.H., 1963, Hierarchical grouping to optimize an objective function; Biometrics, v. **37**, pp. 35-43.

Wrigley, N., 1983, Quantitative methods: On data and diagnostics; Progress in Human Geography, v. **7**, pp. 565-575.

Wrigley, N., 1984, Quantitative methods: Diagnostics revisited; Progress in Human Geography, v. **8**, pp. 525-535.

Wrigley, N. and Dunn, R., 1986. Graphical diagnostics for logistic oil exploration models; Jour. Math. Geol., v. **18**, pp. 355-374.

POISSON REGRESSION ANALYSIS AND ITS APPLICATION.*

C.F. Chung and F.P. Agterberg
Geological Survey of Canada
601 Booth Street
Ottawa, Ontario K1A 0E8
CANADA.

ABSTRACT. In order to detect favorable areas for mineral exploration, a Poisson regression model is applied to the number of the occurrences of mineral deposits and spatially overlapping geological attributes. We obtain the maximum likelihood estimator for this model which is identical to the estimator of the generalized linear model minimizing the deviance.

1. Introduction.

This paper deals with an application of Poisson regression model for detecting favorable areas for mineral exploration under the assumption that the expected number of the occurrences of mineral deposits in an area can be approximated by a linear function of the selected geological attributes in the area.

In an earlier paper (Chung and Agterberg, 1980), we discussed several regression models including Jorgenson's Poisson regression for estimating the number of occurrences of volcanic massive sulphide deposits in 10-km cells from five selected volcanic rock types for the island of Newfoundland. These five rock types are non-overlapping map patterns and we could apply the exact Jorgenson's Poisson model where the number of occurrences in each 10-km cell is assumed to be the sum of five independent Poisson variables and the expected value of each of the five Poisson variables is equal to a constant (occurrence rate) times the subarea covered by the corresponding rock type in the 10-km cell.

However, when we consider several geological attributes which overlap spatially each other, and the occurrences of a given type of mineralization, Jorgenson's model is no longer applicable. In this exposition, we generalize Jorgenson's model into a general class of linear Poisson model where Jorgenson's assumption - the sum of independent Poisson variables - is not required. Also we obtain the maximum likelihood estimators of the parameters in the model as in Jorgenson's model. This Poisson model can be considered as a special

* Geological Survey of Canada Contribution Number 16787 .

C. F. Chung et al. (eds.), Quantitative Analysis of Mineral and Energy Resources, 29–36.

case of the class of the generalized linear models described by McCullagh and Nelder (1983).

We first review Jorgenson's Poisson model and its assumptions. We then discuss the general linear Poisson model and its application to the data for volcanic massive sulphide deposits in Abitibi area of the Canadian Shield (Table 2 of Agterberg elsewhere in this volume). Finally the coefficients and the expected numbers estimated by Poisson regression will be compared to results obtained by the least squares estimators and logistic regression.

2. Jorgenson's Poisson model.

Let $\{ n_{i,j} : i=1, \ldots ,n \text{ and } j=1, \ldots ,m \}$ be independent Poisson random variables (rv) with the expected values:

$$(2.1) \qquad E(n_{i,j}) = \beta_j X_{i,j} ,$$

where β_j is the occurrence rate and $X_{i,j}$ is a explanatory variable for $n_{i,j}$ for $i=1, \ldots ,n$ and $j=1, \ldots ,m$. For example, Jorgenson (1961) assumed that $n_{i,j}$ is the number of failures for a given time period $X_{i,j}$ at the failure rate β_j.

When $\{ n_{i,j} \}$ and $\{ X_{i,j} \}$ are given and $\{ \beta_j \}$ are unknown parameters to be estimated, then the problem of estimating $\{ \beta_j \}$ can be immediately solved by obtaining the maximum likelihood estimators $\{ \hat{\beta}_j \}$ for $\{ \beta_j \}$ by:

$$(2.2) \qquad \hat{\beta}_j = \sum_{i=1}^{n} n_{i,j} \Big/ \sum_{i=1}^{n} X_{i,j} , \text{ for each } j = 1, \ldots , m ,$$

whereas the least squares estimators $\{ \tilde{\beta}_j \}$ of $\{ \beta_j \}$ are obtained by:

$$(2.3) \qquad \tilde{\beta}_j = \sum_{i=1}^{n} n_{i,j} X_{i,j} \Big/ \sum_{i=1}^{n} X_{i,j}^2 , \text{ for each } j = 1, \ldots , m .$$

The maximum likelihood estimators in (2.2) are unbiased and have the minimum variance property. The least squares estimators in (2.3) are also unbiased but have relatively large variances.

Now let, for $i = 1, \ldots , n$,

$$(2.4) \qquad n_i = n_{i,1} + n_{i,2} + \ldots + n_{i,m} .$$

Then n_i is again a Poisson rv with $E(n_i) = \sum_{j=1}^{m} \beta_j X_{i,j}$.

However, when $\{ n_i \}$ and $\{ X_{i,j} \}$ are provided instead of $\{ n_{i,j} \}$ and $\{ X_{i,j} \}$, the problem of the estimation of the $\{ \beta_j \}$ is no longer as simple as before. Jorgenson (1961) proposed the maximum likelihood estimators $\{ \hat{\beta}_j \}$ for $\{ \beta_j \}$ which are obtained as follows:

For a given set of $\{ n_i \}$ and $\{ X_{i,j} \}$, the log-likelihood function L is given by:

$$(2.5) \qquad L = \sum_{i=1}^{n} n_i \ln \left[\sum_{j=1}^{m} \beta_j X_{i,j} \right] - \sum_{i=1}^{n} \ln(n_i!) - \sum_{j=1}^{m} \beta_j \sum_{i=1}^{n} X_{i,j} .$$

Hence the maximum likelihood estimators $\{ \hat{\beta}_j \}$ are obtained by solving the following system of m equations:

$$(2.6) \qquad - \sum_{i=1}^{n} X_{i,j} + \sum_{i=1}^{n} \left[\frac{n_i X_{i,j}}{\sum_{j=1}^{m} \hat{\beta}_j X_{i,j}} \right] = 0 , \; j = 1, 2, \ldots , m .$$

This system of equations can be solved by using the scoring method (cf. Jorgenson, 1961 and Chung, 1978)

However, Jorgenson's model is limited in the sense that, for each i, $\{ n_{i,j} : j = 1, \ldots , m \}$ must be independent, and the conditions (2.1) and (2.4) must be satisfied. This limitation leads to the development in the next section where Jorgenson's model is generalized to a linear class Poisson model which includes Jorgenson's model.

3. Generalized Poisson model.

Let $N_1, \ldots , N_n$ be n independent Poisson rv with, for each i,

$$(3.1) \qquad E(N_i) = X_i'\alpha,$$

where $\alpha' = (\alpha_0, \alpha_1, \ldots , \alpha_m)$ is the (m+1) dimensional vector of unknown parameters to be estimated and $X_i' = (1, X_{i,1}, \ldots , X_{i,m})$ are observed values of m explanatory variables for the i-th variable N_i with $i = 1, \ldots , n$. When $\alpha_0 = 0$, the model in (3.1) includes that in (2.4). We note here that we do not make any assumption on the sum of independent Poisson rv as was done in (2.4). Hence the linear model in (3.1) can be more widely applied than that in (2.4). We also remark that the model in (3.1) is a special case of the class of generalized linear models (cf. McCullagh and Nelder, 1983), since Poisson belongs to the distributions of the exponential family.

For a given set of $\{ N_i \}$ and $\{ X_{i,j} \}$, let us consider methods for estimating $\{ \alpha_j \}$ in (3.1). A natural procedure for estimating $\{ \alpha_j \}$ in (3.1) is to adapt the least squares estimators $\{ \tilde{\alpha}_j \}$ which are based on (3.1), assuming that the variances of all $\{ N_i \}$ are equal to a constant (i.e.$\mathrm{Var}(N_i) = \sigma^2$). Obviously, the least squares estimators are not appropriate under the Poisson assumption about the distribution of N_i. We now obtain maximum likelihood estimators $\{ \hat{\alpha}_j \}$

of { α_j } in (3.1) by following procedures identical to those of Jorgenson (1961).

Given { N_i } and { $X_{i,j}$ }, the log-likelihood function is provided by:

$$(3.2) \qquad L = \sum_{i=1}^{n} N_i \ln \left[\sum_{j=0}^{m} \alpha_j X_{i,j} \right] - \sum_{i=1}^{n} \ln(N_i!) - \sum_{j=0}^{m} \alpha_j \sum_{i=1}^{n} X_{i,j} .$$

Hence the maximum likelihood estimators { $\hat{\alpha}_j$ } are obtained by solving the following system of (m+1) equations:

$$(3.3) \qquad - \sum_{i=1}^{n} X_{i,j} + \sum_{i=1}^{n} \left[\frac{N_i X_{i,j}}{\sum_{j=0}^{m} \hat{\alpha}_j X_{i,j}} \right] = 0 , \; j = 0, 1, \ldots , m .$$

As in (2.6), this system of equations can be solved by the scoring method.

In the setting of the generalized linear models (cf. McCullagh and Nelder, 1983), the negative of the logarithm of a ratio of likelihoods, to be called the deviance $\delta(\alpha)$ is first defined as:

$$(3.4) \qquad \delta(\alpha) = - \sum_{i=1}^{n} \Big(N_i - \sum_{j=1}^{m} \alpha_j X_{i,j} \Big) + \sum_{i=1}^{n} N_i \ln \left[\frac{N_i}{\sum_{j=0}^{m} \alpha_j X_{i,j}} \right] .$$

Then, instead of maximizing L in (3.2), { α_j } is obtained such that $\delta(\alpha)$ in (3.4) is minimized. However maximizing L and minimizing δ are equivalent. $2\delta(\alpha)$ is labelled G^2 by Bishop et al.(1975). $2\delta(\alpha)$ has X^2-distribution with (n-m-1) degrees of freedom and it can be used for testing the models.

Since the { $\hat{\alpha}_j$ } are maximum likelihood estimators of { α_j }, under the usual regularity conditions, { $\hat{\alpha}_j$ } have asymptotic normal distribution and their asymptotic covariance matrix is estimated by the inverse of the information matrix. Hence an asymptotic confidence interval for α_j can be obtained by taking the estimate of the variance from the inverse matrix. In addition we also obtain a confidence interval of $E(N_i)$ by using the fact that the Poisson assumption implies that $Var(N_i) = E(N_i)$.

Another way measuring the goodness of fit of the Poisson linear model in (3.1) is to look at the estimate $\delta(\hat{\alpha})$ of the deviance in (3.4). The other important measure of goodness of fit is the Pearson X^2 statistics, which takes the form:

$$(3.5) \qquad X^2(\alpha) = \sum_{i=1}^{n} \left[\left(N_i - \sum_{j=0}^{m} \alpha_j X_{i,j} \right)^2 \Big/ \sum_{j=0}^{m} \alpha_j X_{i,j} \right],$$

which has the identical distribution function of twice the deviance $2\delta(\alpha)$. That is, $X^2(\alpha)$ also has X^2-distribution with (n-m-1) degrees of freedom

4. Applications.

The generalized Poisson model and its maximum likelihood estimators discussed in the previous section, are applied to the data for volcanic massive sulphide deposits in Abitibi area of the Canadian Shield (Table 2 of Agterberg elsewhere in this volume). We remark here that, since several cross product terms are present in this data set, the original Jorgenson's model is no longer applicable to this data.

The results obtained by the Poisson model (Tables 1 to 3), can be compared to those by the least squares method and the logistic model in Agterberg (Tables 3, 4 and 12, elsewhere in this volume). In all models, the coefficient of the variable x_{45} is the largest and it is significantly greater than zero because its range is relatively small. The predicted values of Table 2 are close to the those obtained by the least squares mèthod and the logistic model.

In the earlier application of Poisson regression for predicting the occurrences to volcanic massive sulphide deposits in Newfoundland, Chung and Agterberg (1980, p.487) have pointed out that the Poisson model failed to provide an accurate prediction of a cluster of 15 separate mineral deposits within two 10-km cells near Buchans. This cluster was located in a large target area of 10-km cells occurring outside the smaller study area consisting of sixty 10-km cells for which the Poisson regression coefficients were estimated. On other hand, the logistic model was able to predict the presence of deposits in the vicinity of Buchans, mainly because it is a qualitative response model in which presence of relatively many mineral deposits in a 10-km cell is weighted in the same manner as presence of a single deposit in that cell. It can be useful to lump separate orebodies that belong to local clusters, and to count them as single units, especially if they are genetically related to one another.

The data set (Agterberg elsewhere in this volume) used in this application, is for a study area of forty 40-km cells. It has not been attempted to use the estimated coefficients for prediction outside this study area. The problem of local clustering was considered by defining two different dependent variables. In addition to using the number of separate deposits per 40-km cell, the presence of one or more deposits in a 10-km cell was counted as a single event per 40-km cell. The number of this newly defined events is less than that of the separate mineral deposits. In this data set for the Abitibi area, the largest local cluster of massive sulphide deposits occurs near Noranda. In this vicinity as many as 19 deposits are situated within one 40-km cell, and 8 of these 19 occur within a single 10-km cell.

Table 1. Poisson regression coefficients and their standard deviations(in brackets) for prediction of volcanic massive sulphide deposits from seven lithological variables in the Abitibi area of the Canadian Shield. The first dependent variable (y_1) represents the number of 10km cells with one or more deposits per 40km cell; the second dependent variable (y_2) is for the number of separate deposits. Both sets of Poisson regression coefficients are preceded by least squares coefficients used as the starting point of the iterative process followed in the scoring method. It also contains the estimates of the deviances and Pearson X^2 statistics for both dependent variables.

(a) Least squares method on y_1.

constant	x_1	x_2	x_{12}	x_{15}	x_{16}	x_{18}	x_{45}
0.4442	-0.6230	-2.5109	1.3792	0.8734	1.3145	0.2513	16.8350

(b) Generalized Poisson model on y_1.

constant	x_1	x_2	x_{12}	x_{15}	x_{16}	x_{18}	x_{45}
0.5260	-0.3899	-1.5200	0.9466	0.7003	0.5119	0.1385	15.2971
(0.236)	(0.117)	(1.184)	(0.526)	(0.456)	(0.581)	(0.041)	(5.109)

$2\delta(\hat{\alpha}) = 30.54$ and $X^2(\hat{\alpha}) = 28.21$.

(c) Least squares method on y_2.

constant	x_1	x_2	x_{12}	x_{15}	x_{16}	x_{18}	x_{45}
0.2582	-0.8988	-2.5887	1.7186	0.9559	1.9600	0.4329	42.7704

(d) Generalized Poisson model on y_2.

constant	x_1	x_2	x_{12}	x_{15}	x_{16}	x_{18}	x_{45}
0.7760	-0.5738	-2.3282	1.4559	0.8976	0.7088	0.2167	34.8317
(0.282)	(0.138)	(1.432)	(0.637)	(0.538)	(0.694)	(0.048)	(7.362)

$2\delta(\hat{\alpha}) = 53.92$ and $X^2(\hat{\alpha}) = 57.87$.

Table 1 also contains the estimated deviances in (3.4) and the estimated Pearson X^2 from the models. These estimates are compared with the X^2-distribution function with 32 degrees of freedom. For example $X^2_{0.95}$ with 32 d.f.≈ 44. For y_1, we have no reason to reject the models using either the deviance or Pearson X^2. However, Poisson model for y_2 is rejected by both statistics at 95% level of significance.

Table 2. Predicted values obtained by generalized Poisson model with observed values(in bracket) for y_1 for the number of 10km cells with one or more deposits per 40km cell. First row and column show cell numbers for west-east and north-south directions, respectively. For geographic location of the 40 cells in this table, see Figure 4 and Table 4 in paper by Agterberg elsewhere in this volume.

	1	2	3	4	5	6	7	8	9	10
1	0.5 (0)	0.1 (0)	0.3 (0)	0.2 (0)	0.1 (0)	0.9 (2)	0.7 (1)	0.0 (0)	0.6 (0)	0.5 (0)
2	0.7 (0)	1.9 (2)	0.8 (1)	0.5 (0)	0.8 (2)	0.2 (0)	0.8 (2)	0.7 (0)	2.2 (3)	0.0 (0)
3	0.8 (0)	0.6 (0)	1.6 (2)	0.8 (0)	1.7 (1)	4.5 (5)	6.8 (7)	0.8 (1)	1.7 (2)	1.3 (1)
4	0.7 (1)	0.5 (0)	0.7 (0)	0.4 (2)	0.8 (3)	1.0 (1)	0.5 (0)	0.5 (0)	0.4 (1)	1.9 (1)

Table 3. Predicted values obtained by generalized Poisson model with observed values(in bracket) for y_2 for the number of separate deposits per 40km cell.

	1	2	3	4	5	6	7	8	9	10
1	0.7 (0)	0.2 (0)	0.5 (0)	0.3 (0)	0.2 (0)	1.3 (2)	1.1 (1)	0.0 (0)	0.9 (0)	0.7 (0)
2	1.0 (0)	2.7 (4)	1.2 (1)	0.7 (0)	1.1 (3)	0.3 (0)	1.1 (2)	1.0 (0)	3.4 (4)	0.0 (0)
3	1.1 (0)	0.8 (0)	2.3 (2)	1.2 (0)	2.3 (1)	8.2 (7)	14.8 (19)	1.6 (2)	3.2 (4)	1.9 (4)
4	1.0 (1)	0.8 (0)	1.1 (0)	0.6 (4)	1.2 (5)	2.2 (1)	0.8 (0)	0.8 (0)	0.6 (1)	4.1 (1)

The predicted frequencies as obtained by Poisson model are close to the observed values for 40-km cells in the vicinity of Noranda as well as elsewhere in the study area. The good results, particularly for y_1, obtained in this paper indicate that the generalized Poisson model is one more useful regression tool for the estimation of frequencies of mineral deposits in a region. Two statistics for the deviance and Pearson X^2 are useful tools for testing the adquacy of Poisson model.

5. Concluding remarks.

The scoring method to solve (m+1) equations in (3.3) is programmed in FORTRAN 77 on IBM PC and the source code is published by Chung(1987) in Computer and Geosciences (Special Issue for this NATO ASI).

6. References.

Agterberg, F.P., 1986. 'Application of recent developments of regression analysis in regional mineral resource evaluation'. Elsewhere in this volume.

Bishop, Y.M.M., Fienberg, S.E. and Holland, P.W., 1975, Discrete Multivariate Analysis, MIT Press, Cambridge, MA.

Chung, C.F., 1978. 'Computer program for the logistic model to estimate the probability of occurrence of discrete events'. Geological Survey of Canada Paper **78-11**, 23p.

Chung, C.F., 1987. 'FORTRAN 77 Program for Poisson Regression'. Computer and Geosciences, Special Issue for NATO ASI on "Statistical Treatment for Estimation of Mineral and Energy Resources".

Chung, C.F. and Agterberg, F.P., 1980. 'Regression models for estimating mineral resources from geological map data'. Mathematical Geology, **12**, p.473-488.

Jorgenson, D.W., 1961. 'Multiple regression analysis of a Poisson process'. Journal of American Statistical Association **56**, p.253-255.

McCullagh, P. and Nelder, J.A., 1983, Generalized Linear Models, Chapman and Hall, London, 261p.

MEASUREMENT, SAMPLING AND INTERPRETATION

John C. Griffiths
Department of Geosciences
310 Deike Building
The Pennsylvania State University
University Park
Pennsylvania 16802

ABSTRACT. Petrogenesis of detrital sediments is a somewhat elusive exercise. Some of the difficulties which arise are concerned with the nature and kind of measurements performed; whereas others are inherent in the genetic processes which generate detrital sedimentary rocks. It is concluded that to elucidate the genesis of a typical detrial sediment it is necessary to obtain a sample set of events (including the tectonic background) which gave rise to the rock.

"In computability theory, the analysis of algorithms, led to the disclosure of incomputable functions - and confirmed the fact that there is no escape from Goedel's undecidability, even for an idealized computing machine."

Bernhard, 1969

1. Introduction

It will be assumed that every investigation has an objective; furthermore, it is also assumed that the investigator not only recognizes this objective but actually defines it in explicit terms at the outset. Now it will be readily recognized that these assumptions are rarely fulfilled in practice; indeed, in most investigations the objective is specified, if at all, in the most vague and usually ambivalent terms, e.g. the investigator desires the "truth" but is usually unwilling to pay the price such an achievement demands. There is, I believe, a technical term for this!

Another feature of investigation which is rarely contemplated is that the objective is just not attainable. It is only comparatively recently that "incomputable functions" have been recognized as existing (see quote). Perhaps the most striking feature along these lines is the Goedel Incompleteness Theorem (Beer, 1964); here a consistent system is shown to be incomplete (Hofstadter, 1979). If a problem is really Goedelian, the question asked is unanswerable within the system in which it is formulated. To solve such a problem it is necessary to enlarge

C. F. Chung et al. (eds.), Quantitative Analysis of Mineral and Energy Resources, 37–56.

the system, create a metalanguage, and find the solution, if any, within the enlarged system (Beer, op. cit.). It is, therefore, important to realize that the succeeding account emphasizes some of the difficulties in problem-solving within a specific domain.

Some attempt is made to formulate general solutions to problems and to illustrate them by taking examples from the analysis of sedimentary rocks, or, by extension, to materials science in the natural world or, in general, to any aggregate composed of (well-defined) constituents. To this end we may commence by defining a population of interest in terms of its elements; this definition is an exercise in defining sets and it is essential to possess criteria for either assigning an element to the set (or population) or for rejecting it. Needless to say this is another requirement which is rarely fulfilled in practice; does, for example, a rock represent a single homogeneous population? I have never seen this question asked let alone answered!

The title of the article is chosen to illustrate the fact that it is necessary to examine the entire problem and that solutions to parts of problems, such as for example, to sampling, are usually inadequate unless actually tied into the entire problem-solving procedure.

2. Properties of populations and their measurement

Since the properties of the elements will be used to characterize the populations to which they belong it is convenient to formulate this as a conceptual relationship as follows:

$$P = f(m,s,sh,o,p) \qquad [1]$$

where P may be looked upon as a unique index representing the population and P is a function of the kinds and proportions of the different elements (m_i, or m_1, m_2,..m_k), their sizes (s_i), shapes (sh_i), and arrangement; arrangement is sub-divided into two surrogate properties of the elements, their individual position in space (o_i), and their position relative to their neighboring elements (p_i). The index i is meant to represent elements of different kinds. I have elsewhere suggested that these five properties are fundamental in defining P as an unique index (Griffiths, 1967, chapter 3). Any other properties are dependent on variation in these five; for example, behavioral properties, such as density, porosity, permeability, electrical resistance, etc. may be substituted for P and the equation then represents these behavioral properties as functions of the five fundamental properties.

The description of any population, then, commences with a definition of the kinds (m) of elements and, usually, a statement of their proportions. This may be followed by an expression of the sizes and shapes of one specific kind of element and these three properties suffice to describe each of the elements in the population. It is, then, usually necessary to locate the elements both individually (o = orientation) and with respect to each other (p = packing). It will be readily recognized that such a conceptual analysis may be applied to any kind of population, be they a collection of people, animals, quartz grains, atoms, subatomic particles or an astronomical population of galactic nebulae.

This conceptual relationship may be reduced to operational form by express-ing it as a multiple regression equation in which the symbols may be replaced by actual measurements on a set of samples from the population and the regression equations may then be solved by standard procedure (Griffiths, 1958). The general nature of such a solution includes transformation of the independent (fundamental) properties (X_i), say by logarithms, so that the linear regressions are extended to include non-linear relationships in the original variables; or the dependent variable ($Y_i = P_i$) may be expressed on a continuous scale in the form of multiple regression equations or as a presence-absence criterion leading to simple discriminant functions; or as an ordered metric when the relationship represents a multiple discriminant. A very large number of problems in a wide range of subject-matter fields is thus succinctly encompassed in this conceptual relationship.

Another advantage of this conceptual function is best seen when the procedure is reduced to a specific measurement operation; for example, suppose it is decided to measure the sizes of elements by some <u>indirect</u> procedure such as sieving, pipette sedimentation, etc. It is usually necessary to "disaggregate" the specimen, or to reduce the systematic arrangement of the elements to a state in which their arrangement does not interfere with the measurement procedure. Disaggregation does not remove arrangement but randomizes the elements in the population whereas in their natural state they usually possess a systematic structure, e.g. a layering of sizes. The equation may be manipulated to express this by reducing it to $P=f(m,s,sh)$ to illustrate the removal of the <u>effects</u> of arrangement; it is not generally recognized that the resulting measure of "size" is some function of the three remaining sources of variation plus their four interactions. This is the reason why this so-called size measure is so difficult to interpret in terms of the genetic processes it is supposed to represent (Pettijohn, 1957, p. 38ff).

Another popular procedure for measuring size of the elements of a population avoids the disaggregation procedure and permits the choice of a single kind of element; in petrography this is represented by analysis of thin sections and, suppose we select the ubiquitous quartz grain, then the equation becomes: $P=f(s,sh,o,p)$ and again this "size" measure is a function of four interacting sources of variation; which means that the variation measured as the variance of P_i is some function of four main factors, six first-order interactions, four second order interactions and one third order interaction (see Griffiths, 1967, op. cit., p. 39). To ignore the possible contributions of these interactions, as is customary in most of the literature, is courting disaster when it comes to interpreting variation in this second measure of "size". Once more variation in the measure can rarely be interpreted in terms of the genetic processes which purportedly produced it. Another rather misleading aspect of this operational procedure of measurement in thin section is that no matter what the measurement is called it will always be a function of these same four interacting variables (i.e. sources of variation) and will therefore tend to contain an unknown degree of redundancy. For example, in point-counting a thin section to determine the proportion of quartz, the outcome is a variable called composition, say m_i= proportion of quartz and variation in m_i = f(m,s,sh,o,p); when the size of quartz is measured on the same thin section as say its longest axis, variation in this "size" measure = f(s,sh,o,p) so that both measures contain contributions to variation from the same four sources, i.e. they contain an unknown degree of redundancy. In most cases when the grain size of quartz increases so does its area. Are these really two different measures or two measures of the same property? An analogous procedure is frequently used to analyze maps or pictures of various kinds (Griffiths, 1974, figure 3, p. 89); in this case, when a grid is applied to a map and the procedure of point-counting is used to estimate the proportions (by area) of the map colored differently, or otherwise represented by different symbols, the procedure is similar to estimating the proportion of minerals under the microscope; the defining equation becomes P = f(m,s,sh,o,p) and the variation in proportion is some function of five main factors and their associated 26 interactions. Manipulation of these data and interpretation of the outcome of analysis is hardly a straight-forward procedure. Standard image analysis is strictly analogous and some of the difficulties associated with attempts to interpret this kind of data are described in subsequent paragraphs.

It became clear that to measure size, as far as possible free from interacting effects, the procedure should be some form of direct measurement; in this case, selecting a single mineral, again for convenience, quartz, and extracting the grains from the rock to remove the effect of arrangement, we arrive at a measure of variation in P = f(s,sh). The actual procedure was devised by Hulbe (1955) and it was found that no further simplification appears feasible; that is, one cannot measure size without shape or vice versa. In fact, size in different directions is one measure of shape. An attempt to compare these three measures on the same rock-specimens showed that while measures of central tendency (means and medians) from all three procedures are very similar, variances differed materially among procedures (Griffiths and Ondrick, 1969).

3. Measurement and Information level of the data

An important aspect of measured data is the level of information it contains; ideally, if direct measurement is used the information level may be considered as a ratio scale (Churchman and Ratoosh, 1959, Stevens, 1959). Conventional parametric statistical analysis is appropriate for this scale of information. It is customary in sedimentary petrography to convert the direct size measure to its logarithms because this serves to normalize the variation. Conceptually, grain size (radius = r) is related to current velocity (V) and the relationships are of the form $V = Cr^2$ so that in logarithms the relationship is linear:

$$\log V = \log C + 2 \log r$$

Again, on the basis of theory, variation in current velocity of wind, water and ice, etc. can be heuristically modelled; currents of high velocity are rare, of low velocity common and the resulting frequency function is positively skew. variation in current velocity is proportional to current velocity so that logarithms tend to normalize the frequency distributions of current velocity. Hence there appears to be a one:one relationship between grain size and current velocity expressed in logarithms. The measure and its logarithm still achieve ratio scale level; this presumes that the measure obtained is a direct measure of size and only direct measurement appears to achieve this requirement.

Similar analytical procedures are used on size measures made in thin section or by indirect measurement, particularly sieving. Here, however, there is no guarantee that the measures, particularly indirect measures, achieve a ratio level. In fact indirect measures, both in concept and operation, probably achieve ordinal level; this is tacitly understood for sieve data by the common use of medians, quartiles and percentiles as summary statistics, i.e. the use of ordinal statistics. The actual level of information of thin section measurement is not, however, obvious and empirical procedure of using log "size" as a suitable approximation to a normal model is difficult to substantiate in concept. The difficulty is best exemplified by observing that the two measures of thin section size, apparent long and short axes of quartz grains, are so highly correlated as to be almost completely redundant (Griffiths, 1967); the ratio of apparent short to long axis (b/a) in microns tends to be a constant (Griffiths, 1967, table 6.6, p. 123, 1978a, table 3, p. 728). This is largely attributable to the fact that both apparent axes measure the same sources of variation, i.e. P = f(s,sh,o,p) for both axes. It should perhaps be emphasized that the same features will also be characteristic of all measures made in thin section and, by analogy, on a map or picture using these procedures; it seems likely that the measure tends to ordinal level. This has important implications when data collected in this manner is analyzed (statistically) and attempts are made to interpret these analytical results.

4. Sampling populations and analysis procedures

It will be convenient to consider the problem of sampling and data analysis of the elements of a population separately for the size, shape measures and the estimates of composition because the information or scale level of the data used to estimate these properties of elements is very different. Also, following practice, sampling for the ratio-interval measures will be treated first.

4a. Sampling variates with ratio or interval scaling

The procedure follows the algorithm displayed in figure 1. The population, element and variate are explicitly defined. The model frequency distribution of the variate in the population is heuristically established; this suffices to set up parameters which the samples are used to estimate. As previously explained, the population is clearly defined, and, for purposes of these experiments, consists of either, a scree on a hillside (stage 1, figure 2, and Griffiths, 1959) or a gravel pit (stage 3, figure 2) at Montoursville, Pennsylvania (Griffiths and Ondrick, 1968); in each case the population is explicitly defined and completely accessible(1). The elements, in both cases, are similar white quartzite boulders (in scree) and white quartzite pebbles, in the gravel, i.e. "m" is a constant. The variates are size and shape or size measured across the elements in three orthogonal dimensions (Griffiths and Ondrick, 1968, p. 10) which, when combined, yields a measure of shape (Griffiths and Smith, 1964). The selection of these two populations is derived from the genetic model of sedimentary processes (see figure 2) but is quite general and may be used for many other populations in different subject matter fields as for example, any process, natural or artificial, which commences with a heterogeneous input (stage 0, fig. 2) and is differentiated into segregated products (stage 3, fig. 2) which, combined, equals the starting state. These processes are examples of increases in negative entropy or structure and a typical example is the entry of crude oil into a refinery which subsequently produces a number of differentiated (cracked) products.

The population, element and variates are therefore, given (figure 1) and the only choice the investigator may exercise is in the selection of what constitutes a sample; this is not as simple as it appears and the alternatives may be illustrated by use of the defining equation [1]. Thus, all samples are similar so "m" becomes a constant and may be omitted from further consideration; we may however, vary size, shape and arrangement of samples or $P = f(s,sh,o,p)$. Following the algorithm "random" samples are taken and measurements performed yielding an observed frequency distribution and its appropriate statistical estimators; these may be tested against the model parameters and if they are consistent, unbiased estimates they will, following the model, also be efficient and sufficient in the sense defined by Fisher (1948). There is, then, no evidence against the hypothesis that these are, indeed, random samples from the postulated distribution. Otherwise, if the hypothesis is rejected, and the estimators are not unbiased, it is assumed that the sampling is not random or equivalently that the sampling arrangement interacts with the element arrangement (= structure) and so the size, shape and arrangement of the samples may be adjusted and the experiment repeated. When the hypothesis tests do not reject the estimators the experiment has been successful in estimating the model parameters and, usually, has led to a definition of the structure - where structure is defined as the pattern of variation of the variate in the population.

The two outcomes illustrate that the initial problem really requires two separate sets of information (from one set of data!) and that until the structure is defined it is extremely difficult, to impossible, to select a sampling design which will yield unbiased

(1) This is not usually a common feature in geological investigations, but, in most cases, it can be achieved if really desired.

Figure 1. Algorithm for evaluation of structure by sampling.

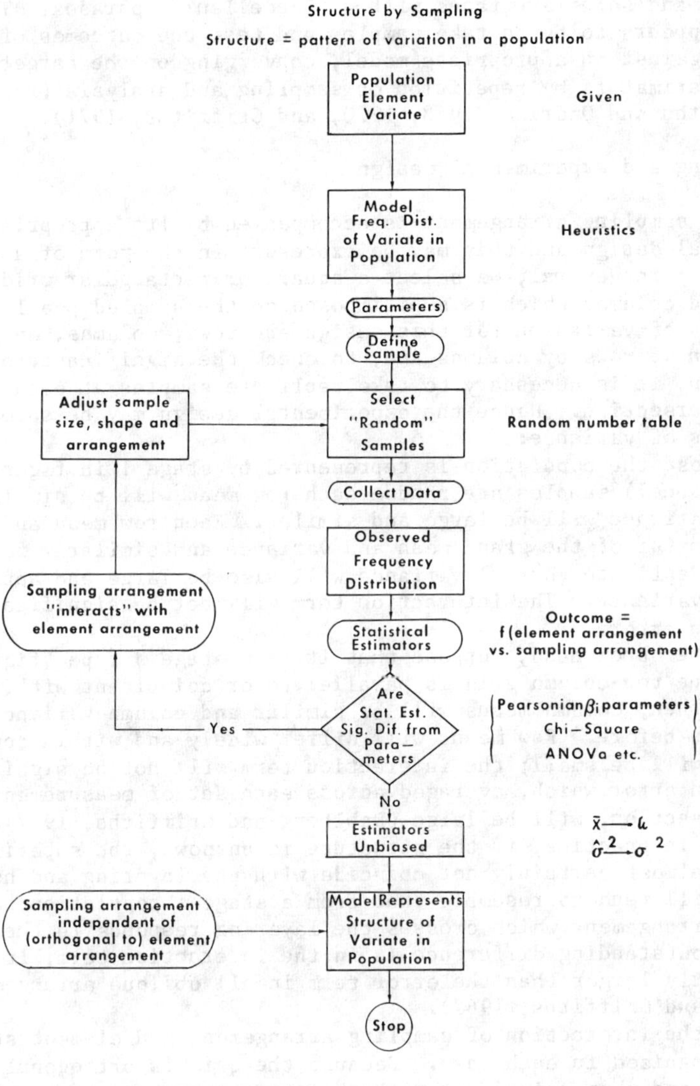

Reproduced, by permission of the Pennsylvania State University Press, from Griffiths and Ondrick (197).

estimates of population parameters. This is also the reason why it is very unusual to be able to resolve the "sampling problem" on the basis of any isolated single experiment. The algorithm of figure 1 deals with this problem by means of the recursive loop; in the beginning, when structure is undefined, it is necessary to take (random) samples to define the structure, but there is no guarantee that the samples taken are random and so we are faced with a (Goedelian?) paradox. The only solution appears to be to take samples and test the outcomes of the analysis against an appropriate model, converging on the target of unbiased estimators by repetition of sampling and analysis (for example, see Griffiths and Ondrick, 1968, 1970, and Griffiths, 1971).

4b. Sampling and experimental design

Each sampling arrangement is accompanied by its appropriate experimental design and this may be expressed in the form of an analysis of variance; in general, we select a square or rectangular grid composed of rows and columns which is superimposed on the sampled population. The sources of variation for this design are rows, columns, an interaction of rows by columns and, to check the significance of this interaction, it is necessary to take replicate samples at each row by column intersection. Hence the experimental design may be summarized as an analysis of variance:

Suppose the population is represented by stage 1 in figure 2; then if the (channel) samples are random each row mean will be similar and each row variance will be large and similar. Each row mean and variance is an estimator of the grand mean and variance and similarly for the columns. Replicate (Error) variance will also be large and approximates the total variance. The interaction term will not be significantly larger than error.

On the other hand, suppose that it is a stage 3 type (figure 2) and that the row-column grid is parallel to or coincident with, the layering; then, column means will be similar and column variances will be large as before. Row means will differ widely and within row variances will be small; the interaction term will not be significantly larger than error which, averaged across each set of measurements at each intersection, will be large (Dahlberg and Griffiths, 1967).

Now, in practice, if the structure is unknown, the superimposed grid will almost certainly not coincide with the layering and hence the response will tend to resemble that from a stage 1 population; any sampling arrangement which crosses the layering responds in the same way. The outstanding difference is in the interaction term; it will be significantly larger than the error term in all oblique arrangements (Dahlberg and Griffiths, 1967).

Thus the interaction of sampling arrangement and element structure can be recognized in each case. Because the grid is orthogonal it may be rotated and when coincident with the structure the interaction variance will no longer be significantly larger than error (Bennion and Griffiths, 1966); if the population is at stage 3, than a mini-max solution of largest difference between row mean and smallest between column means represents the optimum arrangement. If the population is

Figure 2. Relationship between sample arrangement and structure. Variation in the properties of size and shape of quartz grains reflects the process of selective sorting = f(intensity x time).

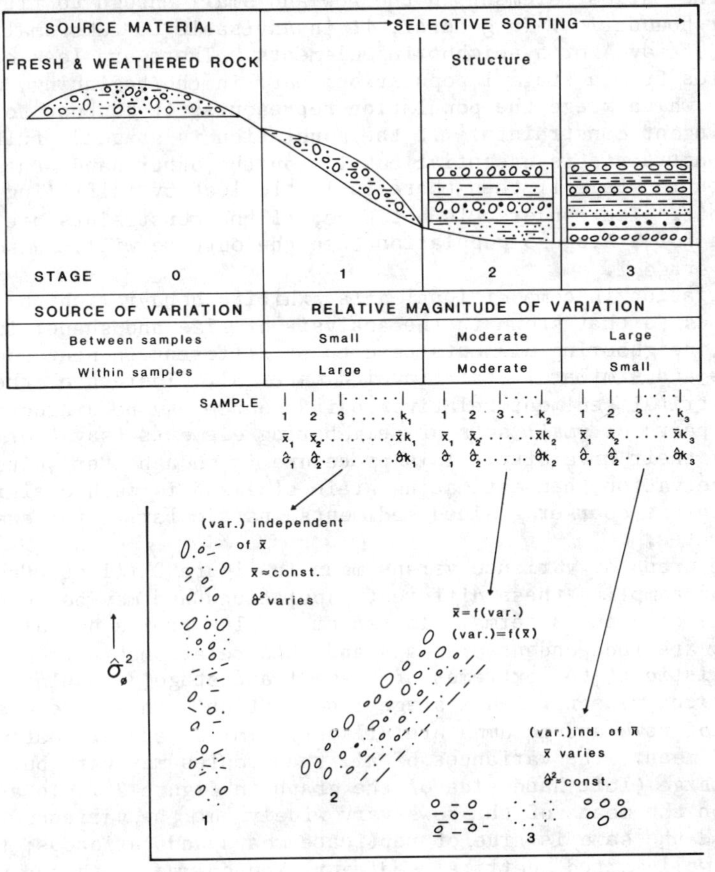

Table 1. Sampling, experimental design and analysis of variance.

Source of variation	Degree of freedom	Expected mean square
Rows	r-1	$\sigma_e^2+k\sigma_{re}^2+kc\sigma_r^2$
Columns	c-1	$\sigma_e^2+k\sigma_{re}^2+kr\sigma_c^2$
Row by column	(r-1)(c-1)	$\sigma_e^2+k\sigma_{re}^2$
Error(Replicates)	rc(k-1)	σ_e^2
Total	rck-1	σ_{tot}^2

stage 1 then the absence of a significant interaction variance is the acceptable criterion; row means and variances and column means and variances will be 'good' estimators of the grand mean and variance.

Each sample should, in the stage 3 population, be large enough to include the largest element in the row and small enough to fit between the layer boundaries. In general, it is advisable to take small samples composed of say 4 or 5 neighboring elements. There are less constraints for samples from a stage 1 population; but, in the beginning, when it is not known which stage the population represents, it is best to apply the most stringent constraints. If the population is stage 3, fulfillment of the constraints is essential but if, on the other hand, the population is stage 1, then there is little lost by fulfilling the constraints. The converse is not true; if the constraints are not fulfilled for a stage 3 population then the outcome will, falsely, indicate stage 1.

Variation in compositional data exhibits behavior which corresponds to that shown in the analysis of size and shape; in terms of figure 2, neighboring elements tend to be different in near source sediments and similar in far-traveled material. Therefore, the location of the detrital sediment relative to its source may be diagnosed by sampling pairs or small sets of neighboring elements (say 4 or 5) and measuring their diversity. This procedure is much better suited to field observation than estimating grain size and is much easier to observe in the coarser grained sediments, particularly, for example, conglomerates.

The graph of variance versus mean in figure 2 illustrates the outcome of sampling these different populations and may be interpreted in genetic or process terms. In random samples from a normal population variances are independent of means and this relationship is characteristic of the extremes of stage 1 and stage 3 populations but for different reasons. In a stage 1 population each sample mean and variance of rows and columns are all very similar and are estimates of the grand mean. The variances of all the samples may vary but are usually large (left-hand side of the graph in figure 2). In a stage 3 population the means of the rows vary widely but the variances vary little and the same is true of replicate means and variances; this is a typical "well-sorted" detrital sediment. The column means and variances estimate the grand mean and variance. Hence by appropriate sampling and analysis it is possible to decide whether the population is near its source (stage 1) or far from it (stage 3). In intermediate populations, variances are some function of their means as shown in figure 2 (stage 2). These relationships were first discovered in a study of sphericity and round-ness of quartz grains from detrital sediments by Curray (Curray and Griffiths, 1955, Griffiths, 1967, figure 6.15, p. 139). It is an important feature which may be used in classifying and naming detrital sediments for various purposes (e.g., Griffiths, 1968, 1978b).

4c. Variates on nominal or ordinal scales

Stratified random sampling, described above, is difficult to achieve for samples taken for less direct measurement procedures; for indirect measures such as sieving, a shovel-full or bagfull represents a spot sample and their statistical estimators fail to converge even when their number "n" gets quite large (Griffiths and Ondrick, 1968, Griffiths, 1967, figure 5.5, p. 96). The estimators suffer, therefore, from varying degrees of instability and cannot be as rigorously analyzed in the same manner as measures which achieve ratio or interval scaling.

Stratified sampling may be achieved in thin section size measurement but, to obtain suitable estimators that converge, the basic sampling unit is the traverse across the thin section; thin section means tend to yield unstable estimators and represent spot samples of layered populations; that is they contain some layers and some parts of layers so that appropriate weighting to yield unbiased estimators is difficult to achieve.

This, unfortunately, also applies to measures of composition derived from point-counting thin sections and, by analogy, from maps, pictures and other images. It is fairly straightforward to heuristically establish model frequency distributions such as the binomial for individual minerals, i.e. elements of one kind, in homogeneous populations; or multinomial models for elements of several different kinds (Griffiths, 1967, chapter 14). However, the resulting estimators show very peculiar behavior and, in particular, variates that sum to a constant, such as proportions or percentages, tend to exhibit negative correlations among pairs because of this constraint (Chayes, 1971). Aitchison has emphasized some of the difficulties which arise in analyzing this kind of data and has recommended a logistic-normal transformation to overcome the "closed" nature of the data (Aitchison, 1984).

One way of treating this kind of data is to use a measure of diversity which is equivalent to a variance and may be used even with nominal data (Pielou, 1975). Sampling such variates shows similar problems of interaction among sampling arrangements and element structures. The recommended practice is, therefore, quite similar; take small samples of neighboring elements and arrange the replicate sampling at intersections of a, preferably, square grid. Whether use of analysis of variance is justified for data suffering these constraints is not clear although, after logarithmic transformations of various kinds, log-linear analysis is available (Bishop et al, 1975).

An alternative approach to collection and analysis of this kind of data is represented by image analysis (Fabbri, 1984). Frequently the data is collected from maps or thin sections by procedures analogous to point-counting and the data are then digitized into a binary state (0/1). Analysis entails the use of Boolean logical operators on the digitized data (Fabbri, op. cit., p. 19). More sophisticated analysis uses the Minkowski operations between sets to transform the images and clarify their structure. An useful description of these various procedures together with a number of geological examples including analysis of maps and thin sections of rocks is contained in the text by Fabbri (1984).

In my opinion 'constant sum' data should be regarded as not better than ordinal scale as the following heuristic argument suggests. One of my distinguished colleagues claimed that the difference between 1 and 2 percent in counting accessory heavy mineral grains is far more important than that between 51 and 52 percent; this tacitly implies that one unit on this scale is of different magnitude depending on where on the scale it occurs. Since "equal" intervals are not equal on this scale the information level cannot be better than ordinal. Many people believe that in counting to establish frequency, rare events (minerals) are more important than commoner ones; this despite the fact that sampling error around the smaller numbers is relatively much larger than around the larger ones. I have elsewhere suggested using the information transformation as one procedure for analyzing such data (Pelto, 1954, Griffiths, 1967, p. 315ff.). This implies that the information level of the data is nominal and suitable analytical procedures should be adopted. One of the more recent approaches to the analysis of these kinds of data is by Atkin's Q-analysis (Atkin, 1976). Since this procedure is designed to analyze information which can be expressed geometrically as simplicial complexes in multidimensional space and, because it is based on the algebra of sets and very elementary matrix operations, it avoids many of the difficulties in the conventional approach; geological examples are described by Gould (1981) and Griffiths (1983).

5. Interpretation

Interpretation of the outcome from measurement, sampling and analysis is usually made through the use of a model; for example, the petrogenesis of detrital sedimentary rocks may be expressed in terms of three stages (figure 3). First, there is a source material which consists of a rock and its (in situ) weathered soil; secondly, this material is eroded, transported and deposited to form detritus and, thirdly, this detritus is modified by diagenetic processes which take place during and after deposition. The product is a typical, often consolidated, detrital sediment. Now, in practice, the second stage is dominated by a single process called selective sorting by the transporting fluids on the basis of density (=composition), size and shape of the elements (minerals or grains). It is generally understood that the kind of source material (both fresh and weathered) may be recognized by studying mineral composition, particularly the "essential and varietal" minerals of the detrital sediment. The detrital process is reflected in variation in the size, size-sorting and shape of the detrital minerals, usually quartz; and, finally, different aspects of the diagenetic processes are reflected in 1. the arrangement of the mineral grains during and after deposition (o and p in equation 1), and, 2. the amount and distribution of the cementitious constituents such as carbonate and silica cement (Griffiths, 1966, Dahlberg and Griffiths, 1967, Griffiths and Ondrick, 1969).

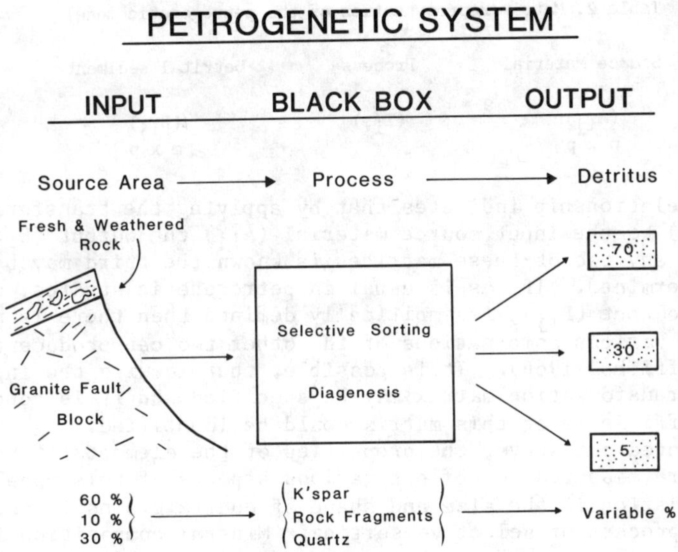

Figure 3. Petrogenetic system, showing variable output from a single input.

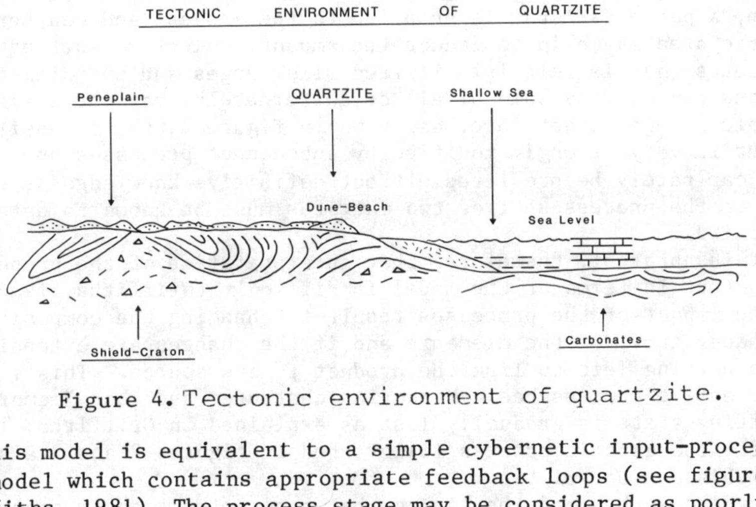

Figure 4. Tectonic environment of quartzite.

This model is equivalent to a simple cybernetic input-process-output model which contains appropriate feedback loops (see figure 4, p. 9, Griffiths, 1981). The process stage may be considered as poorly defined or the process is represented by a black box. Such a model has wide application in many different fields (e.g. in mineral resource assessment, Griffiths, 1978b, 1981, figure 8).

This petrogenetic I-O model may be represented in a matrix relationship as follows (Griffiths, 1966, table 1, p. 655):

Table 2. Matrix representation of the petrogenetic model.

Source Material	Process	Detrital sediment
(A_{ij}) n x p	(T_{ij})	(B_{ij}) m x p

This relationship indicates that by applying the transformation process (T_{ij}) to the input source material (A_{ij}) the output (B_{ij}) is obtained; if any two of these matrices is known the third may be uniquely determined. If, as is usual in petrogenetic studies, only one matrix, the output (B_{ij}), is empirically defined then there is no unique solution and various combinations of the other two can produce the same outcome (Griffiths, 1966). It is possible, that knowing the input and output the transformation matrix may be specified and it is conceivable that the coefficients of this matrix could be identified.

As pointed out above, the properties of the elements of the population are measured to reflect various aspects of this model; for example, variation in the size and shape of quartz grains is used to reflect the process of selective sorting. Mineral composition is usually believed to represent the source material and its weathered products. Ideally, the property measured should be, conceptually and operationally, related to one genetic feature and be independent of all others; for example, the presence of X percent of potassium feldspar should be related solely to its occurrence in the source area and ideally, subsequent processes should not affect this percent. In practice, X percent K'spar is supplied to the process and weathering in the source area may help to reduce the amount; selective sorting concentrates what is left into limited size ranges and and diagenetic conditions may destroy some or all or, alternately, may encourage its authigenic growth. Therefore, as shown in figure 3, the composition of the input is very strongly modified by subsequent processes and the outcome can rarely be predicted without definitive knowledge of the effects of the processes (i.e. two matrices must be known to determine a unique third).

It is apparent, therefore, that interpretation of the properties of the output in terms of the model is difficult (Griffiths, 1966). In fact, the impact of the processes result in changing the composition, size, shape, etc., of the elements and if the changes are extensive there is nothing left to link the product to its source. This implies that the entire process has a Markovian character and that "memory" of its starting state is gradually lost as explained in Griffiths, 1981, figure 2. The genetic process illustrated in figures 2 and 3 also display this characteristic.

Krynine emphasized the linkage between tectonic environment and detrital sediments (1948, 1951) and it seems clear that each of the end-member environments envisaged by Krynine is destined to be reduced eventually to a peneplain. In terms of rocks this implies that no matter what the starting state, whether arkose, low rank or high rank graywacke, the ultimate detrital product is a quartzite (figure 4).

Thus the outcome of this process is equifinal (Griffiths, 1981). There is no way that this output can be uniquely traced back through its history to its starting state. This is a geological example of Goedelian incompleteness; to overcome the difficulty it is necessary to enlarge the system, i.e. to examine a wider range of rocks. Fortunately, since the entire history is represented in an extensive sequence of rocks it is possible to search beyond a quartzite and find an associated stratigraphic neighbor which contains greater compositional diversity and so to identify the rock-sequence to which it belongs. The classification of detrital sediments is facilitated by this association and the fact that any one tectonic environment (and therefore the rock-type which represents it) tends to persist for a long time so that extensive rock-sequences enable the petrographer to assign chemical rocks, shales etc. to the appropriate sequence in which they belong (Griffiths, 1968, figure 3).

If the relationships between sources of variation in the model and variation in the properties of the elements are represented in a rectangular table with properties as rows and sources as columns, then there would be an entry in some cell, say relating occurrence of a phyllite rock-fragment and the source area that supplied it and there would be zeros elsewhere along this row (Griffiths, 1966, table 2, p. 658). This could represent the condition of independence; if, on the other hand, the real situation is portrayed there would be many cells with entries and few with zeros. Because all the measured properties are known to interact interdependently (see equation [1]) it is necessary to manipulate the collected data to obtain the greatest degree of independence (i.e. the greatest number of zeros) in the matrix. Components analysis is an analytical technique suited to this purpose; when it is applied to data matrices(2) collected from detrital sediments it is found that some 5 to 7 components appear to make significant contributions to and account for almost all the variation included in the original data matrix.

Identification of these components in the simplified diagonalized matrix (e.g. table 3, Griffiths, 1966) is more or less straightforward. In the Cow Run Sandstone, a typical example, the larger part of the variation, some 60 percent, is accounted for by the first two components and these are strongly associated with size and sorting respectively. The third component brings the total accounted for to over 75 percent and the dominant contributing variables are proportion of matrix and rock-fragments. Since these two properties possess opposite signs it suggests that there may have been some confusion in assigning points during the point-count to these two constituents, i.e. a finegrained rock-fragment becomes matrix if its external boundary becomes diffuse.

---

(2) A typical n x p data matrix, where n = 39 and p = 12 is shown in table 1, Griffiths, 1958, for the Cow Run Sandstone, a low rank graywacke. The corresponding steps in the analysis of this data matrix (reduced to n = 28 and p = 10) through the correlation matrix, the components matrix and simplification and diagonalization are shown in table 3 and tables A-D in the appendix of Griffiths, 1966.

Proportion of quartz is almost independent of all other petrographic properties and occurs on its own component with an 8 percent contribution to variance accounted for; this is important because it suggests that variation in the size of quartz grains and proportion of quartz may be treated independently. Proportion of silica cement, the last petrographic property and the fifth component brings the total variance accounted for, to 96 percent which is as high as one should expect to achieve. In this example, the process of selective sorting, as reflected in variation in size and size-sorting of quartz grains and diagenesis, (variation in the proportion of silica cement) account for some 70 percent of the total variation; identification of the other components is obscure to say the least. This is a typical outcome of the analysis of detritals some distance from their source. It shows very little "memory" of its starting state. On the other hand, Krynine, in his classic study of the Triassic of Connecticut (1950), could quite appropriately claim that he could stand on the Triassic outcrops and quite literally "see" the source, and vice versa. By way of contrast consider the question "what is the source and from where comes the material, for the detritus in the Mississippi delta? - the answer is the larger part of the North American continent.

7. Discussion

There appears to be two sets of obstacles to resolving the petrogenetic problem described above; the first concerns the nature of the data collected for the specification of mineral composition, the variates suffer from the "closure" constraint. Logarithmic-normal transformation of these data, as recommended by Aitchison (1984), results in a components matrix of the transformed data showing one major component and a few small ones; this may be interpreted to mean that there is only one substantive source of variation in these data. Similar features occur if data of this kind is analyzed by Imbrie's Q-mode factor analysis (1963) i.e., there are few significant eigenvalues (≤ 3) and, usually, the first one is much larger than the others. When data matrices of detrital sediments, such as the Cow Run sandstone, are analyzed using Aitchison's transformation there is again one large eigenvalue and with Imbrie's Q-mode analysis there are generally three, with the first much larger than the others. Now in Imbrie's procedure the transpose of the original matrix is used for analysis and so we meet the problem that R-mode analysis of the data yields 5-7 components whereas its transpose only yields three; since the rank of the rows should equal the rank of the columns this becomes something of a dilemma.

Again, in our procedure, all the petrographic data are collected by measures performed in thin section and although the measures are different conceptually and are given different names, they all contain the same confounded sources of variation; some degree of redundancy is, therefore, expected. Some preliminary attempts to clarify this issue by means of cannonical correlation analysis between compositional and textural petrographic data suggests that the two groups contain the same information. Since the textural measures, essentially size and sorting, do not suffer from closure, it is advantageous to use their contribution and ignore most of the, more or less redundant, compositional data. This leaves the problem of identifying source unresolved.

The second set of obstacles arises from the nature of the dynamic processes; they suggest that the detritus changes its characteristics depending on the time and intensity during which the process persists. Essentially, the outcomes from our investigations have suggested that "memory" of source is gradually lost.

Hence the two sets of obstacles appear to emphasize the same difficulty and, while some alternative measurement and/or analytical procedure may be devised to gather data which avoids the problem of closure, this alone is unlikely to offset the loss of memory which appears to be built into the characteristics of the genetic processes. As proposed earlier, it seems necessary to enlarge the system and adopt a metalanguage to circumvent this genetic obstacle or the problem of the petrogenesis of detrital sediments will remain an example of Goedelian incompleteness, i.e. it may become an example of an insoluble problem. Perhaps analysis of a single geological formation, which represents a limited geological sequence of events, would be expected to reveal only a limited part of the entire story; hence "enlargement", as recommended above, may indicate, that to elucidate the complete sequence of events it is essential to examine a 'complete' sequence of rocks.

8. Acknowledgments

The information upon which this review is based includes a large contribution from graduate student research towards their advanced degrees under my direction. My colleagues, Drs. D.P. Gold and C.P. Thornton, Professors of Geology in the Department of Geosciences at Penn State University, were kind enough to read the manuscript critically and their suggestions are much appreciated. The manuscript was typed by Ms. J. Bailey and the figures were drafted by Ms. T. Mistrick.

9. References

1. Aitchison, J., 1984, 'The Statistical analysis of geochemical compositions', Math. Geol., 16, 531-564.

2. Atkin, R.H., 1976, Mathematical Structure in Human Affairs, Heinemann Educ. Books Ltd., London, 212 p.

3. Beer, S., 1964, Cybernetics and Management, J. Wiley and Sons, N.Y., 214 p.

4. Bennion, D.W., and Griffiths, J.C., 1966, 'A stochastic model for predicting variations in reservoir rock properties', Jnl. Soc. Petrol. Engr., 6, 9-16.

5. Bernhard, R., 1969, 'Uncomputable functions: The Ghost in the computer', Sci. Res., 3, 47-49.

6. Bishop, Y.M.M., Feinberg, S.E., and Holland, P.W., 1975, Discrete Multivariate Analysis: Theory and practice, The M.I.T. Press, Camb., 557 p.

7. Chayes, F., 1971, Ratio Correlation, Univ. Chicago Press, Chicago, 99 p.

8. Churchman, C.W., and Ratoosh, P., 1959, Measurement Definition and Theories, J. Wiley and Sons, N.Y., 274 p.

9. Curray, J.R., and Griffiths, J.C., 1955, 'Sphericity and roundness of quartz grains in sediments', Bull. Geol. Soc. Amer., 66, 1075-1096.

10. Dahlberg, E.C., and Griffiths, J.C., 1967, 'Defining gradients in a sample of sedimentary rock', Sedimentology, 8, 291-309.

11. Fabbri, A.G., 1984, Image Processing of geological Data, Van Nostrand Reinhold Co., N.Y., 244 p.

12. Fisher, R.A., 1948, Statistical Methods for Research Workers, 10th edn., Oliver and Boyd Ltd., London, 354 p. (1st. edn., 1925).

13. Gould, P.R., 1981, 'A structural language of relations', p. 281-312 in Craig, R.C., and Labovitz, M.L. (eds.), Future Trends in Geomathematics, Pion Ltd., London, 318 p.

14. Griffiths, J.C., 1958, 'Petrography and porosity of the Cow Run Sandstone, St. Marys, W. Va.', Jnl. Sediment. Petrol., 28, 15-30.

15. _____, 1959, 'Size and shape of rock fragments in Tuscarora Scree, Fishing Creek, Lamar, Central Pa.', Jnl. Sediment. Petrol., 29, 391-401.

16. _____, 1966, 'A genetic model for the interpretive petrology of detrital sediments', Jnl. Geol., 74, 655-672.

17. _____, 1967, Scientific Method in the Analysis of Sediments, Mcgraw-Hill Book Co., N.Y., 508 p.

18. _____, 1968, 'Geological data for classification', Western Miner., 41, 37-42.

19. _____, 1971, 'Problems of sampling in geoscience', Trans/Sectn. B., Inst. Mng. and Metal., London, B346-356.

20. _____, 1974, 'Quantification and the Future of geoscience', 83-101 in Merriam, D.F., ed., The Impact of Quantification on Geology, Syracuse Univ. Geol. Contribn., No. 2, 104 p.

21. _____, 1978a, 'Sediment Parameters', p. 724-731 in Fairbridge, R.W. and Bourgeois, J., 1978, Encycl. Earth Sci. Series, VI, Encycl. of Sedimentology, Dowden, Hutchinson and Ross, Inc., Pa., 901 p.

22. _____, 1978b, 'Mineral Resource Assessment using the Unit Regional Value Concept', Math. Geol., 10, 441-472.

23. _____, 1981, 'Systems behavior and geoscience problem-solving', p. 1-21, in Merriam, D.F., Computer Applications in the Earth Sciences; an update of the 70's, Plenum Press, N.Y., 385.

24. _____, 1983, 'Geological Similarity by Q analysis', Math. Geol., 15, 85-108.

25. _____, and Ondrick, C.W., 1968, Sampling a Geological population, Comp. Contribn. No. 30, State Geological Survey of Kansas, The Univ. of Kansas, Lawrence, 53 p.

26. _____________________, 1969, 'Modelling the petrology of detrital sediments', p. 73-97 in Merriam, D.F., ed., Computer Applications in the Earth Sciences, Plenum Press, 281 p.

27. _____________________, 1970, 'Structure by sampling in the geosciences', In Random Counts in Physical Science, Patil, G.P., ed., Penn State University Press, 31-55.

28. _____, and Smith, C.M., 1964, 'Relationship between volume and axes of some quartzite pebbles from the Olean Conglomerate, Rock City, N.Y.', Amer. Jnl. Sci., 262, 497-512.

29. Hofstadter, D.R., 1979, Goedel, Escher, Bach; an eternal golden braid, Basic Books, Inc., N.Y., 777 p.

30. Hulbe, C.W.H., 1955, 'Mounting technique for grain size and shape measure-ment', Jnl. Sediment. Petrol., 25, 302-303.

31. Imbrie, J., 1964, Factor and vector analysis programs for analyzing geologic data, Tech. Rept. No. 38-135, Office of Naval Research, Geography Branch, 83 p.

32. Krynine, P.D., 1948, 'The Megascopic and Field Classification of Sedimentary Rocks', Jnl. Geol., 56, 130-165.

33. _____, 1950, Petrology stratigraphy and origin of the Triassic sedimentary rocks of Connecticut, State of Conn. Geol. and Natl. Hist. Sur. Bull., no. 73, 247 p.

34. _____, 1951, 'A critique of geotectonic elements', Trans. Amer. Geophys. Union, 32, 743-748.

35. Pelto, C.R., 1954, 'Mapping of multicomponent systems', Jnl. Geol., 62, 501-511.

36. Pielou, E.C., 1975, Ecological Diversity, J. Wiley and Sons, N.Y., 165 p.

37. Pettijohn, F.J., 1957, Sedimentary Rocks, 2nd edn., Harper and Row Publ. Inc., 718 p.

38. Stevens, S.S., 1959, 'Measurement, psychophysics and utility', p. 18-61 in Churchman, C.W. and Ratoosh, P., eds., Measurement Definition and Theories, J. Wiley and Sons, Inc., N.Y., 274 p.

LEVELS OF INFORMATION AND PROBALITITY MAPPING

Wolfdietrich SKALA
Freie Universität Berlin
Institut für Geologie

ABSTRACT

Exploration is a sequential process, possibly resulting in the discovery of a new deposit. In the early phases of exploration, data and information is generally lacking or insufficient.

Nevertheless, it must be possible to make decisions of economic import, e.g. whether exploration should proceed or not or which methods and strategies should be used to maximize search efficiency. In this regard, the prognostic value of "probabilitiy mapping" cannot be underestimated; it provides an objective basis to develop subsequent strategies and to aid in decision making.

Objective planning data can only be produced when the available information is optimally used. For example, decisions can be facilitated if data from areas with a high level of information is included. An "historical analysis" of training areas can complement standard geologic analysis, furnishing valuable insights into the problem.

The following case studies are used to illustrate the affect which the level of information has on the choice of the methods used and on the probability maps.

1. INTRODUCTION

Exploration is a sequential process, possibly resulting in the discovery of new deposits. It is characterized by distinct levels of information and knowledge, upon which decisions of economic import must be based, e.g., whether exploration should proceed or not or which methods and strategies should be used to maximize search efficiency. Since the resource base, i.e., the total collection of deposit occurrences in an exploration area, is unknown in most cases, the resources - the sum of the amount of raw materials that could be recovered from these deposits - can only be determined by means of resource models as well as estimation procedures. Such models provide an objective basis to develop subsequent strategies and aid in decision making. The goal of such models is to support exploration planning and management; they should therefore not only more or less realistically represent the resource base but should also include some objective and comparable measure of realiability. Furthermore, such resource models have to consider all economical, technological and geological data and information available. Using the resource model classification given by HARRIS 1984, compound models, including geological, economical and geostatistical components are appropriate for the aforementioned situation. These models employ mathematical methods to estimate the unknown endowment as a purely physical event, which is independent of economics and

C. F. Chung et al. (eds.), Quantitative Analysis of Mineral and Energy Resources, 57–77.

technology. Resources and potential supply from the unknown endowment are estimated by interacting economics and technology with this endowment using computer simulation techniques. Hence, the "dynamics of resources" can be considered, which means that a change in product or factor prices or an improvement in the technology of production causes a change in resources: Although resource base is an absolute concept, resources are not absolute; they are a function of economics and technology.

Sufficient data and information are needed to construct resource models, which are dependent on exploration levels. In the initial phases of exploration (e.g. reconnaissance phase) data and information are generally lacking or rarely sufficient: Knowledge of the geology, particularly in grass-roots prospection areas, is often based on small scale, unreliable maps. When available, quantitative observations and measurements are usually widely spaced.

In general, the increase of information during successive exploration levels is controlled by the following facts:

There is an increasing tendency toward more variables. Whereas early prospection stages are confined, par example, to geological map information, detailed exploration often has at its disposal data from different geochemical variables, geophysics or remote sensing activities.

- The information content of these variables increases too. Reconnaissance exploration is limited, in most cases, to subjective and therefore often unreliable qualitative information. Decisions during subsequent exploration can be based, on the other hand, more and more on quantitative measurements.

- Reliability of exploration results as a foundation for decisions concerning further exploration or mining activities is also strongly influenced by the density of information, particularly sampling grids. So, in general, exploration progression is connected with concentrations of sampling grid densities.

Most of exploration quantities and qualities are regionalized variables which have to be mapped or combined into integrated maps. The reliability of exploration results is therefore documented in the reliability of maps. Because of the previously mentioned, significant differences in data and information types, structures and supports, it is very difficult to develop and present a practical oriented measurement of "integrated map goodness" which would allow the quantification of the reliabilities of exploration levels and consequently direct comparisons of decision bases. In Geostatistics a quantity for global reliability evaluation, is the global estimation variance (JOURNEL & HUIJBREGTS 1978), restricted however to constrained conditions, e.g. variables have to be treated separately etc. It will be demonstrated later, that Co-Kriging methods and Co-Kriging Cross-Validation can serve as useful tools in the construction and further development of reliability measurements of exploration maps.

Objective planning data can only be produced when the available information is optimally used. For examples, decisions can be facilitated if data from areas with a high level of information is included. The works of GRIFFITHS 1969 and AGTERBERG 1974 are well known examples of this method of including " training areas". An "historical analysis" of training areas, as documented in oil exploration projects (HAMBLETON, DAVIS & DOVETON 1975) can complement standard geologic analysis, furnishing valuable insight into the problems.

During the last years, members of the working group "Mathematical

Geology" of the Freie Universität Berlin have been working in different practical as well as methodological oriented projects of assessment and appraisal of mineral resources. Since the initial data and information available are obviously different for each project, the following case studies can be well used to illustrate the affect which the level of information has on the choice of the methods used and on the probability maps.

2. COMPUTER ASSISTED EXPLORATION FOR PHOSPHORITES IN THE SINAI PENINSULA, EGYPT

One of the primary goals of subproject C 4 "Geostatistical Studies of Sedimentary Mineral Deposits", of the special research project "geoscientific problems of arid areas" financed by the German Research Foundation, has been to develop a mathematical exploration strategy that can be applied to various mineral deposit types, that is particularly suited to arid areas, and that is versatile enough to fulfill diverse corporate or government purposes, e.g. export, improvement of an underdeveloped area etc. To achieve this goal, the following procedures or theories were explicitly integrated into the exploration process:

- Favorability mapping using a Monte Carlo Simulation
- Search theory, and
- Bayesian statistical decision theory.

The exploration strategy has been formulated for the reconnaissance phase of exploration. Both favorability mapping and search theory have been adapted and incorporated into the exploratin strategy. More detailed descriptions of the results of this project can be found in SHULMAN (1984a, 1984b, 1984c) and SHULMAN & SKALA (1984, 1985).

The probability that the Sinai Peninsula contains exploitable phosphorite deposits is high. The Sinai Peninsula forms a barren area between the Israeli and Jordanian phosphate fields on the one side and the southern phosphate belt in Egypt on the other. In fig. 1, the Israeli phosphate fields are shown along with the Upper Cretaceous rocks and major anticlinal axes in the Negev Desert and in the Sinai Peninsula. The economic deposits in the Negev Desert are restricted to the flanks of synclines. The numerous anticlines which transect Northern Sinai and the Negev Desert trend northeast and form part of the Syrian Arc System. Under the assumption that phosphorite occurs in the Upper Campanian-Maastrichtian rocks, the problem becomes not where can phosphorites be found, but which anticlines should be investigated, and how should the exploration resources be allocated to assure maximum probability of discovery. Probability mapping and search theory furnish answers to both questions.

The exploration strategy designed for the reconnaissance phase consists of the following steps:

- Generate an initial probability map based on all available data
- Allocate resources for geological reconnaissance using search theory
- Geologic reconnaissance
- Update the probability map using information gained during the geologic reconnaissance
- Allocate resources for reconnaissance drilling/trenching using search theory
- Reconaissance drilling/trenching
- Comparison of reconnaissance results using Bayesian statistical

decision theory

2.1 Probability Map Generation

The initial probability map was generated using a Monte Carlo simulaton. Descriptions of the method can be found in SHULMAN 1984a, 1984b and SHULMAN & SKALA 1984. The method consists basically of two steps: 1. A geologic group is selected randomly from a subjective probability distribution. 2. An area belonging to the chosen group is selected randomly from a probability distribution which is based upon weighting functions.

In fig. 1, seventeen favorable areas are shown, each corresponding to an anticline. The geologic group separates the anticlines according to a geologic criteria. For the phosphorite occurences in the Sinai Peninsula, the age of the anticline was chosen as the geologic factor defining the group. Table 1 shows the different groups and the numbers assigned to each group. Each group is assigned a credence reflecting the probability that the scenario contains an economic deposit. The credences are then transformed into a probability distribution which is used in the first step of the simulation. Rather than assigning concrete subjective probabilities to each group and transforming them into a probability distribution, the probability distribution was additionally simulated to meet the following conditions:

$$S_5 > S_4 > S_3 > S_2 > S_1 > 0 \qquad (1)$$

$$S_5 + S_4 + S_3 + S_2 + S_1 = 1 \qquad (2)$$

To differentiate the favorability between areas within a given group and between areas belonging to different groups, weighting functions were used. The weighting function need not be restricted to geologic criteria or information, but may also include other relevant data, e.g. mining, infrastructure, etc.. Among the advantages of the use of weighting functions and of the simulation in general are the capability of explicitly including mining, infrastructure, economic factors, along with geological factors into the probability calculations, the capability of updating the probability map as new information becomes available, and not being dependent on a training or control area. Based upon weighting functions, each area is assigned a weight. The probability distribution of the areas belonging to a group is determined by a transformation of the weight. The following weighting functions were used for the Sinai Peninsula:

$$w_e = A\ 1/d + Bt + Cs \qquad (3)$$

$$w_g = 1 - A_f f_d \qquad (4)$$

$$w = w_e w_g \qquad (5)$$

with

w = area weighting function
w_e = economic weighting function
w_g = geologic weighting function
f_d = fracture density of area
t = elevation above sea level
s = size of area
d = distance to nearest port
A, B, C, A_f = coefficients

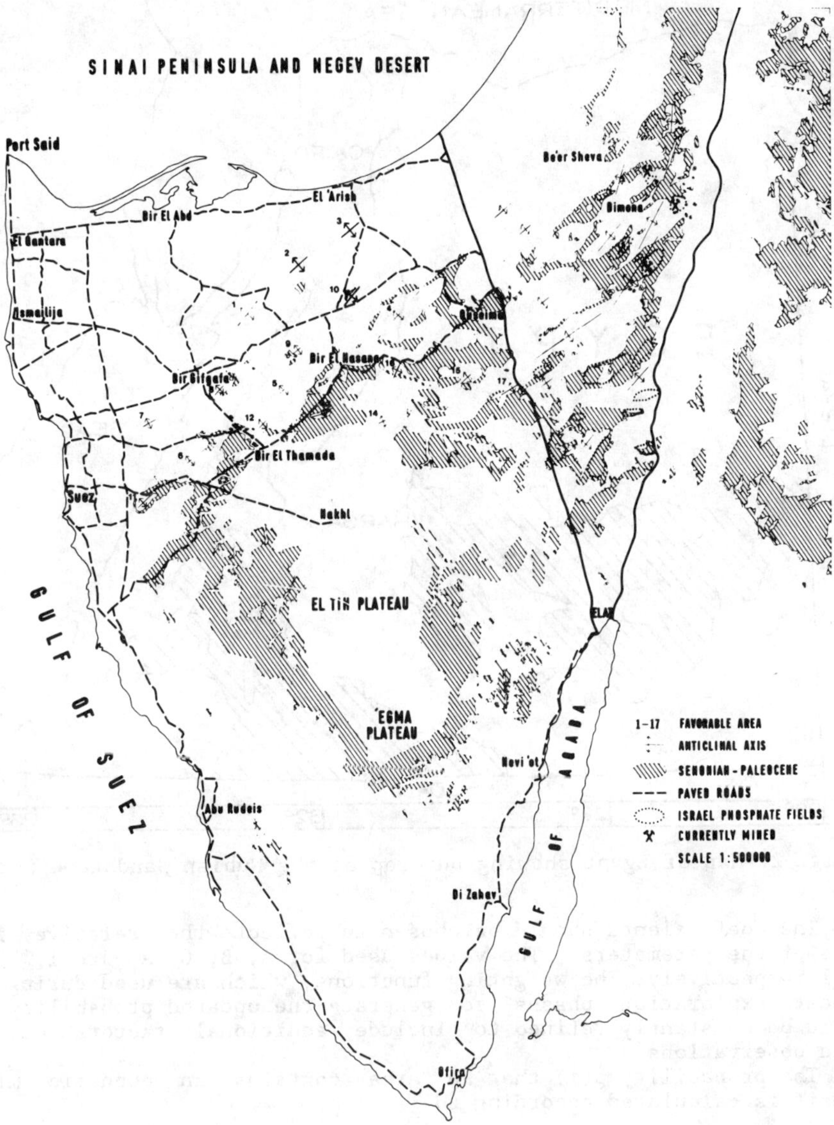

Fig. 1: Map of the Sinai Peninsula and of the Negev Desert
(after SHULMAN & SKALA 1984)

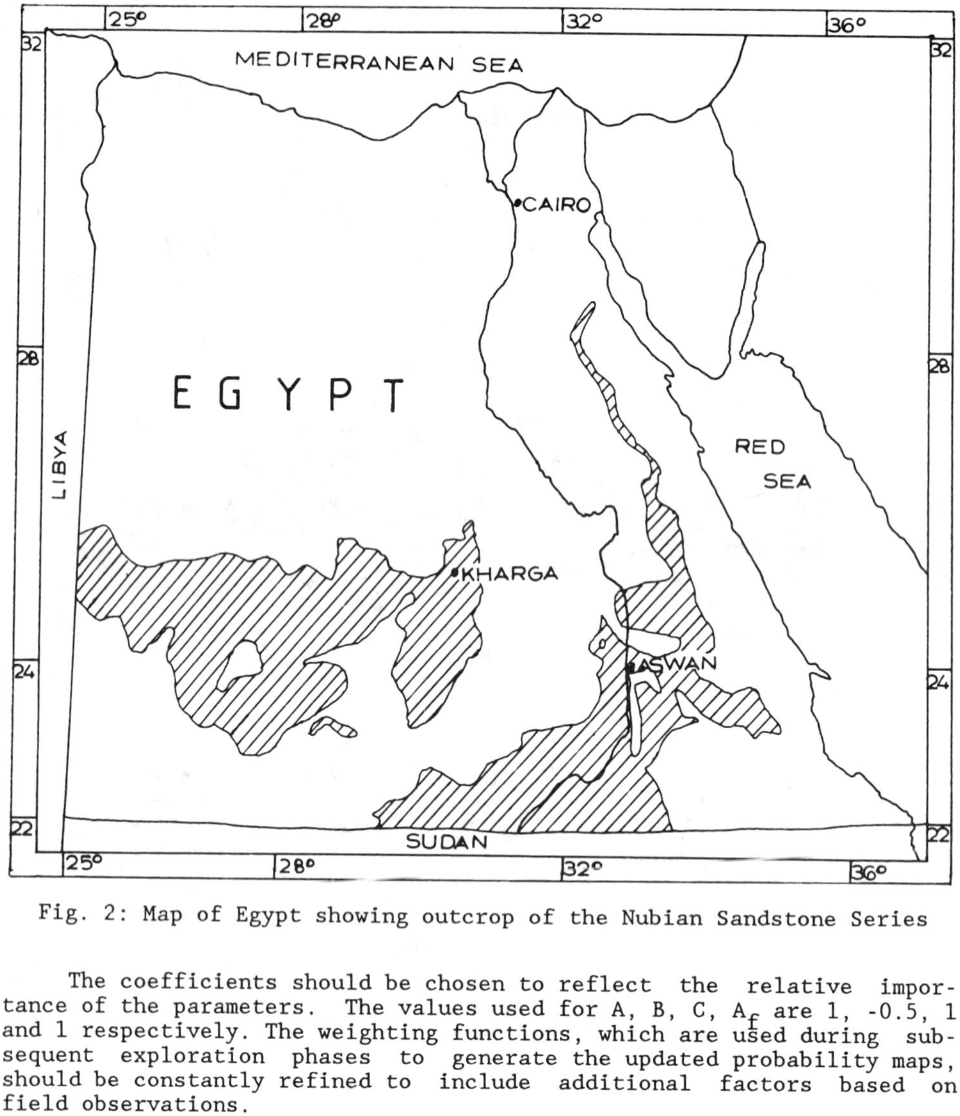

Fig. 2: Map of Egypt showing outcrop of the Nubian Sandstone Series

The coefficients should be chosen to reflect the relative importance of the parameters. The values used for A, B, C, A_f are 1, -0.5, 1 and 1 respectively. The weighting functions, which are used during subsequent exploration phases to generate the updated probability maps, should be constantly refined to include additional factors based on field observations.

The probability p(i) that an area contains an economic mineral deposit is calculated according to

$$p\,(i) = \frac{\sum_{i=1}^{J} w\,(i)}{\sum_{l=1}^{L} w_l} \qquad (6)$$

where

$w(i)$ = weight of area i
J = number of replications that area i was chosen
L = total number of replications
w_l = weight of area chosen during the l'th replication

The denominator is necessary to scale the probabilities between 0 and 1.

In Table 2, the area probabilities calculated using this method are shown. The area numbers correspond to those in fig. 1. To generate the probabilities, the group distribution was simulated 500 times. For each of the group probability distribution simulations, the two step procedure outlined previously was carried out 1000 times.

2.2 Optimal Allocation of Exploration Effort

Search theory was used to optimally allocate exploration effort based on the initial probability map. Under the accessory condition that a predetermined exploration budget is not exceeded, search theory attempts to maximize the probability of discovery. The basic theory behind search theory and an algorithm that may be applicable to mineral exploration is described by SHULMAN 1984c. The algorithm was used to allocate the exploration resources.

Search theory assumes that some knowledge of the whereabouts of the target is available and that a target probability distribution p(i) can be constructed. The p(i) is the probability that the target occurs in cell i. Furthermore, a method to detect the target must be available and therefore a detection function b(i,z) is simply the conditional probability of detecting the target in cell i when effort z (time, money, etc.) is expended provided that the target is in cell i. The probability of discovery P (i,z) is the product of the two functions, i.e. $P(i,z) = p(i)\, b(i,z)$. The search is restricted by the resources allocated for the exploration. A cost function c(i,z) can therefore be defined where c(i,z) is the cost of exploring cell i when effort z is expended. The cost function may be defined in dollars, time, mileage etc. The search problem can be stated as maximizing the probability of discovery under the condition that the total cost allocated for exploration is not exceeded. The problem is solved by introducing a Lagrangian multiplier λ and by maximizing the following equation:

$$l\,(i,\lambda,z) = p(i)\, b(i,z) + \lambda c\,(i,z) \qquad (7)$$

$l(i, \lambda ,z)$ is called the pointwise Lagrangian. Solution of the above equation enables the resources available to be optimally allocated among the various areas. Table 3 shows the optimal allocation of field days per area for field seasons of various lengths. In Table 4, the optimal allocation of drilling costs are shown for various drilling budgets. Drilling costs of $100 per meter and a maximum drilling depth of 30 m were assumed in calculating the drilling costs. It should be noted that the optimal allocation of drilling costs was determined using the initial probability map. Results from the geological field reconnaissance enable an updated version of the probability map to be calculated. Drilling allocations should be based upon the updated version.

Group	Exposed core	No. areas
1	Triassic	1
2	Jurassic	4
3	Lower Cretaceous	5
4	Cenomanian	5
5	Turonian	2

Table 1: Definition of groups and the number of areas assigned to each group (after SHULMAN & SKALA 1984)

Area	Group	Probability
1	2	.0727
2	2	.0180
3	2	.0570
4	3	.0975
5	3	.1109
6	3	.0342
7	4	.0594
8	3	.0056
9	5	.1837
10	4	.0760
11	4	.0796
12	4	.0058
13	2	.0010
14	3	.0017
15	5	.1505
16	4	.0425
17	1	.0036

Table 2: Probabilities calculated using the Monte Carlo simulation (after SHULMAN & SKALA 1984)

	Field seasons in days				
Area	90	120	150	200	250
9	23.96	27.29	30.49	35.32	39.87
15	19.97	23.31	26.51	31.34	35.88
5	13.87	17.20	20.40	25.23	29.77
4	11.29	14.63	17.82	22.65	27.20
11	7.24	10.57	13.77	18.60	23.14
10	6.31	9.64	12.84	17.67	22.22
1	5.42	8.76	11.95	16.78	21.33
7	1.38	4.71	7.91	12.74	17.29
3	.56	3.89	7.09	11.92	16.46
16			1.22	6.05	10.59
6				1.70	6.26

Table 3: Allocations of field days per cell for various field seasons (from SHULMAN & SKALA 1984)

Area	Drilling budget in dollars 75,000	100,000	150,000	200,000	250,000
9	51,090	63,757	81,990	95,520	108,100
15	23,910	36,243	54,810	68,340	80,920
5			13,200	27,000	39,250
4				9,140	21,730

Table 4: Allocation of drilling costs per area for various reconnaissance drilling budgets (after SHULMAN & SKALA 1984)

2.3 Discussion

During this early stage of exploration, quantitive data such as geochemical analyses are not available. Consequently this level of knowledge is not appropriate for detailed planning activities, however sufficient for initial economical and technological appraisals. Therefore the use of probability mapping and search theory alleviates two major pitfalls of reconnaissance exploration: 1) Deciding on an initial workplan and 2) Formulating subsequent work plans contingent on preliminary results. Typical problems of arid environments such as lack of water or lack of infrastructure can be taken into consideration by defining suitable weighting functions during generation of a probability map. Although the economic weighting function used was more export oriented, the weighting function may also be defined to reflect domestic goals. For example, if settlement and agricultural development of the Sinai Peninsula is desired, availability of and proximity to water, distance to nearest town, etc. may be incorporated into the weighting function.

The integration of the two mathematical procedures into the exploration process allows several alternative strategies to be formulated, especially if computer facilities are available in the field. The strategy as formulated is non-adaptive, i.e. the allocations are not influenced by the initial field observations or sampling. The predetermined strategy is carried out regardless of any results obtained. If, on the other hand, the probability map is updated concomitantly with exploration, new allocations of effort can be calculated. Such a search strategy is called adaptive. Various search strategies can be formulated between the two extremes. The allocations of effort may, for instance, be used as a stopping criteria.

3. IRON POTENTIAL IN THE NUBIAN SANDSTONE SERIES

Outcrops of the Nubian Sandstone Series in Southern Egypt cover approximately 180 000 km^2 (see fig. 2). Within the series, oolitic iron ore bands occur. The iron ore bands are restricted to the Middle Group of the Series (ATTIA 1955). Recent attempts at standardizing the nomenclature have placed the oolites within Facies 3^{16} of the Abu Agag Formation (KLITZSCH & LEJAL-NICOL 1984). The iron ore beds vary spatially in the number of iron-bearing horizons, in their thickness, and in their lateral extent. Currently none of the known iron deposits (9 deposits are noted in the Mineral Map of Egypt 1979) are being worked. Several were submerged upon completion of the Aswan High Dam. The largest deposit, with mineable reserves of approximately 30 million tons, is located east of Aswan; the mine was closed in 1983.

Outcrops of the Nubian Sandstone Series are divided into 100 cells, each comprising approximately 1800 km^2. An estimate of the iron potential can be made by assuming that the number of deposits per cell obeys a negative binomial distrubution, i.e.

$$P\ (K=k) = \frac{\Gamma\ (K + r)}{K!\ \Gamma\ (r)}\ p^r\ q^K \qquad (8)$$

$r > 0,\ 0 < p < 1,\ p + q = 1$

with mean and variance

$$E(K) = qr\ /\ p; \qquad \sigma^2\ (K) = qr\ /\ p^2 \qquad (9)$$

K equals the number of deposits, Γ is the gamma function and p, q, and r are parameters. E(K) is the average number of deposits per cell. Estimates of p, q, and r can be made indirectly using the method proposed by AGTERBERG 1977. The method is based on the relationship between the negative binomial distribution and the gamma and poisson distributions.

$$P\ (K=k) = \int_0^{\infty} \frac{\lambda\ K\ e^{-\lambda}\ \theta^{M+1}}{K!\ \ M!}\ X^M e^{-\theta X} d\ \lambda \qquad (10)$$

$X > 0,\ M > -1,\ \theta > 0$

with mean and variance

$$E\ (X) = (M + 1)\ /\ \theta$$

$$\sigma^2\ (X) = (M + 1)\ /\ \theta^2 \qquad (11)$$

X is the percentage of the variable in a cell which is related to the type deposit sought, here the Middle Group, and λ is the average number of deposits per cell. The number of deposits is assumed to be proportional to the percentage of the Middle Group which outcrops, i.e. $\lambda = a\ X$.
It can be shown that

$$\theta = a\ p\ /\ q; \qquad M + 1 = r; \qquad E\ (K) = E\ (X) \qquad (12)$$

Thus if E(K), E(X), and σ^2(X) were known, the parameters p, q, and r could be determined and the negative binomial distribution calculated. For the Nubian Sandstone Series, neither detailed maps showing the Middle Group, thereby allowing E(X) and σ^2 (X) to be calculated, nor a knowledge of the average number of deposits per cell, E(K), are available. The assessment has to be based on estimates by geologists who are familiar with the area. Estimates of the percentages of the Middle Group which outcrops were furnished by geologists from the Remote Sensing and Photogeological Section of the Institut für Angewandte Geologie, Freie Universität Berlin. The mean and variance of the Middle Group were estimated to be 25 and 100 respectively. E(K) remains unknown. The probability of occurence for the number of deposits per cell is shown in fig. 3 as a function of E(K). The total number of deposits is determined by estimating E(K) and multiplying the probability of occurence for the number of deposits by the total number of cells, here 100.

The reliability of the results are dependent on the geologists' estimates. The authors' experience has shown that most geologists overestimate visual parameters, i.e., their perception is distorted since they concentrate on what they are searching for and tend to neglect their surroundings. The resource assessment must therefore be considered optimistic. Furthermore, the size and number of cells are related to the average number of deposits per cell. The economic feasibility of the deposits is implicit in the choice of E(K). An estimate of E(K) is, however, difficult. Investigations have shown that the thickness and iron content of the iron beds in the vicinity of Kharga are much less than those near Aswan. Coupled with the inaccessibility of the major portion of the Nubian Sandstone Series, an estimate of $0 < E(K) < 0.025$ may be realistic. This implies that a maximum of two deposits can be expected to occur. This assessment is speculative. But based on the available information, it furnishes an expeditious assessment whose reliability is approximately commensurate with the data. The method does not furnish information about specific areas, i.e. where the deposits ae located, or about the size of the deposits.

Whereas the above method supplies a global assessment, distributions may also be used to yield an assessment of specific areas whose exploitation may be economically feasible. The following method was used: The area is divided into n cells, with each cell being large enough to contain one deposit. Each cell is then sampled. The probability that the cell contains an iron deposit is governed by the binomal distribution, i.e., there are only two outcomes, yes or no. If yes, its tonnage is estimated from a tonnage distribution. The binomal distribution B(n,p), where n is the number of attempts and p the probability of occurrence, is related the poisson distribution $\Pi(\lambda)$, where λ is the number of occurences.

$$\lim_{n \to \infty} B(n,p) = \Pi(\lambda) \tag{13}$$

By setting $\lambda = n\,p$, Eq. 13 is approximately fulfilled. A Monte Carlo simulation consisting of the following steps was used to duplicate this process:

1. The number of deposits is selected randomly from a Poisson distribution with given λ.
2. Each deposit is assigned a tonnage based a given distribution, either discrete or continuous.

Each replication is comprised of steps 1 and 2 and is repeated several hundred times. The probability that the area contains one or more deposits of a particular size can then be easily calculated.

The average number of deposits λ can be taken from a geologic corresponding and well-defined outcropping of the Middle Group. The rationale behind the Monte Carlo sampling is that the Middle Group need not contain iron beds, and that mineable reserves may be covered by a thin layer of Upper Nubian. The number of discrete deposits is therefore either greater or less than the number delineated on the map. The tonnage distribution may be based on known occurrences, e.g., the area east of Aswan, or on an theoretical distribution. The tonnage distribution used was derived from reserve estimates for the area east of Aswan (ATTIA 1955). A theoretical distribution, which is to be preferred, was not used in order to illustrate how the method may correct historical or training values. The fifteen deposits east of Aswan were divided into three classes: 1. 0-10 million tons, 2. 10-20 million tons, and 3. > 20 million tons. Their probabilities were assumed to be 0.52, 0.42, and 0.06 respectively. In fig 4, the probabilities that 1, 2, or 3 deposits occur are shown as a function of λ . For example, according to the distribution, every 15th deposit should be greater than 20 million tons. The probability that one class 3 deposit occurs in an area containing 15 deposits should therefore be 100 %. But fig. 4 shows that P(k > 1) = 61 %, while P(k = 1) = 45 %.

The extrapolation of historical values to unexplored areas is an unreliable guide for exploration and with risk. The clustering of deposits causes resulting distributions to be distorted. Once a discovery has been made, exploration effort is concentrated in the vicinity of the discovery. This leads to further discoveries and does not mirror the risk inherent in the initial discovery. This also applies to theoretical distributions such as mineral wealth being lognormally distributed. The Monte Carlo simulation lowers the optimistic estimate to a more realistic level. Once an area is chosen for exploration, the method enables an estimate of the probability of occurence to be made.

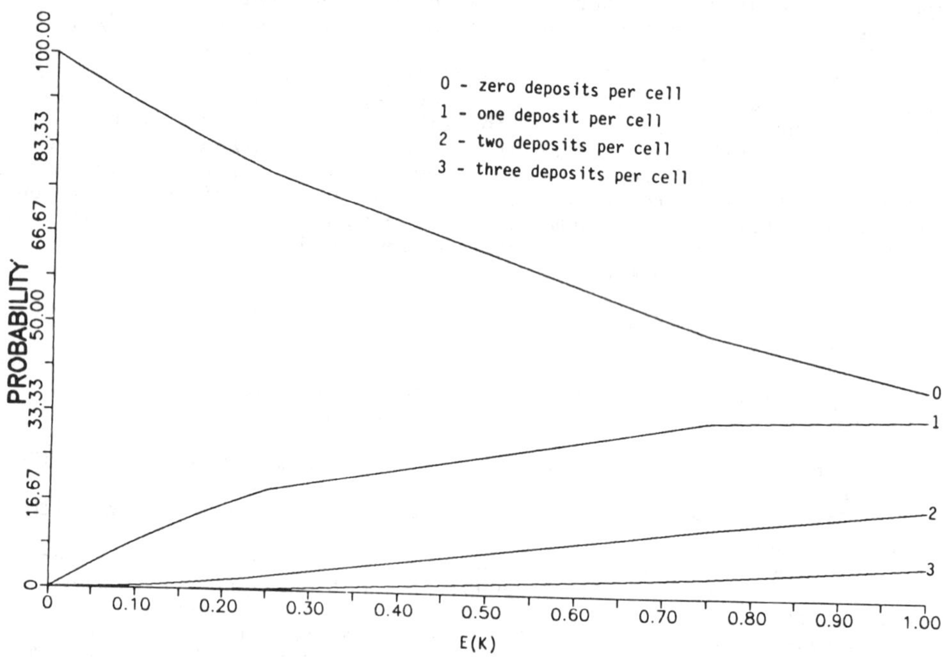

Fig. 3: Probability of occurrence for deposits per cell as a function of the expectation, E (K) of the negative binomial distribution (after SHULMAN & SKALA 1985)

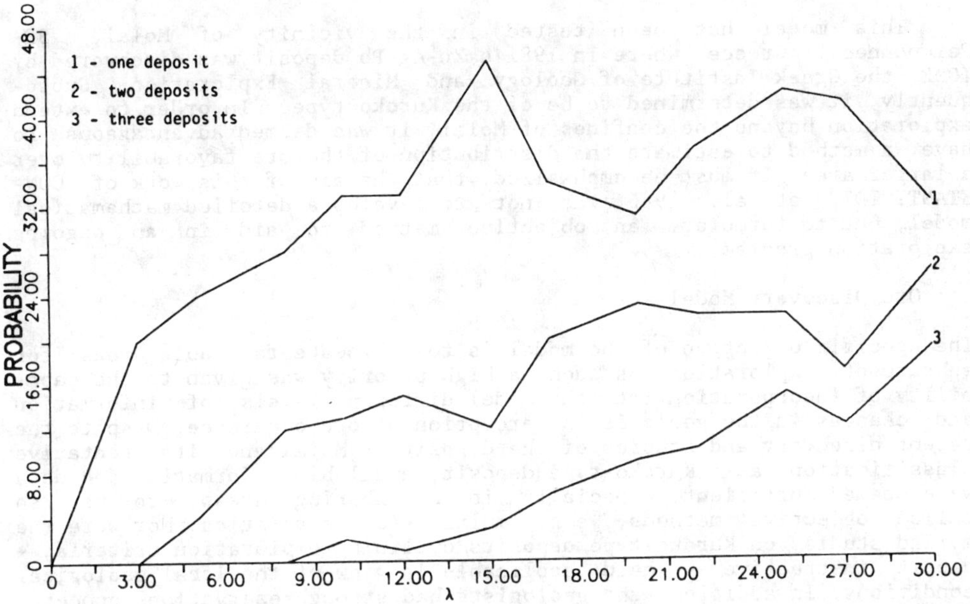

Fig. 4
Probability of occurrence for the number of iron deposits in an are as a function of the number of discrete, Middle Groups outcrops (after SHULMAN & SKALA 1985)

4. THE CLASSIFICATION AND RANKING OF EXPLORATION PROSPECTS: KUROKO TYPE ZN-AG-PB DEPOSITS, GREECE

Both preceding case studies have been determinated by very low information levels, characterized by small geological maps and speculative knowledge of criteria which control the occurence of phosphate and iron deposits (reconnaissance phase). Therefore only general recommendations for further planning activities can be given. In later stages of exploration, when ore favorability has to be considered in more detail, larger scaled geological maps as well as knowledges from training areas or training deposits often are available, from which ore favorability criteria can be deduced. During this exploration stage, PROSPECTOR (DUDA et al., 1978) can be very useful tool, if models of respective classes of ore deposits are available. Another method is "deposit modelling" method, such characteristic analysis (McCAMMON et al., 1981, 1984). Since characteristic analysis is a multiple regression method (of binary data) it requires information from well known deposits. A modified and more pragmatic approach to "favorability mapping" has been proposed by CONSTANTINIDES et al., 1986. This model, developed in our working group, works by Monte Carlo simulations based on qualitative and quantitative input data. Although sufficient geological information is necessary, it doesn't necessarily require any individual training deposit.

This model has been tested in the vicinity of Molai, SE-Peloponnese, Greece, where in 1981 a Zn-Ag-Pb deposit was discovered by IGME, the Greek Institute of Geology and Mineral Exploration. Subsequently, it was determined to be of the Kuroko type. In order to extend exploration beyond the confines of Molai, it was deemed advantageous to have a method to estimate the distribution of the ore favorability over a larger area. It must be emphasized, that the aim of this work of CONSTANTINIDIS et al. 1986 was not to develop a detailed mathematical model, but to formulate an objective method to aid in an ongoing exploration program.

4.1 Ore Discovery Model

The specific objective of the model is to delineate favorable areas for subsequent exploration. As such, a high priority was given to the capability of incorporation into the model different levels of information and changes in the geologist's perception of ore occurence. Despite the recent discovery and studies of the deposit at Molai and its tentative classification as a Kuroko-type deposit, available information and data were deemed sufficient, especially in neighboring areas, to use so called objective methods, e.g. multivariate statistics. Nor were the myriad studies on Kuroko-type deposits outlining exploration criteria - albeit instructive - held applicable in view of the local geological conditions. In addition, the geologists had strong reservations concerning statistical methods which could not be overcome, especially regarding lack of a specific ore characterization model, interpretation of results and its usurpation of their knowledge and responsibility. It was finally decided that a subjective ore discovery model, a simplified expert system, would best fulfill the objective; that such a model could be easily adapted to incorporate different levels of information and changes in the geologists perception and knowledge of ore occurence; and that the exploration geologists, by being brought directly into the initial stages of the assessment and by assisting in the mathematical formulation of the model, would not only accept, but also use the results. Secondary considerations included inculcating and encouraging introspection on the part of the geologists in quantifying the exploration criteria, in resolving differences of opinion and in forming a consensus.

The model decided upon consists of a set of n exploration criteria c_i, weights w_i and an ore favorability index f defined by:

$$f = \sum_{i=1}^{n} w_i P(c_i) \qquad (14)$$

with

$$\sum_{i=1}^{n} w_i = 1 \ : \ \sum_{i=1}^{n} p\ (c_i) \neq 1 \ : \ 0 \leq w_i \leq 1$$

The weights can be interpreted as the favorability of exploration criteria. A Monte Carlo simulation was used to calculate the ore-favorability. The weights remained constant, whereas the criteria obeyed either an uniform, triangular or point distribution. To rank the areas, the favorablity curves must be compared and a specific favorability index derived from each curves. CONSTANTINIDES et al. 1986 used the simulation to represent the curve. Two other, more sophisticated methods were also available.

4.2 Determination of the Favorable Criteria

The first and most critical step is to determine the favorable criteria. Exploring for deposits similar to that found at Molai (Kuroko type) requires:

a) that criteria indicating Kuroko type mineralization as described in the literature and b) that the specific characteristics of the Molai deposit as well as of the regional geology be taken into consideration. The difficulty lies not in preparing such a list, but in assigning a weight to each criterion, which reflects its importance relative to the other criteria as an ore indicator. Differences in opinion among geologists as to the weights naturally result.

Another problem which arises in preparing such lists is related to the "information level" of the area under exploration. For example, if exploration is initiated in a completely unknown area, the favorable criteria may be stratigraphy, petrography, etc. In a partially explored area, however, the criteria may include detailed petrological types, alteration zones, preferred trends of faults, etc. Therefore different sets of criteria based on the information level if the area must be drawn up.

In this case, in order to avoid the above problems, a method similar to "Delphi" (BAXTER et al., 1978) was used. Three groups of experienced economic geologists familiar with the Molai area independently prepared lists of favorable criteria for each level of information. In other words, a list of information or data which a geologist must gather from the field or laboratory to estimate the ore-favorability of an area was determined. Furthermore, the three groups assigned to each criterion a weighting factor (ranging from 0 to 1), reflecting its importance. Since no significant differences among the three lists were ascertained, it wasn't difficult to reach a consensus.

The three levels of information were denoted as A, B, and C. SHULMAN 1986 (table 1) shows the criteria assigned to each level as well as their weighting factors. Since the sum of the weighting factors equals 1, the presence of all the criteria implies on ore-favorability of 1.

4.3 Estimation of Favorability

The Molai area was subdivided into a number of blocks (cells) whose dimensions depended on the information level used. The basic criterion used in the determination of the block dimensions was the possibility or capability of assigning to each block an estimate of the ore-favorability criteria. For Molai, levels A, B and C consisted of 24, 48 and 68 blocks, respectively.

Only the information assumed available at the particular level of information was used to assign values to the criteria. This implies that information gathered subsequent to a level of information was, as far as possible, not incorporated into values. For example, information gained from field studies, thereby increasing the level of information from A to B, was not used in the test of level A.

The probability distribution preferred was triangular, being uniquely determined by assigning maximum, minimum and most probable values to each criterion. Since a minimum and maximum of 0.0 and 1.0, respectively were assumed, only the most probable value had to be given.

The ore-favorability of each block was calculated using the Monte Carlo program - 500 iterations were used. The results are shown in SHULMAN 1986 (fig. 2) for the three information levels of Molai.

4.4 Discussion and Conclusion

The results proved to exceed expectations and were found to be extremely valuable by geologists currently involved in exploring for the Molai type deposits. Regions known or thought to be promising were clearly outlined on the favorability maps, having higher favorability indices. In the case of the Molai region, the area of known ore occurence was estimated to have a favorability of approximately 1, while its probable extensions exhibited favorability indices greater than 0.8. Subsequent drilling in these extensions confirmed the presence of ore bodies. Furthermore, a distinction between areas of Cu mineralization and Zn-Ag-Pb mineralization was apparent, implying that different sets of favorable criteria are valid for each type of mineralization. With an increase of the level of information this fact, as well as, other geological trends, are clarified.

Favorability mapping activities in Greece in cooperation between the Mathematical Geology Working Group, Berlin and IGME, Athens are going to be continued. Therefore, investigations about possible interactions between optimal exploration strategies and policy decisions have to be considered. A first review using multiobjective operation research methods is presented by SHULMAN 1986.

5. BIGA PENINSULA (TURKEY) PROBABILITY MAPPING PROJECT

When this project began in our working group, detailed information concerning the entire area of interest in the Balikesir province (Southern Biga Peninsula, Turkey) already existed, having been gathered during intensive exploration activities by the Turkish and German Geological Surveys. So the actual exploration level of this project can be called advanced in comparison with the preceding projects. Since the Balikesir province includes well known mineral deposits, some dating from antiquity it furnishes an ideal situation to develop and test quantitative techniques for mineral resource assessment. Since data and information are available from geophysical, geochemical and geological exploration as well as from remote sensing, integrated data analyses seem to be successful for the selection of favorable areas of mineralization in Balikesir region (BURGER & KLIEM 1985). Some parts of the exploration area (about 5000 km^2) are characterized by relatively high density of geochemical analyses data. These areas of detailed exploration mostly surround old mining districts or centers of high ore favorability. How these areas can be used for traning purposes will also be shown by BURGER & KLIEM 1985. In connection with the goals of this paper, an investigation of the internal structures of these training areas can be of interest. One of the areas explored in detail is the Handeresi district, an area of about 3 km^2, located between Edremit and Kalkim. This area, mostly underlained by metamorphic schists (Kalabak-Series) and granodiorites contains the currently worked mine of Handeresi. Approximately 1800 soil samples have been taken from a 100 x 50 m (100 x 25 m respectively) grid and analyzed for Pb-Zn-Cu contents by BGR (Bundesanstalt für Geowissenschaften und Rohstoffe, Hannover) and MTA (Maden Tektik ve Arama Enstitüsü, Ankara)

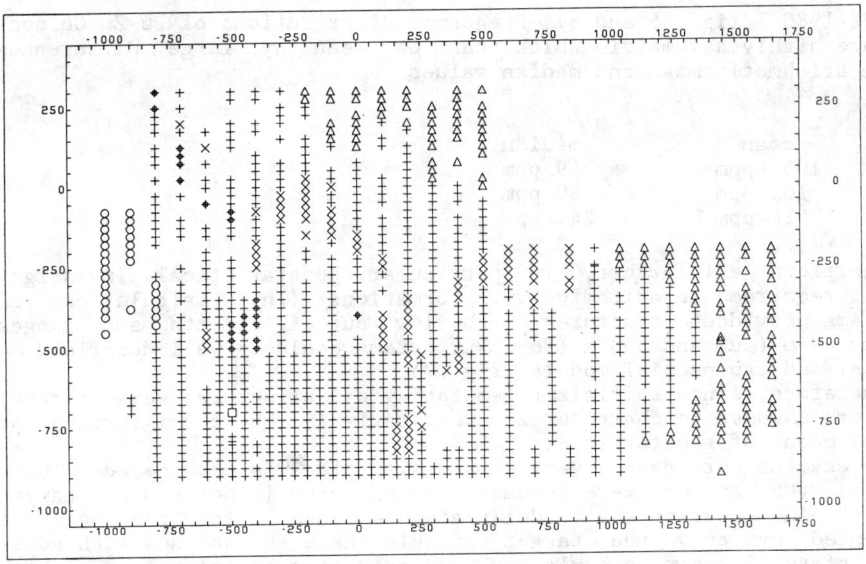

Fig. 5: Handeresi, Biga Peninsula, distribution of soil samples (horizon BC) for different rock units

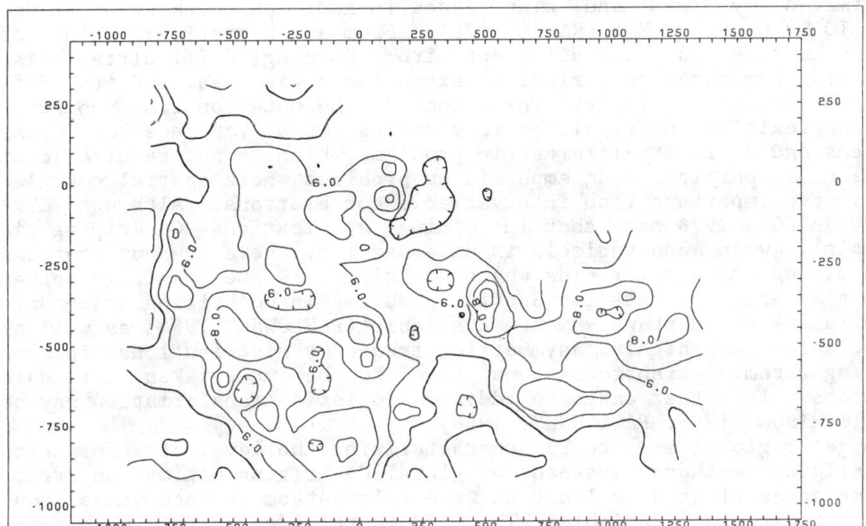

Fig. 6: Handeresi, Biga Peninsula: distribution of Pb-contents (Ln(Pb)) in soil samples (horizon BC)

in 1978-1980 (fig. 5 and 6). Frequency distributions of Pb-Zn-Cu contents are highly asymmetric which can be seen by large differences between arithmetic mean and median values.

	mean	median
Pb	1053 ppm	759 ppm
Cu	103 ppm	60 ppm
Zn	526 ppm	241 ppm

Therefore, all further, in particular geostatistical investigations, required logarithmic transformations (ln). Calculations of variograms provided structures with low nugget effects and ranges between 60 m (Cu) and 500 m (Pb). Variograms could be well described by spherical and exponential models (fig. 7).

Therefore, since sufficient geostatistical structures were present, they were assumed to characterize quantitatively the actual information level by means of geostatistics.

To examine our ideas a very simple situation was considered: Three variables (Pb, Zn, Cu) were assumed, one of which is denoted as "target variable". The question was to delineate, how much information can be represented by only the target variable (here Pb) and how much additional information can be explained by considering additional variables (Zn, Cu). A solution of this problem will be attempted by Co-Kriging methods, which are applied in geostatistics, when one variable was not sampled sufficiently to provide estimates of acceptable precision. In these cases, the precision of this estimation may then be improved by considering the spatial correlations between this variable and other, better sampled variables. For example, the silver grade of a panel can be estimated by lead and zinc grades in addition to those of silver (GUARASCIO 1976, BORGMAN & FRAHME 1976). From a theoretical point of view, Co-Kriging is not different from Kriging; difficulties arise essentially from notations, since an extra index is required to differentiate between variables. These include the notational and computational complexities, difficulties in modeling cross variance or cross variograms and a primary interest in problems which do not require joint estimation, in particular an emphasis on problems where spatial correlation is more important than intervariable correlations. Although JOURNEL & HUIJBREGTS 1978 note that in mining applications Co-Kriging is used mainly when one variable is undersampled, there are applications both in mining and other fields where Co-Kriging is relevant if other difficulties are surmountable. Spatial problems in Co-Kriging arise only when variables of constant sum are available (PAWLOWSKY 1984) as well as for variables of highly asymmetric frequency distributions. In some cases, logarithmic transformations have to be undertaken for data corrections, in other cases an additive-logistic transformation may be used (AITCHISON, 1986; PAWLOWSKY, 1986).

To get a global measure for characterizing the level of information by Co-Kriging methods instead of global Co-Kriging estimation error, global measures of cross-validation were calculated, in each case considering additional variables. The fundamental idea of this method can be compared with the stepwise regression method on the base of cross-validation.

The original goal of cross-validation has been to provide methods of "testing" whether a semivariogram adequately represents the study area. With this approach, the analyst uses a partial data set to estimate values at actual sampled positions. Real (z) and estimated (z^*) values are then compared in such way that the model can be accepted or rejected. The process of cross- validation has been discussed in some detail by CLARK 1986.

For our purpose of global estimation of reliability of the information level, the arithmetic mean of the error $z-z^*$ and its variance were computed.

Comparable with the method of stepwise regression after evaluation of cross-validation expectations of individual variables, cross validations of Co-Kriging results were computed by the stepwise addition of variables. For lead as target variable, this method provides the following results for 1130 used points (results in ln ppm).

	$E(z - z^*)$	$var(z - z^*)$
Pb(individual)	$0.203.10^{-2}$	0.78
Pb(Zn)	$0.600.10^{-3}$	0.57
Pb(Zn,Cu)	$0.397.10^{-4}$	0.80

The results of this very simple case study demonstrate that cross-validation of co-kriged data can provide a useful measure for estimating the reliablity of mapped values by stepwise addition of available variables. These results can be compared immediately as a basis for further exploration ore mining activities.

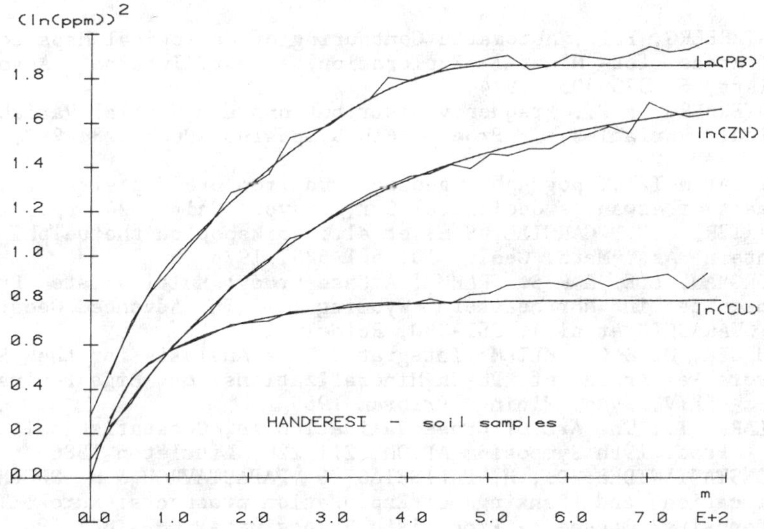

Fig. 7: Variograms of Handeresi soil samples
(Biga Peninsula, Turkey)
Mean variogram of ln (X)

7. CONCLUSIONS

The initial step for the development of a measure to quantify and to compare the reliability of information levels in exploration, as well as to assess their applicablity for planning and decision activities, must be the description and interpretation of all variables, their spatial density and their economical and technological importance. Since data from complete case histories were not available, characterization of individual information levels had to be obtained from different case studies. Comparable measures for reliability of different information levels can only be provided, if a unique target variable can be determined which is chosen to represent significantly all available information. In the Biga Peninsula case study an attempt has been made to develop such a measure by cross validation of co-Kriged values. Further studies have to carried out to decide, if this method can be seen and treated in a more general sense. Because of the differences in target variables as well as available information a more successful solution to this problem should be prepared by presentation of "unit regional values" concepts, as proposed for assessment studies by GRIFFITHS 1969 and SINGER 1971. As a second step, an extension from regional values to global measures to characterize the individual information level as a whole with regard on a specific error measure can be provided by geostatistical methods.

8. REFERENCES

AGTERBERG, F.P.: Automatic Contouring of Geological Maps to Detect Target Areas for Mineral Exploration. - Jour. Internat. Assoc. Math. Geologists, 6, 373-395, 1974

AGTERBERG, F.P.: Frequency Distributions and Spatial Variability of Geological Variables. - Proc. 14th Symposium APCOM, 287-298, New York 1977

ATTIA, M.I.: Topography, geology and iron-ore deposits of the district east of Aswan. - Geological Surv. Egypt, Cairo, 247 pp., 1955

BAXTER, G.G., CARGILL, S.M. et al.: Workshop on the Delphi Method. - J. Intern. Ass. Math. Geol., 10, 581-588, 1978

BORGMAN, L.E. & R.B. FRAHME: A Case Study: Multivariate Properties of Bentonite in Northeastern Wyoming. - In: Advanced Geostatistics (Eds.: GUARASCIO et al.), 381-390, Reidel, 1976

BURGER, H. & Cl. KLIEM: Integrated Data Analysis for the Selection of Favorable Areas of Pb-Zn-Mineralizations on Biga Peninsula, NW-Turkey. - XXIV. Symp. Mining, Pribram 1985

CLARK, I.: The Art of Cross Validation in Geostatistical Applications. - Proc. 19th Symposium APCOM, 211-220, Littleton 1986

CONSTANTINIDES, D., H. FALTSETAS, S. PAPASTAVROU & M. SHULMAN: The Classification and Ranking of Exploration prospects: Kuroko Type, Zn-Ag-Pb Deposits, Greece. - Proc. 19th Symposium APCOM, 687-694, Littleton, 1986

DUDA, R., HART, P., et al.: Development of the Prospector Consulation System for Mineral Exploration,. SRI International, Final Report, 173 S., 1978

GRIFFITHS, J.C.: The Unit Regional-Value Concept and its Application to Kansas: Kansas Geol. Surv. Spec. Distr. Publ., 38, 48 pp., Lawrence, Kansas 1969

GUARASCIO, M.: Improving the Uranium Deposits Estimations (The

Novazza Case). - In: GUARASCIO et al. (Eds.) Advanced Geostatistics in the Mining Industry, 351-368, Reidel, Dordrecht 1976

HAMBLETON, W.W., DAVIS, J.C. & DOVETON, J.H.: Estimating exploration, in HAUN (Ed.) Methods of Estimating the Volume of Undiscovered Oil and Gas Resources: Amer. Ass,. Petrol. Geol., Studies in Geology, 1, 171-185, 1975

HARRIS, D.P.: Mineral resources appraisal. - Oxford Geol. Sci. Ser., 445 S., Carendon, Oxford 1984

JOURNEL, A. & HUIJBREGTS, Ch.: Mining Geostatistics. - Academic Press, London 1978

KLITZSCH, E. & LEJAL-NICOL, A.: Flora and Fauna from Strata in Southern Egypt and Northern Sudan (Nubia and Surrounding Areas). - Berliner Geowiss. Abhandl., A, 50, 47-80, Berlin 1984

McCAMMON, R.B., J.M. BOTBOL, R. SINDING-LARSEN & R.W. BOTBOL: The New Characteristic Analysis / (NCHARAN) Program. - Geol. Survey Bulletin 1621, 27 pp., 1984

Geological Survey of Egypt: Map of Mineral Resources of Egypt. - Cairo 1979

PAWLOWSKY, V.: La Covariance Spatiale des Variables de Somme Constante. - Sci. de la Terre, Paris 1984

SHULMAN, M.: An Exploration Strategy for Phosphates in the Sinai Peninsula, Egypt. - Proc. 18th APCOM Intern. Symposium, London 1984 (a)

SHULMAN, M.: Probability Mapping Using a Monte Carlo Simulation. - Berliner Geowiss. Abh., 50, Berlin 1984 (b)

SHULMAN, M.: Initial Optimal Allocation of Effort in Mineral Exploration. - Transactions SME-AIME, Littleton, 1984 (c)

SHULMAN, M. & W. SKALA: Computer Assisted Exploration: Exploration for Phosphorites in the Sinai Peninsula, Egypt - a Review. - Berliner Geowiss. Abh., A, 50, 399-406, Berlin 1984

SHULMAN, M. & W. SKALA: Predictive Assessment of Mineral Potential in Desert Terrain. - In: Prospecting in Areas of Desert Terrain, 59-69, IMM, London 1985

SINGER, D.A.: Multivariate Statistical Analyisis of the Unit Regional Value of Mineral Resources.: Ph. D., Dept., Geochem. Min., Penn State Univ., 210 pp., State College 1971

REVIEW OF GEOMATHEMATICAL APPLICATIONS FOR MINERAL RESOURCES EVALUATION IN CHINA

Zhao Pengda
Wuhan College of Geology
Wuhan, Hubei
People's Republic of China

ABSTRACT

In the recent decade, the study of geomathematics in China has gained extensive development. Of the numerous research directions, making use of mathematical models in quantitative prediction of mineral deposits and in assessment of mineral resources has been taking the most important position. Five nation-wide academic symposia on geomathematics were held in China during the last ten years. Over a quarter of the papers presented dealt with problems of statistical prediction and resource assessment. This review summarizes the types of geomathematical applications performed.

1. Introduction

The basic approach to statistical applications for mineral resource evaluation in China is illustrated in Figure 1. As shown many different statistical techniques are applied to the data. In addition a great number of feedback mechanisms are built into our approach to improve our statistical models and the results obtained by means of these models.

The methods of statistical prediction for mineral resource evaluation can be classified into three categories according to the following criteria:

1. Prediction of the total amount of the resources in a large region.
2. Quantitative prediction for specific regions within an ore-bearing zone or ore province.
3. Statistical analysis of ore-controlling factors and assessment of ore-bearing environments.

The current status of this work in China is shown in Table 1 and Table 2.

C. F. Chung et al. (eds.), Quantitative Analysis of Mineral and Energy Resources, 79–87.

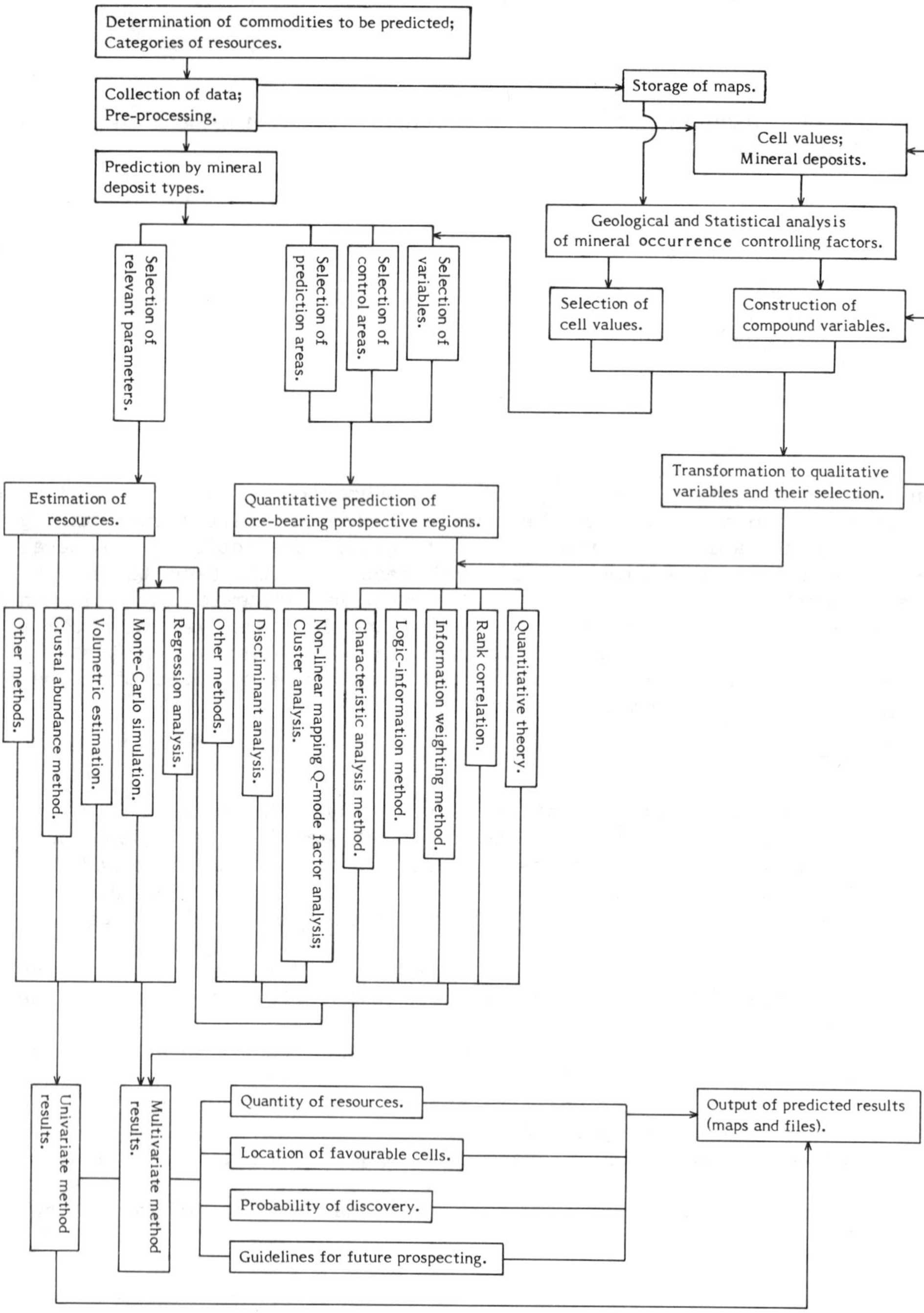

Figure 1. Flowchart of process for statistical prediction of mineral deposits and resources assessment in China.

Table 1. Several selected examples of statistical prediction and assessment of mineral resources in China. The numbers in column "Method used" are referred in Table 2.

Type of prediction and scale	Commodity of prediction	Prediction area	Method used	Date	Organization
Resources Quantity Estimation (RQE) 1:10,000	Ni	Ore field model for ore-bearing parts of Jilin Province; All Chinese Platform regions	6, 23, 29, 5.	1981-1983	Jilin Geology-Mineral Bureau; Changchun College of Geology
RQE 1:200,000	Cu, Fe, S, Alunite	Single map sheet area	23, 25, 30, 27, 26, 7, 8, 5, 6.	1981-1983	Wuhan College of Geology; Computer Centre of Mineral Geology
RQE 1:50,000	Fe, Cu	Southeastern part of Hubei Province	18, 15, 23, 6, 5.	1981-1983	Hubei Geology-Mineral Bureau
RQE 1:500,000	Cu	Regions along Yangzie River in Anhui Province	12, 29, Delphi method estimation, Zipf law method	1981-1983	Anhui Geology-Mineral Bureau
RQE 1:200,000	W	Single map sheet area	6, 25, 23. 17. 27, Zipf law method	1981-1983	Jiangxi Geology-Mineral Bureau
RQE 1:10,000 1:25,000	V, Ti, Fe	Part of Sichuan Province	17, 6, 23.	1981-1983	Sichuan Geology-Mineral Bureau
RQE 1:200,000 1:50,000 1:10,000	Sn	Western part of Yunnan Province	Delphi estimation 17, 32, 22, Favourability index method	1984-1986	Chengdu College of Geology; Yunnan Geology-Mineral Bureau
Prospective Regions Prediction (PRP) 1:50,000 1:25,000 1:10,000	Fe, Cu	Part of Jiangsu, Anhui, Hebei and Neimeng Provinces	7, 8, 22. 23, 25.	1976-1980	Wuhan College of Geology
PRP 1:200,000	Fe	Part of Fujian Province	22, 23, 25.	1977	Institute of Mineral deposits
Statistical Analysis of Factors (SAF) 1:10,000	Pb, Zn	Ore deposit	21, 8.	1982	Geoexploration Co. for nonferrous metals in Hansu Province
PRP 1:10,000	Au	Ore deposit; ore field	23, 25, 27.	1980-1983	Geoexploration Co. for nonferrous metals in Gansu Province

Table 2. Methods used for statistical prediction and assessment of mineral resources in China.

Type of Data	Qualitative Data														Quantitative Data																					
Methods / Purpose	1. Quantitative I	2. Quantitative II	3. Quantitative III	4. Quantitative IV	5. Logic information	6. Characteristic	7. Rank correlation	8. Quantity of information	9. Conditional probability	10. Markov process	11. Probability distribution	12. Subjective probability	13. Deposit modeling	14. Embedded models	15. Correlation coefficient	16. Canonical correlation	17. Crustal abundance est.	18. Volumetric estimation	19. Geometrical probability	20. Time rate	21. Trend surface	22. Discriminant	23. Multiple regression	24. Logistic	25. Cluster	26. Nonlinear mapping	27. Factor	28. Correspondence	29. Monte-Carlo	30. Fuzzy cluster	31. Fuzzy discriminant	32. Geostatistics	33. Bayesian	34. Decision-making	35. Optimal partitioning	36. Step function
1. Correlation of ore-bearing strata										X																									X	X
2. Classification of igneous bodies, gossans and anomalies		X																				X			X	X	X	X		X	X					
3. Identification of ore geohunting indicators					X	X	X	X	X						X						X	X	X		X	X		X							X	
4. Spatial distribution of Mineralization																					X											X				
5. Establishing the relation between mineralization and geological factors	X		X		X	X	X	X	X		X		X		X	X						X	X				X	X								
6. Correlation of ore forming conditions and delineation of prospective regions	X	X	X	X	X	X	X	X	X		X	X	X	X							X	X	X		X	X	X			X	X	X				
7. Estimation of probabilities of known mineral deposits or risk in prospecting	X			X							X			X					X			X	X	X					X				X			
8. Evaluating the scale of mineralization	X	X			X						X	X					X	X					X						X							
9. Evaluation of ore-bearing igneous bodies	X				X							X					X	X					X						X							
10. Prediction of the spatial extent of ore bodies		X										X									X	X														
11. Target selection																				X													X	X		
12. Identification of optimal intervals for ore hunting indicators							X																													

2. Some studies on theory and methods in statistical prediction for mineral resources evaluation in China

2.1 Basic principles

There are different opinions as to whether or not a mineral deposit can be quantitatively predicted. By using the so-called logic-information method we have learned it is generally possible to define one or more indexes that can be used to distinguish between different types of mineral deposits. The larger the number of objects needed to be distinguished, the larger the number of indexes that have to be considered simultaneously. On the other hand, to form a certain kind of mineral deposit, there must have existed a certain number of factors. This is what we call "the principle of determination of the combination of indexes for factors controlling the formation of a deposit". The indexes should represent the factors as much as possible. If index combinations are used in practice, we sometimes find there exists a mineral deposit, sometimes not. This reflects a random characteristic of the deposit-forming process. However, if more essential indexes are added to the combinations, the probability of finding a deposit increases. Ideally, as the number of necessary conditions or indexes increases, the deposit-finding probability gradually approaches one representing certainty of occurrence. If, for an area, only a few indexes can be defined for factors related to mineralization, the deposit-finding probability will remain small. This is "the principle of uncertainty for the deposit-finding probability".

The following three principles underlie statistical prediction for mineral resources evaluation in practice.

(1) Model analogy – "Seeking the common features principle".
(2) Anomaly definition – "Seeking the abnormal situation principle".
(3) Contrast evaluation – "Optimization principle".

The effect of statistical prediction for mineral resource evaluation mainly depends on:

(1) The reliability and availability of the original data.
(2) The suitability and sufficiency of the variables constructed and selected.
(3) The accuracy and efficiency of the predictive models.
(4) The correct selection of the targets.

2.2 Reliability and availability of the original data

In order to enhance the effect of statistical prediction for mineral resources evaluation, it is very important to make use of synthesized information variables. Several aspects have to be considered from the point of view of geological theory, which is used to synthesize as many ore-controlling factors and ore prospecting features is possible. From the point of view of methodology, this leads to the construction of synthesized information maps showing ore-forming regularities.

One of these maps, for example, was made by Wang Yutian in his 1985 thesis "Construction of synthesized information maps and estimation of the total amount of fluorite resources in Zhejiang Province".

2.3 Construction and selection of the variables

Based on the detailed study of geological conditions, the construction of the variables is the foundation of the effective performance of statistical prediction for mineral resource evaluation.

2.3.1 Weighted variables

If a good degree of relationship between a geological factor and the mineralization of the studied area has been established, we can give the factor a certain weight. For instance, in a study on occurrence of copper deposits in the Toagling area, stratigraphic variables were constructed, based on the different favourability degrees of mineralization with regard to the different ages of strata. We adopted variables that represent the occurrence areas of the strata of certain ages within a cell and these were multiplied by mineralization frequencies for the same ages. In this way, the strata with a relatively close relation to copper mineralization were given more weight.

2.3.2 Ratio variables

If the geological meaning of the ratio of two elements is well understood, we can construct a ratio variable related to these two elements. In one of our recent studies we have constructed the following ratio variables in the prediction of a tin deposit:

Ba/Sr Reflecting the degree of acidity of magmatite. The smaller the value of the ratio, the lower the degree of acidity and vice versa.

V/Ti Reflecting the sedimentary environment. This ratio is relatively small in mechanically deposited sediments, and relatively large in chemical sediments.

Pb/Be Reflecting the proximity of a deeply buried ore body. According to the study of zoning of elements in this type of deposit, Pb indicates the frontier halo and Be indicates the tail halo. Then a large ratio value indicates shallow depth of the ore body.

2.3.3 Synthetical variables

In order to reduce the number of the original variables and to create new variables with specific geological significance, we have constructed various synthetical variables. For example, in 1976 we constructed "feature value of rock association" and "feature value of association of country rock alteration" by means of calculation of entropy functions. We also made use of the entropy function to represent "deposit-forming energy" $En = \sum_{i=1}^{n} K_i \ln K_i$ in the prediction of a tin deposit in eastern Yunnan, China, where K_i is the Clark abundance of the ith mineralization element.

2.3.4 Scoring variables

These represent another kind of synthetical variables obtained by means of geological analysis and logical deduction. In the prediction of the Gejiu tin deposits, we constructed structural scoring variables. In the prediction of a copper deposit in one of our recent studies, we constructed structure, magmatite and strata scoring variables. These scoring variables were used to evaluate the favourable ore-bearing degrees of different cells. They represented the various favourable ore-controlling geological conditions and their varous possible combinations. The deposits in the study area were a type of stratigraphically controlled skarn copper deposit. Through geological analysis, we concluded that, East-West directional basement faults controlled magmatic intrusions; the mineralization favourable magmatic intrusions were acidic diorite intrusions; NE-SW directed fold axes within the covering strata determined the position of the ore deposits; the favourable strata and the properties of the country rock provided mineralization with favourable material components and affected the shape and occurrence of the ore bodies. Statistical scores were assigned to these four ore-controlling favourable factors. Then we obtained the scoring variables which indicate favourable degree of mineralization for different cells.

2.4 Study and selection of statistical models

To perform statistical prediction for mineral resource evaluation in an area where the geological and exploration data are relatively poor, we generally have the difficulty that the number of model areas (cells) is not sufficiently large, as there occur relatively few known deposits within the study area. In recent years we have proposed several "zero-control prediction methods".

(i) The "favourable degree of mineralization method" (Luo Junfeng, in 1985) using geostatistical techniques;

(ii) "Anomaly-finding prediction method" (Yuan Ding, in 1985) using cluster analysis and statistical experimental design techniques;

(iii) "Location of probably abnormal cells by simple statistical methods" (Zhao Pengda, Huwangliang, and Li Zijin in 1984). This included use of stepwise regression analysis to predict frequencies of copper deposits in the Anhui Province as well as use of Agterberg's method of comparing a single mineralized cell with all other cells in a region.

2.5 Selection of targets

There is still a lack of strict and effective criteria for optional target selection. Generally we choose:

(1) The area that the majority of the methods indicate as a favourable mineralized area;
(2) The area where the value of the estimation of a parameter connected to mineralization obtains its maximum;
(3) The area where there are more than one mutually overlapping or matching anomalies.

Sometimes the prediction results of different methods differ considerably from each other. For example, in the Tongling area, Anhui Province, we applied three methods based on 16 variables and 914 cells. The numbers of the favourable cells predicted by each method, and the numbers of the favourable cells predicted jointly by every pair of two methods are listed in the following table 3:

	Characteristic analysis	Logic-information	Information weighting calculation
Characteristic analysis	26	4	12
Logic-information		12	1
Information weighting calculation			24

In total, only three cells were predicted simultaneously by all three methods.

3. Characteristics of statistical prediction for mineral resource evaluation in China – Concluding remarks.

Study of applications of geomathematical prediction of mineral deposits began in 1976 in China. The ten year's work has the following characteristics:

(1) In China, the predictive work on total amounts of resources mainly dealt with the disaggregated prediction in local areas at large or medium scales. The level of detail of the data used in the work was relatively high, and so are the reliability and accuracy of the predictions. Consequently, the results can not only provide the resources policy-maker with necessary information, but also provide the deposit exploration planner with a scientific basis.
(2) The prediction of total amount of resources was mostly accomplished on the basis of thorough geological research. On the one hand, it was based on the deep study of the geology of ore deposits to build proper deposit models; and, on the other hand, on extensive analysis of detailed regional geological data in order to make analogous inference. Thus we can overcome the problem of oversimplification.

(3) We devote much attention to the basis of the mathematical model — the study of variables. In this field, we mainly emphasize (1) pre-processing of the variables including the screening of data from a multivariate population, calculation of statistical parameters of data from a single population, and data transformations; (2) representation of the variables, including representation of the model's components, the different vertical and horizontal representations of the characteristics of cells, the inferred continuation of the variables under the surface of the earth; (3) specific characteristics of the variables, including the evaluation of the information value of each variable, and the screening of these information values; (4) synthesis of the variables, i.e. constructing synthesized and composite variables with particular geological meanings; and (5) mutual compatibility of the variables, including the correspondence of the scale of the study area to the scale and spatial extent associated with the variables; compatibility of the numbers of variables of any type incorporated into the model; and relative importance of the same types of variables in controlling the formation or indicating the occurrence of ore deposits.

(4) We have also paid attention to the study of the classification of different types and categories of mineral resources such as the distinctions between resources that belong to different classes of deposits, ore-zones, ore-fields and ore-belts. The assessment of resources and potential resources of new ore deposit belts or new types of deposits have also been considered. The period of study has not been very long, and there remain many problems to be overcome.

NEW APPROACHES TO THE RECOGNITION OF ANOMALIES IN EXPLORATION GEOCHEMISTRY.

J.J. Royer
Centre de Recherches Pétrographiques et Géochimiques
B.P. n°20, 15 rue ND des Pauvres
54501 Vandoeuvre -Les - Nancy
France.

Abstract. Data analysis procedures employed in exploration geochemistry are briefly reviewed. Elementary statistical methods such as histograms or cumulative frequency curves are used to code the data into two characteristic functions related to the *Beginning of Anomalous Grade* (**BAG**), and to the *Significant Anomalous Grade* (**SAG**). Methods for accurately estimating the BAG and the SAG under varying conditions, the limitations of the methods, and interpretation of the data are presented. External data are used to identify the relevant sources of dispersion.

Multivariate *factor proximity analysis* (**FPA**) is discussed in the framework of spatial analysis. Theoretical results are compared to those obtained using *factor kriging analysis* (**FKA**). The method is applied to several different geochemical mineral surveys and to exploration for an oil-bearing formation.

1. INTRODUCTION

Since the early days of mineral exploration, data analysis has formed an integral part of survey activities. This is especially true for geochemistry, geophysics and remote sensing.

The specific features of geochemical data analysis are: the large amounts of data to be processed, the lack of precision of these data, their multivariate nature and their spatial dependency. These characteristics imply adapted methods for the interpretation of the large masses of numerical data in order to extract from them the essential information. I shall briefly review these different features and the techniques used to solve the problems.

In the 1930's, Soviet and Scandinavian geoscientists developed the principles of geochemical exploration for mineral deposits [1], but it was only in the 1950's that the first systematic regional surveys were undertaken in large countries such as the U.S.S.R. and Canada [2,3], followed by other countries, in particular the work of Hawkes in the U.S., Warren in Canada, Webb in the U.K. and Africa, or companies such as BRGM and CEA in France. Most of this work was devoted to the exploration of localized zones, or extensions of already known mineralized districts [There were in fact some pioneer regional surveys at large scale in the early and mid 60's: i.e the Namwala project in Africa, various regional surveys in Sierra Leone, many large U.N. surveys using base metals Cu, Pb, Zn and others in various countries including USA, Canada, Africa and USSR]. But with the recent development of automated analytical techniques (atomic absorption, plasma induction spectrometry, mass spectrometry, neutron activation) which provided fast, cheap and large amount of multielement geochemical data sets, geochemical prospecting methods were

C. F. Chung et al. (eds.), Quantitative Analysis of Mineral and Energy Resources, 89–112.

more commonly used on a large scale in regional surveys. In the 1970's, most countries carried out systematic regional surveys: the National Uranium Resource Evaluation (NURE) and USGS programs in the U.S., the Uranium Reconnaissance Program (URP) in Canada, the geochemical atlas in the U.K., the mineral inventory program in France undertaken by the B.R.G.M. [4], and the United Nations Mineral Explorations Program in many developing countries (Africa, South America,...). A pertinent review of the major geochemical exploration work performed up to the present has recently been given by Garrett [5]. The main objectives of these surveys were primary to locate overburied or blind deposits, but also to elaborate lithological or geological maps. For example, associations of chemical elements were used to determine the nature of the bedrock, which provided information on the occurrences of anomalies possibly related to hidden mineral deposits. The mining industry was actively experimenting with the use of geochemical methods particularly in weathered, altered or glaciated outcrop environments (Africa, Australia and Canada). Moreover, petroleum companies used geochemistry in deep exploration wells in order to identify the lithological sedimentary formations.

This development of highly sensitive and accurate analytical techniques and the interest in regional geochemistry or oil exploration brought about an overabundance of data. This was the main reason why geochemists were among the pioneers in applying computers in geosciences. Their contributions concerned retrieval systems, data base, contour mapping and multivariate analysis. Their main objective was to reduce to a useful form large sets of numbers, which in their raw state were cumbersome and difficult to interpret.

Optimized statistical treatment combined with a graphical representation was first proposed by Lepeltier [6] in 1969, but was also used by Soviet geochemists [7]. The visual study by a graphical representation of the frequency distribution is performed in order to define background and limits above which the anomalies occur, as well as the existence of one or several populations in the surveyed area. Since that time, much work using multivariate techniques such as factor, cluster or correspondence analysis applied to geochemical data analysis was been carried out (see [8],[9], other references can be found in [10],[11]). The fuzzy set theory of Zadeh [12] was also applied in order to take into account the uncertainty or inaccuracy of geochemical data, the idea being to point out potential zones associated with a probability of true occurrence used as a occurrence-measure of reliability [13],[14]. These more or less standard data analysis procedures failed to take into account the spatial disposition of the samples before the stage of graphical representation of the results using spatial plots, contoured maps or elaborate 3D block diagrams.

All samples were processed using the same procedure irrespective to their geographical relations The regionalized aspects of geochemical data were well known by geochemists, but until recently, there was no simple way to incorporate the location of the data points into the multivariate analysis. For this reason, active research work was carried out in the mid 1980's in order to produce geostatistical methods incorporating the structural characteristics of variables within the multivariate aspect. Two methods were explored: *Factor kriging Analysis* [15, 16] and *Proximity Analysis* [17].

This paper presents an overview of several of the different statistical methods used in geochemical data analysis.

The first part is devoted to elementary statistical techniques (histograms, cumulative frequency curve) and their uses for coding the data into two characteristic functions related respectively to the *Beginning Anomalous Grade* (BAG) and to the *Significant Anomalous Grade* (SAG). Methods for accurately estimating the BAG and the SAG under varying conditions, limitations of the methods [which assume a monotone relationship between geochemical level and ore potential], interpretation of the data and a generalization to the multivariate case are discussed. The appropriate use of external data related to the dispersion factors (relief map, direction of the relief slope, ...) is used to indicate the

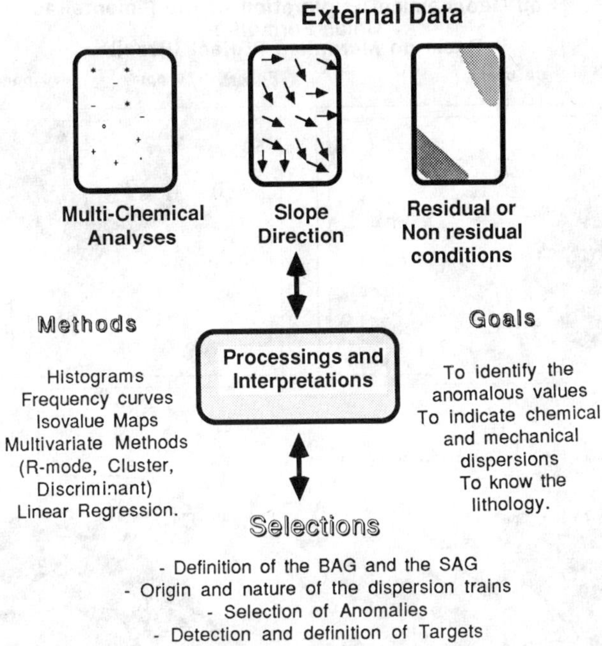

Figure 1 - Typical data necessary for Geochemical Exploration.

relevant sources of geochemical anomalies. A numerical method named the *Cosine Method* is applied to produce a map of the possible source areas for geochemical anomalies in soils, which are useful in recognizing primary halos.

The second part deals with multivariate spatial methods. The principles of proximity analysis are briefly reviewed: this method can be considered as a local discriminant analysis where the between matrix is the local cross-correlogram matrix and the within matrix the usual correlation matrix. To take into account the possible lack of precision of the data, each variable or each sample can be weighted when computing the cross-correlogram or the correlation matrix. Comparisons with Factor Kriging Analysis are discussed in both theoretical and practical terms. The method is applied to different geochemical mineral surveys and to exploration for an oil-bearing formation. Mapping techniques discussed in the first part can be used to produce condensed multivariate geochemical presentations.

2. GEOCHEMICAL DATA ANALYSIS STARTS IN THE FIELD

The first step in geochemical data analysis, as with any other geoscience, is the collection of appropriate information in the field. The most important items of field data are the spatial coordinates of the sample and also its nature (soils, alluvium, eluvium, rocks,...) and a unique identification number [5]. The other data collected in the field depend very much on

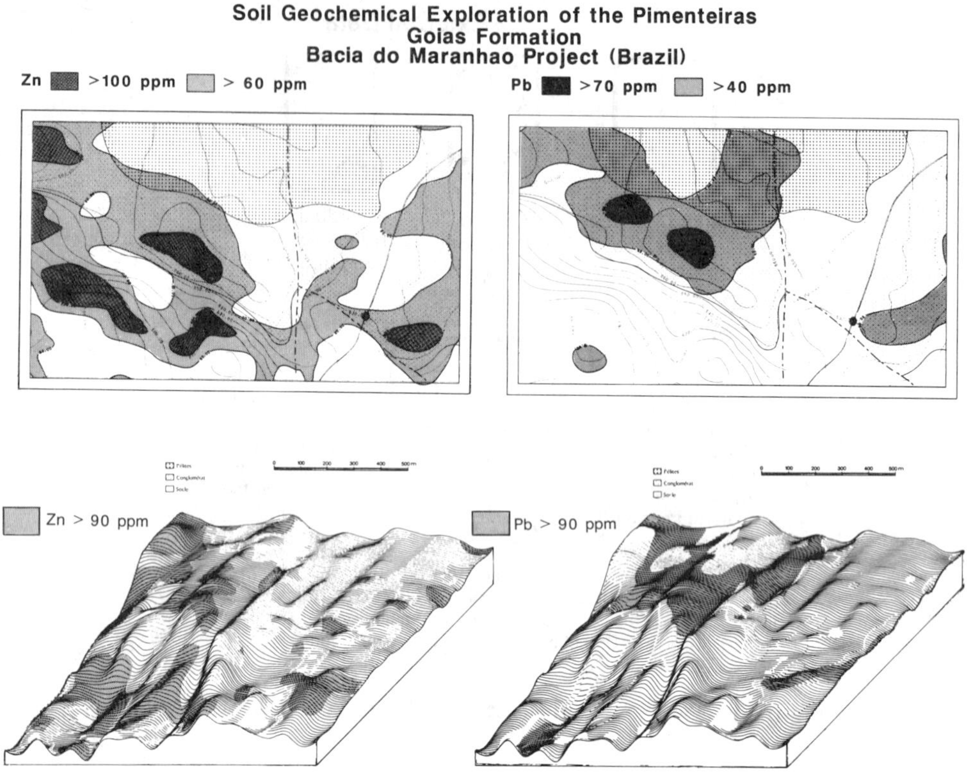

Figure 2 - Pb-Zn secondary dispersions are clearly identified using BAG and SAG isovalues.Two soil maps superimposed onto a topographic block diagram. (i) The BAG and SAG are used to select isovalues for Pb. The resulting map is clear and readable. (ii) Isovalues for Zn are not well selected. The resulting presentation is not therefore useful.

the geological environment, on the exploration goal (ore exploration, petroleum, gas,...). For example, in the case of geochemical prospecting for ore bodies, secondary halos and dispersion trains must be taken into account at the interpretation step. These dispersion factors may either be (i) mechanical during the weathering of ore deposits (solifluxion) - called residual halos - or (ii) chemically produced, for instance by leaching and concentra - tion, or by biogenic accumulation in non-residual halos [2]. Figure 1 represents the typical data necessary for a good interpretation in geochemical exploration. The topographic slope is used to determine the direction of dispersion in residual zones surrounding mineral deposits or primary halos. When it is possible, non-residual and residual conditions (marsh, possible biochemical concentrations, water table, ore horizon) can be mapped after little training by the field geochemist.

There are two main reasons for collecting this additional field information, named external data in the following: firstly, they are essential to aid the geochemical interpreta -

tion, and secondly they can be included in geochemical data processing using disjunctive recoding of ordinal or categorical values. Correspondence analysis applied on such disjunctive tables (coded variables into characteristic functions) is a non-linear multivariate technique well adapted for such data, even if they are some limitation in interpreting the factor loadings [18],[19]. Other techniques based on multivariate or multimapping techniques are suggested in the following section.

3. UNIVARIATE ANALYSIS

In order to check the field data, geochemists, like any other geoscientists, usually use the computer to plot and draw contour maps of elements. These plots are prepared at such a scale that they can be compared with the geological map. At this step, no interpretation is really required. Drawing condensed contour maps is a bit more complicated because the isovalues must be chosen in such a way that they preserve the primarily information. The main goal is to obtain readable synthetic geological maps. Figure 2 represents two soil maps of the Pimenteiras-Goias Formation (Brazil) superimposed on a topographic block diagram . In the first case, the Pb isovalues are adequately chosen, primary and second dispersion halos can be easily identified; on the other the inadequate choice of Zn isovalues produces an non-interpretable presentation. So the question is: *how to determine realistic significant values in a map representation?*

Much work related to the statistical distribution of the data has been done by geochemists. The basic tools are the histogram or the cumulative frequency curve. The main goals of this elementary statistical treatment are to define the background range and the different anomalous thresholds.

3.1. The frequency distribution curve

The geochemical concentration of an element is a random variable. The main characteristics of the frequency distribution curve are used to define the background range and the anomalous threshold values. Sidorenko [7] has defined the average background Z_m as the median of the distribution of the grade Z; all values greater than Z_m are affected to the halo i weighted by p_i which represents the significance of the halo i. The weighting function p_i is defined subjectively according to the geological context, or statistically as the percentage of sample values which belong to the i-th halo (at a regular grid, p_i is proportional to the surface of the i-th halo).

The average background can also be defined as the arithmetic or geometric mean of the data (center of gravity of the distribution), the mode or the median of the distribution. It is common to define an average background for each geological formation (back shales, limestones, carbonates...).

The notion of anomalous values is difficult to conceptualize. Different techniques have been used. Discussing the threshold problem, Hawkes and Webbs [20] recommended to (i) choose it from an orientation survey, (ii) select the top p%, e.g., 2%, of the data for detail study. Beus and Grigorian [2] suggested studying the distribution using a normal law where m is the mean and s the standard deviation. The so-called three sigma rule is based on the normal distribution, where an anomaly corresponds to a value greater than m+3s (can be generalized to m+ns to select the top p% of the data for detail study, see [20]). [Note: the assumption of the normal law is no longer useful if one uses a computer, only (i) and (ii) can be retained].

These different definitions mean that the shape of the distribution must be taken into account. This fact has been used by Lepeltier [6] to define the anomaly threshold. He suggested to take as anomalous values the brake point where the tail of the frequency distribution presents the greater statistical dispersion. The method can be used on normal as well as on lognormal distributions.

In the following, this value is called the *Beginning Anomalous Grade* (**BAG**). The advantages of this procedure are: (i) in the case of a mixed distributions, the BAG respects the individuality of the populations; (ii) this simple method can be performed on a small computer. The disadvantages are: (i) it is limited to monovariable, (ii) the threshold is determined graphically. In practice, one can use a second parameter: the *Significant Anomalous Grade* (**SAG**) which can be data driven (defined graphically) or chosen arbitrary as twice the BAG [21]. If threshold is defined as the upper limit of background variation for the univariate case, this base notions can be extended to several populations, then given g groups/population where the upper limit of background variation is T_k, k=1,g, one can choose BAG = Min(T_k, k=1,g) and SAG = Max(T_k, k=1,g).

3.2. Distribution Mixtures

The concept is that the density $f_Z(z)$ of variable Z can be modelled as a mixture of random variables. Several works have been published on technics to separate distribution mixtures. Sinclair [22] have used a graphic representation on probability plots, Bridges and McCammon [23] have suggested refinements of the method using an interactive approach on a graphic system, recently, an iterative statistical procedure have been developed by Campbell [24]. Suppose we have a bi-distribution mixtures, respectively Z_b for the background ranges or lithological effects, and Z_a for the anomalies. In the method suggested by Lepeltier, it is implicitly supposed that Z_b and Z_a have a similar distribution law (normal or lognormal) with different means and standard deviations:

(1) $f_Z(z) = p_a\, f_a(z) + p_b\, f_b(z)$ where p_a and p_b are such that $p_a + p_b = 1$ and represent respectively the frequency of anomalies and background ranges.

Supposing that $f_a(z)$ and $f_b(z)$ are normally distributed, then the resulting cumulative frequency curve $f_Z(z)$ plotted on a gaussian scale paper can be adjusted to a broken line as shown on Figure 3. The first line is adjusted to the background ranges using the assumption that the distribution of the studied element in the formation is normal. It is also assumed that the anomalous values can suitably be fitted to a normal distribution. The automatic extraction of a normal distribution from a mixed frequency distribution curve can be performed by the technique of residual. The end part of the frequency distribution curve is fitted to a normal distribution, then the same procedure is repeated iteratively on the difference between the originate distribution and the model.The procedure can be easily extended to lognormal or normal mixed distributions.

3.3. Generalization to multivariate data

Mallet [25] developed the idea of a correlation coefficient between data points based on normal fuzzy characteristic functions. The same idea can be used to split a mixed distributions into several components which are normally distributed.

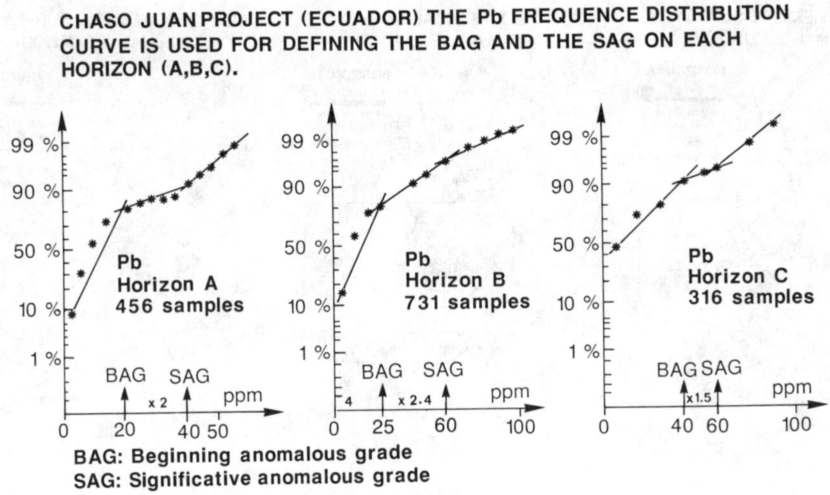

Figure 3 - Definition of the Pb BAG and SAG values for each horizon A,B,C using the frequency distribution curve. Chaso Juan project (Ecuador).

Let $x_1, \ldots, x_n$ be a sample of independent observations x_j of p variables **X**. The basic framework of the technique is the estimation of the multivariate density f(**x**) of **X** by a sum of *fuzzy functions* (Parzen Kernels) centered on N characteristic points x_i:

$$(2)\ f(\mathbf{x}) = (1/N) \sum_{i=1}^{N} \Delta(\mathbf{x} \mid \mathbf{x_i})$$

where :

$$(3)\ \Delta(\mathbf{x} \mid \mathbf{x_i}) = \{1 / [\det(\mathbf{C})^{1/2} (2\pi)^{p/2}]\} \exp\{-1/2\ \|\mathbf{x} - \mathbf{x_i}\|^2/\sigma^2\}$$

and $\|\mathbf{x}\|^2 = \mathbf{x}^t\, \mathbf{C}^{-}\, \mathbf{x}$ represents the variance covariance metric in the variable space $\mathbf{R}^P$. $\Delta(\mathbf{x} \mid \mathbf{x_i})$ is a multivariate normal distribution with $\sigma^2\,\mathbf{C}$ as covariance matrix. The fuzzy canonical characteristic function defined for each sample point s is :

$$(4)\ J_s(\mathbf{x}) = \exp\{-1/2\ \|\mathbf{x} - \mathbf{s}\|^2/\sigma^2\}$$

$J_s(\mathbf{x})$ is maximum for $\mathbf{x} = \mathbf{s}$ and $J_s(\mathbf{s}) = 1$, elsewhere $J_s(\mathbf{x}) \in [0,1]$ which can be interpreted as the probability of sample **x** belonging to the normal distribution centered on s.

The correlation coefficient defined between two sample points s and r is given by:

$$(5)\ R^Z(s,r) = E(J_s^Z \cdot J_r^Z)/[E(J_s^Z.J_s^Z)E(J_r^Z.J_r^Z)]^{1/2}$$

where E denotes the expectation of $J_s^Z = J_s(\mathbf{Z})$ for **Z** a vectorial random variable.

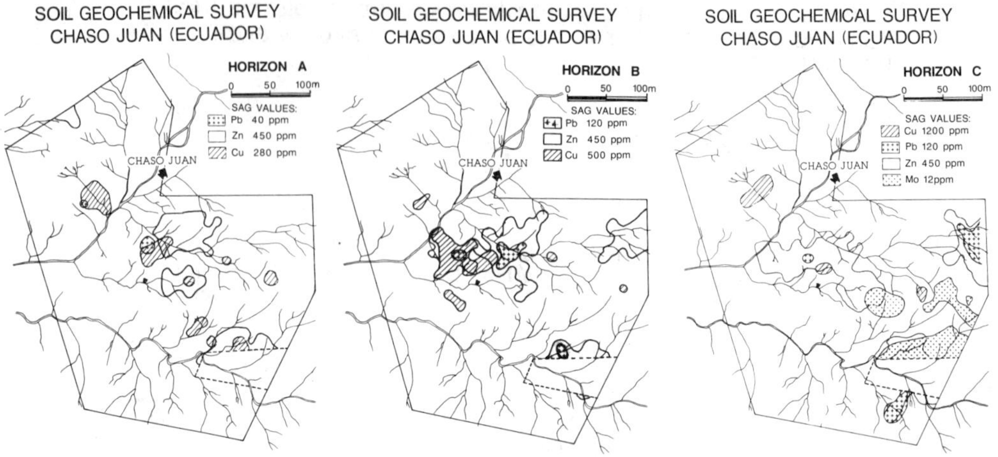

Figure 4 - Typical synthetic geochemical maps obtained by using the SAG and BAG isovalues.- Chaso Juan project (Ecuador).

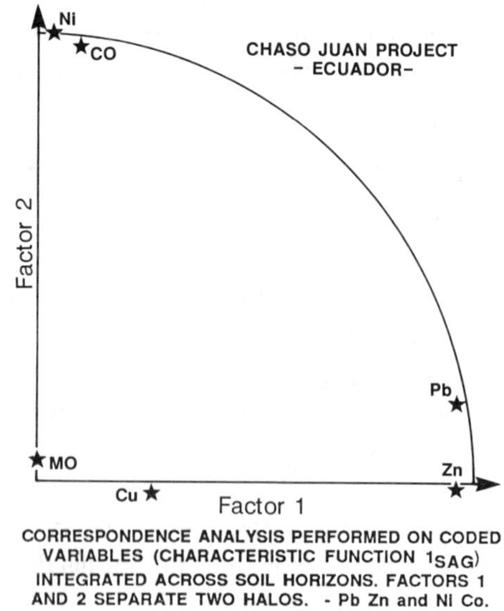

Figure 5 - Correspondence cross-plot of Factors 1 against 2 showing the association and the dispersion of elements in each halo. - Chaso Juan project (Ecuador).

Equation (2) requires the N and σ^2 parameters to be determined by "curve pealing" of the frequency distribution curve in the mono variable case, or more accurately, by cluster or dynamic cluster analysis in the multivariate case. [Note: dynamic CA is better adapted to this problem because this method computes the appropriate number N of clusters necessary to classify the data, N is not fixed as an input parameter]. All data points whose correlation coefficients are greater than a given threshold are included in the same cluster. Then, the mean and standard deviation are recalculated for each cluster. Repeating this procedure iteratively, the N parameters are determined by the number of clusters obtained by the cluster analysis. The velocity of the convergence and the resulting clusters depends on the initial choice of **C** or σ^2 parameters, a discussion of which can be found in [26].

Once the initial data set is split up into N mixed distributions whose coefficients are evaluated, one could define the background population as the distribution whose mean is minimum. Moreover, considering a sample point located at **x**, $J_s(\mathbf{x})$, s=1, ... n indicates the probability of **x** belonging to the s cluster. For instance, the BAG could be defined as the 2nd J_s, the SAG as the 3rd J_s and so on. Therefore, a recoding of the initial variables could be performed using $J_s(\mathbf{x})$. This is a generalization of disjunctive coding [27]. Its advantages stem from the possibility of evaluating the probability of occurrences using fuzzy characteristic functions for cells superimposed on a map.

3.4. Rank coding

Non-linear recoded variable methods have received much attention. The main reason for this is that they provide simple ways of treating field data which are often nominal or or - dinal information. Among the possible coding methods (disjunctive, fuzzy clusters, ...) the rank correlation techniques have been evolved by statisticians because they are independent of any non decreasing function (such as scale, unit, monotone functions). In geochemistry, this technique is simple to use : for a given variable, all sample values are sorted into increasing (or decreasing) order, and then recoded according to their rank. Contoured maps of the recoded values, called *rank maps*, are useful presentations because the recoded variables (normalized) range from 1 to the number of samples. Map comparison can be performed using the same scale on each document. Multivariate techniques (Principal components, proximity, kriging analysis) applied on such recoded variables offer new ways for handling different types of data.

3.5. CASE STUDY n°1: *The Chaso Juan (Ecuador) multi-element soil survey (horizon A,B,C) at a scale of 1:20,000.*

The investigation was performed by the D.G.G.M. (Ecuador) in collaboration with the British in the framework of a United Nation Program [28].

Geological setting: The study area represents the 1/20,000 map of Chaso Juan situated in the Andean Occidental Cordilliera which is characterized by (i) a volcano-sedimentary series of Cretacious age; (ii) intrusive volcanic rocks (granodiorite, dyke, quartz-diorite). This area was chosen according to the results of a geochemical survey in the Central West Ecuador region well known for its metallogenic importance (porphyry copper).

Sampling procedure: Soil was sampled on a regular 200 x 50 m grid at the three horizons A,B, and C. About 1500 samples were analyzed by atomic absorption for: Cu, Pb, Zn, Co, Ni, Mo.

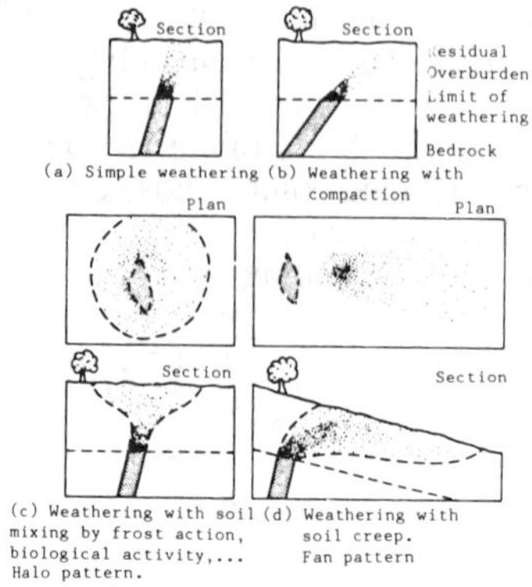

Figure 6 -Different types of surface dis persion patterns in residual overburden. (After Rose,A.W Hawkes,H.E., and Webb,J.S.,[29]).

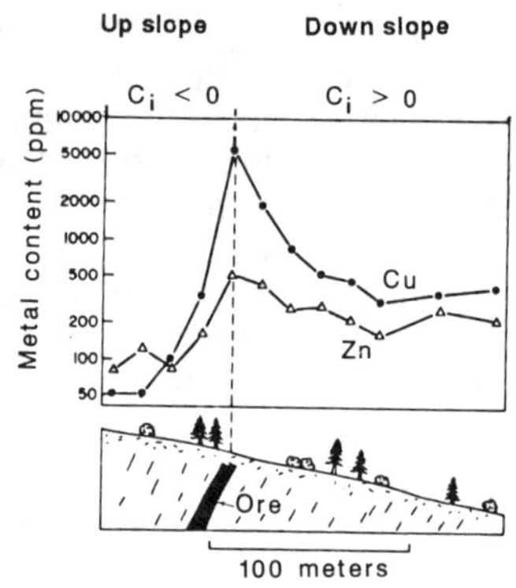

Figure 7 - Interpretation of the C^i coef ficient.C^i is >0 down slope and <0 up slope. The over-burden target is at the limits of these two zones. (Malachite copper mine Colorado - Data from Huff (1952) in Rose,A.W. et al.[29]).

Statistical treatment: The frequency distribution curves have been systematically used to define the BAG and the SAG for each soil horizon in order to obtain the synthetic geo - chemical maps shown in Figure 4. Then a correspondence analysis was performed after a coding of the initial variables using a characteristic function $1_{sag}(x_i)$ defined for each ele - ment, being equal to 1 if the grade at location x_i is greater than the SAG, 0 elsewhere. The factors reported on Figure 5 reflect the association and the dispersion of elements in each halo [Fig. 5 is an integration across soil horizons]. The advantages of this techniques are: (i) it provides a rapid synthesis view of the dispersion avoiding manual map super - position; (ii) it reveals geochemical associations and dispersions of an element depending on the substratum (i.e., Cu in the C horizon, or Pb); (iii) it permits data for the different soil horizons to be integrated.
Conclusion: The coding technique outlined above has been use on each horizon and shows that the theoretical mobilities of Cu, Mo, Pb, Zn, Ni and Co, are respected near the surface whilst at depth Pb is more easily dispersed in this acid equatorial environment, indicating possible hydrothermal primary halos. Selection of anomalies can be optimized according to these criteria. Further geological field surveys have confirmed this hypothesis.

4. LOCALIZATION OF DISPERSION PATTERN SOURCES

In the context of overburden solifluxion, contour maps or plots of the chemical data can be inadequate at identifying relevant potential sources because they emphasize the lowest or the greatest values, which may be dispersed from the relevant sources of anomalies as shown on Figure 6. The steep slope induces a mechanical dispersion of the elements, the up-slope increasing part of the concentration profile is a better indicator of the possible underlying ore than thresholds. The concept of orientation in geochemistry (solifluxion, creeping) is often simply evaluated visually from the maps by the expert. A possible automatic procedure taking into account the external field data [30] (direction of the slope, topography) is described in this section.

4.1. Statement of the problem

Let $Z^i(x)$ be the grade of ith element observed at location x; **S**(x) the unit vector represen - ting the topographic slope; $G^i(x)$ the gradient of the concentration for the ith element.

The gradient vector is oriented toward the maximum concentration and its length is proportional to the first derivative of the concentration curve. (i.e., variability of the grade).

The greater the concentration increases the greater the gradient is. $G^i(x)$ is defined as:

$$(6)\quad G^i(x) = \mathbf{grad}\ Z^i(x) = [\partial Z^i/\partial x\ ,\ \partial Z^i/\partial y]$$

In a similar way, if z(x) is the altitude at location x :

$$(7)\quad S(x) = -\ \mathbf{grad}\ \ z(x) \,/\, \|\ \mathbf{grad}\ \ z(x)\ \|$$

where $\|\ \ \|$ is the euclidean norm for vector **grad** z.

4.2. The Cosine Function

The cosine function is a measure of the geochemical dispersion. Let $\theta^i(x)$ be the angle for - med by vector S(x) and $G^i(x)$: $\theta^i(x)$ = angle $(S(x),G^i(x))$; $C^i(x)$ the cosine of that angle:

$$(8)\quad C^i(x) = \cos\ \theta^i(x) = [S(x)\ .\ G^i(x)] \,/\, [\|S(x)\|\ \ \|G^i(x)|]$$

As shown on Figure 7, an anomaly can be split in two parts (i) the down-slope part of the concentration curve $Z^i(x)$ were $0 \leq C^i \leq 1$; (ii) the up-slope side of $Z^i(x)$ where $-1 \leq C^i \leq 0$, the concentration increases as the slope decreases corresponding to the most favorable zone for locating the sources. The C^i coefficient measures the surface rate of transfer of the diffuse element i through a unit area (Fick's Law) [31].

4.3. Graphic representation

Two possible methods for graphically representing the results are outlined below.

• <u>*The oriented cosine map*</u>: At each sample point, a vector $v^i(x)$ is drawn such that :

$$(9)\quad v^i(x) = [C^i(x)\ .\ S(x)] \ /\ \|S(x)\| = [C^i(x)\ .\ S(x)].S(x) \ /\ \|S(x)\|$$

$v^i(x)$ has the same orientation as the slope vector S(x), but the norm of the vector is proportional to $C^i(x) = \cos\ \theta^i(x)$. The vector map $v^i(x)$ is superimposed on the contoured map of the ith element. The orientations of the vectors $v^i(x)$ converge toward the possible sources of dispersions (see Figure 8).

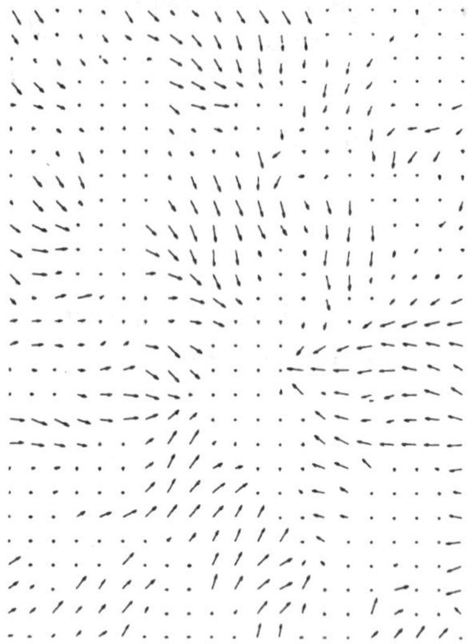

Figure 8 - Oriented cosine map for Pb. Arrows converge toward possible geochemical targets. - Pretoria-Queensland project (Australia).-

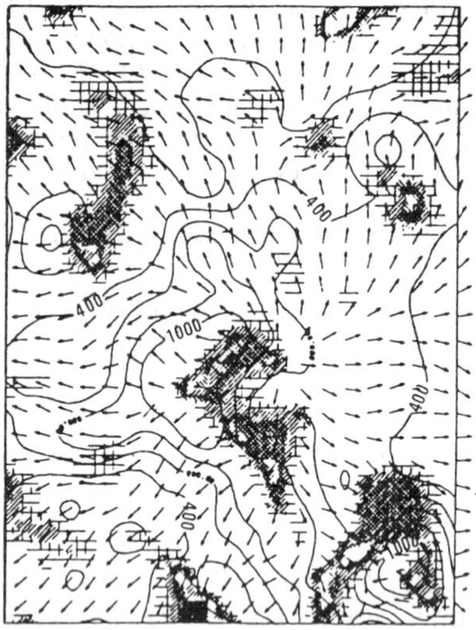

Figure 9 - Synthetic map of Pb constituted by: (i) arrows which indicate the topographic slope; (ii) isovalues of Pb; (iii) zoning area map valuable targets (C^i is < 0)

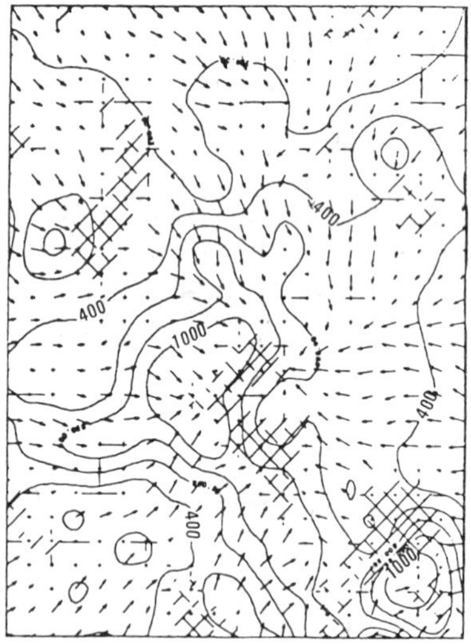

Figure 10 - Synthetic map of Pb constituted by: (i) oriented cosine map; (ii) isovalues of Pb; (iii) zoning area map showing valuable targets (C^i is < 0) - Pretoria-Queensland project (Australia).

•<u>The zoning area map:</u> Negative values of $C^i(x)$ can be outlined using, for instance, a zoning area map. Let ε be a threshold for $C^i(x)$; the set of sample points whose $C^i(x)$ is less than ε would denote potential source for the dispersion of pattern (see Figure 9).

•<u>Discussion:</u> The first method is dynamic in the sense that it underlines the direc - tions of the dispersions. Following vector orientation, one could point out possible sources of the dispersion.

The second is static. Suppose one put a light on each topographic hill lighting the concentration curve superimposed on the surface. The light zones correspond to the zoning area outlined by the second method.

4.4. Future developments

The cosine coefficient defined between the concentration gradient and the slope could be extended to a cosine coefficient between two concentration gradient :

$$(10)\ C^{ij}(x) = [\mathbf{grad}C^i(x).\mathbf{grad}C^j(x)]/[\|\mathbf{grad}C^i(x)\|.\|\mathbf{grad}C^j(x)\|]$$

which is a measure of the similarity between the local features of variable i compared to j at location x. Moreover, $C^{ij}(x)$ can be interpreted as a local correlation coefficient between the variables at location x. Multivariate techniques similar to those applied in characteristic analysis [8] could be used together with the production of composite maps.

4.5. CASE STUDY n°2: *The Pretoria-Queensland (Australia) survey at a scale of 1:10,000.*

The study area comprises the 1:10,000 map of Pretoria, which is located in Queensland (Australia). Pb-Zn mineralization is associated with volcano-sedimentary formations in this region. Systematic soil sampling was undertaken by SMN Penarroya. An isovalue map of Pb is presented in Figure 9. Arrows indicate the direction of the slope at each sampling point. The isovalues have been chosen using the frequency distribution curve (BAG and SAG method). The shaded area corresponds to a region with negative cosine coefficients, high-lighting possible sources of anomalies.

Figure 10 shows the results of the oriented cosine method. Cross patterns are associated with the zone where $C_i^{Pb}(x) > 0$. Arrows are oriented toward the sources of the anomalies. The use of such composite maps, either of uncoded elements or coded ones, is wide spread in geochemical soil surveys. Used in conjunction with other criteria (e.g. geophysics) it improves the selection of potential mineralized zones.

5. MULTIVARIATE CASE

A limiting factor in most traditional geodata multivariate techniques including factor analysis, Q-mode, R-mode, correspondence or characteristics analysis is the quantitative knowledge of the spatial location of samples. These techniques use only the variables themselves and do not take into account their spatial context. These essential characteristics are only included when examining factor maps, sometimes prepared using optimized interpolation techniques (ordinary, universal, intrinsic of order-k kriging).

Computing factor loadings or factor scores makes sense only if one wishes to study linear correlations in the variable space (association of elements in a mineral, identification of chemical reactions,...), but it makes no sense to use regionalized mapping techniques on factors which have been computed independently from the spatial variability. It makes sense only if the sample points are so widely spaced they are effectively independent spatially. At ordinary scale, sample points are spatially correlated. They are related to each other by physical processes (in geochemistry mechanical or chemical dispersion, weathering, biological concentration,...). In the univariate case, this variability may be measured by the covariograms or the semi-variograms, whilst in the multivariate case cross correlograms or cross variograms are used.Thus, there is a need to develop multivariate methods which take into account the spatial variability (underlying structure) and the linear correlation between variables as suggested in recent works [32], [17],[15]. *Factor kriging analysis* (**FKA**) first models the spatial structures using general nested cross covariance or cross variogram functions, and then calculates kriging factors by solving the related cokriging system. The whole procedure is described elsewhere [16], but it involves large computer resources at the structural analysis stage when large number of variables are involved. An alternative technique named *Factor Proximity Analysis* (**FPA**) is described in the following section. This method defines a proximity index between samples and optimizes the ratio of the local variance to the total variance. The method provides linear combinations of variables called *proximity factors*. This section presents the theory and compares the results with factor kriging analysis in the case of two nested structures. Case studies chosen in geochemical and petroleum exploration.are discussed.

5.1. Notion of Proximity.

The coefficient of proximity defined between two samples is of special interest because it can concern either spatial properties (distance between samples, connection between two adjacent regions,...) or qualities (i.e.,, two samples are close together if the distance between them is short and if they have the same geological features). Figure 11 illustrates this notion in a 2D case.

Let us consider N sample points located at x_α, the proximity index is defined between sample α and β by a non oriented graph N by N matrix $\mathbf{W}$, the graph of proximity. The 2nd order neighbours are given by matrix $\mathbf{W}^{(2)}$, the nth order by $\mathbf{W}^{(n)}$. Examples of $\mathbf{W}$ matrices are given below:

(i) $w_{\alpha\beta} = \delta_{\alpha\beta}$ where $\delta_{\alpha\beta} = 1$ if α and β are contiguous
$= 0$ otherwise
in this case $\mathbf{W}^{(n)} = \mathbf{W}^n$.

(ii) $w_{\alpha\beta} = \delta(h)_{\alpha\beta}$ where $\delta(h)_{\alpha\beta} = 1$ if the distance between samples α and β is equal to h in a given direction, anisotropic case (**a**) or in any directions, isotropic case (**b**).

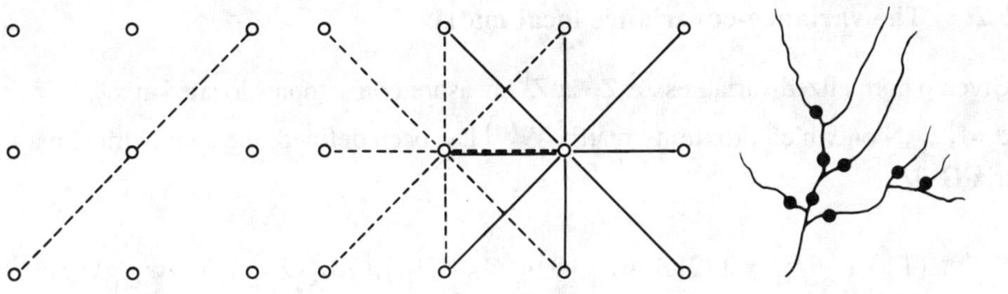

Figure 11 - Definition of proximity. Two samples α and β are contiguous with proximity factors $w_{\alpha\beta}=1$ (otherwise 0) **a)** and **b)** if they belong to the same neighbourhood, **c)** if they are taken along connected streams.

Other more elaborate indices include the followings.

(iii) $w_{\alpha\beta} = p_{\alpha\beta}$ where $p_{\alpha\beta}$ is a transition probability of a Markovian process, ($\mathbf{W}^{(n)} = \mathbf{W}^n$ if homogeneous) or a fuzzy correlation coefficient between samples.

(iv) $w_{\alpha\beta} = s_{\alpha\beta}$ where s is a parameter describing the topography between samples α and β, i.e., $s_{\alpha\beta} = 1$ if α and β on the same slope, 0 otherwise.

(v) $w_{\alpha\beta} = r_{\alpha\beta}^{-p}$ where $r_{\alpha\beta}$ is the distance between α and β along a stream, along a traverse profile or along a well-log (see (c)).

An *Ad hoc* proximity coefficient $w_{\alpha\beta}$ could be defined depending on the problem and on the geodata to be processed. It could take into account local information an expert may wish to consider, but which is difficult to formulate (constraint, discontinuity, nature of the bed rock,...).

For large sets of data, special attention must be paid to the algorithms used to store and compute matrix **W** in order to save space and time. This is a general problem concerning graph theory used in other fields (cluster analysis, segmentation,...). The main thrust of future research could be directed towards formulating more efficient algorithms.

5.2. The variance-covariance local matrix

Given p normalized variables $Z^1, Z^2 \ldots Z^P$ measured on samples located at x_α, $\alpha = 1, \ldots, N$ on which proximity matrix $\mathbf{W}^{(n)}$ has been defined, the symmetrical ma - trix $\mathbf{G}^{(n)}$

$$(11)\ \Gamma^{(n)}_{ij} = 1/2[\Sigma\ w_{\alpha\beta}^{(n)}.(Z^i_\alpha - Z^i_\beta).(Z^j_\alpha - Z^j_\beta)]/[\Sigma\ w_{\alpha\beta}^{(n)}]$$

is called the n-th order local variance-covariance matrix. The diagonal matrix $\Delta^{2(n)}$ whose diagonal elements are Γ_{jj} , represents the n-th order local variance matrix, $\Delta^{(n)}$ is the n-th order local standard deviation matrix. Therefore, the matrix $\mathbf{R}^{(n)} = \Delta^{(n)-1}.\Gamma^{(n)}.\Delta^{(n)-1}$ is the n-th order local correlation matrix.

Most traditional standardization techniques (cross-products matrix, correlation matrix, maximum-minimum row normalization,...) could be defined by a scaling metric in variable space. An exhaustive study of the different distances used in geology, including most of the classical multivariate methods, has been published by Miesch [33]. Moreover, disjunctive coding of the initial variables could provide non-linear results.

The direct study of the n-th order variance-covariance matrix could provide information on spatial variabilities. For instance, when $w_{\alpha\beta} = \delta(h)_{\alpha\beta}$ case (ii), Γ correspond to the cross-variogram matrix defined at step h and for large h, Γ is similar to the usual variance-covariance matrix denoted $\mathbf{C}$ by convention.

5.3. The proximity coefficient

Given any linear combination $\mathbf{u}$ of the p centered variable X^i in variable space $\mathbf{R}^P$, then the variance of $\mathbf{u}$ is given by:

$$(12)\ \sigma^2(\mathbf{u}) = \mathbf{u}^t\ \mathbf{C}\ \mathbf{u}$$ where ^{t} denotes the matrix transposition.

The n-th order local variance of $\mathbf{u}$ could be calculated on the contiguous samples using the same formula:

$$(13)\ \gamma^{(n)}(\mathbf{u}) = \mathbf{u}^t\ \Gamma^{(n)}\ \mathbf{u}$$ which is the variogram of $\mathbf{u}$ in case (ii).

Moreover, the ratio $G^{(n)}(\mathbf{u})$ of the local variance divided by the total variance characterizes the spatial variability analysis of $\mathbf{u}$. This is the n-th order proximity coefficient.

$$(14)\ G^{(n)}(\mathbf{u}) = \gamma^{(n)}(\mathbf{u}) / \sigma^2(\mathbf{u})$$

G(**u**) is equal to 1 if, and only if, **u** is a pure Brownian process over the sampled area **V** (pure nugget effect), in that case $G^{(n)}(\mathbf{u}) = 1 \; \forall$ n. If $G^{(n)}(\mathbf{u}) = 0$, **u** is spatially correla - ted for order n, whilst if $G^{(n)}(\mathbf{u})$ is < 1, **u** has a weak local variability. Note that $G^{(n)}(\mathbf{u})$ could either be > 0 or < 0. Therefore, the proximity coefficient measures the influence of the spatial dependance (range) on which the variable **u** is correlated, and so generalizes the cross variogram tool. The study of $G^{(n)}(\mathbf{u})$ provides information on local and regional causes of variability.

As for simple correlation analysis, it is difficult to analyze relationships between variables using an *a priori* linear combination **u**. In the following section,a multivariate method is defined, similar to the R-mode analysis, which condenses the p initial variables into factors, maximizing and minimizing the proximity coefficient.

5.4. Factorial proximity analysis

The basic idea springs from the observation that: as for any linear combination **u**, G(**u**) is minimal when **u** is a pure nugget effect. The ratio in (14) is optimized by searching for the **u** which maximizes or minimizes G(**u**). This is the classical minimization quadratic ratio problem known in discriminant analysis as Fisher's method (see [34]). One can show that this is equivalent to extracting the p eigen-values Λ and the p eigen-vectors Ω of matrix $\mathbf{C}^{-1}.\Gamma$. More precisely, let $\lambda_1 \geq \lambda_2 \dots \geq \lambda_p$ denote the p eigenvalues of $\mathbf{C}^{-1}.\Gamma$ and the corresponding p eigenvectors $\mathbf{e}^1, \dots, \mathbf{e}^p$ (scaled so that $\mathbf{e}^t\mathbf{C}\mathbf{e} = 1$). Then the vector of coefficient $\mathbf{u}^1 = \mathbf{e}^1$, whose matrix notation is Ω, maximizes the ratio G(**u**). The linear combination $\mathbf{u}^{1t}\mathbf{Z}$, where **Z** denotes the observations, is called the *first local factor*. Continuing, $\mathbf{u}^i = \mathbf{e}^i$ maximizes (14) subject to $\mathrm{cov}(\mathbf{u}^{it}\mathbf{X}, \mathbf{u}^{jt}\mathbf{X}) = 0$, j<i, and is called the ith *local discriminant factor*. Note also that $\mathrm{var}(\mathbf{u}^{it}\mathbf{X}) = 1$, i=1, ... ,p.

According to (14), if the eigenvalue λ is equal to 1 the corresponding eigenvector **u** describes a nugget effect, i.e., local noise contained in the **Z** variables; on the other hand if the eigen-value is less than 1, it corresponds to a regular phenomena. Given the correlation coefficient between $\mathbf{X}^i$ and $\mathbf{u}^j$, one can separate a regional trend from local variability (noise) by plotting the linear combinations $\mathbf{u}^{it}\mathbf{X}$ corresponding respectively to the large or small eigenvalues λ_i. The **u** are constructed such that $\Omega^t\mathbf{C}\Omega = \mathbf{I}$, where **I** is the identity matrix, and consequently the linear combinations $\Omega^t\mathbf{Z}$, say $\mathbf{u}^{1t}\mathbf{Z}, \dots \mathbf{u}^{pt}\mathbf{Z}$, corresponding to nonzero eigenvalues of $\mathbf{C}^{-1}.\Gamma$, form an orthogonal basis in the variable space for the metric associated with **C**. When **C** denotes the correlation matrix or the variance covariance matrix, a set of correlated regionalized variables $Z^i(x)$, i=1, ... p, is decomposed into a set of uncorrelated factors $u^{1t}Z(x), \dots u^{pt}Z(x)$. The procedure can be used to remove background signals from small anomalies and thus to obtain filtered maps. Figure 12 illustrates this effect on a geochemical log sampled from an offshore oil well drilled in Gabon. The proximity factor is interpreted as resulting from analytical errors for large eigenvalues, the medium eigenvalues are related to the lithology (limits of formations), and the small eigenvalues reflects the sedimentary trends.

Additional theoretical results on the proximity factors can be found in [17].

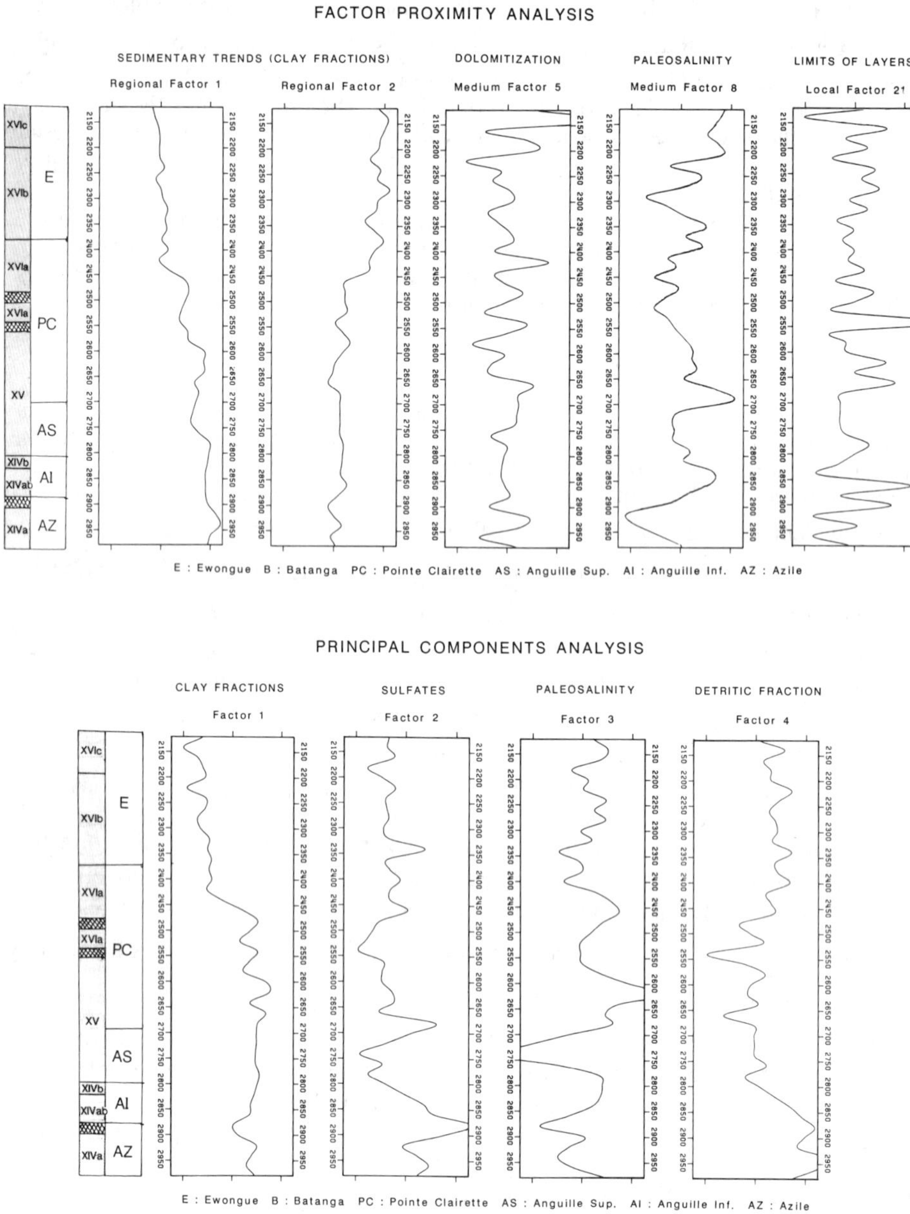

Figure 12 - Typical regional proximity factor reflecting the sedimentary trends in an oil-bearing formation. The local factor is related to the limits of layers. Offshore well in a Senonien formation (Gabon).

Note: (i) Fisher's sample discriminant analysis attempts to obtain a reasonable repre - sentation of a population in which k groups have been observed. Let **B** denotes the between group covariance matrix, and **W** the covariance matrix within groups. To solve this problem, Fisher's method involves only a few linear combinations of the observations by optimizing the ratio:

$$(15)\quad D(\mathbf{u}) = \mathbf{u}^t \mathbf{B} \mathbf{u} / \mathbf{u}^t \mathbf{W} \mathbf{u}$$

Therefore, the proximity analysis proposed in this paper is similar to a Fisher discriminant analysis for which the *between covariance matrix* is the variance-covariance local matrix and the *within covariance matrix*, the variance-covariance matrix.

(ii) Factorial proximity analysis has been described for G(u), i.e., corresponding to the first order contiguity coefficient, evidently there are no limitations applying this technique to $G^{(n)}(\mathbf{u})$ and thus obtaining the nth order local factors.

(iii) As only the first and last eigenvalues are needed for the interpretation of the local factors, optimized algorithms, such as deflation and spectral translation techniques, can be used in order to speed up the calculation of the eigenvectors.

(iv) A portable Fortran version using this procedure is under implementation in the GEOL system [35]. It runs on a HP9000 computer using the Unix system.

5.5. A comparison of proximity analysis in terms of factor kriging analysis

Work on FKA can be found in [32], [15], [16]. Let $Z^i(x)$ be a stationary process whose generalized cross-covariance is defined by $s_{ij}(h)$, the FKA divides $Z^i(x)$ into the sum of p orthogonal stationary processes $Y_{uk}(x)$ (k=1,p; u=0,m) such that:

$$(16)\quad Z^i(x) = \sum_{u=0}^{m} \sum_{k=1}^{p} a_i^{uk}\, Y_{uk}(x)$$

where m is the number of nested structures used to fit the generalized covariance and p the number of variables. Assuming that covariance between Z^i and Z^j is given by a nested structure:

$$(17)\quad \sigma_{ij}(h) = \sum_{u=0}^{m} K^u{}_{ij}\, \sigma_u(h) = \sum_{u=0}^{m} \sum_{k=1}^{p} a_i^{uk} a_j^{uk}\, \sigma_u(h)$$

then for any linear combination **u** of variables $Z^i(x)$, the covariance of **u** is given by:

$$(18)\quad \mathbf{u}^t \sigma(h)\, \mathbf{u} = \sum_{u=0}^{m} (\mathbf{u}^t \mathbf{K}^u \mathbf{u})\, \sigma_u(h)$$

This last relationship is useful in establishing the properties of FPA (using case (ii) $w_{\alpha\beta} = \delta(h)_{\alpha\beta}$) compared to FKA. Using matrix notation,

$\Gamma^{(h)} = 1/2[\sigma(h) + \sigma^t(h)]$ and $\mathbf{C}=\sigma(0)$; then optimizing $G(\mathbf{u})$ is equivalent to

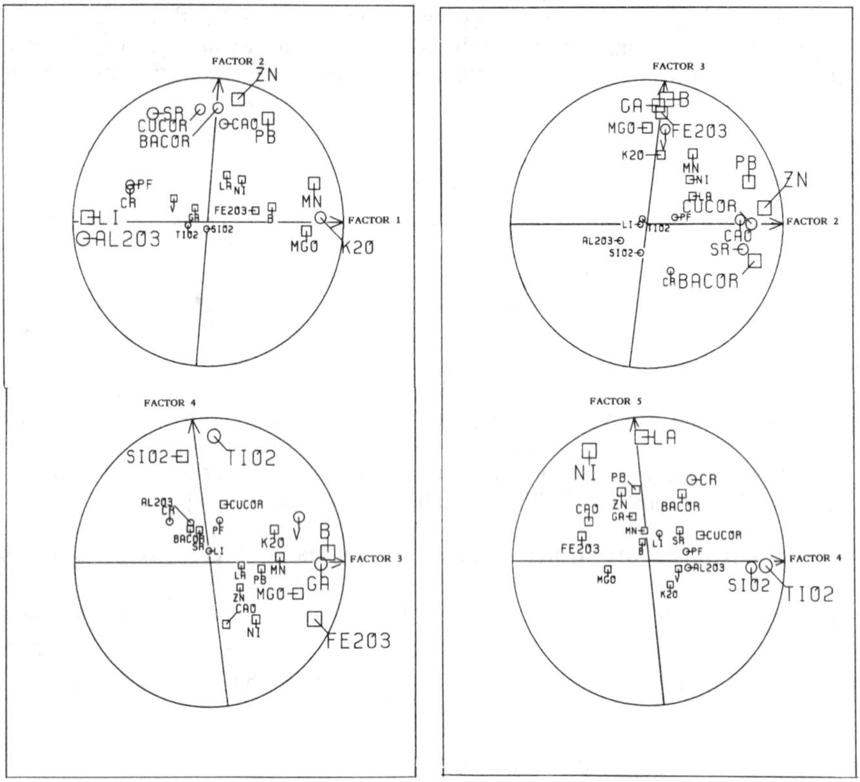

Figure 13 - Factor analysis cross plots using comima oblique rotation. Geochemical survey in oil exploration drilling (Gabon).

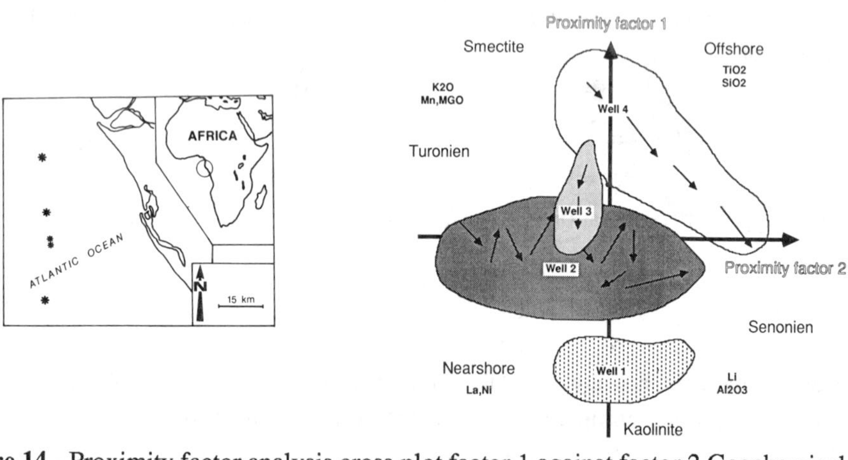

Figure 14 - Proximity factor analysis cross plot factor 1 against factor 2 Geochemical survey in oil exploration drilling (Gabon).

determining λ and **u** such that $\Gamma(h)\mathbf{u} = \lambda\, \sigma(0)\, \mathbf{u}$; by substitution and using the fact that the $Y_{uk}(x)$ are orthogonal,the problem reduces to finding λ and **u** verifying the p linear equations:

$$(19)\ \sum_j \sum_u K^u{}_{ij}\,[\sigma_u(h) - \lambda\sigma_u(0)]\ u_j = 0$$

i) The univariate case: Equation (19) is solved for $\lambda = \sigma_u(h)/\sigma_u(0)$ which is the first local proximity factor.
ii) The intrinsic model case: The intrinsic model case corresponds to a generalized cova - riance fitted by $\mathbf{C}.\sigma(h)$ where **C** is the variance covariance matrix and $\sigma(h)$ is a function. Equation (19) has only one solution $\lambda = \sigma_u(h)/\sigma_u(0)$. The proximity factor analysis performed at different orders h gives the intrinsic covariance.
iii) The case of two nested structures: Supposing that covariance can be written as follow - ings: $\sigma(h) = \mathbf{K}^1\sigma_1(h) + \mathbf{K}^2\sigma_2(h)$, where $\mathbf{K}^1$ and $\mathbf{K}^2$ being positive symmetrical matrices, and that the inverse $(\mathbf{K}^2)^{-1}$ exists; (19) is then simplified to:

$$(20)\ (\mathbf{K}^2)^{-1}\mathbf{K}^1\mathbf{u} = -[\sigma_2(h) - \lambda\sigma_2(0)]/[\sigma_1(h) - \lambda\sigma_1(0)]\mathbf{u} = \beta.\mathbf{u}$$

Then, the eigen-vectors computed by the proximity factor analysis are the same as those of $(\mathbf{K}^2)^{-1}\mathbf{K}^1$; moreover:

$$(21)\ \lambda(h) = [\beta.\sigma_1(h) + \sigma_2(h)]/[\beta.\sigma_1(0) + \sigma_2(0)]$$

(20) and (21) mean that proximity factor analysis is linked to the regionalized structure. (20) is similar to the discriminant analysis of model 1 upon model 2: the proximity factor analysis compares the large spatial structures to the smaller ones, it is therefore a comple - mentary method to factor kriging analysis.

5.6. CASE STUDY n°3: *A geochemical survey in oil exploration drilling (Gabon).*

This survey has been undertaken by SNEAP in the Senonien sedimentary formations situated offshore from Gabon. Twenty-one elements have been determined in samples collected during the systematic logging of deep oil wells [36].

Factor Analysis: the results of a classical factor analysis using comima oblique rotation (minimum maximum constraint [37]) are shown in Figure 13. Five factors explain 78% of the total variance:

- the 1st factor which is related to the environmental conditions of the sedimen - tation, opposing the kaolinite clay fraction with the smectite, montmorillonite, chlorite.
- the 2nd factor indicating the abundance of Zn, Ba, Cu, Pb, Sr and CaO has been interpreted as a *sulfate component* in the series.
- the 3rd factor correlated with the depth, is a measure of the *paleosalinity* (V, B, Ga); it is also correlated with the kaolinite content and to the sulfate factor.

- the 4th factor is related to the detritic fraction (TiO_2, SiO_2) and provides arguments in favor of a *continental origin for the sedimentary material*, whilst the 5th factor (La, Ni) indicates the presence of *biogenetic phosphates*.

Detailed analysis of these results can separate the lithological geochemical effect (formation effect) from biogenetic and diagenesis effects.

Proximity Coefficient The proximity between two samples has been defined using the following principles: two samples are said to be contiguous if they are in a similar sedimentary formation (layer) and within a distance of 20m.

Factor proximity analysis: The regional factors (i.e., corresponding to $G(\mathbf{u}) < 0.3$) reflect the sedimentary conditions of the formation (distance to the coast, depth,...) whilst local regional factors ($0.3 < G(\mathbf{u}) < 0.5$) are related to the environmental conditions of the sedimentation (carbonate content, organo-biochemical fraction,...). For instance, factor 8 reflects the paleosalinity (Figure 12). Local factors ($G(\mathbf{u}) > 0.5$) separate the different layers within a formation.The spatial dispersion for each variable may also be evaluated.

Conclusions: Most of the main geochemical characteristics pointed out by the classical factor analysis have been found by the proximity analysis. But the latter method provides more understandable factors which can be separated into regional trends and local effects. It seems an efficient technique to filter out the analytical error or the local structures.

6. DISCUSSION

Methods of statistical and proximity analysis can be effectively applied to geochemical data. Problems related to the definition of pertinent thresholds, referred to as BAG and SAG, were discussed in the first part of the paper. Spatial aspects were taken into account in the second. All these techniques must be considered as tools useful in summarizing geo - chemical data, the final goal being to obtain the understandable condensed presentations required for decision making for the follow up phases of field work. In that context, the recoding of the initial variables together with the production of composite maps integrating external data should be useful in identifying relevant sources of dispersion patterns...

Proximity analysis has been used to produce factors which can be interpreted in terms of either the lithology of the geological formation, the physical conditions of sedimentation, or diagenesis related factors (salinity, sulfate, phosphate). The method is complementary to factor kriging analysis as it provides a discriminant analysis of small structures contained within larger ones. The factor maps as well as the composite maps described above, are relevant to the principal question asked by the geochemist : how can one differentiate the anomalies from the background by a multivariate approach ?

7. ACKNOWLEDGEMENTS

I am indebted to the various Companies who have provided the geochemical data used in the case-studies, especially the Société Nationale Elf Aquitaine, SNEAP, the SMN Penarroya, the General Direction of Geology and Mines of Ecuador (D.G.M.M.) and the Petrobras Company of Brazil. Part of this research was funded by the DGRST (Délégation Générale de la Recherche Scientifique et Technique) and by the Centre National de la Recherche Scientifique (CNRS). C. Devey is thanked for correcting the English. The contribution of D. Heliot from the Geological School of Nancy, who helped in the computer analysis of the data, is acknowledged.

Contribution CRPG n° 705.

REFERENCES

[1] Ginzberg, I.I., -1960- *Principles of Geochemical Prospecting*, Pergamon Press, New York.

[2] Beus A.A. and Grigorian S.V. -1977- *Geochemical Exploration Methods for Mineral Deposits*. Applied Publ. Ltd, Wilmette,Illinois, U.S.A., 288p.

[3] Levinson A.A. -1974- *Introduction of exploration geochemistry*. Applied Publ. Ltd, Wilmette,Illinois, U.S.A., 612p.

[4] Barbier J., Wilhelm E. -1978- 'Superficial geochemical dispersion around sulphide deposits: some examples in France.' *Journal Geoch. Expl.* (**10**), pp. 1-39.

[5] Garrett, R.G., -1985- 'Geochemical Data - retrospect and prospect.' *in The role of Data in Scientific Progress*, P.S. Glaeser (ed.), Elsevier Science publishers B.V. (North-Holland), CODATA, pp 25-32.

[6] Lepeltier, Cl. , -1969- 'A simplified statistical treatment of geochemical data by graphical representation.' *Econ. Geol.* **64**, pp. 538-530.

[7] Sidorenko, A. , -1965- *Introduction to the data processing of geochemical data.* in Russian , Tchoumakova F.N., Ierousalimskaya E., Kapouskaya E.I., (Eds.), Moscou, 226p.

[8] Botbol, J.M., Sinding-Larsen, R., McCammon, R.B. and Gott, G.B. -1978- 'A regionalized multivariate approach to target selection in geochemical exploration', *Econ. Geol.* **73**, pp.534-546.

[9] Li Yuwei, Yu Jinshen and Xie Xilin -1980- 'Cluster prediction of mineral resources.' *Sci. de la Terre, Série Inf. Géol.* , N°**15**, pp. 75-85

[10] Howarth, R.J. and Sinding-Larsen, R.-1983- 'Multivariate analysis, in Statistics and Data Analysis in Geochemical Prospecting.', Howarth, R.J. (Ed.), Vol. **2** *Handbook of exploration Geochemistry*, Govett, G.J.(Ed.), Elsvier, Amsterdam, pp 207-289.

[11] -1984- *Computers in Earth Sciences for Natural Resources Characterization*, Royer J.J. (Ed.), Sci. de la Terre, Série Inf. Géol. , N°**20**, 2t, 698 p.

[12] Zadeh L.A. -1978- 'Fuzzy sets as a basis for a theory of possibility.' *Fuzzy Sets and Systems*, **1**,(1), pp. 3-28.

[13] Granath, G., - 1984- 'Application of fuzzy clustering and fuzzy classification to evaluate the provenance of glacial till', *Math. Geol.* **16**,pp. 283-303.

[14] Tang S. and Zhao Z. -1984- 'The computer application for regional assessment of uranium potential.' *Sci. de la Terre, Série Inf. Géol.* , N°**20**, Vol. 1, pp. 113-120.

[15] Sandjivy L. -1984- 'The Factorial Kriging Analysis of regionalized data. Its application to geochemical prospection.' *in Verly G. et Al. (eds.), Geostatistics for Natural Characterization*, D. Reidel Pub. Comp., Part **1**, pp. 559-571.

[16] Wackernagel -1985- *L'inférence d'un modèle linéaire en géostatistique multivariable* Ph Thesis, Centre de morphologie Mathématiques, Fontainebleau,

[17] Royer J.J. -1984- 'Proximity Analysis: a method for multivariable geodata processing. Application to geochemical processing'. *Sci. de la Terre, Série Inf. Géol.* , N°**20**, part 1, pp. 223-243.

[18] Royer J.J. and Leymarie P. -1977- 'Use of mathematical methods for metallogenetic investigations on stream sediment data.' *Symp., Matematické Metody v Geologii*, Pribram, Tchecoslovakia, pp. 367-391.

[19] Royer J.J. -1983- 'Closed Data Array and Correspondence Analysis in Geology.' *Sci. de la Terre, Série Inf. Géol.* , N°**16**, pp. 75-83.

[20] Hawkes, H.E. and Webb, J.S. -1962- *Geochemistry in Mineral Exploration.* 1st ed., Harper and Row, New York, 415p.

[21] Granier Cl. -1977- 'Equipement des cartes de géochimie de sols utilisées en prospection tactique.' *Sci. de la Terre, Série Inf. Géol.* , N°**9**, pp. 55-62.
[22] Sinclair, A.J. -1976- 'The Application of Probability Plots to Mineral Exploration', *Association of Exploration Geochemistry*, Special v. **4**, 95p.
[23] Bridges, N.J. and McCammon, R.B. -1980- 'Discrim: a computer program using an interactive approach to dissect a mixture of normal or lognormal distributions.'- *Computer & Geosciences*, v. **6**, n°1, pp 361-396.
[24] Campbell, N.A. -1984- 'Mixture Models and Atypical Values.'- *Mathematical Geology.*, v. **16**, n°5, pp 465-477.
[25] Mallet, J.L. -1983- 'Definition of a correlation coefficient between data points.' *Math. Geol.*, v. **15**, n°2, pp. 213-214.
[26] Mallet, J.L. and Wild, P. -1984- 'An analogue to correspondence analysis with fuzzy characteristic functions.' *Sci. de la Terre, Série Inf. Géol.* , N°**20**, part 1, pp. 121-135.
[27] Wild, P. -1984- *Estimation automatique de la densité multivariée et application à l'analyse des données.* Thèse 3ème cycle, INPL, 100p.
[28] Granda Delgado, B. -1983- *Prospection géochimique sur sol (Horizons A,B et C) multiélément Cu, Pb, Zn, Co, Ni, Mo de la région de Chaso Juan (Equateur)-* Int. Rep., Master, CESEV,Nancy, 90p.
[29] Rose, A.W., Hawkes, H.E., Webb, J.S. -1979- *Geochemistry in Mineral Exploration.* 2nd ed., Academic Press, 658p.
[30] Leymarie, P., Durandau, A. -1985- 'The intregration of geochemical exploration data.' *in The role of Data in Scientific Progress*, P.S. Glaeser (ed.), Elsevier Science publishers B.V. (North-Holland), CODATA, pp 33-37
[31] Crank, J. -1979- *The mathematics of diffusion.* 2nd ed., Clarendon Press Oxford, 414p.
[32] Matheron G. -1982- *Pour une analyse krigeante des données régionalisées.* Note interne, n°**732**, CGMM Fontainebleau, 20 p.
[33] Miesch, A.T. -1980- 'Scaling variables and interpretation of eigenvalues in Principal Component Analysis of Geologic data.' *Math. Geol.*, v. **12**, n°6, pp. 523-538
[34] Johnson, R.A. and Wichern, D.W. -1982- Applied Multivariate Statistical Analysis, Prentice-Hall, Inc., Englewood Cliffs, New Jersey 07632, 594 p.
[35] Jacquemin, P., Mallet, J.L., Royer, J.J., -1985-'Interactive computer aid design in the processing of mining and geological data.' *in The role of Data in Scientific Progress*, P.S. Glaeser (ed.), Elsevier Science publishers B.V. (North-Holland), CODATA, pp 19-24
[36] Brevart, O., Heliot, D. ,Royer, J.J. -1985- *Essai d'Application à l'analyse des proximités en géochimie pétrolière. Exemple du Sénonien du Gabon.* Rap. Etude SNEAP, Nancy, 154p.
[37] Mallet, J.L. -1971- 'Contribution à l'étude des facteurs non orthogonaux en analyse factorielle'- *Rev. Stat. Appl.*, v. **19**, n°1, pp 57-76.

UNIVARIATE PATTERNS IN THE DESIGN OF MULTIVARIATE ANALYSIS TECHNIQUES FOR GEOCHEMICAL DATA EVALUATION

Clifford R. Stanley and Alastair J. Sinclair
Department of Geological Sciences
The University of British Columbia
6339 Stores Rd.
Vancouver, B.C.
V6T 2B4, CANADA.

ABSTRACT. The evaluation of regional geochemical data normally begins with a univariate analysis of each element. Patterns within the data are frequently recognized, but in cases where the patterns are obscure or not immediately understood, or where a large number of elements make interpretation difficult, a multivariate analysis is commonly undertaken. Unfortunately, this analysis is often made without consideration of the univariate results. Ignoring discernable univariate patterns in the design of subsequent multivariate analyses may lead to unneccessary ambiguities and/or complexities in the multivariate results and the subsequent interpretation. Geochemical concentration data from a soil survey in northern British Columbia are used to demonstrate this point. Univariate examination of the data reveals that most elements are bimodally distributed. However, the elements could be divided into two groups differing in the relative contribution of the two component populations (modes) making up their distributions. One group of elements has bimodal distributions with component populations of approximately equal size (55% and 45%); the other has bimodal distributions with component populations of vastly different size (90% and 10%).

This information was used in three subsequent principal component (PC) analyses. The PC loadings and scores from (1) an analysis of all elements in the data set (the standard approach) were compared with the loadings and scores from two other PC analyses. These include (2) an analysis which used only those elements with distributions containing component populations of approximately equal size (55:45), and (3) an analysis which used those elements with distributions containing component populations of vastly unequal size (90:10). Combined results from the two "structured" PC analyses (2 and 3) relate better to the geology and mineral deposits in the survey area than results from the wholesale, "unstructured" PC analysis (1). Furthermore, patterns from the "structured" PC analyses are more distinct than the wholesale PC analysis and lead to a less ambiguous and more complete interpretation.

The "structured" approach to multivariate analysis, which uses information from the univariate analysis in its design, thus represents a logical and superior data evaluation philosophy.

1. INTRODUCTION

C. F. Chung et al. (eds.), Quantitative Analysis of Mineral and Energy Resources, 113–130.

Since the mid 1960's, multivariate statistical data reduction techniques have become widely used in defining geochemically similar geologic samples. This is largely due to the development of rapid and cost effective analytical techniques for the simultaneous determination of multi-element concentrations in geological samples. The purpose of this study is to compare the relative advantages of univariate and multivariate statistical evaluation techniques in grouping geological samples into geochemically similar clusters, and to demonstrate how univariate analysis results can be used to design more effective multivariate statistical evaluation strategies.

Figure 1 - Location Map of the Cassiar Mining District, British Columbia, Canada.

Geochemical data from 114 B-horizon soil samples from a mineral prospect in northern British Columbia were evaluated. The samples were collected on a slightly irregular, 50 m by 200 m, rectangular grid over the Beaver Claims, located within the Cassiar Mining District (Figure 1), approximately 150 km SSW of Watson Lake, Yukon Territory. On the Beaver Claims two principal lithologies occur (Somerville and Smit, 1985), but outcrop is limited to less than 5% of the survey area largely due to locally-derived overburden. The lithologies include the Atan Group, a sequence of autochthonous miogeoclinal sediments, and the structurally superjacent Sylvester Group, an easterly thrusted, allochthonous package of three lithologic assemblages deposited in volcanic arc and marine basin environments.

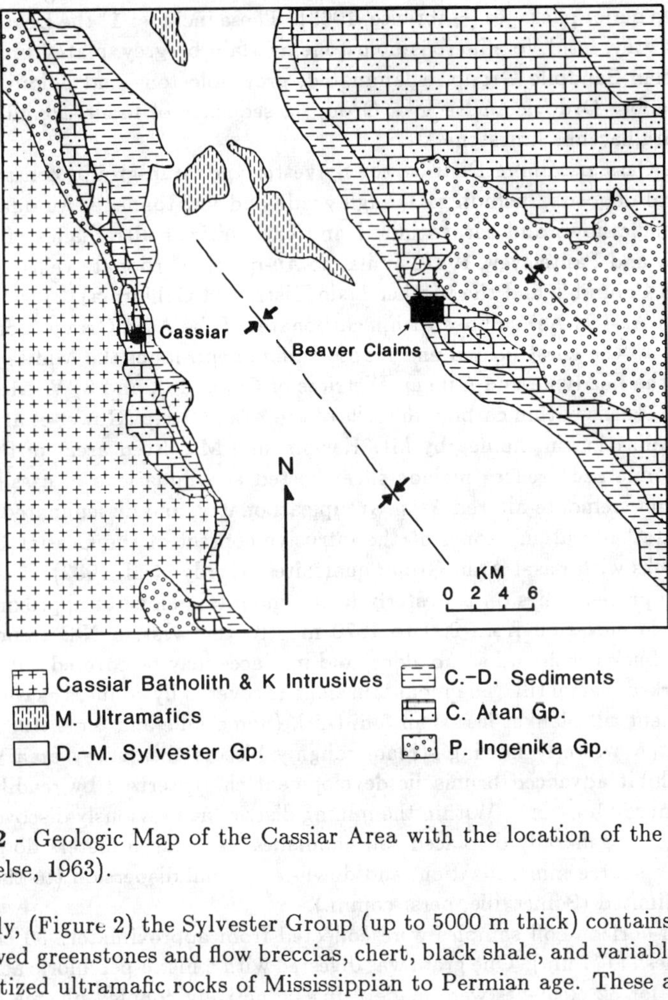

Figure 2 - Geologic Map of the Cassiar Area with the location of the Beaver Claims (after Gabrielse, 1963).

Regionally, (Figure 2) the Sylvester Group (up to 5000 m thick) contains a basal thrust slice of pillowed greenstones and flow breccias, chert, black shale, and variably serpentinized and carbonatized ultramafic rocks of Mississippian to Permian age. These are structurally overlain by a Pennsylvanian to Permian assemblage of basalt, shale, siltstone, chert, and limestone. Both of these assemblages are in steep fault contact with a third assemblage composed of basaltic to andesitic lapilli tuff, quartz diorite, quartz arenite, and shale of uncertain age (Gordy et al, 1982). Graphitic, thin bedded argillites of the Sylvester Group occur in a small outcrop in the SW corner of the Beaver claim block (Somerville and Smit, 1985).

The Atan Group is at least 1000 m thick, lower Cambrian in age, and contains a basal silici-clastic unit of argillite, siltite, and quartz arenite overlain by an upper unit of limestone and dolostone (Gordy et al, 1982). Limited outcrops of Atan Group quartzites, limestones, dolomites, and thin interbeds of tan-colored shale are found in the east-central part of the property, along a westerly flowing stream (Somerville and Smit, 1985).

Locally within the Cassiar Mining District, three additional lithologies may occur between these two major packages (Gabrielse, 1963). These include: 1) the Devonian McDame Group, a dark grey, fetid, fossiliferous dolostone overlain by grey, platey limestone, 2) the Siluro-Ordovician Sandpile Group, a laminated, grey dolostone and dolomitic sandstone, and 3) the Cambro-Ordovician Kechika Group, a sequence of fissile argillite and phyllite underlain by argillaceous limestone.

Within the Cassiar Mining District, the Sylvester and Atan groups contain a variety of metallic mineral deposits, including: 1) high grade and low tonnage, Au-quartz-carbonate veins within mafic flows and serpentinized and carbonitized ultramafics of the Sylvester Group (Diakow and Panteleyev, 1981; similar to Au-quartz-carbonate veins of the Erickson mine of British Columbia and the Mother Lode District of California), 2) irregular replacements of Pb-Zn-Ag-(Sn-As) sulfides within carbonates of the Atan Group (Somerville, pers. comm.; similar to the karst-hosted and manto sulfide deposits of the Midway area, Yukon Territory and the Leadville and Gilman Districts of Colorado), and 3) W-Mo-(Pb-Zn) and Ag-W-(Cu-Pb) skarns within carbonate rocks of the Atan Group (Gower et al, 1985; similar to skarn mineralization in the nearby Mt. Haskins and Mt. Reed areas of British Columbia). Subeconomic occurrences include shear-hosted accumulations of cassiterite and arsenopyrite within tremolite-altered Atan Group carbonates, and disseminated chalcopyrite-pyrite-pyrrhotite "alteration" zones at the intrusive contact of the Cassiar Batholith and Cretaceous stocks with basal Atan Group quartzites (Gordy et al, 1982).

The Beaver property lies on a westerly facing mountain slope of approximately 20 degrees, ranging in elevation from 900 to 1670 m. To the west, a N-S trending valley is vegetated with buck brush and scrub alder, and in places may be covered with up to 100 m of exotic, reworked glacial till. The mountain slope is covered by conifers as well as a locally-derived lodgement till blanket less than 3 m thick (Somerville and Smit, 1985; Somerville, pers. comm.). A westerly flowing stream roughly bisects the survey area. Soils on the claim block exhibit advanced brunisolic development characterized by readily identifiable Ah, Ae, and Bm soil horizons. Within the mining district at previously discovered prospect sites, slumping has generally displaced soil anomalies less than 5 meters down-slope from the subcropping source mineralization, and down-ice glacial dispersion (to the N and NE) appears to be limited (Somerville, pers. comm.).

The 1 kg B-horizon soil samples were collected from approximately 30 cm depth and sieved to −80 mesh (177 um). One gram was digested with a nitric-perchloric acid attack and determinations for 32 elements were made using inductively coupled plasma spectroscopy. In addition, Au was determined using a sequential nitric-perchloric, aqua regia, and methyl-iso-butyl-ketone (MIBK) extraction on a 10 gram sample. Determinations were made with an atomic absorption spectroscopy finish (Somerville and Smit, 1985). Only 25 of the 33 elements analyzed had a sufficient number of determinations above the detection limit and enough statistical dispersion to be considered of value in this interpretation. These elements are Ag, Al, As, Au, Ba, Bi, Ca, Cd, Cu, Fe, K, Li, Mg, Mn, Mo, Na, Ni, P, Pb, Sb, Sn, Sr, V, W, and Zn.

2. DISCUSSION

2.1. Univariate Evaluation

Initially, statistical treatment of the multi-element data consisted of a separate univariate analysis of each element. Histograms and probability plots (Sinclair, 1976) were made for raw and log-transformed data and were evaluated in an attempt to recognize polymodal distributions which might be spatially related to mineralization or lithologic controls. With the exception of B, Be, Co, Ga, Ge, Se, Th, and U, which suffered from major detection limit truncation and/or had values reported at nearly all the same concentration (possibly due to limited dissolution of the element by the nitric-perchloric acid attack), all elements exhibited polymodal distributions. Elements with distributions which were modeled better by a polymodal normal distribution when their concentrations were log-transformed include Au, Bi, Ca, K, Mo, V, W, and Zn. All subsequent evaluation of these elements was performed on their log-transformed values.

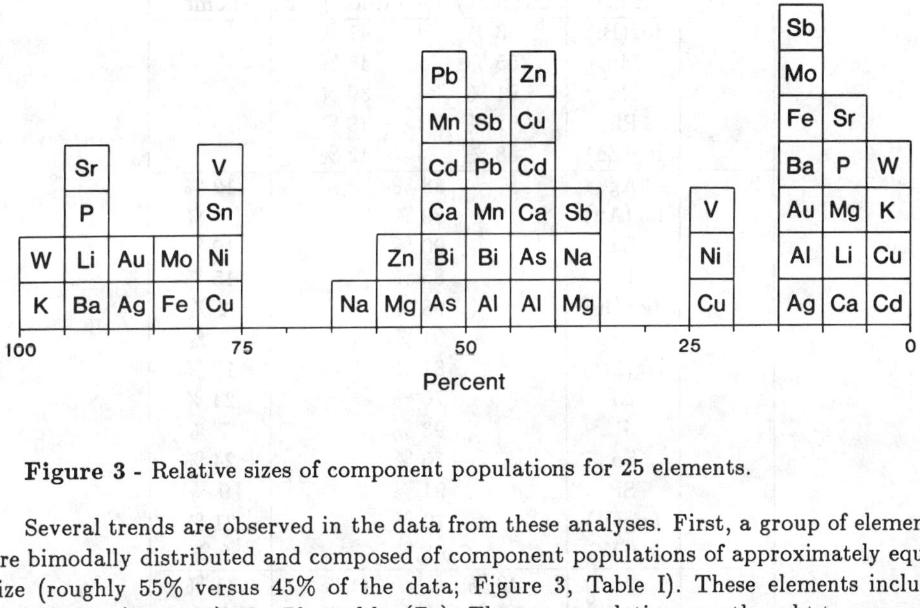

Figure 3 - Relative sizes of component populations for 25 elements.

Several trends are observed in the data from these analyses. First, a group of elements are bimodally distributed and composed of component populations of approximately equal size (roughly 55% versus 45% of the data; Figure 3, Table I). These elements include log(Bi), Mn (Figure 4), Na, Pb, and log(Zn). The two populations are thought to represent different lithologies, each of which underlie about one half of the survey area. For this reason, elements exhibiting this component population structure are classified as part of the "lithology" element suite.

A second group of elements are also bimodally distributed but are composed of component populations of vastly unequal size (roughly 90% versus 10% of the data; Figure 3, Table I). These elements include Ag (Figure 5), log(Au), Ba, Fe, log(K), Li, log(Mo), Ni, P, Sn, Sr, log(V), and log(W). The smaller population observed for these elements is thought to be related to sulfide mineralization on the property, because one would not expect the mineralization to underlie a large proportion of the survey area. Thus, elements exhibiting this component population structure are classified as part of the "mineralization" element suite.

TABLE I. Sizes of Component Populations (in percent)

The frequency distributions of elements evaluated in this study were broken up into component normal populations with the indicated percentage sizes. Group 1 elements are bimodally distributed with component populations of approximately equal sizes; group 2 elements are bimodally distributed with component populations of vastly different size; group 3 elements are trimodally distributed. Elements in groups 1 and 3, combined, comprise the "lithology" element suite; elements in groups 2 and 3, together, combined the "mineralization" element suite; *all* elements together comprise the "total" element suite.

Group	Element	Lithology 1	Lithology 2	Anomaly
1.	log(Bi)	53 %	47 %	
	Mn	55 %	45 %	
	Na	61 %	39 %	
	Pb	52 %	48 %	
	log(Zn)	58 %	42 %	
2.	Ag	88 %		12 %
	log(Au)	89 %		11 %
	Ba	90 %		10 %
	Fe	85 %		15 %
	Log(K)	98 %		2 %
	Li	91 %		9 %
	log(Mo)	81 %		19 %
	Ni	76 %		24 %
	P	93 %		7 %
	Sn	76 %		24 %
	Sr	91 %		9 %
	Log(V)	79 %		21 %
	log(W)	98 %		2 %
3.	Al	48 %	41 %	11 %
	log(Ca)	51 %	43 %	6 %
	Cd	55 %	41 %	4 %
	Cu	76 %	20 %	4 %
	Mg	60 %	35 %	5 %

Finally, a group of elements are trimodally distributed, and have component population proportions which may be a combination of *both* of the previously described structures (that is having two component populations of roughly equal proportion and one component population which is substantially smaller; i.e. 45% versus 45% versus 10% of the data;

Figure 3, Table I). These elements include Al (Figure 6), As, log(Ca), Cd, Cu, Mg, and Sb. Because these elements contain component populations of approximately equal size (similar to both the "lithology" and "mineralization" element suites), they are classified as part of *both* the "lithology" and "mineralization" element suites and may serve to discriminate areas underlain by different lithologies as well as truely anomalous sites.

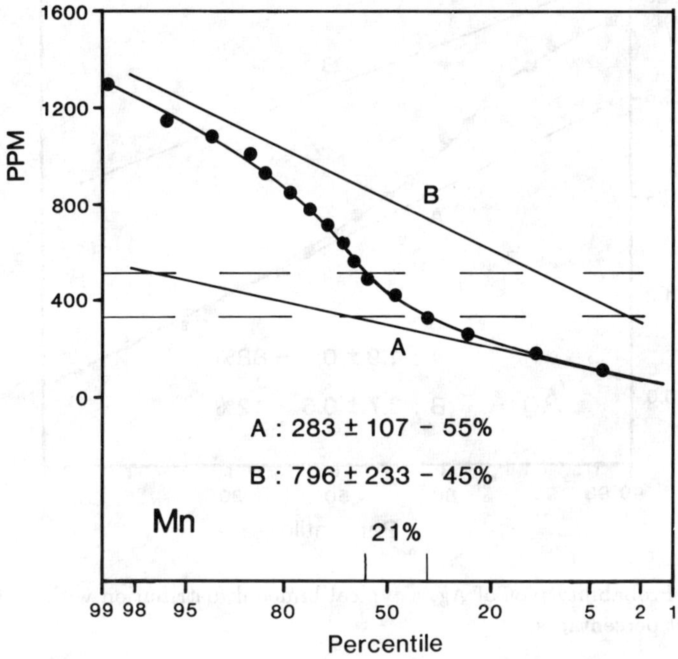

Figure 4 - Probability plot of Mn, a typical bimodal distribution with roughly equal sized component percentages.

Optimal thresholds were selected for each of the polymodally distributed elements. These thresholds were designed to minimize the number of classification errors (both Type 1 (omission) and Type 2 (inclusion)) created by choosing a single threshold in each case where there is a substantial amount of component population overlap. Samples from each classification group were plotted on the claim map with different sized symbols, and geographic clustering of the three previously recognized component populations were apparent, as illustrated in Figure 7.

Two major spatially coherent geochemical domains occur on the map with their contact trending across the center of the survey area in the NW regional strike direction. To the SW, higher concentrations of As, Ba, Cd, Mn, Na, Pb (Figure 7A) and log(Zn) (Figure 7B) are representative, whereas to the NE, higher concentrations of Al, Mg, Sb and Sr occur. These two zones are thought to be related to subcropping Sylvester Group rocks to the SW and subcropping Atan Group rocks to the NE.

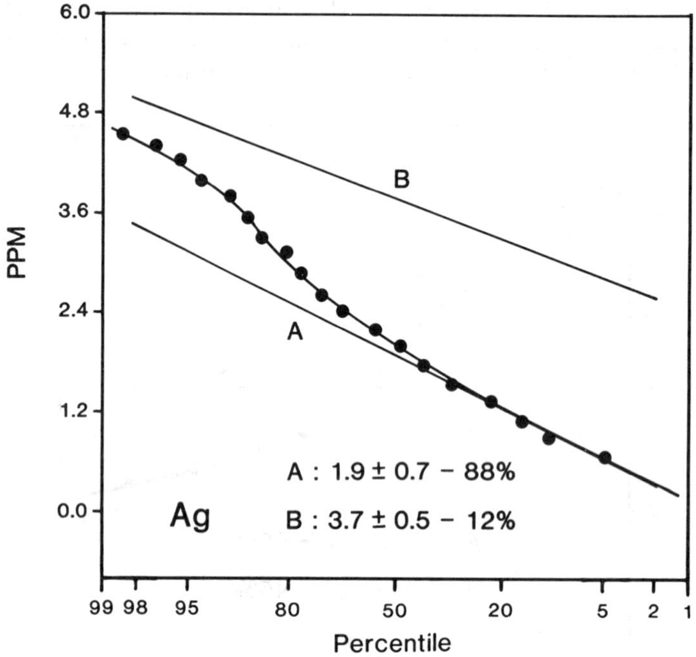

Figure 5 - Probability plot of Ag, a typical bimodal distribution with vastly unequal sized component percentages.

Additionally, a less spatially coherent area in the north-central part of the survey area which is devoid of outcrop has higher concentrations of Ag, As, Ba, Cu, Fe, log(Mo), Pb, Sb, Sn and W (Figure 7C) with lower concentrations of log(Ca). This is an exploration target thought to be underlain by a sulfide occurrence.

Finally, a relatively thin zone trending NW-SE across the center of the survey area contains higher concentrations of Ni (Figure 7D), and lower concentrations of P and log(V). This zone occurs roughly along the geochemical boundary thought to be related to the Sylvester Group/Atan Group contact and may be related to a third, geochemically distinct geologic or geochemical domain.

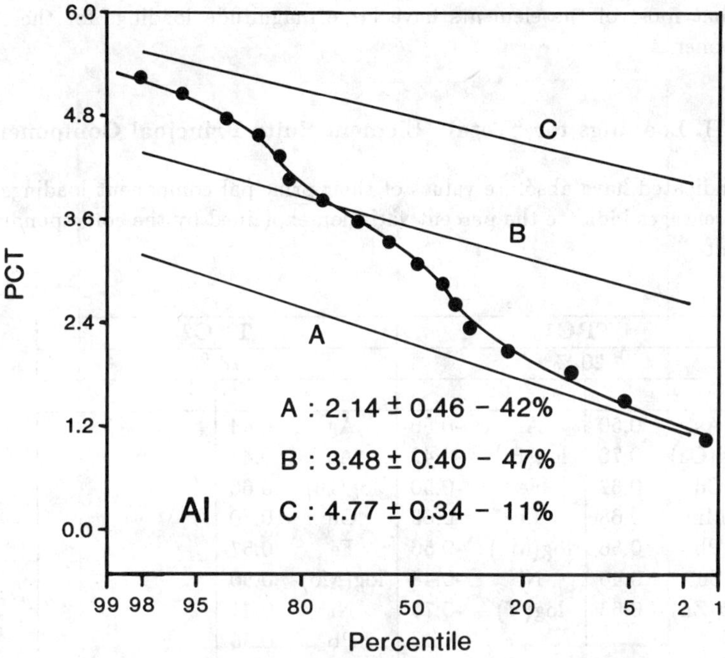

Figure 6 - Probability plot of Al, a typical trimodal distribution.

2.2. Multivariate Evaluation

Due to the large number of variables (elements) analyzed in this soil survey, a multivariate analysis of the data was also undertaken in an attempt to reduce the number of variables for consideration in the interpretation process. The approach used was a principal component analysis because this technique does not require assumptions of normality (clearly many of the elements are polymodally distributed), nor does it impose arbitrary constraints on the analysis which affect the form of the results. In the application presented, the principal component analysis was merely used to rotate the data rigidly in hyperspace for the purpose of viewing the data from a variety of perspectives. In this way, it may be possible to discern and discriminate multiple populations within the data which may be related to geological or geochemical phenomena.

2.2.1. Principal Component Analysis # 1 - "Total" Element Suite

Initially, a principal component analysis was made on the correlation matrix of all 25 elements (the "total" element suite). The loadings greater than 0.40 in absolute magnitude for the first 4 principal components are presented in Table II. Results indicate that 60% of the variation in the data set is described by the first 4 principal components (TPC1-

TPC4), and that most of the elements have large magnitude loadings for the first two principal components.

TABLE II. Loadings for "Total" Element Suite Principal Components

Elements indicated have absolute values of their principal component loadings greater than 0.40. Percentages indicate the percent variation explained by the corresponding principal component.

TPC1				**TPC2**			
20 %				18 %			
+		-		+		-	
As	0.50	Al	-0.56	Ag	0.44		
log(Ca)	0.70	log(Bi)	-0.42	As	0.44		
Cd	0.62	Fe	-0.50	log(Bi)	0.66		
Mn	0.68	Li	-0.63	Cu	0.70		
Pb	0.56	log(Mo)	-0.56	Fe	0.57		
Sn	0.40	Ni	-0.48	log(Mo)	0.50		
log(Zn)	0.66	log(V)	-0.75	Ni	0.41		
				Pb	0.55		
				Sb	0.51		
				log(Zn)	0.42		
				log(W)	0.60		
TPC3				**TPC4**			
11 %				10 %			
+		-		+		-	
Al	0.66	Ni	-0.43	log(Ca)	0.51		
log(K)	0.53	log(V)	-0.43	Mg	0.71		
Na	0.66			Ni	0.43		
				P	0.60		
				Sr	0.70		

Principal component scores from this "total" element suite analysis were extracted and subjected to a histogram and probability plot analysis (Sinclair, 1976). An overlapping polymodal distribution was observed for only the first principal component (TPC1) scores, and an optimal threshold was selected to separate its component populations (Figure 8A). The remaining four principal components were either approximately normally distributed, or bimodally distributed with component population means of roughly equal magnitude but with vastly different standard deviations (intersecting populations similar to a single combined distribution with an apparent highly positive kurtosis). These intersecting distributions are cases in which a set of thresholds is ineffective in partitioning the component populations with a minimum of classification errors. Thus, neither the normal nor intersecting distributions are amenable to threshold selection and as such, no further analysis was made on their principal component scores.

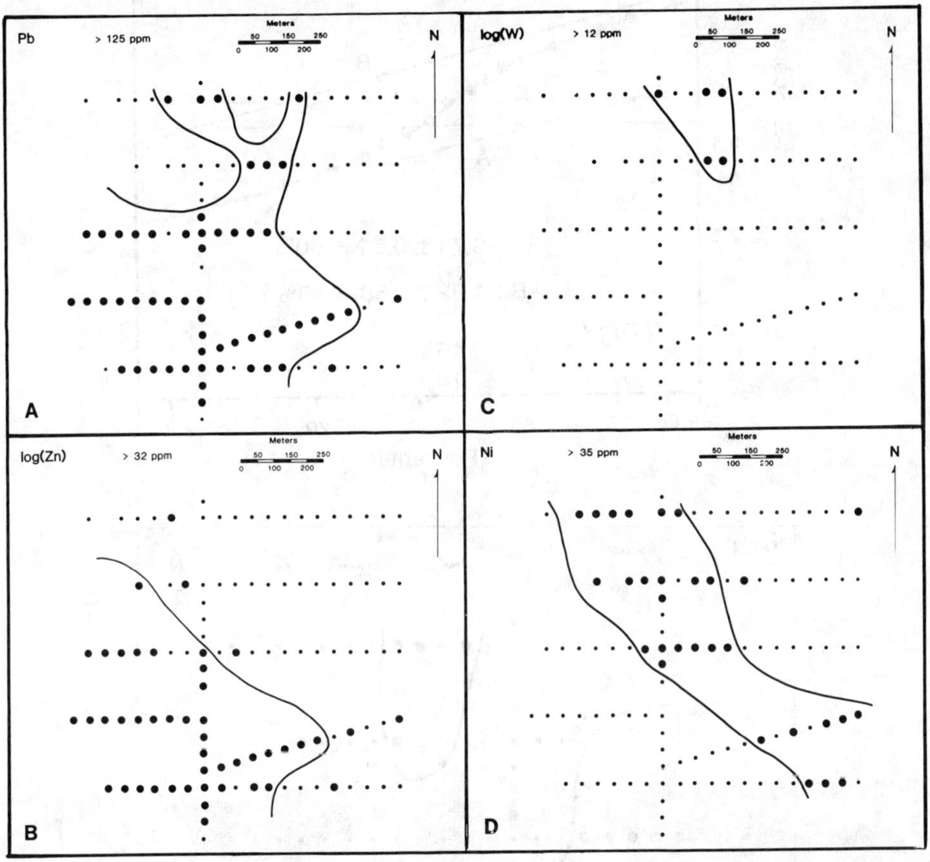

Figure 7 - Symbolic soil grid maps of Pb [A], log(Zn) [B], log(W) [C], and Ni [D].

The first principal component (TPC1) score was plotted on the soil grid with different sized symbols (Figure 8B), and a spatial pattern similar to the pattern defined by elements from the "lithology" element suite was observed. The relative sizes of the component populations from the TPC1 scores are similar to those in elements from the "lithology" suite; thus, this principal component is thought to represent a vector in hyperspace which passes through the corresponding centroids of two lithology-related data clusters. In effect, the probability graphs coupled with the principal component analysis perform the same operation as a discriminant function. What is probably the major structure in the data set can be recognized by this multivariate technique; however, the two other spatial and geochemical trends observed in the univariate analysis results are not apparent in the results of this principal component analysis.

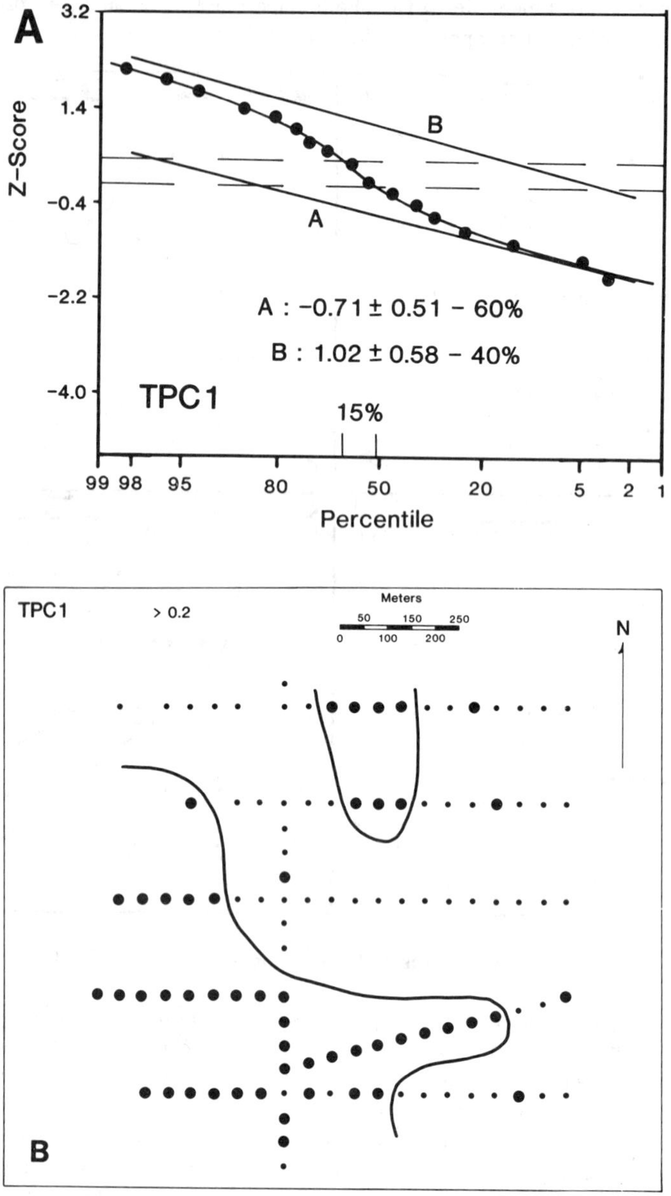

Figure 8 - Probability plot and symbolic soil grid map for TPC1 score.

For this reason, a second approach to the multivariate evaluation of the data set was undertaken. This approach utilized the structure recognized in the data set through univariate analysis to design a multivariate analysis strategy, and involved the evaluation of two separate "structured" principal component analyses.

2.2.2. Principal Component Analysis # 2 - "Lithology" Element Suite

The first of these "structured" principal component analyses used the correlation matrix of only those elements which were classified as part of the "lithology" element suite (Table I), whereas the second used the correlation matrix of those elements which were classified as part of the "mineralization" element suite (Table I). Those principal component loadings with absolute values greater than 0.40 for the "lithology" element suite principal analysis is presented in Table III.

TABLE III. Loadings for "Lithology" Element Suite Principal Components

Elements indicated have absolute values of their principal component loadings greater than 0.40. Percentages indicate the percent variation explained by the corresponding principal component.

LPC1				**LPC2**			
35 %				16 %			
+		-		+		-	
As	0.63	Al	-0.46	Al	0.48	log(Bi)	-0.48
log(Ca)	0.68	Sb	-0.49	log(Ca)	0.55		
Cd	0.65			Na	0.72		
Mn	0.81						
Pb	0.78						
log(Zn)	0.85						

The first two principal components (LPC1-LPC2) from the "lithology" element suite principal component analysis account for more than 50% of the variation in the data set; thus, only scores for these principal components were extracted. Histograms and probability plots of these scores were evaluated (Sinclair, 1976); however, the second one (LPC2) was roughly normally distributed and thus was not considered further. Loadings of high magnitude for the first principal component (LPC1; with absolute values greater than 0.40) have signs which are consistent with the univariate distributions of the bimodally distributed elements over the two lithologic packages (Al and Sb are anomalous to the NE and have negative loadings; As, Cd, Mn, Pb, and log(Zn) are anomalous to the SW and have positive loadings). The first principal component (LPC1) scores are distributed bimodally with roughly equal sized component populations; thus, a threshold was chosen (Figure 9A), and different sized symbols were plotted on the soil grid (Figure 9B). The LPC1 scores also define a spatially coherent geochemical zone in the SW half of the soil grid.

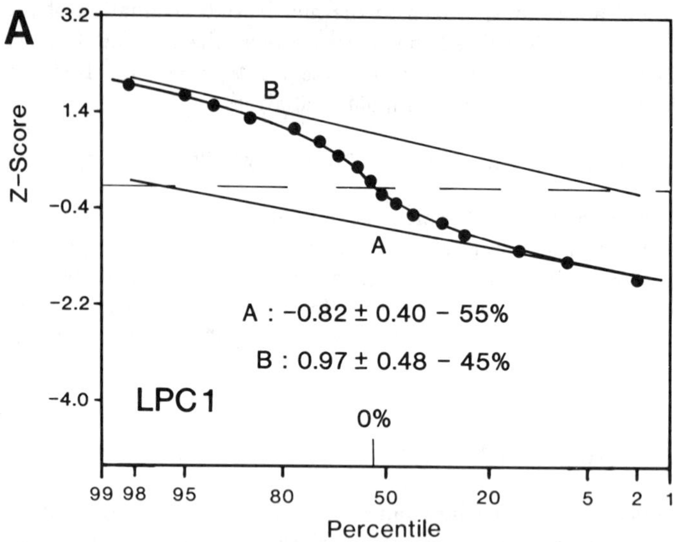

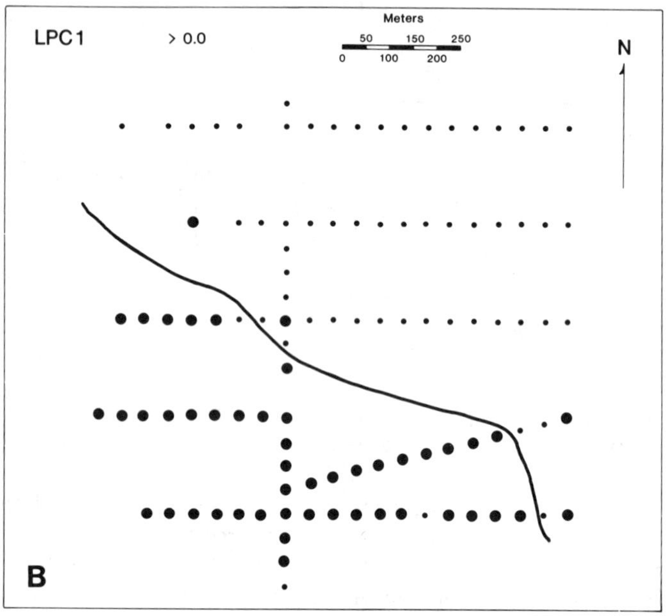

Figure 9 - Probability plot and symbolic soil grid map for LPC1 score.

The zonation defined by the LPC1 scores, the roughly similar contribution to the total data set of each component population, and the LPC1 element loadings suggest that this "lithology" element suite principal component gives results essentially identical to the TPC1 principal component. Thus, the LPC1 scores are also serving as a discriminant score in a manner similar to the TPC1 scores from the "total" principal component analysis.

However, a comparison of the results from these two principal component analyses demonstrates that the clarity of the corresponding results are not identical. Firstly, the amount of component population overlap for LPC1 is much smaller than for TPC1 when measured in terms of both the percentage of data in the zone of overlap and the Mahalanobis Distance statistic (Table IV). In fact, the, TPC1 scores do not significantly decrease the amount of overlap when compared to the relevant single element distributions, whereas the LPC1 scores essentially eliminate the overlapping zone.

TABLE IV. Comparison of Component Population Overlap

Amount of overlap for variables as measured by Mahalanobis distance and "percent overlap" criteria. "Percent overlap" is determined by the percentage of the data which occur between the 'mean plus 2 standard deviations' value of the lower population and the 'mean minus 2 standard deviations' value of the upper population. See figures 4, 8A and 9A for graphical explanation.

Variable	Percent Overlap	Mahalanobis Distance
Pb	4 %	3.71
Zn	19 %	2.82
As	5 %	3.35
Al	10 %	3.02
Cd	18 %	3.07
log(Ca)	25 %	2.70
TPC1	15 %	3.01
LPC1	0 %	4.05

Thus the clarity of the spatial interpretation (Figure 8B versus Figure 9B) is improved by the careful selection of input variables into the principal component analysis, and the anomalous region defined by the TPC1 scores in the north-central portion of the survey area, is not defined by the LPC1 scores. This is due largely to the fact that only those elements which effectively partition the data into these two (possibly lithologically-related) groups were considered. Indiscriminately including all of the elements in the analysis, as in the case of TPC1, appears to introduce random scatter to the principal component scores and has led to more ambiguous and inconsistent results.

2.2.3. Principal Component Analysis # 3 - Mineralization Element Suite

The second of these "structured" principal component analyses, which used the "mineralization" element suite (Table I), was made in an identical manner to the "total" and "lithology" element suite principal component evaluations. The first three principal components (MPC1-MPC3; Table V) account for greater than 50% of the variability in the data. Histograms and probability plots of the extracted scores from these three principal components were evaluated (Sinclair, 1976); thresholds could be chosen because each principal component score was distributed bimodally. Symbolic plots of the soil grid for these principal component scores are presented in Figure 10.

TABLE IV. Loadings for "Mineralization" Element Suite Principal Components

Elements indicated have absolute values of their principal component loadings greater than 0.40. Percentages indicate the percent variation explained by the corresponding principal component.

MPC1			MPC2			
24 %			16 %			
+		-	+		-	
Cu	0.48		Cu	0.59	P	-0.43
Fe	0.71		log(K)	0.43	log(V)	-0.72
Li	0.53					
log(Mo)	0.74					
Ni	0.71					
Sr	0.43					
log(V)	0.54					
log(W)	0.47					
MPC3						
11 %						
+		-				
log(W)	0.43	Fe -0.41				

Three spatial trends are observed. Two of these trends, seen in the MPC1 and MPC3 (Figures 10A and 10C) scores overlie the region in the north-central portion of the survey area defined by the univariate data. This zone may be an exploration target. Slight offsets in the location of high-valued scores from MPC3 relative to MPC1 may indicate that a crude geochemical zoning exists in that region (Figure 10D). Based on the significant element loadings for each score (Table V), this zonation may be related to an Fe enrichment halo around this zone. The region defined by high scores from MPC2 (Figure 10B) occurs along a trend also previously recognized in the univariate analysis. This zone is a highly coherent NW-SE trending region possibly related to a geochemically distinct stratigraphy (possibly the McDame, Sandpile, and/or Kechika Groups) located (but not exposed) between the two major lithologic packages underlying the soil grid, or to geochemical differences of either primary (deuteric alteration-related) or secondary (supergene drainage-related) origin caused by the presence of the thrust fault and associated fractures.

Results from the "mineralization" element suite principal component analysis have defined areas on the soil grid which were not defined by either the "total" or "lithology" element suite principal component analyses. The recognized trends add to the interpretive value of the results and, when considered with the results from the "lithology" element suite principal component analysis (the other "structured" multivariate analysis), are significantly superior in scope and clarity to the "unstructured" principal component analysis results.

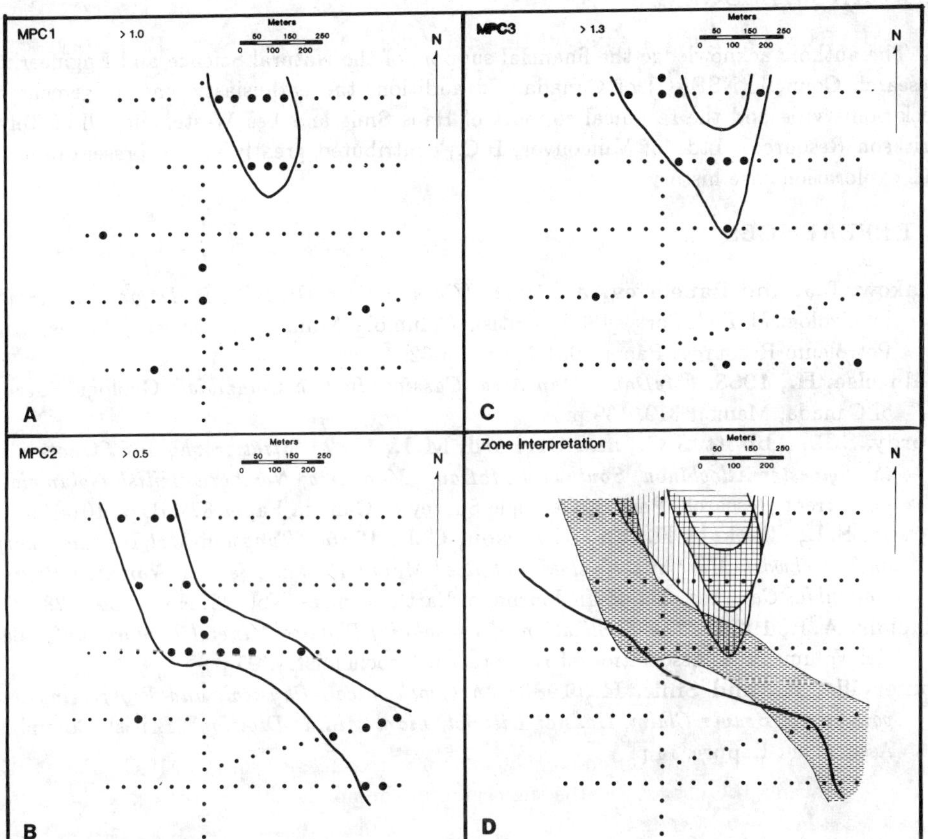

Figure 10 - Symbolic soil grid maps of MPC1 [A], MPC2 [B] and MPC3 [C], and composite interpretation from principal component analyses # 2 and # 3 [D].

3. CONCLUSIONS

The combined evaluation of a "structured" set of principal component analyses (the "lithology" and "mineralization" element suites) offers significant advantages over an evaluation where the elements utilized are indiscriminately input into the analysis ("total" element suite). This "structured" approach did not require the evaluation of more principal components (in this case the "total" principal component analysis and the combined "lithology" and "mineralization" principal component analyses both had a total of 5 significant vectors), and led to a clearer and more comprehensive interpretation of the data.

Thus, the consideration of the univariate numerical structure of a data set in the design of multivariate analysis procedures may allow for easier evaluation of the multivariate results and lead to a more comprehensive interpretation.

4. ACKNOWLEDGEMENTS

The authors acknowledge the financial support of the Natural Science and Engineering Research Council (NSERC) of Canada. In addition, the enthusiastic encouragement of Rick Somerville and the technical support of Hans Smit and Les Westervelt, all of Total Erickson Resources, Ltd. of Vancouver, B.C., contributed greatly to the presentation of this exploration case history.

5. REFERENCES

Diakow, L.J. and Panelayev, A., 1981, *"Cassiar Gold Deposits, McDame Map Area"*, in Geological Fieldwork, 1980, British Columbia Ministry of Energy, Mines, and Petroleum Resources Paper 1981-1, pp. 55-62.

Gabrielse, H., 1963, *"McDame Map Area, Cassiar, British Columbia"*, Geologic Survey of Canada, Memoir 319, 139 p.

Gordy, S.P., Abbott, J.G., and Orchard, M.J., 1982, *"Stratigraphy and Structure of the Sylvester Allochthon, Southwest McDame Map Area, Northern British Columbia"*, in Current Research, Part B, Geologic Survey of Canada Paper 82-1B, p. 101-106.

Gower, S.J., Clark, A.H., and Hodgson, C.J., 1985, *"Tungsten-Molybdenum Skarn and Stockwork Mineralization, Mount Reed-Mount Haskin District, Northern British Columbia, Canada"*, Canadian Journal of Earth Sciences, Vol. 22, No. 5, pp. 728-747.

Sinclair, A.J., 1976, *"The Application of Probability Plots to Mineral Exploration"*, Special Volume # 4, Association of Exploration Geochemistry, 95 p.

Somerville, R., and Smit, H., 1985, *"A Geochemical, Physical, and Prospecting Report on the Beaver Claim, Cassiar District, Liard Mining Division"*, British Columbia Assessment Report, 14 p.

APPLICATION OF ROBUST STATISTICS IN THE ANALYSIS OF GEOCHEMICAL DATA

F. Wurzer
Joanneum Research Society
Mineral Resources Research Division
Roseggerstr. 15
8700 Leoben, Austria

ABSTRACT. In mineral exploration, regional geochemical surveys are widley used, and these often benefit from a statistical analysis of the resulting data. However, classical statistics often fail to describe the data adequately. In this study robust and resistant methods which improve the characterisation of uni- and multivariate element behaviour are presented. For the univariate case these are: resistant data summaries (e.g., to determine class boundaries for coded maps), and the robust estimation of parameters. In the multivariate case robust estimaties of the variance-covariance matrix are used in principal components and canonical correlation analysis. For the determination of outliers not corresponding to multivariate patterns an approach based on an influence function is applied.

0. INTRODUCTION

This paper presents examples of a case study based on more than 700 soil samples covering an area of about 90 km^2 in the Austrian Alps. For each sample the concentrations of 27 chemical elements has been determined (Na, Cr, Fe, Co, Ni, As, Sb, Ba, La, Hf, Ta, W, Au, Th, U, Mg, Al, K, Ca, Sr, Mn, Cu, Zn, Pb, S, Si, P). In addition CO_2, LOI, pH, Eh and conductivity were determined. Classical statistical analysis based on the (log-)normal law often fails to produce satisfactory results. The reasons for this include: highly skewed and outlier disturbed distributions, and sampling and analytical errors resulting in very heterogeneous data. These problems provided the motivation to adopt robust and resistant statistical methods for the analysis of regional geochemical data. This study does not include procedures and methods used to reduce or measure sampling and analytical errors (see Reimann and Wurzer, 1987). For the application of the methods described below the data have to be reliable.

C. F. Chung et al. (eds.), Quantitative Analysis of Mineral and Energy Resources, 131–143.

1. THE UNIVARIATE CASE

1.1. Resistant Data Summaries

At the beginning of a data analysis univariate statistical procedures have to be applied to each variable. For an initial inspection of the data, resistant data summaries like Tukey's boxplot (Tukey, 1979) can be used. These provide a resistant description of the data, specifying location, spread, skewness, taillength, and outlying data points. Because the measured variables areregionalized, the best way to present the observations is to map them in a coded way. For coding we use class boundaries corresponding to thesummary points of the boxplot: the median, the upper and the lower hinge (H_U, H_L) which can be seen as quartiles, and the adjacent values which are the smallest respective highest values falling within the interval of the outlier cutoffs (see also Velleman and Hoaglin, 1981). These cutoffs are given by

$$H_U + 1.5 * h \quad \text{and} \quad H_L - 1.5 * h.$$

where h is called the h-spread and

$$h = H_U - H_L.$$

In the case of the normal distribution this interval contains approximately 99.3 % of the data. To finish, the boxplot, outliers (values falling beyond the mentioned cutoffs) are plotted separately.

For mapping we code the observations according to the following classes corresponding to the boxplot: lower respective upper outliers, values between the lower respective upper hinge and the corresponding adjacent value, and the middle 50%-batch (values between the hinges). Figure 1 shows an example of such a coded map.

For exploration applications additional information is presented for the most interesting class of the upper outliers by making the size of the plotted symbol proportional to the measured data value.

When using classical statistics, coding is often on the basis of arbitrarily chosen histogram classes or the empirical percentiles. Too many classes will result in mapping sampling and analytical errors. With the boxplot, however, class boundaries and outlier definition are based on the data, and different regions can be compared directly. No subjective data handling is necessary.

1.2. Robust Location and Scale

To estimate the parameters of location and scale, we consider the M-estimator proposed by Huber (1981). Suppose that we have n univariate observations $x_1,\ldots,x_n$. According to Huber's Proposal 2, we estimate location and scale by the solution of the system of equations:

$$\sum_{i=1}^{n} \psi\left(\frac{x_i - \mu}{\sigma}\right) = 0$$

$$\sum_{i=1}^{n} \left[\psi\left(\frac{x_i - \mu}{\sigma}\right)\right]^2 = \beta$$

where β is the expected value in the normal case

$$\beta = E_{\Phi}[\psi(X)]^2$$

and ψ is a robustifying function. We use two such functions. The first is

$$\begin{aligned} \psi(t) &= -c \quad \text{for } t < -c \\ &= t \quad \text{for } |t| \leq c \\ &= c \quad \text{for } t > c \end{aligned}$$

where c = 1.345 is chosen corresponding to 95% efficiency in the normal case.
The second is the piecewise linear function proposed by Hampel

$$\begin{aligned} \psi(t) = -\psi(-t) &= t, && \text{for } 0 \leq t < a \\ &= a, && \text{for } a \leq t < b \\ &= ((c-t)/(c-b))*a, && \text{for } b \leq t < c \\ &= 0, && \text{for } t \geq c \end{aligned}$$

where a = 1.7, b = 3.4, and c = 8.5 (see Afifi and Azen, 1979).

Fig. 2 shows a typical strongly skewed empirical distribution; Table I the corresponding statistics. It is clear that, compared to the arithmetic mean and the standard deviation, the robust estimates give better results describing the bulk of the data.

Table I:

VARIABLE: Pb [ppm]
NUMBER OF OBSERVATIONS: 709

ARITHMETIC MEAN:	37.69	STAND. DEVIATION:	47.16
ROB. LOCATION (HUBER):	28.07	ROB. ESTIMATE OF	
ROB. LOCATION (HAMPEL):	27.11	SCALE (HUBER):	19.02

Tab. I: Univariate statistics corresponding to the empirical distribution shown in Fig. 2.

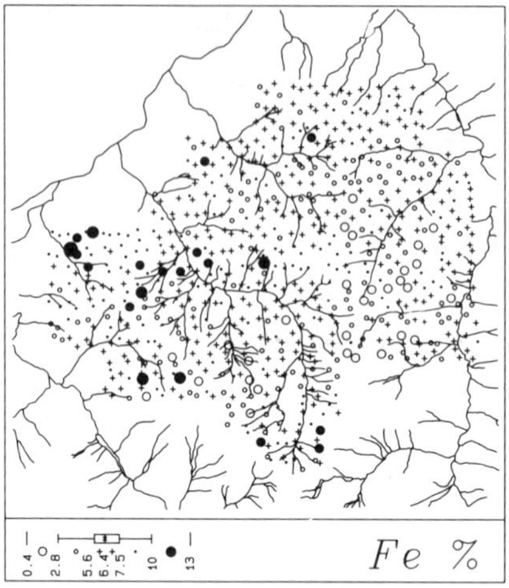

Fig. 1: Coded map for the element iron based on classes determined by the boxplot.

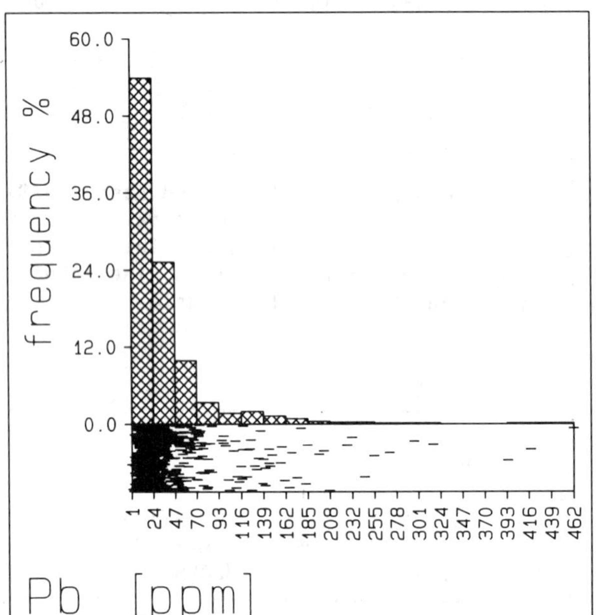

Fig. 2: Histogram and jittered one-dimensional scatterplot for the element Pb

2. MULTIVARIATE ANALYSIS

2.1. Robust Covariances

Most of the parametric multivariate statistical methods are based on the analysis of the covariance structure, e.g., principal components and canonical correlation analysis. It is well known that the empirical covariance matrix is very sensitive to outlying data values. The effects of a proportion ε of outlying data are biased estimates and a sometimes enormously increased statistical variability. If ε becomes sufficiency large estimation may even become meaningless. For the classical empirical covariance matrix, this critical amount of outlying data is equal to zero because one single point lying far away from the bulk of the data spoils the whole estimation of the covariance structure. In exploration geochemistry, however, outliers almost always have to be expected.

Therefore, to improve the effectiveness of multivariate statistical methods, robust estimates for multivariate location and scatter are introduced. The algorithm used is discussed briefly; for details see Dutter (1983) and Karnel (1985).

Suppose that we have a complete, multivariate matrix of observations (x_{ij}), $i = 1,\ldots,n$ observations, $j = 1,\ldots,p$ variables. $\underset{\sim}{x}_i = (x_{i1},\ldots,x_{ip})$ represents a p-dimensional observation. The aim is to estimate the location and the covariance between the elements of $\underset{\sim}{x}_i$.

First, try to find a transformation $(A,\underset{\sim}{b})$ where A is a matrix of dimension (pxp) and $\underset{\sim}{b}$ is a (px1) vector and the data $\underset{\sim}{x}_i$ will be transformed to

$$\underset{\sim}{y}_i = A(\underset{\sim}{x}_i - \underset{\sim}{b}),\ i = 1,\ldots,n,$$

such that

$$\operatorname*{ave}_i \{w(\|\underset{\sim}{y}_i\|)\underset{\sim}{y}_i\} = \underset{\sim}{0},$$

$$\operatorname*{ave}_i \{u(\|\underset{\sim}{y}_i\|)\frac{\underset{\sim}{y}_i\underset{\sim}{y}_i^T}{\|\underset{\sim}{y}_i\|^2} - v(\|\underset{\sim}{y}_i\|)\ I\} = 0.$$

u, v, and w are certain weight functions. The definition of a robust estimate of location is then given by b, and a pseudo-covariance matrix by

$$C = (A^TA)^{-1}\tau^2.$$

τ denotes a correction factor to standardize the estimation approximately for the case of a normal distribution. As weight functions we use

$$\begin{aligned} u(t) &= a^2 && \text{if } t < a\\ &= t^2 && \text{if } a \le t \le b\\ &= b^2 && \text{if } b < t,\end{aligned}$$

$$\begin{aligned} w(t) &= 1 && \text{if } t < c\\ &= c/t && \text{if } c \le t,\end{aligned}$$

The Function v,

$$v(t) = 1/p \; E_{\Phi}[u(||X||)],$$

is constant and is given by the condition of Fisher-consistency. To determine the constants a, b, and c, we use the model assumption that we have a certain portion of non-normal data values ε. By specifying ε, these constants can be computed by the minimax solution using the gross-error model.

The parameter ε is of special interest for the practical application to geochemical data. The gross-error model gives a mathematical formulation of the underlying data model consisting of samples belonging to the background population in contrast to those reflecting local influence.

2.2. Robust Principal Component Analysis

Principal components analysis (PCA) is a multivariate statistical method directly based on the covariance (or correlation) matrix. Thus it is very sensitive to the estimation of the variance-covariance structure. To robustify this method, the covariance matrix C is calculated in the robust way described above.

Principal components are orthogonal linear transformations of the data matrix Y = XA, where X and Y are (nxp) matrices and A is a (pxp) matrix, with the first principal component having the largest variance, down to the pth which has the smallest variance. It can be shown that the variances of the principal components are the eigenvalues of C and the vectors $\tilde{a}_j = (a_{ij},\ldots,a_{pj})^T$ of the transformation matrix A are the corresponding eigenvectors (Srivastava and Carter, 1983).

In our studies, PCA is used to obtain a reduction in dimensionality. This is achieved by considering only the first k principal components, and discarding those linear combinations with the smallest variances. Also of interest for data analysis in exploration geochemistry is the search for significant outliers in the distribution of principal components which can be interpreted according to certain dominant element associations, e.g., regional variability or specific mineralizations.

Table II presents the PCA results for 709 soil samples, the elements Na, Cr, Fe, Sb, Ba, Mg, Al, K, Ca, Sr, Mn, Cu, Zn, and Pb are considered. The computed principal components are based on a robustly estimated covariance matrix with a proportion ε = 5 % of non-normal data values. Because the elements are analyzed in different measurement units (ppm and %), the computation is based on robustly standardized data, i.e., standardization by robustly estimated locations and variances.

To demonstrate the effect of robustification, Table III presents the corresponding results based on the empirical variance-covariance matrix.

Important in the quest for a reduction in dimensionality is the fact that the cumulative percent of total variation exhibited by the first few principal components increases for the robust approach. Another effect is apparent which improves the interpretability of the computed results. For the analysis based on robust covariances, a better contrast between high and low variable loadings is achieved See for example Pb and Sb in the third eigenvector or Cu, Zn, Al, K, and Ba in the second one.

P C A - Results (robust)
****

E I G E N V E C T O R S

	1	2	3	4	5	6	7	8	9	10	11	12	13	14
Na !	0.242	-0.160	-0.208	0.475	0.158	-0.463	0.216	0.258	-0.339	-0.281	0.184	-0.077	-0.246	-0.027
Cr !	0.333	0.156	0.064	0.094	-0.442	-0.203	-0.391	0.265	0.320	-0.030	-0.327	-0.321	-0.185	0.211
Fe !	0.345	0.073	0.043	-0.158	-0.464	0.082	0.423	0.037	-0.204	0.494	0.377	-0.117	-0.060	0.035
Sb !	0.083	0.029	0.681	0.174	-0.136	-0.046	0.031	0.150	0.018	-0.049	-0.034	0.628	-0.163	-0.171
Ba !	-0.178	0.414	-0.102	0.309	0.204	-0.387	-0.174	0.020	0.066	0.585	0.098	0.199	0.196	0.186
Mg !	0.395	0.161	-0.075	0.068	-0.069	0.049	-0.476	-0.159	-0.247	-0.072	0.174	0.007	0.370	-0.562
Al !	-0.181	0.413	-0.133	0.263	-0.313	0.364	0.083	0.208	-0.221	-0.384	0.073	0.133	0.298	0.348
K !	-0.326	0.360	-0.030	0.202	0.014	0.291	-0.101	0.022	0.064	0.051	0.216	-0.223	-0.620	-0.374
Ca !	0.367	-0.167	-0.057	0.115	0.230	0.366	-0.357	-0.231	-0.027	0.054	0.266	0.213	-0.303	0.493
Sr !	0.257	-0.123	-0.190	0.526	0.073	0.375	0.330	-0.019	0.377	0.192	-0.324	0.033	0.152	-0.207
Mn !	0.249	0.212	0.157	-0.264	0.516	0.188	0.045	0.646	0.137	-0.008	0.150	-0.135	0.135	-0.010
Cu !	0.222	0.403	0.122	-0.031	0.096	-0.217	0.289	-0.474	0.445	-0.358	0.267	-0.067	0.032	0.065
Zn !	0.235	0.430	0.007	-0.147	0.224	0.034	0.155	-0.207	-0.462	0.050	-0.596	0.015	-0.216	0.043
Pb !	-0.103	-0.106	0.612	0.345	0.132	0.089	-0.025	-0.176	-0.222	0.072	0.011	-0.553	0.215	0.147

	EIGENVALUES	PERCENT OF VARIATION	CUM. PERCENT
1	4.1882	29.916	29.916
2	2.5162	17.973	47.889
3	1.7163	12.260	60.148
4	1.1711	8.365	68.513
5	0.9074	6.482	74.995
6	0.8090	5.779	80.774
7	0.6659	4.756	85.530
8	0.5046	3.604	89.134
9	0.4113	2.938	92.072
10	0.3020	2.157	94.229
11	0.2746	1.961	96.190
12	0.2507	1.791	97.981
13	0.1496	1.068	99.049
14	0.1331	0.951	100.000

Tab. II: Results of a principal components analysis based on the robust variance-covariance estimation.

P C A - Results
****

E I G E N V E C T O R S

	1	2	3	4	5	6	7	8	9	10	11	12	13	14
Na !	0.128	0.029	-0.482	-0.282	0.361	-0.118	-0.363	0.198	0.293	0.301	0.323	0.152	-0.229	0.013
Cr !	0.240	0.325	-0.210	-0.045	-0.342	0.424	-0.079	0.425	-0.125	-0.431	0.112	0.178	-0.055	-0.242
Fe !	0.274	0.314	-0.264	-0.036	-0.298	-0.153	0.355	-0.076	0.246	0.377	-0.482	0.265	0.079	-0.015
Sb !	0.083	0.313	0.345	-0.507	-0.053	0.131	0.044	0.099	0.046	0.183	-0.093	-0.563	-0.354	0.035
Ba !	-0.345	0.221	-0.010	0.192	0.331	0.201	-0.367	0.319	0.105	0.065	-0.561	-0.104	0.251	-0.072
Mg !	0.407	-0.015	0.159	0.278	0.107	0.430	0.042	0.107	0.198	0.132	0.129	-0.011	0.158	0.651
Al !	-0.358	0.271	-0.165	0.079	0.062	0.142	0.471	0.090	0.125	0.211	0.470	-0.239	0.373	-0.182
K !	-0.429	0.140	0.133	0.160	0.120	0.148	0.285	0.130	-0.093	0.100	-0.001	0.442	-0.611	0.177
Ca !	0.349	-0.222	0.290	0.225	0.256	0.273	0.084	-0.036	0.091	0.273	0.003	0.078	-0.131	-0.663
Sr !	0.214	-0.036	-0.315	-0.209	0.591	0.086	0.429	-0.049	-0.359	-0.272	-0.227	-0.087	0.010	0.063
Mn !	0.237	0.162	0.269	0.202	0.116	-0.600	0.088	0.579	-0.262	0.062	0.098	0.001	0.077	0.009
Cu !	0.100	0.497	-0.017	0.141	0.063	0.070	-0.309	-0.433	-0.567	0.290	0.138	0.060	0.048	0.004
Zn !	0.106	0.462	0.099	0.265	0.238	-0.223	-0.002	-0.312	0.475	-0.478	0.067	-0.063	-0.164	-0.056
Pb !	-0.046	0.134	0.446	-0.537	0.182	0.037	-0.003	-0.052	0.092	-0.084	0.073	0.523	0.400	-0.007

	EIGENVALUES	PERCENT OF VARIATION	CUM. PERCENT
1	3.6386	25.990	25.990
2	2.3805	17.004	42.994
3	2.0029	14.306	57.300
4	1.4532	10.380	67.680
5	0.9794	6.996	74.675
6	0.8173	5.838	80.513
7	0.6450	4.607	85.120
8	0.5413	3.866	88.986
9	0.4328	3.091	92.078
10	0.3362	2.401	94.479
11	0.3006	2.147	96.626
12	0.2222	1.587	98.213
13	0.1733	1.238	99.451
14	0.0769	0.549	100.000

Tab. III: Results of a principal components analysis based on the classical empirical variance-covariance matrix.

2.3. Robust Canonical Correlation Analysis

Canonical correlation analysis is a multivariate statistical method to analyse the dependence of two groups of variables. The aim is to find a linear combination

$$Z_X = u_1X_1 + \ldots + u_pX_p$$

of the first group of variables $(X_1,\ldots,X_p)$ and also a linear combination

$$Z_Y = v_1Y_1 + \ldots + v_qY_q$$

of the second group of variables $(Y_1,\ldots Y_q)$ such that their correlation

$$r(Z_X,Z_Y) = \frac{\Sigma\,(Z_X-\bar{Z}_X)(Z_Y-\bar{Z}_Y)}{\sqrt{\Sigma\,(Z_X-\bar{Z}_X)^2\,\Sigma\,(Z_Y-\bar{Z}_Y)^2}}$$

is maximized. The centered values $Z_X-\bar{Z}_X$ and $Z_Y-\bar{Z}_Y$ are called canonical variates.

A mathematical equivalent formulation for the correlation coefficient $r(Z_X,Z_Y)$ can be given by

$$r(Z_X,Z_Y) = \frac{\underset{\sim}{u}^T S_{XY} \underset{\sim}{v}}{(\underset{\sim}{u}^T S_{XX} \underset{\sim}{u})(\underset{\sim}{v}^T S_{YY} \underset{\sim}{v})},$$

where

$$\underset{\sim}{u}^T = (u_1,\ldots,u_p),$$
$$\underset{\sim}{v}^T = (v_1,\ldots,v_q),$$

and S_{XY}, S_{XX}, and S_{YY} are the sample variance-covariance matrices for the corresponding groups of variables.

Thus, to robustify this procedure we use - just as in principal components analysis - robust covariances computed by the algorithm presented in chapter 2.1.

The best way to interpret the bivariate structure of the canonical variates is to present them on a scatter plot.

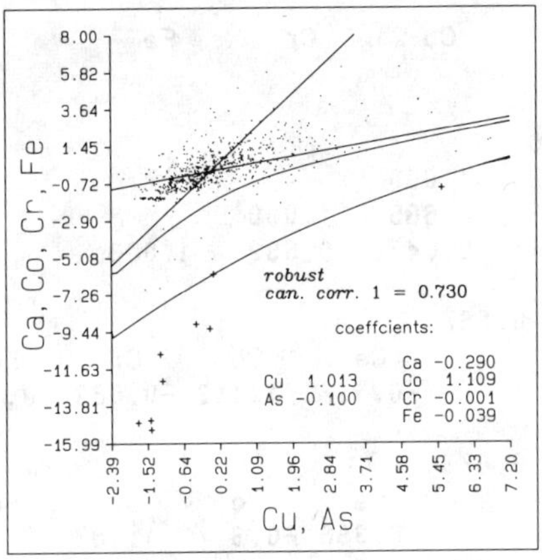

Fig. 3: Scatterplot of canonical variates (robust); the asymptotes and hyperbolic isolines of the influence function for the definition of multivariate outliers are plotted.

In Fig. 3 an example of the application of canonical correlation analysis to geochemical data is given. Considered are the variables Cu and As as a mineralization element-association group. To reduce the influence of greenschists that generally tend to have elevated Cu- and As-values within this area under investigation, the second group of variables consists of Co, Cr, Fe and Ca. Table IV contains the robustly computed canonical correlations, and the corresponding coefficients for the linear combinations.

CORRELATIONS :

	Cu	As	Ca	Co	Cr	Fe
Cu	1.000					
As	0.178	1.000				
Ca	0.112	-0.054	1.000			
Co	0.698	0.043	0.419	1.000		
Cr	0.389	-0.051	0.313	0.593	1.000	
Fe	0.392	0.122	0.314	0.594	0.509	1.000

CANONICAL CORRELATION 1 = 0.730

coefficients :	Cu	As	Ca	Co	Cr	Fe
	1.013	-0.100	-0.290	1.109	-0.001	-0.039

CANONICAL CORRELATION 2 = 0.190

coefficients :	Cu	As	Ca	Co	Cr	Fe
	0.081	-1.011	0.392	0.040	0.825	-1.048

Tab. IV: Results computed by robust canonical correlation analysis for the element groups (Cu, As) and (Ca, Co, Cr, Fe).

CORRELATIONS :

	Cu	As	Ca	Co	Cr	Fe
Cu	1.000					
As	0.331	1.000				
Ca	-0.072	-0.026	1.000			
Co	0.622	0.018	0.042	1.000		
Cr	0.384	-0.028	-0.035	0.685	1.000	
Fe	0.353	0.034	-0.024	0.647	0.532	1.000

CANONICAL CORRELATION 1 = 0.667

coefficients :	Cu	As	Ca	Co	Cr	Fe
	1.059	-0.318	-0.158	1.142	-0.083	-0.153

CANONICAL CORRELATION 2 = 0.073

coefficients :	Cu	As	Ca	Co	Cr	Fe
	0.034	-1.011	0.358	-0.317	1.187	-0.721

Tab. V: Results computed by classical canonical correlation analysis for the element groups (Cu, As) and (Ca, Co, Cr, Fe).

The effect of robustification can be seen by comparison to Table V, which shows the canonical correlation results based on classical variance-covariance estimation. Note the higher first canonical correlation for the robust procedure, which is exactly the objective of canonical correlation, i.e., to obtain two linear combinations of the groups of variables with the maximum possible correlation between them.

The question in mineral exploration is which observations deviate most from the general multivariate behaviour. For a discussion of that topic see also Howarth and Sinding-Larsen (1983).
A new approach for the definition of outliers in canonical correlation analysis shall be discussed here.

For simplicity, let us consider the case of correlation analysis. The influence of a data point upon the correlation coefficient depends on the position of that point in the bivariate space spanned by the two variables. A point lying at 45 degrees to the data clusters major axis increases the correlation, otherwise a point lying on the perpendicular line through the bivariate center decreases it. Thus, somewhere in between points have no effect at all. This property can be measured by influence functions.

Such an empirical influence function which determines the effect of each data point upon the correlation coefficient is given by:

$$I(y_{i1},y_{i2};r) = (n-1)(r-r_{-i}) \overset{\text{appr.}}{=} (1-r^2)\, u_{i1}u_{i2},$$

where (y_{i1},y_{i2}) are n bivariate observations and r_{-i} denotes the correlation coefficient based on n-1 observations (discarding observation with index i). Further notations:

$$u_{i1} = 1/2\,\sqrt{n}\,\{(d_{i1}+d_{i2})/\sqrt{1+r} + (d_{i1}-d_{i2})/\sqrt{1-r}\},$$
$$u_{i2} = 1/2\,\sqrt{n}\,\{(d_{i1}+d_{i2})/\sqrt{1+r} - (d_{i1}-d_{i2})/\sqrt{1-r}\},$$

where

$$d_{ij} = (y_{ij} - m_j)/\sqrt{a_{jj}},$$
$$m_j = (\sum y_{ij})/n, \text{ and}$$
$$a_{jj} = \sum (y_{ij} - m_j)^2$$

(see Devlin et al., 1975).

The data can then be ranked according to their influence. Values with the highest influence are marked as outliers. The shape of this influence function (over the whole bivariate space) is hyperbolic. Points lying on the asymptotes do not have any effect on the correlation. Outlier cutoffs in a scatterplot are given as hyperbolic curves.

The application of this method to canonical correlation analysis is demonstrated in Fig. 3. For mineral exploration purposes, outliers having the high values in the first group of variables are of the greatest interest. Therefore hyperbolas for the definition of outliers are drawn only in the appropriate section at the 80th and 95th percentiles. To determine the geographical position of the samples, a map is plotted with outlying observations marked separately. This resulting map can guide the geologist in his interpretation of the data.

Another improvement of robust computation is that the outliers are displayed more clearly than for classical canonical correlation analysis, Fig. 4. The bivariate cloud of the canonical variates is more scattered and outlying data values cannot be seen as obviously.

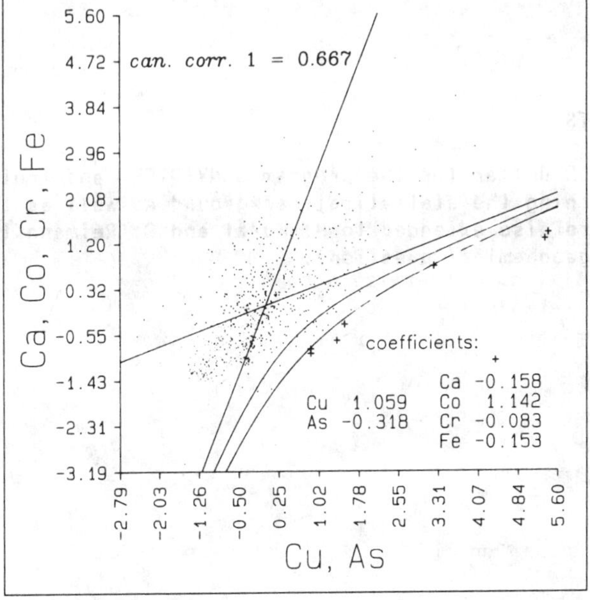

Fig. 4: Scatterplot of canonical variates based on classical covariance estimation; asymptotes and hyperbolic isolines of the influence function are plotted as in Fig. 3.

3. CONCLUSIONS

The presence of anomalies (outliers in the statistical sense) is a basic assumption of geochemical exploration, resulting in highly skewed and outlier disturbed data. In addition, geochemical samples are usually taken from multiple populations. Therefore classical statistical analysis will often fail to produce results which can be successfully interpreted. The aims of robust statistics are to describe the structure best fitting the bulk of the data and to enhace the identification of outliers and unusual deviating substructures. Thus they are well suited to treat geochemical data.

As the examples show, resistant data summaries and robust estimation of location and scale perform well in the univariate case. For the multivariate procedures, it is even more important to apply robust estimates for multivariate location and the variance-covariance structure. General data behaviour can be described more accurately, providing a better contrast to hidden substructures or deviating data points. Based on robust estimates of covariance (correlation) Principal Components and Canonical Correlation Analysis end up with more reliable results giving a detailed insight to geochemical processes. The influence function used in Canonical Correlation Analysis proofed to be very effective to detect multivariate outliers. A value of influence on the estimation of canonical correlation is assigned to each data point, thus giving a clear criterion to decide whether a sample is deviating from the multivarite data structure or not.

4. ACKNOWLEDGEMENTS

Thanks are due to R.Dutter for the program CONVINTER and fruitful discussions concearning the statistical background as well as to G.Karnel. Thanks are also extended to H. Kürzl and C. Reimann for the discussion of geochemical questions.

5. REFERENCES

Afifi A.A. and Azen S.P. (1979): *Statistical Analysis, A Computer Oriented Approach*. Academic Press, New York.

Devlin S.J., Gnanadesikan R., and Kettenring J.R. (1975): 'Robust Estimation and Outlier Detection with Correlation Coefficients.' Biometrika, 62,3, p531.

Dutter R. (1983): 'COVINTER: A Computer Program For Computing Robust Covariances and For Plotting Tolerance Ellipses.' Res. Rep. No. 10, Institute for Statistics, Techn. Univ. Graz, Austria.

Howarth R.J. and Sinding-Larsen R. (1983): 'Multivariate Analysis.' In Handbook of Exploration Geochemistry. Vol. 2: Statistics and Data Analysis in Geochemical Prospecting. Elsevier, Amsterdam-Oxford-New York.

Huber P.J. (1981): *Robust Statistics*. J. Wiley, New York.

Karnel G. (1985): 'Robuste Kovarianzmatrizen und kanonische Korrelation.' Unpublished diploma thesis, Techn. Univ. Graz, Austria.

Reimann C., Kürzl H. and Wurzer F. (1987): 'Monitoring Accuracy and Precision - Improvements by Introducing Robust and Resistant Statistics.' COBAC IV, September 15-19, 1986 - Graz, Austria; Microchemica Acta (in press.).

Srivastava M.S. and Carter E.M. (1983): *An Introduction to Applied Multivariate Statistics*. North Holland, New York.

Tukey J.W. (1979): *Exploratory Data Analysis*. Addison Wesley.

Velleman P.F. and Hoaglin D.C. (1981): *Application, Basics, and Computing of Exploratory Data Analysis*. Duxbury Press, Boston.

U.S. GEOLOGICAL SURVEY ASSESSMENT METHODOLOGY FOR ESTIMATION OF UNDISCOVERED PETROLEUM RESOURCES IN PLAY ANALYSIS OF THE ARCTIC NATIONAL WILDLIFE REFUGE

Robert A. Crovelli
U.S. Geological Survey
P.O. Box 25046, MS-971
Denver Federal Center
Denver, Colorado 80225, U.S.A.

ABSTRACT. The objective of this investigation was to develop a geostochastic system for estimation of undiscovered petroleum resources in play analysis. A resource appraisal system was designed for play analysis using a reservoir-engineering model and an analytic probabilistic methodology. The system resulted in a package of computer programs, called FASP, for play analysis, subplay analysis, mixed-lithology analysis, and aggregations. Resource estimates of crude oil, nonassociated gas, dissolved gas, and total gas are calculated in terms of probability distributions as tables and graphs. The FASP computer package is a large complex system that is very flexible and efficient, but which can easily be adapted to most microcomputers. This system was used by the U.S. Geological Survey to assess the in-place petroleum potential of the Arctic National Wildlife Refuge located in northeastern Alaska.

1. INTRODUCTION

An analytic method using probability theory was developed for petroleum resource appraisal of undiscovered oil and gas in the Arctic National Wildlife Refuge (ANWR) in Alaska (U.S. Department of the Interior, 1987). An objective was to replace an existing Monte Carlo simulation method in order to increase the efficiency of the petroleum resource appraisal process. The resulting analytic method is a geostochastic system for estimating the undiscovered oil and gas potential of a geologic play and an aggregation of plays. Play analysis is a general term for various geologic models and probabilistic methods for analyzing a geologic play.

A geologic model for the quantity of undiscovered petroleum resources in a play involves uncertainty due to the incomplete or fragmentary geologic information generally available. The basic geologic model used in this study was developed by the U.S. Department of the Interior and applied by the U.S. Geological Survey in petroleum assessments of the National Petroleum Reserve in Alaska (U.S. Department of the Interior, 1979; White, 1979) and the Arctic National Wildlife Refuge (Mast et al., 1980). The probabilistic methodology used in those two

C. F. Chung et al. (eds.), Quantitative Analysis of Mineral and Energy Resources, 145–160.

assessments was a Monte Carlo simulation method.

2. GEOLOGIC MODEL

In play analysis a petroleum assessment area is partitioned into geologic plays, and the individual plays are analyzed. A play consists of a collection of prospects having a relatively homogeneous geologic setting. A prospect is a potential hydrocarbon accumulation. A hydrocarbon accumulation is a discrete oil or gas deposit, which may consist of one or more pools depending upon the specific play concept. A prospect is modeled by separately considering the uncertainty as to the presence of a hydrocarbon accumulation, and its size if present. An accumulation of hydrocarbon is modeled as either crude oil with its dissolved gas or nonassociated gas. The amount of dissolved gas present in an accumulation of oil is calculated from a gas-oil ratio. Because gas refers to either nonassociated gas or dissolved gas, the amount of gas in a play is the sum of the two types of gas from the prospects. There are three sets of geologic attributes or random variables involved in this play-analysis approach; these are for the play, the prospect, and the hydrocarbon volume. The play and prospect attributes are concerned with the presence or absence of certain geologic characteristics at the play and prospect levels, respectively. The hydrocarbon-volume attributes are concerned with the size of the hydrocarbon accumulation.

The play attributes are (1) existence of a hydrocarbon source, (2) favorable timing for migration of hydrocarbons from source to trap, (3) potential migration paths, and (4) existence of potential reservoir facies. The presence of all four play attributes (in which case the play is said to be "favorable") is a necessary, but not sufficient, condition for the existence of oil or gas deposits in the play. Thus, if one or more of these attributes is not present, all the prospects within the play are dry. Subjective judgments are made by experts for estimating the probability of the presence of each play attribute. The product of these four probabilities is the probability that the play is favorable for the existence of hydrocarbon accumulations and is called the marginal play probability.

The prospect attributes are (1) trapping mechanism, (2) effective porosity, and (3) hydrocarbon accumulation. Given a favorable play, the presence of all three prospect attributes is a necessary and sufficient condition for the existence of a hydrocarbon accumulation in the prospect. Subjective judgments are made by experts for estimating the probability of the presence of each prospect attribute. The product of these three probabilities is the probability that a prospect is a hydrocarbon accumulation, given the play is favorable, and is called the conditional deposit probability.

The hydrocarbon-volume attributes are (1) area of closure, (2) thickness of reservoir rock, (3) effective porosity, (4) trap fill, (5) depth to reservoir, and (6) hydrocarbon saturation. The hydrocarbon-volume attributes jointly determine the volume of the hydrocarbon accumulation within the prospect. The following reservoir engineering equations are used to calculate the in-place volumes of oil (barrels) and nonassociated gas (thousand cubic feet), respectively:

$$\text{Oil in place} = 7{,}758*1{,}000*A*F*H*P*S_h/B_o$$

$$\text{Nonassociated gas in place} = 1{,}537.8*1{,}000*A*F*H*P*S_h*(P_e/T)*(1/Z)$$

where A = area of closure (1,000 acres)
F = trap fill (decimal fraction)
H = reservoir thickness (feet)
P = effective porosity (decimal fraction)
S_h = hydrocarbon saturation (decimal fraction)
B_o = oil formation volume factor
P_e = original reservoir pressure (psi)
T = reservoir temperature (degrees Rankine)
Z = gas compressibility factor

Other mathematical functions are:

$$B_o = 0.96513 + 0.37186 * 10^{-3} * (CN) + 0.13039 * 10^{-6} * (CN)^2 - 0.23326 * 10^{-10} * (CN)^3$$

$$CN = 1000 * R_s\left(\frac{G_{sq}}{G_o}\right)^{0.5} + 1.25(T - 460.0)$$

$$R_s = \frac{G_{sq}}{1000} * \left[\left(\frac{P_e}{20}\right) * \frac{10^{(0.0125*API)}}{10^{0.00091(T-460)}}\right]^{\frac{1}{0.83}}$$

where API = API gravity of oil

= 7.473 + 0.00182 * D with a maximum of 48.5

D = reservoir depth (feet)

$$G_o = \text{specific gravity of oil} = \frac{141.5}{131.5 + API}$$

G_{sq} = specific gravity of solution gas

$$= 0.5595 + 0.002725*(API) + 0.0001325*(API)^2$$

$$P_e = 0.4643099 * D + 3.74494791$$

T = 0.016 * D + 495.0
Z = a non-linear function of depth in tabular form.

Both equations for oil in place and nonassociated gas in place consist of a product of factors that are functions of the hydrocarbon-volume attributes. The attributes are treated as continuous independent random variables, with the exception of effective porosity which possesses near perfect positive correlation with hydrocarbon saturation. The probability distribution for an attribute is determined from subjective judgments made by experts, usually geologists, based either on actual geological and geophysical data, when available, or on the experience and knowledge of the experts using analog data and geologic extrapolations when data is unavailable. The probability distribution for each attribute is described by a complementary cumulative distribution function determined from seven estimated fractiles (100th, 95th, 75th, 50th, 25th, 5th, 0th). (The 5th fractile, for example, is an attribute value such that there is a 5% chance of at least that value). In each play analyzed the seven fractiles are estimated for all of the hydrocarbon-volume attributes, except hydrocarbon saturation whose seven fractiles are one of two possible sets of fixed values depending upon the expected reservoir lithology, which is estimated by the geologists as either sandstone or carbonate. The experts also estimate the hydrocarbon-type probabilities which are the respective probabilities of a given accumulation being either oil or nonassociated gas; however, if the reservoir depth is greater than a specified depth, namely 22,500 feet, the accumulation is always assumed to be nonassociated gas.

The number of drillable prospects in the play is treated as a discrete random variable, and seven fractiles are estimated.

Probability judgments concerning each of the three sets of attributes are developed by experts familiar with the geology of the area of interest. The experts first review all existing data relevant to the appraisal, identify the major plays within the assessment area (e.g., basin or province), and then assess each identified play. All of the geologic data required by this model for a play is entered on an oil and gas appraisal data form (Figure 1). Information from the data form is entered into computer data files as the input for a computer program based upon an analytic method.

The 1987 ANWR study also included several other geologic models, some of which were very complex. A probability histogram model was designed to aid in the modeling of the input distributions for the hydrocarbon-volume attributes. A subplay model was developed for estimating resources in a fraction of the play from estimates of the entire play. A mixed-lithology model was necessary for estimating resources in dependent sandstone and carbonate reservoirs. An aggregation model was needed to aggregate individual play and subplay estimates of oil and gas, respectively, in order to estimate the petroleum potential of the entire assessment area and blocks of subplays. The geologic modeling that was done in the 1987 ANWR study was considerably more complex than in the 1980 ANWR study.

3. ANALYTIC METHOD OF PLAY ANALYSIS

An analytic method using probability theory was developed as a more efficient alternative to the costly and time-consuming Monte Carlo

Oil and Gas Appraisal Data Form

Evaluator : U.S.G.S./B.L.M. Play Name Turbidite

Date Evaluated: Oct. 23, 1985

	Attribute	Probability of Favorable or Present	Comments
Play Attributes	Hydrocarbon Source	1.0	
	Timing	1.0	
	Migration	1.0	
	Potential Reservoir Facies	1.0	
	Marginal Play Probability	1.0	
Prospect Attributes	Trapping Mechanism	0.8	
	Effective Porosity (>3%)	0.9	
	Hydrocarbon Accumulation	0.7	
	Conditional Deposit Probability	0.504	

	Attribute		
Hydrocarbon Volume Parameters	Reservoir Lithology	Sand	✓
		Carbonate	
	Hydrocarbon	Gas	0.3
		Oil	0.7

	Attribute \ Fractiles: Probability of equal to or greater than	100	95	75	50	25	5	0
Hydrocarbon Volume Parameters	Area of Closure ($x10^3$ Acres)	0.6	1.5	3.5	7	14	25.1	35
	Reservoir Thickness/vertical closure (Ft)	5	28.3	51.7	75	131.3	225	300
	Effective Porosity %	6	9	12	15	18.8	25	30
	Trap Fill (%)	1	14	27	40	55	80	100
	Reservoir Depth ($x10^3$ Ft)	4	6.8	9.7	12.5	15.7	18.8	22
	No. of drillable prospects (a play characteristic)	5	13	22	30	40	50	60

Figure 1. Oil and gas appraisal data form used in appraisal of Turbidite play, Alaska, for the 1987 Arctic National Wildlife Refuge Study.

simulation method for petroleum play analysis (Crovelli, 1987). The analytic method is based upon the same geologic model, same type data, and same probability assumptions as the simulation method.

The analytic method was developed by the application of many laws of expectation and variance in probability theory. The analytic method systematically tracks through the geologic model, computes all of the means and variances of the appropriate random variables, and calculates all of the probabilities of occurrence. The lognormal distribution is used as a model for two unknown distributions in order to arrive at probability fractiles (Crovelli, 1984). Oil, nonassociated gas, dissolved gas, and gas resources are each assessed in turn. Separate methodologies have been developed for analyzing individual plays and for aggregating the plays.

The basic steps of the analytic method of play analysis are:

1. Select the play.
2. Select oil as the first resource to be assessed.
3. Compute the mean and variance of each of the following hydrocarbon-volume attributes: (1) area of closure, (2) thickness of reservoir rock, (3) effective porosity, (4) trap fill, (5) depth to reservoir, and (6) hydrocarbon saturation. Determine the mean and variance from the estimated seven fractiles, assuming a uniform distribution between fractiles, that is, a piecewise uniform probability density function (as was done in the case of the simulation method). Recall that the hydrocarbon saturation distribution depends on whether the estimated reservoir lithology is sandstone or carbonate. Calculate the mean and variance of the product of effective porosity and hydrocarbon saturation, assuming they possess near perfect positive correlation. Also compute the mean and variance for the reciprocal of the oil formation volume factor, which is a function of reservoir depth through a series of formulas.
4. Compute the mean and variance of the accumulation size of oil in place using a reservoir engineering equation. The equation involves the product of a constant, area of closure, reservoir thickness, trap fill, effective porosity, hydrocarbon saturation, and the reciprocal of the oil formation volume factor. Various laws of expectation and variance are involved in the calculations.
5. Model the accumulation-size distribution by the lognormal probability distribution with mean and variance from step 4. Calculate various lognormal fractiles of the accumulation size for oil.
6. Compute the probability that a prospect has an oil accumulation, given the play is favorable. This is called the conditional prospect probability of oil. This probability is the product of the conditional deposit probability, the probability that the reservoir depth is less than 22,500 feet, and the hydrocarbon-type probability of oil.
7. Compute the mean and variance of the conditional prospect potential for oil, which is the quantity of oil in a prospect, given the play is favorable. They are arrived at by applying the conditional prospect probability of oil to the mean and variance of the accumulation size of oil.
8. Compute various fractiles of the conditional prospect potential for oil by a transformation to appropriate lognormal fractiles of the

accumulation size of oil using the conditional prospect probability of oil.

9. Compute the mean and variance of the number of prospects from the estimated seven fractiles, assuming a uniform distribution between fractiles (as is also the case in the simulation method).

10. Compute the mean and variance of the number of oil accumulations, given the play is favorable. They are arrived at by applying the conditional prospect probability of oil to the mean and variance of the number of prospects.

11. Compute the mean and variance of the conditional (A) play potential for oil, which is the quantity of oil in the play, given the play is favorable. They are determined from the probability theory of the expectation and variance of a random number (number of prospects) of random variables (conditional prospect potential).

12. Compute the conditional play probability of oil, which is the probability that a favorable play has at least one oil accumulation, and is a function of the conditional prospect probability of oil and the number-of-prospects distribution.

13. Compute the mean and variance of the conditional (B) play potential for oil, which is the quantity of oil in the play, given the play is favorable and there is at least one oil accumulation within the play. They are arrived at by applying the conditional play probability of oil to the mean and variance of the conditional (A) play potential for oil.

14. Compute the unconditional play probability of oil, which is the probability that the play has at least one oil accumulation, and is the product of the conditional play probability of oil and the marginal play probability.

15. Compute the mean and variance of the unconditional play potential for oil, which is the quantity of oil in the play. They are arrived at by applying the unconditional play probability of oil to the mean and variance of the conditional (B) play potential for oil.

16. Model the probability distribution of the conditional (B) play potential for oil by the lognormal distribution with mean and variance from step 13. Calculate various lognormal fractiles.

17. Compute various fractiles of the conditional (A) play potential for oil by a transformation to appropriate lognormal fractiles of the conditional (B) play potential for oil using the conditional play probability of oil.

18. Compute various fractiles of the unconditional play potential for oil by a transformation to appropriate lognormal fractiles of the conditional (B) play potential for oil using the unconditional play probability of oil.

19. Select nonassociated gas as the second resource to be assessed. Repeat steps 3 through 18, substituting nonassociated gas for oil, with two basic modifications as follows. A reservoir engineering equation is used to calculate the accumulation size of nonassociated gas in place. The conditional prospect probability of nonassociated gas is equal to the conditional deposit probability minus the conditional prospect probability of oil.

20. Select dissolved gas as the third resource to be assessed.

Repeat steps 3 through 18, substituting dissolved gas for oil, with two basic modifications as follows. The reservoir engineering equation for the accumulation size of oil in-place is multiplied by a gas-oil ratio which is a function of reservoir depth. The conditional prospect probability of dissolved gas is the same as the conditional prospect probability of oil.

21. Select gas as the fourth resource to be assessed. Repeat steps 4 through 18, substituting gas for oil, with two basic modifications as follows. Replace step 4 to compute the mean and variance of the accumulation size of gas in-place by using conditional probability theory and conditioning on the type of gas. The conditional prospect probability of gas is the same as the conditional deposit probability.

A simplified flowchart of the analytic method of play analysis is presented in Figure 2.

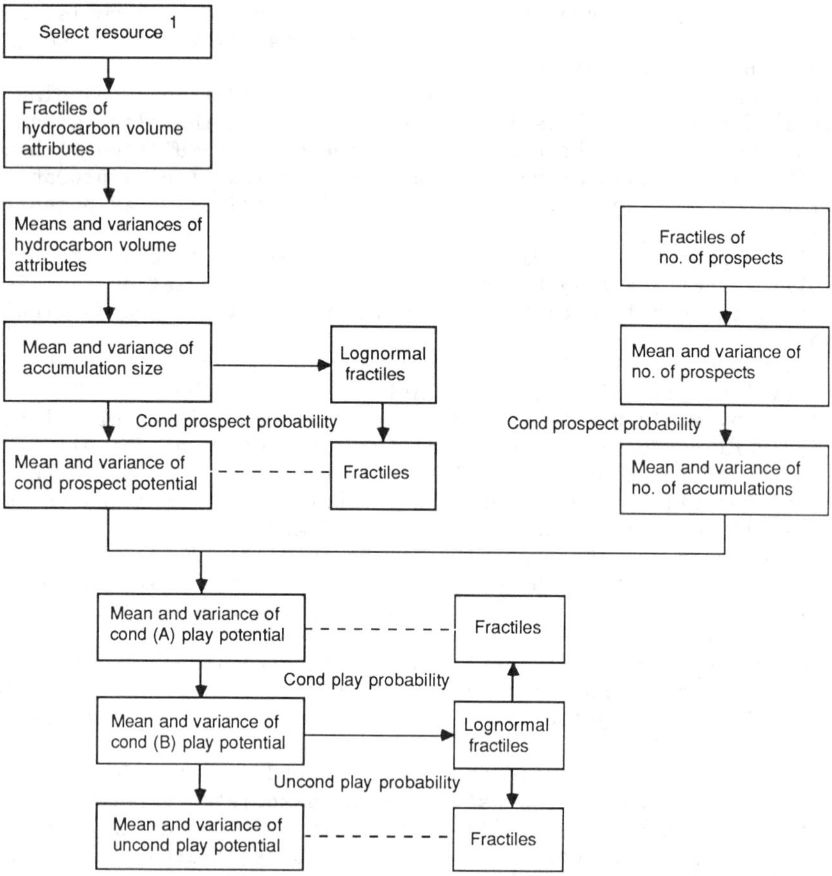

[1] Oil, nonassociated gas, dissolved gas, and gas resources are each assessed in turn.

Figure 2. Flowchart of analytic method of play analysis.

On the basis of the analytic method, a computer program was designed and called Fast Appraisal System for Petroleum, Reservoir Engineering (FASPRE). An early version of FASPRE was called FASP (Crovelli and Balay, 1986). Because both cost and running time are negligible, FASPRE allows for quick feedback evaluation of geologic input data.

3.1. Numerical Example of Play Analysis

The Arctic National Wildlife Refuge study (U.S. Department of the Interior, 1987) assessed undiscovered in-place oil and gas resources of geologic plays in the North Slope of Alaska using the reservoir-engineering model described in section 2. The geologic model is similar to the reservoir-engineering model in the 1980 ANWR assessment except some of the parameter values in the reservoir-engineering equations were changed. The oil-floor depth was also changed from 15,000 feet to 22,500 feet. The geologic plays were completely redefined in the new 1987 ANWR assessment.

A numerical example of the application of the analytic method of play analysis is illustrated using the resource assessment of an individual play, called Turbidite, from the 1987 ANWR study. The completed oil and gas appraisal data form for the Turbidite play is given in Figure 1 as filled out by the U.S. Geological Survey resource appraisal team of experts for that study. Estimates of undiscovered in-place petroleum resources for the Turbidite play are presented in FASPRE output form in Table I.

4. ANALYTIC METHOD OF PLAY AGGREGATION

A separate methodology was developed for estimating the aggregation of a set of plays. In this method the resource estimates of the individual plays from the analytic method of play analysis using the FASPRE program are aggregated by means of probability theory. Oil, nonassociated gas, dissolved gas, and gas resources are each aggregated in turn.

The basic steps of the analytic method of play aggregation are:

1. Select plays to aggregate.
2. Select oil as the first resource to be aggregated.
3. Compute the mean and variance of the unconditional aggregate potential for oil, which is the quantity of oil in the assessment area. They are determined by adding all of the individual play means and variances of the unconditional play potential for oil, respectively, assuming independence among the plays.
4. Compute the unconditional aggregate probability of oil, which is the probability that the assessment area has at least one play with oil, and is a function of the individual unconditional play probabilities of oil.
5. Compute the mean and variance of the conditional aggregate potential for oil, which is the quantity of oil in the assessment area, given the assessment area has at least one play with oil. These are

TABLE I. Estimated in-place oil and gas resources for the Turbidite play

	Mean	S.D.	F95	F75	F50	F25	F05
OIL (Millions of BBLs)							
Number of Accumulations	10.7	5.0	3	7	10	14	20
Accumulation Size	228.8	386.0	17.3	53.3	116.7	255.3	787.3
Cond. (B) Play Potential	2463.6	1705.5	723.7	1328.2	2025.6	3089.2	5669.9
Uncond. Play Potential	2458.3	1707.5	713.9	1323.8	2022.2	3088.5	5676.2
NONASSOCIATED GAS (Billions of CuFt)							
Number of Accumulations	4.6	2.7	1	3	4	6	9
Accumulation Size	571.3	928.5	46.1	139.1	299.4	644.5	1942.7
Cond. (B) Play Potential	2717.2	2511.4	547.9	1174.5	1995.5	3390.2	7268.2
Uncond. Play Potential	2630.3	2516.7	387.7	1102.7	1932.9	3327.2	7192.7
DISSOLVED GAS (Billions of CuFt)							
Number of Accumulations	10.7	5.0	3	7	10	14	20
Accumulation Size	255.2	435.8	18.9	58.7	129.0	283.6	881.4
Cond. (B) Play Potential	2747.7	1915.1	800.7	1474.5	2254.1	3445.9	6346.3
Uncond. Play Potential	2741.7	1917.3	789.8	1469.6	2250.4	3445.2	6353.4
GAS (Billions of CuFt)							
Number of Accumulations	15.3	6.7	5	10	15	20	27
Accumulation Size	350.0	642.3	22.7	73.9	167.5	379.9	1234.3
Cond. (B) Play Potential	5374.0	3435.1	1728.8	3051.1	4528.0	6719.9	11859.0
Uncond. Play Potential	5372.0	3436.0	1724.0	3049.2	4526.8	6723.8	11877.0

arrived at by applying the unconditional aggregate probability of oil to the mean and variance of the unconditional aggregate potential for oil.

6. Model the probability distribution of the conditional aggregate potential for oil by the lognormal distribution with mean and variance from step 5. Calculate various lognormal fractiles.

7. Compute various fractiles of the unconditional aggregate potential for oil by a transformation to appropriate lognormal fractiles of the conditional aggregate potential for oil using the unconditional aggregate probability of oil.

8. Select nonassociated gas as the second resource to be aggregated. Repeat steps 3 through 7 using play analysis estimates of nonassociated gas, namely the individual play means and variances of the unconditional play potential for nonassociated gas, along with the individual unconditional play probabilities of nonassociated gas.

9. Select dissolved gas as the third resource to be aggregated. Repeat steps 3 through 7 using play analysis estimates of dissolved gas, namely the individual play means and variances of the unconditional play potential for dissolved gas, along with the individual unconditional play probabilities of dissolved gas.

10. Select gas as the fourth resource to be aggregated. Repeat steps 3 through 7 using play-analysis estimates of gas, namely the individual play means and variances of the unconditional play potential for gas, along with the individual unconditional play probabilities of gas.

A simplified flowchart of the analytic method of play aggregation is presented in Figure 3.

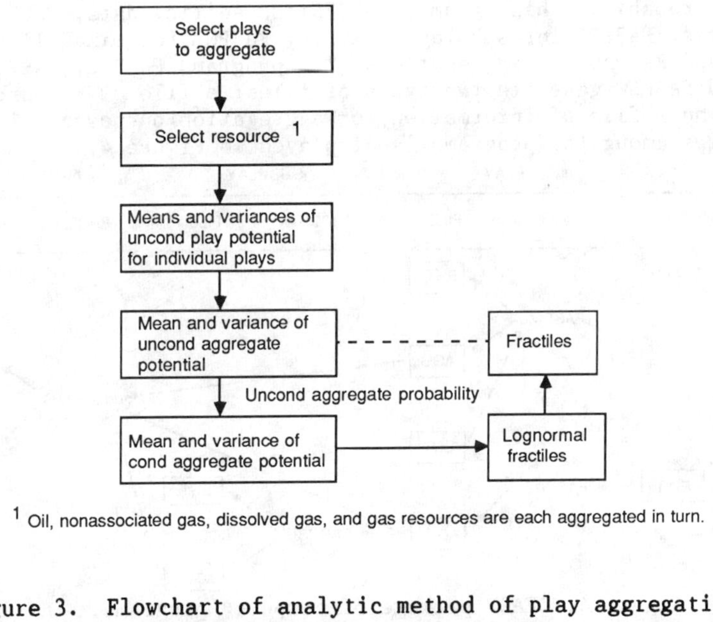

Figure 3. Flowchart of analytic method of play aggregation.

A computer program was designed on the basis of the analytic method for the aggregation of plays and called the Fast Appraisal System for Petroleum Aggregation (FASPAG). FASPAG interfaces with FASPRE as follows. FASPRE not only generates a file of resource estimates for an individual play, but also outputs a second file of results which consists of the mean and standard deviation of the unconditional play potential for each of the four resources, along with the corresponding unconditional play probabilities. The second file is needed for an aggregation of plays and forms an input file for FASPAG. Therefore, after FASPRE is run on each play in a set of plays, any subset of plays can be aggregated by running FASPAG on the corresponding subset of aggregation input files. FASPAG not only generates a file of resource estimates for an aggregation of plays, but also outputs a second file of results needed for an aggregation of aggregations, which forms yet another input file for FASPAG. Hence, after FASPAG is run on each aggregation in a set of aggregations, any subset of aggregations can be aggregated at once. FASPAG also possesses the capacity of aggregating a set of plays under a dependency assumption. Under the assumption of perfect positive correlation, all of the individual play standard deviations (instead of the variances) of the unconditional play potential for a resource are added together. In terms of the standard deviation of the unconditional aggregate potential, any degree of dependency is possible between 0 and 1, where 0 corresponds to independence and 1 denotes perfect positive correlation.

Computer programs were also written on the basis of the analytic methodologies that were mathematically derived from the other geologic models established in the 1987 ANWR study. A computer package, called FASP, was created and consisted of six computer programs: FASPPH for developing probability histograms, FASPED for editing data, FASPRE for play analysis, FASPSP for subplay analysis, FASPML for mixed-lithology analysis, and FASPAG for aggregations. The programs FASPRE, FASPSP, FASPML, and FASPAG generate two types of files: a file of resource estimates and a file of information for aggregation purposes. The relationships among the programs is displayed in Figure 4.

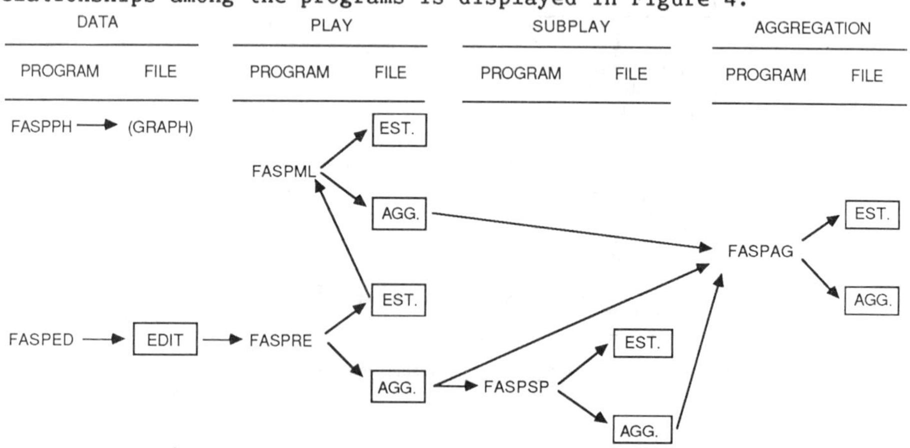

Figure 4. FASP package of computer programs.

4.1. Numerical Example of Play Aggregation

A numerical example of the application of the analytic method of play aggregation is illustrated using the assessment of total in-place petroleum resources in the Arctic National Wildlife Refuge. The assessment area had been partitioned into seven geologic plays: (1) Topset, (2) Turbidite, (3) Thomson/Kemik, (4) Undeformed Pre-Mississippian, (5) Imbricate Fold Belt, (6) Folded Ellesmerian/Pre-Mississippian, and (7) Undeformed Ellesmerian. An oil and gas appraisal data form (Figure 1) was filled out for each of the seven plays by USGS resource appraisal experts. The data forms for the seven individual plays were first analyzed by the analytic method of play analysis using the computer program FASPRE. The seven aggregation files generated by FASPRE were then aggregated by FASPAG to produce the total ANWR estimates. Estimates of undiscovered in-place petroleum resources for the aggregation of the seven plays, assuming independence among the plays, are presented in FASPAG output form in Table II. The corresponding graphs of the complementary cumulative distribution functions for the estimated total in-place oil and gas resources, respectively, of the ANWR area are displayed, again in direct computer output form, in Figure 5.

5. CONCLUSIONS

The analytic method using probability theory is a practical alternative to the simulation method for petroleum play analysis. The computer package FASP based on the analytic method operates thousands of times faster than a computer program based on the simulation method. Because the cost and running time are negligible, FASP allows for quick feedback evaluation of the estimated geologic data, a feature which was invaluable during the Arctic National Wildlife Refuge resource assessment meetings. Moreover, FASP can be adapted to most microcomputers; FASP needs no system-dependent subroutines or unusual library functions. The analytic method produces not only numerical estimates of petroleum resources, but also mathematical equations of probabilistic relationships involving these resources; whereas the simulation method produces no such equations.

A tremendous savings of time and cost can result using FASPRE, especially when analyzing hundreds of individual plays. However, the greater advantage of the analytic method might lie in the aggregation of a set of plays, especially if the set is large and there are many combinations of aggregations required. The computer program FASPAG based on the analytic method can aggregate any subset of plays almost instantly, and it can aggregate aggregations. FASPAG has considerable flexibility, even aggregating under a dependency assumption.

ACKNOWLEDGMENT

The author wishes to thank Richard H. Balay for computer programming.

TABLE II. Summary of 1987 ANWR assessment of total in-place petroleum resources for aggregation of all seven plays

	Mean	S.D.	F95	F75	F50	F25	F05
OIL (Billions of BBLs)							
Accumulation Size	0.3	1.3	0.0	0.0	0.1	0.3	1.3
Cond. Aggregate Potential	13.8	8.2	4.8	8.2	11.9	17.2	29.4
Uncond. Aggregate Potential	13.8	8.2	4.8	8.2	11.9	17.2	29.4
NONASSOCIATED GAS (Trillions of CuFt)							
Accumulation Size	0.9	3.4	0.0	0.1	0.2	0.7	3.5
Cond. Aggregate Potential	16.7	14.6	3.7	7.6	12.6	20.9	43.5
Uncond. Aggregate Potential	16.7	14.6	3.7	7.6	12.6	21.0	43.6
DISSOLVED GAS (Trillions of CuFt)							
Accumulation Size	0.4	1.6	0.0	0.0	0.1	0.3	1.4
Cond. Aggregate Potential	14.5	9.9	4.4	7.9	12.0	18.2	33.1
Uncond. Aggregate Potential	14.5	9.9	4.4	7.9	12.0	18.2	33.1
GAS (Trillions of CuFt)							
Accumulation Size	0.6	2.4	0.0	0.0	0.1	0.4	2.1
Cond. Aggregate Potential	31.3	17.5	11.5	19.2	27.3	38.8	64.5
Uncond. Aggregate Potential	31.3	17.5	11.5	19.2	27.3	38.8	64.5

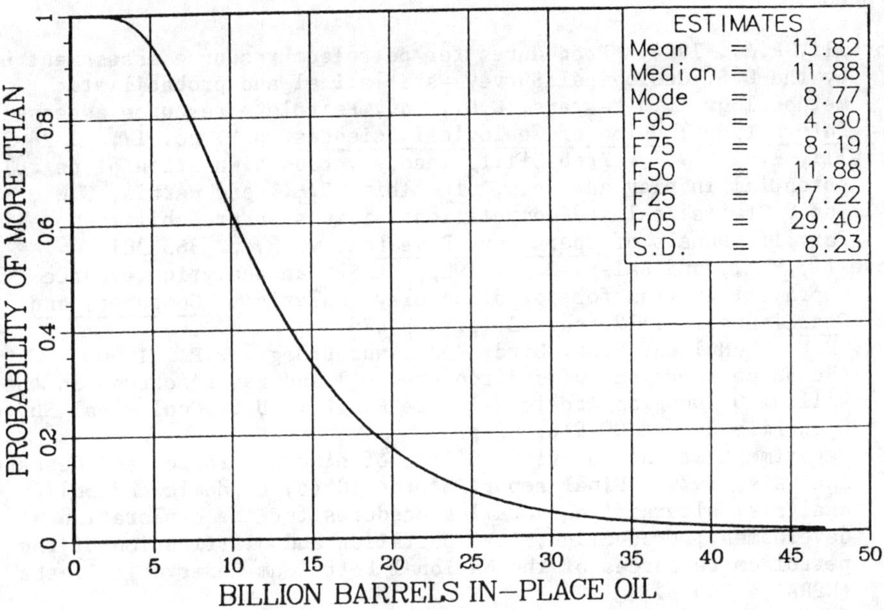

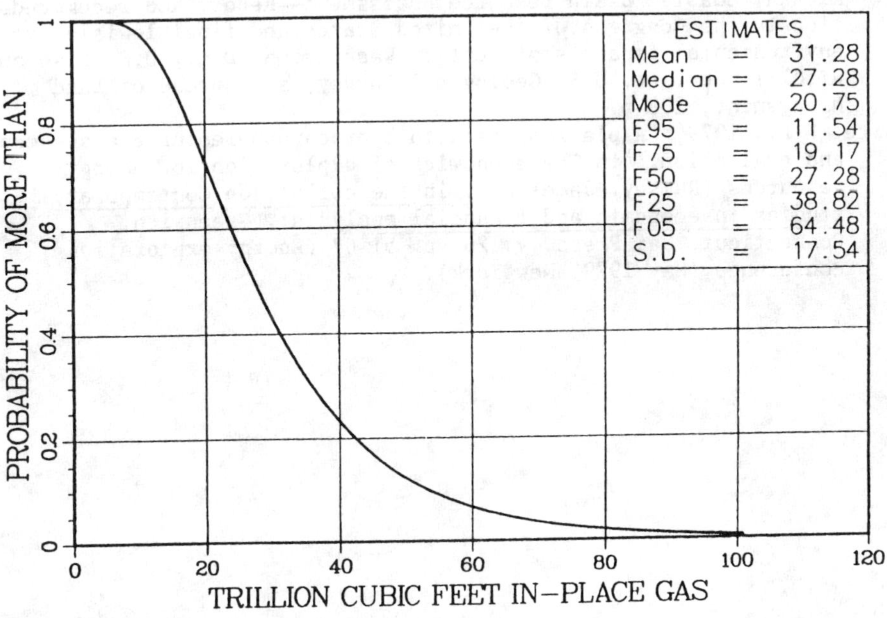

Figure 5. Complementary cumulative probability distribution graphs showing the estimated total in-place oil and gas resources of the ANWR area. F, fractile.

REFERENCES

Crovelli, R.A., 1984, 'Procedures for petroleum resource assessment used by the U.S. Geological Survey--statistical and probabilistic methodology,' in Masters, C.D., ed., Petroleum resource assessment: International Union of Geological Sciences, pub. no. 17, p. 24-38.

Crovelli, R.A., 1987, 'Probability theory versus simulation of petroleum potential in play analysis,' in Albin, S.L., and Harris, C.M., eds., Statistical and computational issues in probability modeling, Part 1: Annals of Operations Research, v. **8**, p. 363-381.

Crovelli, R.A., and Balay, R.H., 1986, 'FASP, an analytic resource appraisal program for petroleum play analysis': Computers and Geosciences, v. **12**, no. 4B, p. 423-475.

Mast, R.F., McMullin, R.H., Bird, K.J., and Brosge, W.P., 1980, 'Resource appraisal of undiscovered oil and gas resources in the William O. Douglas Arctic Wildlife Range': U.S. Geological Survey Open-File Report **80-916,** 62 p.

U.S. Department of the Interior, Office of Minerals Policy and Research Analysis, 1979, 'Final report of the 105(b) economic and policy analysis, alternative overall procedures for the exploration, development, production, transportation and distribution of the petroleum resources of the National Petroleum Reserve in Alaska (NPRA),' 145 p.

U.S. Department of the Interior, 1987, 'Arctic National Wildlife Refuge, Alaska, coastal plain resource assessment--Report and recommendation to the Congress of the United States and final legislative environmental impact statement': Washington, D.C., U.S. Fish and Wildlife Service, U.S. Geological Survey, and Bureau of Land Management, 208 p.

White, L.P., 1979, 'A play approach to hydrocarbon resource assessment and evaluation,' in The economics of exploration for energy resources (Ramsey, James B.), in the collection Contemporary studies in economic and financial analysis: Greenwich, Connecticut, Jai Press, v. 26, p. 51-67 (Energy exploration conference, May 1979, New York).

STATISTICAL EVALUATION OF PETROLEUM DEPOSITS BEFORE DISCOVERY

John C. Davis
Kansas Geological Survey
1930 Constant Avenue
Lawrence, Kansas 66046
USA

ABSTRACT. Pre-discovery evaluations of prospects are conditional on perceptions of trap geometry and assumptions about regional endowment. These conditional relationships must be determined explicitly in order to assess the expected worth of petroleum deposits, and to achieve realistic resource and reserve estimates.

INTRODUCTION

The objective of a petroleum exploration program is to define areas favorable for the discovery of commercial quantities of oil and gas. Exploration is a decision-making process, in which the outcomes of decisions cannot be known in advance. Because of this uncertainty, petroleum exploration has a pronounced probabilistic character, best expressed in Bayesian terms.

The uncertainty associated with exploration does not end with the successful discovery of hydrocarbons, although it is greatly reduced. The reserves in a newly discovered pool must be estimated, in order to plan for the field's economic development. Since a large initial capital investment is required to bring a new field on stream, particularly in remote areas or hostile environments, detailed assessments of the inplace reserves are essential. Throughout a field's lifetime, additional investment decisions must be made: whether to drill additional wells, install more equipment, begin enhanced recovery procedures. All of these decisions hinge on economic considerations, which in turn are based on predictions of the remaining productive life of the field. As exploitation continues, uncertainty about the true magnitude of a reservoir diminishes, but it never completely vanishes.

A producing area has a continuous commercial history, from the initial definition of a prospect, to discovery of a field, through development, to eventual depletion. Actions taken during this historical sequence are based on probabilities as perceived by decision-makers at the time. The perceived distribution of probable outcomes first leads explorationists to drill the wildcat well that results in discovery. The outcome of that action leads to a distribution of probable amounts of reserves, which guides development of a field. Continued production leads to a distribution of probable remaining reserves, which in turn leads to continued development decisions. Thus we see a continuously repeating pattern of prior probabilities, decisions, and outcomes which leads to a distribution of posterior probabilities that becomes the prior probability distribution for a new round of decisions.

It is more than this classical pattern of changing probabilities that gives the oil business its Bayesian flavor. The probabilities themselves, particularly those used in the initial, exploratory

C. F. Chung et al. (eds.), Quantitative Analysis of Mineral and Energy Resources, 161–186.

phase, are highly subjective. The use of subjective probabilities, personal probabilities, or belief functions has always been a hallmark of the Bayesian approach. In spite of pleas for the systematic collection of data that could be used to calculate relative-frequency estimates of probabilities in petroleum exploration (Harbaugh, Doveton, and Davis, 1977, p. 38-42), very little information has been systematically gathered from the millions of exploratory holes that have been drilled. Instead, explorationists base their probabilistic assessments on experience, intuition, and "gut feelings." This is the way the oil business has operated historically, the way it operates today, and undoubtedly is the way it will operate in the future.

WILDCAT PROBABILITIES CONDITIONAL ON REGIONAL ENDOWMENT

The simplest circumstance in exploratory drilling arises when the outcome is regarded as conditional only on basinwide or regional considerations. That is, local geologic factors play no role in determining the location of a well, either because prior information about local geology is unavailable, or there are no discernible conditional relationships between geologic details and drilling outcomes. This situation would be unusual, to say the least, but would correspond to a grid drilling program or one in which exploratory wells were drilled at random. In addition, it is a useful hypothetical concept to calculate a "background performance" level against which more sophisticated exploration strategies can be judged.

Grid-drilling schemes have been thoroughly investigated (Drew and Griffiths, 1965; Griffiths and Drew, 1966, Drew, 1967; Griffiths and Singer, 1971); a summary of these studies is given by Harbaugh, Doveton, and Davis (1977, p. 47-57). Most of the studies have concentrated on estimating the regional success ratio using geometric probability (Singer and Wickman, 1969; Singer, 1972), or upon estimating the aggregate amount of petroleum that could have been discovered by grid drilling. Instead, attention is focused here on estimating the likely outcome from drilling an individual exploratory hole.

In order to estimate the probable outcome of a wildcat well, it is necessary to know something about the spectrum of possibilities. One possible outcome, of course, is that nothing will be discovered—the hole will be dry. The alternative is that oil or gas will be encountered; but to predict the magnitude of the discovery, it is necessary to know (or assume) a probability distribution for field sizes in the region.

The pioneering work by Arps and Roberts (1958) on field sizes in the Denver-Julesburg Basin of Colorado established the concept that volumes of discovered fields in a maturely explored region tend to follow a lognormal distribution. (Interestingly enough, Arps and Roberts declined to explicitly invoke lognormality in their own analysis.) Subsequent studies (Kaufman, 1963, 1965; McCrossan, 1969; Haun, 1971; Drew, 1972; Kaufman, Balcer, and Kruyt, 1975; Meisner and Demirmen, 1981; Lee and Wang, 1983; Forman and Hinde, 1985) have tended to confirm (or to assume) that field sizes are, indeed, lognormally distributed (Fig. 1). However, it must be remembered that the lognormal is a model of the posterior distribution of fields that have already been discovered, and the relevant distribution in evaluating a prospect is the prior distribution which describes sizes of undiscovered fields. In the early stages of exploration, the prior distribution is essentially the same as the original distribution of sizes of pools initially in place in the region. As exploration proceeds, the prior distribution is the difference between this initial distribution and the posterior distribution of discovered fields.

Three factors make it difficult to deduce the form of the prior field-size distribution from the posterior. One of these is definitional, the other two reflect biases in sampling. The size distribution of discovered fields is relevant only for those exploration wells that were successes; a far larger proportion of wildcats do not result in commercial discoveries and the discovered field-size distribution provides no information about them. Lumped together without distinction in the

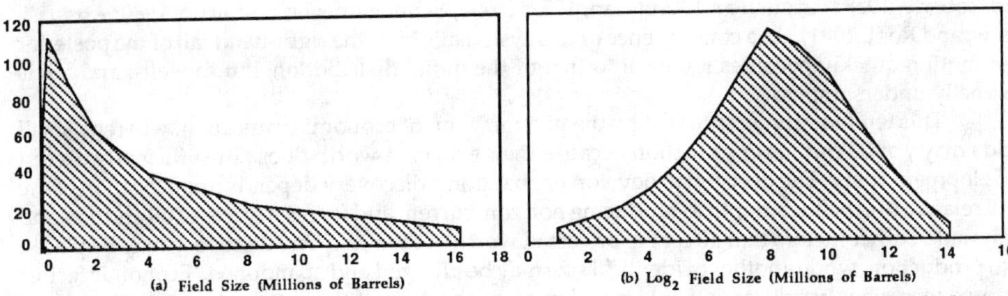

Figure 1. Field-size distribution for Denver-Julesberg Basin, Colorado, showing all fields discovered through 1970. Distribution shows typical lognormal shape. Vertical axis is number of fields. (a) Plotted on linear scale. (b) Plotted on logarithmic (log_2) scale. Data from Haun (1971).

"dry" category are holes that lacked the slightest hint of hydrocarbons; holes that produced "shows" of oil and gas; and holes that discovered significant quantities of oil and gas, but were abandoned because of economic, technical, or other considerations.

It is interesting to speculate, as did Newendorp (1975, p. 325), that oil may accumulate in a spectrum of volumes, from teacup-sized or smaller "pools" to multibillion-barrel giants. There may be an almost infinite number of infinitesimally small accumulations, and decreasing numbers of increasingly larger pools forming a J-shaped distribution. Even if we sample at random from such a population, we will define the vast majority of outcomes as "dry holes," because they do not satisfy an economic or conceptual idea of what constitutes a discovery. It would be difficult to determine the shape of such a distribution, because no information is available for the left-hand tail. Fortunately, it is not necessary to make size distinctions in this part of the curve since these all constitute uneconomic deposits, and a noncontinuous prior distribution can be used (Fig. 2).

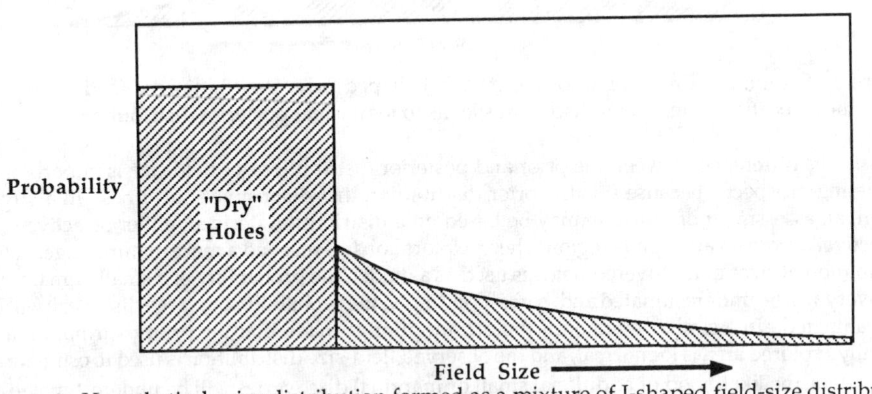

Figure 2. Hypothetical prior distribution formed as a mixture of J-shaped field-size distribution and uniform distribution of outcomes classed as "dry."

Two forms of sampling bias occur in petroleum exploration that cause the posterior distribution of sizes of discovered fields to deviate from the initial size distribution. Large accumulations of oil and gas generally are associated with large prospects (principally structural closures) which are easier to find than smaller features. Therefore, big accumulations are found early in exploration, and are selectively removed from the prior distribution of undiscovered pools. Many authors have simulated the discovery process by sampling from a hypothetical initial distribution of fields, assuming that the probability of discovery is proportional to field size (Arps and Roberts, 1958; Barouch and Kaufman, 1977; Drew, Schuenemeyer, and Root, 1980; Attanasi, Drew, and Root, 1981). As a consequence of this systematic bias, the right-hand tail of the posterior distribution quickly becomes identical to that of the initial distribution, but the left-hand tail is distinctly undersampled.

This tendency is accentuated by the phenomenon of economic truncation, whereby small fields may not be put into production because their expected worth does not sufficiently exceed development costs. The decision to develop or abandon a discovery depends upon location of the well relative to pipelines, depth to producing horizon, current and forecast prices of oil and gas, and a myriad of other factors that affect the projected worth of the find. One discovery may be placed into production, while another of identical size may be plugged and abandoned. Economic factors become increasingly critical as field sizes diminish, so fewer of the extremely small discoveries enter the posterior distribution. The economic censoring of marginal-sized fields leads to the abrupt left-hand tail of the observed lognormal size distribution of discovered fields (Fig. 3). This characteristic would emerge regardless of the form of the initial distribution.

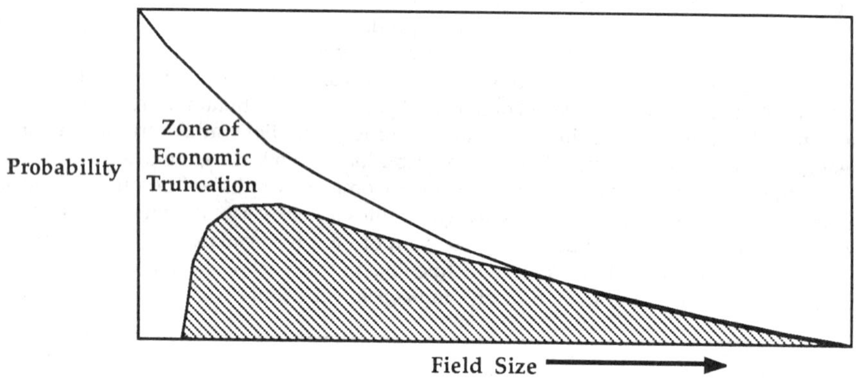

Figure 3. Sampling proportional to size from a J-shaped prior distribution of field sizes plus economic truncation of marginal discoveries leads to lognormal posterior distribution.

The difference between the prior and posterior distributions of field size is important in evaluating prospects, because it is the prior distribution that determines outcomes. In a virgin region, an assessment of outcome may be based on a distribution derived, either objectively or subjectively, from an analogous region where exploration has reached a more mature stage. If the distribution of sizes of discovered fields is used as a model, the probability of a small or marginal discovery will be underestimated and, correspondingly, the probability of a large discovery will be overestimated. In a similar manner, if the presumption is made that the prior distribution in a partially explored area is lognormal, and the observed field-size distribution is used to estimate its parameters, the likelihood of additional small or marginal discoveries will be underestimated.

Schuenemeyer and Drew (1983) have developed a procedure for estimating the initial (or parent) field-size distribution for a region which has a sufficient history of exploration to establish the upper tail. Because of size-biased sampling, the observed sample distribution can be presumed equal to the population of field sizes in this upper range. The method involves fitting a J-shaped distribution to the observed field sizes, using only those fields which are equal to or larger than the modal size. The upper tail is then extrapolated to estimate the initial form of the distribution. Perhaps the most obvious J-shaped model that could be used is the exponential,

$$f(V) = \lambda e^{-\lambda V} \tag{1}$$

where V is field size. However, based on their experiences with this method using data from the Denver Basin of Colorado and the Permian Basin of Texas, these authors conclude that the exponential does not provide a good fit to field sizes. They state that it tends to underestimate the probability of very large and very small fields, and to overestimate the probability of medium-sized fields. That is, the exponential distribution is regarded as insufficiently concave.

Of the many alternatives, Schuenemeyer and Drew (1983) selected the Pareto, an empirical distribution widely used in economics to describe sizes of income. The well-known "Zipf's Law," which has been used to model the size distributions of mineral resources, is a special case of the Pareto (Arnold, 1985, p. 569). Although Howarth, White, and Koch (1980), in a study of Stafford County, Kansas, found that a Zipf's Law model overpredicted large fields, the generalized Pareto distribution is more flexible in form, and presumably would provide a better fit.

The probability function for the generalized form of the Pareto distribution is

$$f(V) = \theta a^{\theta} V^{-(\theta+1)} \qquad (2)$$

where V again is field size. The left side of the distribution is set by a, which is a truncation parameter constrained so $V \geq a > 0$. The parameter θ sets the decline rate. Schuenemeyer and Drew (1983) set alternative values of a and estimate θ by nonlinear least squares; their experiments indicate that $0.58 < \theta < 1$ for oil and gas fields. (Because their study is based on discrete categories of field sizes, the authors actually use a log-geometric distribution, which is a discrete equivalent of a truncated Pareto.) The method was subsequently applied to field-size distributions such as those in the "Miocene-Pliocene Trend" of the U.S. Gulf Coast (Attanasi and Drew, 1985), with the results shown in Figure 4.

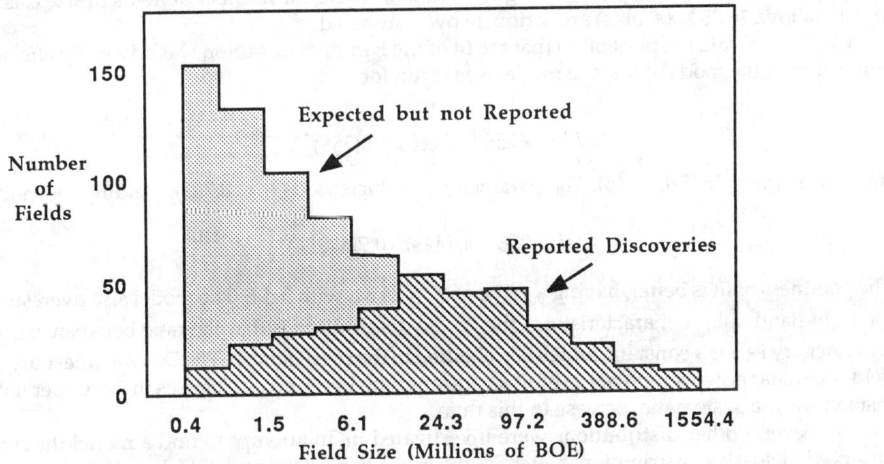

Figure 4. Observed (posterior) and expected (prior) field-size distribution modeled by truncated Pareto, for "Miocene-Pliocene Trend" of U.S. Gulf Coast (from Attanasi and Drew, 1985). Field size in millions of barrels of oil equivalent.

To model field-size distributions, the Pareto must be truncated on the left at an economic limit set by computing the expected point of no economic return from small fields at several hypothetical prices per barrel for petroleum. Schuenemeyer and Drew (1983) also found it necessary to truncate the right-hand tail.

There is no theoretical reason for using the Pareto distribution; even in econometrics where the Pareto was first applied, it is regarded as empirical. The criteria for selecting a specific model must be its goodness of fit, its behavior beyond the region of fit, and computational convenience. Both the Pareto and alternative J-shaped distributions were fitted to the field-size data from the Denver Basin given in Table 1 to examine their relative performances.

Since only observations for V > 128,000 bbls (the lower limit of the modal class of field sizes) were used, the fitted Pareto function f(V | V > 128,000 bbls) does not depend upon the truncation parameter a. Under these conditions Eq. 2 becomes

$$f(V \mid V > 128{,}000\ \text{bbls}) = \theta(128)^{\theta}\, V^{-(\theta+1)} \tag{3}$$

This model was fitted by the minimum χ^2 criterion. This iterative procedure was chosen because it works on grouped data, and it can be applied to *any* assumed parameterized distribution. [The nonlinear regression procedure used by Schuenemeyer and Drew (1983) is limited when using grouped data to those distributions which have discrete equivalents.] An important theoretical advantage is that the χ^2 goodness-of-fit measure is known to be asymptotically correct for large samples only if the model is fitted by the minimum χ^2 procedure.

Table 1(a) lists the observed and expected number of fields according to the Pareto model in each size category in the Denver Basin. The parameter $\hat{\theta}$ = 0.584, so the fitted model is

$$f(V \mid V > 128{,}000\ \text{bbls}) = 0.584(128)^{0.584}\, V^{-1.584}$$

The χ^2 value for the fitted Pareto distribution is $\chi^2 = 48.76$, with 6 d.f., indicating significant lack of fit. Major discrepancies occur in the right-hand tail, where the number of fields in size classes 15 and 16 (above 16,384,000 bbls) are seriously overestimated.

To evaluate the possibility that the fit of the Pareto distribution was adversely affected by truncation at the modal class, the model was rerun for

$$f(V \mid V > 256{,}000\ \text{bbls}) = \theta(256)^{\theta}\, V^{-(\theta+1)} \tag{4}$$

Results are given in Table 1(b). The parameter $\hat{\theta}$ is increased ($\hat{\theta}$ = 0.709), yielding the model

$$f(V \mid V > 256{,}000\ \text{bbls}) = 0.709(256)^{0.709}\, V^{-1.709}$$

The goodness of fit is better, having a value of $\chi^2 = 24.81$, with 5 d.f. The model also overestimates the right-hand tail. A characteristic of the Pareto distribution is that the ratio between category i and category i + 1 is a constant, which is embodied in the coefficient θ. The Denver-Julesburg Basin field-size data, however, do not exhibit a constant ratio between size classes in the upper tail, but instead show a systematic increase in this ratio.

Several other distributions were investigated in an attempt to find a model that fits the observed field-size distribution better in the right-hand tail, and which extrapolates in the hypothesized manner on the left. Only J-shaped distributions with well-known integrals were checked. These included the hyperbolic distribution, widely used to model oil well production decline curves (Arps, 1945); the mean-shifted lognormal distribution; and the mean-shifted Γ distribution, which includes as a subcase the Pareto, and asymptotically approaches the lognormal. Unfortunately, with these models it was not possible to simultaneously fit the observed distribution of large fields better than the Pareto, and to extrapolate large numbers of undiscovered small fields. The tested distributions had essentially the same defects as the Pareto model.

It is possible that continued investigation will lead to a family of J-shaped distributions having the desired characteristics, perhaps Pearson Type I or Type III curves. Of course, it is also possible that the initial field-size distribution is not monotonically declining at all, but rather could be adequately modeled by one of the distributions tested.

TABLE 1. Numbers of fields discovered through 1970 in logarithmic size classes in the Denver–Julesburg Basin, Colorado (from Haun, 1971). Total number of wildcat wells is 7939, of which 703 resulted in discoveries. Expected values predicted by truncated Pareto distribution fitted to (a) size classes 8 and above, (b) size classes 9 and above; (c) expected values predicted by hyperbolic distribution fitted to size classes 8 and above; (d) expected values predicted by lognormal distribution fitted to size classes 9 and above. Expected values for smaller classes are extrapolations.

Field-size Class (2^n x 1000 bbls)	Field-size Class Limits (1000 bbls)	Observed No. Fields	Expected No. Fields (a)	Expected No. Fields (b)	Expected No. Fields (c)	Expected No. Fields (d)
1	1 - 2	9				1
2	2 - 4	14				2
3	4 - 8	20				7
4	8 - 16	28				17
5	16 - 32	42				34
6	32 - 64	63				59
7	64 - 128	89				84
8*	128 - 256	112	146		104	102
9	256 - 512	104	97	127	112	103
10	512 - 1024	88	65	77	93	88
11	1024 - 2048	63	43	47	62	64
12	2048 - 4096	38	29	29	35	38
13	4096 - 8192	21	19	18	17	20
14	8192 - 16,384	8	13	11	8	8
15**	16,384 - 32,768	3	26	17	7	4
16**	32,768 - 65,536	1				
*Modal class						
**Class combined for χ^2 tests			$\chi^2 = 48.76$	$\chi^2 = 24.81$	$\chi^2 = 3.57$	$\chi^2 = 0.15$

The hyperbolic and lognormal distributions have characteristics that are interesting in this context. We can briefly examine these. The general form of the hyperbolic is

$$f(V) = (\alpha + \beta V)^{-\gamma} \tag{5}$$

Since the function must integrate to 1.0 to serve as a probability distribution, the coefficients α, β, and γ must be interdependent. Taking this into account, Eq. 5 may be rewritten as the probability density function

$$f(V) = \frac{c}{c+1} \, \frac{\left(a + \frac{1}{1+c} V\right)^{-(1+c)}}{a^{-c}} \tag{6}$$

The relationship between Eq. 5 and Eq. 6 is given by the following equivalencies:

$$\alpha = a \left| \frac{c}{(c+1)} a^{-c} \right|^{-1/1+c}$$

$$\beta = \frac{\alpha}{a(1+c)}$$

$$\gamma = 1 + c$$

Table 1(c) gives the results of fitting the Denver Basin field-size data by a hyperbolic model. The parameters are a = 167.61 and c = 1.18, so the model is

$$f(V) = 228.39(167.61 + 0.4587\,V)^{-2.18}$$

The value for goodness of fit is $\chi^2 = 3.57$ with 5 d.f., a major improvement over the Pareto. The hyperbolic distribution slightly overestimates the right-hand tail and slightly underestimates the modal class. Although the hyperbolic distribution is J-shaped on an arithmetic scale, when expressed in logarithmic class sizes the distribution is skewed and unimodal. This is expressed as smaller expected class frequencies for smaller fields. Figures 5, 6, and 7 compare the two fitted Pareto distributions and the hyperbolic distribution to the observed field-size distribution over the range of observed field sizes, and projects the distributions through the region of economic truncation.

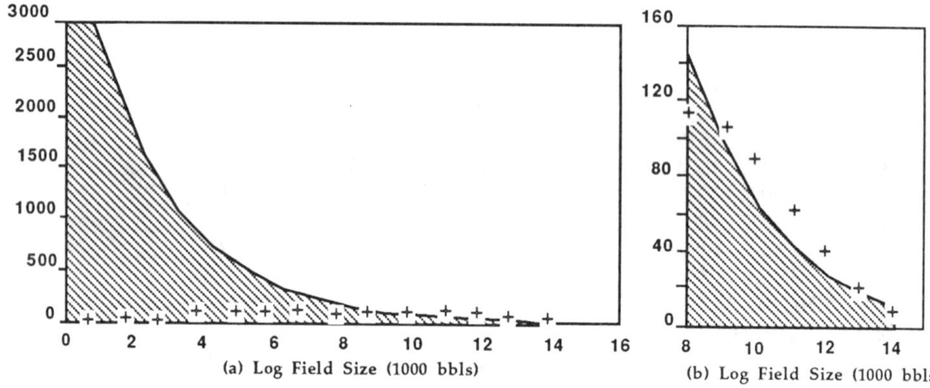

Figure 5. Prior distribution of field sizes estimated by Pareto model fitted to field classes larger than 128,000 bbls (class size 8) for the Denver-Julesberg Basin, Colorado. χ^2 value is 48.76. (a)Distribution extrapolated into the region of economic truncation. (b) Fit of Pareto to observed field-size data in right tail; note expanded vertical scale.

The lognormal distribution can be converted to mean-shifted form by introducing the constraint $\mu \leq 0$. Then, only the right-hand tail of the distribution, which is monotonically declining, will be fitted to the data. The fits achieved are similar to those given by the Pareto, with severe overestimation of the largest size classes and corresponding underestimation of size classes near the mode.

When the constraint on the mean is removed, the lognormal distribution perfectly fits that part of the observed field-size data which is beyond the mode ($\chi^2 = 0.14$, with 5 d.f.). Extrapolation to the left of the mode conforms closely to the observed distribution of smaller fields, but there is a small but consistent underestimation of the numbers of these fields [Table 1(d)]. If the upper tail of the field-size distribution indeed represents a complete sampling of all pools larger than the mode, this suggests that the initial population does contain more small accumulations than the lognormal predicts.

Since the true nature of the prior distribution where it is truncated is not known, the choice of model must be based upon preference and degree of fit in the region where sampling is believed complete. The search for possible models of the prior field-size distribution should be continued.

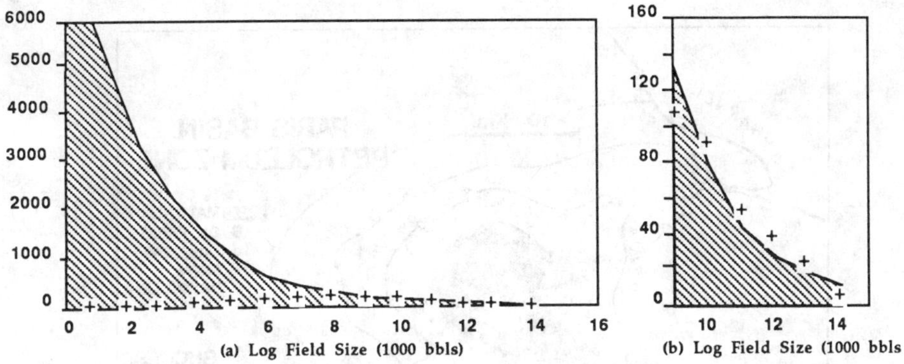

Figure 6. Prior distribution of field sizes estimated by Pareto model fitted to field classes larger than 256,000 bbls (class size 9) for the Denver-Julesberg Basin, Colorado. χ^2 value is 24.81. (a) Distribution extrapolated into the region of economic truncation. (b) Fit of Pareto to observed field-size data in right tail; note expanded vertical scale.

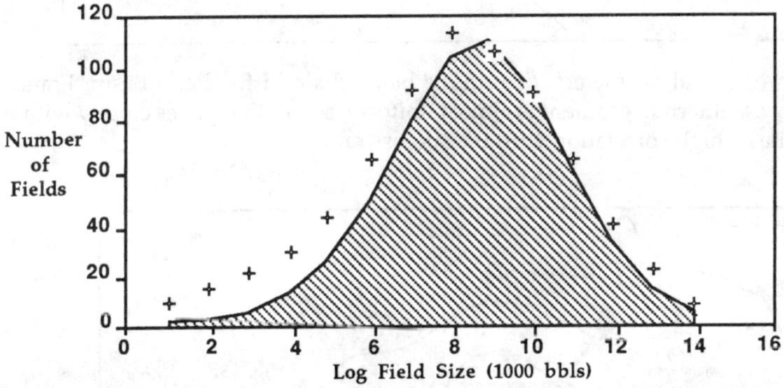

Figure 7. Prior distribution of field sizes estimated by hyperbolic model fitted to field classes larger than 128,000 bbls for the Denver-Julesberg Basin, Colorado. χ^2 value is 3.57. (a) Distribution extrapolated into the region of economic truncation. (b) Fit of hyperbolic to observed field-size data in right tail.

The probability distribution describing field sizes in a region forms only part of the prior distribution necessary to forecast drilling outcomes. If X is the number of barrels of oil that may be discovered, the complete prior distribution is given by

$$\Pr(X \leq b) = (1-p) + p \cdot \Pr(F \leq b)$$

where $(1-p)$ is the dry hole probability and $\Pr(F \leq b)$ is the probability that a discovered field is equal to or smaller than a specified volume b. This probability $\Pr(F \leq b)$ may be derived from the exponential, Pareto, hyperbolic, or other model used to characterize the population field-size distribution. The resulting mixture distribution describes the prior probabilities associated with all possible outcomes of a wildcat well, from a dry hole to a giant discovery. The general form of this prior distribution is shown in Figure 2.

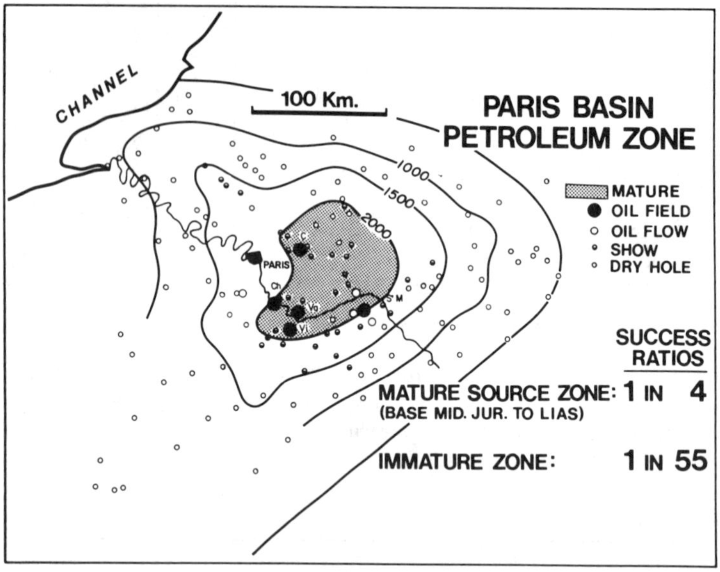

Figure 8. Depth of burial, in meters, of Jurassic black shale in the Paris Basin, France (from Demaison, 1984). Geothermal gradient is almost uniform, so depth equates closely with thermal maturity, which has a high correlation with the success ratio.

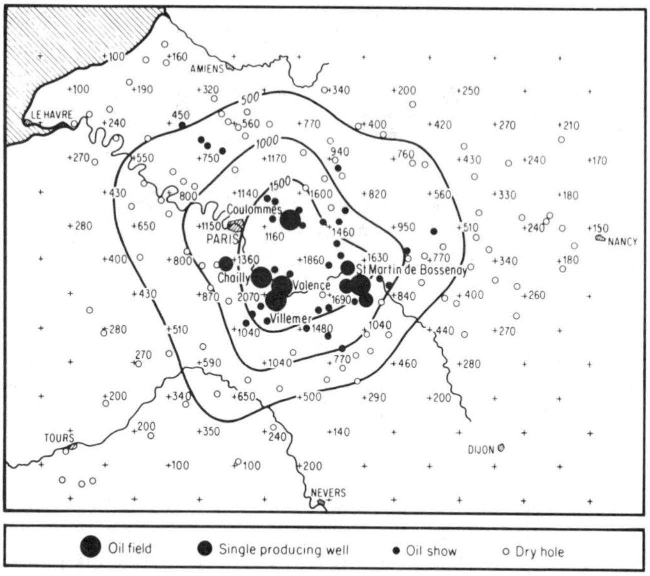

Figure 9. Grams of petroleum per ton of source rock in the Paris Basin (from Tissot and Welte, 1984). The petroleum yield correlates highly with the success ratio and field size.

data (Fig. 12), but interrelationships are not shown. These, of course, are critical in the development of conditional probabilities that can be used to assess prospects.

The variable to be estimated in the Shell procedure is called the "charge volume," the amount of oil and gas in place in a trap. If a trap is filled to the spillpoint, then the amount present is the minimum quantity of hydrocarbon that may have originally moved into the trap. The charge volume is considered to be related to the volume of source rock that potentially contributed to the trap, the potential yield of the source rock, the fraction of the yield that was generated and expelled, and the fraction that migrated to the trap. Each of these independent, hypothetical factors is in turn composed of several real variables that can be measured. The volume of source rock, for example, is the product of the area of source beds that potentially drain into the trap and which have reached critical thermal maturity, and the thickness of the source beds.

A discriminant function analysis is first used to identify those variables effective in classifying traps as productive or barren. A second discriminant function distinguishes between oil-bearing and gas-bearing members of the productive class. In the Shell scheme, the dry hole probability (or rather its complement, the probability of commercial quantities of hydrocarbons) is assessed independently of the prediction of hydrocarbon volume. Estimates of the volumes of oil and gas in place, conditional on the trap being productive, are made by linear regression. According

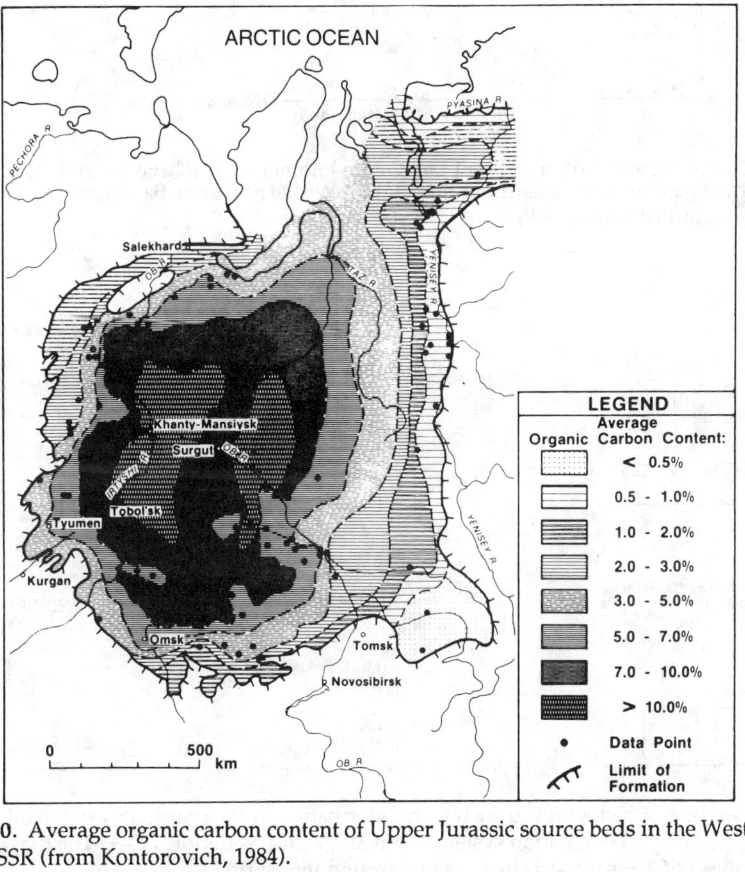

Figure 10. Average organic carbon content of Upper Jurassic source beds in the Western Siberia Basin, USSR (from Kontorovich, 1984).

to Sluijk and Nederlof, the effective variables for the conditional relationship, $p(V \mid X_1, X_2, ..., X_n)$, are the effective drainage area, source rock thickness, potential ultimate yield, yield fraction for maturity, and the yield fraction for timing. (Potential ultimate yield is based on organic carbon analyses by pyrolysis, yield fraction for maturity is based on vitrinite reflectance, and yield fraction for timing involves assessments of the time of formation of the trap.)

A similar type of analysis is performed to determine the critical factors that allow a trap to retain hydrocarbons. The dependent variable is the difference in formation pressure within a reservoir and in the rock immediately above the sealing unit. The independent variables are seal thickness, seal lithology, depth, and degree of fracturing. The authors do not disclose the relative effectiveness of these variables.

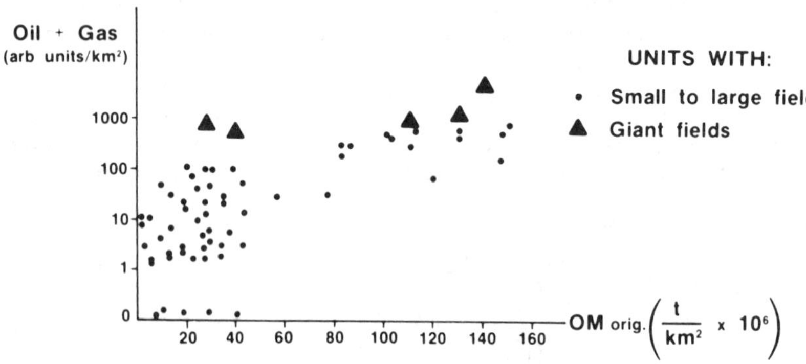

Figure 11. Plot of organic carbon content per square kilometer (in 10^6 metric tons/km^2) versus volume of oil and gas fields per square kilometer for the Western Siberian Basin (from Kontorovich, 1984). Triangles indicate giant fields.

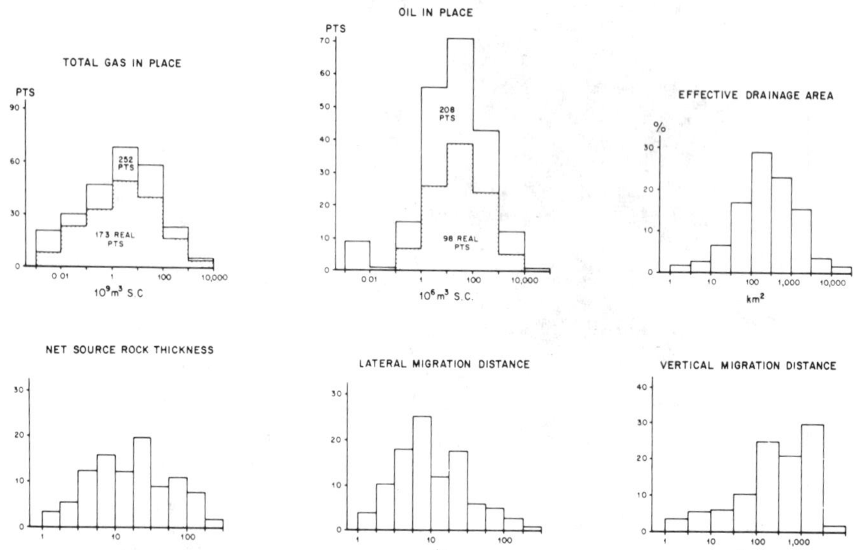

Figure 12. Samples of histograms drawn from information in the proprietary databank used to calibrate Shell's prospect evaluation system (from Sluijk and Nederlof, 1984). The diagrams are based on analyses of 346 prospects in 37 basins around the world.

From the relationships observed between the variables in their model, Shell has constructed a set of linear equations relating geological and geochemical properties that can be estimated prior to drilling to the success ratio and the volume of oil and gas. The equations are solved by Monte Carlo simulation, so the input parameters are not simply point estimates but rather sample distributions. Similarly, the coefficients of the estimating equations are represented by normal distributions based on the standard errors of the regression parameters. Output is in the form of a probability distribution, the volume of oil or gas present, conditional on the specified geological and geochemical conditions.

ASSESSMENT BASED ON SEISMIC CHARACTERISTICS OF PROSPECTS

The detailed characteristics of a prospect are usually established by seismic methods, because these provide the only information that is specific to a particular location short of drilling itself. From reflection seismography, the geometry of a trap can be inferred, and its area and height of closure measured with a degree of precision. In some limited circumstances, the presence of hydrocarbons can be detected directly, but usually the best that can be done is to establish that a structure having a geometry suitable for hydrocarbon entrapment is present. This is critical, because explorationists know that while the presence of a trap does not guarantee the presence of oil, the absence of a trap essentially precludes the possibility of oil.

Seismic information is especially critical in high-cost areas such as offshore and in remote, hostile environments like the North Slope of Alaska or the Canadian Arctic. Only prospects potentially capable of a very large return can justify the expense of drilling. Seismic surveying is the only way to establish the geometry of a prospect with any degree of assurance, and to determine if it contains sufficient gross volume to hold an economic quantity of oil.

In more mature areas such as the U.S. Midcontinent, the extremely high density of drilling permits the delineation of small prospects by structural mapping of formation tops picked on well logs. It also is possible to explore for more subtle hydrocarbon traps by regional mapping of updip pinchouts, facies changes, and the like that cannot yet be detected with precision by seismic means. These map data are interpreted and analyzed in a manner similar to that applied to seismic structural data. However, the economics of exploration determine that prospects defined by non-seismic techniques will only be pursued in relatively low-cost circumstances, when the price of failure in what is a high-risk situation is not prohibitive.

Seismic information is highly proprietary, and not often available for research purposes. However, in Federal waters of the offshore United States, all seismic data must be shared with the U.S. government. In a study for the U.S. Geological Survey, Davis and Harbaugh (1981, 1983) were given access to some of this data. The study involved an appraisal of 68 major seismic prospects in part of the "Pleistocene Trend," which includes about 3 million acres (1.2 million hectares) in the Louisiana-Texas Outer Continental Shelf (OCS) region of the Gulf of Mexico.

Most seismically detected structures in this area are the result of diaperic salt or shale movement and many possess shapes that are exceedingly complex. These structures, which had constituted exploration prospects, were classified by four parameters: area of closure, height of closure, length of bounding faults (BFD), and drainage distance. Structural properties were measured on regional seismic reflection-time maps of "Horizon II," a deep Pleistocene reflecting horizon. In addition, engineering estimates of recoverable oil and gas were available for those structures in which discoveries had been made. These are expressed as barrels or metric tons of oil equivalent, combining oil and gas.

Within the area of study, all closed structures were tabulated. These range from simple unfaulted anticlines to complexly faulted structures which may have piercement cores of salt or shale. Individual fault blocks that form the components of a structure were aggregated for statistical purposes. Although it is possible to estimate structural attributes of individual fault blocks, the information required to estimate reserves in individual reservoirs within these blocks is not readily available.

Area of Closure is defined as the area of an anticlinal structure bounded by the lowest closing reflection-time contour. In a faulted anticlinal structure, one or more faults may also form boundaries to the area of closure, but the deepest part within a fault block is necessarily defined by the lowest closing reflection-time contour. The lowest closing contour value may differ from one fault block to another, as shown in Figure 13. The total area of closure of a faulted structure is the sum of the areas of closure of all closed fault blocks. The structure shown has about 11,900 acres (4800 hectares) of closure.

Height of closure is defined as the vertical distance in milliseconds between the lowest bounding contour and the highest part of the anticlinal structure or fault block. There must be no faults of sufficient length to disrupt the continuous slope within the area being measured.

It is generally a simple matter to determine the height of closure within a single fault block, or within an unfaulted dome. However, individual fault blocks on a piercement dome may exhibit different heights of closure. In this study, the maximum height of closure obtained on any part of a complexly faulted structure was assigned to the entire structure. The maximum height of closure of the structure shown in Figure 13 is about 700 milliseconds, or roughly 3500 feet (1070 meters).

Bounding fault distance (BFD) is defined as the distance along faults which bound areas of closure within individual fault blocks, as illustrated in Figure 13. By this definition, a bounding fault is that part of a fault within or forming the bounds of an area of closure. If the fault separates two fault blocks, each with closure, the BFD is measured along each side of the fault; otherwise, the bounding distance is measured only once.

Drainage distance is defined as the straight-line distance, generally perpendicular to slope, which lies between the highest point of structural closure and the lowest point along a continuous slope. The maximum drainage distance for the structure illustrated in Figure 13 is 4 miles (6.4 kilometers).

The distribution of the sizes of fields discovered in the 68 seismic prospects is shown as a log-probability graph in Figure 14(a). Areas of closure of tested structures are shown plotted against cumulative percent in Figure 14(b); producing and dry structures are distinguished, and can be seen to follow different distributions. In each of these graphs, the distributions of points approximate straight lines, and so it may be concluded that oilfield volumes and areas of structures, both producing and dry, are lognormally distributed. This relationship is expected, because studies in many other regions show that these properties generally are distributed lognormally (Drew and Griffiths, 1965; Harbaugh, Doveton, and Davis, 1977; Harbaugh and Ducastaing, 1981).

Figure 14(a) is based on 59 fields and represents the population of structures known to contain commercial reservoirs of hydrocarbons in the study area. The median field size is about 16 million barrels (2.2 million metric tons). For a virgin area believed to be analogous to this part of the OCS, Figure 14(a) provides an estimate of the prior unconditional probability distribution of the hydrocarbon volume in those structures that contain oil or gas. In *this* OCS region, Figure 14(a) represents the posterior distribution of discovered fields. The prior distribution appropriate for further exploration on undrilled structures must be found as the difference between this distribution and an (unknown) initial distribution of hydrocarbons in place.

It is important to note that the frequency distributions of sizes of producing structures and of dry structures that exhibit closure and that have been tested by drilling have different medians and different standard deviations. The median size of structures that contain producible hydrocarbons [Fig. 14(b)] is about 5000 acres (2000 hectares), with a logarithmic standard deviation of about 0.34. The median size of nonproducing tested structures is about 3200 acres (1300 hectares), but has a greater logarithmic standard deviation of about 0.40. This is reflected in the difference in the fitted lines in the figure.

Figure 15(a) is a plot of log BOE versus log area of closure for the productive structures. (Because zero values cannot be plotted on logarithmic scales, only points that represent structures with measurable closure and that contain recoverable hydrocarbons can be shown.) Note that there

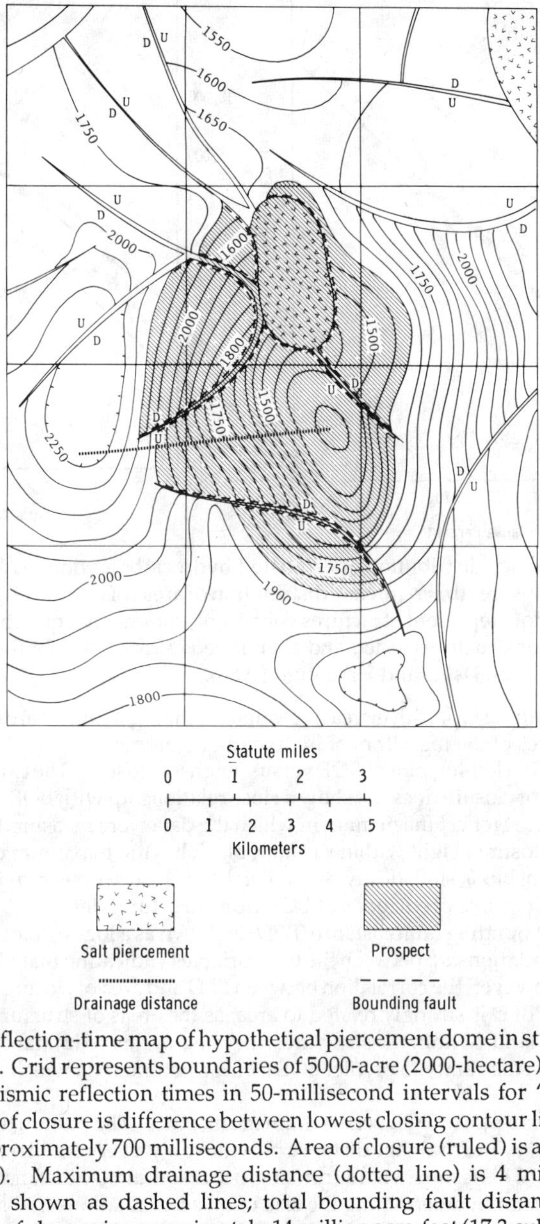

Figure 13. Seismic reflection-time map of hypothetical piercement dome in study area (from Davis and Harbaugh, 1981). Grid represents boundaries of 5000-acre (2000-hectare) offshore lease tracts. Contour lines are seismic reflection times in 50-millisecond intervals for "Horizon II" (Lower Pleistocene). Height of closure is difference between lowest closing contour line and highest point within structure, approximately 700 milliseconds. Area of closure (ruled) is approximately 11,900 acres (4800 hectares). Maximum drainage distance (dotted line) is 4 miles (6.4 kilometers). Bounding faults are shown as dashed lines; total bounding fault distance is 24 miles (38.6 kilometers). Volume of closure is approximately 14 million acre-feet (17.3 cubic kilometers). Area occupied by salt piercement is included within area of closure.

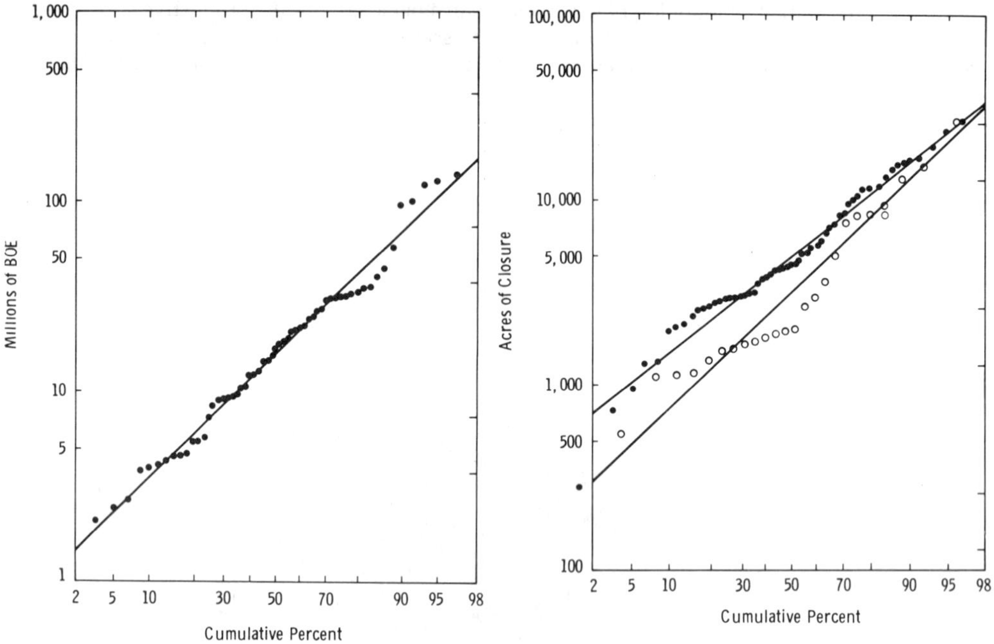

Figure 14. (a) Frequency distribution of estimated hydrocarbon volumes associated with structures which exhibit closure. (b) Frequency distribution of areas of closure of structures which have been drilled. Solid dots represent structures containing commercial quantities of hydrocarbons. Open circles represent structures tested and abandoned as dry. Distributions are plotted as log-probability graphs. From Davis and Harbaugh (1983).

is a marked linear relationship, with increasing values of the logarithm of area of closure associated with increasing values of the logarithm of hydrocarbon volume.

Figure 15(b) is a log-log plot of BOE versus height of closure. The extreme scatter of points indicates that height of closure does not show a close relationship with BOE. This apparent lack of relationship may in part reflect the manner in which the data were measured, as on highly faulted structures only the closure height within the fault block having maximum closure was recorded. Although the relationship is statistically weak, height of closure does provide some information that can marginally improve predictions of BOE from structural data.

BOE versus bounding fault distance (BFD) is shown as a log-log plot in Figure 16(a). There is a relatively strong relationship between the two variables, indicating that BFD may be an efficient predictor of BOE. However, the correlation between BFD and area of closure is very high ($r = 0.88$), which indicates that BFD is strongly related to area; as the areas of structures increase, so do the lengths of bounding faults within the structures. Because of this redundancy, BFD provides only a marginal increase in the ability to predict BOE from structural data which also include measurements of area.

Figure 16(b) is a plot of BOE versus drainage distance. The pattern of points indicates that the relationship is weak, and drainage distance provides only a very small marginal increase in the ability to predict BOE. This is somewhat surprising, because drainage distance is popularly reputed to be an important factor in determining the size of oil accumulations in this area.

The correlation coefficients between the five basic logarithmically transformed variables, BOE, area of closure, height of closure, BFD, and drainage distance, are presented in matrix form in Table 2.

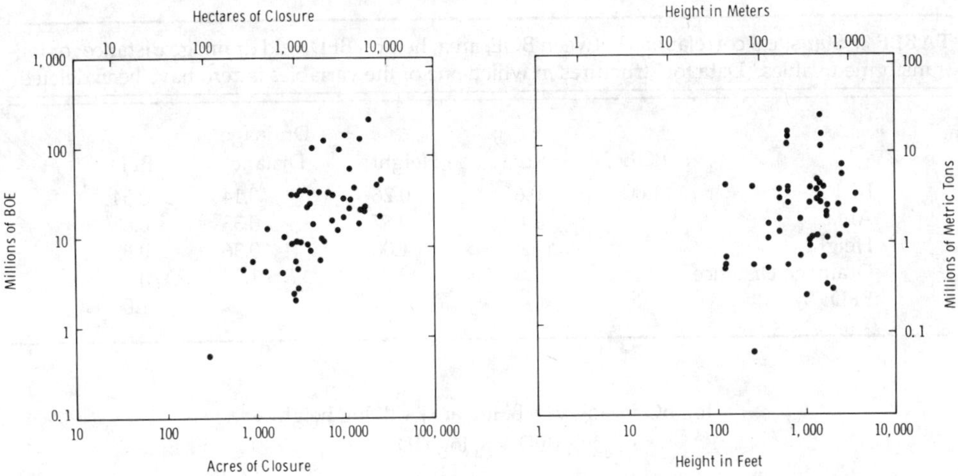

Figure 15. (a) Logarithmic scale plot of estimated volumes of hydrocarbons versus areas of structural closure for all structures for which an estimate of recoverable hydrocarbons is available. (b) Logarithmic scale plot of estimated volumes of hydrocarbons versus maximum heights of closure. From Davis and Harbaugh (1983).

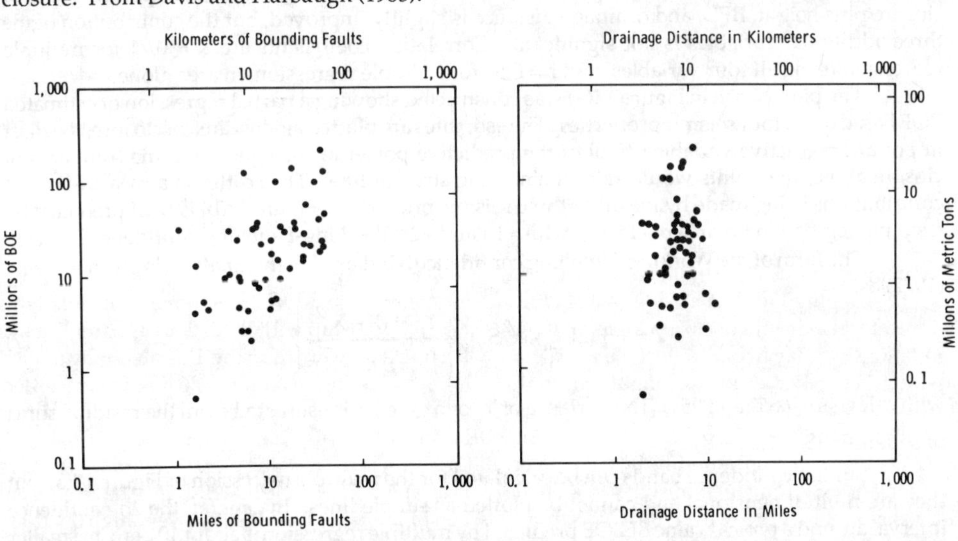

Figure 16. (a) Logarithmic scale plot of estimated volumes of hydrocarbons versus bounding fault distances (BFD) for tested structures containing one or more bounding faults. (b) Logarithmic scale plot of estimated volumes of hydrocarbons versus drainage distances. From Davis and Harbaugh (1983).

Predicted values of BOE can be found by multiple regression of the log of estimates of recoverable BOE on the logs of seismic properties of area, height of closure, bounding fault distance, and drainage distance:

TABLE 2. Statistical correlations between BOE, area, height, BFD, and drainage distance for log-transformed values. Data for structures in which one of the variables is zero have been deleted.

	BOE	Area	Height	Drainage Distance	BFD
BOE	1.00	0.66	0.26	0.14	0.54
Area		1.00	0.67	0.33	0.88
Height			1.00	0.36	0.63
Drainage Distance				1.00	0.38
BFD					1.00

$$\log \mathrm{BOE} = \log \alpha_0 + \beta_1 \log \mathrm{area} + \beta_2 \log \mathrm{height} + \\ + \beta_3 \log \mathrm{BFD} + \beta_4 \log \mathrm{DD} + \log \varepsilon_i$$

The procedure used to predict BOE from seismic structure is a multiple regression in which all geologic variables are assumed to contribute to the prediction. However, it is apparent from the plots that almost all of the predictive power comes only from the area of closure. The fit of a linear regression of area alone is highly significant; the fit of a multiple regression incorporating area of closure plus height, BFD, and drainage distance is slightly improved, but the contribution of the three additional variables is not significant. Correlation coefficients are $R = 0.71$ for multiple regression using all four variables, and $r = 0.66$ for a simple regression on area alone.

The plot shown in Figure 17(a) was constructed showing a partial regression of estimated BOE based on all four seismic properties. The estimates are plotted against area of closure, the most important predictive variable. If all of the predictive power of the equation came from area of closure alone, the points would fall on a sloping straight line. The scatter is a measure of the contributions being made by the other three seismic properties. Figure 17(b) is BOE predicted by a simple regression on area of closure with $\approx 1\sigma$ and $\approx 2\sigma$ Working-Hotelling confidence bands.

The form of the Working-Hotelling confidence band used in this analysis is given by Seber (1977) as

$$Y = \hat{Y} \pm (2F^{\alpha}_{2,n-2})^{1/2} S_{y|x} \left\{ \frac{1}{n} + \frac{(X-\bar{X})^2}{\Sigma(X_i-\bar{X})^2} \right\}^{1/2} \tag{7}$$

with α-levels of 68% and 95%. The variance of Y for a given X is estimated from the residual sums of squares, $SS_r = \Sigma(Y_i - \hat{Y})^2$.

Similar confidence bands can be calculated for the multiple regression of Figure 17(a), but they are multidimensional and cannot be plotted as simple lines. In general, the 2σ confidence interval around a typical value of BOE predicted by multiple regression is about 10 percent smaller than the interval around a value of BOE predicted by simple linear regression on seismic area of closure alone.

The regression and associated confidence limits in Figure 17(b) can be used to estimate the form of the probability distribution of possible field sizes associated with a seismically defined structural prospect in the "Pleistocene Trend," if it is assumed that the error distribution is lognormal with mean zero and variance $\sigma^2_{y|x}$.

Following the suggestions of Draper and Smith (1981), various residual plots were examined for indications of nonconstant variance, bias or drift, and lack of fit. Figure 18(a), for

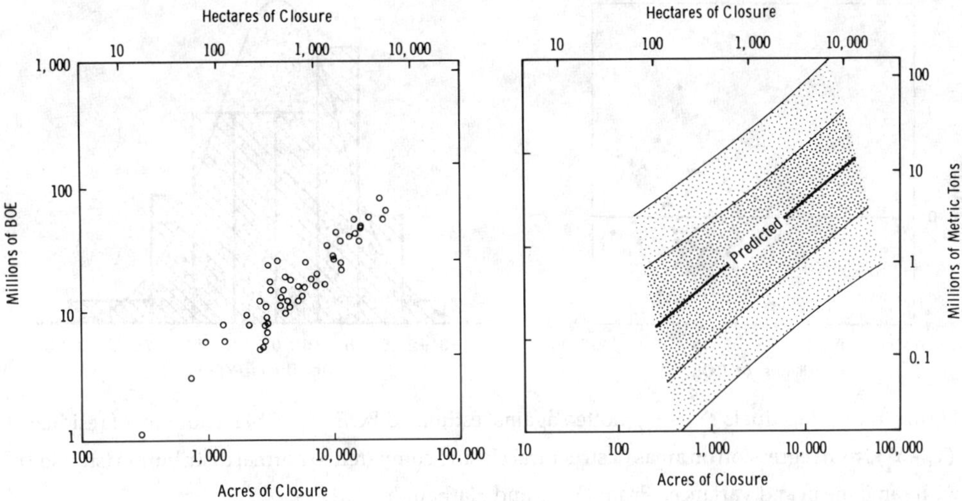

Figure 17. (a) Logarithmic scale plot of volumes of hydrocarbons (estimated by four-variable multiple regression) versus areas of structural closure. (b) Regression of estimated volumes of hydrocarbons on areas of structural closure. Dark shaded area is contained within approximately 1 standard deviation confidence limits; lighter area is contained within approximately 2 standard deviation confidence limits. From Davis and Harbaugh (1983).

example, shows residuals $(Y_i - \hat{Y}_i)$ plotted against estimated BOE $(\hat{Y}_i)$. The plot shows limited heteroscedasticity (variance increasing with increasing estimated BOE), but the effect is not severe. A histogram of residuals alone is shown in Figure 18(b). The residuals are approximately symmetrical in distribution, although they are significantly platykurtic. These aberrations in the residuals apparently result from the extreme scatter in the large values of BOE, which is not completely removed by the log transformation. As a result, the fitted regressions tend to underestimate BOE when the reserve figures are extremely large. Since the regression is constrained to pass through the bivariate mean, this implies that the regression also tends to overestimate BOE when it is very small, but because of the logarithmic transformation this error is negligible. Heteroscedasticity tends to inflate the confidence bands around the predictions. The particular pattern of deviation of residuals shown by these data from that assumed by the regression model is unfortunate but seems unimportant in practice (Neter and Wasserman, 1974). The general results will be a slight exaggeration of the uncertainty in the estimates of BOE over most of its range, and a tendency to underestimate extremely large prospects.

As an example of the use of confidence belts to construct a prior distribution of BOE conditional on prospect size, Figure 19 shows the probability distribution for a structure having 1000 acres (400 hectares) of closure on the seismic map of Horizon II. The center of the distribution and other points along the curve are values of the regression line and confidence belts corresponding to a prospect area of 1000 acres (400 hectares). Although the mean estimate of the volume of recoverable oil contained in the structure is 4 million BOE (0.5 million metric tons), the probability that the prospect contains less than 1 million BOE (0.1 million metric tons) is equal to 1/6, the proportion of the probability distribution which lies to the left of –1 standard deviation. Similarly, the probability is 1/6 that the prospect may contain over 18 million BOE (2.4 million metric tons), because this is the proportion of the distribution that exceeds +1 standard deviation.

Other discovery sizes and their associated probabilities can be estimated, using the characteristics of the lognormal distribution. The field size associated with the upper 5 percent of

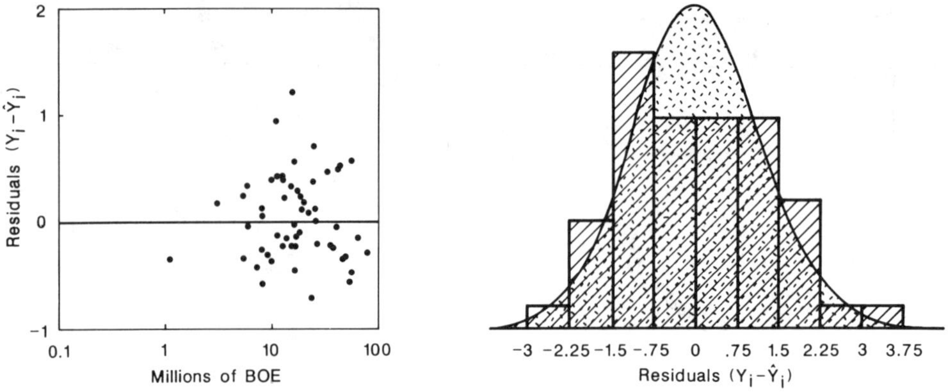

Figure 18. (a) Residuals $(Y_i - \hat{Y}_i)$ plotted against estimated BOE $(\hat{Y}_i)$. (b) Histogram of residuals $(Y_i - \hat{Y}_i)$ from regression on areas of structural closure compared to normal distribution (stippled) with same mean and variance. From Davis and Harbaugh (1981).

the probability distribution is often considered critical, because it is taken as the upper limit on the size of field that may be discovered. In this example, the uppermost field size is 82 million BOE (11 million metric tons); the probability that a larger field will be associated with a structure having 1000 acres (400 hectares) of closure is less than 5 percent.

Figure 20 shows a series of prior probability distributions for estimated BOE in seismic prospects conditional upon different areas of closure. These curves are derived from Figure 17(b) in the same manner as Figure 19, but are plotted in cumulative form as a log-probability graph. The estimated field sizes associated with any specified levels of probability can be read directly from the curves. By interpolating between the curves, prior probability distributions can be obtained for any size of prospect.

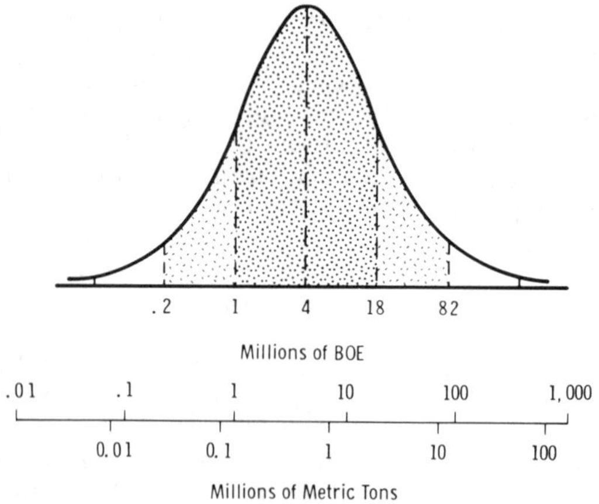

Figure 19. Probability distribution of field sizes conditional on a prospect having an area of 1000 acres (400 hectares). Shaded areas correspond to values of approximately ±1 and 2 standard deviations, read from Figure 17(b). From Davis and Harbaugh (1983).

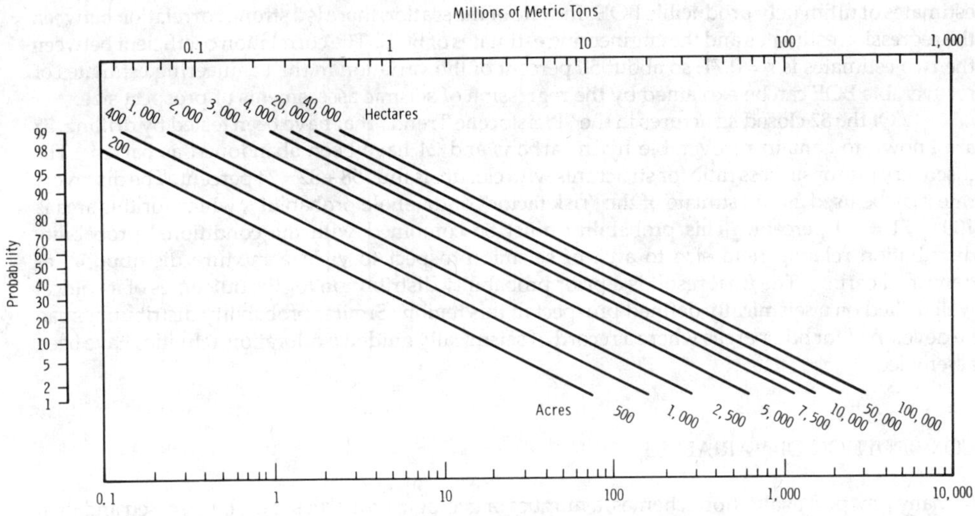

Figure 20. Graph showing probability distribution of field sizes for various areas of closure on seismic prospects (from Davis and Harbaugh, 1983). Vertical axis is probability of occurrence of a field whose size is specified on horizontal axis. Diagonal lines represent area of seismic prospect.

Figure 21 is a plot of estimated ultimately producible BOE (based on reservoir evaluations) versus BOE estimated by regression from seismic structural attributes. The failure of the points to lie on a straight line is a measure of the error in the regression, plus any errors in the engineering

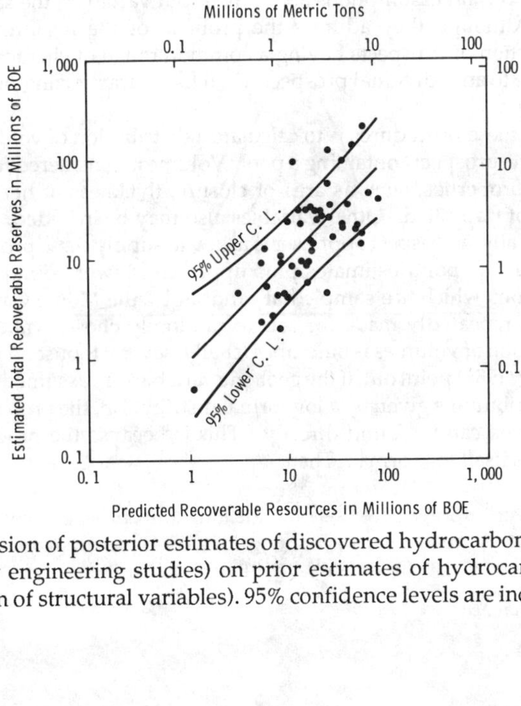

Figure 21. Regression of posterior estimates of discovered hydrocarbon volumes (based on U.S. Geological Survey engineering studies) on prior estimates of hydrocarbon volumes (based on multiple regression of structural variables). 95% confidence levels are indicated. From Davis and Harbaugh (1983).

estimates of ultimately producible BOE. In spite of the scatter, there is a strong correlation between the regression estimates and the engineering estimates of BOE. The correlation coefficient between the two estimates is r = 0.71, so about 50 percent of the variation in the engineering estimates of recoverable BOE can be explained by the regression of seismic assessments of prospect size.

Of the 82 closed structures in the "Pleistocene Trend" that have been tested by drilling, 58 are known to contain recoverable hydrocarbons and 24 have been abandoned as barren. The discovery rate or success ratio for structures with closure is thus 58 ÷ 82 = 71 percent. The discovery rate may be used as an estimate of the "risk factor" or dry-hole probability, which for this area is 100 – 71 = 29 percent. This probability must be combined with the conditional probability distribution relating field size to area of seismic prospect to yield a mixture distribution, as described earlier. The final result is a prior probability distribution for the outcomes of a wildcat well drilled on a seismically defined prospect in this region. Similar probability distributions can be developed for other areas where a record of seismically guided exploration activities have been assembled.

COMBINATION OF VARIABLES

In many prospect evaluation schemes, a number of geologic properties must be assessed and then combined to yield a prediction of the outcome of drilling the prospect. Since the geologic properties are seldom known exactly, they commonly are provided as probability distributions. Typically, these are combined using a Monte Carlo procedure; the methodology is described in detail in several texts (e.g., Newendorp, 1975, p. 369-490; McCray, 1975, p. 190-221). The U.S. Geological Survey developed a Monte Carlo simulation program called RVP (Range-of-Values Program) to assess offshore prospects in Federal waters. A similar Monte Carlo procedure has been used by the Canadian Geological Survey to combine geologic information on a regional basis for resource appraisal (Roy, Proctor, and McCrossan, 1975). Lee and Wang (1983) discuss the Canadian program and point out that, given certain assumptions about the input variables, the same results can be obtained analytically. Although they address the problem of the resource assessment of a petroleum play (or collection of prospects having a common range of characteristics), the same procedures are applicable to an individual prospect when the characterizing properties are given as probability distributions.

The basic step in these procedures is to calculate a distribution of volume of oil or gas in place, conditional upon the prospect containing a pool. Volume is considered to be the product of a number of geological properties, such as area of closure, thickness of net pay, porosity, oil saturation, and fraction of trap filled. Other variables also may be included, provided they are quotients of volume. Usually, a prospect appraiser is asked to supply these properties in the form of distributions, or at least as point estimates plus upper and lower bounds. These are then converted into distributions which are sampled at random by the Monte Carlo program. The volumetric calculation is repeatedly made for sets of randomly chosen values of the geologic variables, until a distribution of volumes is built up. Usually, several thousand trials are required.

As Lee and Wang (1983) point out, if the geologic variables are assumed to be independent and their probability distributions given by a lognormal distribution, the probability distribution describing prospect volume can be found directly. This is because the product of a series of lognormal distributions is itself lognormal. Then,

$$\mu_{\log V} = \Sigma\mu_{\log X}$$

and

$$\sigma^2_{\log V} = \Sigma\sigma^2_{\log X} \qquad (8)$$

Furthermore, for lognormal X:

$$\mu_X = e^{\mu_{\log X} + 1/2\,\sigma^2_{\log X}}$$

and

$$\sigma^2_X = (\mu_X)^2 (e^{\sigma^2_{\log X}} - 1)$$

where

V	=	Volume
X	=	Geologic variable
μ_X, σ^2_X	=	Means, variances of geologic variables
$\mu_{\log X}, \sigma^2_{\log X}$	=	Means, variances of logarithms of geologic variables
$\mu_{\log V}, \sigma^2_{\log V}$	=	Means, variances of logarithms of volume.

Since the lognormal distribution is completely specified by the mean and variance, a conditional probability distribution for volume can be constructed from these estimated values.

In an example, Lee and Wang (1983) show that geologists' subjective estimates of distributions of four geological variables are close to lognormal in form (Fig. 22). They calculated a

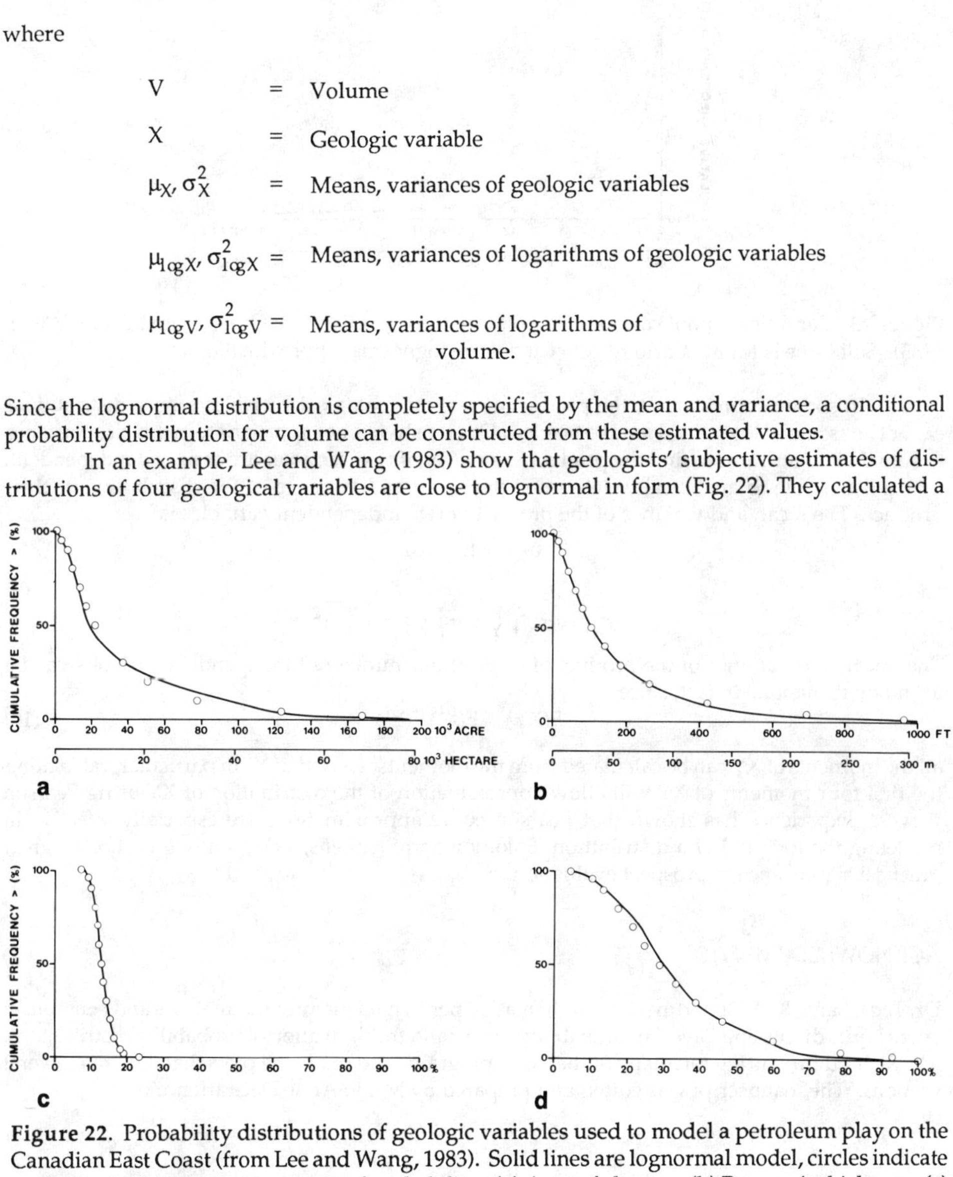

Figure 22. Probability distributions of geologic variables used to model a petroleum play on the Canadian East Coast (from Lee and Wang, 1983). Solid lines are lognormal model, circles indicate geologists' subjective assessments of probability. (a) Area of closure. (b) Reservoir thickness. (c) Percent porosity. (d) Percent trap fill.

lognormal field size distribution (Fig. 23) from these four variables, regarding oil saturation as a constant. The conditional probability distribution shown in Figure 23 can be combined with an estimate of the dry hole probability to yield a mixture distribution, which is then in the appropriate form for the prior distribution needed for prospect evaluation. The resulting distribution would resemble that shown in Figure 2.

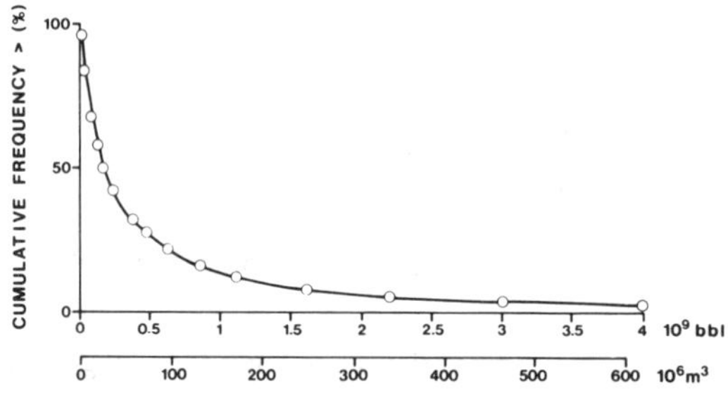

Figure 23. Conditional pool volumes for a play on the Canadian East Coast (from Lee and Wang, 1983). Solid line is Monte Carlo result. Circles are lognormal approximations.

In some circumstances, the product of several geological variables whose distributions cannot be assumed lognormal may be needed. Obviously, it is not appropriate in this circumstance to sum their logarithms to find the product. However, if two variables are assumed independent, it is still possible to analytically determine the mean and variance of the distribution of their product. The mean and variance of the product of two independent variables is

$$\mu_{XY} = \mu_X \cdot \mu_Y$$

and (9)

$$\sigma_{XY}^2 = \sigma_X^2 \mu_Y^2 + \sigma_Y^2 \mu_X^2 + \sigma_X^2 \sigma_Y^2$$

The mean and variance of the product of an arbitrary number of independent variables can be found by recursion. In fact, since

$$E(XY)^r = E(X^r)\,E(Y^r) \tag{10}$$

all the moments of XY can be calculated from the momentsof X and of Y. In particular, calculating the first four moments of XY will allow approximation of the distribution of XY using Pearson curves. Experience has shown that Pearson curve approximations are especially effective in modeling the long tail of a distribution (Solomon and Stephens, 1978), an observation of great practical significance in prospect evaluation.

ACKNOWLEDGMENTS

Dr. Ted Chang, K.U. Department of Mathematics, performed many of the analyses and generously spent hours discussing Bayesian probability theory and the derivation of probability distributions. Mr. Richard Brownrigg ran many of the IMSL programs and made the plots that illustrate several sections. The manuscript was edited and prepared by Ms. Jo Anne DeGraffenreid.

REFERENCES

Arnold, B. C., 1985, 'Pareto distribution,' *in* Kotz, S., Johnson, N. L., and Read, C. B., eds., *Encyclopedia of Statistical Sciences*, 6: John Wiley & Sons, Inc., New York, 758 p.

Arps, J. J., 1945, 'Analysis of decline curves': *Transactions of the AIME*, 160, p. 228-247.

Arps, J. J. and Roberts, T. G., 1958, 'Economics of drilling for Cretaceous oil on the east flank of the Denver-Julesburg basin': *Am. Assoc. Petroleum Geologists Bull.*, 42, p. 2549-2566.

Attanasi, E. D. and Drew, L. J., 1985, 'Lognormal field size distributions as a consequence of economic truncation': *Mathematical Geology*, 17, no. 4, p. 335-351.

Attanasi, E. D., Drew, L. J., and Root, D. H., 1981, 'Physical variables and the petroleum discovery process,' *in* Ramsey, J. B, ed., *Economics of Exploration for Energy Resources*: JAI Press, Greenwich, Ct., p. 3-18.

Barouch E. and Kaufman, G. M., 1977, 'Estimation of undiscovered oil and gas,' *in Proceedings of the Symp. in Applied Mathematics*: Amer. Math. Soc., XXI, p. 77-91.

Davis, J. C. and Harbaugh, J. W., 1981, A method for rapid evaluation of oil and gas prospects in the OCS regions: Report, U.S. Dept. Interior, Geological Survey, No. 14-08-0001-18785, 62 p.

Davis, J. C. and Harbaugh, J. W., 1983, 'Statistical appraisal of seismic prospects in Louisiana–Texas Outer Continental Shelf': *Am. Assoc. Petroleum Geologists Bull.*, 67, no. 3, p. 349-358.

Demaison, G., 1984, 'The generative basin concept,' *in* Demaison, G. and Murris, R. J., eds., *Petroleum Geochemistry and Basin Evaluation*: Am. Assoc. Petroleum Geologists, Mem. 35, p. 1-14.

Demaison, G. and Murris, R. J., eds., 1984, *Petroleum Geochemistry and Basin Evaluation*: Am. Assoc. Petroleum Geologists, Mem. 35, 426 p.

Draper, N. R. and Smith, H., 1981, *Applied Regression Analysis*: John Wiley & Sons, Inc., New York, 709 p.

Drew, L. J., 1967, 'Grid-drilling exploration and its application to the search for petroleum': *Economic Geology*, 62, no. 5, p. 698-710.

Drew, L. J., 1972, 'Spatial distribution of the probability of occurrence and the value of petroleum; Kansas, an example': *Mathematical Geology*, 4, no. 2, p. 155-171.

Drew, L. J. and Griffiths, J. C., 1965, Size, shape and arrangement of some oilfields in the USA: Symp. on Computer Applications in the Mineral Industries, Pennsylvania State Univ., p. FF1-FF31.

Drew, L. J., Schuenemeyer, J. H., and Root, D. H., 1980, Petroleum-resource appraisal and discovery rate forecasting in partially explored regions—An application to the Denver Basin: U.S. Geological Survey Prof. Paper 1138-A, 11 p.

Forman, D. J. and Hinde, A. L., 1985, 'Improved statistical method for assessment of undiscovered petroleum resources': *Am. Assoc. Petroleum Geologists Bull.*, 69, p. 106-118.

Griffiths, J. C. and Drew, L. J., 1966, Grid spacing and success ratios in exploration for natural resources: 6th Ann. Symp. and Short Course on Computers and Operations Research in Mineral Industries, Pennsylvania State Univ., 17 p.

Griffiths, J. C. and Singer, D. A., 1971, A first generation simulation model for selecting among exploration programs, with special application to the search for uranium ore bodies: *Geocom Programs*, no. 2, London, 42 p.

Harbaugh, J. W., Doveton, J. H., and Davis, J. C., 1977, *Probability Methods in Oil Exploration*: John Wiley & Sons, New York, 269 p.

Harbaugh, J. W. and Ducastaing, M., 1981, Historical changes in oilfield populations as a method of forecasting field sizes of undiscovered populations: A comparison of Kansas, Wyoming, and California: Kansas Geol. Survey, Subsurface Geology Series 5, Univ. Kansas, 56 p.

Haun, J. D., 1971, Potential oil and gas resources: Unpub. Progress Rept., Colorado School Mines, Golden, Colo., 18 p.

Hitchon, B. and Horn, M. K., 1974, 'Petroleum indicators in formation waters from Alberta, Canada': *Am. Assoc. Petroleum Geologists Bull.*, 58, p. 464-473.

Howarth, R. J., White, C. M., and Koch, G. S., 1980, 'On Zipf's law applied to resource prediction': *Inst. Min. Met. Trans.* (Sect. B: Applied Earth Sciences), 89, p. B182-B190.

Kaufman, G. M., 1963, *Statistical Decision and Related Techniques in Oil and Gas Exploration*: Prentice-Hall, Inc., Englewood Cliffs, N.J., 307 p.

Kaufman, G. M., 1965, 'Statistical analysis of the size distribution of oil and gas fields', *in* Symposium on Petroleum Economics and Evaluation: *Transactions of the AIME*, 224, p. 109-124.

Kaufman, G. M., Balcer, Y., and Kruyt, D., 1975, 'A probabilistic model of oil and gas discovery,' *in* Haun, J. D., ed., *Methods of Estimating the Volume of Undiscovered Oil and Gas Resources*: Am. Assoc. Petroleum Geologists, Studies in Geology No. 1, p. 113-142.

Kontorovich, A. E., 1984, 'Geochemical methods for the quantitative evaluation of the petroleum potential of sedimentary basins,' *in* Demaison, G. and Murris, R. J., eds., *Petroleum Geochemistry and Basin Evaluation*: Am. Assoc. Petroleum Geologists, Mem. 35, p. 79-109.

Lee, P. J. and Wang, P.C.C., 1983, 'Probabilistic formulation of a method for the evaluation of petroleum resources': *Mathematical Geology*, 15, no. 1, p. 163-181.

McCray, A. W., 1975, *Petroleum Evaluations and Economic Decisions*: Prentice-Hall, Inc., Englewood Cliffs, N.J., 448 p.

McCrossan, R. G., 1969, 'An analysis of size frequency distribution of oil and gas reserves of western Canada': *Canadian Jour. Earth Sci.*, 6, no. 2, p. 201-211.

Meisner, J. and Demirmen, F., 1981, 'The creaming method: A Bayesian procedure to forecast future oil and gas discoveries in mature exploration provinces': *Jour. Royal Statistical Society*, Series A (General), 144, Part 1, p. 1-31.

Neter, J. and Wasserman, W., 1974, *Applied Linear Statistical Models*: Richard D. Irwin, Inc., Homewood, Ill., 842 p.

Newendorp, P. D., 1975, *Decision Analysis for Petroleum Exploration*: Petroleum Publ. Co., Tulsa, Okla., 750 p.

Roy, K. J., Proctor, R. M., and McCrossan, R. G., 1975, Hydrocarbon assessment using subjective probability, *in Probability Methods in Oil Exploration*: Proc. Am. Assoc. Petroleum Geologists Research Symposium, Stanford Univ., Stanford, Calif.

Schuenemeyer, J. H. and Drew, L. J., 1983, 'A procedure to estimate the parent population of the size of oil and gas fields as revealed by a study of economic truncation': *Mathematical Geology*, 15, no. 1, p. 145-161.

Seber, G.A.F., 1977, *Linear Regression Analysis*: John Wiley & Sons, Inc., New York, 465 p.

Singer, D. A., 1972, Elipgrid, a FORTRAN IV program for calculating the probability of success in locating elliptical targets with square, rectangular and hexagonal grids: *Geocom Programs*, no. 4, London, 16 p.

Singer, D. A. and Wickman, F. E., 1969, Probability tables for locating elliptical targets with square, rectangular and hexagonal point-nets: Pennsylvania State Univ. Mineral Science Experiment Station Spec. Pub. 1-69, 100 p.

Sluijk, D. and Nederlof, M. H., 1984, 'Worldwide geological experience as a systematic basis for prospect appraisal,' *in* Demaison, G. and Murris, R. J., eds., *Petroleum geochemistry and Basin Evaluation*: Am. Assoc. Petroleum Geologists, Mem. 35, p. 15-26.

Solomon, H. and Stephens, M. A., 1978, 'Approximations to density functions using Pearson curves': *Jour. American Statistical Assoc.*, 73, p. 153-160.

Tissot, B. P. and Welte, D. H., 1984, *Petroleum Formation and Occurrence*, 2nd ed.: Springer-Verlag, Berlin, 699 p.

Waples, D., 1981, *Organic Geochemistry for Exploration Geologists*: Burgess Publ. Co., Minneapolis, Minn., 151 p.

ON ASSESSING DRY PROBABILITIES IN OFFSHORE OIL AND GAS EXPLORATION: AN APPLICATION OF BAYES'S THEOREM

Andrew R. Solow
Woods Hole Oceanographic Institution
Woods Hole, Massachusetts 02543

ABSTRACT

United States Government estimates of offshore oil resources have been based largely on subjective information. Through time, results of exploratory drilling become available, and the estimates are updated. Although this situation strongly suggests the use of Bayes's Theorem, this approach has not been taken. In this paper, the problem is explicitly framed in terms of Bayes's Theorem, and two simple exercises are performed on actual results to demonstrate the usefulness of the approach.

INTRODUCTION

A perennial problem in statistical estimation and prediction is how to combine subjective information with objective measurements. This problem arises in many fields. It is particularly important in the earth sciences, where there is often considerable subjective information and objective measurements, due to their great cost, are often scarce. In certain applications, the combination of subjective and objective information is done informally by either "fixing up" results or introducing fictitious measurements which are taken to represent the quantitative expression of subjective information (e.g., Kulkarni, 1983; Journel, 1986). The purpose of this paper is to describe a situation in which subjective information is quite important, and in which we will be able to frame the problem in such a way that this information can be combined with objective measurements via Bayes's Theorem. This situation concerns the estimation of undiscovered offshore oil resources in the United States.

In 1980, the United States Geological Survey (USGS) undertook an appraisal of undiscovered recoverable oil and gas resources in the United States. In 1984, the Minerals Management Service (MMS) undertook a similar exercise for the offshore United States. A report of the USGS assessment appears in USGS (1981a), with further details in USGS (1981b). A report of the MMS assessment appears in MMS (1985). We will focus on the assessment of undiscovered offshore oil resources. In Table 1, we report the results of these two studies. What is most striking about these results is the enormous drop in the estimates, particularly for the Alaska and Atlantic regions. In monetary terms, this constitutes a downward revision on the order of hundreds of billions of dollars in the asset value of the nation's undiscovered offshore oil resources

C. F. Chung et al. (eds.), Quantitative Analysis of Mineral and Energy Resources, 187–198.

(Boskin, et al., 1985). The work presented in this paper represents the beginning of an attempt to explain these results.

In order to investigate these results, it is necessary to understand the methodology under which the assessments were carried out. A brief description is presented in the next section. The problem is extremely large in scope, and we will focus on one aspect: estimating the probability that a region is dry (i.e., contains no oil resources). We will see that this estimation was based largely on subjective information, although between 1980 and 1984 the results of some exploratory drilling became available. This sequence of events -- initial subjective estimation, gathering of objective measurements, updating the estimation -- is highly suggestive of an application of Bayes's Theorem. This approach, however, was not explicitly taken by either the USGS or the MMS.

In the third section, we formulate this problem in Bayesian terms. In this way, prior dry probabilities based on subjective information may be combined with drilling results via Bayes's Theorem to provide posterior dry probabilities. Under this formulation, we perform two exercises. First, treating the USGS results as prior, and the MMS results as posterior, to the drilling results, we solve for the implicit likelihood of the drilling results. Second, using a simple model for the occurrence of oil targets and the drilling strategy, we update the USGS prior dry probabilities to find the corresponding posterior dry probabilities. In the final section, we discuss ways in which the study can be made more realistic.

There is a fairly substantial literature on the estimation of oil resources. The review article by Wiorkowski (1981) and the comments following it list many references. Important contributions to this area have been made by Attanasi, Drew, Root, Schuenemeyer, and Singer. Many of their articles appear in the references of Wiorkowski (1981). We also wish to note the discovery model of Barouch and Kaufman (1976a,b). A more complete Bayesian estimation of oil resources than that presented here would certainly make use of this and related models. Another source of statistical work in this area is the Princeton University Resource Estimation and Validation Project (e.g., Bloomfield, et al., 1979).

CURRENT METHODOLOGY

Before turning to a brief description of the assessment methodology, we make the following remark. Recoverable resources refer to quantities of oil which may be economically recovered. Thus, as economic and technological conditions change, marginal oil resources will change in status from recoverable to non-recoverable, or vice versa. In other words, the definition of the estimand in the assessment exercise may change with time. We will not deal with this problem here beyond making two comments. First, broadly speaking, oil prices were rising at the time of the USGS assessment, but had levelled off at the time of the MMS assessment (MMS, 1985). Presumably, an effect of this was a reduction in status of some oil prospects from recoverable to non-recoverable. Second, while there is considerable uncertainty regarding the characteristics of oil in the ground, there is also uncertainty regarding future economic conditions which bear on the recoverability of oil resources. A comprehensive statistical treatment of the assessment problem should take this into account.

Estimating undiscovered oil resources is necessarily attended by a high level of uncertainty. To see this, it is sufficient to note that as soon as information confirming the presence of oil becomes available, the resource is no longer undiscovered. As a consequence of this, the assessment of undiscovered resources is carried out in frontier regions, where little or no exploratory drilling has occurred. In order to overcome the lack of drilling information, both the USGS and the MMS based their assessments on subjective estimates of key parameters given by teams of petroleum geologists and other experts. The details of the methods are given in USGS (1981a) and MMS (1985), respectively. The general approach, which has been called the Delphi method, is discussed in Harris (1984).

To begin with, the overall area under consideration (i.e., the entire United States for the USGS and the offshore United States for the MMS) was divided into provinces on the basis of rough geological homogeneity. Although there have been changes in the delineation of provinces, to give an idea of their size, the size distribution for the ten offshore Alaska provinces given in MMS (1984) had mean, median, minimum, and maximum of 56, 49, 8, and 133 million acres, respectively. The assessment proceeded province by province. For a single province, the assessment was broken into two parts. First, because the province was virtually unexplored, a subjective estimate of the dry probability was given. Again, to give an idea of the magnitudes of these probabilities, for the offshore Alaska provinces mentioned above the distribution of USGS dry probabilities had mean, median, minimum, and maximum of 0.68, 0.72, 0.00, and 0.90, respectively (USGS, 1981b). Second, conditioning on the complementary event that the province was not dry, subjective estimates of key parameters determining the quantity of the resource were given. The two steps are then combined to give an overall estimate of the quantity of oil. We will focus exclusively on the first step: namely, assessing the dry probability.

Between 1980 and 1984, a certain amount of exploratory drilling took place in some of the offshore provinces. The results of that drilling were disappointing (MMS, 1985). Partly as a consequence of this, the dry probabilities given by the MMS were generally larger than those given by the USGS: in fact, as we point out in the next section, as further drilling results become available we expect this probability to either increase or fall to zero. This was particularly true for offshore Alaska, although similar results were recorded for the Atlantic provinces. The MMS (1985) cites this increase in dry probabilities as a major cause of the sharp drop in total estimated resources. In the next section, we formulate the process of estimating dry probabilities in Bayesian terms. In this way, the results of exploratory drilling may be combined with previous estimates in a rigorous way according to Bayes's Theorem to provide updated dry probabilities. We apply this Bayesian approach in two ways to some of the results published by the USGS and the MMS.

A BAYESIAN FRAMEWORK

As we described in the previous section, the estimation of undiscovered oil resources as performed by the USGS and the MMS requires the estimation of the probability that a region is dry. For unexplored regions, this estimation must be based on scientific judgement. Over time, exploratory drilling will occur in some of these regions. As the results of this drilling become available, it will be important to update the estimated dry probability. Of course, if the exploratory drilling results include one or more successes, the dry probability falls to zero.

Let d be the event that the region under consideration is dry, and let I denote the results of exploratory drilling. Then, Bayes's Theorem states that the way in which the prior probability of d should be updated to reflect I is:

$$pr(d/I) = pr(I/d)pr(d)/pr(I) \qquad (1)$$

where:

pr(d/I) is the posterior dry probability
pr(d) is the prior dry probability
pr(I/d) is the likelihood of I given d
pr(I) is the probability of I

We make three observations. First, as we have said, if the exploratory drilling includes at least one success then:

$$pr(I/d) = 0$$

and consequently:

$$pr(d/I) = 0$$

Thus, the only interesting case is when the results of drilling include no successes. Let N denote this event. Second, note that:

$$pr(N/d) = 1$$

since it is impossible to have a drilling success in a dry region. Third, we may express the denominator in (1) as:

$$\begin{aligned} pr(N) &= pr(N/d)pr(d) + pr(N/d^c)pr(d^c) \\ &= pr(d) + pr(N/d^c)(1-pr(d)) \end{aligned}$$

where c denotes complement (i.e., d^c is the event that the region contains oil). Taking account of these points, the interesting form of (1) is:

$$pr(d/N) = pr(d)/(pr(d) + pr(N/d^c)(1-pr(d))) \quad (2)$$

In order to apply (2), it is necessary to specify $pr(N/d^c)$ -- that is, the likelihood of having no drilling successes given the presence of oil. This probability depends on the way in which the oil is distributed in the region and the number and location strategy of the exploratory wells. We will illustrate this using a simple model in our second exercise below.

It is worth pointing out that the notion that exploratory drilling results are binary (hit or miss) is certainly a simplification. For example, an exploratory well may encounter an economically unrecoverable oil pool, whose presence virtually guarantees the existence of recoverable resources nearby.

Exercise One: Solving for the Implicit Likelihood

As our first exercise, we solve (2) for $pr(N/d^c)$ to find:

$$pr(N/d^c) = pr(d)(1-pr(d/N))/pr(d/N)(1-pr(d)) \quad (3)$$

(which is the ratio of the prior dry odds to the posterior dry odds). Then, using the USGS (1981b) dry probabilities as priors and the MMS (1985) dry probabilities as posteriors, we find the implicit likelihoods of having no successes in the presence of oil. We note that the treatment of USGS probabilities as prior and MMS probabilities as posterior is a construction of this paper, and was not shared by the USGS or the MMS. This exercise was performed for four provinces in the Bering Sea region of Alaska. The results, which appear in Table 2, consist of the following:

s, the size (in millions of acres) of the provinces given in MMS (1984)
$pr(d)$, the dry probability given in USGS (1981b)
n, the number of dry wells given in MMS (1985)
$pr(d/N)$, the dry probability given in MMS (1985)
$pr(N/d^c)$, the likelihood of having no successes given the presence of oil, found from (3)

Also, as a crude way of correcting $pr(N/d^c)$ for n, we report:

$$(pr(N/d^c))^{1/n}$$

which is the probability of having a single dry well in the presence of oil, if the outcomes of the n wells are independent and identically distributed. That is, if the outcomes of the n wells are independent and the conditional dry probabilities for the n wells are all equal, and the overall conditional dry probability for the region is $pr(N/d^c)$, then the dry probabilities for the individual wells are all equal to $pr(N/d^c)^{1/n}$.

We make the following two remarks.

First, the case of the Bristol Basin provides a form of control. No wells were drilled in the Bristol Basin between 1980 and 1984. We would expect, ceteris paribus, the prior and posterior probabilities to be the same. In fact, there is a slight change. This change may be due to factors such as a change in economic conditions which affects recoverability or the implicit treatment of the provinces as non-independent (in which case, failure to find oil in one province may increase the dry probability for another). Within the putative Bayesian framework, a possible explanation is that the USGS prior was not the same as that used by the MMS. In any case, the result for the Bristol Basin suggests that drilling information, and not other factors, accounts for the larger differences between prior and posterior dry probabilities in other provinces.

Second, we turn to the values of $pr(N/d^c)^{1/n}$. As we have said, this quantity represents a crude attempt to correct $pr(N/d^c)$ for the number of dry wells, n. It is interesting to note how similar these values are for the three provinces, despite the substantial disparities in province size. Under certain circumstances, this result is reasonable, while under others it is not. For example, suppose that the event d^c corresponds to the existence of a single oil pool of size S. If S is constant across provinces, we would generally expect $pr(N/d^c)^{1/n}$ to be higher for larger provinces. On the other hand, suppose the event d^c corresponds to the existence of K identical targets each of size S. If S is again constant across provinces, but K is proportional to province size, we would generally expect $pr(N/d^c)^{1/n}$ to be constant across provinces. In the next exercise, we consider a simple model similar to this second example. Finally, it is interesting to note that $pr(N/d^c)^{1/n}$, which may be thought of as the failure rate for a single well, is quite high.

It would be interesting to extend this exercise to several more offshore provinces. Unfortunately, the information contained in the USGS and MMS reports is insufficient to do this. One problem is that the USGS and MMS provinces do not match except in a few cases. A second problem is that drilling information is only sporadically reported.

Exercise Two: Using a Model to Find Posterior Dry Probabilities

As a second exercise, we will introduce a simple model which provides the likelihood, $pr(N/d^c)$. We will use this model in conjunction with the prior dry probabilities given in USGS (1981b) to find posterior dry probabilities for the four Bering Sea provinces. Again, note that $pr(N/d^c)$ depends on both the distribution of oil in the region and the way in which locations for exploratory wells are chosen. Thus, a model which provides $pr(N/d^c)$ must account for both factors.

We adopt the following model for the distribution of oil. Oil occurs in circular pools, each with known radius r. The number of pools, T, has a prior Poisson distribution, so that:

$$pr(T = t) = e^{-a} a^t / t! \quad t=0,1,\ldots \qquad (4)$$

Note that:

$$pr(d) = pr(T = 0) = e^{-a}$$

Conditional on T = t, the pool centers are independently uniformly distributed over the region, R. For simplicity, we take R to have unit area. This model is called the bombing model (Solomon (1953), Matern (1960)). Note that it is possible for pools to overlap. Throughout the analysis, we will ignore border effects. This is reasonable if the pools are small relative to R.

For the well-siting part of the model, we will suppose that the first well is located at x_1, which is chosen at random (i.e., uniformly) in R. The second well is located at x_2, which is chosen at random in:

$$R \cap (d_2(x_1))^c$$

where $d_2(x_1)$ is the circle with radius 2r centered at x_1. Similarly, the third well is located at x_3, which is chosen at random in:

$$R \cap (d_2(x_1) \cup d_2(x_2))^c$$

and so on. It is easy to show that this strategy, which avoids the possibility that a pool is discovered more than once, maximizes the probability of discovery under the bombing model.

We are now in a position to find the posterior dry probability given that n out of n exploratory wells fail to hit oil -- an event which we will denote by N(n). It is easy to show that under the bombing model:

$$pr(d/N(n)) = pr(T=0/N(n)) = \exp(-a(1-np)) \qquad (5)$$

where:

$$p = \pi r^2$$

is the area of a single oil pool. In fact, it can be shown that for the prior distribution for T given by (4), the posterior distribution for T is:

$$pr(T=t/N(n)) = e^{-a(1-np)}(a(1-np))^t/t! \qquad t=0,1,\ldots$$

That is, the posterior distribution is also Poisson.

In order to apply (5), it is necessary to specify a and p. For the Bering Sea provinces, we will choose values of a to identify the prior dry probabilities given in Table 2. That is:

$$a = -\log pr(d) \qquad (6)$$

To choose p, we proceed as follows. Note that:

$$pr(N(1)/d^c) = (e^{-ap}-e^{-a})/(1-e^{-a}) \qquad (7)$$

Recall that for the Navarin Basin we were able to solve directly for this probability in the previous exercise. We found the implicit value to be 0.93. Also, pr(d) is 0.71 for the Navarin Basin, corresponding to a = 0.3425. We will, therefore, solve (7) for p using these values to find p = 0.06. Recall that p is expressed in terms relative to the size of the province. We will assume that absolute target size is constant across provinces. By taking account of the province sizes given in Table 2, we find that this value of p for the Navarin Basin corresponds to p = 0.088 for the Norton Basin and p = 0.032 for the St. George Basin.

The results of this exercise are presented in Table 3, where we report:

pr(d), given in Table 2
a, found from (6)
p, given above
n, given in Table 2
pr(d/N(n)), given in Table 2
pr(d/N(n)), found from (5) for the values of p given above

It is interesting to note that the values of pr(d/N(n)) given by (5) are in fairly close agreement with those given in Table 2.
an illustration of how this Bayesian approach might be used, suppose that a stopping rule of the form:

$$\text{Terminate exploration if } pr(d/N(n)) > 0.95$$

is used. The question arises: How many dry wells does it take for termination to occur? For the simple model described above it is easy to show from (5) that this number is given by:

$$n = (1 + (\log 0.95/a))/p$$

Using the values of a and p given in Table 3, the critical number of dry wells is 9 for the Norton Basin, 26 for the St. George Basin, and 14 for the Navarin Basin. Note that we are able to perform this calculation for the Bristol Basin, for which a = 0.236 and p = 0.07. The critical number of dry wells for the Bristol Basin is 11. Clearly, a stopping rule of this form, which does not take into account the substantial unit cost of sampling, is unrealistically simple. Nevertheless, as an informal procedure it may be useful.

DISCUSSION

We began this paper by pointing out the need for methods which combine qualitative information with quantitative measurements. Bayes's Theorem is an appealing way of doing this. There are, however, three problems with this approach. First, the problem must be formulated in a way which allows the application of Bayes's Theorem. Second, a prior distribution must be specified which reflects the subjective information. Third, a model must be specified which provides the likelihood of the objective measurements.

For the problem of estimating dry probability, the Bayesian formulation is quite natural. Also, the summarization of subjective information in terms of prior probability does not appear to be a problem: in any case, both the USGS and the MMS seem comfortable with this step. It is in the third area -- that of modelling -- where considerable improvement is possible. An improved bombing model might include the following two features:

(1) The size of each oil pool is considered random, possibly following a lognormal distribution.

(2) The pool centers follow an inhomogeneous Poisson process (Cox and Isham, 1980). The intensity function could be made to depend on covariates such as porosity.

The model for the drilling strategy could also be made more realistic. For example, under the inhomogeneous Poisson model for pool centers, the first exploratory well would be located where the prior intensity is greatest. Conditional on a miss at the first well, the second well would be located where the posterior intensity is greatest, and so on. As we have said, the simple analysis presented in this paper ignores the problem of recoverability. Apart from economic factors, recoverability is affected by water depth, depth of the resource below the seafloor, distance to port, etc. It should not be too difficult to introduce these kinds of factors into the analysis. Of course, as the problem becomes more complicated, with the introduction of more complex models and a larger number of parameters, attention will have to be paid to numerical methods.

A more fundamental question arises: Is the parameterization in terms of numbers of discrete pools reasonable? As we have seen, the USGS and the MMS do not parameterize the problem in this way. Instead, they break the problem into two parts: assessing the dry probability and, conditional on the presence of oil, assessing the quantity. Although no justification for this parameterization is given, a possible explananation is the following. The quantity of oil in a region depends on two independent factors: a source of hydrocarbon and the geological structure of the region (e.g., the potential for traps). The question of parameterization is a fairly important one. A model may be very appealing in terms of mathematical convenience, yet it may be unrealistic to the point of uselessness. It is the role of the geologist to guide the model-building stage of the analysis.

The first exercise which we performed might be useful for checking, a posteriori, the reasonableness of the MMS assessment of dry probabilities, using the USGS dry probabilities as a starting point. For example, once the MMS experts give their assessments, the implicit likelihood could be found from (5). The experts could then

region	USGS 1981	MMS 1985	% change
Alaska	12.2	3.3	-73
Atlantic	5.4	0.7	-87
Gulf of Mexico	6.2	6.0	-3
Pacific	3.2	2.2	-31
Total	27.0	12.2	-55

Table 1: Comparison of estimated undiscovered offshore oil resources from 1981 and 1985. Values are in billions of barrels. Source: USGS (1981a), MMS (1985).

be asked to assess the reasonableness of this likelihood. If the likelihood is felt to be unreasonable, then the MMS results would require further explanation. An explanation might be that the provinces are viewed implicitly as non-independent, so that drilling misses in one province increase the dry probabilities in a number of provinces. The second exercise which we performed is simply an example of how Bayes's Theorem would be applied directly. We have stressed the great simplicity of the bombing model in this context. However, for frontier provinces where very little is known, even such a simple model might be useful as a starting point.

ACKNOWLEDGEMENTS

This research was supported by the J.N. Pew, Jr. Charitable Trust and the Marine Policy and Ocean Management Center of the Woods Hole Oceanographic Institution. Support was also given by the National Science Foundation and the Organizing Committee of the NATO Advanced Seminar Institute. The author would like to thank Dr. James M. Broadus of the Woods Hole Oceanographic Institution for his help.

Woods Hole Oceanographic Institution Contribution Number 6218.

province	s	pr(d)	n	pr(d/N)	pr(N/d^c)	pr(N/d^c)$^{1/n}$
Norton Basin	25	0.78	5	0.85	0.63	0.91
St. George Basin	70	0.72	4	0.78	0.73	0.92
Navarin Basin	37	0.71*	1	0.73	0.93	0.93
Bristol Basin**	32	0.79	0	0.80	----	----

*average of probabilities for shelf and slope

**also called the North Aleutian Basin in MMS (1985)

Table 2: Results from Exercise One.

					pr(d/N(n))	
province	pr(d)	a	p	n	from Table 2	from (5)
Norton Basin	0.78	0.249	0.088	5	0.85	0.87
St. George Basin	0.72	0.329	0.032	4	0.78	0.75
Navarin Basin	0.71	0.343	0.060	1	0.73	0.73

Table 3: Results from Exercise Two.

REFERENCES

Barouch, E. and G.M. Kaufman (1976), Probabilistic modelling of oil and gas discovery, in Energy: Mathematics and Models, F. Roberts, ed., SIAM, Philadelphia.

Barouch, E. and G.M. Kaufman (1976), Oil and gas discovery modelled as sampling proportional to random size, Working Paper 888-76, Sloan School of Management, Massachusetts Institute of Technology.

Bloomfield, P. et al. (1979), Volume and Area of Oilfields and Their Impact on Order of Discovery, Princeton University Resource Estimation and Validation Project, Princeton, New Jersey.

Boskin, M.J., et al. (1985), New estimates of the value of Federal mineral rights and land, The American Economic Review, 75(5), pp. 923-936.

Cox, D.R. and V. Isham (1980) Point Processes, Chapman and Hall, London.

Harris, D.P. (1984), Mineral Resources Appraisal, Clarendon, Oxford.

Journel, A.G. (1986), Constrained interpolation and qualitative information - the soft kriging approach, Math. Geol., 18(3), p. 269-286.

Kulkarni, R. (1984), Bayesian kriging in geotechnical problems, in Geostatistics for Natural Resources Characterization, G. Verly, et al., eds., Reidel, Dordrecht.

Minerals Management Service (1984), Federal Offshore Statistics, Report 84-0071.

Minerals Management Service (1985), Estimates of Undiscovered Economically Recoverable Oil and Gas Resources for the Outer Continental Shelf as of July 1984, Report 85-0012.

Matern, B. (1960), Spatial Variation, Meddelanden Fran Statens Skogsforskningsinstitut, Band 49, No. 5.

Solomon, H. (1953), Distribution of the measure of a random two-dimensional set, Ann. Math. Stat., 24, p. 650-656.

United States Geological Survey (1981), Estimates of Undiscovered Recoverable Conventional Resources of Oil and Gas in the United States, Circular 860.

United States Geological Survey (1981), Conditional estimates and marginal probabilities for undiscovered recoverable oil and gas resources by province, Open File Report 82-666A.

STATISTICAL ANALYSIS OF OIL AND GAS DISCOVERY DATA

Paul C. C. Wang[1] and Vijayan N. Nair[2]

1 - Department of Statistics
University of Calgary
Calgary, Alberta, TN2 1N4
Canada

2 - Mathematical Sciences Research Center
AT&T Bell Laboratories
Murray Hill, NJ, 07974
USA

ABSTRACT. We provide a systematic approach for analyzing oil and gas discovery data based on a successive sampling model for the discovery process. First, the size distribution of deposits is estimated nonparametrically. Graphical goodness-of-fit procedures are then used to select an appropriate parametric form for the distribution. The parameters of this distribution are estimated using maximum likelihood techniques. From this, various quantities of interest such as the total remaining potential and the size of the largest undiscovered deposit can be predicted. The estimation of the number of deposits that could exist in a petroleum play is also discussed. The methods are illustrated on discovery data from the Rimbey-Meadowbrook reef play in Central Alberta, Western Canada.

1. INTRODUCTION

A central problem in petroleum resource evaluation is the estimation of the size distribution of oil and gas deposits based on discovery data that are incomplete and not representative of the population from which they are drawn. Indeed, in many petroleum regions, larger deposits tend to be discovered early during exploration. In some areas, a combination of attributes such as depth, area and distance to transportation centers influence the order of discovery. In offshore exploration, distance to shore or ocean depth have significant impact on which prospects are to be tested first. When factors such as these have influenced the order of discovery, the discovered deposit data cannot be considered a random sample from the population. The biased nature of the sample must be taken into account in making predictions about the undiscovered oil and gas resources in a petroleum region.

A knowledge of the size distribution and the number of deposits that could exist in a petroleum region are the fundamental objectives in petroleum resource evaluation. From these, quantities such as the total remaining potential, the number of deposits in each size class and the size of the largest of the undiscovered deposits can be predicted. These provide the basic ingredients for petroleum economic analysis and the planning of exploration strategies.

In order to address the size-biased phenomenon, Arps and Roberts (1958) proposed a

C. F. Chung et al. (eds.), Quantitative Analysis of Mineral and Energy Resources, 199–214.

discovery model that postulates that the probability of finding one more field with an area between y and $y+dy$ in the basin for each additional wildcat well drilled is proportional to:

i) the area y of such fields,

and

ii) the remaining number of undiscovered fields of that size.

Let n denote the number of fields of area y that have been found after drilling w exploratory wells, N denote the total number of fields of that size area existing in the basin, and B denote the total basin area. Then, assuming the area searched is small relative to B, they arrived at the model

$$n = N(1 - \exp\{-cwy / B\}) \tag{1}$$

where c is interpreted as the exploration efficiency with respect to the basin area B. For random drilling, the constant c would be 1. Drilling based on geological or geophysical information would in principle yield a larger constant c. Arps and Roberts (AR) used the value $c=2$ in their analysis of the Lower Cretaceous fields in the east flank of the Denver-Julesburg basin, which covered an area of 5.7 million acres. They did not, however, indicate how the exploration efficiency might be estimated from data. A detailed statistical description of the AR model is given by Kaufman (1986).

Drew, Schuenemeyer and Root (1980) extended the AR model by using a geometric approach based on the concept of the area of influence of wells to estimate search effort instead of the cumulative number of exploratory wells. Both the historical drilling and discovery data are used to estimate the parameters of their generalized discovery model — the DSR model. Although both the AR and the DSR models are probabilistic, they do not lend themselves easily to standard statistical treatment for estimation of unknown parameters. These type of models are essentially constrained by equation (1), and it is not clear how to extend them to incorporate additional factors for explaining more of the variation in the discovery data.

Meisner and Demirmen (1981) describe the declining trend of field size with advancing exploration as the "creaming phenomenon". They developed a statistical model for forecasting the distribution of the total recoverable petroleum as a function of the cumulative number of exploratory wells. This model assumes that the underlying size distribution is log-normal and postulates that i) the mean of log field size is a linear function of the corresponding exploratory well number, and ii) the probability of success is a linear logistic function of the cumulative number of exploratory wells. The parameters of the underlying size distribution are not identifiable under this model. Moreover, the finiteness of the population of deposits, a very important feature in petroleum resource evaluation, is not captured here.

In a completely different attack on the problem, Kaufman et al. (1975) proposed a statistical sampling approach which was further developed in Barouch and Kaufman (1976, 1977). The key idea in this approach is to model the discovery process in the petroleum region as sampling successively from a finite population of deposits without replacement and with probability proportional to deposit magnitude. Following Kaufman (1986), we shall refer to this model as the successive sampling (SS) discovery model. Although this model was developed with the intention of evaluating petroleum plays, it can also be applied, in principle, to provinces or basins. The only restriction is that the region under study should be sufficiently homogeneous geologically. Methods of estimation for the finite population *SS* model have been considered by Gordon (1981,1983) and Andreatta and Kaufman (1986). Application of these methods to the Western Gulf Miocene-Pliocene discovery data and a North Sea discovery data are given in Kaufman (1986). Some very useful comparisons between the *AR* and the *SS* discovery models are also discussed in this paper.

The successive sampling discovery model considers the region under study, say a play, as a statistical population consisting of N oil and gas deposits with magnitudes $W_1, ..., W_N$. Given all such magnitudes and the number of discoveries $n < N$, the model assumes that the probability of discovering deposits with magnitudes $W_1, ..., W_n$ *first and in that order* is

$$\prod_{j=1}^{n} \frac{W_j}{W_j + \cdots + W_N} . \tag{2}$$

The magnitude is usually taken to be some function of the area or the volume of a deposit. Bloomfield et al. (1979) defined magnitude in the form $W_j = Y_j^{\beta}$, where Y_j is the volume or the area of a deposit and β is some unknown parameter to be estimated from the discovery data. This definition of magnitude was used by Smith and Ward (1981), and Lee and Wang (1985,1986) for estimating undiscovered petroleum resources in a North Sea oil play and two Leduc oil plays in Central Alberta, Western Canada. More generally, the magnitude may be a function of deposit attributes

$$W_j = W(Y_{j1}, ..., Y_{jk}), \tag{3}$$

where the attributes $(Y_{j1}, ..., Y_{jk})$, associated with a deposit are observable once the deposit is discovered, and $W(\cdot)$ is a function to be modeled from the data. For example, a magnitude may be defined by

$$W_j = \text{Volume}^{\beta_1} \times \ \text{Area}^{\beta_2} \times \ \text{Depth}^{\beta_3} \tag{4}$$

and β_1, β_2 and β_3 are unknown parameters. If in fact the deposits with shallower depths tend to be found early during exploration, then β_3 would be negative. If it has no impact on the order of discovery, β_3 would be close to zero. In principle, many different functional forms can be entertained for the magnitude. Time-dependent factors such as the price of oil, land availability and the level of geologic knowledge about the play can also be included. A "learning period" during which the field size tends to increase may be modeled by a piecewise function of the form

$$W_j = \begin{cases} Y_j^{\beta_1}, & j < J \\ Y_j^{\beta_2}, & j \geq J \end{cases} \tag{5}$$

with $\beta_1 < 0$ and $\beta_2 > 0$, Y_j is a measure of field size and J is the turning point. In practice we need to obtain a compromise between a realistic representation of the sampling process and a model which is both conceptually and computationally simple to use as a working tool.

In this paper, we consider a superpopulation successive sampling discovery model, i.e., in addition to the SS discovery model (2), we assume that the magnitudes $W_1, ..., W_N$ were generated independently and identically according to a distribution function (df) F. Throughout, a deposit is taken to be a pool (a natural underground reservoir containing an accumulation of oil or gas or both separated or appearing to be separated from any other such accumulation). In general, the observed data would consist of deposit magnitudes *in order of discovery* and the values of the associated attributes. In this paper, we take deposit magnitude to be the attribute of interest itself -- pool size measured in millions of barrels of oil-in-place. See, however, Nair and Wang (1986) for a more general formulation. So we observe $\mathbf{x}=(x_1, ..., x_n)$ with x_j denoting the magnitude of the jth discovery. Note that the discovered deposit magnitudes are not independent. The log-likelihood function under this model is given by (Nair and Wang, 1986)

$$\log L = \text{Constant} + \sum_{j=1}^{n} \log dF(x_j) + \log \int_0^{\infty} [\phi(t;F)]^{N-n} g_n(t)dt \tag{6}$$

where $\phi(t;F)$ is the Laplace transform of W, $b_j=x_j+\cdots+x_n$, and $g_n(t)$ is the general gamma or Erlang density with parameters $b_1, ..., b_n$ (Johnson and Kotz, 1970, p.222).

This paper provides a systematic approach for analyzing oil and gas discovery data based on some recently developed methods for estimating the underlying size distribution function. The paper is organized as follows. In Section 2, we give the nonparametric maximum likelihood estimator of the distribution function F. We then use a graphical goodness-of-fit procedure, known as the quantile-quantile plot, to select an appropriate parametric form for the distribution. This parametric distribution is estimated using maximum likelihood techniques. We illustrate the methods for the log-normal distribution. The simultaneous estimation of the parametric distribution and the population size N are discussed in Section 4.

Table I. Summary statistics of the Rimbey-Meadowbrook reef discovered pool sizes (measured in 10^6 barrels of oil)

sample size	=	23	minimum $x_{(1)}$	=	0.294
mean $\bar{x}$	=	157.82	5 percentile	=	0.587
s.d. s	=	308.81	25 percentile	=	3.403
geometric mean	=	20.91	median	=	14.655
mean of log's	=	3.04	75 percentile	=	169.61
s.d. of log's	=	5.56	95 percentile	=	765.89
skew. coeff.	=	1.40	maximum $x_{(n)}$	=	1295.4

The techniques will be illustrated on discovery data from the Rimbey-Meadowbrook (R-M) reef play in Central Alberta. See Lee and Wang (1986) for a full discussion of this data set. The summary statistics of the R-M data given in Table I. It can be seen from these that the discovered pool data are highly skewed to the right. The largest discovered pool is over 4000 times bigger than the smallest discovered pool. The box plot of the logarithm of the pool sizes in Figure 1 shows no outlying values in the log-scale. It also suggests that the tails of the logged data are roughly symmetrical, but that the median is still low relative to the quartiles. This could be the result of the size biased sampling. Figure 2 shows a plot of field sizes against the discovery date (the *spud date* of the discovery well). The smooth LOWESS curve (Cleveland, 1979) in the figure shows the general declining trend in discovered sizes with advancing exploration. Figure 3 illustrates the same phenomenon in terms of cumulative discovery against discovery numbers. The steep rise at the beginning of the cumulative discovery curve illustrates that larger pools tend to be found early during exploration.

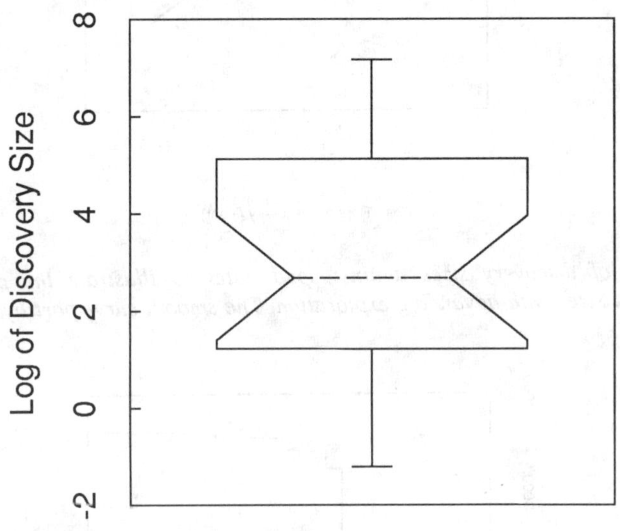

Figure 1. A notched boxplot of the logarithm of the discovered pool sizes of the Rimbey-Meadowbrook reef play.

The major implication of the latter two figures is that this set of data cannot be considered as a random sample. Consequently, summary statistics contained in Table I must not be interpreted as estimates of corresponding values for the underlying population. If the underlying size distribution is log-normal, the random sampling estimates of μ and σ^2 are 3.04 and 5.56 (Table I). These would, however, overestimate the true population parameter μ and would underestimate σ^2. In the sequel, we illustrate how they can be estimated in the presence of the biased sampling phenomenon prescribed by the SS discovery model (2).

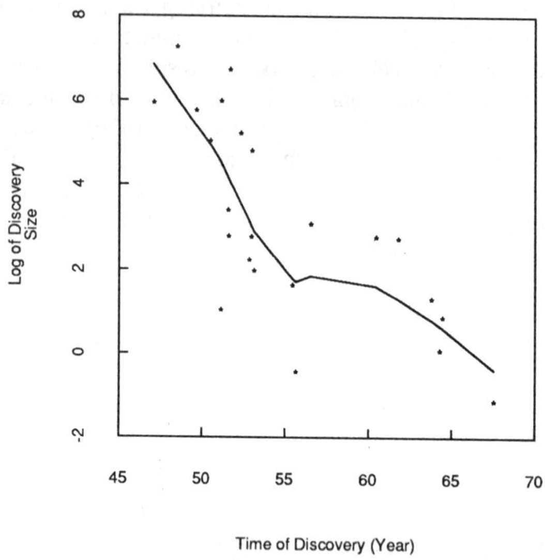

Figure 2. Plot of discovery sizes against spud dates to illustrate the phenomenon of diminishing pool sizes with advancing exploration. The smooth curve portrays the decline in pool sizes.

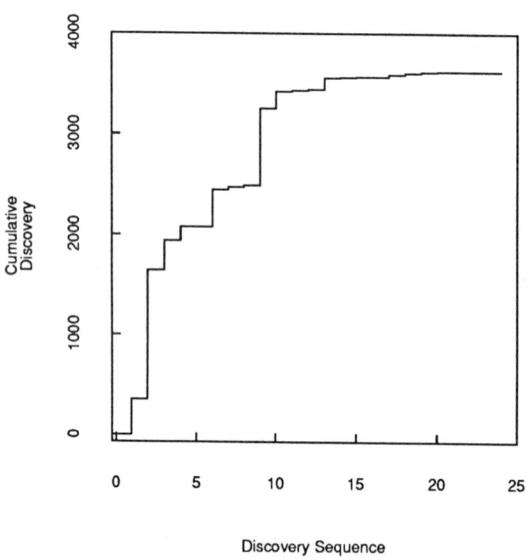

Figure 3. Plot of cumulative discovery against discovery number to illustrate that large pools tend to be discovered early during exploration. The large pools are represented by large discontinuous upward jumps.

2. NONPARAMETRIC ESTIMATION

To determine the shape of the underlying size distribution F, we first need to estimate it without making any parametric assumptions. The standard empirical distribution function (edf) defined by

$$F_n(y) = \frac{\text{no. of pool sizes} \leq y}{n} \tag{7}$$

would not be appropriate since we have a biased sample.

Let the n observed pool sizes $x_1, \ldots, x_n$ be distinct. The nonparametric estimation of F involves estimating the probabilities $\{p_i\}_{i=1}^n$ associated with $\{x_i\}_{i=1}^n$ with $\sum_{i=1}^n p_i = 1$. The maximum likelihood estimator of F would then be given by

$$\hat{F}_n(y) = \sum_{i: x_i \leq y} \hat{p}_i \tag{8}$$

where $\{\hat{p}_i\}_{i=1}^n$ with $\sum_{i=1}^n \hat{p}_i = 1$, maximize the log-likelihood

$$\log L = C + \sum_{j=1}^n \log p_j + \log \int_0^\infty [\sum_{j=1}^n p_j e^{-tx_j}]^{N-n} g_n(t)dt\,, \tag{9}$$

and C in a constant. The values $\hat{p}_1, \ldots, \hat{p}_n$ that maximize (9) have to be determined numerically. We provide below a simple algorithm, which is an application of the EM algorithm (Dempster, Laird and Rubin, 1977), for determining the $\hat{p}_i$'s. Details can be found in Nair and Wang (1987).

Step 1. Start with a set of initial values $\{p_i^{(0)}\}_{i=1}^n$ with $\sum_{i=1}^n p_i^{(0)} = 1$;

Step 2. For any $j \geq 0$, given $\{p_i^{(j)}\}_{i=1}^n$ compute $\{p_i^{(j+1)}\}_{i=1}^n$ as

$$p_i^{(j+1)} = \frac{1}{N} + \left(1 - \frac{n}{N}\right) \frac{\int_0^\infty p_i^{(j)} e^{-tx_i} \left[\sum_{k=1}^n p_k^{(j)} e^{-tx_k}\right]^{N-n-1} g_n(t)dt}{\int_0^\infty \left[\sum_{k=1}^n p_k^{(j)} e^{-tx_k}\right]^{N-n} g_n(t)dt}\,. \tag{10}$$

Step 3. Repeat Step 2 until the likelihood stops increasing.

The above iteration leads to a maximizer of the likelihood function (9). If L has a unique maximum, then the EM algorithm will converge to the unique MLE $\{\hat{p}_i\}_{i=1}^n$.

Table II gives the values of the nonparametric MLE (NPMLE) of F for the R-M reef data for $N = 23$, 40, 100, and $+\infty$. The NPMLE for $N = 23$ is given by the standard edf (7). The NPMLE for the infinite population case ($N = \infty$) is

$$\hat{F}_{\infty}(y) = \sum_{i:x_i \leq y} \hat{q}_i \tag{11}$$

where $\hat{q}_i = x_i^{-1} / \sum_{k=1}^{n} x_k^{-1}$ (Cox, 1969 and Vardi, 1982).

Table II. Nonparametric pool size distributions for $N=23$, 40 100 and $+\infty$ for the Rimbey-Meadowbrook discovery data.

Pool Size	$\hat{F}_{23}(x)$	$\hat{F}_{40}(x)$	$\hat{F}_{100}(x)$	$\hat{F}_{\infty}(x)$
0.294	.043	.227	.344	.416
0.587	.087	.352	.524	.625
0.954	.130	.435	.637	.753
2.150	.174	.482	.691	.810
2.612	.217	.524	.737	.857
3.403	.261	.560	.773	.893
4.658	.304	.592	.802	.919
6.580	.348	.621	.824	.938
8.596	.391	.648	.842	.952
14.222	.435	.673	.856	.960
14.653	.478	.699	.870	.969
14.655	.522	.724	.883	.977
15.039	.565	.750	.897	.985
20.086	.609	.775	.909	.991
27.562	.652	.800	.920	.996
110.960	.696	.825	.930	.997
142.900	.739	.850	.940	.998
169.610	.783	.875	.950	.9985
290.830	.826	.900	.960	.9989
351.490	.870	.925	.970	.9992
366.660	.913	.950	.980	.9996
765.890	.9565	.975	.990	.9997
1295.400	1.0000	1.000	1.000	1.0000

Note from Table II that as N increases, the NPMLE's put more mass at smaller values. Also, note that the $\hat{p}_i$'s associated with the larger x_i's are close to $1/N$. This is intuitively reasonable since the larger pool sizes are more likely to have been discovered and the remaining $N-n$ undiscovered pools are likely to be the smaller ones. A grouped size distribution based on the NPMLE for $N=40$ is given in Table III.

Table III. A grouped size distribution based on NPMLE for $N=40$

Size Interval (MM bbl)	No. of Pools Discovered	Undiscovered	Total
0 - 2	3	14	17
2 - 4	3	2	5
4 - 8	2	1	3
8 - 16	5	0	5
16 - 64	2	0	2
64 - 128	1	0	1
128 - 256	2	0	2
256 - 512	3	0	3
512 - 1024	1	0	1
1024 and over	1	0	1
total	23	17	40

This is obtained by computing the total probability mass of each class and multiplying by 40 to get the total number of pools in each class. The numbers are rounded so that the total number of pools adds up to 40. This table suggests that the largest size of the remaining pools is in the size class of 4 to 8 million barrels of oil in-place.

The advantage of the NPMLE is that it does not require any distributional assumptions. However, it assigns mass only to the observed data. It assumes, for example, that the smallest undiscovered pool is no smaller than the smallest discovered one — a clearly unrealistic situation. It is preferable to choose an appropriate parametric distribution, estimate the parameters and use this to predict the various quantities of interest. The usual methods for calculating the critical values for formal goodness-of-fit test procedures such as chi-square or Kolmogorov tests do not apply here since the data observed under the successive sampling model (2) do not constitute a random sample from F. However, informal graphical goodness-of-fit procedures such as the quantile-quantile (Q-Q) plot can still be used provided we have a nonparametric estimate of F. See Chambers et al. (1983) for a discussion and interpretation of Q-Q plots. We illustrate the construction of a log-normal Q-Q plot since it is the most commonly used parametric distribution in petroleum resource evaluation. Similar plots can be constructed for assessing the adequacy of fit for other distributions such as the Weibull.

Let $x_{(i)}$ be the ith ordered observation, and for $i=1, ..., n$, let

$$u_i = \frac{\hat{F}_n(x_{(i)}) + \hat{F}_n(x_{(i-1)})}{2} . \tag{12}$$

Define $\hat{F}_n(x_{(0)})=0$ in (12). Then the log-normal Q-Q plot is a plot of $\Phi^{-1}(u_i)$ against $\log x_{(i)}$, $i=1, ..., n$, where Φ is the standard normal cumulative distribution function. If the underlying distribution is log-normal with parameters μ and σ^2, the plot will be approximately linear with slope σ and intercept μ. A nonlinear configuration would suggest that a log-normal model is unreasonable. In this case, the shape of the plot can be used to infer what alternative model may be appropriate (see Chambers et al., 1983). Note that the usual Q-Q plot is based on the standard edf (7) with $u_i=(i-\frac{1}{2})/n$, $i=1, ..., n$. However, this plot is inappropriate when the data are generated by the successive sampling model (2).

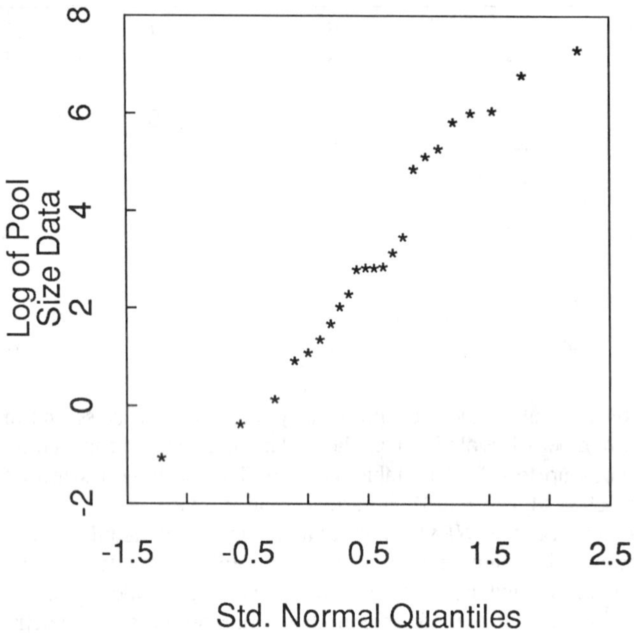

Figure 4. Log-normal quantile-quantile (Q-Q) plot of the Rimbey-Meadowbrook discovery data using the nonparametric maximum likelihood estimates with N = 40.

Figure 4 shows the log-normal Q-Q plot for the R-M reef data, and it suggests that the log-normal assumption for the underlying pool size distribution is reasonable. Quick-and-easy estimates of μ and σ^2 can now be obtained from this plot by fitting a line, say by the method of least squares, and using the slope and intercept to estimate μ and σ respectively. Nair (1984) has examined the properties of such estimators in the random sampling situation. For $N=40$, these estimates are given by 1.30 and 8.79, respectively. Note that, as mentioned earlier, the estimate of μ is much smaller than the sample average of the data in Table I the estimate of σ^2 is much larger than the sample variance.

3. PARAMETRIC INFERENCE

In this section, we illustrate maximum likelihood estimation of the parameters of F for the log-normal case. We consider the conditional estimation of θ assuming that the population size N is known. The log-likelihood for this case is obtained by substituting in (6) the log-normal distribution with density

$$f(y \mid \theta) = (2\pi)^{-1/2} (\sigma y)^{-1} \exp\left\{-\frac{1}{2\sigma^2} (\log y - \mu)^2\right\}, \quad y > 0, \tag{13}$$

where $\theta = (\mu, \sigma^2)$. The maximum likelihood estimates of μ and σ^2 can be obtained by numerically maximizing the likelihood using the Newton-Raphson procedure. However, this procedure is computationally expensive, and it requires good initial values. A simpler alternative is obtained by using the E-M algorithm. See Barouch, Kaufman and Nelligan (1983) and Nair and Wang (1986). For the log-normal case, this algorithm finds the fixed points of the pair of equations

$$\mu = \frac{n}{N} g + \left\{1 - \frac{n}{N}\right\} E\left[\log Z \mid \text{data};\ \mu, \sigma^2\right]$$

and

$$\sigma^2 = \frac{n}{N} S^2(\mu) + \left\{1 - \frac{n}{N}\right\} E\left[(\log Z - \mu)^2 \mid \text{data};\ \mu, \sigma^2\right], \tag{14}$$

where Z denotes anyone of the $(N-n)$ undiscovered pool sizes $Y_{n+1}, \ldots, Y_N$, and

$$g = \frac{1}{n} \sum_{j=1}^{n} \log x_j \quad \text{and} \quad S^2(\mu) = \frac{1}{n} \sum_{j=1}^{n} (\log x_j - \mu)^2 . \tag{15}$$

Note that if $n/N = 1$, i.e., all the pools have been discovered, the equations in (14) gives the usual MLE of μ and σ^2 under random sampling. On the other hand, if n/N is close to 0, i.e., the number of pools in the population is very large and we observe only a tiny fraction of it, the MLE's can be approximated by

$$\hat{\mu} = g - S^2(g) \quad \text{and} \quad \hat{\sigma}^2 = S^2(g). \tag{16}$$

This follows from the fact that in the infinite population case (n fixed and $N \rightarrow \infty$) the size-biased distribution itself is log-normal with parameters $\mu + \sigma^2$ and σ^2 (Meisner and Demirmen, 1981). For the intermediate cases with $0 < n/N < 1$, the MLE's have to be evaluated numerically. The computational aspects of this procedure are much more complex than the simple appearance exhibited by the pair of equations (14). The numerical difficulty associated with the superpopulation SS discovery model has always been the main obstacle in using this model as a standard working tool in petroleum resource evaluation.

See Nair and Wang (1986) for maximum likelihood estimation of the parameters for general F and for the multivariate case where magnitudes are defined as a function of various deposit attributes such as the one given by (4). Computational issues in deriving the MLE's are also discussed there. The random sampling estimates for μ and σ^2 provide good initial estimates for the EM algorithm. Also, the EM algorithm provides a substantial reduction in the amount of computations required for each iteration. However, the speed of convergence is still slower than the NR algorithm. This is not surprising since the NR algorithm has a quadratic convergence rate whereas the EM is only a linear procedure. By combining the two algorithms, greater improvement may be obtained.

Table IV. Maximum likelihood estimates of μ and σ^2 for different values of N of the Rimbey-Meadowbrook reef play

N	$\hat{\mu}$	$\hat{\sigma}^2$	$\log L$
23	3.04	5.32	-145.35
25	2.66	6.63	-145.40
30	1.95	8.60	-145.68
35	1.40	9.91	-145.83
40	0.93	10.93	-145.93
50	0.17	12.49	-146.04
60	-0.43	13.69	-146.11
70	-0.94	14.67	-146.15
80	-1.38	15.48	-146.18

For the R-M reef data, Table IV gives the MLE's for different values of N. Note that as N increases, $\hat{\mu}$ decreases and $\hat{\sigma}^2$ increases. These trends are intuitively reasonable because with a large value of N, the SS model implies that more smaller pools remain to be discovered. Consequently, the average pool size would be smaller and their variability would be larger. Note that the $\hat{\mu}$ and $\hat{\sigma}^2$ values for $N=40$ are quite different from those reported in Section 2 obtained by fitting a LS-line to the Q-Q plot in Figure 4. The initial estimates used for all the EM iterations shown in Table IV are $\tilde{\mu}=5$ and $\tilde{\sigma}^2=1$. These are quite different from the MLE's and, with these estimates, the NR algorithm would fail to converge in many of the cases considered. The convergence criteria were $||\theta_{new} - \theta_{old}|| \leq .0001$ and $||\log L_{new} - \log L_{old}|| \leq .00001$. Additional maximum likelihood analyses of the R-M reef data can be found in Lee and Wang (1986).

4. SIMULTANEOUS ESTIMATION AND PREDICTION

In the previous section, we illustrated how the MLE's of μ and σ^2 for the log-normal distribution can be obtained when the number of deposits N in the play is known. In this section, we consider the simultaneous estimation of μ, σ^2 and N. As before, the method can be extended to other distributions. We also show how to predict i) the total remaining potential and ii) the number of deposits remaining to be discovered in different size classes for the R-M reef population in Central Alberta.

There are many approaches that utilize the discovery data to estimate $\theta=(\mu, \sigma^2)$ and N simultaneously. One approach would be a 3-dimensional grid search of the likelihood function. Alternatively, we can compute $\hat{\theta}(N)$ for different values of N and then determine the $\hat{N}$ that maximizes the likelihood profile of N. For our discovery data, we get $\hat{N}$ as 23 (see column 4 of Table IV), and $\hat{\theta}=\hat{\theta}(23)=(3.04, 5.56)$, which is unrealistic and is known to be incorrect as there have been further discoveries since. The above approaches are computationally expensive, and the likelihood profile is usually uninformative as the profile is rather flat as a function of N. It frequently produces an estimate $\hat{N}=n$, that is all the pools have been discovered. Kaufman (1986) proposes a moment-matching method to estimate N, and then estimate θ conditional on this N. This method requires a splitting of the discovery data into an "early" part and a "late" part and the estimator appears to be sensitive to how the data are split.

An alternative to the above approaches is to postulate that the number of pools N in the play also has a superpopulation distribution $P(N \mid \gamma)$, indexed by a parameter γ, and that conditional on N, the magnitudes are distributed as described in Sections 2 and 3. We can then simultaneously estimate the parameters $\theta=(\mu, \sigma^2)$ and γ. The superpopulation distribution for N may be interpreted as a model describing a random mechanism as to how N is generated or it may be thought of as a prior distribution describing the subjective beliefs of the geologists. Nelligan (1980) assumed N is a Poisson random variable with mean γ and computed the MLE's by a 3-dimensional grid search of the likelihood function. Nair and Wang (1987) show how the MLE's can be obtained in general using the E-M algorithm. For the log-normal — Poisson superpopulation distribution, the algorithm finds the fixed point of three equations consisting of the pair of equations given by (14) with N replaced by γ and the following equation

$$\gamma = n + \gamma \frac{\int_0^\infty \phi(t;\theta)\, e^{-\gamma(1-\phi(t;\theta))}\, g_n(t)dt}{\int_0^\infty e^{-\gamma(1-\phi(t;\theta))}\, g_n(t)dt} \tag{17}$$

with $\phi(t;\theta)$ being the Laplace transform of the Y_i's. Clearly, the MLE $\hat{\gamma}$ satisfies $\hat{\gamma}>n$. By substituting the MLE's $(\hat{\theta}, \hat{\gamma})$ into the conditional distribution of N given the data, prediction intervals for N can be obtained. The total remaining potential can be estimated by

$$\hat{R} = (\hat{\gamma}-n)\, E[Z \mid \text{data}; \hat{\mu}, \hat{\sigma}^2, \hat{\gamma}]\,. \tag{18}$$

Similarly, the remaining number of deposits in the size class $(a, b]$ can be estimated by

$$\hat{N}(a, b) = (\hat{\gamma}-n)\, P(a<Z\leq b \mid \text{data}; \hat{\mu}, \hat{\sigma}^2, \hat{\gamma})\,. \tag{19}$$

and the largest undiscovered deposit can be estimated by

$$\hat{Y}_{(1)} = E[\max(Y_{n+1}, \ldots, Y_N) \mid \text{data}; \hat{\mu}, \hat{\sigma}^2, \hat{\gamma}]\,. \tag{20}$$

Each of the quantities (18), (19) and (20) are readily computable once the MLE's of θ and γ have been obtained. Estimating the standard errors of these estimators appears to be a difficult problem.

For the R-M data, the MLE's are $\hat{\mu}=1.86$, $\hat{\sigma}^2=8.66$ and $\hat{\gamma}=31.2$. The 99% prediction interval for N is (23, 41] with a posterior mode of 30 pools in play. The total remaining potential is estimated to be $\hat{R}=6.11$ million barrels of oil in-place. Table V gives the MLE's of the numbers of pools remaining to be found in different size classes. We see that there are about 7 more pools to be found in the range of 0 to 2 million barrels and possibly another one in the size class between 2 to 8 million barrels of oil in-place. These estimates are quite conservative compared to those found in Lee and Wang (1986). Our purpose here, however, is not to re-assess the potential of the R-M play but only to illustrate the basic techniques.

Table V. A grouped size distribution based on the MLE's $\hat{\mu}=1.86$ $\hat{\sigma}^2=8.66$ and $\hat{\gamma}=31.2$ for the Rimbey-Meadowbrook discovery data

Size Interval (MM bbl)	Expected No. of Pools		
	Discovered	Undiscovered	Total
0 - 2	3	7.43	10.43
2 - 4	3	.53	3.53
4 - 8	2	.20	2.20
8 - 16	5	.04	5.04
16 - 64	2	.01	2.01
64 - 128	1	0	1.00
128 - 256	2	0	2.00
256 - 512	3	0	3.00
512 and over	2	0	2.00
total	23	8.21	31.21

5. CONCLUDING REMARKS

We have illustrated some procedures for predicting the undiscovered oil and gas resources under a superpopulation successive sampling model for the discovery data. We have taken deposit magnitude to be the pool size itself in this paper. Generalizations as indicated in (3) are possible and are discussed in Nair and Wang (1986). The multivariate SS discovery model would provide a more comprehensive analysis of a set of discovery data including the associated attribute values. Although we have given explicit algorithms for computing the maximum likelihood estimates, their numerical evaluation is still involved. Therefore, further research on simple and reasonably efficient methods of estimation would be worthwhile. Results on interval estimation, especially prediction intervals, are also needed.

REFERENCES

1. Andreatta, G. and Kaufman, G. M. (1986), ''Estimation of Finite Population Properties when Sampling is Without Replacement and Proportional to Magnitude'', *J. Amer. Stat. Assoc.*, 81, 657-666.
2. Arps, J. J. and Roberts, T. G. (1958), ''Economics of Drilling for Cretaceous Oil Production on the East Flank of the Denver-Julesberg Basin'', *Bull. AAPG*, 42, 2549-2566.
3. Bloomfield, P., Deffeyes, K. S., Watson, G. S., Benjamini, Y., and Stine, R. A. (1979), ''Volume and Area of Oil Fields and Their Impact on Order of Discovery'', Resource Estimation and Validation Project, Department of Statistics and Geology, Princeton University.
4. Barouch, E. and Kaufman, G. M. (1976), ''Probabilistic Modelling of Oil and Gas Discovery'', in *Energy: Mathematics and Models*, ed. F. Roberts, Philadelphia: SIAM, 248-260.
5. Barouch, E. and Kaufman, G. M. (1977), ''Estimation of Undiscovered Oil and Gas'', in *Proceedings of Symposia in Applied Mathematics*, 21, 77-91.
6. Barouch, E., Kaufman, G. M. and Nelligan, J. (1983), ''Estimation of Parameters of Oil and Gas Discovery Process Models Using the Expectation — Maximization Algorithm'', in *Energy Modelling and Simulation*, eds. A. S. Kydes et al., North-Holland, Amsterdam, 109-117.
7. Cleveland, W. S. (1979), ''Robust Locally Weighted Regression and Smoothing Scatterplots'', *J. Amer. Stat. Assoc.*, 74, 829-836.
8. Cox, D. R. (1969), ''Some Sampling Problems in Technology'', in *New Developments in Survey Sampling*, eds. N. L. Johnson and H. Smith Jr., Wiley-Interscience, New York, 506-527.
9. Chambers, J. M., Cleveland, W. S., Kleiner, B. and Tukey, P. A. (1983), *Graphical Methods for Data Analysis*, Duxbury Press, Boston.
10. Dempster, A. P., Laird, N. M. and Rubin, D. B. (1977), ''Maximum Likelihood from Incomplete Data via the EM Algorithm (with discussion)'', *J. Roy. Statist. Soc.*, *B*, 39, 1-38.
11. Drew, L. J., Schuenemeyer, J. H. and Root, D. H. (1980), ''Petroleum Resource Appraisal and Discovery-Rate Forecasting in Partially Explored Regions: Part A, An Application to the Denver Basin'', U.S. Geological Survey Professional Paper no. 1138.
12. Gordon, L. (1981), ''Successive Sampling in Large Finite Populations'', *Annals of Statistics*, v. 11, no. 2, 702-706.
13. Gordon, L. (1983), ''Estimation for large successive samples with unknown inclusion probabilities'', Preprint.

14. Johnson, N. L., and Kotz, S. (1970), *Continuous Univariate Distributions - 1* , Wiley, New York.

15. Kaufman, G. M., Balcer, Y. and Kruyt, D. (1975), "A Probabilistic Model of Oil and Gas Discovery", in Methods of Estimating the Volume of Undiscovered Oil and Gas Reserves, Studies in Geology #1, ed. I. D. Haun, Amer. Assoc. Pet. Geol., 113-142.

16. Kaufman, G. M. (1986), "Finite Population Sampling Methods for Oil and Gas Resource Estimation", in Oil and Gas Assessment — Methods and Applications, Studies in Geology #21, ed. D. D. Rice, Amer. Assoc. Pet. Geol., 43-53.

17. Lee, P. J. and Wang, P. C. C. (1985), "Prediction of Undiscovered Oil and Gas Pool Sizes When Discovery Record is Available", *Jour. Math. Geol.*, 17, no. 2, 95-113.

18. Lee, P. J. and Wang, P. C. C. (1986), "Evaluation of Petroleum Resources from Pool Size Distributions", in Oil and Gas Assessment — Methods and Applications, Studies in Geology #21, ed. D. D. Rice, *Amer. Assoc. Pet. Geol.*, 33-42.

19. Meisner, J. and Demirmen, F. (1981), "The Creaming Method: A Bayesian Procedure to Forecast Future Oil and Gas Discoveries in Mature Exploration Provinces", *J. Roy. Statist. Soc.*, *A*, 144, 1-31.

20. Nair, V. N. (1984), "On the Behavior of Some Estimators from Probability Plots", *J. Amer. Stat. Assoc.*, 79, 823-836.

21. Nair, V. N. and Wang, P. C. C. (1986), "Estimating the Superpopulation Parameters of a Successively Sampled Finite Population", Draft.

22. Nair, V. N. and Wang, P. C. C. (1987), "Nonparametric Estimation of the Superpopulation Distribution Under Biased Sampling from a Finite Population", in prep.

23. Nelligan, J. D. (1980), Petroleum Resources Analysis Within Geological Homogeneous Classes, Ph.D. Thesis, Department of Mathematics and Computer Sciences, Clarkson College of Technology, Potsdam, New York (1980).

24. Smith, J. L. and Ward, G. L. (1981), "Maximum Likelihood Estimates of the Size Distribution of North Sea Oil Fields", *Jour. Math. Geol.*, 13, no. 5, 399-413.

25. Vardi, Y. (1982), "Nonparametric Estimation in the Presence of Length Bias", *Ann. Statist.*, 10, 616-620.

AUTOMATED INTEGRATION OF MINERAL RESOURCE DATA BY IMAGE PROCESSING AND ARTIFICIAL INTELLIGENCE

Andrea G. Fabbri and Tonis Kasvand*
Canada Centre for Remote Sensing, Ottawa, Canada K1A 0Y7
*National Research Council of Canada, Ottawa, Canada K1A 0R8

ABSTRACT. Resource assessment involves an overwhelming amount of information of diverse nature and format. One of the main challenges is trying to represent a problem in a computational form either qualitatively or quantitatively. This paper describes the application of image processing and artificial intelligence for the integration of data from maps and other forms of digital imagery.

Particular attention is given to the interface between image processing and expert systems. The computer representation and processing of images, expert-system rules and their associated weights, are exemplified to introduce the requirements of more comprehensive and "humane" systems in which high-level queries bring integration from the simple standardization of data formats to intelligent interaction between multiple data sets. Model-driven and data-driven approaches are discussed.

1. INTRODUCTION

The problem of resource assessment characteristically involves overwhelming amounts of information of diverse nature and format. Geology, geophysics, geochemistry, geographical data, aerial photographs, archival information on mineral prospects and drilling sites, the annotations of surveys for sampling lithologies, and structures, are examples of the most common data in economic geology work. Hand drawn maps and sketches, contour maps and their original digital data, computer files with location of samples and of their synthetic description either from field notes or from coding forms, are typical forms of data.

In addition, much data is provided by remote sensing, from satellites and from aerial surveys with optical and radar sensors. Satellites such as SPOT can be activated at will to scan a desired area. This is done only if information is needed for that area, otherwise it would be impossible to analyze all the information received. The resolution of satellite imagery can range between 1.4 km and 10 m per pixel. In airborne radar and aerial imagery, resolutions below 1 m can

C. F. Chung et al. (eds.), Quantitative Analysis of Mineral and Energy Resources, 215–236.

be easily achieved. Obviously, so much information cannot be handled without the usage of computers. Even then, the information has to be integrated in physical format, resolution, spatial correspondence and logical relationship in terms of natural processes. Furthermore, the main challenge to solving any problem is to represent it in a computational form, both quantitatively and qualitatively.

The following quotation from Varnes (1974) points at the problem complexities of map making and analysis in the earth sciences. Such problems are even greater if we consider remotely sensed imagery as ancillary to maps.

> "A map is a spatial classification that transmits information about features at or near the earth's surface for a defined purpose.... One of the most fundamental problems in the construction and use of maps is the isolation and identification of those attributes that are essential to the definition of map units.
>
> Maps are both prepared and modified through four principal types of operations: generalization, selection, addition or superposition, and transformation.... Examination of the logic, or lack of it, in maps is aided by various kinds of plot and graphical analyses. Among the more useful and easily constructed are the data matrix, tree of logic, table of logical division, and three-dimensional map unit matrix.... A look at the future suggests an increasing need for precise information and growing sophistication in acquiring and processing the data" (Varnes, 1974, p. 1).

Many different methods have been used for systematically quantifying and integrating spatially-distributed data for multivariate statistical analysis. Agterberg et al. (1972), Fabbri, Divi and Wong (1975) and Fabbri (1975) exemplified how systematically quantified regional geological data can be useful for a variety of statistical applications. Those authors used square grid patterns (equal-area cells) produced according to the Universal Transverse Mercator (UTM) geographic projection. Special-purpose geological maps were prepared that were designed to reveal metallogenic features. Transparencies of the grids were placed over the maps to be coded, and counts were made to express the areal proportions of all rock units occurring in each square cell of the grid. Similar procedures were employed for other ancillary maps such as geophysical contour maps and mineral-occurrence distribution maps.

The target of those studies was the spatial analysis and it became clear that much data preparation and manipulation (such as getting acquainted with the data after digitization, transformation of the variables coded, combination of variables and creation of new variables) was required by the statistical methods. Recent tools for data manipulation and statistical analysis are interactive systems such as CHARAN (McCammon et al. 1983, 1984) and SIMSAG (Chung, 1983).

Kasvand, Fabbri and Nel (1981) and Agterberg et al. (1981) used image processing methods in a more systematic procedure to digitize map data. The map-unit boundary-lines were digitized into thin lines consisting of continuous chains of black pixels or picture elements of desired resolution. All areas enclosed by such lines were identified and assigned map unit labels corresponding to the legends. Images of up to one million pixels were produced from different maps covering the same area on the ground. Pixel-to-pixel correspondence was assured for each image. Either image processing methods could then be used for analysis or, by regridding, new data sets could be computed for cells that were then used for multivariate statistical analysis. Interactive systems for image processing such as GIAPP (Fabbri, 1980; 1984) or for statistical analysis such as CHARAN and SIMSAG, mentioned earlier, were developed to derive "optimal" exploration strategies interactively. Other software and hardware tools for image analysis in geology have been reviewed by Fabbri (1986).

The expert system PROSPECTOR is an approach to data integration developed at Stanford Research Institute International (SRI) during 1976-1983. It had several methods for taking into account spatial locations of geological field data and ore deposit models. In its drilling-site-selection mode of interaction a facility for using graphical input was described by Duda et al. 1978 and Duda, 1980. More recent developments about this system speak in favour of its application to regional geologic environments and at district and outcrop levels (McCammon, 1984). PROSPECTOR was programmed to classify spatial data in the form of square cells for images of up to 128 x 128 pixels. The use of artificial intelligence, AI, in computer vision and the related fields of remote sensing and geographic information systems was reviewed by Fabbri, Fung and Yatabe (1986).

Can all these experiences and tools be brought together and made usable to the economic geologist rather than being just attractive toys for computer fans? The answer is yes, but it naturally requires much work and problem understanding. In addition, the user must be able to communicate with the computer in human language, via images, maps, plots, photos, tables, diagrams, etc., and not in some unfriendly computer jargon. Table 1 shows examples of computer interactions in their historical development of increasing user friendliness and sophistication.

2. WHY TO USE IMAGES AND HOW TO REPRESENT THEM IN THE COMPUTER

In many instances data come in the form of images, maps, aerial photographs, remotely sensed data, including both optical and radar sensors, that have an original pictoral representation. Some systematically collected data such as gravity measurements, aeromagnetic recordings, and geochemical samples are often represented as contour maps. The contouring procedure is analogous to the one needed to compute a digital image representation. Thus, much of the data is available in image form (i.e., 2D). Other spatially distributed data, such as tabulated survey

Table 1: Examples of Computer Interaction.

[A] Simplest interaction: 1-Routines for entering: integer, real and character parameters. 2-Routines for calling separate subprograms once the parameters have been selected. 3-Prompts requesting entry of mathematical formulas in alphanumeric characters to be interpreted and compiled (online compilation).
[B] Interaction with menus: 1-Menus requesting a numeric or alphanumeric choice, or 2-Menus with line prompts to point at with a cursor for choosing the processing algorithms.
[C] Graphic menus: 1-Graphic displays with boxes and prompts to point at with cursor (with or without subsequent entry of keyboard characters), 2-Mouse activated windows with menus and symbols (icons) to point at.
[D] Icon assisted processing: 1-Icon assisted navigation through processing algorithms by an image processing language. 2-Development of macroinstructions for complex icons and image algebra. 3-Parsing of natural language queries into picture operations.

information for population density, land use or any point data related to a geographical location, can be converted to image form if it is wanted.

Ideally, one would like to put a map or a prepared mineral sample in front of a scanning device connected to the computer and push a button to initiate the process. Given adequate equipment, the scanning aspect is feasible, but the resultant digitized image is not of much use for a computer if the original input material is not "simplified and interpreted before digitizing" and allows registration with all other data.

If the map is traced by an operator, the computer records a sequence of X-Y coordinates as the point of the stylus or the crosshair of the cursor is moved along some trajectory on the surface of the tablet. Thus, this method basically produces one-dimensional sequences (strings) of X-Y coordinate pairs or "vectors". Whatever other information is entered together with the vectors is at the moment immaterial.

Figure 1. A very generalized flowchart of the problem of representation.

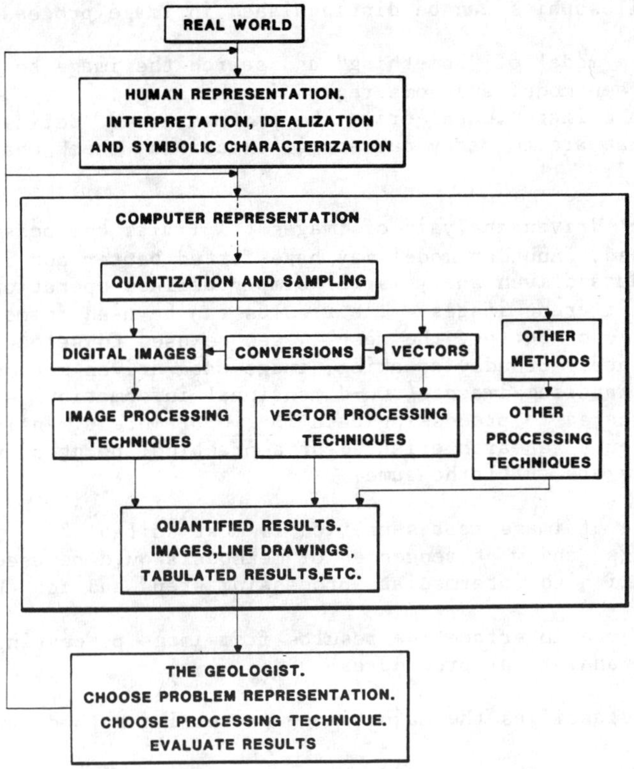

It is to be noted that the computer receives predefined data in a predefined form. The process should be made as convenient as possible from the human point of view, but there is much tedious work to be done. Entering the data introduces the possibility of different kinds of errors and the digitized data, have to be carefully checked and corrected. Consequently, the data-entry process must allow for human interaction for error correction. To the computer, raster images are only arrays of numbers: one number per pixel. The numbers are "individuals", since no relations have yet been established between pixels, if any exist at all. The main purpose of subsequent image processing is to detect or establish relationships between the pixels within one image or among images containing different information.

Whether the data are stored in the computer as (binary) images, a list of run-length codes, as "quadtrees" (Samet et al. 1984), or as sets of vector strings are in principle equivalent if registration and resolution are suitable. One form of data base can be converted to the other and back again, but the conversion may not be simple and significant information loss may be expected due to changes of spatial resolution. Data resolution cannot be increased during conversion.

3. HOW TO COMPUTE WITH IMAGES: ELEMENTS OF IMAGE PROCESSING

Two basic philosophies can be distinguished in image processing:

(1) We have a model of "something" and search the image to find best fit between model and some region in the image.
(2) We let the image (data) drive the process, i.e., define only operations that are to carry out specific actions, check what was found, and use it.

The model-driven analysis of images (1) limits the possibilities to the models used. Another model may have fitted better but it was not tried. The data-driven analysis (2) simply defines operations within and between different images. The results can be used in models and also as in (1), or better, the results can be used to select models that seem appropriate. Broadly speaking, image data-driven methods are more suitable for handling images (two-dimensional information) since these methods are geared to processing data in two or more dimensions.

For either a research-oriented or a practical point of view, the basic problems are still the same.

(a) What type of image representation is most suitable?
(b) What images and what sequences of methods should be used to obtain the solution to intermediate processing steps and for the whole problem?
(c) How does one interface the results from image processing to existing analytical procedures?

Table 2 classifies the major image processing methods and provides some examples.

How to select suitable sequences of operations is the major problem to the user not versed in image processing techniques. He does not even know what can be done, far less how to establish the necessary string of operations. This is a general problem of human understanding of any field. The best we can do is to let the computer advise and assist. Kasvand and Fabbri (1985) provided examples of image memories and processing loops of importance in geological data integration. Figure 2 shows some computational aspects in terms of types of processing required and the form of the input and output data for preprocessing and processing of binary and labeled images. The P in the illustration refers to "pipelineable" processes, i.e., computations that are particularly well suitable for special computers (pipeline and parallel machines) designed to perform at high speed for spatially distributed data.

Table 2. Major processing methods for digital images.

Establish "global" characteristics of an image or between images.	->Fourier transforms to detect spatial periodicities. ->Auto- and cross-correlation to establish spatial dependencies. ->Morphological transforms: global case (Serra, 1982). ->Statistics: one dimensional histogram, two dimensional and n-dimensional relationships, co-occurrences, contingency table analysis, higher order statistics. ->Pixel counts giving area measures and contour length
Local spatial operations within an image: single operations or iterative	->Detection of edges of areas. ->Erosions, dilatations, opening, closing, "hit-or-miss" transforms to filter for global processing (Serra, 1982, Fabbri, 1984). ->Distance computations: measures of distance between selected elements. ->Thinning to obtain thin line representations of boundaries, e.g., faults, dikes lineaments, etc. ->Error detections such as broken elevation contours or segments. ->Detection of junctions where lines cross or touch. ->Processing of thin lines on maps.
Labeling and relabeling operations within an image	->Coherence operations (topological continuity). ->Relabeling of pixels after classification (local operations that produce global results. ->Unique labels for areas, edges, junctions, phases, etc., which allow measurements of individual components in the image (faults, dikes, contacts, unconformities) to be used for global methods, statistics, object counting, etc.

4. THE "INSIDE" OF EXPERT SYSTEMS

Expert systems are, in their simplest form, based on the "if X and Y then Z" class of rules (production rules), the consequences of which are actions performed by the computer. The set of these rules is called an "inference network" that is used by programs termed "inference engines".

The variables (or predicates) X, Y, etc., in these rules can be classical logical variables, i.e., they are expressions of first order predicate calculus or Boolean algebra consisting of "AND" (conjunction), "OR" (disjunction), and "NOT" (negation). For example, "Z is true if X is true and Y is true or Q is false" or

"Z <<== X .AND. Y .OR. .NOT. Q".

This "hard" logic is too unrealistic in most practical circumstances. For instance, we may have a far more common situation when X is mostly true, Q is frequently false, etc. The classical approaches to such problems are termed "hypothesis testing" and "Bayesian probability calculation". The testing of hypotheses for and against a particular

Table 2. Continued.

Operations between images	->Boolean algebra: .AND., .OR., .NOT. and combinations. ->Fuzzy logic (non binary). ->Expert system type of logic with weights: fuzzy .AND.'ing and .OR.'ing (e.g., PROSPECTOR).
Processing from image to tabulated data as interface to existing statistical programs or packages	->Lists of areas sizes, shapes. ->Particle or object counts. ->Lists of edge lengths.
Mixed operations: image --> list --> image --> list --> etc.	->Establish sequential relationships between elements in image, n-step adjacencies. ->Detection of Markovian properties (e.g. in crystalline textures, Fabbri and Kretz, 1986).
Other types of processing	->Noise cleaning. ->Enhanching. ->Displays. ->Vector-to-raster conversion. ->Raster-to-vector conversion. ->Other type of image data encoding or decoding (also multiresolution).
Pattern recognition	->Multivariate statistics (clustering, discrimination, classification.
Computer vision	->AI-based techniques and expert systems.

conjecture requires the knowledge of, or estimation of the underlying probability distributions of the variables. Of course, in most situations, these distributions are not available and even their estimation is problematic. These will be described later for a particular expert system.

To "soften" the logic and to include statements about uncertainty, beliefs, and so on, numerous methods have been introduced. They have been discussed by Wise and Henrion (1985). In expert systems, those concepts are represented as weights for "presence" and "absence" of a particular piece of evidence which supports or denies the particular hypothesis. The composite of the "pros" and "cons" of all the evidence for a hypothesis is computed from certain formulas given by the same authors. The evidence consists of some form of "ground truth" given to the expert system by the user, or the evidence is a computed hypothesis from a lower level in the chain of reasoning. The evidence also carries a weight representing its significance.

Let us briefly consider the expert system for mineral exploration PROSPECTOR (Hart, Duda and Einaudi, 1978). In it, judgemental knowledge is represented by rules that can have the form of simple implicational statements. Rarely can any of the implications be made with complete certainty. For this reason the rules must specify the evidences, the

Figure 2. Computational aspects of processing binary and labeled images for geological applications. P indicates pipelineable processes.

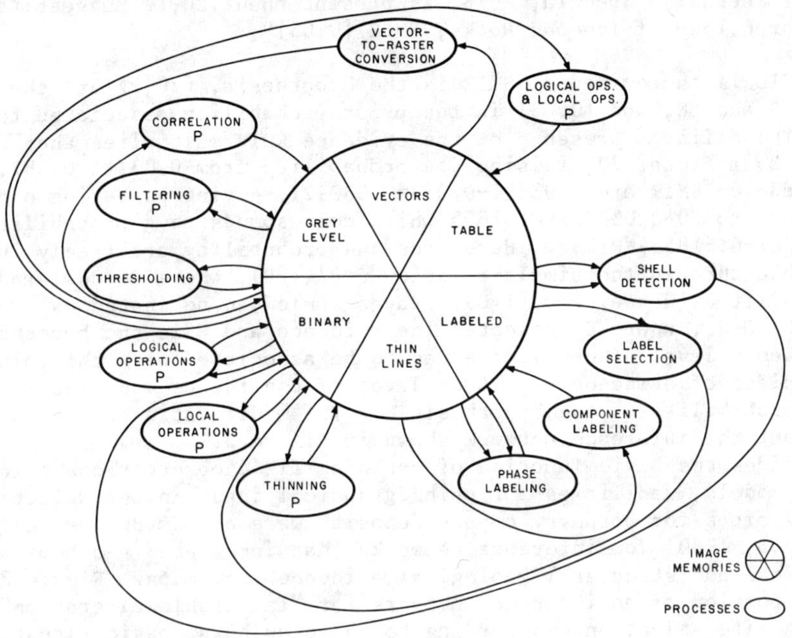

hypotheses, the implications and their strengths: a priori probability values are associated with evidences and with hypotheses that represent the state of knowledge when the implication was made. In PROSPECTOR, to account for uncertainty in both the evidences and the rules, a subjective probability can be associated with every relation and a pair of strength values with every rule (one for the observed presence of the evidence in that rule and one for the observed absence). A Bayesian inference scheme used in developing PROSPECTOR has been discussed by Duda, Hart and Nilsson (1976) and it was demonstrated in mineral-deposit model development by Duda, Gaschnig and Hart (1979). Rules for (a) multiple evidence, (b) logical combinations of evidences, and (c) their contextual relations were established.

In addition, the volunteering of new evidence with different weights updates the a priori probability of the consequent hypothesis by amounts that depend on the strengths of the rules. The implicational statements assume the form:

"if E then (to degree LS, LN) H",

where E is the evidence, H is the hypothesis, and LS, LN are the weights associated to the observed presence (likelihood sufficiency) and to the observed absence (likelihood necessity) of the evidence.

Let us consider the following statement in PROSPECTOR:

"If Intrusive Breccia, RCIB, is present then (20,1) Suggestive Morphology of Igneous Rocks, SMIR [0.03]",

where RCIB is the evidence, SMIR is the hypothesis, (20,1) are the values LS and LN, and [0.03] is the prior probability associated to SMIR. The definite presence of the evidence RCIB multiplies the "odds" of SMIR by a factor 20, raising its probability from 0.03 to 0.382. The prior odds on SMIR are 0.03/(1-0.03)=0.030927, giving posterior odds on SMIR equal to 20*0.030927=0.61855 which corresponds to a probability of 0.61855/(1+0.61855)=0.382. Here odds and probability are freely interchangeable through the simple relation O=P/(1-P), where O=odds, and P=probability. Hence, P=O/(1+O). Bayes' rule can be stated as $O(H|E)=LS*O(H)$, where E indicates the evidence and H is the hypothesis.

Often a level of confidence has to be associated with the volunteered piece of evidence E'. Such level of confidence modifies the a priori probability of the hypothesis H. The modification can propagate throughout the inference network shown in Figure 3.

Besides the basic function of matching field observations to ore-deposit models, facilities for using graphical input in the selection of drilling sites for porphyry copper deposits were added (Duda et al. 1978; Duda, 1980) for inference networks that integrated geochemical, geophysical and structural geological evidence from maps. Figure 3 shows a portion of an inference network for "favourable alteration" for drilling site selection. According to those authors, basic situations from digitized map-samples or cells are: (i) being within some region, or (ii) near some contour. For transition regions, a piecewise "favourability function" was used to gradually reduce the probability from 1 to 0 according to distances within a predefined range. Figures 4a and 4b show the digitized data for structures and for alteration for the Island Copper deposit in British Columbia, Canada (resolution is 128 cells by 128 cells). Figure 4c shows that the favourable structures, i.e., the zones of influence from fractures and from their intersections, have been identified by computing simple spatial transforms. Figure 4d shows a final favourability digital map of the area.

5. EXPERT SYSTEMS AND IMAGE PROCESSING

As illustrated above, the central idea of expert systems is to combine evidence for and against a particular hypothesis. The combinations may follow the rules of Bayesian probabilities, fuzzy logic, Dempster-Schafer belief functions, measures of belief MB and disbelief DB, etc., and merge these into certainty factors CF's. No unanimity has been reached on how to combine uncertainties, beliefs, or disbeliefs, as became apparent in a recent workshop on uncertainty and probability in artificial intelligence (see UCLA, 1985).

For digital image-processing, the uncertainties can be represented as 2-D mappings. The value of a pixel or its "gray level", i.e.,

Figure 3. Inference network for drilling site selection (simplified after Duda et al., 1978, p. 109). The hypothesis of favourable alteration for deposit types M1 or M2 is represented. In the boxes (termed spaces) are evidences and hypotheses. Arrows represent rules, broken arrows represent context-constraint rules (i.e., one evidence or hypothesis must be assessed before another one can be assessed), and values beside the arrows represent the weights for LS and LN, respectively. The value of a priori probability assigned to the hypothesis FAM1+M3 is 0.05. Intermediate spaces are for logical implications.

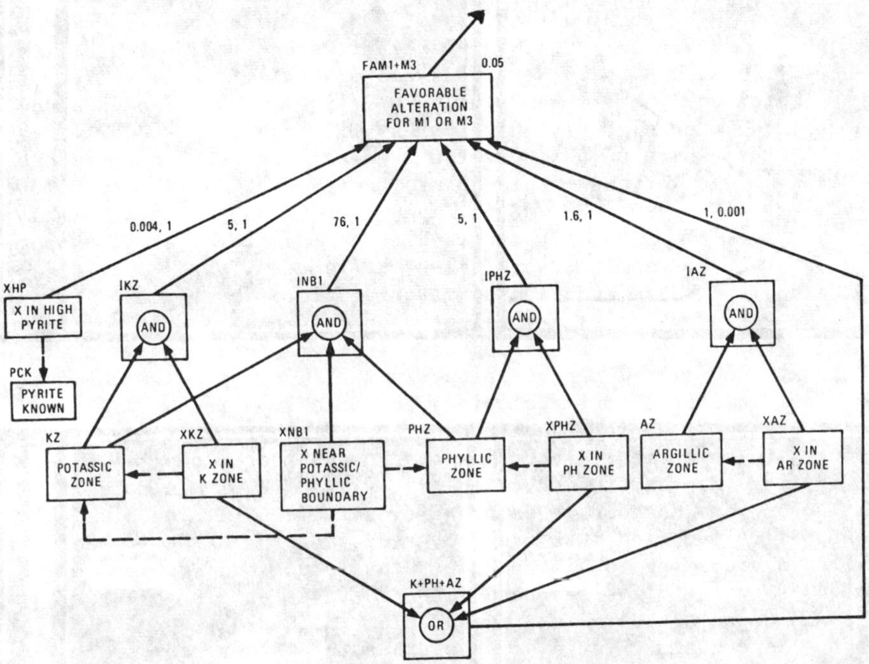

Ik(i,j) may be interpreted as evidence for or against a particular hypothesis: Ik(i,j) may be "the degree of membership of the pixel" according to fuzzy logic, or it may represent any quantitative or qualitative measure, where, k=1,2...., are the images, i=1,2...Im; j1,2,...,-Jm; and Im,Jm= horizontal and vertical dimensions of the image in columns and rows, respectively. The acquisition of the actual numerical values for the matrix Ik(i,j) is problem dependent.

The functions that weigh evidence, may be looked upon as special pixel-based operations between several registered images, which create a new image S(i,j). Thus, S(i,j)= S(Ika(i,j),Ikb(i,j),...), where, for instance, Ika = image of supporting evidence, Ikb = image of contrary evidence, s = image of certainty factors for a particular hypothesis, and i=1,2,..., j=1,2,..., k=1,2,.... In fuzzy logic, given the "fuzzy" variables f1 and f2, with membership functions F1 and F2, fuzzy 'AND' = min[F1(f1),F2(f2)], and fuzzy 'OR' = max[F1(f1),F2(f2)]. A fuzzy

Figure 4. Graphic displays for Island Copper study area, British Columbia, Canada. (a) the structures, (b) the alterations, (c) the favourable structures, and (d) the final favourability map (after Duda et al. 1978), pgs 119-122). The values between -5 and +5 represent the certainties as gray levels in the illustrations of monitor displays.

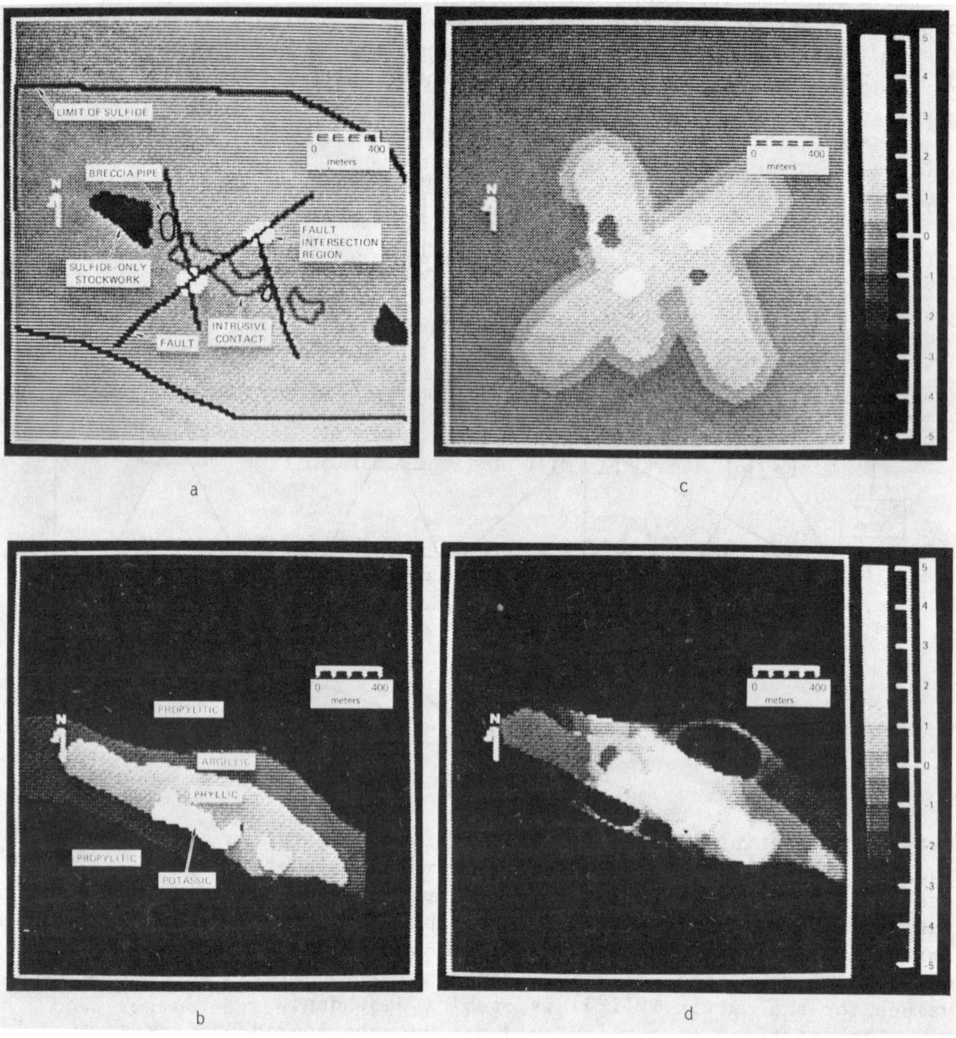

logical operation between two registered "gray level" images I1(i,j) and I2(i,j) becomes a pixel-by-pixel operation. Thus, fuzzy 'AND' between two images I1 and I2 is Ie(i,j)= min[I1(i,j), I2(i,j)], i=1,2,..., j-=1,2,... . Boolean operations ('AND', 'OR', etc.) on binary (0,1) images are a special case of fuzzy operations. Hence, conceptually there is no difference between the processing of single numbers of evidence and processing matrices of evidence. These calculations may be considered as special image processing operations. Weights for evidence can be represented as matrices of multipliers. Of course, the problem of determining the original probabilities and weights remains the same as in all expert systems.

In practice there are several problems to overcome. Firstly, the images in remote sensing are very large, for example, one LANDSAT image consists of approximately 3000 by 4000 elements (Im=3000, Jm=4000). Many inputs to the inference network (the "leaves") are images and the total number of input images can become quite large if we consider remote sensing and ancillary data. To simplify the processing, the images have to be registered and rectified, i.e., to be in the same scale, resolution, map position and orientation. Secondly, at each node of the inference network (hypothesis) an image of hypothesis is created. The storage requirements in the computer become large, even though the processing could be synchronized so that the images can be read sequentially from mass storage devices, e.g., one row of pixels at a time. Thirdly, large images will take a long time to process.

However, images usually contain large areas where the contents or the information is the same or similar. Furthermore, the information in an image is always quantized in amplitude as well as spatially. For example, the areas between any two contour lines in a contour map of elevations can be considered to be one of the same altitude, even though they seldom are in reality. To reduce computation time and to reduce the amount of data that has to be processed, the input images should be quantized to acceptable levels for a given problem. The following procedure is suggested to overcome some of the problems discussed.

(a) Quantize only the amplitudes ("gray levels") in the registered input images. For simplicity, consider each quantization level (q=1,2,...) as an independent binary image. This creates a set of images Ik'q(i,j), i,j,k as above.

(b) Label or tag each pixel in Ik'q with a symbol that gives the quantization level. Usually the labels are numbers and the quantization levels themselves serve as the labels.

(c) Form an image of intersections Ir(i,j) of all the labeled input images Ik'q. Assign a specific label to each pixel in Ir which corresponds to each unique combination of labels in, Ik'q(i,j), i.e.,

$$I2'(i,j) \wedge I2'(imj) \wedge \ldots \wedge Ik'(i,j) \wedge \ldots \; ; \; [\wedge = \text{conjunction}].$$

(d) Form a list or table T of labels in Ir, including the images processed, their labels, and other information, such as size of the areas involved.

(e) The number of distinct labels in Table T corresponds to the number of times the expert system has to be run, since each table entry contains a different combination of input parameters. The results of the hypotheses may also be stored in T.

Figure 5. (a) labeled image of intersections between images of soil class S in (b), ground-water condition I in (c) and lithology A,B,C in (d), modified after Varnes (1974, p. 3, Fig. 1). Labels in (a) are: S, A, B, C, SA, SB, SC, SI, AI, CI, #=SAI, *=SBI and +=SCI. Picture coordinates are added on top and to the left of the image in (a). The input images are of size Im=38, Jm=34, i.e., 1292 pixels, requiring 1292 runs of the expert system. However, by using the "tabular approach" (see Table 3), only 12 runs are required.

```
      1 2 3 4 5 6 7 8 91011121314151617181920212223242526272829303132333435363738
     ----------------------------------------------------------------------------
  1| . . . . . . . . . . . . . . . . . . . . . . . . . . . . . . . . . . . . . .
  2| . . . . . . . . . . . . . . . . . . . . . . . . . . . . . . . . . . . . . .
  3| . . . . . . . . . . . . . . . . . A A . . . . . . . . . . . . . . . . . . .
  4| . . . . . . . . . . . . . . . . A A A . . . . . . . . . . . . . . . . . . .
  5| . . . . . . . S . . . . . . . A A A A A . . . . . . . . . . . . . . . . . .
  6| . . . . . . S S S S . . . . A A A A A A . . . . . . . . . . . . . . . . . .
  7| . . . . . S S S S S . . . A A A A A A A . . . . . . . . . . . . . . . . . .
  8| . . . . . S S S S S S . A A A A A A A A . . . . . . . . . . . . . . . . . .
  9| . . . . . S S S S S S A A A A A A A A AAI I I I I . . . . . . . . . . . . .
 10| . . . . . S S S S S SSA A A A A A AAIAIAI I I I I I . . . . . . . . . . . .
 11| . . . . . S S S S SSASA A A AAIAIAIAIAIAIAI I I I I . . . . . . . . . . . .
 12| . . . . . S S S S SSASASA AAIAIAIAIAIAIAIAIAI I I . . . . . . . . . . . . .
 13| . . . . . S S S S SSASA # #AIAIAIAIAIAIAIAIAIAI I . . . . . . . . . . . . .
 14| . . . . . S S S S SSA # # #AIAIAIAIAIAIAIAIAIAI A . . . . . . . . . . . . .
 15| . . . . . S S S S S # # # # #AIAIAIAIAIAIAIAIAI A A . . . . . . . . . . . .
 16| . . . . . S S S SSISI # * * * *BIBIBIBIBIBIBI B B B B . . . . . . . . . . .
 17| . . . . . . S S SSISI * * * * * *BIBIBIBIBI B B B B B B . . . . . . . . . .
 18| . . . . . . S S SSISI * * * * * * * *BIBI B B B B B B B B . . . . . . . . .
 19| . . . . . . S S SSISISI * * * * * * *SBSB B B B B B B B B C . . . . . . . .
 20| . . . . . . . S SSISISISI * * * * *SBSBSBSBSB B B B B C C C . . . . . . . .
 21| . . . . . . . S S SSISISI * * * *SBSBSBSBSBSBSB B C C C C C C . . . . . . .
 22| . . . . . . . S S S SSISISISBSBSBSBSBSBSBSBSCSCSC CCICICICICI I I . . . . .
 23| . . . . . . . . S S S S S S SSBSBSBSBSBSCSC + + + +CICICICICI I I I I . . .
 24| . . . . . . . . S S S S S S S SSBSBSCSCSC + + + + + +CICICICI I I I I . . .
 25| . . . . . . . . . S S S S S S S SSCSCSC + + + + + + +CICICICI I I I I . . .
 26| . . . . . . . . . . S S S S S S S SSC + + + + + + + +CICICICI I I I . . . .
 27| . . . . . . . . . . S S S S S S S S SSC + + + + + + +CICICI I I I I . . . .
 28| . . . . . . . . . . . S S S S S S S SSISISISISI + + +CI I I I I . . . . . .
 29| . . . . . . . . . . . . S S S S S S SSISISISISISISI I I I I . . . . . . . .
 30| . . . . . . . . . . . . . S S S S S S S S S SSISI . . . . . . . . . . . . .
 31| . . . . . . . . . . . . . . S S S S S S S S S S . . . . . . . . . . . . . .
 32| . . . . . . . . . . . . . . . . . . S S . . . . . . . . . . . . . . . . . .
 33| . . . . . . . . . . . . . . . . . . . . . . . . . . . . . . . . . . . . . .
 34| . . . . . . . . . . . . . . . . . . . . . . . . . . . . . . . . . . . . . .
     ----------------------------------------------------------------------------

                          #=SAI, *=SBI, +=SCI
```

(a)

(f) When the expert system runs are completed, table T and the labeled input images Ik' are used to create an output image of results.

The above procedure can be further refined in several ways, for example, by varying the spatial resolution of the images (multi-resolution), i.e., first a coarse analysis, followed by finer spatial resolution in areas of special interest. If only various "bits and pieces" of evidence are available at distinct map locations, image processing operations may be used to "spread" this information over likely areas. Distances from various features in a map, such as faults, can be represented in image form, and many other image processing operations are available to transform the input information into a form considered suitable by the operator.

Figure 5 exemplifies the labeling of overlap of attributes (binary images) of lithology, A, B, and C, slope, S, and ground-water conditions, I. A list of labels (T) is shown in Table 3.

An expert system may have several modes of operation. PROSPECTOR has two modes:

Figure 5. Continued.

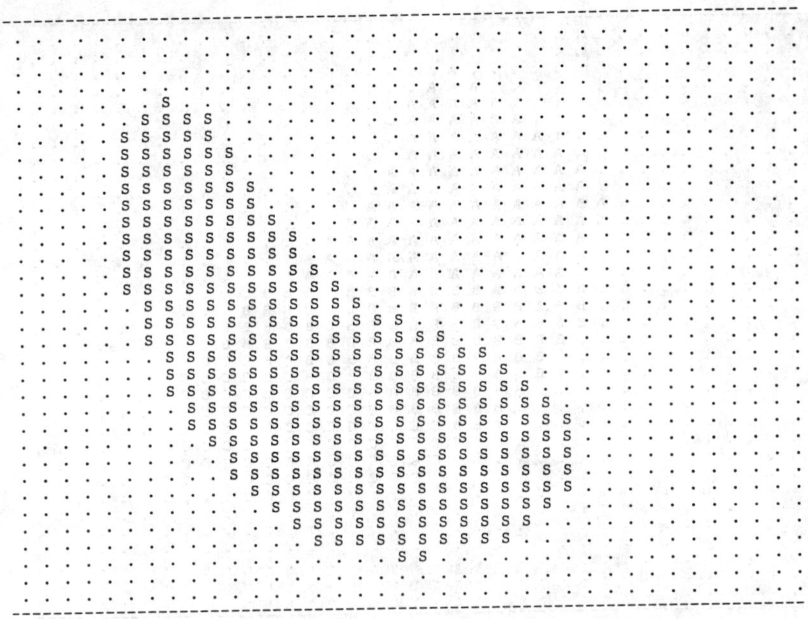

(b)

S = 316, . = 886

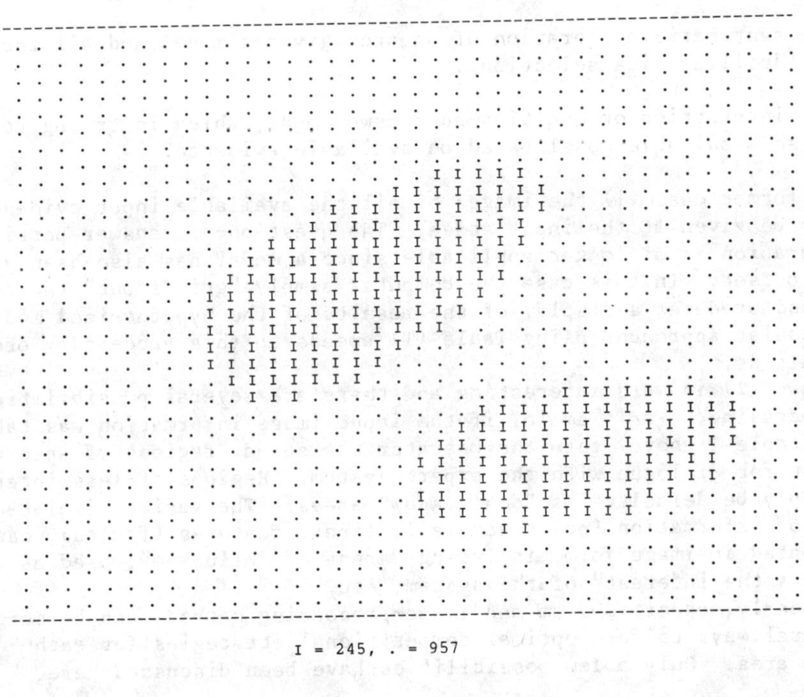

(c)

I = 245, . = 957

Figure 5. Continued.

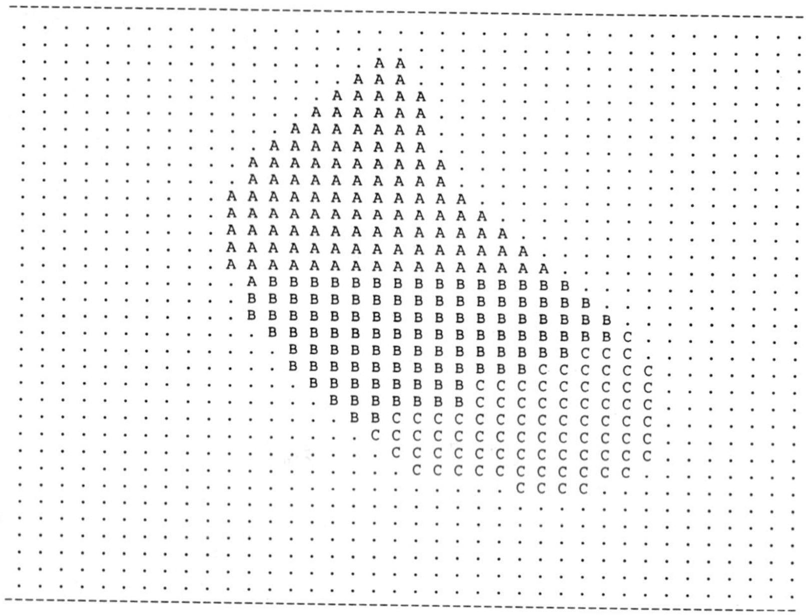

(d)

A = 122, B = 110, C = 82

(1) The systematic exploration of an area given a model and all the inputs (drilling site selection).

(2) The interaction or question-and-answer mode, which is trying to establish a suitable model based on available evidence.

In the former case (1) the images of all the available input evidence have to be given at the input stage. The question-and-answer portion of the operation is no longer applicable since a model has also been given at the outset. In this case the computer simply "grinds out" the output image and produces a display of the results of the hypothesized model. The "tabular approach" using Table (T) speeds up this process by orders of magnitude.

Case (2) is more interesting and there are several possibilities for processing. For example, if the input image information was tabulated (table T above) then the operator may select regions of special interest for dialogue with the expert system. Regions of less interest may simply be defaulted as "don't know" areas. The various isolated pieces of information (point samples), linear features (faults), can be represented in image form, including "areas of influence", used as means to "focus the interest" of the system, etc.

Clearly, expert-system and image-processing methods can be merged in several ways to form optimal computational strategies for each problem area. Only a few possibilities have been discussed here.

Table 3. Table of labels computed from the images in Figure 5. The row numbers in the left column represent the table address. From the pixel counts in the 7th column, a contingency table can be extracted.

	S	A	B	C	I	no.	.	.	etc. ...
1	S	.	.	.	.	150			
2	.	A	.	.	.	53			
3	.	.	B	.	.	31			
4	.	.	.	C	.	11			
5	.	.	.	.	I	29			
6	S	A	.	.	.	9			
7	S	.	B	.	.	29			
8	S	.	.	C	.	13			
9	S	.	.	.	I	33			
10	S	A	.	.	I	11			
11	S	.	B	.	I	34			
12	S	.	.	C	I	35			
13									
14									
15									

6. REQUIREMENTS FOR MORE COMPREHENSIVE AND "HUMANE" SYSTEMS

Many well established mathematical analysis techniques exist in statistics as well as in digital image processing. These require input data which has to be supplied in a predefined form. However, the user has to know algorithms and sequence of processing steps to operate the system and also has to know image processing. Such systems offer no "precanned" advice to user, leaving the user free to select whatever processing sequences he/she desires: meaningful as well as meaningless sequences are allowed. Expert systems, on the contrary, force the user to follow a rigid protocol (e.g., inference rules) from which one cannot deviate unless provision is made for "learning". What should we do to make the computer a better tool and assistant? What kind of questions could we expect the computer to answer with quantitative and qualitative replies? These are difficult questions to answer, but one attempt will be made.

Any system that is to provide information to a user is only useful if a "symbiosis" can be established between the computer system with the information and the user with the question. If the system were intelligent in a human sense, then usually a question and answer situation could arise. In the human case, the medium of communication is the natural language, mathematic, graphs, maps, images, and so on. Our present computer systems are still exceedingly primitive. Computer "understanding" of spoken language is very limited; "understanding" of typed-in text is also rather limited; "understanding" of formulas, graphs, and images are only research topics at present. On the opposite side, human understanding of computer languages and procedures to obtain the desired results takes too long to learn. The expert in some field is not going to learn the computer methods and the computer experts cannot become experts in fields from which the query may originate. Simply stated, there is a "communication gap" between man and machine.

From the computer programming point of view, the only feasible approach at present is to limit the "field of interaction" between the user and the machine to rather narrow domains both linguistically and in the fields of "expertise". The machine can be programmed to "understand" typed input with constrained vocabulary and grammar, and it is able to reply in stereotyped sentences. Computer "expertise" has to be limited to quite narrow areas and its "understanding" of a request from a user is only revealed by a display of the processing steps used.

Let us consider two examples of queries to illustrate the possibilities and to clarify what may be required from the designers of such computer systems.

Given a model to relate overlaps of features extracted from maps with uranium mineralization in northwestern Manitoba, Canada, the following evidence was described by Fabbri (1984, pp. 101- 104).

(i) The geological map shows a promising contact between pelitic and psammitic rocks. Marble and calc-silicate rocks occur along this contact.

(ii) The occurrence of gravity and aeromagnetic anomaly lows may relate to areas where the sedimentary belt is thickest.

(iii) The occurrence of eU/eTH highs in the area seem to relate to the development of white anatectic granite and pegmatite within the pelitic rocks.

The interactively extracted pattern obtained by Fabbri (1984) required a number of spatial transforms and logical operations within and between binary images. From the contact in (i) a part was isolated for further exploration because of its compliance with evidences (ii) and (iii).

A question for our expert system to produce the same pattern could be:

"FIND AREAS WHERE URANIUM MINERALIZATION MAY OCCUR NEAR 1/2 KM CONTACT BETWEEN PELITIC AND PSAMMITIC ROCK UNITS IN THE NORTHWESTERN MANITOBA DATA BASE"

The following action by machine would be initiated:

[1] Parse question and find model(s) and corresponding images from data base;

[2] Reply describing model(s) and related images available;

[3] Ask questions related to modification of data base (e.g., addition of proprietary information to available images);

[4] Ask questions about which variant or modification of model to use;

[5] Compute transformation of images that relate to implications in model (i.e., run model);

[6] Display output results in numeric/graphic form;

[7] Go back to beginning of action and prompt for modification of question or stop session.

Another question that would involve remote sensing could be the following:

"EXTEND LENSES OF IRON FORMATION IN TURBIDITIC SEDIMENTS IN BATHURST INLET AREA USING KNOWN MAPS AND REMOTE SENSING IMAGERY".

The evaluation of this query follows roughly the same pattern except that remotely sensed images are used. In this instance, a search is required through archives of such images for a coverage of the area indicated. Search methods will have to consider some of the following facts: (a) cloud-free images are to be preferred, (b) several periods might have to be analyzed, and (c) up to four images could be needed to cover the area if it is located at the center of a mosaic. Also needed, among many other things, are the properties of the image in terms of the resolution of the objects to be extended (or mapped further).

The interpretation of the questions for creating a meaningful response requires:

(1) Parsing of the command. Standard packages are available and the command can be verified with "counter-questions" from the machine.

(2) Establishing a dictionary of available images and how to find them according to the question.

(3) Creating the necessary command string of image processing operations. This requires very detailed knowledge about image processing and what programs are available. Some research on this problem is being done in Japan.

(4) The creation of a response to the user in several forms:
(a) Natural language output created from the question.
(b) Printing of resultant maps, graphs, etc.

Operations (1) and (4a) are intensively studied by AI researchers and packages are available. Operation (2) depends on present organization of image data. Operation (3) is feasible, based on special expert systems (stereotyped replies for (4a) are used now). Operation (4b) is standard.

8. CONCLUDING REMARKS

Given the very large amounts of data available from various sources, the use of computers is now critical for integrating and analyzing the information. If the possible interactions between the data sets are properly handled, they can provide answers to many questions about resources and environmental impact. Our concept of automatic integration stresses that various spatially-distributed data-sets must be properly prepared.

If it is known in advance what data are needed, the initial problem is only an archival search that could be greatly assisted by an "intelligent" data retrieval system. If the user does not know what data must be processed for his problem, then the need for an expert or an expert system is unavoidable, unless specially trained personnel are available. The next step is to select adequate computational techniques and to determine if all the necessary data are at hand. If they are not, it is necessary to determine what is missing, to format the data, and so on. The assistance of an expert system to represent a given problem in a computationally suitable form, to select the necessary algorithms, and to advise the user on the path to follow towards a reasonable solution, would greatly benefit both the experienced user and is indispensable for the uninitiated user.

We have attempted to show that a comprehensive integration of multivariate statistical analysis, of image processing, and expert-system-based analysis is feasible. To obtain results, the user does not need to know the particulars of the analytical procedures: the details should only be made available upon request. No unsurmountable obstacles have been encountered.

9. REFERENCES

Agterberg, F.P., Chung, C.F., Fabbri, A.G., Kelly, A.M. and Springer, J.S., 1972, 'Geomathematical evaluation of copper and zinc potential of the Abitibi Area, Ontario and Quebec'. Geol. Surv. Canada, Paper **71-45**, 55 p.

Agterberg, F.P., Chung, C.F., Eade, K.E. and Fabbri, A.G., 1981, 'Preliminary geomathematical analysis of geological, mineral occurrence and geophysical data, southern District of Keewatin, Northwest Territories'. Geol. Surv. Canada, Open File **718**, 31 p.

Chung, C.F., 1983, SIMSAG: 'Integrated computer system for use in evaluation of mineral and energy resources'. Mathematical Geology, v. **15**, p. 47-58.

Duda, R.O., 1980, 'The Prospector system for mineral exploration', Stanford Research Institute, Final Report April 1980, 120 p.

Duda, R.O., Gaschnig, J. and Hart, P.E., 1978, 'Model design in the Prospector consultation system for mineral exploration'. In, Donald Michie, Ed., Expert Systems in the Microelectronic Age. Edinburgh Univ. Press, 1979, p. 153-167.

Duda, R.O., Hart, P.E., Barrett, P., Gaschnig, J.G., Konolige, K., Reboh, R. and Slocum, J., 1978, 'Development of the Prospector consultation system for mineral exploration'. Stanford Research Institute International, Final Report, October 1978, 193 p.

Duda, R.O., Hart, P.E. and Nilsson, N.J., 1976, 'Subjective Bayesian methods for rule-based inference systems'. Proc. National Computer Conference AIFIPS Conference Proc. v. **45**, p. 1075-1082.

Fabbri, A.G., 1985, 'Design and structure of geological data banks for regional mineral potential evaluation'. Can. Inst. Mining. Metall. Bull., v. **68**, p.91-98.

Fabbri, A.G., 1980, 'GIAPP: Geological Image Analysis Program Package for estimating geometrical probabilities'. Computers and Geosciences, v. **6**, p. 153-161.

Fabbri, A.G., 1984, Image Processing of Geological Data. Van Nostrand Reinhold, New York, 244 p.

Fabbri, A.G., 1986, 'Promising aspects of geological image analysis'. In, J.T. Hanley and D.F. Merriam, Eds., Microcomputer Applications in Geology, New York, Pergamon Press, p. 199-213.

Fabbri, A.G., Divi, S.R. and Wong, A.S., 1975, 'A data base for mineral potential estimation in the Appalachian region of Canada'. In, Report of Activities, Geol. Surv. Canada Paper **75-1C**, p. 123-132.

Fabbri, A.G., Fung, K.B. and Yatabe, S.M., 1986, 'The usage of artificial intelligence in remote sensing: a review of applications and current research'. In, Chung, C.F., Fabbri, A.G. and Sinding-Larsen R., Eds., Proc. of Nato ASI Statistical Treatments for Estimation of Mineral and Energy Resources, Lucca, Italy, June 22-July 4, 1986 (Reidel, Dordrecht, in press).

Fabbri, A.G. and Kretz, R., 1986, 'A study of two-dimensional grain sequences in rocks'. In, Merriam, D.F., Ed., book in press.

Hart, P.E., Duda, R.O., and Einaudi, M.T., 1978, 'PROSPECTOR - computer-based consultation system for mineral exploration'. Mathematical Geology, v. **10**, n. 5, p. 589-610.

Kasvand, T. and Fabbri, A.G., 1985, 'Considerations on pyramidal pipelines for spatial analysis of geoscience data'. In, H. Freeman and G.G. Pieroni, eds., Computer Architectures for Spatially Distributed Data, NATO ASI Series, Vol. **F18**, Berlin, Springer-Verlag, p. 295-311.

Kasvand, T., Fabbri, A.G. and Nel, L.D., 1981, 'Digitization and processing of large regional geological maps'. Nat. Res. Council Canada, Electr. Eng. Divs. Report, **ERB-938**, 91 p.

McCammon, R.B., 1984, 'Recent developments in Prospector and future expert systems in regional resource evaluation'. Proc. 9th W.T. Pecora Symposium on Remote Sensing, Sioux Falls, SD, p. 243-248.

McCammon, R.B., Botbol, J.M., Sinding-Larsen, R. and Bowen, R.W., 1983, 'Characteristics Analysis-1981: Final program and possible discovery', Mathematical Geology, v. **15**, p. 59-84.

McCammon, R.B., Botbol, J.M., Sinding-Larsen, R. and Bowen, R.W., 1984, 'The New Characteristic Analysis (NCHARAN) Program", U.S. Geological Survey, Bulletin **1621**, 27 p.

Samet, H.A. Rosenfeld, A., Shaffer, C.A. and Webber, B.R.E., 'Use of hierarchical data structures in geographical information systems'. Proc. Intl. Symp. on Spatial Data Handling, Zurich, Aug. 20-24, 1984, p. 392-430, 1984.

Serra, J., 1982, Image Analysis and Mathematical Morphology. New York, Academic Press, 610 p.

UCLA, 1985, Uncertainty and Probability in Artificial Intelligence. Workshop, UCLA, Aug. 14-16, 1985, sponsored by RCA and AAAI, 282 p.

Varnes, D.J., 1974, 'The logic of geological maps, with reference to their interpretation and use for engineering purposes'. U.S. Geol. Survey Prof. Paper 837, 48 p.

Wise, B.P. and Henrion, M, 1985, 'A framework for comparing uncertain inference systems to probability'. In, Proc. Workshop on Uncertainty and Probability in Artificial Intelligence, UCLA, Aug. 14-16, 1985, p. 99-108.

PRINCIPLES OF MATHEMATICAL MORPHOLOGY

Jean SERRA
School of Mines of Paris
Centre of Mathematical Morphology
35, rue St Honoré
77305 Fontainebleau
France

ABSTRACT. Mathematical Morphology is introduced as a method Image Analysis based on set theory. Two examples are treated in detail, in order to exhibit sequences of operations. Emphasis is put on the progressive loss of information. The concept of a morphological filter is then presented. After a discussion on the current equipments, the conclusion compares Mathematical Morphology with the other methods for Image Analysis.

1. MORPHOLOGICAL OPTICS DERIVED FROM VISION

In vision, a distinction should be made between geometric and morphological optics. It is well-known that the theories of geometric optics depend initially on the postulate that a system is identified when it is possible to predict the image of a luminous <u>point</u>. The complete field of vision is then deduced by superposition of elementary transforms. Often, there is an even stronger hypothesis, namely : "the image of the point is itself a point". Obviously, in such a process the convolution and point transforms of Euclidean space (homothetics, rotations, affinities, etc...) play a major role. The usefulness of the linearity property need not be demonstrated : when an image is taken, one sums the images to attenuate the background noise ; short-sighted correction lenses deconvolute the retinal image, etc...

Linearity also occurs in acoustics. Indeed, the intensity of sounds, when one leaves aside considerations of phase, combine arithmetically. When several sources emit sound at the same time, the hearing process accommodates all the vibrations, and, to a certain extent, isolates and compares them. If this were not the case, there would be no orchestras. Since preserving the ratios among the sound sources is necessary for an intelligent understanding of the

C. F. Chung et al. (eds.), Quantitative Analysis of Mineral and Energy Resources, 237–254.

sound scene, all amplifiers (or transmitters) are required to comply with the relative proportions of the source origins, i.e., in mathematical terms, they must be <u>linear</u>.

However, visual signals combine differently. Objects in space generally have three dimensions, which are reduced to two dimensions in the photograph or on the retina. In this projection, the luminances of the points located along a line oriented directly away from the viewer are not summed, because most physical objects are not translucent to light rays, in the same way that they would be to X-rays, but are opaque. Consequently, any object which is seen hides those that are placed beyond it with respect to the viewer : this self-evident property is a basic one.

In fact, it serves as a starting point for Mathematical Morphology (SERRA [1986]), for, whenever we wish to describe quantitatively phenomena in this domain, a set approach must be used. Stating that A is in front of B is equivalent to asserting that we see the visible contour of B minus, in the set sense, that of A. Stating that A hides B is equivalent to saying that the contour of B is included in the contour of A, etc...etc... A morphological description, i.e. a description of their shapes, must primarily use <u>portions</u> of space. When transformations are involved, they must apply globally, i.e., they cannot be reduced to simple juxtapositions of point transformations (just like *gestalt*-psychology when it deals with human vision).

Now, the set $P(R^n)$ of subsets of Euclidean space is equipped with an order relation, called inclusion, such that any family X_i of elements of $P(R^n)$ admits a smallest upper bound, called the union, and denoted $\cup X_i$ and a greatest lower bound, called the intersection (the dual of the union), and denoted $\cap X_i$. In the same way that the theory of geometric optics puts its emphasis on the transformations which commute with addition, morphology naturally stresses the transformations, or mappings $\psi : P(R^n) \rightarrow P(R^n)$ that are related to the basic structures of $P(R^n)$. Thus they will be developed :

- either from those which preserve inclusion, i.e.

$$X \subset Y \Rightarrow \psi(X) \subset \psi(Y) \qquad X, Y \in P(R^n)$$

these transformations are designated as <u>increasing</u> transformations ;

- or from those which commute with the union, i.e.,

$$\psi(\bigcup_i X_i) = \bigcup_i \psi(X_i) \qquad X_i \in P(R^n)$$

which are called <u>dilations</u>. (The dual operation which commutes with the intersection is called an <u>erosion</u>). It will be noted, moreover, that the three classes are not independent : each of the latter two can generate the first class.

2. IRREVERSIBILITY AND MORPHOLOGICAL DISCOURSE

The parallel that could be drawn with linear methodologies ceases at this point, insofar as there is a major problem. When we say that in geometric optics we improve a fuzzy picture, making it sharp, we are expressing the point of view of the spectator. The physicist would tend to feel that nothing had been gained since it is always possible to revert from the sharp to the fuzzy picture : both of them contain exactly the same amount of information. The implied linear process is <u>reversible.</u> We are well aware, since the works of SHANNON [1949] and of WIENER [1949], of the emphasis that signal processing lays on the notion of information content of a message. This high level of interest is all the more justified when one considers questions related to transmission (amplifiers, broadcasting, etc...).

In computer vision, what are we seeking - to transmit information or rather to assimilate it ? Reversibility is acceptable when we improve the images that provide the input to a system, as with the case of spectacles for shortsighted people. It is also acceptable to encode images for transfer to processing devices, as is the case when the retinal image is transferred to visual cortex of the occipital lobe. But beyond this point ? The brain does not add a third "eye", which would then look at the visual zones and be observed itself. The chain stops there, and with it the notion of reversibility.

Recognition of an object simply means that all the rest has been eliminated from the scene. This is a definitive irreversible operation.

The tool created within the framework of mathematical morphology satisfies this property. The simplest dilation, the union of a set X and its translate X_h can only lose information : from $X \cup X_h$, one cannot backtrack to identify X. The question then arises as to how we can spread out successive losses among a series made up of dozens of transformations, so that the result converges to a single aim ? This is the central question of morphology ; in various forms and in particular expressions this question occurs frequently. Thus :

i) since we no longer have the structure of a group, which proved to be useful in the case of geometric optics (the usual similarities and convolutions), what are the conditions required in order that the composition product of two morphological operators remains in the same class as (at least) one of them ? If the answer is no such conditions exist, does this mean that the composition takes on a new meaning ? If such conditions exist, how can we interpret the composition ? For example, the product of two dilations is yet another dilation, but a dilation followed by an erosion leads to a product which has the characteristics of neither. Thus, as the various possible combinations between dilations and their complementary operations take place, there arises an infinite world of <u>new</u> mappings which can then be concatenated with each other : openings, closings, size distributions, morphological filters, ultimate erosion, skeletons, conditional bisector, and many others.

ii) Among all the possible "new species", how do we go about selecting them ? The answer is to have a set of reference properties on which to base the assessment. Thus, since there are sequences of operations, it is advisable to apply the first set of criteria to the choice of comparison between inputs and outputs, and to direct one's interest to the mappings, which may be :

- extensive, i.e., such that $\psi(X) \supset X$ (or anti-extensive if $\psi(X) \subset X$),
- over or underpotent, depending on whether $\psi\psi \supset \psi$ or $\psi\psi \subset \psi$,
- and especially idempotent, when $\psi\psi = \psi$.

The last-mentioned concept is a particularly powerful one. Its meaning is that a chain of processes, symbolized by ψ has acted as far as was possible, and that there is no point in pursuing the iteration. Idempotence acts as a buffer to the propensity for ψ to lose information. Filters, skeletons and ultimate erosions are operators that fall into this class, and which have a very high degree of internal consistency.

The first set of criteria, although useful, are not the only ones. It may be required of ψ that it be isotropic, or that it satisfy a given anisotropy, or be self-dual, or preserve connectivity or homotopy, or that it act, or not act, selectively on a given class of sets known in advance, etc...

3. FROM SENTENCES TO WORDS

The fact that we know all the "good" mappings does not ensure that we have worked our way through the problem, and does not give any indication as to how we should manage a sequence of morphological operations with a given aim in view. In a sense, it is a lexicon in a language the words of which are the key basic transformations and in which sentences are the sequences we are trying to construct. We now raise the question as to how the sentences are constructed. Let us approach this question via two specific examples.

3.1. First example : extraction of roads in remote sensing

In this first example, we start from an aerial photograph which simulates a satellite spot image (Fig. 1-a). It has been taken over the region of Beauce, in France. The morphologist is asked to implement an automatic procedure to extract the roads from the image. This problem, proposed to I. DESTIVAL [1985], led her to observe that the objects of interest to the case are long, narrow, connected, and their clear aspect contrast against a dark background. Hence the following strategy :

a) open image 1a by a structuring element (in this case, an hexagonal prisma) not much larger than the roads (Fig. 1-b);

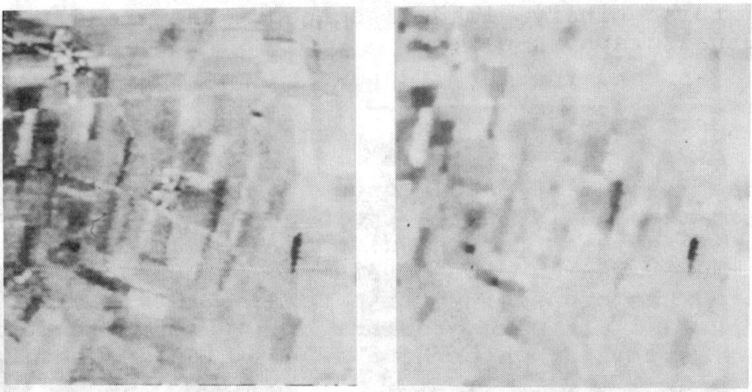

1-a 1-b

b) perform the arithmetic difference between the initial picture and its opening 2b. All bands that are larger than the roads are eliminated, as well as the bands that are darker than their surroundings. This first filtering was based both on contrast on width parameters (Fig. 1-c) ;

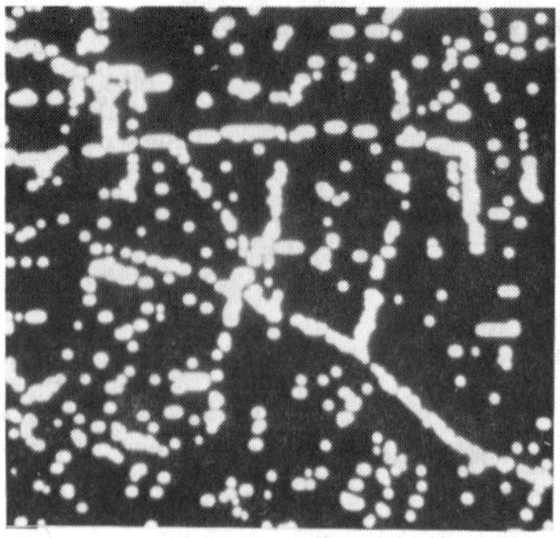

1-c

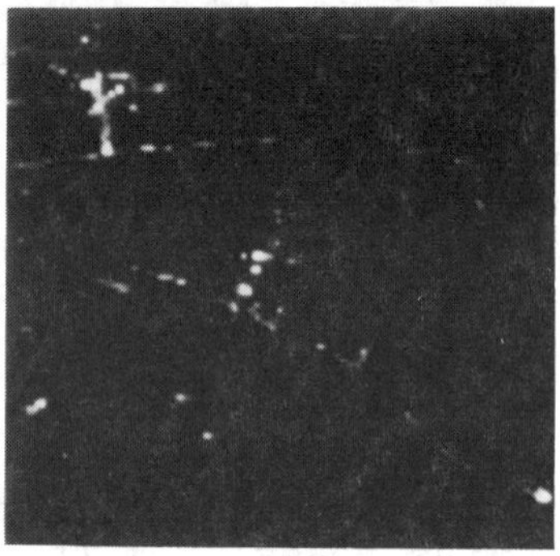

1-d

c) place a threshold on the transformation 1-c. The result appears in black on white in the picture (1-d) ;
d) In this binary image, connect the road portions by perceptual graph (1-e) ;

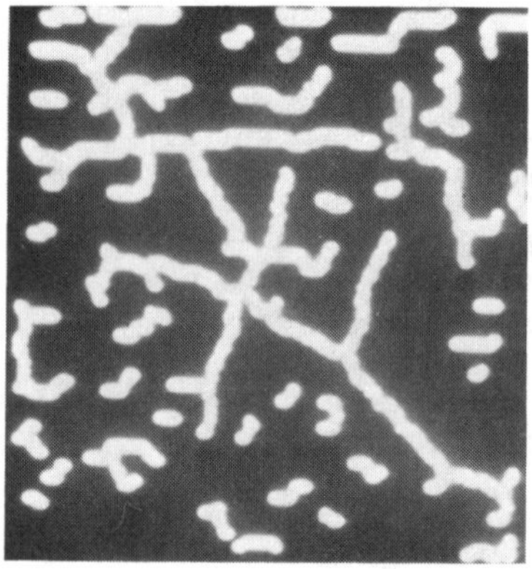

1-e

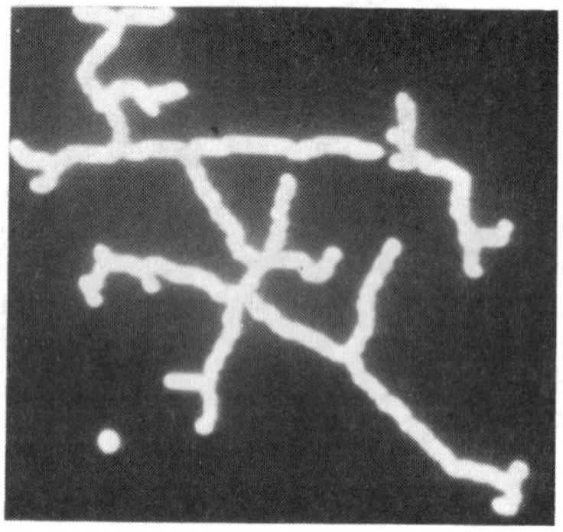

1-f

e) identify and retain only the connected components the skeleton of which have a length greater than a given length (1-f) ;
f) clip image (1-f). This final step is shown in Fig. 1-g.

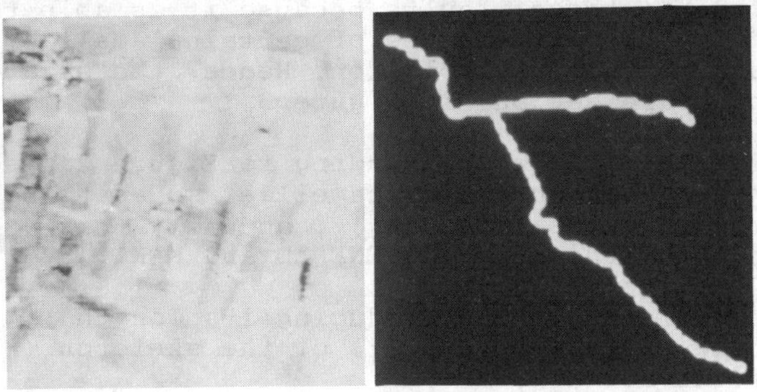

1-g

3.2. Second example : quantitative microscopy

We now move over from macro- to micro-scopy, from grey-tone image analysis to binary structures, and from robotics to a study of composite materials.

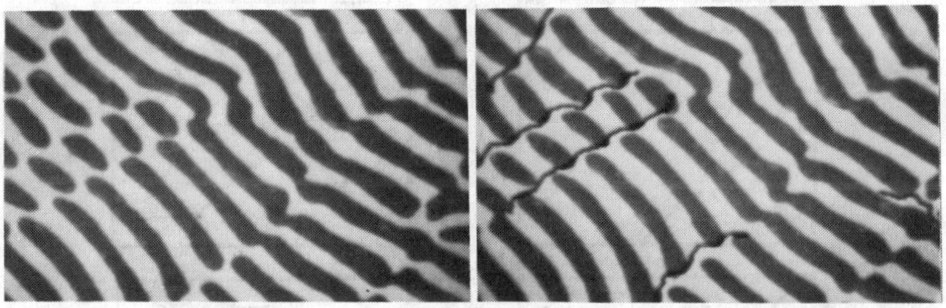

2-a 2-b

Fig. 2 : (a) Defect lines in an oriented eutectic material (GX200). A typical situation where no geometric primitive exists.
(b) Result of the treatment shown in fig. 3.

What we are observing in Fig. 2 is a cross-section perpendicular to its axis of a lamellar eutectic with an oriented solidification. It was produced by C.E.N.G. in Grenoble ; they wanted to estimate the number and the size distribution of fault lines per unit area of the surface. In order to solve this problem, LANTUEJOUL [1978] noted that the eye "sees" these faults because it is able to detect alignments in the extremeties of certain lamellae, and the brain makes a mental interpolation. Hence, the strategy, the steps of which are set out in Figure 3 :

a) threshold the image corresponding to Figure 2a (set X) ;
b) skeletonize the eutectic lamellae so as to be able to identify their extremities (X_1) ; actually this "skeleton" is a homotopic thinning (ROSENFELD and KAK [1976]) (SERRA [1882]) ;
c) clip the skeleton without reducing its length (X_3) ;
d) extract the extreme points (Y) of the skeleton (X_3), from the rest of the set X_4 ;
e) determine by homotopic thickening the zones of influence of the connected components of the set $Y \cup X_4$ (i.e., X_5) ;
f) connect all zones of influence associated with a point, and which are adjacent to each other (X_6) ;
g) skeletonize X_6, with the result that appears in Figure 2b, superimposed on the original picture.

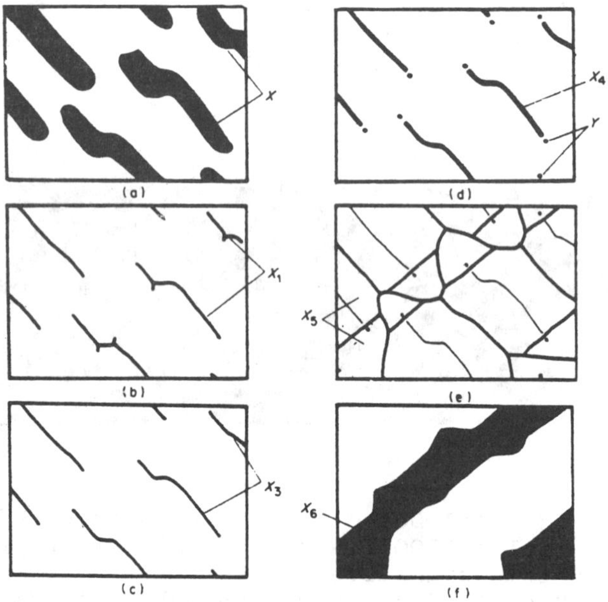

Fig. 3 : The successive steps between fig. 3a and the final result 3b.

3.3. Primitives or gestalt

Can we qualify the small brackets or the fault lines as primitives, in the same way that we can qualify the phoneme "a" in the expression "a drink !" as being a primitive, or that of automobiles on a motorway as seen from an aircraft ? As far as the motor is concerned, there may be dozens of small dark marks that are similar in size and shape to the safety brackets. The second example is even worse : in essence the fault lines are merely an absence of crystals : how could we threshold a lack of set ?.. When analysed carefully, we see that perception is more a modification of the environment : pipes are "broken" in the middle, or dendrites stop at one point and take up a little further away. It is this full context, as described in Gestalt Psychology (KOHLER [1929], KOFFKA [1935]), that defines an object in terms of a gap and there is no reasonable need to invoke the notion of primitive. However, in the first example we made use of a triangle to identify a zone, which had exactly the same shape as it (stage e). Unlike the other openings, this was a template matching operation, i.e., a search for a geometric primitive. Note that human perception also oscillates between the two terms of :

primitives <——————————> gestalt

in which each move to one of the poles improves knowledge acquired in the preceding move. Even if the expression "a drink !" is pronounced so "badly" that no distinct phonemes can be distinguished, the context, the tone used, etc., will point in the right direction. Then one mentally recalls the sounds, the phonemes are mentally restored, and the call is recognized..

In this dialectic, images and sounds do not oscillate round the same mean point. The laws of visual combination and perspective are such that near objects can hide distant objects, thus breaking up their elementary shapes, whereas the sounds of a voice speaking can follow each other with no interference. Different voices can mix simultaneously and each voice will retain its individuality. Thus it is seen that vision tends to operate more in gestalt mode. The same arguments apply to computer vision. For example, when particles in a thin section overlap too much, it proves impossible to separate them. It is precisely in this sort of situation that the gestalt hypotheses of Boolean models (G. MATHERON [1967]) enable a very simple, rugged, method to be used for particle counting. It can happen, on the contrary, that the very organisation of elements in a scene can lead to the generation of new primitives, artificially. Examples are the circular trinity of Roublov, or the diagonal cross in Gericault's painting of the "Raft of the Méduse" ;

indeed, painters often make use of this ambiguity to create an esthetic touch.

If such a dialectic exists, it is doubtless because the "gestalts" are not absolute, and that they are not totally disorganized when a few elements are removed. The main guideline of the two processes we examined earlier (as is the case in all other analyses in Mathematical Morphology) is based on this hypothesis. Actually, all strategies used in a morphological process tend to subdivide total perception into a number of sub structures, which are then individually apprehended in turn, the sequence of which is equivalent to a global perception. In a sense, they are primitives, but primitives of the processing and not of the image. The residual of the opening will extract all light shaded contrasts with respect to darker backgrounds, to the extent that they have a given amplitude and that they have a given width. The binary skeleton, when it is correctly pruned of its spurs, can be used to extract all extremities of long objects. These are morphological primitives. Although they are structural, they are still elementary enough to be called primitives. But above all, they are no longer isomorphic to the objects analysed (in an opening you don't go about moving circles in an image because you are looking for circles), and this makes them of general application. How many of them are there ? Close to 200, if we refer to the routines of B. LAY's Morpholog Package [1983], but doubtless less. (Some routines of Morpholog are sophisticated sentences).

At this point, we are beginning to measure the art of the morphologist. With a set of tools as defined above, he must reduce any visual process to a series of morphological primitives, each of which is known, and he must use them at the right moment, and with the right parameters. This reformulation is sometimes a difficult operation and requires a perfect knowledge of the situation.

4. FROM WORDS TO LETTERS

Let us pursue our progress from the "complex" towards the "simple", reviewing the syntax of morphological filters, and that of elementary operations.

4.1. Morphological filters

There is a temptation to group together the two most important properties of morphological operations and to study in particular the class of transformations which are simultaneously increasing and idempotent, i.e., the morphological filters. A detailed version of the theoretical aspects is set out in G. MATHERON [1983], J. SERRA

[1982][1986] ; thus, for the present, we shall concentrate our attention on the main syntax rules that are needed to build the operators.

a) all known filters are derivatives of two classes, extensive filters, better known as algebraic closings, and anti-extensive filters, the algebraic openings ; the filters may be obtained by composition products, union, intersection and limits.
b) these two classes are dual with respect to extensivity and therefore we need only examine one of them closely : e.g., the opening.

In a lattice $\mathcal{P}$ any algebraic opening γ can be decomposed into a union of morphological openings (i.e., products of an erosion by dual dilation (SERRA [1986]) ; if moreover, $\mathcal{P} = \mathcal{P}(E)$, is the set of parts of the space E in which a translation is defined (Euclidean space, digital grids), then any opening γ over $\mathcal{P}(E)$ which is invariant by translation, can be written as union of <u>morphological</u> openings, i.e., product of an erosion by the dual dilation (SERRA [1982]), where B describes the class $\mathcal{B}$ of the invariants of (2)

$$(2) \qquad \gamma(X) = \cup \{X_B, B \in \mathcal{B}\} \qquad \text{for all } X \subset E$$

From a syntactic point of view, this relationship must be read from right to left, with the implied meaning that <u>any</u> union of morphological openings will produce an anti-extensive filter, and enables the investigator to ascertain the class of invariants (this being an excellent aid to build customised operators). An illustration of this was given earlier in the example dealing with the motor, in which the opening using a ball could be replaced by a union of openings using horizontal cylinders.

As far as openings that are non-invariant by translation are concerned, a case must be made for the <u>connected opening</u> $\gamma_x(X)$ which is relative to a given point x and which, for any set $X \subset E$ has as an associate either the empty set $\emptyset$ (if $x \in X$) or the connected componenet of X which contains x. The example of the motor illustrates this process of a union of such openings, when, in stage d, we extracted the lengthwise shapes.

c) absorption laws and semi-groups : up to now, we have left aside the possibility that the process stages involve one or several digital parameters. In the case of filters, these operations can be organized by observing that the less images a filter acts on, the more active the filter. If we choose to study a family ψ_λ which depends on a positive parameter λ, their power will be greater or lesser to the extent that $\lambda \geqslant \mu$ implies $\mathcal{B}_\lambda \subset \mathcal{B}_\mu$, where $\mathcal{B}$ designates the class of invariants of the filter ψ.

(3) $$\lambda \geqslant \mu \Rightarrow \mathcal{B}_\lambda \subset \mathcal{B}_\mu$$

Also the family $\{\psi_\lambda\}$ will be consistent if the successive invariant classes derive from each other, i.e., if

(4) $$\lambda \geqslant \mu \Rightarrow \psi_\lambda \quad \psi_\mu = \psi\lambda \quad \Leftrightarrow \mathcal{B}_\lambda = \psi_\lambda (\mathcal{B}_\mu)$$

Now, the two conditions (3) and (4) are equivalent to :

(5) $$\psi_\lambda \psi_\mu = \psi_\mu \psi_\lambda = \psi_{\sup(\lambda,\mu)}$$

Expressed literally, when we iterate a first filter with a second filter, the resulting action is that of the more powerful of the two. This composition law symbolises a simple yet powerful syntax, to the extent that it also represents a principle of classification by invariants. This is called a commutative semi-group (SERRA [1986]).

If we are dealing with openings (or closings), conditions (3) and (4) are equivalent, and any family satisfying (3) is called a size distribution (or anti-size distribution) (MATHERON [1975]). In fact, it is easy to see that in any sizing operation, in the physical acceptance of the term, there is involvement of a family of openings with a positive parameter which satisfies relationship (5). It should also be noted that syntactic structure (5) is not the only feature of morphological transforms. In linear signal processing, the iteration of two low-pass frequency filters with limits λ and μ, satisfies the composition law (5).

4.2. Syntax of elementary operations

Relationship (2) takes us to the area of dilations and erosions (i.e., to those transforms which commute with union and intersection respectively, anc which are invariant by translation. Let us pursue the reasoning with respect to the dilation (the case of erosion can be deduced from what follows by duality). These two constraints : commuting with union and translation invariance together define a set-oriented operation which is well known as Minkowski's addition. Let B be the transform of origin O, and B_x be its translate by vector O_x. Set B, which we designate as the structuring element, will characterize the transformation, and the Minkowski sum of X according to B will be :

$$X \oplus B = \bigcup_{x \in X} B_x = \bigcup_{x \in X} \bigcup_{b \in B} (x+b)$$

Each point x of set X spreads out to occupy the zone B_x ; we then proceed with the union of these zones (SERRA [1982]).

This particular formula devised by MINKOWSKI provides some syntactic structures that have been known now for a long time (HADWIGER [1957]) which also prove to be most useful.

In effect we see that

(6) $$(X \oplus B_1) \cup (X \oplus B_2) = X \oplus (B_1 \cup B_2)$$

and that

(7) $$(X \oplus B_1) \oplus B_2 = X \oplus (B_1 \oplus B_2)$$

The first of these relationships, which states that Minkowski's addition distributes the union, enables us to obtain dilations using the structuring elements B of any size whatsoever, by producing a union of elementary dilations.

5 - WHAT IS A MORPHOLOGICAL PROCESSOR ?

Since the onset of the 80's, we have witnessed a spectacular development of morphological processors. Doubtless for commercial reasons, certain manufacturers announce a strong filiation with the mother theory, while others prefer to ignore the connexion. Anyway, the theory does exist and implies a strong common denominator for all these systems.

In fact, we can always decompose any morphological process into a series of dilations, and complementations. Indeed, these very dilations can be reduced to a smaller number in different ways, that one need only take as basic material. We can therefore define a morphological processor as being any computer-based system which provides :

i) a set of dilation primitives, which act as the letters, on the basis of which
ii) other modules can elaborate "words", or even complete composite processes (i.e., "sentences").

Such a system can be designed in terms of software (e.g., "Morpholog" program) or, on the other hand can be entirely hard-wired, i.e., integrated into circuits (specific to a given application, for instance) ; all intermediate mixtures of software and hardware can also be envisioned.

The choice of a set of basic dilations, from which all other dilations will be developed, followed by complex morphological operations, is obviously highly important. Up to the present time the manufacturers have always limited the primitives they present to translation invariant dilations, characterized by structuring elements. For Leitz texture analysers, the size of the basic elements varies from 1 to 100 pixels ; in the Morpholog software package or in the Numelec system, they are obtained as sub-sets of the elementary 7 pixel hexagon ; in the Machine Vision Inc. systems, they are reduced to a pair of points that may be considerably separated from each other. It is fairly easy to imagine other structuring alphabets (e.g., square ones, 3-D ones, ...) which would be optimal for other technological domains. Note that some of these alphabets are based on <u>cellular logics</u> (i.e., comparisons between one pixel and its immediate neighbours, PRESTON and DUFF [1975]) but not all of them. In association, several techniques are able to accelerate the implementation of the dilation primitives : "operator pipelines", simultaneous processing of various sectors of the image, and for the convex structuring elements, recursive algorithms.

The quality factors of a morphological processor can be assessed also in terms of the <u>kinds of dilations</u> that the processor can easily perform. Now, in the simplest case, i.e., the 2D-binary case, there are three of which are used very frequently. There are the translation invariant operators, or structuring elements, the operators which are translation invariant in a given zone (the eroded part of the image field) and identically null beyond them. Also, there are the disks for geodesic operations (LANTUEJOUL, BEUCHER [1981]). Moreover, for any given space E (digital grids, Euclidean space, etc) there is the associate lattice of the partitions of this space which becomes important whenever we wish to simplify an over-segmented scene, and finally, the lattice of the sub-graphs of the digital functions defined over E. Computer systems should be evaluated not only in terms of the beauty of their architecture, but also in terms of the capability of the system to tackle all these lattices and dilations. If one makes the assessments from the latter point of view, it will be seen that not all morphological processors are equivalent.

One must then add on, the transformation sequences. From this point of view, the real degree of the power of a morphological processor is not to be attached so much to what we termed the words, such as skeleton, ultimate erosions, etc... but rather in the panoply of application software packages (sentences). This is only natural and logical. Geometric optical devices are mostly used to transmit images, i.e., images which are to be analysed, and reduced by morphological optical devices. As for all robots, their value is linked to hereditary or acquired memory, with respect to the cases they can effectively resolve, or the scenes they can "understand".

6 - CONCLUSION

To conclude, we shall examine the classification of image analysis techniques into four categories (Fig.4).

LINEAR	DATA TREATMENT
signal processing	for stereology
convolution, Fourier analysis	for multivariate analysis
tomography	for taxonomy
MORPHOLOGICAL	**SYNTACTICAL**
morph. filtering	extraction of primitives
integral geometry	grammars of contours
random sets	Hough transform
skeleton	

Fig. 4 : The four major approaches in Image Analysis

First there is the class of linear methods. As the name suggests, these methods develop image transformations which preserve addition, and beyond addition, the notion of group structure, thereby making reversibility an important feature. Their scope of application is enormous. We touched on them earlier when we discussed geometric optics, but they will also be found in other signal processing domains, in radiological tomography, in cartography (by kriging, for example). We have already seen how morphology, insofar as it puts the emphasis on irreversibility, takes the opposite course from that of the linear approach.

The second group emphasizes the measurement approach. Here, we consider that a picture is well described when there is a large number descriptors (volumes, histograms, diameters,...). These numbers are then reduced to a set of essential values, which, for example, have a stereological meaning, or which identify factor axes, or which may be grouped together by some taxonometric method. In contrast to this, morphology is more directed to mappings of a set E onto itself than onto R or onto R^+ ; it is, moreover, much more sequential : mapping ψ_3 follows ψ_2 which follows ψ_1, etc... whereas, the measurements usually derive in parallel from the initial image.

Lastly, we will place under the heading "syntactic" those methods which decompose the picture to be analysed into a family of simple shapes, such as disks, convex polyhedra, contour arcs, and then go on to describe each object as a particular organization of these primitive templates, using more or less deterministic rules of proximity or succession. These approaches are much more global than those undertaken in mathematical morphology, which, on the contrary, seek sequences of local operators. They are also more discrete, and finally, they use the calculation of optima more than morphology does.

The fact that these 4 methods differ vastly from each other and are sometimes contradictory in no way prevents them from co-existing. Experimenters have known this for a long time. It is never illegal to perform an image subtraction when applying morphological techniques, nor, once a specific object has been extracted by an ultimate erosion, to subject it to measurements for statistical or stereological computation. It would be more precise to state that the first stages of an image processing operation tend to be linear (summing, auto-focus, image enhancement by deconvolution,...) ; then followed by morphological processing, and finally by statistical computations or syntactic recognition processes. This particular approach may be reduced to one or two stages. Human vision follows this procedure exactly : the lens, the retinal nerve endings, and the various brain stages are assigned to distinct operations, which follow one another (this does not exclude flash backs and logical loops).

However, there is a more profound link between the methods. All of them restore explicitly and sometimes visually, a sort of "object" which is more or less hidden and more or less spatial. The scanner 'slice' in tomography, shows something that was invisible ; a stereological calculation estimates the specific three-dimensional surface of a porous medium using just a few plane slices (in this case the restoring function relates to a number). The Boolean random model which is invoked to deduce and explain a given crystalline growth interprets structures seen...

Finally, in these situations, more than in the result, it is the procedure followed that was made intelligible, thus enabling it to be "taught" to an image analysis device.

REFERENCES

- CHERMANT J-L., COSTER M. (1985) : Précis d'analyse d'image C.N.R.S., Paris, 521 p.

- HADWIGER H. (1957) : Vorslesungen èber inhalt Oberfléche, und Isoperimetric, Springer, Berlin, 312 p.

- KOFFKA K. (1935) : Principles of Gestalt Psychology, Harcourt, N.Y., 720 p.

- KOHLER W. (1929) Gestalt Psychology, Liveright (N.Y.), 403 p.

- LANTUEJOUL Ch. (1978) : 'Détection automatique des lignes de défaut dans les eutectiques lamellaires', rapport interne, 22 p.

- LANTUEJOUL Ch., BEUCHER S. (1981) : 'On the use of the Geodesic metrics in Image Analysis', j. of Micr. Vol. 121, pt. 1, pp. 39-49.

- LAY B., LANTUEJOUL (1983) : 'Morpholog User's Guide', Ecole des Mines de Paris, 84 p.

- MATHERON G. (1967) : Eléments pour une théorie des Milieux Poreux, Masson, Paris, 166 p.

- MATHERON G. (1975) : Random sets and Integral Geometry, Wiley, 275 p.

- MATHERON G. (1983) : 'Filters and lattices', rapport interne Ecole des Mines, 36 p.

- MEYER F. (1979) : 'Iterative Image Transformations for an Automatic Screening of Cervical Smears' (Journal of Histochemistry and Cytochemistry, Vol. 27, n° 1, pp. 128-135).

- MEYER F. (1980) : 'Empiricism or Idealism...' ("Pattern Recognition in Practice", E.S. Geselma and L.N. Kanal, Editors, North-Holland Publishing Co ; pp. 21-33).

- MEYER F. (1982) : 'The perceptual graph : a new algorithm' (ICASSP 82. Proc., IEEE Internat. Conf. on Acoustics, Speech and Signal Processing, Paris, 3-5 May 1982, pp. 1932-1935).

- MEYER F., VAN DRIEL A. (1980) : 'Automatic Screening of Papanicolaou Stained Cervical Smear with the T.A.S.', Microscopica Acta, Suppl. n° 4, pp. 82-94.

- PRESTON K., DUFF M. (1985) : Modern Cellular Automata Plenum Press, N.Y., 300 p.

- PRETEUX F. (1984) : 'Détection automatisée de différents organes en coupes scanner' (Colloque Image, GRESTI CESTA), 6 p.

- ROSENFELD A., KAK A.C. (1976) : Digital Picture Processing, Ac. Press, N.Y., London.

- SERRA J. (1982) : Image Analysis and Mathematical Morphology. Ac. Press, 600 p.

- SERRA J. (1983) : 'Quelques semi-groupes de filtrages morphologiques', rapport interne, Ecole des Mines, 30 p.

- SERRA J. (1986) : Elements de théorie pour l'optique morphologique Doctor Thesis, Edn. of Paris, 217 p.,

- SERRA J. (Ed.) : Advances in Mathematical Morphology (to be published in 1987).

- SHANNON E., WEAVER W. (1949) : A Mathematical Theory of Communication, Univ. of Illinois Press.

- STERNBERG S.R. (1982) : 'Cellular Computers and Biochemical Image Processing', Proc. US-France Seminar Biochemical Image Processing, Springer Verlag, New York.

- WEIBEL E.R. (1980) : Stereological methods (Vol. 1), Ac. Press, London, 416 p.

- WIENER N. (1949) : The extrapolation, interpolation and smoothing of stationarity time series with engineering applications, Wiley, N.Y.

THE USE OF GEOLOGICAL IMAGE ANALYSIS IN REMOTE SENSING AND SEISMIC ATTRIBUTE INTEGRATION FOR RESOURCE ASSESSMENT

R. Sinding-Larsen(*), Per R. Stokke(**)

(*) Department of Geology, Norwegian Institute of Technology, 7034 Trondheim-NTH, Norway
(**)Statoil A/S, Postboks 300, Forus, 4001 Stavanger, Norway

ABSTRACT. Geological images whether it be a map, a Landsat scene or a seismic cross-section need to be preprocessed before the image can be interpreted and integrated with other images. In this paper image analysis techniques for preprocessing data are illustrated by three examples. In the first example edge detection techniques are used on Meteosat- and Landsat Multispectral (MSS) data to define and compare lineaments related to different geographic scales. In the second example a multidimensional scaling method is used on airborne MSS data to remove shadow and other proportional effects. The third example shows how characteristic analysis can be used to integrate seismic attribute data.

INTRODUCTION

Better data interpretation techniques in mineral and petroleum exploration are needed in order to reduce exploration costs.

The success of exploration efforts worth several million dollars depend on the interpreter's skill in analysing multiple data sets and in identifying favorable areas where further exploration investments are justified.

Different exploration scenarios necessitate different preprocessing methods in order to extract the desired information from the raw data collected.

Each of the three preprocessing techniques illustrate different aspects of the process of extracting information: Edge detection helps enhance linear features; Multidimensional scaling suppresses proportional amplitude variations; and the ternary coding in characteristic analysis permits the expression of favorability dependent on the values of the attributes.

C. F. Chung et al. (eds.), Quantitative Analysis of Mineral and Energy Resources, 255–269.

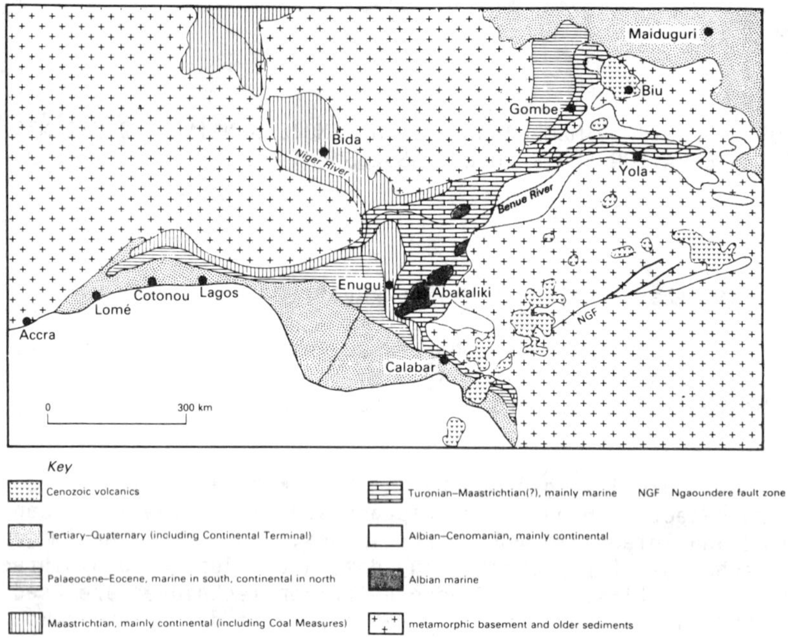

Fig. 1. Simplified geological map of the Benue Trough. (From Wright, J. B. 1985)

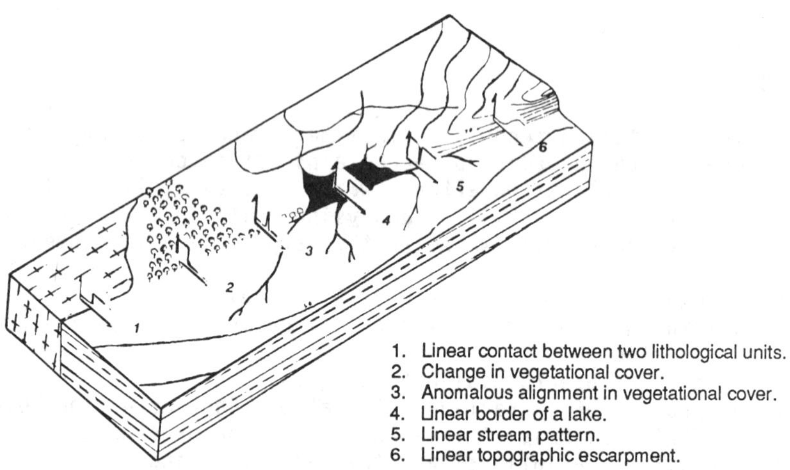

Fig. 2. Diagramatic representation of variations in radiometric reflection across a lineament. (After Fraipont and Hirsch 1984)

1. EDGE DETECTION FOR LINEAMENT ANALYSIS

The following example illustrates how Meteosat and Landsat MSS data can be analysed using an edge detection technique as a basis for extracting lineaments. This example is not presented as a complete lineament analysis but as an illustration of the advantage of using a preprocessing sequence extracting edges prior to traditional lineament analysis.

Geological Setting.

The study area used in this illustration is located in the northern part of the Benoue Trough centered around the Gombe area (see Figure 1). The area was selected because Meteosat and Landsat data was available and petroleum accumulations are known to be present both in the Niger Delta to the southwest and in the north towards lake Chad. The Benoue Trough is an elongated rifted depression which probably provided the major link between the Mediterranean and the Gulf of Guinea during upper Cretaceous time. Sediments are over 5000 m in thickness in places and have been strongly folded, probably by later adjustments along faults in the underlying basement.

Intensity discontinuities in an image can be regarded as edges. Such discontinuities can form lineaments or linear features in a spectral image as shown in Figure 2.

The basic conflict encountered in edge detection is the one of eliminating noise and insignificant edges without distorting the shape of the edges. By blurring a grey level image we remove noise and unnecessary detail, but at the cost of poorer localization.

A simple edge detector can, logically, be divided into two parts; a) blurring the image by using a two-dimensional average operator, b) registration of maxima in the first derivative in the gradient direction.

The theoretical discussion in recent years has centered around the choice of average and difference operators. The basic conflict prevails no matter what average operator we use, but some average operators like the Gaussian average operator do better than others (Schreiber and Troxel 1985).

The degree of blurring (smoothing) is essentially determined by the size of the average operator. So, the user chooses a desired level of resolution in the grey level image by blurring with an operator, and gets as a result, after having performed a search for maxima in the gradient direction, an edge image at a higher (=a lot of noise and detail) or lower resolution.

Let us assume that we wish to detect the most important edges in an image, and at the same time obtain good localization. We will briefly discuss our approach to this issue using the lineament analysis example from Benoue.

The detection of edges can broadly be classified into three methods; 1) the method based on differential operators based on spacial differenciation; 2) the zone fitting method which assumes an edge zone in a small region and seeks the zone parameters and 3) the statistical approaches including zones of noise.

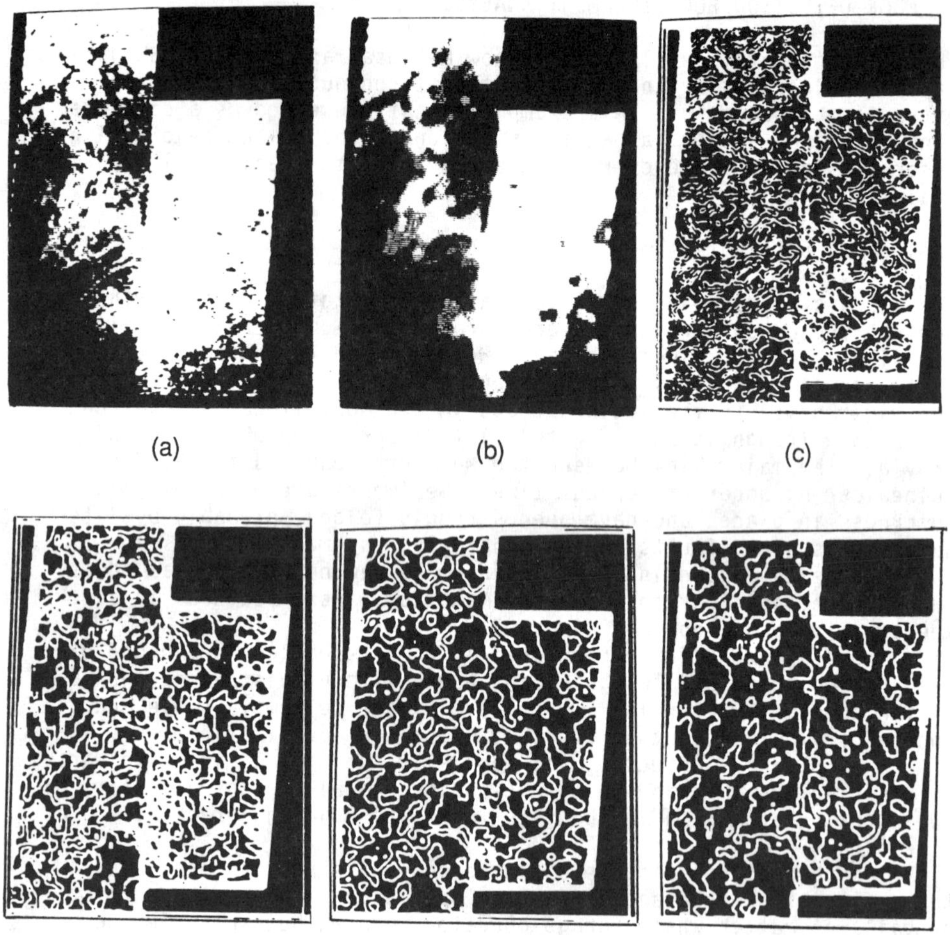

(a) (b) (c)

(d) (e) (f)

Fig. 3. (a) Composite Landsat scene channel 7.
(b) Image 3.a. smoothed with the Gaussian sharpening filter.
(c) Sobel edge detection on image 3.b.
(d) Sobel edge detection on image 3.b. with 10% increase in the smothing parameter.
(e) Sobel edge detection on image 3.b. with 15% increase in the smothing parameter.
(f) Sobel edge detection on image 3.b. with 20% increase in the smothing parameter.

Fig. 4. Meteosat image preprocessed with Gaussian sharpening filter with white contours indicating edges.

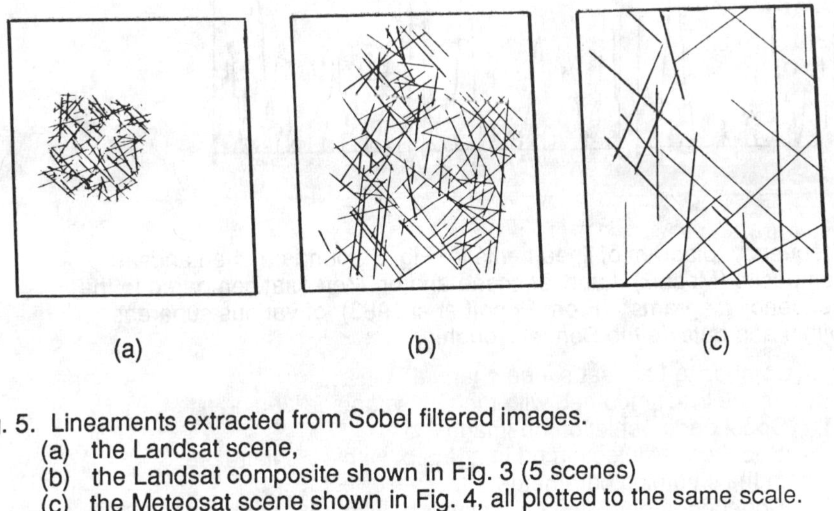

(a) (b) (c)

Fig. 5. Lineaments extracted from Sobel filtered images.
(a) the Landsat scene,
(b) the Landsat composite shown in Fig. 3 (5 scenes)
(c) the Meteosat scene shown in Fig. 4, all plotted to the same scale.
(From Sinding-Larsen et al. 1986).

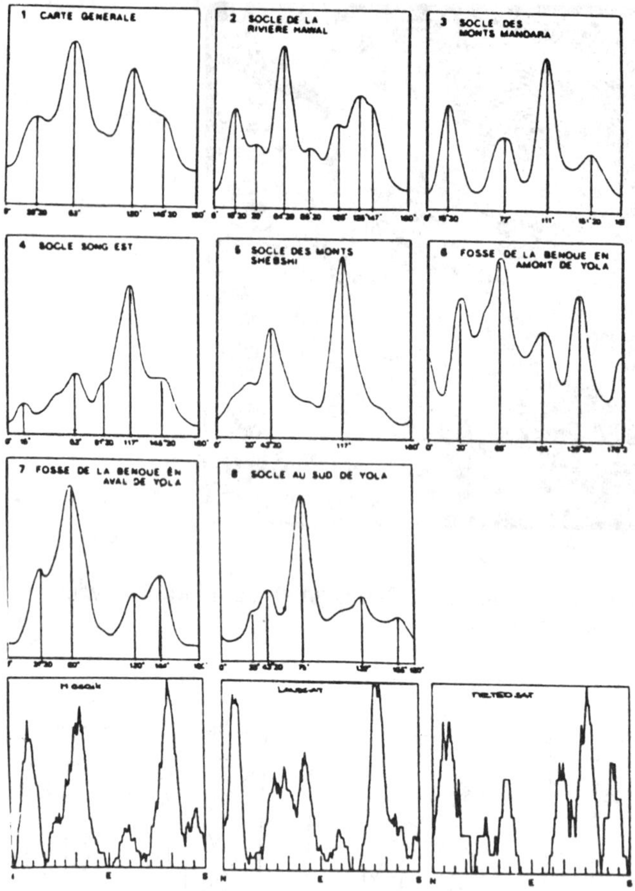

Fig. 6 Frequency diagram of lineaments in Fig. 5 found on the Landsat composite (Mosaik), Landsat scene and on Meteosat compared to the frequency diagrams (From Popoff et al.1983) of various subareas within and outside the Benue Trough.

In this paper the first type of methods using a differential operator by Sobel (Grimson and Hildreth 1985) was used. The Landsat and Meteosat images were pretreated by convolving the image with a sharpening filter (Schreiber and Troxel, 1985)

$$\begin{matrix} 1 & -7 & 1 \\ -7 & 49 & -7 \\ 1 & -7 & 1 \end{matrix}$$

and thereafter interpolated first horizontally and then vertically with a central Gaussian filter.

(0 1 3 6 11 16 20 21 20 16 11 6 3 1 0)/135

This procedure smooths out short discontinuities and increases the contrast between the larger structures. A Sobel filtering of this picture will give edges which can be convenientely related to geological lineaments. In Figure 3 we see a Landsat composite of five Landsat scenes and the result of the application of the Gaussian sharpening filter as well as the use of Sobel edge detection on increasingly blurred (smoothed) images. The Sobel edges corresponding to the second most blurred image was used for setting the smoothing parameters for the Meteosat and Landsat treatments.

In Figure 4 we see the Meteosat picture after being treated in a similar way, giving in Figure 7 edges which are more or less similar to the type of edges we had for the Landsat composite covering a much smaller area.

These Sobel filtered images are then used as a basis for visual extraction of lineaments. Because of their standardized form the Sobel filtered images ensure a similar interpretation procedure both for the Landsat and the Meteosat scenes (Figure 5). As we see from the diagram (Figure 6) the lineaments found on the Landsat, the Landsat mosaic and the Meteosat do in fact correspond. This shows that lineament analysis can start on a small scale image, e.g. Meteosat, and then subsequently be performed in more detail using larger scale satelite images.

2. MULTIDIMENSIONAL SCALING OF REMOTELY SENSED MSS DATA

In travelling from its initial source to the remote sensor, the electromagnetic radiation in the visible and near infrared undergoes absorption, reradiation, reflection, scattering, polarization and spectrum redistribution. The interpretation of the relationship between multispectral scanner (MSS) data and geological features of interest for mineral exploration is greatly complicated both by the number and the magnitude of the factors which affect the recorded signal. Spatial variability in the intensity of spectral reflection from pixels that are thematically similar, can be caused by shadow effects or effects due to varying topographic inclination.

The removal of these effects is consequently a useful preprocessing step in the interpretation of multispectral scanner data for mineral exploration or thematic mapping in general (Sinding-Larsen, 1983).

Two pixels that are thematically similar and with shadow falling on one of them will show a proportional relation between the intensitites measured in each of the spectral bands. We will therefore try to define a multidimensional scaling procedure which adjusts pixels to equality if they have proportional spectral intensities.

A multidimensional scaling procedure with these properties (Hill, M., 1974)) is equivalent to correspondence analysis (Benzécri, J.P. 1973)). This procedure can be described as follows with reference to the MSS data matrix X (nxp), where the n rows represent n p-dimensional pixel vectors. Let us consider the _n_ pixels within our remote sensing study area as a "cloud" of _n_ points in _p_-dimensional vector space where _p_ is the number of spectral bands measured within each pixel. Similarity in spectral intensities between pixels in this p-dimensional vector space will be represented by the interdistance between pixel points.
If the elements of the data matrix X(nxp) are divided by the sum of all the elements in the matrix, we obtain a matrix B with element:

$$b_{ij} = x_{ij} / \sum_{i=1}^{n} \sum_{j=1}^{p} x_{ij}$$

If each element of B is scaled according to

$$b_{ij}/b_{i.} \sqrt{b_{.j}}$$

resulting in a streching of the spectral band axes by a factor of

$1/\sqrt{b_{.j}}$ then the distance between pixels g and r in this vector

space will be given by:

$$d^2_{gr} = \sum_{j=1}^{p} \left(\frac{b_{gj}}{b_{g.} \sqrt{b_{.j}}} - \frac{b_{rj}}{b_{r.} \sqrt{b_{.j}}}\right)^2$$

- where $b_{r.}$ is the row sum or albedo of the ith pixel relative to the total albedo from the n pixels of the study and $b_{.j}$ is the proportion of the total albedo coming from the reflected intensity in the jth spectral band. Such a vector space will automatically compensate for any proportional reduction in the reflected intensities e.g. due to a shadow effect.

Let us verify the algorithm by computing the distance between pixel g and pixel r where the spectral intensitites in pixel g, are proportional by a factor of k to those in pixel r, i.e.

$$x_{rj} = k x_{gj}$$

In matrix B the following relations are found:

$$b_{rj} = k\, b_{gj} \qquad b_{r.} = k\, b_{g.}$$

$$d^2_{gr} = \sum_{j=1}^{p} \left(\frac{b_{gj}}{b_{g.} \sqrt{b_{.j}}} - \frac{b_{rj}}{b_{r.} \sqrt{b_{.j}}} \right)^2$$

$$= \sum_{j=1}^{p} \left(\frac{b_{gj}}{b_{g.} \sqrt{b_{.j}}} - \frac{k b_{gj}}{k b_{g} \sqrt{b_{.j}}} \right)^2 = 0$$

Thus if pixel points in this space are close to each other it will signify that the "spectral profiles" of the pixels have similar shapes.

The interdistances between pixels will remain unchanged when pixels or spectral bands are referred to new coordinate axes defined by principal components.

The next step in the processing is the computation of a cross-product matrix from which principal components are extracted. The pixel points are then projected onto these axes. Pixels with proportional spectral intensities having zero interdistances are equal in this scaled multidimensional space and are projected at the same place onto the first principal component.

In Figure 7a the albedo from raw MSS data is shown within a 3 km^2 area in Bamble, in the south of Norway. We can observe shadows from trees falling onto the highway (E18) near the upper left border. In Figure 7b the same area is shown with MSS data processed according to the multidimensional scaling procedure described. This picture shows clearly that reduced spectral reflection due to shadow on the highway is compensated for. It can further be seen that shadows on the flanks of small hills in Figure 7a have equally been removed in Figure 7b.

Application to mineral exploration

The multispectral 10 (Daedalus) air-borne survey over Bamble in the south of Norway was conducted to test the efficiency of the MSS method for identifying limonitic alteration related to mineralization.

A study area was chosen near an old nickel mine where limonitic alteration can be observed. In order to compensate for proportional effects, the spectral axes were scaled as described. A principal component analysis was then performed on a model area selected so that half of the pixels would fall within the zone of limonitic alteration. The first principal component will in this case be computed so that there is maximum separation between the projections of limonitic and non-limonitic pixels.

The effect of this linear classifier is shown in Figure 7b where limonitic alteration shows up in light near the lower right corner.

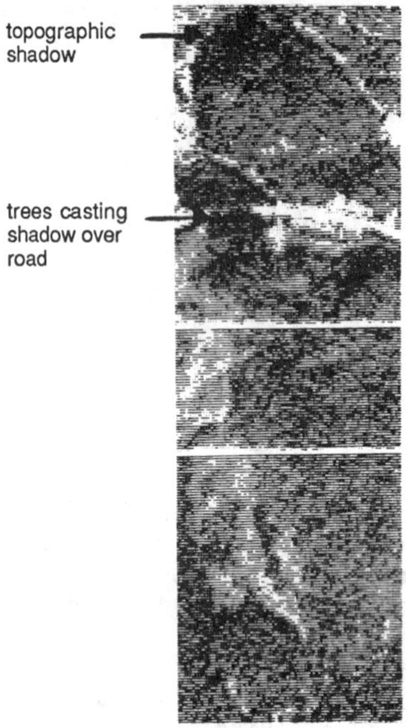

Fig. 7.a. MSS albedo image

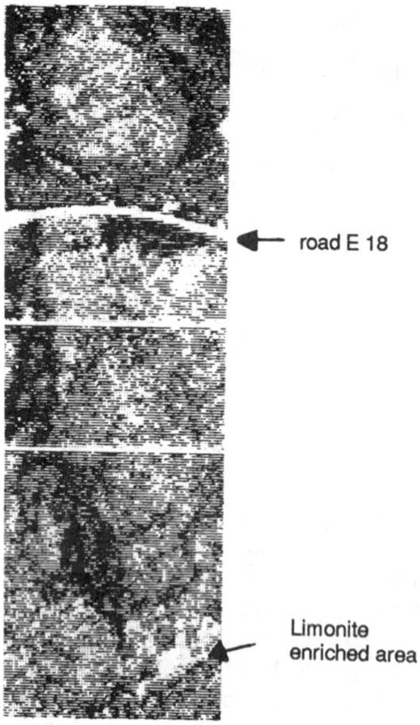

Fig. 7.b. MSS albedo multi-dimensionally scaled.

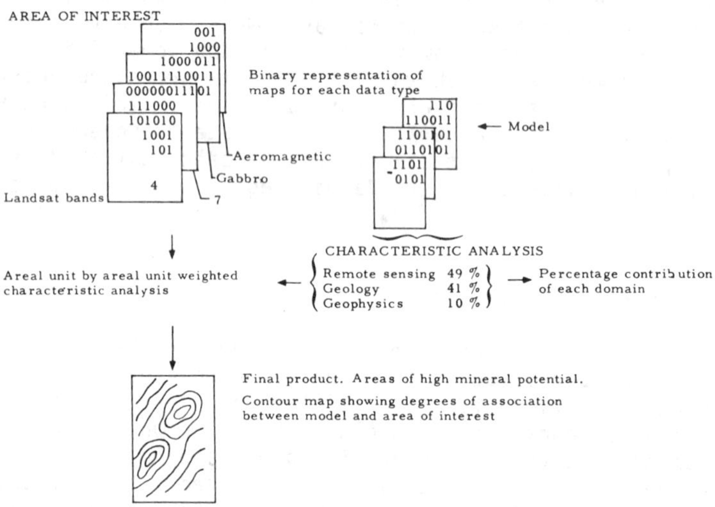

Fig. 8. Example of characteristic analysis process for mineral exploration.

The efficiency of this method in separating limonitically altered from unaltered rock exposures, was verified by field checking. Use of this procedure allows the recognition of limonite altered pixels (2.5 x 5 m) within a densely forested area and will thereby help to speed up field work.

3. CHARACTERISTIC ANALYSIS OF SEISMIC ATTRIBUTES

A diagrammatic representation of the weighted characteristic process applied to mineral exploration is shown in Figure 8.

The stacked and migrated seismic section shown in Figure 9 was used as input to characteristic analysis (Sinding-Larsen, R. 1974, et.al. 1979). The X axis (columns) represent receiver positions placed at regularly spaced intervals of 25 meters. The Y axis (rows) represent the two way travel time for the acoustic signal. The Y axis is sampled every four milliseconds.

Each seismic trace shown vertically in Figure 9 can be represented as complex numbers which are functions of time (Taner, Koehler, Sheriff, 1979). Real and imaginary parts of a time sample defines a vector with magnitude and direction. Different seismic attributes can be described to show how this vector varies with time. Attributes are instantaneous functions of time and are represented by the envelope amplitude, phase and frequency. These attributes are evaluated individually for each data point in the seismic section and transformed into ternary form by assigning the value of 1 meaning favorable, the value of -1 meaning unfavourable and the value of 0 meaning indeterminant (McCammon et al. 1981) .

The reason for using a ternary representation of the data is that in many instances the interpreter can assess the favourability of a given stratigraphic facies in a zone with respect to particular seismic attributes. How thin beds affect the attributes have been computed by Robertson and Nogami (1984). Thus for a given stratigraphic situation, the researcher assumes that in characteristic analysis, the value +1, 0 or -1 can be assigned to each of the seismic attributes at each data point. Characteristic analysis was used in this example to characterize zones of interest within different stratigraphic units.

The rational behind characteristic analysis of a seismic section is that the different attributes will jointly to some degree "characterize" different stratigraphic zones, and that the weight of each seismic attribute can be interpreted as an estimate of the importance of this seismic attribute in characterizing a zone with e.g. a desirable sandstone-shale ratio.

If the degree of common occurrences between the attributes within a zone is tabulated in an array, the array will represent the total of common occurrences between all the variables of this zone. In mathematical terms, this array is a "product matrix" because it is obtained by multiplying the original binary matrix for the attributes within the zone by its transpose.

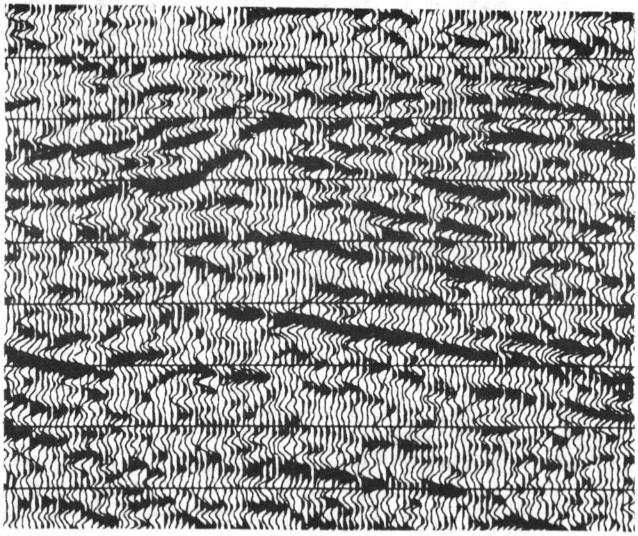

Fig. 9. Final stack, migrated amplitude seismic section.

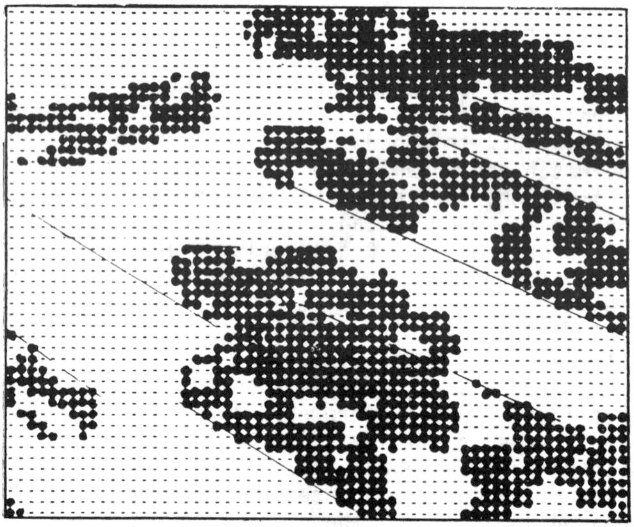

Fig.10. Lithofacies plot from characteristic analysis of seismic attributes. Areas with dominant shale embedded in a sandstone sequence are indicated by clusters of black dots.

Each row of the product matrix represents the degree of common occurrences of each attribute with itself and the other attributes in the zone. If one considers the attributes as vectors in a space with the datapoints in a given zone as axes, then the single vector which maximizes the projections of all the attribute vectors is the eigenvector associated with the largest characteristic root of the product matrix. For convenience, the length of this eigenvector is set arbitrarily to unity. The coefficients of this eigenvector are used to define the weights associated with the attributes chosen within a particular zone.

Multivariate comparison of the seismic section with a specific stratigraphic zone

When the weights associated with the attributes for a specific stratigraphic zone has been constructed and established as "criterion". Then all that remains is the computation of the degree of association between each datapoint and the zone. This is accomplished by multiplying the binary vector that represents the attributes of each region datapoint with the weights representing the equivalent variables within the zone. The resultant vector magnitude will thus be a function of the computed contribution of each attribute. In Figure 10 an example is shown where areas with dominant shale embedded in a sandstone sequence are indicated by black dots.

CONCLUSIONS

By blurring a satellite grey level image with a Gaussian filter followed by a sharpening filter noise and unnecessary detail are removed without significantly distorting the image patterns. A Sobel edge detector used on the satellite images preprocessed by a Gaussian and a sharpening filter provide a standardized basis for lineament extraction from multisatellite images. The example presented from the Benoue Trough showed that the structural pattern identified on small scale scenes (e.g. Meteosat) do reflect the same major structural trends as those found on larger scale remote sensing images like the Landsat or the Spot system.

The use of a multidimensional scaling procedure which stretches differentially the spectral reflectivity vectors by the reciprocal of the square root of their pixel sums and strech, differentially, the pixel vectors by the reciprocal of the square root of their albedo permits the compensation of effects causing proportional spectral intensities between pixels. By processing the data in this way and performing principal component analysis, an efficient linear classifier was obtained separating pixels containing limonitically altered from unaltered rock exposures.

Weighted characteristic analysis allows the combination and use of seismic attributes expressed in a ternary form to characterize a zone on the seismic section and to evaluate the similarity between a region and the zone according to predetermined weights calculated for each attribute within the zone. The selected zone may be part of the seismic section, may be from a different section, or may be a non-existing hypothetical seismic zone produced from syntetic seismograms. Used by an experienced interpreter this method should be a valuable tool in seismic facies analysis as the interactive nature of the analytic process enhances the interpreter's capability to appraise different lithofacies. Future development of the method lies in the optimal selection of attribute values as well as in the incorporation of filtering procedures and other pre-treatments of the variables.

ACKNOWLEDGEMENTS

The cooperation of J.M. Monget, B. Rindstad and T. Rousselin is gratefully acknowledged. Thanks are also due to the Department of Geology, The Norwegian Institute of Technology (NTH) and the Norwegian State Oil Company (STATOIL) for their support and the permission to publish this paper.

BIBLIOGRAPHY

Benzécri, J.P., 1973: 'L'Analyse des Données 2, l'analyse des correspondances' Dunod, Paris.

Fraipont, P., Hirsch, J., 1984b: 'Analyse linéamentaire: procédurede traitement automatique de donneés télédétectées. C.R. Coll. Int. "Informatique en Sciences de la Terre", Nancy, April 1984.

Grimson, W.E.L., Hildreth, E.C., 1985: 'Comments on "Digital step edges from zero crossings of second directional derivatives"'. ISSN 0162-8828, Vol. PAMI-7, No.1.

Hill, M., 1974: 'Correspondence analysis: a neglected multivariate method'. J.R. Statist. Soc., Ser. C, 23, 340-354.

McCammon, R.B., Botbol, J.M., Sinding-Larsen, R., Bowen, R.W. 1983: 'Characteristic Analysis-1981: Final Program and a Possible Discovery.' Math. Geology, Vol. 15, No. 1.

Popoff M., Benkhelil, J., Simon, B., Motte, J.-J., 1983: Approche géodynamique du fossé de la Bénoué (NE Nigeria) à partir des données de terrain et de télédétection. Bull. Centres Rech. Explor.-Prod. Elf-Aquitaine, 7, 1, 323-337.

Robertson, J.D., Nogami, H.H., 1984: 'Complex seismic trace analysis of thin beds.' Geophysics, Vol. 49,344-352.

Schreiber, W.F., Troxel, D.E., 1985: 'Transformation between continuous and discrete representation of images: A perceptual approach.' ISSN 0162-8828, Vol. PAMI-7, No. 2.

Sinding-Larsen, R., Botbol, J.M., McCammon, R.B., 1979: 'Use of weighted characteristic analysis as a tool in resource assessment: Evaluation of Uranium Resources Proceedings Advisory Group Meeting. International Atomic Energy, Vienna, p. 275-285. IAEA-AG/64-15.

Sinding-Larsen, R., 1983: 'A method for shadow removal from multispectral scanner data and its use in mineral exploration.' Data for Science and Technology. P.S. Glaeser (ed.) North-Holland Publ. Co., Codata.

Sinding-Larsen, R., Stokke, P., Rindstad, B., Monget, M.J., Rousselin, T., 1986: 'Multisatellite interpretation of the Benoue Trough area' (Manuscript in preparation).

Taner, M.T., Koehler, F., Sheriff, R.E., 1979: 'Complex seismic trace analysis.' Geophysics. Vol. 44, 1041-1063.

Wright, J.B., 1985: Geology and mineral resources of West Africa. George Allen and Unwin Publishers, London.

STATISTICAL IMAGE PROCESSING

Paul Switzer
Department of Statistics
Stanford University
Stanford, California 94305

ABSTRACT. This is a personal overview of a subject whose literature is largely outside the domain of widely read statistical publications. For this statistician much of the literature is unsatisfying because of weak articulation of image processing problems and fragmentary discussions of the properties of image processing procedures. While in no sense a survey, it is hoped that this brief article, by means of example, will draw attention to strengths and weaknesses of some common approaches. It also suggests alternatives and makes proposals.

The article is divided into six sections which deal respectively with definitions of univariate and multivariate images, geometric and other properties of images, pixellation and observational data grids, probabilistic descriptions of images, local and global criteria of image estimation quality, and a classification of estimation procedures.

1. IMAGES

A sufficiently broad definition of an image would be a function $X(t)$ defined for a geographically meaningful index t taking values in a bounded region R. We distinguish between an image, regarded as truth, and the data which are available to represent or estimate the image.

1.1. Geographic Support

Usually we think of an image as two-dimensional as in the case of photographic style imagery. In this case the location t will be a vector with a horizontal and vertical coordinate. In some applications the image could be three-dimensional such as an ore body, or the evolution of a two-dimensional image over some time interval. It is also useful to consider single transects, paths, bore holes, etc., to be one-dimensional images, as well as the evolution in time at a single geographic point.

In this presentation the geographic support of an image will always be a compact region in a Euclidean space of one, two, or three dimensions, although further generalizations are possible. Even if the available data are on a discrete lattice, as is typical, it is important not to lose sight of the fact that a true continuous image exists apart from the observation lattice.

C. F. Chung et al. (eds.), Quantitative Analysis of Mineral and Energy Resources, 271–282.

1.2. Types of Images

The type of image is characterized by the space of possible values of the image function $X(t)$. Some examples are:

(i) Dichotomous images: {present, absent}, $\{0, 1\}$, {black, white}

(ii) Non-negative images: {intensity}, {concentration}, {grey level}

(iii) Polychotomous images: {colors}, {rock types}, {species}

(iv) Multivariate combinations of the above: {multi-band intensities}, {multi-element assays}, {hue & intensity}

We commonly will want photographic-like representations of the images which have been estimated from available data. For two-dimensional grey-level or dichotomous images this representation is fairly obvious. For three-dimensional or multivariate images, representations will commonly take the form of multiple two-dimensional grey-level images. Three-variate images could be represented by intensity-weighted mixtures of the three primary colors to produce false-color photographic imagery.

1.3. Image Size

Image size is an elusive concept. It could refer to the extent of the geographic domain which one wishes to consider as a whole for a particular application. Or it may be determined by physical limitations on display capabilities or by mapping conventions. Even when there are conventions it would seem reasonable to choose image sizes related to the spatial scales appropriate to applications, even when the full resolution of the physical display are not exploited. The effect of image size on estimation procedures is discussed briefly later.

2. IMAGE PROPERTIES

The image properties here discussed could be regarded as summary statistics reflecting global textural characteristics. A fairly general notion of a global property is that of a spatial frequency distribution function, FQ, of a locally defined pointwise property, $Q(t)$. Specifically, for a real-valued property defined on images of unit area,

$$\begin{aligned} FQ(q) &= \text{ image area where } Q(t) < q, \\ &= \text{Prob}[Q(t') < q] \text{ where } t' \text{ is a random location in the image domain } R. \end{aligned}$$

From the global distribution FQ one can calculate average values, extrema, etc., of the local property $Q(t)$. Generalizations are possible to joint global frequency distributions of two or more locally defined image properties. Global distributions of image properties clearly depend on the choice of image domain and subimages of R will differ from one another with regard to specific global properties. Different image estimating procedures will characteristically differ with respect to global properties of the image estimate.

2.1. Elementary Properties

The elementary global properties are those for which the local defining function $Q(t)$ depends on the underlying image only through $X(t)$ at the single point t. Thus any reshuffling of the image will not change the values of elementary global properties. Some examples are:

(i) For univariate intensity images take $Q(t) = X(t)$; then the global property FQ is just the frequency distribution of intensity values over the image. A variant would be a recentered and/or rescaled distribution, where the transformation may depend on other elementary global statistics. For bivariate (or multivariate) intensity images we could have bivariate frequency distributions of intensity values.

(ii) For bivariate (or multivariate) intensity images take $Q(t) = X1t) \star X2(t)$ where $X1(t)$ and $X2(t)$ are two standardized (zero mean, unit variance) components of the image vector at t. The frequency distributions, FQ of such products of image components can then be averaged to get pairwise correlation coefficients between variables.

(iii) For dichotomous images $Qh(t) = 0$ if $X(t) = X(t+h)$, and $Qh(t) = 1$ otherwise, where h is a lag vector. The global FQh is then essentially the relative frequency of same-color point pairs separated by lag h. If h is varied then the plot of FQh against h is comparable to the variogram. For polychotomous images one could define a variogram for each color pair.

(iv) For differentiable intensity images let $Qm(t)$ be the magnitude of the image gradient at location t. The global distribution function, FQm, contains rich textural information. For example, the variance of this global distribution, normalized by the image intensity variance, may be interpreted as a measure of image patchiness; it is related to the derivative at the origin of the isotropic variogram described above.

(v) For line-segment images $Qh(t) = 0$ if no segment in the image crosses the segment $(t, t+h)$, and $Qh(t) = 1$ otherwise. The resulting global frequencies may once again be plotted as functions of h; the local derivatives may be used to find segment-length distributions within directional class intervals.

3. DATA

We assume that the data are not definitive of the image $X(t)$. Either the variable X is not directly observed, or it is not observed at all locations t, or both. The set of data locations is denoted R' and is a subset of the image domain R.

3.1. Pixels

Image data, typically obtained from scanning devices, are most often measurements recorded on a geographic rectangular grid. Each small rectangle in the grid is called a pixel (picture element). By convention, the pixel located at t means the pixel whose central point is t. Pixels are a creation of the data reporting procedure rather than an intrinsic property of the underlying image.

Usually we think of the data grid as two-dimensional as in the case of photographic style imagery such as satellite imagery. In some applications the grid would be three-dimensional such as data from parallel sections of a rock or satellite imagery repeated at equal time intervals. It may also be useful to think of time series data or data recorded along geographic transects (lines, curves) as pixellations of a one-dimensional image.

Image data not obtained from scanning devices may not be geographically gridded, as is often the case with mineral assay data and other geologic field data. In such cases the data will not be associated with a regular lattice but they will still be spatially and/or temporally indexed by t.

3.2. Measurements

The measurement recorded for the pixel at t will be denoted by $Y(t)$. The types of possible measurements are very diverse. Furthermore, the space of possible values of $Y(t)$ may or may not coincide with the space of possible values of $X(t)$ for the underlying image. Some examples are given below:

(i) $Y(t)$ is a multivariate intensity measurement, as in multi-band satellite data, while the underlying image $X(t)$ is a dichotomous or polychotomous field corresponding to rock types, vegetation categories, land-use classes.

(ii) $Y(t)$ and $X(t)$ are both multivariate intensities, as in multi-band satellite data, where $X(t)$ is interpreted as the true reflectance at t and $Y(t)$ is the degraded and integrated reflectance as actually observed.

(iv) $Y(t)$ is dichotomous, as in very low quality measurements of an intensity field $X(t)$.

The measurement $Y(t)$ for the pixel at location t will commonly be the result of some integration over the t domain, particularly for intensity data. The region of integration is not necessarily coincident with the pixel. There are several possibilities:

(i) The region of integration is small relative to the grid spacing. This is the common situation for field data such as monitoring data, mineral assay data, soil sampling, etc., using small sample volume.

(ii) The region of integration is comparable to the grid spacing. This is a common situation for satellite data where scanners are designed to provide this level of spatial integration. Sometimes, but not as often as supposed, the region of integration exactly corresponds to the rectangular pixel boundary. More commonly there is overlap across pixel boundaries.

(iii) The region of integration is large relative to the grid spacing. This is a situation typical of reflection seismology data.

3.3. Image Size and Data Sparseness

Image size is an elusive concept. For example, there are certain conventions in satellite data analysis which seem to prescribe square images which are 512×512 pixels. Such conventions, which may be related to scanner specifications, data storage and retrieval modes, or to capabilities of CRT display devices, are useful for image identification purposes. However, conventional image sizes may not be related to the sizes of image features which may be of interest.

The related question, whether the image data are sparse or dense, should not be gauged by image size conventions. Usually it is possible to make more or less ambitious demands on the data so that the same data may be viewed as either sparse or dense. Statistical processing of image data may use parameters reflecting both a local spatial scaling and a global spatial scaling related respectively to feature size and image size. It may be said that the most interesting problems for statistical approaches to image processing are those where the data are somewhat sparse rather than very dense.

3.4 Data Models

In the derivation of a formal Bayes image estimation procedure corresponding to a specified estimation error criterion, one combines a probability model for the class of possible images (the 'prior distribution') with a model for the observable data $Y(t)$ conditional on the actual image $X(t)$.

For typical multi-band gridded satellite data, for example, a fairly general representation of such a conditional distribution is

$$Y(t) = \int y[X(t+u)]dK(u) + N(t) \quad \text{for} \quad t \in R',$$

where $N(t)$ is a spatial lattice of unwanted noise vector random variables which model atmospheric effects, measurement error, illumination effects, surface texture variation, scan line effects, quantization, etc. The integrating kernel $K(u)$ represents the spatial resolving capability of the measuring device and the effects of pixellation.

The transformation $y(x)$ represents the conversion of the underlying object image property to a vector of light intensities corresponding to the frequency bands of the measuring device. For example, the object image may represent a patchwork of rock or vegetation species, i.e. a polychotomous image. Sometimes the object image is itself a multi-band intensity image so that the transformation $y(x)$ is trivial. The degree to which the transformation $y(x)$ is specified as well as the integrating kernel $K(u)$ and the distribution of the noise field $N(t)$ will depend on the physical situation as well as the formal requirements of the estimation procedure. For sparse data lattices the $N(t)$ may sometimes be treated as mutually independent random variables. However, typically the $N(t)$ would have some spatial continuity due to the persistence of atmospheric effects, for example. But the success of many estimation procedures depends on having an image, $X(t)$, which changes more slowly spatially than the unwanted noise, $N(t)$.

4. IMAGE MODELS

4.1. Image Models and Bayes Estimation

An image model is used to express global prior information about the true but unknown image. Models may be used to constrain the class of possible estimated images derived from available data. Or, via a Bayesian analysis, they may be used to assign posterior probabilities to image estimates, to find 'most probable' estimates which 'minimize' expected values of estimation error criteria, or more informally to guide estimation procedures.

For a proper Bayes estimation procedure, a unique probability measure needs to be associated with the set of all possible images on the domain of interest R. Since these prior measures describe global properties, it is reasonable to suppose that they could and should depend on R. In particular, if we have information specific to subregions of R, then we might use different models for each subregion, all of which are different from the model we would use for R taken as a whole.

Image models, in the form of complete prior probability measures, are most usefully described constructively in ways which facilitate Monte Carlo image generation. Sample images, i.e., simulated images, are sometimes evaluated visually to assess the appropriateness of a selected image model. Of course, many very different image models may all provide visually satisfying simulations but have different implications for image estimation.

A parametric class of image models is often specified as a means of introducing flexibility into the choice of a prior measure; examples are described a little later. To select a unique measure for a Bayesian calculation, values of the parameters must be assigned. It might seem reasonable to infer parameter values from the available data so that the prior measure might somehow be adapted to the specific image, even though one no longer has a genuine Bayesian analysis.

It is desirable if parameters of an image model can be interpreted as tangible global statistics of the image. Examples are transition frequencies for specified spatial lag, class boundary perimeter, mean size of 'objects', class frequencies, mean gradient length and direction, etc. If such textural properties of the image are allowed to vary slowly over the image then additional parameters could describe this slow variation.

While the model-dependent optimality of a Bayes estimation procedure should not be taken too seriously, such procedures nevertheless do incorporate information which could establish, for example, the degree of localization of the data used for estimation. Further uses of image models such as calculations of average estimation errors can be quite speculative. However, it will always be interesting to use image models for evaluation of the kind of image properties described in Section 2.

We distinguish between models for the true image and models for the observable data, given the true image. Both are needed for a Bayes estimation procedure. The latter have been mentioned briefly in Section 3; examples of the former are described below.

4.2. An Image Model Example for Univariate Intensity Images

The following example of an image model allows for slowly changing average intensities and intensity variability. The model is usefully written as the sum of two components:

$$X(t) = m(t) + r(t) \quad \text{for} \quad t \in R, \quad \text{where}$$

$m(t)$ is the slowly changing smooth part of the intensity image,
$r(t)$ is a zero-mean non-stationary gaussian random function

$$\mathrm{Cov}[r(t), r(t')] = \sigma(t)\sigma(t')\rho(t - t') \quad \text{where}$$

$\sigma(t)$ is a slowly varying intensity standard deviation and
$\rho(h)$ is a stationary autocorrelation function.

The specified functions m, σ, ρ may each have adjustable parameters and these become the global parameters of the image model. The richness of the intensity model depends then on the richness of the m, σ, ρ specification. Of course, even for very parsimonious specifications, the class of model realizations is large because $r(t)$ is a random process.

The gaussian property allows one to calculate some global textural measures. For example the marginal frequency distribution of intensity values will be a mixture of gaussian distributions. The mean of this mixture distribution is the spatial average value of $m(t)$ and its variance is te spatial average of $\sigma^2(t)$ plus the spatial variance of $m(t)$. The average magnitude of intensity changes for lag h could also be calculated in a straightforward way as a function of the global parameters of the model.

4.3. Models for Dichotomous Images

A probability model for a dichotomous image is the same as a model of a random partition of the region R into two subregions, say black and white. Such partitions may be generated, for example, in one of the following ways:

(i) Poisson centers. Points are located at random within R; these serve as the centers of regions of influence; these regions of influence may have random or fixed sizes and shapes; the union of the regions of influence is the black subset of the partition and its complement is the white subset (1).

(ii) Cell structure. The region R is first partitioned into cells, say by means of random lines or the Voronoi polygons of a point process; the cells are then independently colored black or white with either fixed probabilities or spatially varying probabilities. Cell structure models generalize in a straightforward manner to polychotomous images (2).

(iii) Dichotomization of continuous intensity images. The black subregion corresponds to the subregion where intensity exceeds a threshold.

Properties of dichotomous images may be related to parameters of the generating model. For example, the mean length of the boundary of the black subregion is proportional to

$$\frac{\partial p(h)}{\partial h}\Big|_{h=0}$$

where $p(t, t')$ = probability that the image is black at locations t and t'

$p(h) =$ average $\{p(t, t')\}$ with $|t - t'| = h$.

Sometimes models are used which generate image values only on a discrete lattice. Often this lattice is made to correspond to the data observation lattice, but this can be both confusing and unnecessary. The underlying true image lattice should be a refinement of the data lattice, say with a spacing of one-half of the data lattice. Properties of lattice models for a given spacing do not typically carry over to other spacings so lattice models must usually be regarded only as models of convenience.

For dichotomous images on a finite region R, a lattice model corresponds to a discrete probability assignment to each of the finitely many possible images. Lattice models may be obtained by starting with a continuous space model and restricting it to the chosen lattice. Alternatively, one may have a direct specification such as the one described below:

Let bb = the proportion of all pairs of adjacent lattice points which are both black

ww = the proportion of all pairs of adjacent lattice ponts which are both white

Assign probability $G(bb, ww)$ to the image, where G is a specified non-negative function on the unit square.

If G is an increasing function for both arguments, then roughly images with less black-white boundary will be more probable than images with more black-white boundary. If we specify a model of this type for a given lattice then generally no such model would be implied for other lattice spacing. The above model is an example of a first-order spatial Markovian lattice process; see (3) and references therein.

5. ESTIMATION CRITERIA

Available data are used to obtain an estimate $\hat{X}(t)$ of the underlying image field $X(t)$ for all points t in the spatial domain of interest. Goals of estimation are

(i) undo the effects of spatial integration of the measurement process

(ii) interpolate between measurement locations

(iii) filter noise components and other irrelevancies in the data.

Assuming the data are not definitive, we want estimating procedures which try to minimize meaningful error criteria. Here we consider mainly criteria which are local in that we wish to minimize a measure of error at each geographic point t of the image domain. Of course, an estimation procedure which is 'optimum' for local estimation criteria may produce estimated images whose global properties do not match well the global properties of the true image; more on this point later.

5.1. Examples of Pointwise Error Criteria

Let $e(t)$ denote the measure of error at t; a more explicit notation is

$$e(t) = e(\hat{X}(t), X(t)) \quad \text{for} \quad t \in R.$$

It is important to note that an error can be defined at all points of the region R and not just at the set of data points R'. A standard example for univariate intensity images is the squared error criterion $[X(t) - X(t)]^2$. However, it is often the case that the absolute intensity at each location is less critical for interpretation than the intensity relative to some standard scaling for the image as a whole. For example, we may agree that the average brightness and contrast of any estimated image will always be adjusted to standard levels. If errors are calculated from globally standardized true images and standardized estimates, then this could substantially affect the choice of estimation procedures.

The squared error criterion, even in its globally standardized form, is an example of a non-contextual criterion in that it pays no attention to neighboring points. By contrast, a contextual criterion is one which might do local standardization of the image and its estimate before calculation of the error $e(t)$. Hence, an intensity error of a given magnitude would be penalized more heavily in a locality of low variability and less heavily in a locality of high variability.

The pointwise errors, $e(t)$, will generate a frequency distribution as the location index t ranges over the region of the image R. The mean of this error frequency distribution (or median, maximum, etc.) might serve as a measure of the overall quality of the restored image. This spatially averaged error we denote $e(R)$. It is clear that the spatially averaged error depends on the choice of averaging region R.

5.2. Comments on Optimality

Explicit image restoration quality criteria serve several useful ends such as

(i) judging whether the available data are adequate or bounteous

(ii) assisting the design of further data collection or data compression

(iii) permitting comparisons between competing estimating procedures.

Of course the direct calculation of errors is not possible except for experimental situations or simulations where the true image is known. Therefore, errors must themselves be estimated. Such estimated errors will necessarily involve some form of data modelling, for example a probabilistic description of how the observable data depend on the underlying image.

Probabilistic models which relate data to the underlying image might also be used to obtain estimation procedures wich 'optimize' restored image quality according to selected error criteria. However, the optimality property is not only tied to the selected error criteria and the adopted data model, but is also tied to a particular image domain: restoration algorithms which try to minimize the spatially averaged error $e(R)$ may not be the same as those which try to minimize $e(R^{\sim})$ where $R^{\sim}$ is some subregion of R, say. Basically, the statistical properties descriptive of the region R may be very different from those which are descriptive of subregions.

Estimation procedures which optimize a specified error criterion may necessarily produce estimated images which have properties quite different from the underlying true image or image probability model. As an extreme example consider the case of a polychotomous image with uninformative data; then the most probable estimate is the image with the highest prior probability which might well be a trivial single-color image. A more typical phenomenon is the reduction in local contrast when trying to optimize estimates of pointwise intensity, especially seen as a blurring of edges. One may wish to introduce side conditions to the optimization of the image estimate which try to preserve important local and global properties of the type described in Section 2, for example.

6. IMAGE ESTIMATION PROCEDURES

Image estimation procedures may be roughly classified into three types which are described more fully later:

(i) Bayes estimates;

(ii) estimates which minimize an empirical function of the data;

(iii) ad hoc estimates which have certain desirable properties.

In 'easy' problems many competing estimation procedures will perform similarly so comparisons should be made on challenging problems. This does not only mean low nominal signal-to-noise ratios since the effects of high noise levels are quite easily mitigated if the image signal is changing slowly relative to the noise. A challenging problem is one, for example, in which the noise and signal power spectra are not grossly different or in which the noise is correlated with the signal.

6.1. Bayes Estimates

Bayes procedures make use of a prior probability measure on the class of all possible images, a conditional probability specification of the observable data for each possible underlying image, and an explicit estimation loss criterion. The estimated image is the one which minimizes the expected value of the loss criterion; expectation is taken with respect to the calculated posterior distribution on the class of all possible images, given the data. In principle, a complete probability specification is needed for true Bayesian inference. This principle is often compromised by allowing for a parametric specification with parameter values somehow extracted from the data themselves.

The estimated image with maximum posterior probability or probability density corresponds to the use of an all-or-nothing error criterion. Another criterion is the sum of pointwise errors which requires the calculation of the posterior marginal probability distribution at each point of the image domain. Since a Bayes procedure depends so heavily on a complete probability specification it becomes important to examine the robustness of Bayes estimates to alternative specifications. There is no particular reason why a Bayes estimated image should have the same statistical properties as the true underlying image, so a Bayes estimate might be visually misleading even if the probability specification is 'correct'. Examples of prior probability specifications are described briefly in Section 4 for a dichotomous image and for an intensity image.

6.2. Image Estimates Which Minimize Empirical Functions of the Data

The notion that the underlying image to be estimated, $X(t)$, is spatially less changeable than the observable data $Y(t)$ leads to various estimation criteria based on regularization ideas; examples are smoothing splines (4), maximum entropy estimates (5), and principal component projections (6). The first two have been primarily developed for univariate images while the last of these is for multivariate images. As in the case of Bayes procedures, interesting properties of the underlying true image, such as those described in Section 2 may be systematically distorted by the estimate.

In the case of splines the idea is to minimize average curvature of the intensity field for reasons good or bad. If the data are error-free observations of the field on a discrete lattice then the spline estimate can be used to interpolate the field. When the observations are subject to error then allowing for the data points themselves to be reestimated can further reduce the curvature of the estimated intensity image. Smoothing splines are a compromise between image curvature and fidelity to the observation points.

In the case of maximum entropy estimates of an image, it is only the data lattice itself which is reestimated and the procedure is not well adapted for interpolation. A compromise is made between fidelity to the observation points and minimization of the entropy of the marginal intensity frequency distribution. Unlike the spine curvature criterion, there is nothing spatial about the entropy criterion so it will not perform efficiently where the spacing of the observation grid is moderately close relative to changes in the intensity of the image. Indeed, if the data were permuted then the resulting estimation algorithm would be unaffected.

For multivariate gridded data such as multi-band satellite observations, a common estimation procedure to recover a noise-contaminated underlying multivariate image is to use principal components in the following manner. The modulation or overall spatial variation of the underlying true image is assumed to be much greater than the added noise so that the directions in space of variables corresponding to minimum variation are identified as noise. Hence, the projection of the multivariate data on the subspace orthogonal to the low-variance subspace is thought to suppress the noise while sacrificing some dimensionality.

The problem with the principal components procedure described above is the assumption that small global variability corresponds to noise. This criterion is non-spatial as it does not pay attention to spatial integrity and is invariant to spatial permutations of the data. Instead one may use a criterion explicitly related to spatial structure such as unit lag autocorrelation. Then one looks for directions in the space of variables which have minimum autocorrelation, rather than minimum variance, and identifies those as noise (7).

6.3. Ad Hoc Image Estimation Procedures

These procedures do not necessarily differ from those described above nor are they necessarily less effective in certain classes of problems. Rather ad hoc image estimates are distinguished in that they are not derived explicitly as solutions to optimality problems. Instead they are perceived to have desirable properties when applied to selected examples.

Certain ad hoc procedures are motivate by apparent inadequacies of 'optimal' procedures. For example, pointwise optimal procedures will not accurately reproduce derivative behavior of an intensity image. In the case of images with sharp gradients, optimal linear filters will induce blurring. To remedy such effects one might consider median-type filters (8) or similarity type filters (9).

A general description of a median type filter begins with the assigning of weights to observations at locations t' in relation to estimation at location t. Denote these weights by $w(t', t)$. A cumulative frequency distribution of the observations is formed with jumps proportional to the assigned weights. The usual linear filter would take the mean of this distribution as the estimated intensity at fixed location t; instead we take the median of this distribution to be the estimated intensity. The assignment of weights will reflect proximity to t so for each estimating location a new set of weights is needed, for example a moving window with fixed weights inside the window and zero weights outside the window.

A similarity type filter allows the weights $w(t'; t)$ to depend on the similarity of the observations at t' and t assuming the image is being estimated only at data grid points. If $Y(t')$ and $Y(t)$ are the respective observed values, then $w(t'; t)$ would be taken to be a nonincreasing function of the difference $|Y(t) - Y(t')|$. In this way, if there is a steep intensity gradient such as a fault line in the vicinity of the estimation point t, then little weight will be attached to data values on the far side of the fault.

7. REFERENCES

1. Matérn, B., 1960, Spatial variation, *Medd. Statens Bkogsforskningsinstitut*, **49**, 144 p.
2. Switzer, P., 1965, Reconstructing patterns from sample data, *Ann. Math. Statist.*, **36**, 138–154.
3. Besag. J., 1987, On the statistical analysis of dirty pictures, *J. Roy. Statist. Soc. B*, to appear.
4. Wahba, G., 1984, Cross-validated spline methods for direct and indirect sensing experiments, *Statist. Signal Processing*, 179–190.
5. Titterington, D. M., 1984, The maximum entropy method for data analysis (with reply), *Nature* **312**, 381–382.
6. Rosenfeld, A., and Kak, A. C., 1982, *Digital Picture Processing*, second ed., 2 vols., Academic Press.
7. Switzer, P., and Green, A. A., 1985, Min/max autocorrelation factors for multivariate spatial imagery, Stanford University Technical Report No. 9.

SPATIAL ANALYSIS OF PATTERNS OF LAND-BASED AND OCEAN-FLOOR ORE DEPOSITS*

F.P. Agterberg
Geological Survey of Canada
601 Booth Street
Ottawa, Canada K1A 0E8

ABSTRACT. In this paper, the spatial pattern of massive sulfide deposits in the Bathurst area, New Brunswick, is studied by application of a Neyman-Scott cluster process model. Next, the relationship between youngest sheet and pillow flows, which host hydrothermal vents on the East Pacific Rise near 21°N is considered in detail. Finally, the spatial cross-covariance function is derived for the more complex spatial patterns of acidic and mixed volcanic rocks in the Tetagouche Group of Middle-Late Ordovician age which contains the massive sulfide deposits in the Bathurst area, New Brunswick. The purpose of these case studies is to help determine how geostatistical methods can be used to model the variances and covariances of parameters quantified for patterns of mineral deposits and their environments.

1. NEYMAN-SCOTT CLUSTER PROCESS MODEL APPLIED TO MASSIVE SULFIDE DEPOSITS IN BATHURST AREA, NEW BRUNSWICK

The starting point for the application in this section is a map with the geology of the Bathurst mining district and 40 massive sulfide deposits compiled by Franklin et al. (1981, Fig. 45, p. 524) from other published maps. Table I shows the u (west-east) and v (north-south) co-ordinates of these deposits in km measured from a point located in the northwestern corner of the map.

Table II shows frequency distributions for number of deposits per cell for three different cell sizes using the location data of Table I. For larger cells, the variance is somewhat greater than the mean, suggesting relatively weak clustering. Franklin et al. (1981, p. 525) stated that the deposits occur in two groups more or less corresponding to two volcanic centers originally defined by Helmstaed (1973). The purpose of this section is to investigate ways to model this clustering of ore deposits by methods of spatial

* Geological Survey of Canada Contribution 36686.

C. F. Chung et al. (eds.), Quantitative Analysis of Mineral and Energy Resources, 283–299.

TABLE I Geological co-ordinates measured in units of 10 km for 40 deposits shown in diagram by Franklin et al. (1981, p. 524); No. is deposit number as in Franklin et al.; u and v represent west to east and north to south directions respectively.

No.	u	v	No.	u	v	No.	u	v	No.	u	v
1	5.96	3.28	11	2.12	4.35	17c	4.11	1.01	26	1.52	2.96
2	5.38	2.46	12	1.86	3.26	18	3.75	3.21	27	5.27	2.83
3	2.34	1.46	13	2.83	0.96	19	4.09	0.71	28	3.64	4.44
4	3.94	4.54	14	4.37	3.64	20	3.58	0.77	29	2.81	2.68
5	3.58	3.39	15a	3.75	4.29	20a	3.49	0.90	30	2.91	6.21
6	2.79	6.69	15b	3.51	4.41	21	5.94	3.41	31	2.23	4.29
7	6.94	3.00	15c	3.66	4.35	22	3.56	2.74	32	5.55	3.58
8	5.49	4.69	16	4.14	1.29	23	5.12	3.47	33	0.13	2.40
9	0.36	2.16	17a	4.29	1.09	24	3.04	2.57	34	5.59	4.22
10	1.31	1.93	17b	4.26	0.92	25	5.14	2.79	37	4.78	5.14

TABLE II Frequency (f) in number of cells containing k deposits. Boundaries of (10 km x 10 km) cells were chosen at u = 0, 1, ..., 7 and v = 0, 1, ... 7 (see Table I). The 10 km cells were subdivided into equal parts to obtain the smaller cells.

(10 km x 10 km)		(5 km x 5 km)		(2.5 km x 2.5 km)	
k	f	k	f	k	f
0	28	1	167	0	748
1	9	1	21	1	33
2	8	2	6	2	2
3	2	3	1	3	1
4	1	4	1	4	0
5	1	5	0	5	0
Ave = 0.8163		Ave = 0.2041		Ave = 0.0510	
Var = 1.4031		Var = 0.3171		Var = 0.0613	

statistics. For recent discussions of Neyman-Scott models, the reader is referred to Ripley (1981) and Diggle (1983). Earlier, Agterberg (1976) has attempted to apply this type of cluster process model to another set of deposits in the same region. The method followed in this paper yields better results.

The probability generating function (p.g.f.) of a Neyman-Scott model satisfies

$$g_{N_1,N_2}(s_1,s_2) = \exp\{\lambda\int(g_k\{1-\Sigma(1-s_i)\cdot p_i(u)\}-1)\,du\} \qquad (1)$$

where N_1 and N_2 are discrete random variables for numbers of points in cells A_1 and A_2 originating from all clusters. Cluster centers are assumed to be randomly distributed in the plane according to a simple Poisson process with p.g.f.

$$g_N(t) = \exp\{\lambda(t-1)\} \qquad (2)$$

where λ is the expected value of N representing number of cluster centers per cell A_i. Let $x|u$ represent location of the cell A_i with respect to a cluster center at location u; $f(x|u)$ represents a 2-dimensional density function so that a point falls in A_i with probability

$$p_i(u) = \int_{A_i} f(x|u)\,dx \qquad (3)$$

Any event has probability $p_i = P(\alpha_i = 1) = p_i(u)$ of falling in A_i; α_i denotes a Bernoulli-type random variable. The p.g.f. for two cells, A_1 and A_2, then becomes

$$\begin{aligned} g_{\alpha_1,\alpha_2}(s_1,s_2) &= E(s_1^{\alpha_1} s_2^{\alpha_2}) \\ &= 1-(1-s_1)p_1-(1-s_2)p_2 \\ &= 1-\sum(1-s_i)p_i(u) \end{aligned}$$

If K events are deposited independently around a cluster center with $p_k = P(K = k)$ and p.g.f. g_k, k_i of these may fall in cell A_i. K is a discrete random variable, e.g. with a negative binomial distribution. However, in the application of this section, it is not necessary to assume a specific frequency distribution for K. Consequently, for two cells:

$$g_{k_1,k_2}(s_1,s_2) = \sum p_k\{g_{\alpha_1,\alpha_2}(s_1,s_2)\}^k$$

or

$$g_{k_1,k_2}(s_1,s_2) = g_k\{1-\Sigma(1-s_i)p_i(u)\} \qquad (4)$$

Eq. (4) applies to a single cluster. Addition of events originating from all clusters consists of combining Eqs. (4) and (2). This leads to Eq. (1). Suppose that

$$Q_{ij} = \int p_1^i\, p_2^j\, du \qquad (5)$$

Then Eq. (1) yields the following parameters:

$$E(N) = \lambda\nu_1 Q_{10}; \; \sigma^2(N) = \lambda\nu_1 Q_{10} + \lambda(\nu_2-\nu_1) Q_{20};$$
$$\sigma(N_1,N_2) = \lambda(\nu_2-\nu_1) Q_{11} \tag{6}$$

where ν_1 and ν_2 represent the mean and second moment of K, respectively.

Suppose that $f(x|u)$ can be represented by

$$f(x,y) = \frac{1}{2\pi\sigma^2} \exp\{-\frac{1}{2\sigma^2}(x^2+y^2)\} \tag{7}$$

where x and y are distances measured eastward and southward from a cluster center. If the cell A_i of Eq. (3) is small, we have approximately,

$$p_1(u,v) = \frac{1}{2\pi\sigma^2} \exp\{-\frac{1}{2\sigma^2}(u^2+v^2)\}$$

where the subscript i = 1 indicates that A_1 is the first cell of a pair (A_1,A_2) with for the second cell:

$$p_2(u+k, v+\ell) = \frac{1}{2\pi\sigma^2} \exp\{-\frac{1}{2\sigma^2}[(u+k)^2+(v+\ell)^2]\}$$

From Eq. (5), it follows that

$$Q_{10} = 1; \; Q_{20} = \frac{1}{4\pi\sigma^2}; \; Q_{11}(k,l) = Q_{20} \exp\{-\frac{1}{4\sigma^2}(k^2+\ell^2)\} \tag{8}$$

According to Eq. (6), the covariance $\sigma(N_1,N_2)$ is proportional to Q_{11} of Eq. (8). Consequently, an isotropic 2-dimensional Gaussian

TABLE III Frequency of other deposits per 10-km cell measured by successively centering 10-km grid on each of the 40 deposits of Table I. Lower half of symmetrical array is shown only. Numbers in horizontal direction represent 10-km cells in west to east direction; those in vertical direction are for north to south direction; origin (not shown) is situated between 1 and -1 in both directions.

	-7	-6	-5	-4	-3	-2	-1	1	2	3	4	5	6	7
1	0	1	0	16	16	47	44	47	39	26	16	9	6	2
2	0	0	5	22	21	37	32	41	34	21	11	9	7	0
3	0	0	0	2	12	8	22	21	35	18	15	5	3	0
4	0	0	2	2	12	9	33	16	14	7	2	0	0	0
5	0	0	0	0	1	1	2	5	5	3	0	0	0	0
6	0	0	0	0	0	9	5	2	0	0	0	0	0	0

TABLE IV Array similar to Table III for 5-km cells.

	-11	-10	-9	-8	-7	-6	-5	-4	-3	-2	-1	1	2	3	4	5	6	7	8	9	10	11
1	1	0	0	2	6	1	4	16	8	10	20	15	13	8	17	9	4	5	2	5	0	5
2	0	0	0	5	3	5	6	14	9	8	6	8	11	7	7	7	6	9	5	2	2	0
3	0	2	3	7	7	4	11	4	7	1	8	13	4	6	9	7	3	3	1	3	4	2
4	0	0	0	1	7	5	1	9	17	16	7	10	14	12	7	4	7	1	6	1	1	1
5	0	0	0	0	0	4	5	3	3	12	3	5	8	6	12	6	4	11	3	2	0	2
6	0	0	0	0	2	1	2	1	1	3	4	5	3	10	7	6	2	1	0	2	1	1
7	0	0	1	0	2	1	8	3	2	15	7	2	5	5	1	2	1	2	0	0	0	0
8	0	0	1	0	0	0	3	2	2	3	8	8	1	6	2	2	2	0	0	0	0	0
9	0	0	0	0	0	1	0	0	0	0	2	1	3	2	2	0	2	0	0	0	0	0
10	0	0	0	0	0	0	0	0	1	0	0	0	1	1	0	1	0	0	0	0	0	0
11	0	0	0	0	0	0	0	0	4	2	0	2	0	0	0	0	0	0	0	0	0	0

frequency distribution fitted to the original data, would be proportional to Q_{11} and allow the estimation of σ.

Although the product $\lambda\nu_1$ can be estimated, it is not possible to estimate λ and ν_1 separately by the method followed here, and we will concentrate on the estimation of σ. Suppose that a grid with spacing a is successively placed on all 40 deposits of Table I and that frequencies of other deposits in the 10-km cells belonging to this grid are determined. Tables III and IV show such frequencies for square cells measuring 10 km and 5 km on a side, respectively.

Suppose that the deposits were randomly distributed in the area according to a simple Poisson process, then the frequencies of Tables III and IV would all be equal to one another except for random fluctuations around a constant value and an edge effect arising from the fact that the study area (70 km x 70 km) has boundaries outside of which there are no other deposits. Approximations of $\sigma(N_1,N_2)$ were obtained for 10 km and 5 km cells, by correcting for these edge effects and centering with respect to the mean (cf. Agterberg, 1980, p. 1828). The values were also divided by total number of deposits and multiplied by 100. These estimated antocovariance functions are shown in Tables V and VI, respectively. The random fluctuations in both tables are considerable and negative values occur. The latter were replaced by a relatively small positive constant (5 for 10 km cells, and 10 for 5 km cells, respectively), logarithms (base 10) were taken and a quadratic trend surface with its maximum of the origin was fitted to the values shown in bold print in Tables V and VI. Values at the origin were set equal to 72 and 55 for 10 km and 5 km cells, respectively, using the statistics shown in Table II, to account for the fact that extrapolation of $\sigma(N_1,N_2)$ to the origin results in a value less than $\sigma^2(N)$ as can be seen from Eq. (6). The best-fitting quadratic surfaces were, for 10 km cells:

$$\log_{10} \sigma(N_1,N_2) = 1.7645-0.0567k^2-0.0139\, k\ell-0.0508\ell^2$$

and for 5-km cells:

$$\log_{10} \sigma(N_1,N_2) = 1.6753-0.0437k^2-0.0388k\ell-0.0430\ell^2$$

TABLE V Two-dimensional autocovariance function for 10-km cells obtained by counting frequencies in central part of Table III for edge effects and centering them with respect to the mean. The corrected frequencies were divided by total number of deposits and multiplied by 100. Negative numbers are shown as –. Numbers used for surface fitting are shown in bold print.

	-5	-4	-3	-2	-1	1	2	3	4	5
1	-	**7**	-	**81**	**48**	**57**	**54**	**29**	**7**	-
2	-	**60**	**24**	**70**	**30**	**61**	**58**	**24**	-	1
3	-	-	-	-	**13**	**8**	**94**	**29**	37	-
4	-	-	14	-	**98**	**7**	**9**	-	-	-
5	-	-	-	-	-	-	-	-	-	-

TABLE VI Array similar to Table V for 5-km cells.

	-5	-4	-3	-2	-1	1	2	3	4	5
1	**29**	-	-	**118**	-	**82**	**71**	**21**	**155**	**58**
2	-	**80**	**196**	**159**	**17**	**13**	**58**	**16**	**25**	**36**
3	**118**	-	**24**	-	**21**	**85**	-	**9**	**66**	**46**
4	19	**129**	**43**	**21**	-	**59**	**129**	**115**	**45**	-
5	118	-	**24**	-	**21**	-	**52**	**28**	156	51

These results differ from one another but not by much. The solution based on 10 km cells is probably more accurate because the random fluctuations in Table VI exceed those in Table V. The contours for the 10 km solution are slightly elliptical. Replacing them by circular contours with the same areas gives:

$$\log_{10} \sigma\,(N_1,N_2) = 1.7645 - 0.0538\,(k^2+\ell^2)$$

From Eq. (8), it follows that

$$(2\sigma)^{-2} = 0.0538 \log_e 10 = 0.1239, \quad \text{and } \sigma = 1.42 \text{ (= 14.2 km)}.$$

If R represents the distance of a point to its cluster center, R^2 is a random variable distributed as $\sigma^2\chi^2(2)$ where $\chi^2(2)$ represents the chi-squared distribution for 2 degrees of freedom. A few selected probabilities for this distribution are:
$P\,(R<\sigma) = 0.393$; $P(R<2\sigma) = 0.865$; and $P(R<3\sigma) = 0.989$. Because σ = 14.2 km, we would expect most deposits associated with a cluster

center to be contained within a circle with radius equal to about 30 km. The distance between Helmstaed's (1973) two volcanic centers in the Bathurst area is approximately 40 km, suggesting 2 clusters with radii of the order of 20 km. The results from our application of the Neyman-Scott cluster process model are therefore compatible with the hypothesis of existence of two overlapping volcanic centers in the Bathurst area, each with an associated cluster of about 20 ore deposits.

Although this is an encouraging result, it is also apparent that estimation problems for this type of clustering model are considerable. It is easier to use the negative binomial model in combination with geostatistical estimation of the variance for number of deposits per cell of variable size (Agterberg, 1980) although an interpretation in terms of cluster centers is then not possible.

2. AUTO- AND CROSS-COVARIANCE OF VOLCANIC FLOWS ALONG EAST PACIFIC RISE AT 21°N

As shown in Figure 1, Agterberg and Franklin (in press) have quantified lithostratigraphic units represented in map-form by Ballard et al. (1981) for an area on the East Pacific Rise at 21°N. The purpose of this quantification was to characterize the environment of hydrothermal vents in terms of variables genetically related to the vents so that they can be used for estimating probabilities of occurrence. Initially, multiple regression was employed for this purpose; the results were refined by using the non-linear method of logistic regression. As shown in Figure 1, the coding for rock units was performed for subcells. Variables were defined for two sizes of cells: Small cells measuring (4 x 4) subcells and large cells consisting of (8 x 8) subcells. Regression analysis is based on product-moment correlation coefficients or the corresponding covariances. For variables defined for lithostratigraphic units, the latter are related to the geometrical shapes of the boundaries of the units on the map. In this section, this relationship will be studied for the variables x_1 and x_2, representing youngest pillow flows and youngest sheet flows, respectively. The occurrence of vents is strongly correlated with these youngest volcanics. Ballard et al. (1981) have pointed out that the ratio of sheet flows to pillow flows, increases in the southwesterly direction, together with other variables including topographic elevation of the ocean floor, and occurrence of hydrothermal vents and their exiting temperatures. In Agterberg and Franklin's (in press) study, the variable among all other variables considered for small cells, that is most strongly correlated with occurrence of vents is the product variable x_1x_2. The mean of this variable is linearly related to the covariance of x_1 and x_2. Consequently, it is important to investigate the change of this covariance as a function of size of the cells used for quantification of the variables x_1 and x_2.

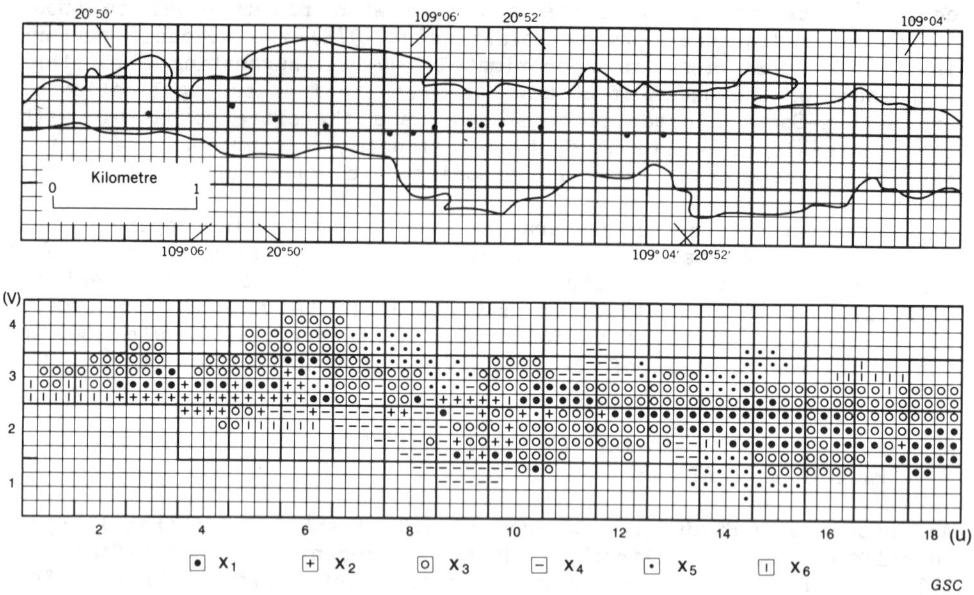

Figure 1: Coded version of area on East Pacific Rise at 21°N studied by Ballard et al. (1981). Upper part shows outline of young volcanic flows and occurrences of hydrothermal vents. Lower diagram shows coded for variables x_1-x_6 in subcells; x_1, x_3 and x_5 represent pillow flows, x_2, x_4 and x_6 are sheet flows; age groups are (x_1, x_2): youngest, (x_3, x_4): intermediate, (x_5, x_6): oldest. There are (4 x 4) subcells in a 'small' cell and (8 x 8) subcells in a 'large' cell. Small cells are shown by heavier boundaries and numbered in horizontal (U) and vertical (V) directions. Large cells consist of (2 x 2) cells each with one odd and one even number for U and V.

TABLE VII Frequencies (f) of pairs of adjacent subcell codes for x_1 and x_2 in lower half of Figure 1. For transformation of f into covariance (cov) and first autocorrelation coefficient (r_1), see text.

	f	cov	r_1
x_1x_1 (SW-NE)	68	0.0739	0.7252
x_1x_1 (SE-NW)	41	0.0418	0.4102
x_2x_2 (SW-NE)	25	0.0292	0.5725
x_2x_2 (SE-NW)	12	0.0132	0.2588
x_1x_2 (SW-NE)	4	0.0008	0.0117
x_2x_1 (SW-NE)	7		
x_1x_2 (SE-NW)	16	0.0079	0.1096
x_2x_1 (SE-NW)	5		

Table VII contains frequencies of pairs of adjacent subcells x_1x_1, x_2x_2, x_1x_2 and x_2x_1, in the horizontal (SW-NE) and vertical (SE-NW) directions of Figure 1. These frequencies can be interpreted as geometrical auto- and cross-covariances, respectively. Definition of a test area T and use of mean values of x_1 and x_2 in this test area allows the transformation from a geometrical covariance into a statistical covariance. The test area T for Figure 1 was taken to consist of the 51 small cells and contains 816 subcells. The transformation of the first geometrical covariance (= 68) for x_1x_1 in Table VII, is done as follows: First the geometrical covariance of T in the horizontal direction is computed. It amounts to $816-9 \times 4 = 780$. The ratio of the two geometrical covariances is $68/780 = 0.0872$. The mean value for x_1 in T amounts to $\bar{p}_1 = 0.1152$. This is the proportion of T underlain by subcells labelled x_1. It also represents the expected proportion of x_1 for a cell of any size and shape placed at random within T. The subcell variance of x_1 amounts to $s_1^2 = \bar{p}_1 - \bar{p}_1{}^2 = 0.1152-0.0133 = 0.1019$. Statistical covariances are obtained by subtracting $\bar{p}_1^2$ from the ratios of the geometrical covariances. The autocovariance for the horizontal direction is $0.0872-0.0133 = 0.0739$ (see Table VII). Division of this value by the variance s_1^2 yield the autocorrelation coefficient $r_1 = 0.7252$ for a single shift in the horizontal direction. The covariances and autocorrelation coefficients in Table VII were obtained in the same way. For this purpose, a mean value $\bar{p}_2$ and subcell variance s_3^2 were defined for x_2 as previously for x_1.

The procedure of calculating cross-covariances and cross-correlation coefficients is similar. The frequencies 4 and 7 for the horizontal direction were combined by averaging them. The ratio $5.5/780 = 0.00705$ was reduced by $\bar{p}_1\bar{p}_2 = 0.0621$. This gave the small cross-covariance 0.0008 for the horizontal direction shown in Table VII. Division by $s_1s_2 = 0.07209$ yields the corresponding first cross-correlation coefficient (=0.0117).

The values of x_1 and x_2 for small and large cells can be expressed as proportions of cell areas underlain by x_1 and x_2, respectively. The means remain equal to p_1 and p_2; the corresponding

TABLE VIII Comparison of observed (o) to estimated (e) variances and covariance for variables x_1 and x_2 in 'small' (4 x 4) and 'large' (8 x 8) cells. For further explanation, see text.

	r_1(ave)	p	s_e^2/s^2	s_e^2	s_o^2
x_1x_1 (4 x 4)	0.5677	0.566	0.36	0.0367	0.0362
x_2x_2 (4 x 4)	0.4157	0.878	0.22	0.0112	0.0098
x_1x_1 (8 x 8)	0.5677	0.566	0.17	0.0134	0.0093
x_2x_2 (8 x 8)	0.4157	0.878	0.09^5	0.0037	0.0027
	r_1(ave)	p	$s_e(x_1, x_2)/s_1s_2$	$s_e(x_1, x_2)$	$s_o(x_1, x_2)$
x_1x_2 (4 x 4)	0.0606	2.804	0.04	0.0029	0.0023

variances depend on cell size. The latter are shown as s_o^2 in Table VIII for small (4 x 4) and large (8 x 8) cells, respectively. The observed covariance of x_1 and x_2 in small cells amounts to $s(x_1,x_2) = 0.0023$. Theoretically, the results in the last column of Table VIII which are used in multivariate analysis can be obtained from the frequencies of adjacency data in Table VII. The following relatively simple technique illustrates this. Correlation coefficients for the horizontal and vertical directions in Figure 1 were averaged. Next, it was assumed that the resulting r_1(ave) belongs to a negative exponential model of the type $r_d = \exp(-pd)$ where d represents distance of shift. The coefficient p is obtained from r_1 with d = 1 (see Table VIII). The same type of model is assumed to hold true for the cross-covariance of subcells for x_1 and x_2. For square cells with edge a, the ratio of the variance of its proportion value s_e^2 to the subcell variance satisfies

$$s_e^2/s^2 = a^{-4}\int_s dx \int_s \exp(-pd)\, dy \tag{9}$$

In this expression x and y denote vectors in the plane and $d = ||x-y||$.

Suppose that the cell covariance of x_1 and x_2 is written as $s_e(x_1,x_2)$, then the ratio $s_e(x_1,x_2)/s_1s_2$ is equal to the right-hand side of Eq. (9) if the value of p for the subcell cross-covariance function is used.

Comparison of the estimated auto- and cross-covariances in Table VIII shows that the agreement is good for the (4 x 4) cells but poor for the (8 x 8) cells. The main reason that this approach did not yield good results for large cells is that the test area for large cells contained (17 x 64 =) 1088 subcells instead of the original test area T with 816 subcells. Because of edge effects related to enlargement of the test area for large cells by including more older volcanic flows, the estimated variances s_e^2 are probably better than the observed variances s_o^2 in Table VIII.

The frequency distribution of cell data can be modelled as a function of cell size by methods originally proposed by Matheron (1984). The observed frequencies of x_1 in (4 x 4) cells are given in the second column of Table IX. In total, there are 51 observations. Dividing the scores of x_1 by 16 and multiplication by 100 gives the cell proportion values p (in percent) also shown in Table IX. The values of f were converted into cumulative frequencies F_o (in percent). As much as 58.8 percent of the 51 cells are without youngest pillow sheets in Table IX. A subcell was assigned to a rock type in Figure 1 only if its cell center is underlain by this rock type on the map. Consequently, up to approximately half of a subcell area may be underlain by a rock type to which it was not assigned. For this reason, the first value of p in Table IX was set equal to 0.5/16 = 3.13 percent instead of zero. A continuous distribution model suitable for fitting to cell data with relatively many zeros was used in Agterberg and Fabbri (1978). It is fully determined by two parameters. For x_1, the mean $\bar{p}_1 = 0.1152$ and geostatistically determined variance $s_e^2 = 0.0367$ of Table VIII can be used. Suppose

TABLE IX Frequency (f) of scores for x_1 (= numbers of x_1 subcells per 4 x 4 cell). For further explanation, see text.

x_1	f	F_o (%)	p (%)	z_p	x_F	F_e (%)
0	30	58.8	3.13	-1.86	0.11	54.4
1	3	64.7	6.25	-1.53	0.35	63.7
2	5	74.5	12.50	-1.15	0.64	73.9
3	4	82.4	18.75	-0.89	0.83	80.0
4	1	84.3	25.00	-0.67	1.00	84.1
6	3	90.2	37.50	-0.32	1.26	89.6
7	2	94.1	43.75	-0.16	1.38	91.6
10	1	96.1	62.50	0.32	1.74	95.9
11	1	98.0	68.75	0.49	1.87	96.9
12	1					

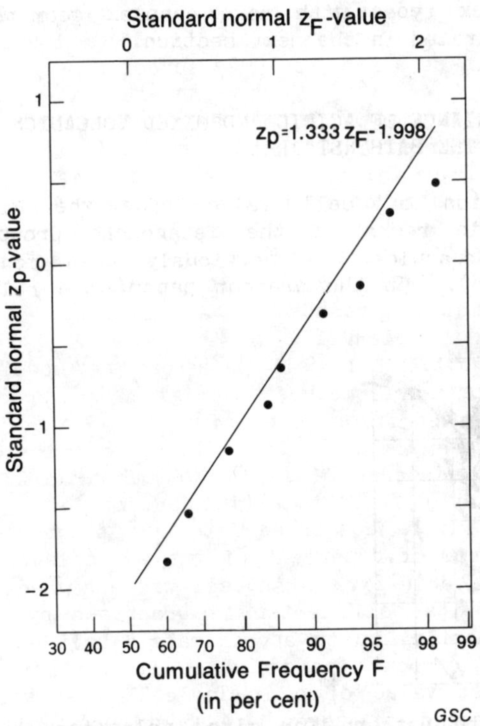

Figure 2: Frequencies of variable x_1 in small cells were transformed into standard normal z_p-values and plotted on normal probability paper. Theoretical straight line is based on geostatistical estimate of variance.

that z_F represents the z-value of the standard normal cumulative frequency $F = \Phi(z_F)$ and that z_p is the z-value of the proportion $p = \Phi(z_p)$. The model predicts a straight-line relationship between z_F and z_p with 2 coefficients which are determined by $\bar{p}_1$ and s_e^2. This straight line as well as the observed values in Table IX are shown in Figure 2. They occur slightly below the straight line reflecting the fact that the geostatistically derived variance s_e^2 on which the straight line is based is slightly greater than the observed variance s_o^2 (cf. Table VIII). The expected frequencies F in the last column of Table IX correspond to the theoretical model and are slightly less than the observed frequencies.

The method of frequency distribution analysis illustrated for x_1 in (4 x 4) cells in Figure 2 and Table IX can also be applied to x_2 and to both variables in (8 x 8) cells. The fit is also good in these instances but fewer observed frequencies are available. It may be concluded that the variables x_1 and x_2, after probit transformations, satisfy approximately a bivariate normal distribution.

The model for the relation between two lithostratigraphic units (x_1 an x_2) developed in this section and the preceding one is relatively simple. The same approach can be applied to study the relationships between rock types with more complex geometrical map patterns as will be illustrated in the next section.

3. AUTO- AND CROSS-COVARIANCE OF ACIDIC AND MIXED VOLCANICS OF ORDOVICIAN AGE IN THE BATHURST AREA

The frequency distribution of cell values for the Middle-Late Ordovician acidic volcanic rocks of the Tetagouche Group in the Bathurst area of New Brunswick has previously been modelled by Agterberg and Fabbri (1978). In the present paper, the relationship

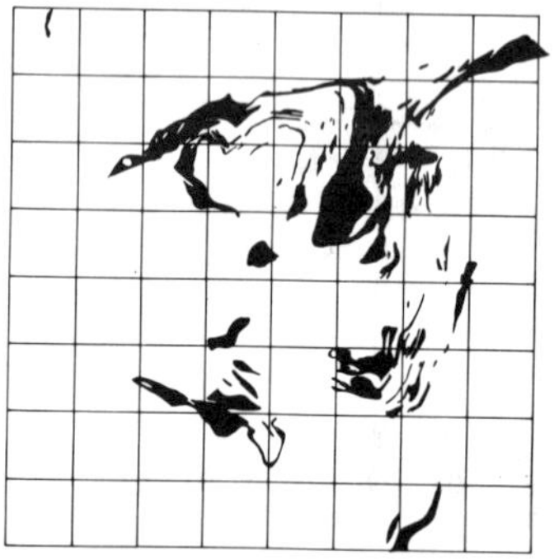

Figure 3: Black and white pattern for mixed volcanics in Bathurst area used with pattern of acadia volcanics (see Agterberg and Fabbri, 1978) to obtain measurements on Quantimet shown in Figure 4. North points upward. Grid of (10 km x 10 km) cells used to obtain point-counting data of Table X is superimposed.

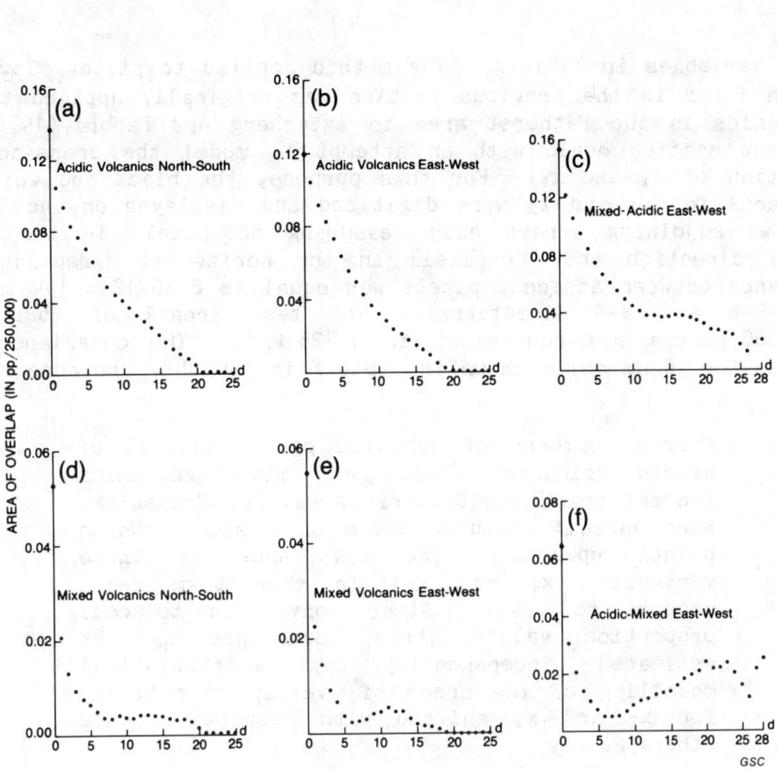

Figure 4: Measurements on Quantimet for patterns of acidic volcanics and mixed volcanics; (a), (b), (d) and (e) represent geochemical autocovariances; (c) one (f) are geometrical cross-covariances botained by moving other pattern across stationary pattern from the west. Unit of shift (d) is 2.33 km. pp. represents picture points or pixels.

between acidic volcanics and mixed volcanics is studied for this same area. 'Mixed volcanics' (see Figure 3) is a rock unit stratigraphically above the acidic volcanics and consisting primarily mafic volcanics mixed with minor acidic volcanics.

The geometrical covariance of the acidic and mixed volcanics was measured on an image analyzer (Quantimet 720 with linear correlator module). Results are shown for the N-S and E-W directions in Figure 4. The unit of shift (d) is 12 pixels or 2.33 km. The pattern for mixed volcanics in the E-W direction shows a secondary peak for d = 11 to 13 (or at about 28 km) which is related to the fact that mixed volcanics mainly occur in two N-S elongated zones (see Figure 3). The two variables x_1 (acidic volcanics) and x_2 (mixed volcanics) were also point-counted for (10 km x 10 km) cells belonging to the Universal Transverse Mercator grid. The square subcells used for this digitizing measure 500 m on a side so that the maximum score of a variable in a cell amounts to 400. These scores are shown for both variables in Table X. The method applied to pillow flows on the ocean floor in the previous section was originally applied to acidic volcanics in the Bathurst area in Agterberg and Fabbri (1978). The present section deals with an attempt to model the cross-covariance function of x_1 and x_2. For this purpose, the black and white input patterns for x_1 and x_2 were digitized and displayed on the Quantimet as two adjoining arrays each measuring 625 pixels in the vertical (N-S) direction and 400 pixels in the horizontal direction. Real distance between adjacent pixels was equal to 2330/12 = 194 meters in the E-W and N-S directions. The test area T of (625 x 400 =) 250,000 pixels was equivalent to 9,425 km^2. The covariance of the scores for 10 km cells shown in Table X is 1393.83. Because there are

TABLE X Scores (numbers of subcells per 10 km-cell of acidic volcanics (bold) and mixed volcanics (normal print) in Bathurst area, New Brunswick. Each subcell measured 500 m on a side. North points upward. The covariance of these variables (x_1 and x_2) in the 10-km cells amounts to 1393. After conversion to cell proportion values, this covariance can be estimated independently by geostatistical modelling of the areas of overlap of patterns for x_1 and x_2 shifted with respect to each other.

0	**0**	**0**	**0**	**0**	**0**	**0**	**0**
0	0	0	0	0	3	24	111
0	**0**	**10**	**172**	**294**	**51**	**0**	**0**
0	0	84	114	50	200	38	0
0	**28**	**211**	**378**	**296**	**8**	**52**	**0**
0	24	112	20	101	272	221	0
0	**0**	**175**	**350**	**253**	**70**	**229**	**52**
0	0	0	47	67	162	16	10
0	**0**	**116**	**333**	**382**	**342**	**267**	**6**
0	0	0	50	0	19	41	0
0	**30**	**12**	**222**	**301**	**198**	**56**	**3**
0	4	65	119	6	153	30	0
0	**0**	**0**	**85**	**17**	**10**	**12**	**0**
0	0	2	66	13	0	0	0
0	**0**	**0**	**4**	**1**	**0**	**24**	**0**
0	0	0	0	0	14	61	0

80 cells measuring 10 km on a side in Table X, the total area is 6,400 km^2 which is smaller than the test area T used on the Quantimet. For this reason, an adjusted covariance equal to 1569.60 was estimated by enlarging the area of the array in Table X to that of T using blank cells. Division of the adjusted covariance by (400 x 400) yields the corresponding covariance $s_o(x_1,x_2) = 0.0098$ of the proportions for x_1 and x_2 in 10 km cells. The corresponding covariance for 20 km cells amounts to 0.0116. The purpose of the application in this section is to estimate these covariances independently from the geometrical cross-covariances measured on the Quantimet (Figures 4c and 4f).

The latter measurements apply to area of overlap within the test area T between the patterns of acidic volcanics and mixed volcanics moved across one another in the east-west direction. For no shift (d = 0), one of the original patterns (acidic volcanics in Figure 4c; mixed volcanics in Figure 4f) is reproduced. For increasing amount of shift, the other pattern moves across the first pattern until the patterns approximately coincide for d = 26 (= 26 x 2.33 = 60.58 km) where the geometrical cross-covariances show local minimum values in Figure 4. The input for the Quantimet was prepared so that the patterns would approximately be in registration after a shift of 60 km. The theoretical amount of overlap is zero at this point because rock types on the geological maps are mutually exclusive. The geometrical cross-covariance is not necessarily symmetrical with respect to this origin.

The measurements for d ranging from 19 to 28 in Figures 4c and 4f were replotted in Figure 5 as follows. The patterns of Figures 4c

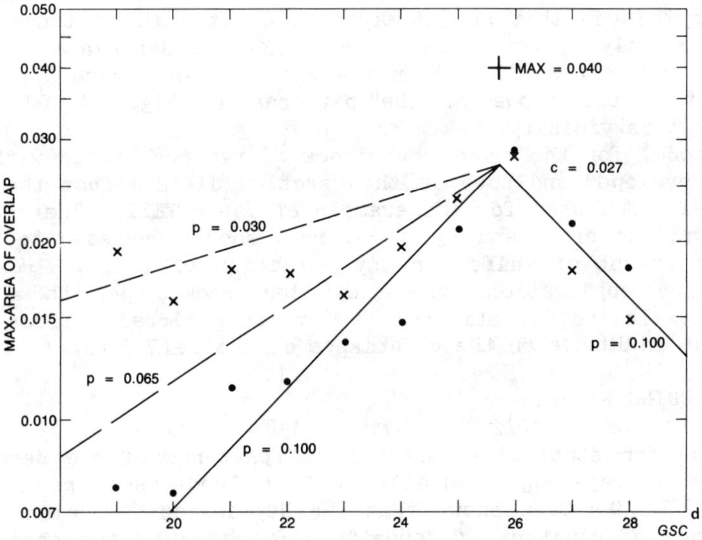

Figure 5: Measurements of Figure 4c (solid dots) and Figure 4f (crosses) replotted for d=19, 20, ..., 28. For further explanation, see text.

and 4f suggest that, in the vicinity of the origin, the geometrical cross-covariance can be approximated by a doubly-exponential curve with its origin at d = 60/2.33 = 25.75 asymptotically approaching the value MAX = 0.040 in both directions. The vertical axis in Figure 5, for MAX — area of overlap, has a logarithmic scale. The dots represent the measurements of Figure 4c and the crosses those of Figure 4f. Although the geometrical cross-covariance function is not necessarily symmetrical with respect to the origin at d = 25.75, the pattern of the crosses at one side of the origin should be symmetrical with respect to the pattern of the dots at the other side, and vice versa.

For geostatistical modelling similar to that for the application of Table VII, a symmetrical model of the type 0.04-0.027exp(-pd') with d' = d - 25.75 was assumed to hold true for the geometrical covariance with p set equal to 0.100, 0.065 and 0.030, respectively. In order to obtain the statistical cross-covariance, it is necessary to subtract $p_1 \cdot p_2 = 0.0103$ from the geometrical cross-covariance where $p_1 = 0.158$ and $p_2 = 0.065$ represent the proportions of the test area T underlain by x_1 and x_2, respectively. Consequently, the point covariance 0.03-0.027 exp(-pd') can be used to replace exp(-pd) in the previous example of Table VIII. For 10 km cells, this cross-covariance model results in estimates $s_e(x_1, x_2)$ equal to 0.0135, 0.0106 and 0.0068, for p = 0.100, 0.065 and 0.030, respectively (cf. Figure 4). The observed covariance $s_o(x_1, x_2)$ is 0.0098 indicating that p = 0.065 provides the best model for 10 km cells. On the other hand, $s_e(x_1, x_2)$ for 20 km cells is equal to 0.0195, 0.0152, and 0.0103, for the three models, respectively. Because $s_o(x_1, x_2) = 0.0116$ for 20 km cells, p = 0.030 is to be preferred. The reason for this is that any straight line fit such as those shown in Figure 4 is only approximately correct in the immediate vicinity of the point with d = 25.75. From the patterns in Figure 3, it can be inferred that the slope of the patterns in Figure 4 becomes less outside of this vicinity.

The model for the cross-covariance of two rock types with complex patterns developed and used in this section differs from the negative exponential model used for the example of Table VIII. The geometrical cross-covariance of x_1 and x_2 in Figure 1 should decrease to zero with decreasing amount of whift for any direction of shift. However, in this earlier application, the fact that rock types on a map are mutually exclusive constitutes a strictly local effect without significant influence on the covariance of the cell data.

5. CONCLUDING REMARKS

The methods for statistical analysis of patterns of ore deposits and rock types on maps apply equally well to land-based and ocean-floor deposits. It has been shown that the Neyman-Scott model can be used in the study of clusters of deposits. By means of two case studies, both dealing with volcanic rocks of different types, it was also shown that covariances of cell proportion data can be obtained by geostatistical modelling of cross-covariance functions.

6. REFERENCES

Agterberg, F.P., 1976, New problems at the interface between geostatistics and geology; in M. Guarascio, M. David and C. Huijbregts, eds., Advanced Geostatistics in the Mining Industry, Reidel, Dordreckt, Holland, pp. 403-421.

Agterberg, F.P., 1980, Geochemical crustal abundance models: Soc. Mining Engineers of AIME, Transactions, vol. 268, pp. 1823-1830.

Agterberg, F.P., and Fabbri, A.G., 1978, Spatial correlation of stratigraphic units quantified from geological maps: (computers and Geosciences, vol. 4, pp. 285-294.

Agterberg, F.P. and Franklin, J.M., in press, Estimation of the probability of occurrence of polymetallic massive sulfide deposits on the ocean floor; in P. Teleki, ed., Marine Minerals: Resource Assessment Strategies, Reidel, Dordrecht, Holland.

Ballard, R.D., Francheteau, J., Juteau, T., Rangan, C., and Normark, W., 1981, East Pacific Rise at 21°N: the volcanic, tectonic and hydrothermal processes of the central axis: Earth and Planetary Sc. Letters, vol. 55, pp. 1-10.

Diggle, P.J., 1983, Statistical analysis of spatial point patterns: Academic Press, London, 148 p.

Franklin, J.M., Lydon, J.W., and Sangster, D.F., 1981, Volcanic-associated massive sulfide deposits: Economic Geology 75th Anniversary Volume, pp. 485-627.

Helmstaedt, H., 1973. Structural geology of the Bathurst – Newcastle district: New England Intercollegiate Geol. Conf., Geology of New Brunswick, Univ. New Brunswick, Field Guide to Excursion, Trip 5-A, pp. 34-46.

Matheron, G., 1984, Isofactorial models and change of support; in G. Verly, M. David, A.G. Journel and A. Marechal, eds., Geostatistics for Natural Resources Characterization: Reidel, Dordrecht, Holland, pp. 449-467.

Ripley, B.D., 1981, Spatial Statistics: Wiley, New York, N.Y., 252 p.

POINT PROCESSES FOR THE EARTH SCIENCES

B. D. Ripley
Department of Mathematics
University of Strathclyde
26 Richmond Street
Glasgow G1 1XH
U.K.

ABSTRACT. Point processes are models for point-like objects such as trees in a forest and earthquake occurrences in space-time. They are used to produce precise descriptions of point patterns. The basic types of pattern are described together with informative graphical summaries.

1. WHAT IS A POINT PROCESS?

Point processes are statistical models for point-like objects or for objects which can be represented as points in some space. The history of their study started with points in time (calls to a telephone exchange) and in two-dimensional space (trees in a forest). Examples in more complicated spaces include earthquakes in three-dimensional space plus time. The theory for points in time is rather distinct because time is ordered and one can make good use of the observation that a point divides time into past and future.

This paper is principally concerned with spatial point processes in a plane. The methods all generalize to three-dimensional space and to space-time. However, two-dimensional point patterns are much easier to illustrate and have been by far the most common application. Before turning to the earth sciences it is illuminating to show some of the many other applications of point processes. Forestry has already been mentioned, and the location of plants has concerned ecologists throughout this century. Animal ecologists also study point patterns when studying territorial behaviour, for instance birds' nesting patterns and patterns of insects on the surface of a pond. Astronomers have studied patterns of stars and galaxies both on the celestial sphere and in three-dimensional space. Cell walls contain various small particles and craters whose pattern in muscle cells may be diagnostic of muscle diseases. Patterns and alignments of ancient monuments and burial places have been studied to assess various unconventional archaeological theories (of 'ley lines').

The study of earthquakes has been, to the author, the most convincing application of point processes in the earth sciences to date. This has been done on several scales, from the data in years over

C. F. Chung et al. (eds.), Quantitative Analysis of Mineral and Energy Resources, 301–322.

several centuries in Iceland (Gudmundsson and Saemundsson, 1980) to the location in space-time of microearthquakes (e.g. Vere-Jones, 1978). This example introduces the idea of a marked point process. As well as a location in space-time, earthquakes have a magnitude, which is associated as a mark to each point. (Other examples of marked point processes are trees marked by size and stars marked by brightness.) A completely different example is the locations of magnetite crystals in a 80cm x 100cm slab of anorthosite rock shown by Davis (1973, pp.308-311) whose pattern is analysed by him and by Ripley (1981, pp.175-180). Other potential examples are point patterns of bore holes, oil fields and mineral seams, all viewed on a large scale.

These are all rather natural examples of representing small objects as points. A less obvious example is that of small geological faults (Stoyan, 1983). Here the faults are considered as short line segments, and can be represented as points in a four-dimensional space giving the centre and direction, with length of the fault as the mark. This device is rather powerful, and is the basis of the subject of stochastic geometry. Another example is from studies of alignments of points. Points are considered in triples. Three points form a triangle: if the points are nearly aligned the largest angle of the triangle will be nearly 180°, and conversely. Thus one could represent the three points as a point in a two dimensional space showing the largest angle (a measure of shape) and the longest side (a measure of size). Further details of this and other representations are given by Broadbent (1980), Kendall and Kendall (1980) and Kendall (1981,5). Deffeyes et al (1982) give other applications of stochastic geometry, to petroleum geology.

2. TYPES OF POINT PATTERN

The aim of statistical point process theory is to describe patterns of points and to indicate a plausible mechanism by which they might have been generated. Many "indices" characterizing point patterns have been proposed, particularly in ecology, but these are all too simplistic and so fail in their aim of characterization.

The baseline for studies of point patterns is no interaction at all between the points. This is often referred to as a "completely random" pattern, and is illustrated in figure 1a. The points are put down independently and uniformly throughout the region of study.

There is still no interaction between the points in figure 1b. However, the density of points varies from place to place in the square. Such a pattern is called heterogeneous. Figure 1c is also heterogeneous but shows a clear directionality. Directional patterns are called anisotropic.

The examples of figure 2 are all homogeneous and isotropic but do exhibit interactions between the points. Figures 2a and 2b are both examples of clustering, in which parent points are laid down at random and then daughter points are distributed around the parents. The pattern in figure 2c is a rather extreme form of regularity, with figure 2d exhibiting mild regularity.

Figure 3 shows some more complicated patterns. If interaction is

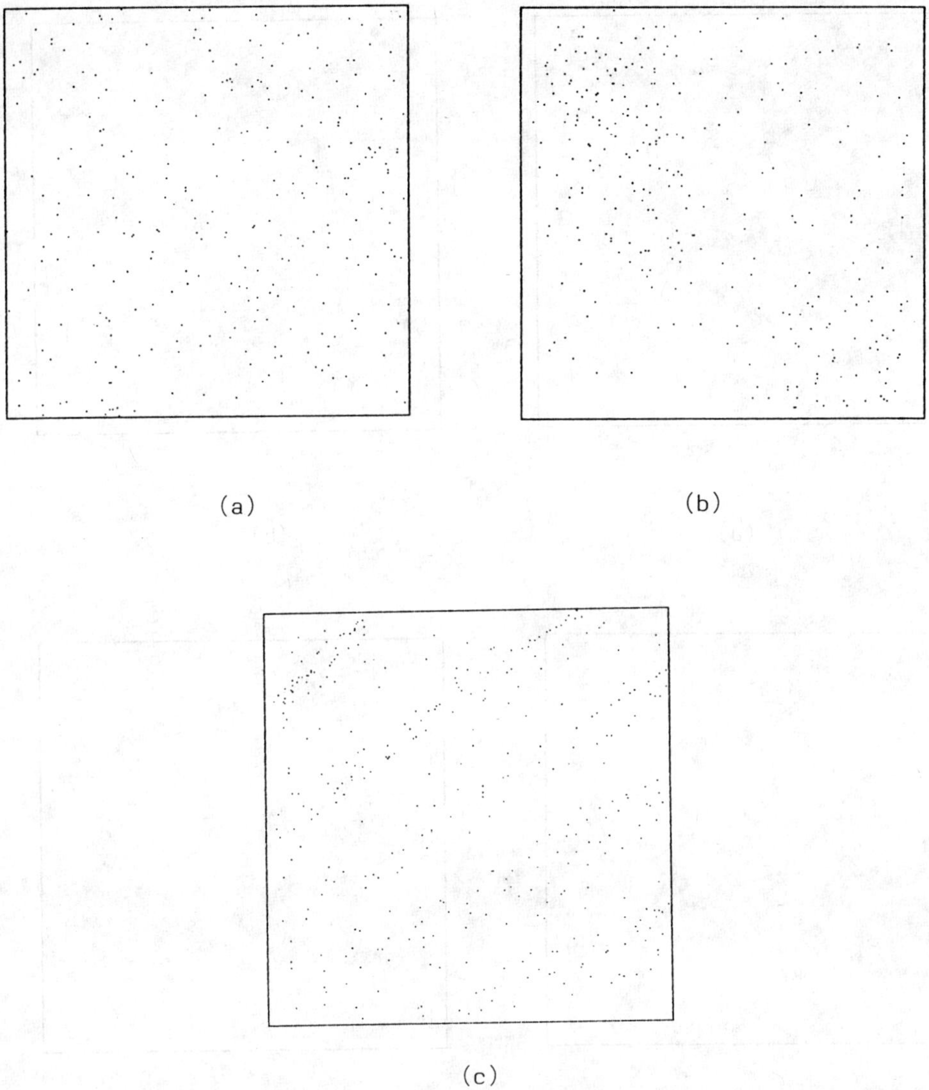

Figure 1. (a) A "completely random" pattern of 300 points.
(b) Heterogeneity: a plot of 314 points in a "patchy" pattern. (c) A banded pattern of 367 points.

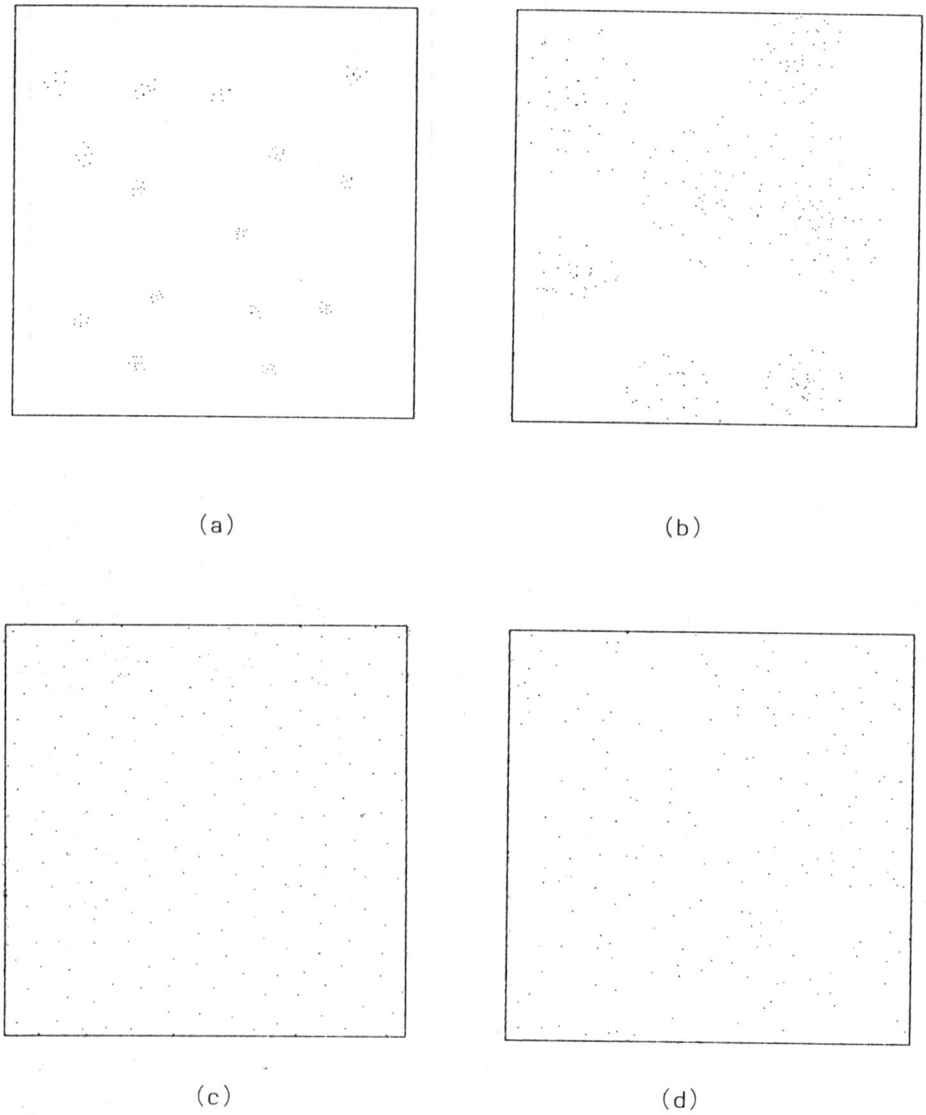

Figure 2. (a,b) Two clustered patterns. (c,d) Two regular patterns of 200 points each.

(a)

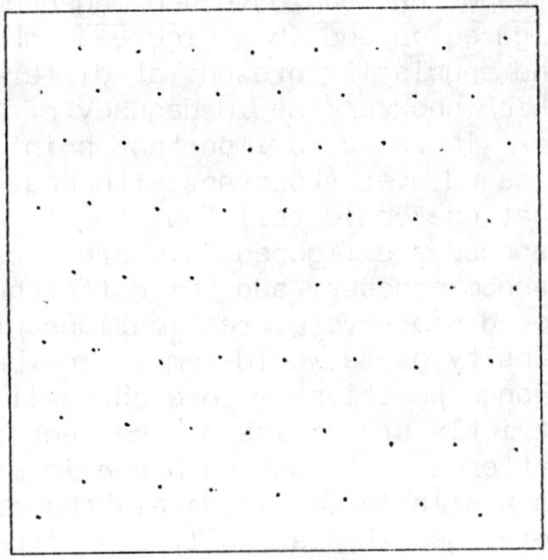

(b)

(c)

Figure 3. (a) An anisotropic clustered pattern. (b) A strongly anisotropic regular pattern. (c) A pattern in the unit square which shows regularity at a scale of 0.02, clustering at a scale of 0.08 and regularity at a scale of 0.2.

present one can have a homogeneous anisotropic pattern, as shown in figures 3a and 3b. Figure 3c shows that one can have both clustering and regularity present at different scales. It is examples like this which show up the inadequacy of "indices".

There is an important point which the above description has glossed over (together with most of the literature). There is no way that one could tell from the figures that figures 1b,c are heterogeneous and figures 2a,b are clustered. Each is a realization of a random process, and the difference between them will only emerge if we could see several realizations of each process. In figure 1 the high density parts would remain in the same areas, whereas in figure 2 they change positions with each realization. All the terms we have used are strictly properties of the generating processes rather than of the patterns. It must be borne in mind that when only one sample pattern is available (as is usual) the distinction between clustering and heterogeneity is really one of convenience. We use heterogeneity where there is a simple description of the density variation (as in figure 1c) and clustered where the pattern appears "patchy".

There is an underlying philosophical problem. In fact one cannot tell from figure 1a that it is random at all. It might be that the objects would always have occurred at those particular points. Nevertheless the simplest description of such a pattern is as a realization of a random process. Similarly, the simplest descriptions of figures 1c and 2b are heterogeneity and clustering respectively. There is a considerable debate in the geographical literature about whether stationarity (of which homogeneity and isotropy are particular forms) is ever a useful assumption. It is a modelling assumption. From a large enough point of view, including the geological processes, all earth science is stationary. Sometimes it is more convenient to include in the model the particular background structure. For point patterns stationarity is very useful. (For more of this debate see Bennett and Haining, 1985, especially the discussion.)

3. BASIC THEORY

A point pattern can be described by giving the coordinates of each of the points, $\mathbf{x}_1,\ldots,\mathbf{x}_n$. Both the number of points n and the coordinates are thought of as random. The order of labelling the points is irrelevant so $\{\mathbf{x}_1,\ldots,\mathbf{x}_n\}$ should be thought of as an unordered set. It turns out to be easier to use an equivalent description. Let N(A) denote the number of points within the set A. If this is known for all sets A we can certainly locate the points. (Take $A_1 = \{\mathbf{x}_1,\ldots,\mathbf{x}_n\}$, $A_2 = A_1^c$. Then $N(A_1) = n$ and $N(A_2) = 0$ tells us that the points are $\mathbf{x}_1,\ldots,\mathbf{x}_n$.) Thus we replace the point pattern by the counting process N().

Thus far we have allowed only a finite number of points. It is difficult to define homogeneity and isotropy carefully for a process defined on a bounded region, so we change our point of view, and suppose that the point pattern extends throughout space and is viewed within a window E. This is normally more realistic in that it allows the interaction of unobserved points outside the window with points inside the window. This introduces the problem of edge effects. Unless E is a very simple shape and n is very large, most of the points will be near the edge and so be interacting with unobserved points. (This can be significant even asymptotically as $n \to \infty$.) In spatial point patterns edge effects can not be ignored but have to be circumvented or compensated for.

The mean value, $\Lambda(A)$, of $N(A)$ is known as the intensity measure. For a homogeneous pattern the average density of points is constant throughout space, so

$$\Lambda(A) = \lambda \times \text{area}(A)$$

and λ is known as the intensity. More generally, it is usually reasonable to assume that the intensity has a smooth density $\lambda(\mathbf{x})$, so

$$P(\text{point is in } A_{\mathbf{x}}) \approx E[N(A_{\mathbf{x}})] \approx \lambda(\mathbf{x})\text{area}(A_{\mathbf{x}})$$

where $A_{\mathbf{x}}$ is a very small set containing $\mathbf{x}$. For a heterogeneous point pattern the most informative description is often a map of $\lambda()$ (section 6).

To consider interactions we have to consider two or more points simultaneously. For a homogeneous isotropic point process we will have

$$P(\text{point in } A_{\mathbf{x}} \text{ and one in } A_{\mathbf{y}}) \approx \lambda^2 g[d(\mathbf{x},\mathbf{y})]\text{area}(A_{\mathbf{x}})\text{area}(A_{\mathbf{y}})$$

where $d(\mathbf{x},\mathbf{y})$ is the distance between $\mathbf{x}$ and $\mathbf{y}$. Thus the function g will be identically one if there is no interaction between the points, less than one if a point at that distance has an inhibitory effect, and greater than one if the effect is positive. Thus g-1 is sometimes known as a pair correlation function. It proved to be difficult to estimate, so an equivalent form is used. Consider

$$\begin{aligned} &E(\text{number of points within distance t of point in } A_{\mathbf{x}}) \\ &= \lambda^2 \text{ area}(A_{\mathbf{x}}) \int_0^t g(r)\, 2\pi r dr \\ &= \lambda^2 \text{ area}(A_{\mathbf{x}})K(t), \text{ say.} \end{aligned}$$

Thus

$$E(\text{number of points within distance } t \mid \text{point at } \mathbf{x}) = \lambda K(t) \quad (1)$$

Estimates of the function K have proved the most informative descriptions of interaction between points. From K one can find cov{N(A), N(B)} for any pair of sets A. Thus we use first moments to detect and describe heterogeneity but second moments for interaction.

However, first and second moments do not characterize a point process. [This has been known for some time but perhaps insufficiently appreciated; Baddeley and Silverman (1984) give an explicit counter-example.] Thus other summaries can also be useful. A simple descriptive statistic is

$$p(t) = P[N(\{\mathbf{x} | d(\mathbf{x},\mathbf{0}) \le t\}) > 0] \tag{2}$$

$$= P(\text{there is a point within distance } t \text{ of the origin})$$

introduced by Ripley (1977). The complementary function

$$1-p(t) = P(\text{disc radius } t \text{ is empty})$$

has subsequently been called an "empty-space statistic" by Lotwick and Silverman (1982).

The function p(t) is the cumulative distribution function of the distance from a fixed point (the origin) to the nearest point of the pattern. This must not be confused with methods based on nearest-neighbour distances from point to point. Let d_i be the distance from point $\mathbf{x}_i$ to its nearest neighbour. Then if we select a point at random we select a sample from $\{d_1,\dots,d_n\}$. Let F(t) be the empirical distribution function of $\{d_i\}$, that is

$$F(t) = \frac{1}{n}(\text{number of } d_i \le t)$$

Then F measures something of the local interaction of the pattern and has been advocated by Diggle (1979, 1983). It has proved less popular than K, for four reasons. To use F one needs to record all small distances between pairs of points. However, F retains only a small proportion of these measurements whereas K uses them all. Second, F is necessarily limited to short-range interactions and (third) is not easy to correct for edge-effects. Finally, the theory for estimating K is much better understood than that for F.

The theory for marked point processes is a simple extension; just confine attention to a single mark (for a finite set of marks) or to a range of marks. Thus, for example, one gets

$$E(\text{number of b-points within distance } t \,|\, \text{a-point at } \mathbf{x}) = \lambda K_{ab}(t)$$

Details of the theory are given by Hanisch and Stoyan (1979), Lotwick and Silverman (1982), Diggle (1983, Chapter 6) and Upton and Fingleton (1985, pp. 255-258).

The "empty-space" ideas can be extended to objects of finite size; Guild and Silverman (1978), Ripley (1986).

4. ESTIMATION FOR HOMOGENEOUS PROCESSES

The main problem in estimation is that of edge effects. One obvious way to circumvent edge effects is to artificially remove the edge. This can be done by using a "border" or "guard area" within E. Suppose we can find a set E* within E so that

$$\{\mathbf{y} | d(\mathbf{x},\mathbf{y}) \leq t \text{ for some } \mathbf{x} \in E^*\} \subset E$$

We will then be able to account for all points within distance t of a point in E*. Although a simple idea, this is usually impracticable. If E is a complex shape there may be no such set E*. If E* is non-empty it may be so much smaller than E that most of the observed points are not in E*. Finally, the number of points in E* is random, which introduces unnecessary uncertainty into our estimation. The main attraction of the border method is its simplicity.

Consider first estimating p(t). From (2), by homogeneity,

$$p(t) = P[N(\{\mathbf{x}|d(\mathbf{x},\mathbf{y}) < t\}) > 0]$$

for any point **y**. This suggests the unbiased estimator

$$\hat{p}(t) = \frac{1}{m}[\text{number of } \mathbf{y}_i \text{ within t of a point}]$$

where $\{\mathbf{y}_1,\ldots,\mathbf{y}_m\}$ is a fixed set of sampling points within E*. A slight change of point of view enables us to estimate p() for all t, not just one distance. Let d_i be the distance from $\mathbf{y}_i$ to the nearest point, and r_i the distance to the boundary of E. Then

$$\hat{p}(t) = [\text{number of } d_i \leq t \leq r_i]/[\text{number of } r_i \geq t]$$

This is again unbiased for any set of points $\{\mathbf{y}_1,\ldots,\mathbf{y}_m\}$ from E; it automatically defines

$$\{\mathbf{y}_i\} \cap E^* = \{\mathbf{y}_i | r_i > t\}$$

Ideally m should be very large, so that $\{\mathbf{y}_i,\ldots,\mathbf{y}_m\}$ densely covers E. In practice m = min(10n, 1000) seems sufficient (Ripley, 1981, p.154).

To use the border method when estimating K(t) we return to definition (1). For each point $\mathbf{x}_i$ in E* we count the number T_i of other points within distance t. Let the total of these counts be T, from N ($\leq$ n) points in E*. Then

$$ET = E[E(T|N)]$$

$$= E[\sum_1^N E(T_i|N)] = E[NE(T_1|N)]$$

$$= E[N\lambda K(t)] = \lambda K(t)EN = \lambda K(t)(\lambda a^*)$$

$$= \lambda^2 a^* K(t)$$

where a^* is the area of the window E^*. Thus T/a^* is an unbiased estimator of $\lambda^2 K(t)$. To estimate $K(t)$, we estimate λ by $\hat{\lambda} = n/a$ (where a = area(E)), so

$$\hat{K}_{border}(t) = a^2 T/a^* n^2$$

This is not unbiased, but its bias is negligible unless n is rather small (less than 20, say).

Ideally we would like to use all pairs of points to estimate K(). Several methods have been suggested to do so. If in the procedure described above we take E^* equal to E, T is the number of (ordered) pairs of points at most distance t apart, and $\hat{K}(t)$ is then a times the proportion of all pairs of points $(\mathbf{x}_i, \mathbf{x}_j)$ with $0 < d(\mathbf{x}_i, \mathbf{x}_j) \leq t$. This will be biased downwards because we ignore points closer than t which are outside E. The idea of the edge corrections is to estimate the effect of those points.

The correction of Ripley (1976, 1977) is based on isotropy. Suppose we rotate the pair $(\mathbf{x}_i, \mathbf{x}_j)$ with $d(\mathbf{x}_i, \mathbf{x}_j) \leq t$ about $\mathbf{x}_i$. Let ϑ be the proportion of the rotation that $\mathbf{x}_j$ is within E. Now unless $\mathbf{x}_i$ is within distance t of the boundary, $\vartheta = 1$. However, if $\vartheta < 1$ we might miss other points at distance $d(\mathbf{x}_i, \mathbf{x}_j)$ from $\mathbf{x}_i$. Thus we count the pair $(\mathbf{x}_i, \mathbf{x}_j)$ $1/\vartheta$ times to compensate. With this method of counting we get T_R. Then

$$ET_R = \lambda^2 a K(t)$$

provided there are two points of E distance 2t apart. [Thus for a rectangle the limit on t is half the length of the diagonal. Ohser (1983) lifted even this restriction.]

The edge-correction of Ohser and Stoyan (1981) is based on homogeneity. Again the pair $(\mathbf{x}_i, \mathbf{x}_j)$ is weighted in the count. This time $\mathbf{x}_i$ is moved about in E and the proportion ϑ of the motion that $\mathbf{x}_j$ is also within E is recorded. (This motion can either be translation or translation plus rotation. Figure 4 illustrates the computation of ϑ.) Notice that in this edge correction ϑ is always less than one, so a computation has to be done for each pair of points, whereas for Ripley's for most pairs $\vartheta = 1$ for small t. One again gets an unbiased estimator of $\lambda^2 K(t)$ and a negligibly biased estimator of K(t).

Some very intricate theoretical work has been done to compare these three estimators of K(t), (Ripley, 1984a; Ohser, Ripley and Stoyan, 1986). This shows that the border-corrected estimator is much more variable than the other two with a slight overall advantage to Ripley's form.

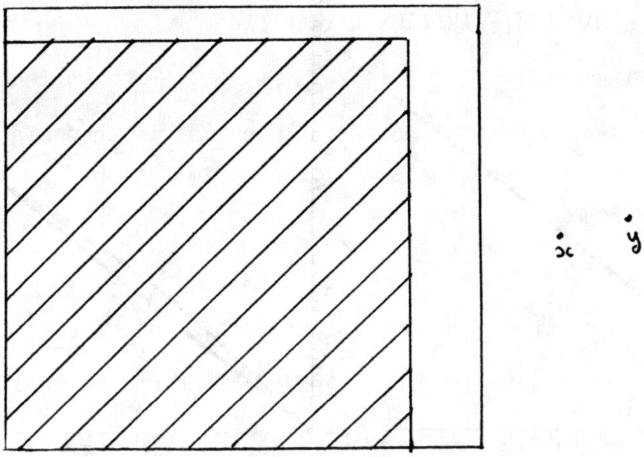

Figure 4. The proportion of the hatched area is ϑ for the Ohser-Stoyan edge correction.

Both $\hat{K}()$ and $\hat{p}()$ provide graphical summaries of the interactions present at various scales. Figures 5 and 6 illustrate $\hat{K}()$ for our examples; $\hat{p}()$ was not too illuminating for these datasets. Quite a lot of thought has been given to computing them efficiently, and they are now rapid to compute even on personal computers. [Ripley (1981, Chapter 8) gives some computational details; Appleyard et al (1985) discuss a study with $n \approx 500$ done on an early Commodore PET, a very modest computer.]

One might worry how much these techniques are affected by inaccuracies in the recording of the points. In general small measurement errors seem to have little effect; Ripley and Silverman (1978) suggest that one can still detect regularity reliably with inaccuracies up to one third of the critical point for L_m given below.

5. MODELS FOR HOMOGENEOUS PROCESSES

In order to interpret our summaries we need to know what they will look like under various models. We can see immediately that $\hat{K}(t)$ will be low at short distances for a regular pattern, and high for a clustered pattern. Similarly, $\hat{p}(t)$ will indicate less empty space for a regular pattern than for a clustered pattern and so give a curve that rises to one more rapidly. However, to make use of these observations we need to know what to expect for a random pattern.

The baseline model, for "complete randomness", is the binomial process. A fixed number n of points are laid down independently and uniformly within E. For a Poisson process the number of points n is

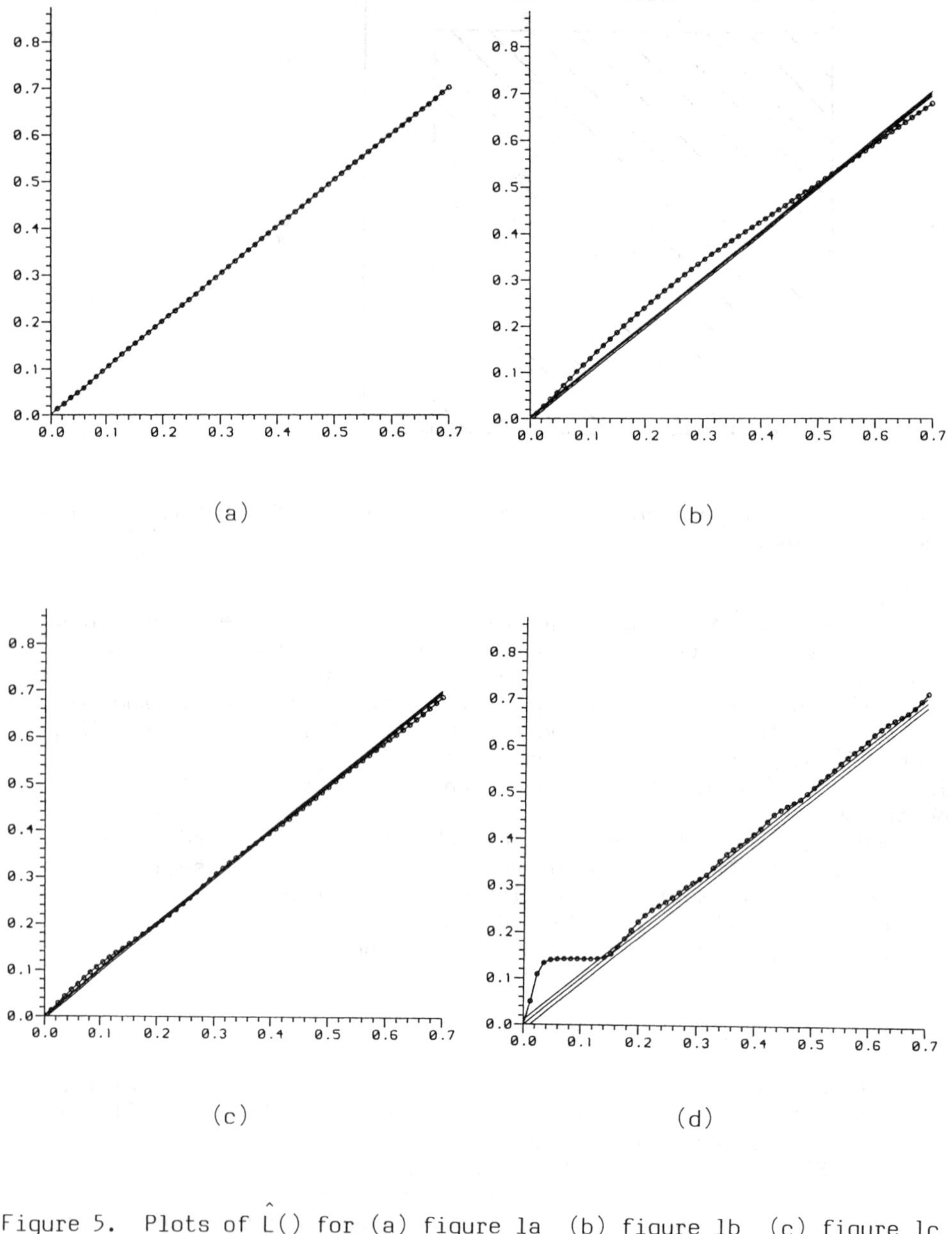

Figure 5. Plots of $\hat{L}()$ for (a) figure 1a (b) figure 1b (c) figure 1c (d) figure 2a (e) figure 2b (f) figure 2c (g) figure 2d (h) figure 3c. The diagonal lines are $y = t$ and $y = t \pm 1.42\sqrt{a}/n$.

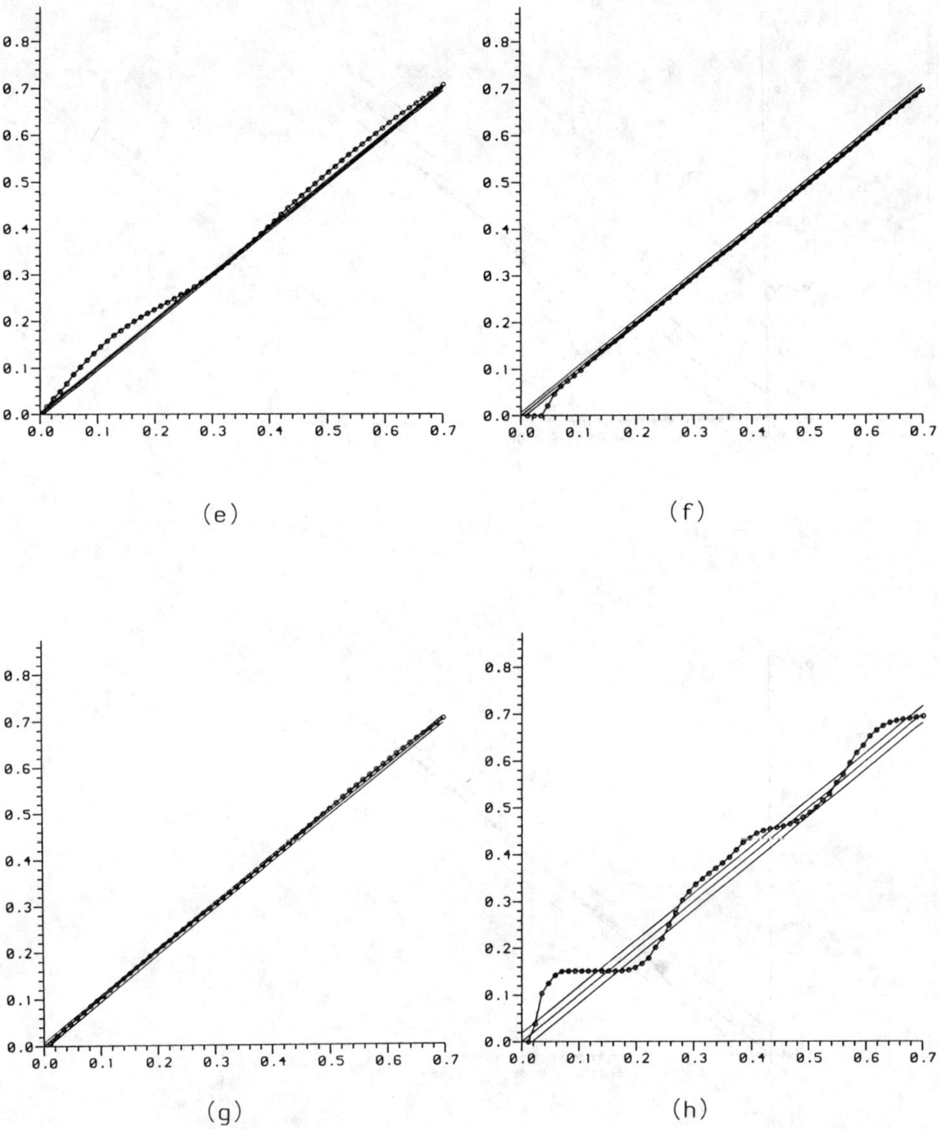

Figure 5 continued

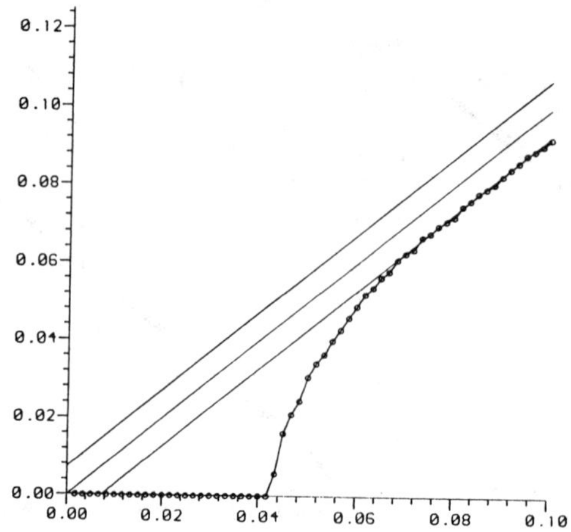

(a)

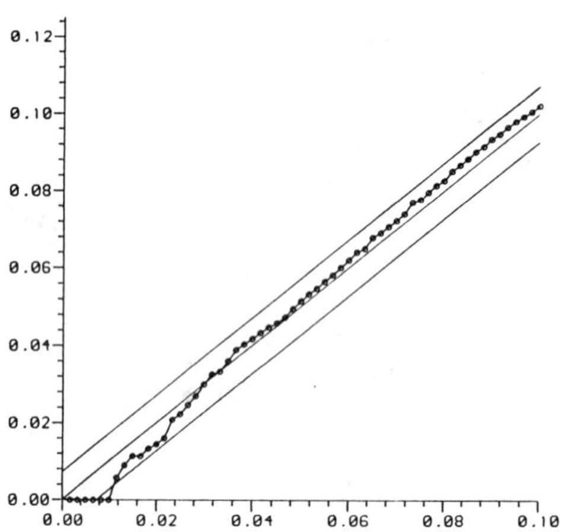

(b)

Figure 6. Enlarged plots of $\hat{L}()$ for small t for (a) figure 2c and (b) figure 2d. Note that L_m rejects randomness of figure 2d at a 5% level.

selected from a Poisson (λa) distribution, and the points distributed according to a binomial process. The major characteristic of a Poisson process is that parts of the pattern in non-overlapping regions are independent. Thus whereas a binomial process can only be defined within a specific window E (and so can not be stationary), a Poisson process can be defined throughout space by building up its realization from bounded non-overlapping regions. Thus a "completely random" or "random" process is always interpreted as meaning a Poisson process. If we take a Poisson process, observe it within the window E and condition on the number of points we return to a binomial process (by the definition). (In technical terms, the number of points n is approximately ancillary to the interaction, so statistical theory recommends conditioning on n.)

We can find p() and K() for a Poisson process. Notice that

$$1-p(t) = P[N(\text{disc centre } 0, \text{ radius } t) = 0]$$

and now N(A) has a Poisson distribution, mean λarea(A). Thus

$$1-p(t) = P[N(A) = 0] = e^{-\lambda \text{area}(A)} = e^{-\pi\lambda t^2}$$

so

$$p(t) = 1-e^{-\pi\lambda t^2} \qquad (3)$$

If λ is unknown (as usual) we will need to estimate it by $\hat{\lambda} = n/a$.

From the definition of K(t),

$$\lambda K(t) = E(\text{number of points within distance } t \mid \text{point at } \mathbf{x})$$

$$= E(\text{number of points in a disc radius } t)$$

by the absence of interaction,

$$= \lambda \times \text{area of disc} = \lambda\pi t^2$$

using the Poisson distribution. Thus for a Poisson process

$$K(t) = \pi t^2$$

(Alternatively, $g() \equiv 1$, whence $K(t) = \pi t^2$.) These results are specific to two dimensions.

The expected form of K(t) suggested to a number of people that it should be plotted on square-root scale. Define

$$L(t) = \sqrt{[K(t)/\pi]}, \quad \hat{L}(t) = \sqrt{[\hat{K}(t)/\pi]}$$

Then for a random process L(t) = t, and we can interpret deviations of $\hat{L}(t)$ from a diagonal line as departures from randomness. This transformation has a (coincidental) side benefit. Remember that $\hat{K}(t)$

is proportional to a (pseudo-) count T, and that statisticians like to plot square roots of counts to stabilize the variance. Thus (in two dimensions only) $\hat{L}()$ has the desirable features of a linear expected shape and more-or-less constant variability about its mean. All that remains to be stated is the extent of that variability.

Suppose we measure the deviation of $\hat{L}(t)$ from L = t by

$$L_m = \max_{t \le t_o} |\hat{L}(t)-t|$$

where t_o is the maximum interaction distance of interest. Then to a rather good approximation

$$P(L_m \le 1.42 \sqrt{a}/n) \approx 5\%$$

$$P(L_m \le 1.68 \sqrt{a}/n) \approx 1\%$$

for moderate t_o. (The virtual independence of t_o results from the fortunate observation that $\hat{L}(t)$ is in fact slightly more variable at very small distances, from the discreteness of the count.) This enables us to say not only what we mean by K(t) being "high" or "low" but to indicate whether we have significant evidence of a departure from randomness.

Many other portmanteau tests of non-randomness have been proposed, but the test based on L_m is overall the most powerful known. It has the advantage of being able to detect non-randomness at a variety of scales (and being unlikely to confuse them), and if the null hypothesis is rejected of giving a graphical explanation of what is wrong.

We can obtain some simple models of clustered patterns by starting with a Poisson process and replacing each point with a cluster. (Such models are, not unreasonably, called Poisson cluster processes.) The clusters are independent copies of a daughter process, a finite point process which is heterogeneous and may be anisotropic. One of the simplest descriptions of a daughter process is to distribute a fixed number of points independently around the parent, with a rotationally symmetric distribution. These are Neyman-Scott cluster processes, for which K() and p() are known (Ripley, 1981, p.164). The important form is that of K(t),

$$K(t) = \pi t^2 + \text{const} \times f(t)$$

and f(t) is the cumulative distribution function of the distance between two points of the daughter process. We can conclude that K(t) increases monotonically above expectation (under randomness) and that the distances over which it does so measure the size of the daughter cluster.

Formal models for regular patterns cause much greater mathematical difficulty. We can borrow ideas from statistical physics but exact calculations are notoriously difficult. The basic idea is to reduce the probability of patterns with points close together. Consider just n points within E. We can modify the density of $\{\mathbf{x}_i\}$ from uniformity in E^n to

$$p(\mathbf{x}_1,\ldots,\mathbf{x}_n) = \text{const} \prod_{i<j} h[d(\mathbf{x}_i,\mathbf{x}_j)]$$

Then $h < 1$ inhibits points at that distance apart. Such models are called pair-potential models and will generally have $h(t) < 1$ for $t < R$, $h(t) = 1$ for $t \geq R$. An extreme example is $h(t) = 0$ for $t < R$. This corresponds to deleting all realizations of a Poisson process with any pair of points closer than R apart. Less drastic is the Strauss process of Kelly and Ripley (1976), which takes $h(t) = c < 1$ for $t < R$.

Unfortunately K() and p() are unknown for such models. They can be estimated by simulation and there are some approximations from statistical physics which work only when interactions are very slight. Simulations of the Strauss model (Ripley, 1977, p.188) show that K() is low for $t \leq R$ and rises rapidly to near expectation by $t = 2R$, thus confirming that regular processes do behave as expected.

Formal statistical theory would estimate parameters in these models, say by maximum likelihood, and so fit the models to data and test their goodness-of-fit. For these models this is well beyond our current understanding and *ad hoc* methods have to be used. Even for a Neyman-Scott cluster process the likelihood cannot be computed, since it involves allocating observed points to clusters. For pair-potential models some approximate techniques are available (Ogata and Tanemura, 1981, 1984; Penttinen, 1984; Glotzl and Rauchenschwandtner, 1981; Fiksel, 1984).

6. ESTIMATING HETEROGENEITY

In studying interaction we have worked mainly with the second-moment statistic K(). Here we return to the first moment and the estimation of $\lambda()$ as a smooth function of space. There are some problems here, since we no longer have stationarity and are unable to deduce from the data how smooth $\lambda()$ actually is. Yet we do need to assume smoothness, otherwise we may as well just report where the points fell!

Similar problems occur when considering how to estimate probability density functions both in one and in more dimensions. Silverman (1981, 1985) gives some two-dimensional examples. Perhaps the most widely-used method in that field is kernel density estimation. Here we use

$$\hat{\lambda}(\mathbf{x}) = \frac{1}{ah^2} \Sigma K\left(\frac{\mathbf{x}-\mathbf{x}_i}{h}\right) \tag{4}$$

for a kernel $K \geq 0$ with $\int K(\mathbf{x})d\mathbf{x} = 1$. Think of $K(\mathbf{x}-\mathbf{y})$ as a smooth "blip" around $\mathbf{y}$; an explicit example is

$$K(\mathbf{y}) = (2\pi)^{-1} \exp\{-\|\mathbf{y}\|^2/2\}$$

the bivariate normal density. The scalar h is the smoothing constant. Large h gives a smooth estimate, small h a rough one; see figure 7. The problem is how to choose h. Some ideas are available for estimating densities (Silverman, 1985) but they depend essentially on the independence of $\mathbf{x}_1,\ldots,\mathbf{x}_n$. We can meet this condition for a heterogeneous Poisson process. This has a random number n of points within E, and conditional on n they are independent with pdf

$$f(\mathbf{x}) = \lambda(\mathbf{x})/\Lambda(E)$$

Thus we can estimate λ by $\hat{\Lambda}(E)\hat{f}$, where $\hat{\Lambda}(E) = n/a$. In fact (4) is just a kernel density estimate for f

$$\hat{f}(\mathbf{x}) = \frac{1}{nh^2} \Sigma K(\frac{\mathbf{x}-\mathbf{x}_i}{h})$$

multiplied by n/a.

A problem arises at the edges, where $K(\mathbf{x}-\mathbf{y})$ is partially outside E. This can be overcome in various ways; in figure 7 the scaling was adjusted so that

$$\hat{\lambda}(x) = \frac{1}{a} \Sigma K(\frac{\mathbf{x}-\mathbf{x}_i}{h}) / \int_E K(\frac{\mathbf{y}-\mathbf{x}}{h}) d\mathbf{y}$$

The denominator is h^2 unless $\mathbf{x}_i$ is near the boundary of E.

If interaction and heterogeneity are both present then there is greater difficulty. The most common example seems to be local regularity with large-scale heterogeneity, for which the lack of independence causes no problem. A combination of clustering and heterogeneity would, of course, be very difficult to unravel.

Some examples of intensity estimation are given by Cox (1979) and Diggle (1981, 1983).

7. ANISOTROPY AND OTHER EXTENSIONS

The methods of sections 3 to 5 can be extended to investigate the anisotropy of homogeneous point patterns. The generalization of the empty-space statistics is almost obvious. Recall that

$$1-p(t) = P(\text{disc radius } t \text{ is empty})$$

Suppose we replace the disc with an anisotropic figure F, such as an ellipse or rectangle. Define

$$q(F) = P(\text{set } F \text{ is empty})$$

and estimate it by

$$\hat{q}(F) = \frac{\#\{i \mid F \text{ centred on } \mathbf{y}_i \text{ is empty}\}}{\#\{i \mid F \text{ centred on } \mathbf{y}_i \text{ is within } E\}}$$

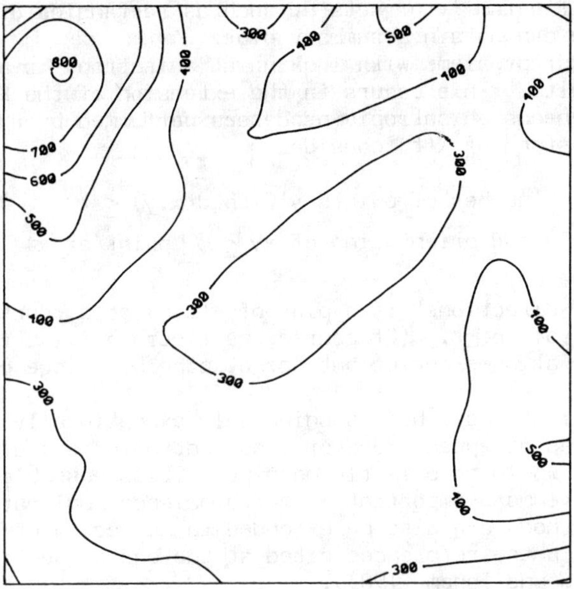

(a)

(b)

Figure 7. Estimates of λ() for the pattern of figure 1c, with the bivariate normal kernel. (a) h = 0.10, (b) h = 0.05.

By taking a parametric family of sets F, such as rectangles of width w and height h, one can obtain a summary plot.

One of the major problems with looking at anisotropy is actually displaying the results. This occurs in the extension of the K() statistic to homogeneous anisotropic processes performed by Ohser and Stoyan (1981). Instead of K(t) consider

$$\lambda K(t,\varphi) = E(\text{number of points } \mathbf{y} \text{ with } d(\mathbf{x},\mathbf{y}) \leq t \text{ and orientation of } \mathbf{xy} \leq \varphi \mid \text{point at } \mathbf{x})$$

Then their "rose of directions" is a plot of $K(t,\varphi)$ as a function of φ averaged over intervals of t. (Of course, as figure 3 illustrates, one can have anisotropy at some scales but not at others.) Edge correction is still possible.

Anisotropy seems to have been studied only exceptionally, probably because most examples do appear more or less isotropic, and often a preferred direction would have little meaning. It is possible that anisotropy will prove more important in earth science applications.

Many of the methods can also be extended to marked point processes. Examples are given in the references cited at the end of section 3, and also in Harkness and Isham (1983).

8. SUMMARY

We have seen that a quite sophisticated theory is available to study point patterns, perhaps with marked points. This has superceded many of the earlier simpler statistics which were based on a false trichotomy regular/random/clustered and enable interactions between the points to be studied on a wide range of scales (for a large enough sample). Further, the statistical sampling theory of these methods is now quite well-understood. Many of the earlier methods (notably the statistic of Clark and Evans, 1954) do not have the sampling distributions claimed because of defects in their developers' understanding of the basic theory. Simulation has played a large part in developing these techniques and is invaluable for exploratory work with point patterns.

A technical reference for most of the material given above is Ripley (1981). This is periodically updated by review articles in the International Statistical Review, of which the first is Ripley (1984b). Two texts at a more elementary mathematical level are Diggle (1983) and Upton and Fingleton (1985) (but see my reviews in Mathematical Geology, **18**, pp. 353-4 and **18**, pp. 863-4).

REFERENCES

Appleyard, S.T., Witkowski, J.A., Ripley, B.D., Shotton, D.M. and Dubowitz, V. (1985). 'A novel procedure for pattern analysis of features present on freeze-fractured plasma membranes'.

J. Cell Sci. **74**, 105-117.

Baddeley, A.J. and Silverman, B.W. (1984). 'A cautionary example on the use of second-order methods for analyzing point patterns'. Biometrics, **40**, 1089-1093.

Bennett, R.J. and Haining, R.P. (1985). 'Spatial structure and spatial interaction : modelling approaches to the statistical analysis of geographical data (with discussion)'. J. Roy. Statist. Soc. **A 148**, 1-36.

Broadbent, S. (1980). 'Simulating the ley hunter (with discussion)'. J. Roy. Statist. Soc. **A 143**, 109-140.

Clark, P.J. and Evans, F.C. (1954). 'Distance to nearest neighbour as a measure of spatial relationships in populations'. Ecology **35**, 445-453.

Cox, T.F. (1979). 'A method for mapping the dense and sparse regions of a forest stand'. Applied Statistics **28**, 14-19.

Davis, J.C. (1973). Statistics and Data Analysis in Geology. Wiley, New York.

Deffeyes, K.S., Ripley, B.D. and Watson, G.S. (1982). 'Stochastic geometry in petroleum geology'. Math. Geol. **14**, 419-432.

Diggle, P.J. (1979). 'On parameter estimation and goodness-of-fit testing for spatial point patterns'. Biometrics **35**, 87-101.

Diggle, P.J. (1981). Some graphical methods in the analysis of spatial point patterns. In Interpreting Multivariate Data (ed. V. Barnett), Wiley, Chichester, 55-73.

Diggle, P.J. (1983). Statistical Analysis of Spatial Point Patterns. Academic Press, London.

Fiksel, T. (1984). 'Estimation of parameterized pair potentials of marked and non-marked Gibbsian point processes'. Elekt. Inf. Kybern. **20**, 270-278.

Glotzl, E. and Rauchenschwandtner, B. (1981). 'On the statistics of Gibbsian processes'. Lect. Notes Statistics **8**.

Gudmundsson, G. and Saemundsson, K. (1980). 'Statistical analysis of damaging earthquakes and volcanic eruptions in Iceland from 1550-1978'. J. Geophys. **47**, 99-109.

Guild, F.J. and Silverman, B.W. (1978). 'The microstructure of glass fibre reinforced polyester resin composites'. J. Microscopy **114**, 131-141.

Hanisch, K.-H. and Stoyan, D. (1979). 'Formulas for the second-order analysis of marked point processes'. Math. Oper. Stat. ser Stat. **10**, 555-560.

Harkness, R.D. and Isham, V. (1983). 'A bivariate spatial point pattern of ants' nests'. Applied Statistics **32**, 293-303.

Kelly, F.P. and Ripley, B.D. (1976). 'A note on Strauss' model for clustering'. Biometrika **63**, 357-360.

Kendall, D.G. (1981). 'The statistics of shape'. In Interpreting Multivariate Data (ed. V. Barnett), Wiley, Chichester.

Kendall, D.G. (1985). 'Exact distributions for shapes of random triangles in convex sets'. Adv. Appl. Prob. **17**, 308-329.

Kendall, D.G. and Kendall, W.S. (1981). 'Alignments in 2-dimensional random sets of points'. Adv. Appl. Prob. **12**, 380-424.

Lotwick, H.W. and Silverman, B.W. (1982). 'Methods for analysing spatial processes of several types of points'. J. Roy. Statist. Soc. **B 44**, 406-413.

Ogata, Y. and Tanemura, M. (1981). 'Estimation of interaction potentials of spatial point patterns through the maximum likelihood procedure'. Ann. Inst. Statist. Math. **B 33**, 315-338.

Ogata, Y. and Tanemura, M. (1984). 'Likelihood analysis of spatial point patterns'. J. Roy. Statist. Soc. **B 46**, 496-518.

Ohser, J. (1983). 'On estimators of the reduced second moment measure of point processes'. Math. Oper. Statist. ser. Statist. **14**, 63-71.

Ohser, J. and Stoyan, D. (1981). 'On the second-order and orientation analysis of planar stationary point processes'. Biom. J. **23**, 523-533.

Ohser, J., Ripley, B.D. and Stoyan, D. (1986). 'On the efficiency of edge corrections for spatial point processes'. Unpublished ms.

Penttinen, A. (1984). 'Modelling interactions in spatial point patterns : parameter estimation by the maximum likelihood method'. Jyväskylä Studies in Computer Science, Economics and Statistics **7**.

Ripley, B.D. (1976). 'The second-order analysis of stationary point processes'. J. Appl. Prob. **13**, 255-266.

Ripley, B.D. (1977). 'Modelling spatial patterns (with discussion)'. J. Roy. Statist. Soc. **B 39**, 172-212.

Ripley, B.D. (1981). Spatial Statistics. Wiley; New York.

Ripley, B.D. (1984a). 'Edge corrections for spatial processes'. In Stochastic Geometry, Geometrical Statistics, Stereology (eds. R.V. Ambartzumian and W. Weil). Teubner-Texte; Leipzig, 144-153.

Ripley, B.D. (1984b). 'Spatial statistics : developments 1980-3'. Int. Statist. Rev. **52**, 141-150.

Ripley, B.D. (1986). 'Statistics, images and pattern recognition'. Canad. J. Statist. **14**, 83-111.

Ripley, B.D. and Silverman, B.W. (1978). 'Quick tests for spatial regularity'. Biometrika **65**, 641-642.

Silverman, B.W. (1981). 'Density estimation for univariate and bivariate data'. In Interpreting Multivariate Data (ed. V. Barnett), Wiley, Chichester, 37-53.

Silverman, B.W. (1985). Density Estimation. Chapman and Hall, London.

Stoyan, D. (1983). 'Quantification of correlations of geological line systems, with applications in the interpretation of remote sensing images'. In Methods of Instrumental and Digital Image Analysis for Geological Problems, Z. Physik der Erde, Potsdam.

Upton, J.G. and Fingleton, B. (1985). Spatial Data Analysis by Example. Volume 1. Point Pattern and Quantitative Data. Wiley, Chichester.

Vere-Jones, D. (1978). 'Space time correlations for micro earthquakes - a pilot study'. Suppl. Adv. Appl. Prob. **10**, 73-87.

AN UNCONVENTIONAL APPROACH TO GEOSTATISTICAL ESTIMATION

A. Grassia
CSIRO Institute of Energy and Earth Resources
Division of Groundwater Research
P O Wembley 6014
Western Australia

ABSTRACT. After a cursory discussion on sample requirements for estimation purposes for uncorrelated data and spatially correlated data for which an experimental semivariogram can be obtained at a reasonable cost and for which no doubts arise about the semivariogram model, I will dwell at a greater length on an alternative "ad hoc" procedure for the latter case. Full information of the semivariogram is not required with this procedure, which would provide good approximation of the estimation variance more directly and at a relatively low sampling and computing cost, and without the uncertainties of the choice of the semivariogram model.

Keywords: Plot, block, square, AOV, semivariogram, estimation, extension, variance, correlation, sill, range, estimate, trend.

1. INTRODUCTION

It is fairly common for a statistician to be asked by a geologist, hydrologist etc. the sample size required to estimate the mean of a regionalized variable for a block, whole deposit or catchment etc. with a given level of precision with a preassigned probability. Hydrologists studying hydrological properties of a particular area are, perhaps, more concerned about sample size than mineralogists because their financial resources, usually from public funds, are often limited, and little immediate profit would result from the investigation.

The answer to the above question would be simple for independently distributed data, but, complications arise when spatial correlation is present. It is mainly the latter aspect that I would like to discuss here. Emphasis is placed on cases in which obtaining the experimental semivariogram would be costly and/or doubts may arise about the appropriate semivariogram model when an alternative "ad hoc" estimation procedure is proposed.

C. F. Chung et al. (eds.), Quantitative Analysis of Mineral and Energy Resources, 323–339.

2. SAMPLE SIZE

2.1 Uncorrelated-Data

With uncorrelated data and random sampling the variance of the sample mean is given by σ^2/n. Then the sample size requirement is simply given by

$$n = \frac{t^2\sigma^2}{(\varepsilon\mu)^2} \tag{1}$$

where μ, σ^2 are the mean and variance of the population under study, t is the value of the normal or t-distribution, and ε the proportional error allowed.

2.2 Spatially Correlated Data

Expression (1) would not apply when data are spatially correlated. While the variance of a sample mean for independently distributed data is expected to be σ^2/n (used already in (1)), for a spatially correlated variable z the variance of the sample mean $\bar{z}$ is given by

$$\mathrm{Var}(\bar{z}) = \frac{1}{n^2}\{\sum_{i=1}^{n} \mathrm{Var}(z) + 2\sum_{i=1}^{n}\sum_{j=i+1}^{n} \mathrm{Cov}(z_i, z_j)\} \tag{2}$$

or

$$\mathrm{Var}(\bar{z}) = \frac{\sigma^2}{n} + \frac{2}{n^2}\sum_{i=1}^{n}\sum_{j=i+1}^{n} C_{ij} \tag{2'}$$

where C_{ij} is the expected covariance between values of z at the ith and jth sites, and

$$C_{ij} = \sigma^2 \rho_{ij}$$

where ρ_{ij} is the expected correlation.

We usually do not know the variance of z, and in classical statistics it is estimated from the sample as

$$s^2 = \frac{1}{n-1}\sum_{i=1}^{n}(z_i - \bar{z})^2 \tag{3}$$

Before proceeding any further, I would like to digress a little and mention here the following identity, which will become relevant later,

$$\frac{1}{n-1}\sum_{i=1}^{n}(z_i - \bar{z})^2 = \frac{1}{2n(n-1)}\sum_{i=1}^{n}\sum_{\substack{j=1\\ i\neq j}}^{n}(z_i - z_j)^2 \qquad (4)$$

(See Appendix B for proof).

2.2.1 Geostatistical Notation

Let us assume now that we are dealing with spatially correlated data. If we make a (weak) second order stationarity assumption for the random function $Z(x)$, the covariance exists and depends only on the separation distance, h,

$$C(h) = E\{Z(x + h) \,.\, Z(x)\} - \mu^2 \qquad (5)$$

where $\mu = E\{Z(x)\}$ and the variance is taken to be

$$C(0) = E\{[Z(x) - \mu]^2\} = \sigma^2$$

(see Jc and Huijbregts, 1978, p.32 and ff

The semivariogram is given by

$$\gamma(h) = \tfrac{1}{2} E\{[Z(x + h) - Z(x)]^2\} \qquad (6)$$

$$= C(0) - C(h)$$

Let us introduce the concept of a "si d assume that on exists. This is indicated by $\gamma(\infty)$. The sill is reached when h becomes "sufficiently" large. Then

$$\gamma(h) = \gamma(\infty) - C(h) \qquad (7)$$

implying

$$C(0) = \gamma(\infty) = \sigma^2 \qquad (8)$$

Also

$$\gamma(h) = \gamma(\infty)\ (1 - \rho(h)) \qquad (9)$$

In the following $\bar{\gamma}$ indicates the average semivariogram and $\hat{\bar{\gamma}}$ its sample estimate obtained as a weighted average over all sample semivariogram values (the '^' stands for 'estimate of').

From (9)

$$\bar{\gamma} = \gamma(\infty)\ (1 - \bar{\rho}) \qquad (10)$$

where $\bar{\rho}$ is a weighted average of the spatial correlations, and from (4) and (6)

$$\hat{\bar{\gamma}} = s^2 \quad (11)$$

Notice that as $\bar{\rho} \geqq 0$, from (10)

$$E\ \{s^2\} = \bar{\gamma} \leqq \gamma(\infty) \quad (12)$$

Expression (12) should sound a warning to those who automatically assume the sill (if it exists) is estimated by s^2, because, especially when the range is large, this can lead to an underestimate of the sill value. More will be said on this at a later stage.

2.2.2 Sample Size Requirement

If the deposit estimation is carried out on a block basis, from (2) and (10) the variance of the block mean from a sample of n cores would be

$$\mathrm{Var}(\bar{z}) = \mathrm{Var}(z)\ \{\frac{1}{n} + \frac{n-1}{n}\ \bar{\rho}\} \quad (13)$$

or

$$\mathrm{Var}(\bar{z}) = \gamma(\infty)\ \{\frac{1}{n} + \frac{n-1}{n}\ \bar{\rho}\} \quad (14)$$

(Cochran, 1977, p.209). And

$$\mathrm{Var}(\bar{z}) = \gamma(\infty)\ \{\frac{1}{n} + \frac{n-1}{n}\ [1 - \frac{\bar{\gamma}}{\gamma(\infty)}]\} \quad (15)$$

(assuming the sample uniformly distributed).

We can rewrite (15) as

$$\mathrm{Var}(\bar{z}) = \gamma(\infty)\ \{1 - \frac{n-1}{n}\ \frac{\bar{\gamma}}{\gamma(\infty)}\} \quad (16)$$

Expression (15) is very instructive. The part within the curly brackets provides a warning not to sample very intensively domains in which the spatial correlation is very high as the variance of the block mean would decrease very slowly for $\bar{\gamma}$ somewhat smaller than $\gamma(\infty)$ (high correlation) as n increases.

So far we have confined ourselves to the artificial situation of a discrete case. For the continuous case and availability of information on the semivariogram we can use, instead of $\bar{\gamma}/\gamma(\infty)$, the auxiliary function F(L), where L is the block width (Journel and Huijbregts, 1978, Section II. E.2).

Then (16) can be rewritten as

$$\mathrm{Var}(\bar{z}) = \gamma(\infty)\ \{1 - F(L)\} \tag{17}$$

as $\hat{F}(L) = \frac{n-1}{n}\ \hat{\bar{\gamma}}/\hat{\gamma}(\infty)$, for L relatively small (substituting $F(l_1, l_2)$ or $F(l_1, l_2, l_3)$ for F(L) in two or three dimensions). See also David (1977, p.252).

With the availability of information on the semivariogram, expression (15) or (17) should provide the experimenter with the means of optimizing the block sizes of the deposit so that an optimum sample will be determined for a given precision level and a chosen probability level. We want then that the block size be such that, if N is the total number of blocks for the deposit the block and sample sizes should be those which minimize Nn (=Nb).

If equal block sizes are used and μ, ε and t are as before, we require the block size and within block sample size such that the total sample size for the whole deposit, Nn, is a minimum and the condition

$$\frac{\gamma(\infty)}{n}\ \{1 + (n - 1)\ [1 - \frac{\bar{\gamma}}{\gamma(\infty)}]\} = \frac{(\varepsilon\mu)^2}{Nt^2} \tag{18}$$

is satisfied.

Or from (16)

$$\gamma(\infty)\ \{1 - \frac{n-1}{n}\ \frac{\bar{\gamma}}{\gamma(\infty)}\} = \frac{(\varepsilon\mu)^2}{Nt^2} \tag{19}$$

Evidently the best block size would be that for which n = 1. This implies that blocks should be as small as it is feasible to minimize the total sample size. Each block should have one core sample, located in its centre. Hence the total sample size would be equal to the number of blocks required.

The subdivision of the deposit in unequal block sizes would not lead to the optimum, although with an appropriate choice of the number of cores per block the total sample size could approach the optimum.

3. SOME "AD HOC" METHODS

If no information on the semivariogram is available, $\bar{\gamma}$ and $\bar{\rho}$, and hence $\sigma^2 = \gamma(\infty)$ and other parameters can be estimated from a random sample of equal size blocks into which the deposit has been subdivided. In turn a selected block will be subdivided into $b = a^d$ equal "plots" with $\underline{d}$ being the dimension of the domain and $\underline{a} = 3$. For a linear domain $\underline{a} = 5$ could be considered. One sample core should be at the centre of the plot.

An analysis of variance in which the total variability has been subdivided into between and within blocks as shown below would provide the estimates of $\bar{\gamma}$ and $\bar{\rho}$, and

AOV

Source of Variation	DF	SSq	MSq	EMSq
Between Blocks	r-1	SSB	MSB	$\sigma_w^2+b\sigma_b^2$
Within Blocks	r(b-1)	SSW	MSW	σ_w^2
Total	rb-1	SSt	MSt	

where b is used instead of n used earlier. In the AOV,

$$\sigma_w^2 = \sigma^2 (1 - \bar{\rho}) \ (= \bar{\gamma}) \qquad (20)$$

and

$$\sigma_b^2 = \sigma^2 \bar{\rho}, \qquad (21)$$

with σ_b^2 being estimated by

$$\hat{\sigma}_b^2 = \frac{\text{MSB} - \text{MSW}}{b} \qquad (22)$$

From (20) and (21)

$$\sigma^2 = \sigma_w^2 + \sigma_b^2 \qquad (23)$$

and

$$E\ \{\text{MSB}\} = \sigma^2 \{1 + (b - 1)\ \bar{\rho}\} \qquad (24)$$

Notice that MSt gives only an approximate estimate of σ^2. In fact

$$E\ \{\text{MSt}\} = [(rb - 1)\ \sigma_w^2 + (r - 1)b\ \sigma_b^2]/(rb - 1)$$

$$= [(rb - 1)\ \sigma^2\ (1 - \bar{\rho}) + (r - 1)b\ \sigma^2\ \bar{\rho}]/(rb - 1)$$

or

$$E\{\text{MSt}\} = \sigma^2\ [1 - (b - 1)\bar{\rho}/(rb - 1)] \leqq \sigma^2 \qquad (25)$$

as $\bar{\rho} \geqq 0$, confirming (12). From (21) and (23)

$$\bar{\rho} \doteq = \sigma_b^2/(\sigma_w^2 + \sigma_b^2) \qquad (26)$$

By the way

$$\text{Var}\ (\bar{z}) \doteq \sigma^2\ [1+ (b - 1)\ \bar{\rho}]/rb \qquad (26')$$

(See also Cochran, 1977, section 9.4).

4. SAMPLE DESIGN - TWO DIMENSIONS

In order to determine the sample requirements for a deposit, the whole deposit is to be or assumed to be subdivided into a square grid, with the squares of the grid constituting the blocks (or primary sample units). The grid should be such that each square should be approximately between one and nine times the desirable (or optimum) sample size. A number of squares (blocks) say 10-12, will be selected in a random fashion. Each selected block would be divided into 9 plots and a "core" sample taken at the centre of each plot. A between and within block analysis of variance will be carried out as in the previous section. This would provide the essential information for the choice of appropriate plot size, and hence the number of sample sites which would provide an estimate of the variable of interest for whole deposit with a chosen percentage error as discussed above.

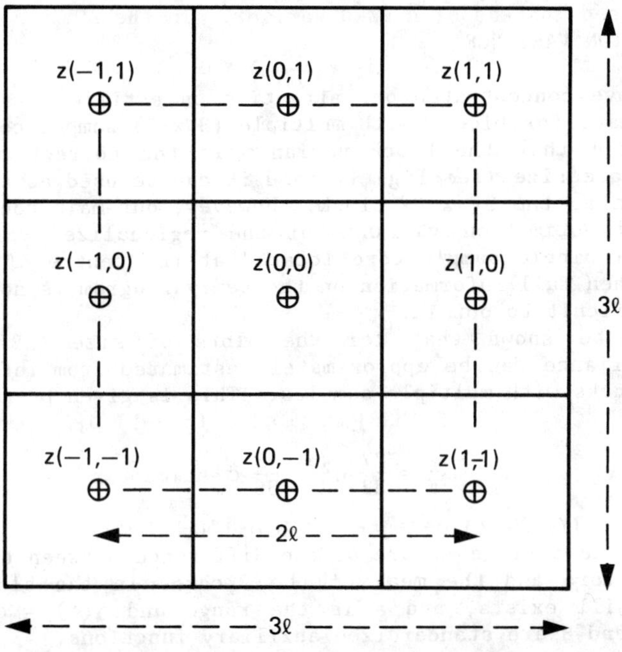

Figure 1. A typical 3 x 3 block with sample cores at the centre of each plot ($\ell = 1$).

If the block size is 3ℓ x 3ℓ, the 9 sample cores are in a 2ℓ x 2ℓ square (see Fig. 1) therefore semivariogram values can be obtained for five distances up to $\ell\sqrt{8}$. The distances between sample sites in each block are

Distance	Number
ℓ	12
$\ell\sqrt{2}$	8
2ℓ	6
$\ell\sqrt{5}$	8
$\ell\sqrt{8}$	2

The ideal situation would be to have blocks such that the semivariogram range would be between 2ℓ and 3ℓ.

The supplementary information of the semivariogram values on the above distances (and different directions, if desirable) may help in the determination of the optimum plot size, and hence the appropriate grid. A core sample will be taken at the centre of the plot or grid square. It would be helpful if the optimum sample plot size is not larger than the block size used in the preliminary investigation, as it would be easier to determine the extension variance for the plot. A $2\ell \times 2\ell$ plot size would be the ideal one. The estimation for the deposit would be obtained by averaging the grades over the (new) plots.

5. ESTIMATION VARIANCE

So far we have concentrated our attention on estimation and variance of the estimate for blocks with multiple (3 x 3) sample cores, and we have indicated that the block variance is the correct one for the $(2\ell \times 2\ell)$ block size (see Fig. 1), and it can be used as a reasonable approximation to the $3\ell \times 3\ell$ block. However, our main concern is the estimate and estimation variance of the regionalized variable from plots with a single sample core located at the centre of the plots, especially when full information on the semivariogram is not available or it is difficult to obtain.

It can be shown that for the plots of size $(2\ell \times 2\ell)$ the extension variance can be approximately estimated from information on the 10-12 blocks with multiple samples. This is given by

$$\hat{\sigma}^2_E = \frac{27}{32}\,\hat{\sigma}^2_d - \frac{5}{36}\,\hat{\sigma}^2_w \tag{27}$$

where $\hat{\sigma}^2_d$ is the average square of the difference between the value at the central core and the mean of the 9 cores in the block. If we assume the sill exists, and $\underline{a}$ is the range and $\gamma(\infty) = \sigma^2$. In the following F and H are standardized auxiliary functions.

$$E\,\{\hat{\sigma}^2_E\} = \sigma^2\,\{2H(\frac{\ell}{a}, \frac{\ell}{a}) - F(\frac{2\ell}{a}, \frac{2\ell}{a})\} \tag{28}$$

where

$$2\,\sigma^2\,\hat{H}\,(\frac{\ell}{a}, \frac{\ell}{a}) \doteq \frac{3}{4}\,\{\hat{\sigma}^2_w + \frac{9}{8}\,\hat{\sigma}^2_d\} \tag{29}$$

and

$$\sigma^2\,\hat{F}\,(\frac{2\ell}{a}, \frac{2\ell}{a}) = \frac{8}{9}\,\sigma^2_w \tag{30}$$

(see Appendix A, from Grassia, 1984, unpublished). Similar formulae can be obtained for one or tri-dimensional cases.

For plots of size smaller than $2\ell \times 2\ell$, as an approximation, some linear interpolation can be used for H and F. For example, for plots of size $\ell \times \ell$ we can use

$$2\,\sigma^2\,H(\frac{\ell}{2a}, \frac{\ell}{2a}) \doteq \sigma^2\,H(\frac{\ell}{a}, \frac{\ell}{a}) = \frac{3}{8}\,\{\hat{\sigma}^2_w + \frac{9}{8}\,\hat{\sigma}^2_d\} \tag{31}$$

and

$$\sigma^2\,F(\frac{\ell}{a}, \frac{\ell}{a}) \doteq \frac{1}{2}\,\sigma^2\,F(\frac{2\ell}{a}, \frac{2\ell}{a}) = \frac{4}{9}\,\hat{\sigma}^2_w \tag{32}$$

and

$$\hat{\sigma}^2_E \doteq \frac{27}{64}\,\hat{\sigma}^2_d - \frac{5}{72}\,\hat{\sigma}^2_w \tag{33}$$

The variance of the combined estimate would be given by a weighted average of the variance for the blocks given by expression (16) and that for single plots given by (27) (or 32), using the area size as weights.

6. TREND

So far we have assumed no trend (or drift). The presence of a trend would affect mainly the Between Blocks component of variance, as the average distance between the 3 x 3 blocks is likely to be large.

If preliminary investigations have indicated the presence of a trend, to overcome its effects on the Between Blocks variance we could select, say, 10 random sets of pairs of neighbouring 3 x 3 blocks, with the two blocks preferably arranged in the direction of the trend. The contrast between blocks within set can be used as an estimate of a detrended Between Blocks variance, and then a detrended estimate of the sill (σ^2) from (23). Alternatively, the Within Block sample sites can be integrated as well as possible into the major grid system of single site plots, then leading to a single component of estimation variance. The latter may lead to some loss of information.

7. ILLUSTRATIVE EXAMPLES

7.1 Porosity Data

For illustration purposes information on Porosity Data in Delfiner "Intrinsic Model of Order k" Tables I and II is used. In Table I a grid of 18 x 18 values is shown and in Table II three subtables of 2 x 2, 3 x 3 and 6 x 6 block means are tabulated.

Table 1 (from Grassia, 1984, unpublished) herewith shows the analyses of variance for the three block sizes (a notation slightly different from the text is used). However, that for 3 x 3 blocks concerns us here, for which $\hat{\sigma}^2_d = 15.86$.

Table 1. Porosity Data - From Delfiner's Geostatistics Course

Analysis of Variance

Source of Variation	DF	Sums of squares	Mean squares	E (Mean squares)
		(a) 2x2 Blocks		
Between Blocks	80	2080.08	26.00	$\sigma_2^2+4\sigma_{B_2}^2$
Within Block	243	5258.36	21.19	σ_2^2
Total	323	7228.44	22.37	
		$\frac{1}{4}\ \sigma_2^2=\underline{5.30}$	$\sigma_{B_2}^2=1.20$	$\frac{1}{4}\sigma_{B_2}^2=0.300$
		(b) 3x3 Blocks		
Between Blocks	35	1655.64	47.30	$\sigma_3^2+9\sigma_{B_3}^2$
Within Block	288	5572.80	19.35	σ_3^2
Total	323	7228.44	22.37	
		$1/9\ \sigma_3^2=\underline{2.15}$	$\sigma_{B_3}^2=3.10$	$1/9\sigma_{B_3}^2=0.345$
		(c) 6x6 Blocks		
Between Blocks	8	662.25	82.78	$\sigma_6^2+36\sigma_{B_6}^2$
Within Block	288	6566.19	20.85	σ_6^2
Total	323	7228.44	22.37	
		$\frac{1}{36}\ \sigma_6^2 = \underline{0.58}$	$\sigma_{B_6}^2 = 1.72$	$\frac{1}{36}\ \sigma_{B_6}^2 = 0.048$

Using the Delfiner semivariogram model with two components, one with sill = 8 and range = 4ℓ and the other with sill = 14.31 and range = 1ℓ, for blocks of size $2\ell \times 2\ell$ (ℓ = core distance) with a single core in the centre we obtain for the estimation (extension) variance $\hat{\sigma}_E^2 = 10.22$. For blocks of size $3\ell \times 3\ell$ $\hat{\sigma}_E^2 = 12.17$.

From formula (27) using all 36 3 x 3 blocks we obtain

$$\hat{\sigma}_E^2 = \frac{27}{32} \times 15.86 - \frac{5}{36} \times 19.35$$

$$= 10.69$$

On the other hand, a random sample of 12 3 x 3 blocks which led to the following analysis of variance table

AOV

Source of Variation	DF	SSq	MSq
Between Blocks	11	514.12	46.74
Within Blocks	96	1679.60	17.50
Total	107	2193.72	20.50

and $\hat{\sigma}_d^2 = 17.31$, gave an estimated extension of variance of

$$\hat{\sigma}_E^2 = \frac{27}{32} \times 17.31 - \frac{5}{36} \times 17.50$$

$$= 12.17 \text{ (>10.69, an overestimate?)}$$

By chance this is equal to that for $3\ell \times 3\ell$ blocks, obtained from the semivariogram information.

If the semivariogram for small values is almost linear we can take

$$H\left(\frac{\ell}{2}, \frac{\ell}{2}\right) = \tfrac{1}{2} H(\ell,\ell)$$

and

$$F(\ell,\ell) \doteq \tfrac{1}{2} F(2\ell, 2\ell)$$

and

$$\sigma_E^2(\ell,\ell) \doteq \tfrac{1}{2} \sigma_E^2(2\ell, 2\ell) \qquad (34)$$

From the semivariogram and using the tables of auxiliary functions we obtain a per cent standard error of SE% = 0.57.

Using the above estimate of variance based on 36 blocks, from (34) we obtain for 36 blocks (plots) with a single core in the centre

$$\text{SE \%} = \left(\frac{5.35}{324}\right)^{\frac{1}{2}} \times \frac{100}{22.8} = 0.56$$

While using the estimate of variance based on 12 3 x 3 blocks SE% = 0.60.

7.2 Phosphorus Content

The data in this example are from McIntyre (1967), with some rearrangement so as to obtain twelve 3 x 3 blocks. The original data consisted of a 10 x 11 grid. For the sake of the excercise they were converted into a 9 x 12 grid by turning row 10 into column 12 and discarding two (outlying) values.

A spherical model was fitted to the data by eye, with $\gamma(\infty) = 90.0$ and range = 3. From the twelve blocks the following estimates were obtained

$$\hat{\sigma}_d^2 = 28.93; \quad \hat{\sigma}_w^2 = 72.29$$

The following estimation variances were obtained for $\ell \times \ell$ blocks (plots) with a single core in the centre. Using semivariogram:

$$\hat{\sigma}_E^2 = 90 \; \{2 \; H \; (\frac{\ell}{6}, \frac{\ell}{6}) \; - \; F \; (\frac{\ell}{3}, \frac{\ell}{3})\}$$

$$= 90 \; (2 \times 0.190 - 0.260)$$

$$= 10.8.$$

Using expression (33):

$$\hat{\sigma}_E^2 = 9.1$$

The agreement is fairly good.

8. COMMENTS AND CONCLUSIONS

It should be clear that the estimation method proposed here is not affected by anisotropy. Trend can be accounted for, at least in part, as suggested above. With it the user has no difficult judgement to make as it is in some cases with the choice of appropriate semi-variogram models. Although an approximate procedure it could be useful in block or deposit estimation.

The semivariogram is required in point estimation (for interpolation, contouring purposes, etc.) with the use of Kriging. However, the possibility of estimating the semivariogram from short distances ($h < 3\ell$), the analysis of variance and sample estimates of H and F, or using the inverse approach with or without assuming a model, is not to be excluded.

The method requires further investigations concerning robustness and generality of application. It needs to be tested on other data.

9. ACKNOWLEDGEMENT

I wish to thank Kathy McGuckin for her excellent typing of the manuscript.

10. REFERENCES

1. Cochran, W.G. (1977). *Sampling Techniques*. Wiley, New York.
2. David, M. (1977). *Geostatistical Ore Reserve Estimation*. Elsevier, New York.
3. Delfiner, P. (1979). *Geostatistics Course*. Seattle.
4. Grassia, A. (1984, unpublished). Spatially Correlated Data. Some Identities and the Variance-Sill Relationship. Presented at a "Spatial Process Workshop", University of Western Australia, Perth.
5. Journel, C.L.J. and Huijbregts, A.G. (1978). *Mining Geostatistics*. Academic Press, London.
6. McIntyre, G.A. (1967). 'Soil sampling for soil testing'. *J. Aust. Inst. Agr. Res.* **33**, 309-320.

APPENDIX A

TWO-DIMENSIONAL CASE

Here the semivariogram function is assumed nearly linear for distance values $\leqq 3\ell$, and the auxiliary functions F and H are assumed to be non-standardized.

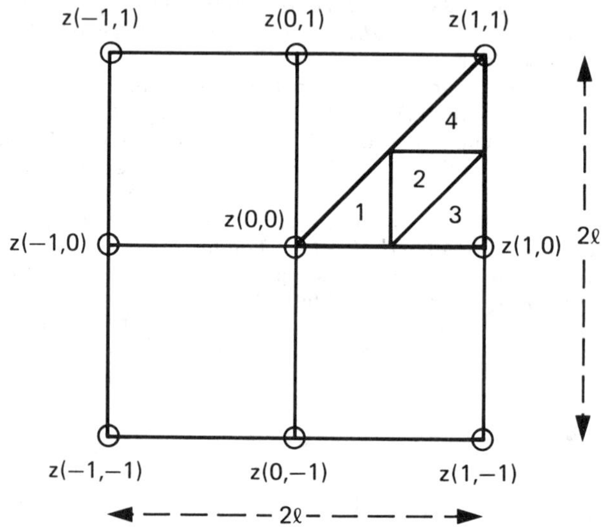

Figure A1 One of possible 8 triangles of a 2ℓ x 2ℓ square subdivided into four equal triangles (ℓ = 1).

(a) Show that

$$F(2\ell,2\ell) \doteq \frac{8}{9}\,\sigma_w^2$$

Let b = 9 be the block size (see above square). We require the average of the semivariances for any pair of points in the 2ℓ x 2ℓ square. For a single block we have

$$\frac{1}{b^2}\,\{\tfrac{1}{2}\sum_i \sum_j \sum_{i'} \sum_{j'} [z(i,j)-z(i',j')]^2\}$$

$$= \frac{b-1}{b}\,\frac{1}{2b(b-1)}\,\{\sum_i \sum_j \sum_{i'} \sum_{j'} [z(i,j)-z(i',j')]^2\} \quad (A1)$$

$$= \frac{b-1}{b} \left\{ \frac{1}{b-1} \sum_i \sum_j [z(i,j)-\bar{z}]^2 \right\}$$

(from expression (4)). Notice there are 9 zero values in (A1) (in terms of expected value of variances)

$$F(2\ell,2\ell) \doteq \frac{b-1}{b} \sigma^2_w$$

$$\doteq \frac{8}{9} \sigma^2_w \qquad \text{(A2)}$$

QED

(b) Show that

$$H(\ell,\ell) \doteq \frac{3}{8} \{\sigma^2_w + \frac{9}{8} \sigma^2_d\}$$

Let us consider in Fig. 1 the triangle with heavy lined perimeter, subdivided into four small triangles numbered 1 to 4. The average semivariance of z(0,0) with itself twice with weight ¼, proportional to the area of triangle No. 1, and of z(0,1) and z(1,1) with z(0,0) with a joint weight of 3/4, proportional to the area of triangles 2, 3 and 4, can be used as an approximation

$$H(\ell,\ell) \doteq \tfrac{1}{4}\{2 \times \tfrac{1}{2} [z(0,0) -z(0,0)]^2\} + \frac{3}{4} \{\tfrac{1}{2}[z(0,1) - z(0,0)]^2$$
$$+ \tfrac{1}{2} [z(1,1) - z(0,0)]^2\}/2$$

But this triangle is one of the 8 contained in 2ℓ x 2ℓ (or 3x3) square. Averaging over the 8 triangles and considering that each difference occurs twice because common to two triangles

$$H(\ell,\ell) \doteq \tfrac{2}{8} \{\tfrac{1}{4}\, 8 \times 0 + \frac{3}{8} \tfrac{1}{2} \sum_i \sum_j [z(i,j) - z(0,0)]^2\}$$

$$\doteq \frac{3}{8} \{\tfrac{1}{8} \sum_i \sum_j [z(i,j) - z(0,0)]^2\}$$

or

$$H(\ell,\ell) \doteq \frac{3}{8}\{\frac{1}{8} \sum_i \sum_j [z(i,j) - \bar{z}]^2 - \frac{9}{8} [z(0,0)-\bar{z}]^2\}$$

$$\doteq \frac{3}{8}\{\sigma_w^2 + \frac{9}{8} \sigma_d^2\} \qquad \text{(A3)}$$

QED

From (A2) and (A3)

$$2\ H\ (\ell,\ell) - F\ (2\ell,\ 2\ell) \doteq \frac{27}{32}\sigma_d^2 - \frac{5}{36}\sigma_w^2 . \qquad \text{(A4)}$$

Considering that

$$F(2\ell,2\ell) \doteq 2F(\ell,\ell)$$

and

$$H(\ell,\ell) \doteq 2\ H(\ell/2,\ \ell/2)$$

The extension variance for a square of size $\ell x \ell$ estimated by the central point would be

$$2\ H(\ell/2,\ell/2) - F(\ell,\ell) \doteq H(\ell,\ell) - \tfrac{1}{2}\ F(2\ell,2\ell).$$

APPENDIX B

PROOF FOR EXPRESSION 4

If $z_1, z_2, \ldots, z_n$ is a random sample with sample mean $\bar{z}$, we have (summing from 1 to n)

$$\sum_i \sum_j (z_i - z_j)^2 = \sum_i \sum_j [(z_i - \bar{z}) - (z_j - \bar{z})]^2$$

$$= \sum_i \sum_j (z_i - \bar{z})^2 + \sum_i \sum_j (z_j - \bar{z})^2 - 2 \sum_i \sum_j (z_i - \bar{z})(z_j - \bar{z})$$

$$= 2n \sum_i (z_i - \bar{z})^2 - 2\sum_i (z_i - \bar{z}) \sum_j (z_j - \bar{z})$$

$$= 2n \sum_i (z_i - \bar{z})^2 \qquad \text{(B1)}$$

as

$$\sum_i (z_i - \bar{z}) = \sum_j (z_j - \bar{z}) = 0$$

Considering that $z_i - z_j = 0$ for $i = j$

$$\sum_i \sum_j (z_i - z_j)^2 = \sum_i \sum_{j,\, i \neq j} (z_i - z_j)^2$$

from (B1)

$$\sum_i (z_i - \bar{z})^2 = \frac{1}{2n} \sum_i \sum_{j,\, i \neq j} (z_i - z_j)^2 \qquad \text{(B2)}$$

On dividing both sides of (B2) by n-1 expression (4) follows.

CASE STUDIES ON MODELLING COMPLEX SULPHIDE OREBODIES FOR ORE RESERVE ESTIMATION USING GEOSTATISTICAL METHODS

F.H.MUGE
Centro de Valorização de Recursos Minerais
Technical University of Lisbon
Instituto Superior Técnico, Av. Rovisco Pais
1096 LISBOA CODEX

ABSTRACT. On modelling complex sulphide orebodies some problems may ari se concerning the spatial definition of ore types based on geological criteria. The definition of the morphology of each ore type can be achieved through the estimation of an indicator function per ore type. In what concerns the estimation of ore reserves per type, the heterogeneity of the spatial variability of the ore within each ore type, can put some problems in the modelling phase.

In this paper, some methods are proposed to deal with these kind of problems and typical case studies are presented concerning some complex sulphide orebodies located in the South of Portugal.

1. GLOBAL ORE RESERVE ESTIMATION

The evaluation of the global and in-situ reserves of an ore deposit consists usually on the estimation of the tonnage (T), quantity of metal(Q) and mean grade (m). In order to estimate these values we must choose, at first, the auxiliary variables we are going to use in the procedure.

1.1. Stratiform Orebodies

If we are dealing with stratiform orebodies we can use the mineralized thickness defined in each drill hole (j) as the sum of the lenghts ($\ell(x_j)$) of the mineralized drill cores (x_j):

$$h_j = \sum_{x_j} i(x_j)\, \ell(x_j) \qquad (1)$$

where: $i(x_j)$ is an indicator variable defined as:

$$i(x_j) = \begin{cases} 1, & \text{if the core sample } x_j \text{ is mineralized} \\ 0, & \text{otherwise} \end{cases} \qquad (2)$$

C. F. Chung et al. (eds.), Quantitative Analysis of Mineral and Energy Resources, 341–358.

The other usual auxiliary variable is the accumulation defined as:

$$a_j = \sum_{x_j} i(x_j)\ \ell(x_j)\ Z(x_j) \tag{3}$$

where: $Z(x_j)$ is the average grade of the core sample x_j.

Using these variables, the estimation of the ore reserves can be achieved by:

$$T^* = s^* \bar{h}_s^* \bar{\rho}_s^* \qquad Q^* = s^* \bar{a}_s^* \bar{\rho}_s^* \qquad m^* = \frac{Q^*}{T^*} \tag{4}$$

where: s^* is the estimated value of the surface occupied by the mineralization

$\bar{h}_s^*$ and $\bar{a}_s^*$ are the estimated average thickness and accumulation in the surface s^*

$\bar{\rho}_s^*$ is the average rock density in the surface s^*

The values $\bar{h}_s^*$ and $\bar{a}_s^*$ can be estimated by kriging, using the variogram models of the mineralized thickness h_j and accumulation a_j.

The error associated to the rock density is usually less than the other components of error and can be estimated by its average sample value in the surface s^*.

In what concerns the mineralized surface s, if it is geologically well defined or if it is imposed by technological criteria, the associated estimation error can be considered negligible. In both cases the ore reserves are estimated within a predefined surface. In other cases, the mineralized surface s is not well defined and can be estimated using an indicator auxiliary variable, defined for each drill hole:

$$i(j) = \begin{cases} 1, & \text{if the drill hole } j \text{ is mineralized} \\ 0, & \text{otherwise} \end{cases} \tag{5}$$

The estimated value of the mineralized surface s^* can thus be written:

$$s^* = S\ \bar{i}_S^* \tag{6}$$

where:

S is a predefined surface that contains the mineralized surface s

$\bar{i}_S^*$ is the estimated average of the indicator variable i(j) in the surface S.

The value $\bar{i}_s^*$ can also be taken as the estimated probability of success of a random drill hole within S and can be calculated by indicator kriging, using the variogram model of the indicator variable i(j).

Using this methodology, an estimation of the mineralized surface s^* is obtained, but its localization in S is lost. Being so, the expression (4) cannot be used directly because the average values $\bar{h}_s^*$ and $\bar{a}_s^*$ must be estimated within the mineralized surface s^*.

One possible solution is to use the composite auxiliary variables:

$$h_j\, i(j) = \begin{cases} h_j, & \text{if the drill hole } j \text{ is mineralized} \\ 0, & \text{otherwise} \end{cases} \tag{7}$$

$$a_j\, i(j) = \begin{cases} a_j, & \text{if the drill hole } j \text{ is mineralized} \\ 0, & \text{otherwise} \end{cases} \tag{8}$$

Using these variables, the estimation of the ore reserves can be obtained by:

$$T^* = S\, \bar{h}^*_S\, \bar{\rho}^*_S \qquad Q^* = S\, \bar{a}^*_S\, \bar{\rho}^*_S \qquad m^* = \frac{Q^*}{T^*} \tag{9}$$

where: $\bar{h}^*_S$ and $\bar{a}^*_S$ are respectively the estimated average values of the variables $h_j i(j)$ and $a_j i(j)$ in S.

The values $\bar{h}^*_S$ and $\bar{a}^*_S$ can also be taken as the estimated average thickness and accumulation in the surface S and can be calculated by kriging, using the variogram models of the composite variables $h_j i(j)$ and $a_j i(j)$.

1.2. Massive Orebodies

If the orebody is of the massive type we can use the average grade $Z(x_j)$ of each core sample x_j as auxiliary variable. Using this variable, the estimation of the ore reserves can be obtained by:

$$T^* = v^*\, \bar{\rho}^*_v \qquad m^* = \bar{Z}^*_v \qquad Q^* = m^*\, T^* \tag{10}$$

where:

- v^* is the estimated value of the volume occupied by the mineralization
- $\bar{Z}^*_v$ is the estimated average grade in the volume v^*
- $\bar{\rho}^*_v$ is the average rock density in v^*.

The value $\bar{Z}^*_v$ can be estimated by kriging, using the variogram model of the average grade $Z(x_j)$.

If the mineralized volume v is geologically well defined or if it is imposed by technological criteria, the associated estimation error can be considered negligible. In these cases, the ore reserves are estimated within a predefined volume and the estimation error of the tonnage is disregarded. On the other hand the estimation error of the quantity of metal equals the estimation error of the mean grade. In other cases, the mineralized volume v is not well defined and can be estimated using the indicator auxiliary variable $i(x_j)$ (2).

The estimated value of the mineralized volume v^* can be obtained by:

$$v^* = V\, \bar{i}^*_V \tag{11}$$

where: V is a predefined volume that contains the mineralized volume v.
$\bar{i}_V^*$ is the estimated average of the indicator variable $i(x_j)$ in the Volume V.

The value $\bar{i}_V^*$ can be calculated by indicator kriging, using the variogram model of the indicator variable $i(x_j)$.

Using this methodology, the localization of v^* within V is lost and the expressions (10) cannot be used directly because the average value $\bar{z}_V^*$ must be estimated within the mineralized volume v^*.

One solution is to use the composite auxiliary variable:

$$Z(x_j)\ i(x_j) = \begin{cases} Z(x_j), & \text{if the core sample } x_j \text{ is mineralized} \\ 0, & \text{otherwise} \end{cases} \tag{12}$$

Using this auxiliary variable, the estimation of the ore reserves can be obtained by:

$$T^* = V\ \bar{i}_V^*\ \bar{\rho}_V^* \qquad Q^* = V\ \bar{z}_V^*\ \bar{\rho}_V^* \qquad m^* = \frac{Q^*}{T^*} \tag{13}$$

where: $\bar{z}_V^*$ is the estimated average value of the variable $Z(x_j)\ i(x_j)$ in V.

The value $\bar{z}_V^*$ can also be taken as the estimated mean grade in the volume V, and can be calculated by kriging, using the variogram model of the composite auxiliary variable $Z(x_j)\ i(x_j)$.

Table I shows the auxiliary variables and the variogram models needed in each case.

	STRATIFORM OREBODIES		MASSIVE OREBODIES	
	well-defined min.surface	not well-defined min.surface	well-defined min.volume	not well-defined min. volume
Tonnage	2-dimensional variogram of thickness h_j	2-dimensional variogram of $h_j\ i(j)$	-	3-dimensional indicator variogram of $i(x_j)$
Quantity of Metal	2-dimensional variogram of accumulation a_j	2-dimensional variogram of $a_j\ i(j)$	-	3-dimensional variogram of $Z(x_j)\ i(x_j)$
Mean Grade	-	-	3-dimensional variogram of $Z(x_j)$	-

TABLE I : Variogram models for ore reserve estimation

2. RESERVE ESTIMATION PER ORE-TYPE

The methodology used for global ore reserve estimation can be applied for ore type reserve estimation.

In this case, a definition of indicator auxiliary variables per ore--type is needed:

$$i_k(x_j) = \begin{cases} 1, & \text{if the core sample } x_j \text{ belongs to ore-type } k \\ 0, & \text{otherwise} \end{cases} \qquad (14)$$

$$i_k(j) = \begin{cases} 1, & \text{if the drill hole } j \text{ is mineralized in ore-type } K \\ 0, & \text{otherwise} \end{cases} \qquad (15)$$

Whatever the methodology used (2 or 3 - dimensional auxiliary variables), order problems may arise due to the fact that each ore-type kriging system is solved independently of the others. One solution is suggeseed by Sullivan (1983) and Journel (1983)through the use of quadratic programming. To obtain the estimated value of the tonnage per ore-type(T_k^{**}), the formulation of the problem well be as follows:

$$\text{Min} \quad : \sum_{k=1}^{n} W_k (T_k^{*} - T_k^{**})^2$$

$$\text{subject to:} \quad \sum_{k=1}^{n} T_k^{**} = T^{*} \qquad (16)$$

$$T_k^{**} > 0$$

where: T_k^{**} (k=1,n) is the final estimate of the tonnage for each ore-type
T_k^{*} (k=1,n) is the initial estimate of the tonnage for each ore-type
W_k (k=1,n) is a weight given to each ore-type
T^{*} is the estimated value of the global tonnage.

In what concerns the mean grade of each ore-type (m_k), the formulation is as follows:

$$\text{Min} \quad : \sum_{k=1}^{n} W_k (m_k^{*} - m_k^{**})^2$$

$$\text{Subject to:} \quad \sum_{k=1}^{n} T_k^{**} m_k^{**} = m^{*} T^{*} \qquad (17)$$

$$m_k^{**} > 0$$

where: m_k^{**} (k=1,n) is the final estimate of the mean grade of each ore type
m_k^{*} (k=1,n) is the initial estimate of the mean grade of each ore-type
m^{*} is the estimated value of the global mean grade.

3. CASE STUDY

3.1. Statistics

In a sulphide orebody located in the South of Portugal, (CORVO orebody) several ore-types were established according to geological and cut-off criteria based on Cu and Zn grades.

The global available data consists of 62 drill holes with 1176 2m core samples assayed for Cu and classified into 5 ore-types.

Table II shows the statistics of the samples (*) for each ore-type (labeled 2 to 6).

TABLE II - Statistics of the samples for each ore-type

		Cu grade (%)		
Ore Type	Total # of samples	Mean (%)	Variance $(\%)^2$	Coefficient of Variation
2	616	8.45	45.42	0.80
3	177	9.00	25.80	0.56
4	136	0.41	0.33	1.40
5	78	4.96	16.02	0.81
6	169	5.20	23.60	0.93

3.2. Histograms

Figures 1A - 1E show the histograms of Cu grade for all five ore-types. The histograms of ore-types 5 and 6 are very much alike in shape. Histograms of ore-types 2, 3 and 4 are well differentiated.

(*) For proprietary reasons, all values have been multiplied by an arbitrary constant.

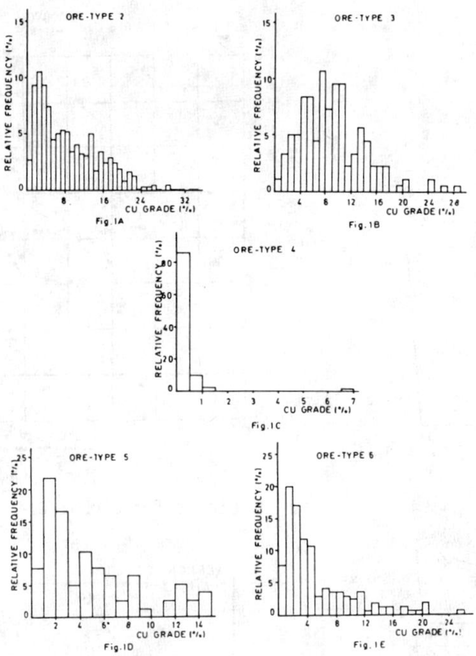

Figure 1: Histograms of Copper grade

3.3. Spatial Statistics

In order to obtain a quick look at the spatial variability of the data, a series of contour plots were drawn. Figure 2 shows a schematic overlay of isopach maps outlining the thickness allocation to each ore-type. The geometrical complexity of the orebody comes up from Figure 2 - several intermingled and overlapping subpopulations appear and it is difficult to split the deposit into spatially distinct zones assignable to each ore-type.

3.4. Global Ore Reserve Estimation

The Global Ore Reserve Estimation was performed within the limits shown in Figure 3.

In order to apply the methodology exposed in this paper, initial attempts to model global experimental variograms of the 3-dimensional auxiliary variable Cu grade were done and the results are shown in Figure 4. The vertical variogram displays a complex pattern of nested structures, revealing a "mixture" of populations. The same cannot be said for the horizontal variograms where the structure is not clear.

Figure 5 shows the experimental variograms and the adjusted theoretical models for the 2-dimensional auxiliary variables: total thickness and total copper accumulation. Based on these variogram models, global ore reserves were estimated, within the fixed surface limits of Figure 3, by kriging the mean total thickness and copper accumulation according to(4).

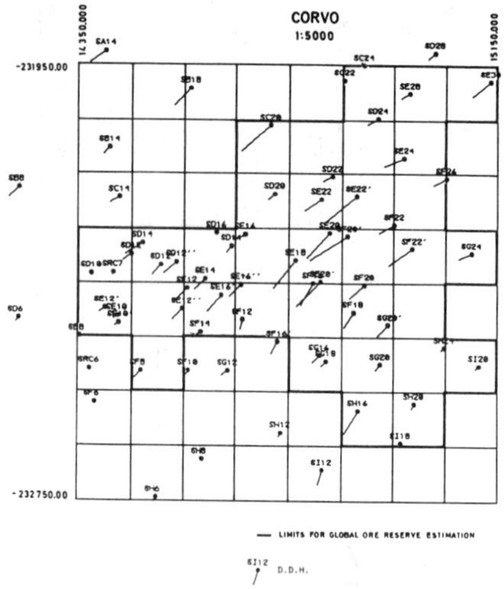

Fig. 3: Limits for Global Ore Reserve Estimation

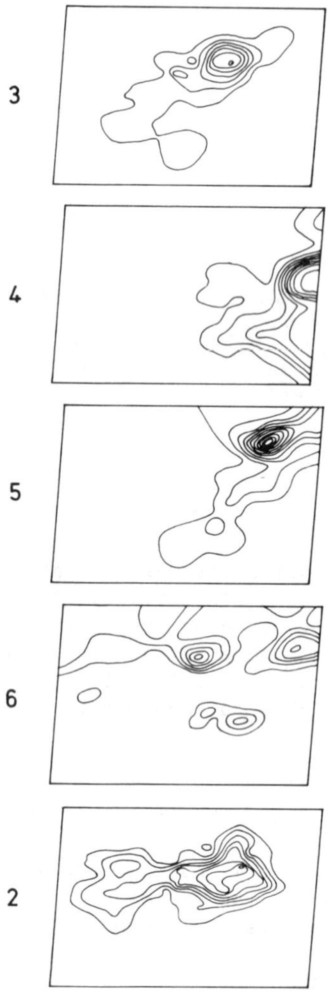

Fig.2: Isopach maps for the 5 Types of Ore

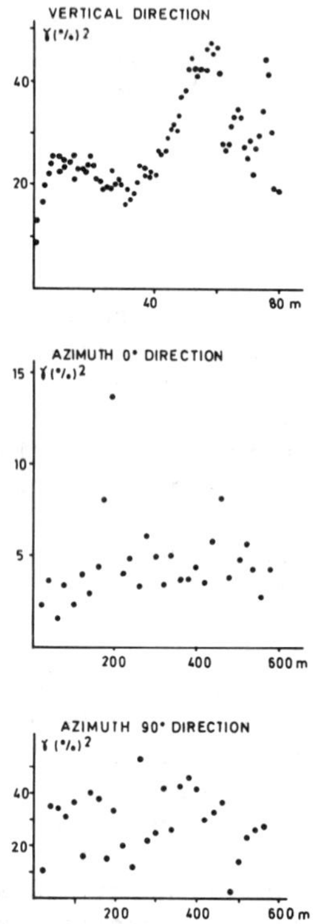

Fig. 4: Global Experimental Variograms of Cu Grade.

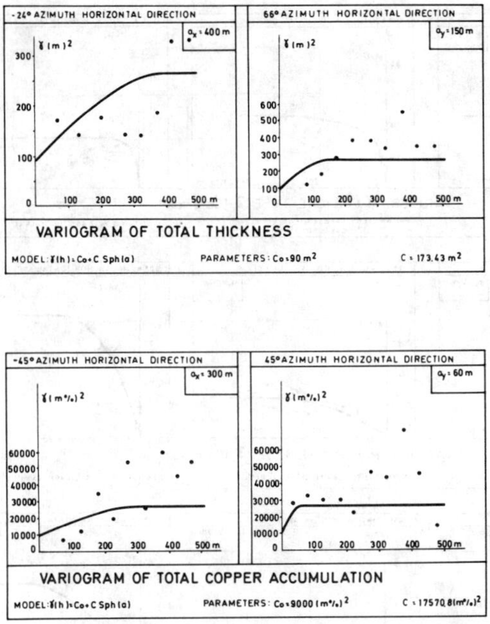

Figure 5: Variograms of Total Thickness and Copper Accumulation

3.5. Reserve Estimation per Ore-type

For the purpose of local estimation per ore-type, separate 3-dimensional variograms of Cu grade were calculated for each subpopulation, as it was defined by the geologist. The results of that calculation are summarized in Figure 6. The modelling of the experimental variograms was performed using spherical schemes but, for several ore-types, especially in horizontal directions, a poor fitting was obtained.

Experimental variograms of the 2-dimensional auxiliary variables: thickness and copper accumulation were calculated for ore-type 2(Figure 8) within the well defined limits of Figure 7. This methodology cannot be applied for the other ore-types due to two main reasons:

i) The number of mineralized holes in the ore-type was too low to justify variogram modelling.

ii) The limits of the mineralized surface were not geologically well defined.

For these ore-types, variograms of the auxiliary composite variables defined in (7) and (8) were calculated and the ore reserves were estimated according to (9) by kriging the mean thickness and accumulation within the global surface of the deposit (e.g. Fig. 4). Figure 9 shows the variograms of these auxiliary variables for ore-type 3.

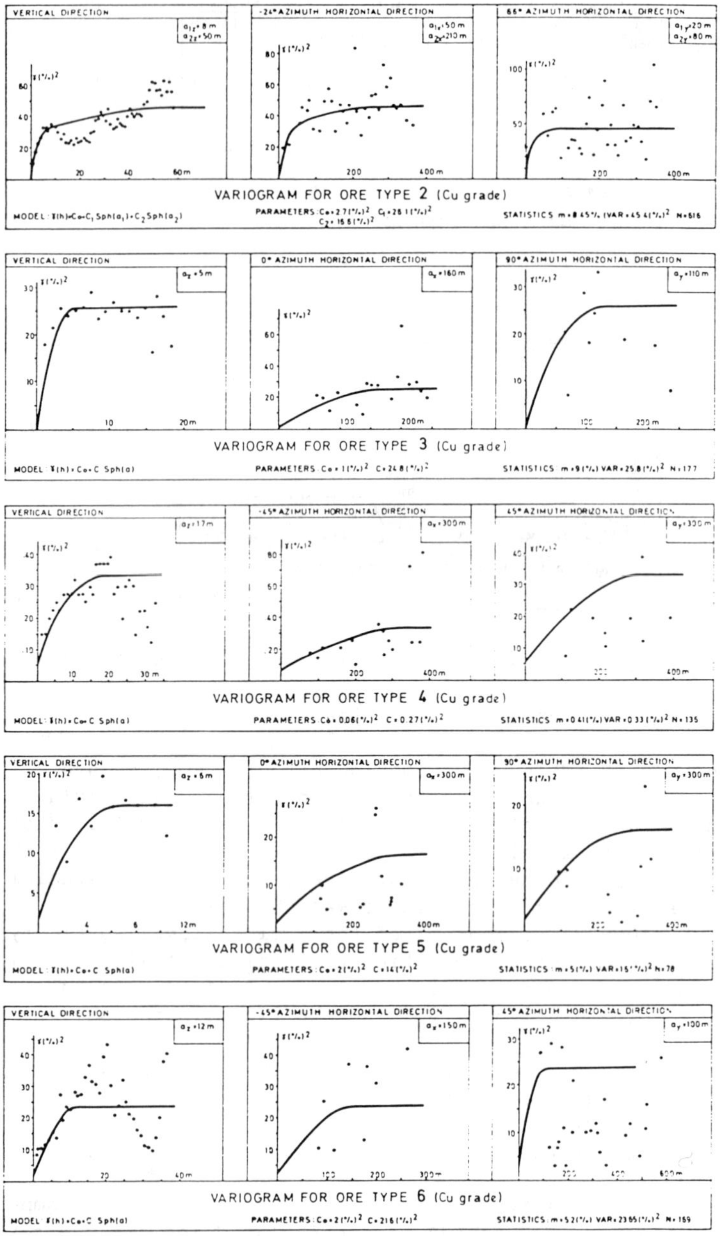

Figure 6: 3-dimensional Variograms of Cu Grade per Ore-type

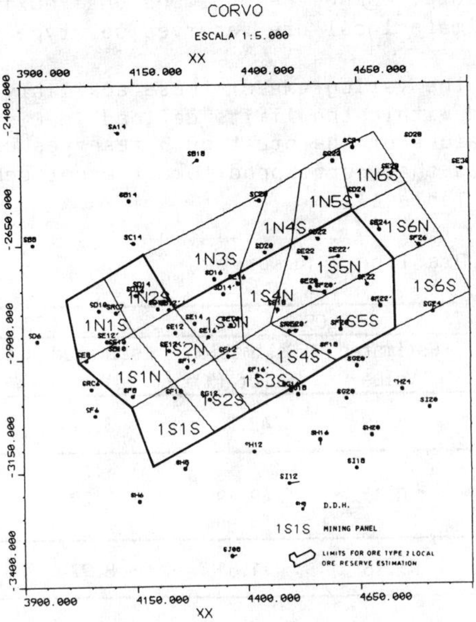

Figure 7 : Limits for Ore-type 2 Reserve Estimation

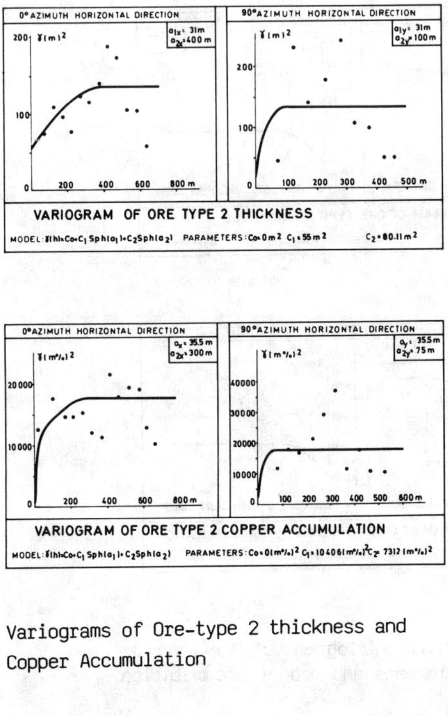

Figure 8: Variograms of Ore-type 2 thickness and Copper Accumulation

It is also possible, using the 3-dimensional auxiliary variables (12) and (14), to estimate local ore reserves per type within certain zones in the orebody.

Figure 11 shows the variograms of these auxiliary variables for ore-type 3, calculated within the limits defined in Figure 10.

The estimated values of the ore-type 3 reserves using the two different methodologies that correspond to different concepts of ore reserves are shown in Table III.

TABLE III - Ore Type 3 Reserves

	GLOBAL		LOCAL	
	Estimated Value	Estimation Error (%)	Estimated Value	Estimation Error (%)
Tonnage (10^6 Ton.)	3.79	42.38	3.22	32.45
Quantity of Copper (10^6 Ton.)	0.35	49.79	0.29	34.87
Mean Copper Grade (%)	9.78	11.81	8.97	19.30

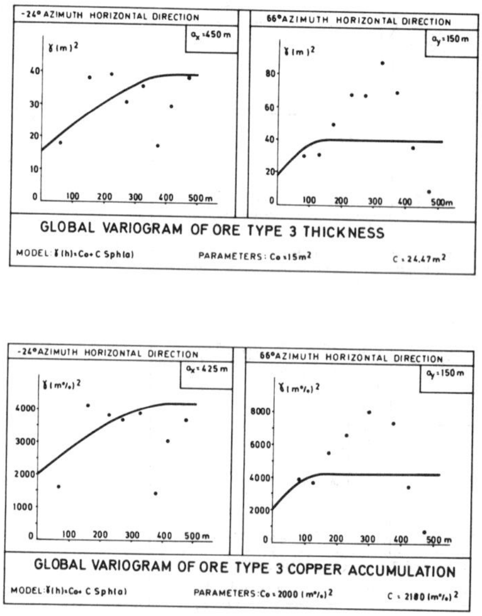

Figure 9: Global Variograms of Ore-type 3 thickness and copper accumulation

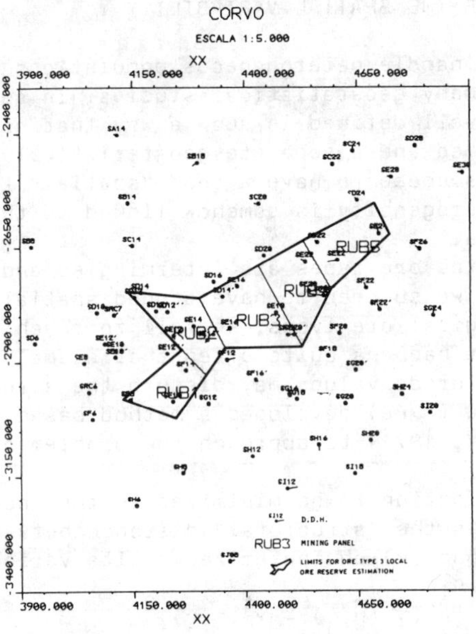

Figure 10: Limits for Ore-type 3 Local Ore Reserve Estimation

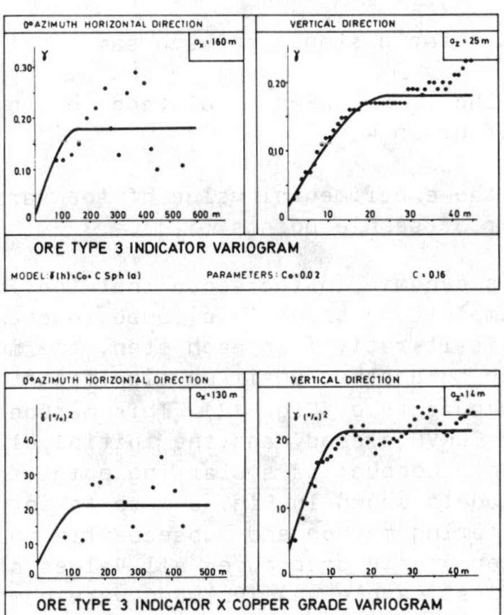

Figure 11: Local Indicator and Indicator x Cu Grade Variograms

4. HETEROGENEITY OF THE SPATIAL VARIABILITY

The problem of how to handle heterogeneous populations in an ore reserve estimation arises in many geostatistical studies. In some cases the ore--types are spatially well defined in such a way that non-overlapping sub-populations are obtained and a separate geostatistical study is possible. In this case we only succeed to have a good "spatial" typology if the source of spatial heterogeneity is somehow linked to the typology based on geological criteria.

In other cases the ore types are intermingled and problems become worst because even if we succeed to have a good spatial typology based on the original geological ore-types, we fail to reach spatial continuity.

In both cases it happens quite often that a small proportion of very high or very low grade values may disturb the structural analysis.

Muge and Pereira (1986) developed a method based on the dynamic clusters method (DIDAY, 1971) to approach the problem of improving the "spatial" typology.

The objective function to be minimized in the course of the algorithm is defined as the "structural" distance between a sample (i) and a "structural" group (k) characterized by its variogram model:

$$D_{ik} = \frac{1}{n_{ik}} \sum_{j=1}^{n_{ik}} (\gamma_k(\vec{h}_{ij}) - \gamma_k^*(\vec{h}_{ij}))^2 \qquad (18)$$

where: n_{ik} is the number of samples that fall within the neighbourhood of sample i defined for the "structural" type k.

$\vec{h}_{ij}$ is the vector distance between samples i and j.

$\gamma_k(\vec{h}_{ij})$ is the value taken at distance h_{ij} by the variogram model of group k.

$\gamma_k^*(\vec{h}_{ij})$ is the experimental value of the variogram function for the pair of sample points (i,j).

The procedure is dynamic in the sense that the initial preliminary allocation of each sample to a group is allowed to change the course of the algorithm and it is iterative: in each step, the method transfers samples from one group to another until no significant transfer of samples from group to group occurs (v.g. Fig. 12). This method was applied to the ore-types defined for CORVO orebody and the initial classification into ore-types was taken into account as a starting point for the procedure. Hence the variogram models shown in Fig. 6 were taken as nuclei for the non-hierarchical clustering method and subsequently updated.

In the first step of the procedure, all values of the experimental variogram where the first sample was included were compared to the correspondent values given by the models depicted in Fig. 6 according to (18) and the sample was reassigned to the ore-type with the lower "distance". The procedure continues for all samples. As an output of the first step, a new classification of samples was obtained and a model was fitted to the experimental variogram relative to each new group of samples.

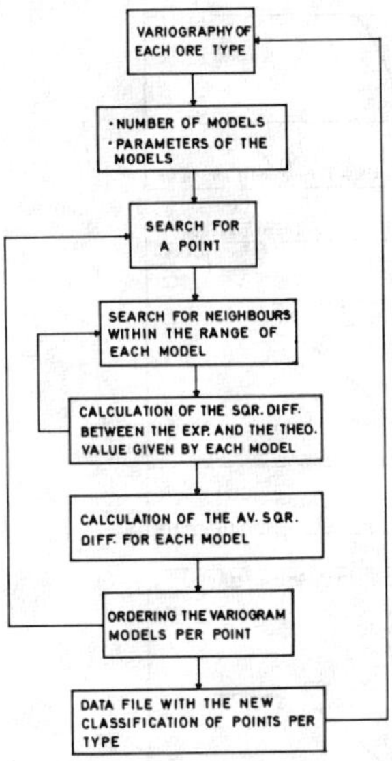

Figure 12: Organigram of the method.

The models fitted in the first step were taken as nuclei for the second step, which leads to a new reassignement of samples to groups. For each group, a new set of models were fitted and the procedure continues up to the convergence of the algorithm.

Applying the algorithm, 5 ore types were obtained. The main statistics of these types (labeled 2' to 6') are shown in Table IV. The variogram models for these final ore-types are shown in Fig. 13.

The composition of the final types in terms of the initial ones is shown in Table V. The initial ore-type 2 was spead into all final types; on the other hand, two final types have a mean grade higher than the global value (2' and 6'); the other final types have almost the same variance but different spatial structures.

Figure 14 shows a schematic comparative overlay of isopach maps outlining the thickness allocation to each initial and final ore-type. As it can be seen ore-types 2', 3' and 4' maintain the spatial localization of the initial ore-types 2, 3 and 4.

Ore-types 5 and 6 were spread into all final ore-types. On the other hand, the experimental variograms obtained show quite a good fit in the horizontal directions, comparing with those of Figure 6.

TABLE IV-Statistics for the final Ore-types

Ore Type	Average (%)	Variance (%)	Number of Samples
2'	10.22	49.50	173
3'	6.72	20.62	239
4'	4.81	26.67	527
5'	8.40	22.55	195
6'	20.91	68.16	42
Global	7.12	40.25	1176

TABLE V - Transition Matrix from the Initial to the Final Types (percentage)

FINAL \ INITIAL	2	3	4	5	6	TOTAL
2'	70	9	2	9	10	100
3'	60	22	3	3	12	100
4'	38	13	23	8	18	100
5'	61	18	3	7	11	100
6'	71	17	0	2	10	100

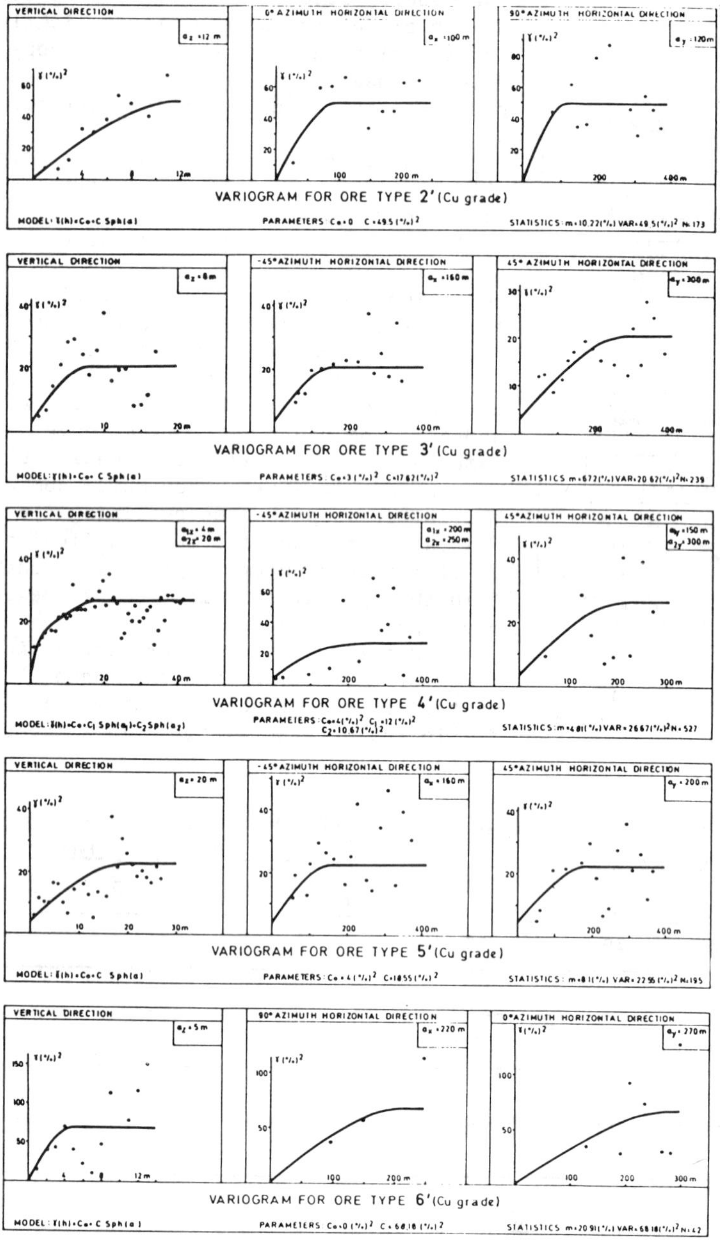

Figure 13: Variogram Models for the Final Ore-types (Cu grade)

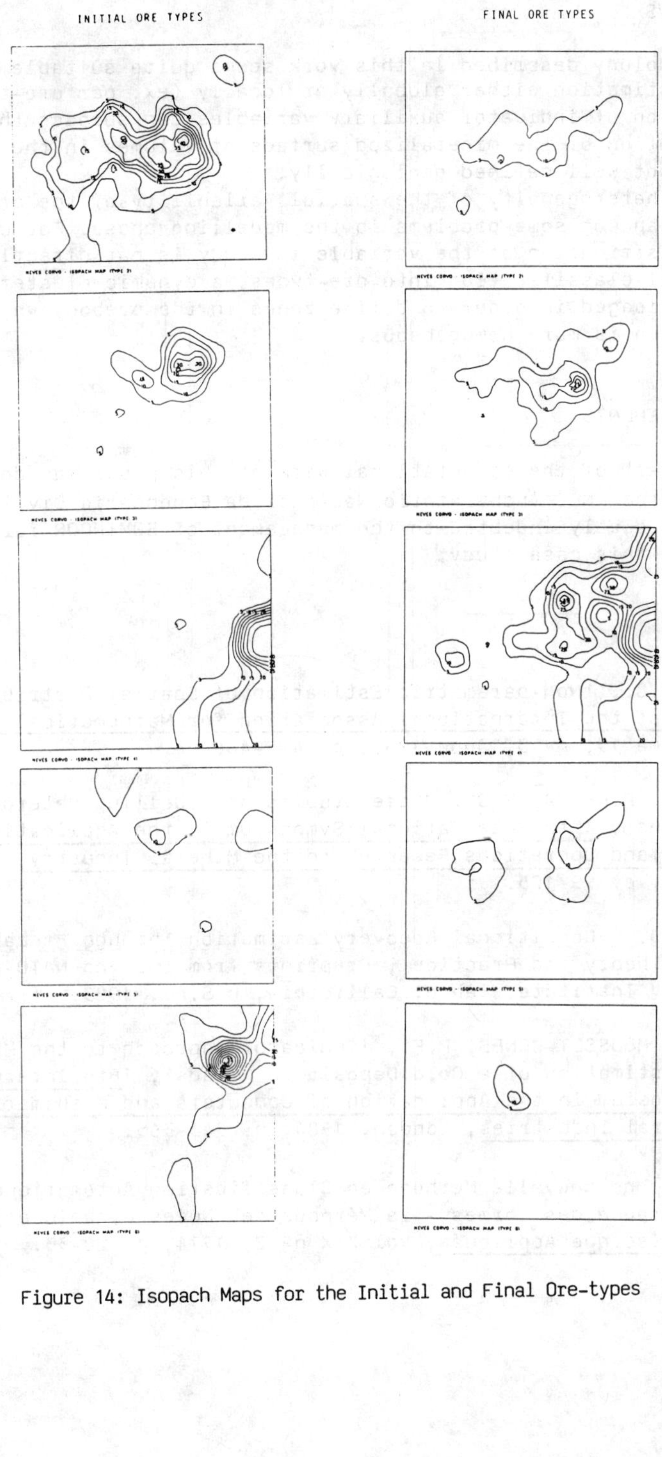

Figure 14: Isopach Maps for the Initial and Final Ore-types

CONCLUSIONS

The methodology described in this work seems quite suitable for ore reserve estimation either globally or locally (ex. per ore-type). The introduction of indicator auxiliary variables can surpass the problem of the estimation of the mineralized surface or volumes in the cases where they are not well defined geologically.

The heterogeneity of the spatial variability of the ore within each ore-type can put some problems in the modelling phase. For cases when the spatial distribution of the variable in study is not directly linked to the initial classification into ore-types, a dynamic clustering method can be envisaged in order to define zones in the orebody where the spatial distribution is more homogeneous.

ACKNOWLEDGEMENTS

The most part of the computational work of this paper was done using the DEC-10 of the LNEC (Laboratório Nacional de Engenharia Civil).

I am deeply indebted to the management of SOMINCOR for permission to publish this case study.

REFERENCES

JOURNEL, A.G., 'Non-parametric Estimation of Spatial Distributions', Journal of the International Association for Mathematical Geology, Volume **15**, nº 3, June 1983, p. 445-468

MUGE, F.H., PEREIRA, H.G., 'Case Studies on Modelling Heterogeneous Populations', 19th International Symposium in the Application of Computers and Operations Research in the Mineral Industry, Pennsylvania, 1986, p. 99-105.

SULLIVAN, J., 'Conditional Recovery Estimation Through Probability Kriging-Theory and Practice', Preprints from the 2nd NATO Advanced Study Institute, Tahoe, California, U.S.A., 1983.

KWA, B.L., MOUSSET-JONES, P.F., 'Indicator Approach to the Mineral Reserve Estimation of a Gold Deposit in Nevada', 18th International Symposium in the Application of Computers and Mathematics in the Mineral Industries, London, 1984, p. 343-365.

DIDAY, E. 'Une Nouvelle Méthode en Classification Automatique et Reconnaissance des Formes - la Méthode des Nuées Dynamiques' Revue de Statistique Appliquée, Vol XIX nº 2, 1971, p. 20-33.

A BAYESIAN APPROACH TO SURFACE ESTIMATION

by

Henning Omre
Norwegian Computing Center
Postbox 335 - Blindern,
0314 Oslo 3
Norway

ABSTRACT

The motive for working with Bayesian approaches to surface estimation is the lack of flexibility in choosing expectation functions in the kriging methods most frequently used. In the method presented, the user can provide qualified guesses for the expectation function and specify the uncertainty associated with this guess. Predictions will be based on the qualified guess and the available observations. A couple of examples are presented.

Key words: Surface estimation, kriging, Bayesian statistics.

1. INTRODUCTION

In many applications the variables to be studied are associated with references to the geographical space. Examples are found in petroleum reservoir evaluation, mining, pollution mapping, meteorology, agriculture, etc. The variables can often be classified as either point processes or continuous processes. In this paper the latter will be discussed.

Many approaches to analysis of continuous processes have been suggested. Methods based on random function theory, among these the kriging techniques, seem to be in frequent use, see Matheron (1965), Journel and Huijbregts (1978) and Vanmarcke (1983). In this paper another approached linked to the family of linear kriging methods will be presented. Practical use of methodology entails establishing a link between reality, and the model underlying one of the available methods. The users are responsible for justifying this link. By extending the family of methods, the chances will increase for the user to find a suitable model formulation for the problem at hand.

C. F. Chung et al. (eds.), Quantitative Analysis of Mineral and Energy Resources, 359–373.

Define the problem at hand by:

Let the regionalized variable of interest be denoted

$$\{ z(x);\ x \in A \}$$

with

x a reference variable in one, two or higher dimensions

A the domain of interest in the reference space

The regionalized variable is not completely known, but it is observed, in the locations x_i; $i=1,..,N$, within A, giving the following set of observations:

$$\{ z(x_i);\ x_i \in A;\ i=1,..,N \}$$

The problem is to predict the value of the regionalized variable in an arbitrary location x_o within A; ie predict $z(x_o)$.

Using random function theory implies representing the regionalized variable by a random function:

$$\{ Z(x);\ x \in A \}$$

The linear kriging approach to evaluation of random functions is based on second order analysis, hence, the form of;

$$E \{ Z(x) \} \qquad x \in A$$

$$Var \{ Z(x) - Z(x') \} \qquad x, x' \in A$$

should be specified. The form of these moments can not be chosen arbitrarily because inference about them have to be made from the observations or other available sources of information.

The linear kriging method will represent a gaussian random function exactly, but usually kriging is considered a second order approximation to an arbritrary random function. The model formulation above allows for evaluation of any linear operator on the random function.

As a consequence of the random function approach, the set of observations will be a set of random variables:

$$\{ Z(x_i)\ ;\ x_i \in A\ ;\ i=1,...,N \}$$

The prediction of the value in an arbitrary location x_o within A is based on a linear predictor:

$$Z^*(x_o) = \sum_{i=1}^{N} \alpha_i Z(x_i)$$

with

α: { α_i; i=1,..,N } a set of unknown weights, dependent on x_o, to be determined.

In linear kriging, the criterion for determining the weights is the best unbiased predictor in the expected squared sense. That implies determining the set of weights, α, from:

$$\underset{\alpha}{\text{Min}} \text{ Var } \{ Z(x_o) - Z^*(x_o) \}$$

$$\text{constrained by } \; E \{ Z(x_o) - Z^*(x_o) \} = 0.$$

The properties of this class of estimators are thoroughly discussed in Journel and Huijbregts (1978).

The application of the kriging methods has often shown that the flexibility in the choice of expectation function, E { Z(x) }, has been too limited. This lack of flexibility appears clearly whenever extrapolations beyond the locations of the observations are required. To compensate for this, the user often introduces fake data in order to force the predictions in the direction of the expectation of the user. This ad hoc solution to the problem is seen as improper. It would be better if the expectations of the users were a part of the model. It should be added that this problem occurs, often to an even higher degree, in other approaches than kriging also.

In the following, a short review of linear kriging methods, emphasizing the flexibility in choice of expectation function, will be given. The three methods familiar to all geostatisticians, simple, ordinary and universal kriging, will be outlined. Thereafter two Bayesian approaches to kriging will be presented. In the next section a more thorough discussion of one of the Bayesian models will be given. Thereafter some examples will be presented.

2. MODELS FOR SURFACE ESTIMATION

Simple kriging was introduced by Dr. G. Matheron in the early sixties, and the following model formulation was used:

$$E\{Z(x)\} = m(x) \qquad ; \quad x \in A$$

(1) $$\mathrm{Var}\{Z(x) - Z(x')\} = 2\,\gamma(x - x') \quad ; \quad x, x' \in A$$

with

$m(x)$; $x \in A$ a user specified function

$\gamma(x - x')$ the variogram function which is assumed known or estimated from the observations.

This entails that the random function is assumed to fluctuate centered around a known function, ie. $m(x)$; $x \in A$. The spatial variogram function is dependent on the relative locations only, ie. $x - x'$. In simple kriging the evaluation should be performed on the centered random function $\{ (Z(x) - m(x)) \ ; \ x \in A \}$. Consequently no bias problems occurs.

In the applications of simple kriging, however, the user found it hard to specify the expectation function exactly. In some cases, guidance can be given by deterministic models of the phenomenon. Examples may be found in pollution mapping and meteorology.

Ordinary kriging was introduced at the same time as simple kriging. The following forms for the two first moments were suggested:

$$E\{Z(x)\} = m \qquad x \in A$$

(2) $$\mathrm{Var}\{Z(x) - Z(x')\} = 2\,\gamma(x - x') \qquad x, x' \in A$$

with

m an unknown constant

$\gamma(x - x')$ the variogram function which is assumed known or estimated from the observations.

This entails that the random function is assumed to fluctuate centered around an unknown constant level. The spatial variogram function is dependent on the relative locations only. The ordinary kriging method proved helpful in many problems, but often the assumption of constant expectations was hard to justify. In petroleum reservoir description interests are primarily linked to anticlines of structures, in the border areas of deposits the decline in mineralization is definite and the vertical direction in many phenomenon show strong trends. An interpretation of ordinary kriging in a local stationarity setting helped relaxing the assumptions. This interpretation was complicated to understand for many users.

In order to meet these needs universal kriging was introduced by Dr. G. Matheron in the seventies. One justification of universal kriging came through an approximation to a random function with a random drift, see Matheron (1971). The first two moments were given the form:

$$E\{ Z(x) \} = a_o + \sum_{l=1}^{L} a_l f_l(x) \qquad x \in A$$

$$(3) \qquad Var\{ Z(x) - Z(x')\} = 2\gamma(x - x') \qquad x,x' \in A$$

with

L	a user specified integer constant
a_l ; $l=0,..,L$	unknown constants
$f_l(x)$; $x \in A$; $l=1,..,L$	user specified functions
$\gamma(x - x')$	the variogram function which is assumed known or estimated from the observations.

This entails that the random function is assumed to fluctuate around a function represented by a linear combination of user specified functions. The spatial variogram function is dependent on the relative locations only. Originally the estimation of this variogram function was a big problem, but new approaches to variogram estimation presented in Switzer (1984) and Stein (1984) may simplify this. Most use of universal kriging assumed the user specified functions to be polynomials in the location coordinates and assumed some local validity of the model. This extension of the kriging model was well suited to solve local deviations from the ordinary kriging assumption. When it came to extrapolation and global aspects had to be considered, the polynomial form was too rough to be useful. By letting the user specified functions be the users expectation of the shape of the regionalized variable, improvements in the extrapolations can be obtained.

Many users, however, found it hard to provide an exact specification of the drift terms, $m(x)$; $x \in A$ or $f(x)$; $x \in A$; $l=1,..,L$. It seemed more natural to consider them as qualified guesses for the expectation function, and to associate uncertainty with this guess. The link to Bayesian theory was obvious.

In a recent paper, Omre (1986), one Bayesian approach to kriging is presented. The model formulation is:

$$E\{ Z(x') \mid M(x) ; x \in A \} = a_o + M(x') \quad ; \quad x' \in A$$

$$(4)$$

$$Var\{ Z(x') - Z(x'') \mid M(x) ; x \in A \} = 2\gamma(x' - x'') \quad ; \quad x' x'' \in A$$

with

a_o	an unknown constant
$\{ M(x) \; ; \; x \in A \}$	a random function, apriori specified by its two first moments;
	$E \{ M(x) \} = \mu_M(x) \qquad x \in A$
	$Var \{ M(x) - M(x')\} = 2 \gamma_M(x,x') \qquad x,x' \in A$
$\gamma(x' - x")$	the conditional variogram function which is assumed known or estimated from the observations.

This entails that the random function of interest is assumed to fluctuate around another random function which represents the users qualified guess for the expectation function. The conditional variogram function is independent of the conditioning and depends on the relative locations only.

The qualified guess for the expectation function should be apriori specified. The function $\{ \mu_M(x) \; ; \; x \in A \}$ corresponds to the best guess while $\gamma_M(x,x')$ represents the uncertainty associated with the guess. Note that this variogram function is location dependent. It can be shown that the corresponding covariance function could take the form;

$$(5) \qquad Cov \{ M(x) , M(x') \} = \sigma_M(x)\sigma_M(x') \, \varrho_M(x - x')$$

with

$\sigma_M(x) \; ; \; x \in A$	an arbitrary function
$\varrho_M(x - x')$	a positive definite correlation function

and ensure conditionally, positive definiteness in the variogram function. Consequently the uncertainty in the qualified guess should be specified by a correlation function which is dependendent on the relative locations only, ie $\varrho_M(x - x')$, and a value in each location representing the standard deviation of the guess in that location, ie $\sigma_M(x)$. The Bayesian approach provides the user with the possibility to include a qualified guess into the prediction procedure. The guess may be based on several sources of qualitative information. In applications with few observations, experienced users and a need for extrapolation beyond the locations of the observations, this method is expected to be useful. The introduction of this approach should make the use of fake data unneccessary. Examples of applications could be mapping of petrophysical variables in description of reservoirs, geotechnical engineering and meteorology.

An alternative Bayesian approach was suggested by Solow (1985):

$$E \{ Z(x) \mid A_l;\ l=0,..,L \} = A_o + \sum_{l=1}^{L} A_l f_l(x) \qquad x \in A$$

$$(6) \qquad Var \{ Z(x) - Z(x') \mid A_l;\ l=0,..,L \} = 2\,\gamma(x - x') \qquad x,x' \in A$$

with

$A_l;\ l=0,..,L$ — a set of random variables, a priori specified by its two first moments:

$$E \{ A_l \} = \mu_l \qquad l=0,..,L$$

$$Cov \{ A_l, A_k \} = \sigma_{lk} \qquad l,k=0,...,L$$

$f_l(x);\ x \in A;\ l=1,..,L$ — user specified functions

$\gamma(x - x')$ — the variogram function which is assumed known or estimated from the observations.

This entails that the random function is assumed to fluctuate around a function represented by a linear combination of user specified functions. The coefficients of the combination are known through apriori specified distributions. The variogram function is dependent on the relative locations only. The predictors resulting from this model are at present being developed by A. R. Solow and the author.

This model can be useful whenever the user specified functions can be fairly precisely given and the coefficients can be interpreted by the user. One example would be depth conversion of seismic reflection time data. Then $\{ f_1(x);\ x \in A \}$ is identical to the reflection time function and the coëfficient a_1, is the corresponding velocity. The observationsare the depth measurements in well locations.

3. OUTLINE OF BAYESIAN APPROACH

The form of the model was presented in expression (4) in the previous section. The discussion in this section is based on Omre (1986). In linear Bayesian analysis, see Hartigan (1969), the expressions for conditional expectation and covariance replace the familiar Bayesian relations. Recall the expressions:

$$E \{ Y_1 \} = E \{ E \{ Y_1 \mid Y_3 \}\}$$

$$Var \{ Y_1 - Y_2 \} = E \{ Var \{ Y_1 - Y_2 \mid Y_3 \}\} + $$
$$+ Var \{ E \{ Y_1 \mid Y_3 \} - E \{ Y_2 \mid Y_3 \}\}$$

with Y_1, Y_2 and Y_3 three arbitrary random variables.

A natural predictor for the value of $Z(x_o)$ is

$$(8) \qquad Z^*(x_o) = \sum_{i=1}^{N} \alpha_i (Z(x_i) - \mu_M(x_i)) + \mu_M(x_o)$$

since

$$E\{ Z(x) \} = a_o + \mu_M(x).$$

Unbiasedness requires that:

$$E\{ Z(x_o) - Z^*(x_o)\} = 0$$

$$<=>$$

$$a_o = a_o \sum_{i=1}^{N} \alpha_i$$

Hence a neccessary and sufficient condition is

$$(9) \qquad \sum_i \alpha_i = 1$$

The prediction variance is:

$$\begin{aligned} &\mathrm{Var}\{ Z(x_o) - Z^*(x_o) \} \\ &\quad = \mathrm{Var}\{ Z(x_o) \} + \\ (10) \qquad &\quad + 2 \sum_i \alpha_i \,\mathrm{Cov}\{ Z(x_i) , Z(x_o)\} - \\ &\quad - \sum_i \sum_j \alpha_i \alpha_j \,\mathrm{Cov}\{ Z(x_i) , Z(x_j) \} \end{aligned}$$

From expression (7) and (9) it can be shown that:

$$\begin{aligned} &\mathrm{Var}\{ Z(x_o) - Z^*(x_o) \} \\ (11) \qquad &\quad = 2 \Sigma_i \alpha_i [\gamma(x_i - x_o) + \gamma_M(x_i, x_o)] - \\ &\quad - \sum_i \sum_j \alpha_i \alpha_j [\gamma(x_i - x_j) + \gamma_M(x_i, x_j)] \end{aligned}$$

Assume, for now, the variogram expressions to be known and to be conditionally, positive definite. Later in the section the conditions on $\gamma_M(x,x)$ for ensuring conditionally, positive definiteness and the estimation of $\gamma(x - x')$ will be discussed.

The set of weights α: $\{ \alpha_i ; i=1,..N \}$ is determined by minimizing the prediction variance under the unbiasedness constraint:

$$\min_{\alpha} \operatorname{Var}\{ Z(x_o) - Z^*(x_o) \}$$
$$\sum_i \alpha_i = 1$$

The solution of the minimization can be determined by using Lagrangean multipliers. Denote the solution $\tilde{\alpha} :\{ \tilde{\alpha}_i \ ; \ i=1,\ldots,N \}$. The resulting predictor and prediction variance are:

$$Z^*(x_o) = \sum_i \tilde{\alpha}_i \ (Z(x_i) - \mu_M(x_i)) + \mu_M(x_o)$$

$$\text{(12)} \qquad \operatorname{Var}\{ Z(x_o) - Z^*(x_o) \} = 2 \sum \tilde{\alpha}_i[\gamma(x_o - x_i) + \gamma_M(x_o, x_i)] - \sum_i \sum_j \tilde{\alpha}_i \tilde{\alpha}_j \ [\gamma(x_i - x_j) + \gamma_M(x_i, x_j)]$$

It is worth noting that the predictor will be dependent on the following:

. structural distance through the variogram function $\gamma(x - x')$

. structural distance between locations of prediction and observations; ie $\gamma(x_o - x_i)$; $i=1,..,N$

. structural distances inbetween locations of observations ; ie $\gamma(x_i - x_j)$; $i,j=1,..,N$

. qualified guess for expectation function, ie $\mu_M(x)$

. precision in qualified guess, ie $\gamma_M(x,x')$.

Consequently the predictions from this Bayesian approach will not coinside with the ones from ordinary kriging. The properties of the predictions are:

. the predictor is exact in the sense that it reproduces the observations. From this follows that decreasing confidence is put in the qualified guess with increasing number of observations.

. in areas without observations, particularly in extrapolations, the qualified guess has large influence on the predictor.

So far the conditions on the variogram expressions for being conditionally, positive definite, and hence ensuring positive prediction variances, have not been discussed. The conditional variogram function $\gamma(x - x')$, should be chosen among the family of functions usually applied in kriging. The variogram function associated with the qualified guess, $\gamma_M(x,x')$, however, are dependent on the absolute locations. It can be shown that variogram functions with associated co-variance functions as in expression (5) will be conditionally, positive definite. The fact that the sum of two conditionally, positive definite functions will be conditionally, positive definite ensures a positive prediction variance.

As mentioned previously, the variogram function $\gamma(x - x')$ is not apriori known, and has to be estimated from the observations and the qualified guess. It can be shown that one unbiased estimator is:

$$\hat{\gamma}(h) = \frac{1}{2N_h} \sum_{(i,j)\in D_h} [(Z(x_i) - Z(x_j))^2 - (\mu_M(x_i) - (\mu_M(x_j))^2 - 2\,\gamma_M(x_i,x_j)]$$

with

$$D_h : \{ (i,j) \mid x_i - x_j = h \text{ or } x_j - x_j = h \ ; \ i,j=1,\ldots,N \}$$

$$N_h = \# D_h$$

Extensions of the method to treat predictions of block averages and observations with errors can easily be made.

4. EXAMPLES

In this sections some examples of applications of the Bayesian approach are presented. The examples should not be considered a complete study, it is merely a demonstration of the consequences of applying the Bayesian approach. The observations represent thickness of geologic layer and they originate from the North Sea.

The locations of the 30 observations are presented in figure 1. The area of interest is rectangular with size 20000 x 10000. The observations are located close to one of the diagonals in the area. This is caused by preferential sampling in areas believed to have large thickness by the geologists. In later evaluation, the geologists need prediction of the thickness in areas without observations.

In the study, four alternative geostatistical models are compared. Let the thickness over the area be denoted $\{ z(x);\ x\in A \}$, and let its underlying random function be $\{ Z(x);\ x\in A \}$. The notation in expression (4) and (5) will be used in the proceeding presentation. The models are denoted case 1A and 1B, and 2A and 2B, because they go in pairs:

Case 1A - is based on the following model:

$$\mu_M(x) = 0.0 \qquad ;\ \text{all } x\in A$$

$$\sigma_M(x) = 8.11 \qquad ;\ \text{all } x\in A$$

$$\rho_M(h) = \exp\{-(h^2/10000.)\}$$

$$\gamma(h) = 10.0\,[1.5(h/2000.) - 0.5(h/2000.)^3]$$

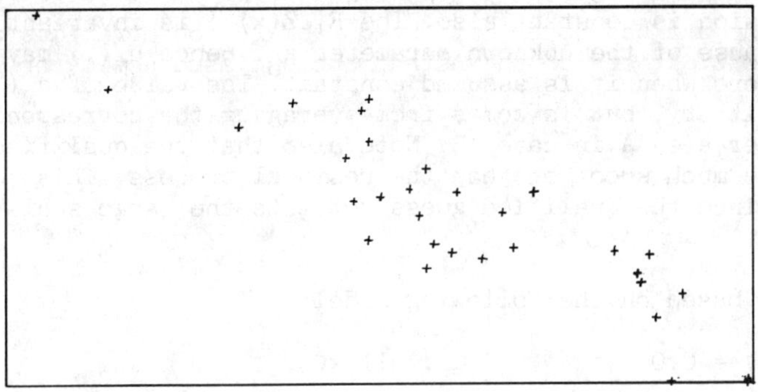

Figure 1. Locations of the observations

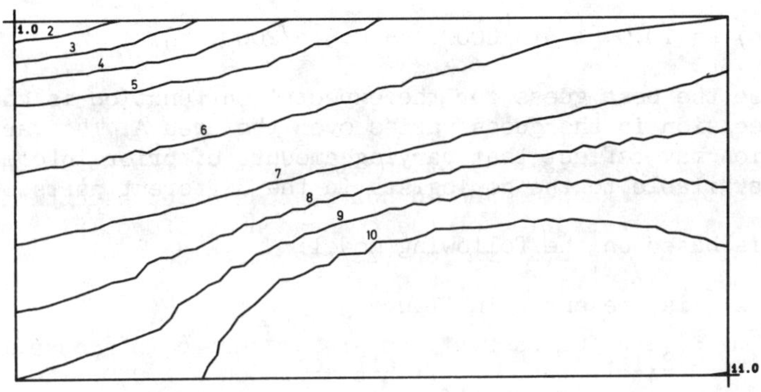

Figure 2. Contour map of the precision in the qualified guess - $\{ \sigma_M(x); x \in A \}$

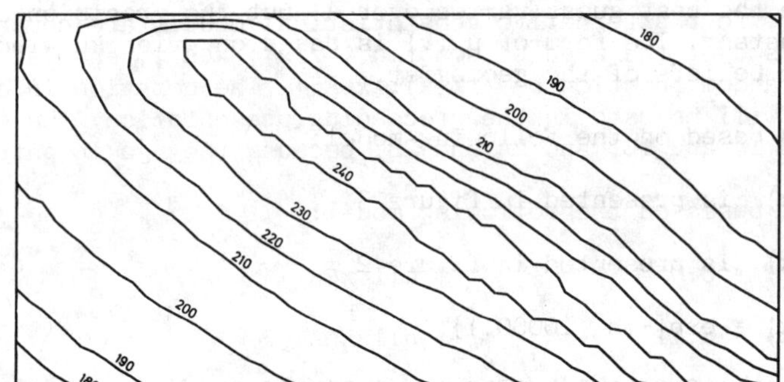

Figure 3. Contour map of the best guess - $\{ \mu_M(x); x \in A \}$

In this case the best guess for the expectation function is constant, and it precision is constant also. The E{ Z(x) } is invariant to shift in $\mu_M(.)$ because of the unknown parameter a_o. Hence $\mu_M(.)$ may be set exactly to zero when it is assumed constant. The value of $\sigma_M(.)$ looks somewhat arbitrary, but is comes from averaging the corresponding variances over area A in case 1B. Note also that the qualified guess is assumed to be much smoother than the residual process. This seems reasonable since the qualified guess reflects the large scale beliefs of the user.

Case 1B - is based on the following model:

$\mu_M(x) = 0.0$; all $x \in A$

$\sigma_M(x)$ is presented in figure 2

$\varrho_M(h) = \exp\{-(h^2/10000.)\}$

$\gamma(h) = 10.0\ [1.5(h/2000.) - 0.5(h/2000.)^3]$

In this case the best guess for the expectation function is constant, but the precision in the guess varies over the area A. The variation in the precision may reflect that varying amounts of prior information have been available to the geologists in the different parts of area A.

Case 2A - is based on the following model:

$\mu_M(x)$ is presented in figure 3

$\sigma_M(x) = 8.11$; all $x \in A$

$\varrho_M(h) = \exp\{-(h^2/10000.)\}$

$\gamma(h) = 10.0\ [1.5(h/2000.) - 0.5(h/2000.)^3]$

In this case the best guess varies over A, but the precision in the guess is constant. The form of $\mu_M(.)$ is based on prior knowledge and reflects the beliefs of the geologist.

Case 2B - is based on the following model:

$\mu_M(x)$ is presented in figure 3

$\sigma_M(x)$ is presented in figure 2

$\varrho_M(x) = \exp\{-(h^2/10000.)\}$

$\gamma(h) = 10.0\ [1.5(h/2000.) - 0.5(h/2000.)^3]$

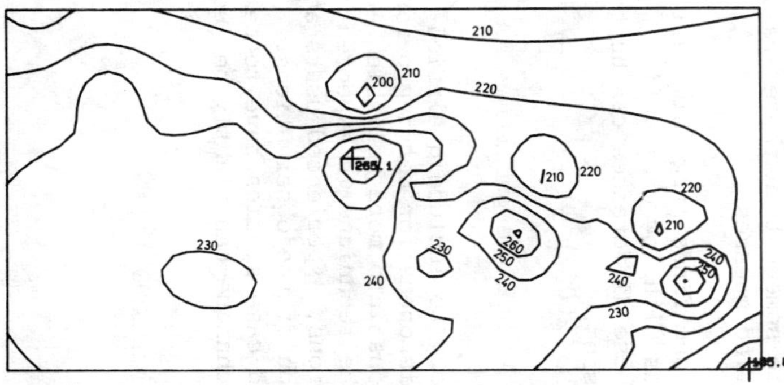

Figure 4A. Contour map of predictions based on case 1A.

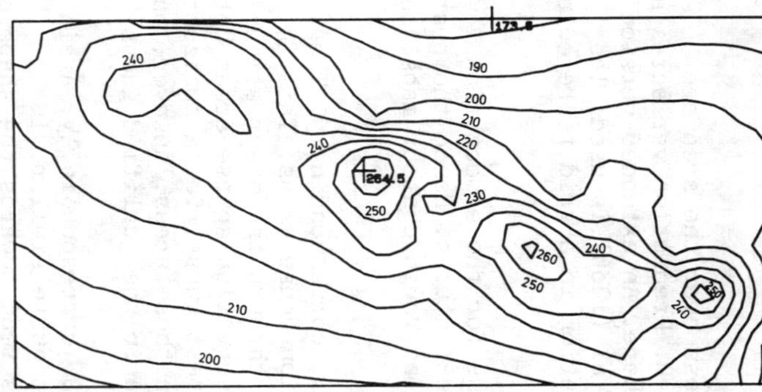

Figure 5A Contour map of predictions based on case 2A

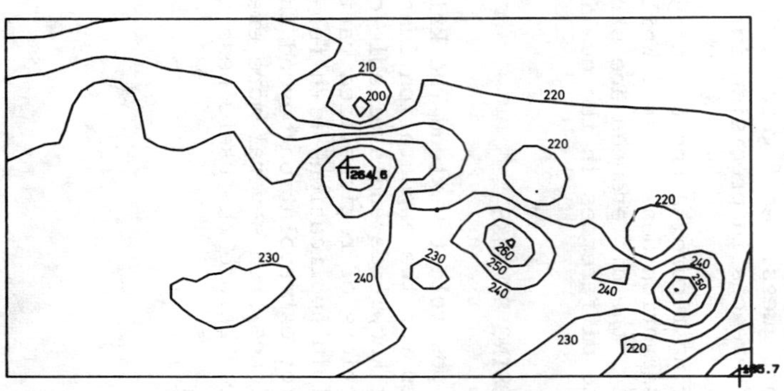

Figure 4B. Contour map of predictions based on case 1B

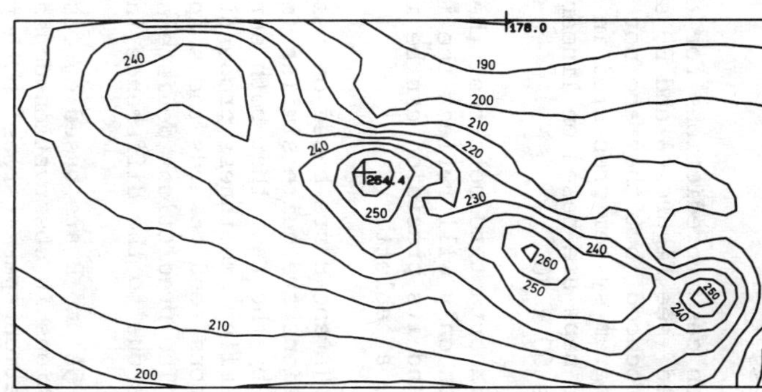

Figure 5B. Contour map of predictions based on case 2B.

In this case the best guess varies over A, and so does the precision in the guess.

For each case, predictions for the thickness over the area are produced, see figure 4A and B, and 5A and B. Prediction variances are also produced, but they are not reported here. An extended version of the program system presented in Bjerch et al (1986) is used. The contour maps are based on linear interpolation in a grid representation of size 60 x 30.

It is important to note that the best guess for the expectation function only will reflect the large scale features. Hence the best guess and its precision can be introduced as rough contour maps, on for example a tablett.

Figure 4A and B are based on case 1A and B, correspondingly. The predictions are quite similar in observation dense areas. This is caused by the fact that both surfaces run through the observations, and have similar smoothness properties. In observation sparse areas both predictons tend towards the shape of the best guess, ie. a horizontal plane. The predictions deviate somewhat in these areas, however, and that is due to the difference in precision in the qualified guess.

Figure 5A and B are based on case 2A and B, correspondingly. Again the predictions in observation dense areas are quite similar. In observation sparse areas, both predictions tend towards the shape of the best guess, see figure 3. They deviate somewhat in these areas due to differences in precision in the qualified guesses.

Comparing figure 4A and B, and 5A and B, one finds large similarities in observation dense areas. While in areas with low observation density, the differences are striking. These differences can be traced back to differences in the qualified guesses.

5. CLOSING REMARKS

It is possible to extend the kriging theory to include qualified guesses for the expectation function. This opens for merging prior expert knowledge and available observations into more reliable predictions of regionalized variables. The technique is expected to be useful in applications with few observations, experienced users and a need for extrapolation beyond the location of the observations. Experience from several more examples and case studies have to be gathered before the usefulness of the technique can rightly be judged.

6. ACKNOWLEDGEMENT

Thanks to Thomas Bjerch for providing valuable assistance in working out the examples. The research is financially supported by Norwegian Computing Center and the national Norwegian petroleum company, Statoil a.s.

7. LIST OF REFERENCES

Bjerch, T., Homleid, M., Omre, H. and Sollie, B.H., 1986: Surface Estimation Based on Observations and General Trends; NCC-Note, SAND/2/1986 (in Norwegian).

Hartigan, J.A.,1969: Linear Bayesian Methods, Journ. of the Royal Stat. Soc., Ser. B, Vol. 31, pp 446 - 454.

Journel, A.G. and Ch. J. Huijbregts, 1978: Mining Geostatistics, Academic Press, 600 p.

Matheron; G., 1965: Les Variables Regionalisees et leur Estimation; Masson & Cie; Paris.

Matheron; G., 1971: The Theory of Regionalized Variables and its Applictions; Le Cahiers du Centre de Morphologie Mathematique, No. 5, Fontainebleau, France; 211 p.

Omre, H, 1986: Bayesian Kriging - Merging Observations and Qualified Guesses in Kriging, to appear in Math. Geol.

Solow, A.R., 1985: Personal communication.

Stein, M., 1984; Estimation of Spatial Variability Part I: Non-Parametric Variogram Estimation, Tech. Report no. 75, Dept. of Statistics, Stanford University, 183 p.

Switzer, P., 1984: Inference for Spatial Autocorrelation Functions in Geostatistics for Natural Resources Characterization, Part I; Ed. G. Verly et al., D. Reidel Publ. Co.; pp 127 - 140.

Vanmarcke, E., 1983: Random fields; The MIT Press, 382 p.

CONDITIONAL SIMULATION OF INDICATOR DATA. CASE STUDY OF A MULTISEAM COAL DEPOSIT

A.O.SOARES
Centro de Valorização de Recursos Minerais
Technical University of Lisbon
Instituto Superior Técnico, Av. Rovisco Pais
1096 LISBOA CODEX

ABSTRACT. When planning the production schedulling resulting from an existing mining method, numerical models of the spatial variability of the grades and geometry of mineralized phases of a deposit are required previous to the simulation of the mining methods.

This paper presents a simulation of the geometry of lignite seams, which are intermingled with barren, using the conditional simulation of indicator data.

Inside the simulated geometry of lignite seams, the classical Conditional Simulation is applied for the lignite heat values.

1. INTRODUCTION

The selectivity characteristics of a mining method combined with the ore grades spatial variability, determines the time variation of the mined raw material that have to satisfy the quality and quantity specifications of the stockyards and processing plants.

When planning that production schedulling, numerical models of the spatial variability of the geometry of the mineralized phases and inside grades are required, previous to the simulation of the mining methods.

The technique of Conditional Simulation of random functions (cf. Journel, 1974) provides such models where the spatial variability of grades is characterized.

In mineral deposits with differentiated mineralized phases the spatial variability of the relative geometry of each phase must be modelled before the grades. To simulate the relative geometry of such deposits Journel and Isaaks (1984), propose to simulate independently each phase by using the Conditional Indicator Simulation. The process assumes that all phases are mineralized and consequently an indicator variable is defined by a cutoff grade.

This paper presents the simulation of the relative geometry of a deposit where mineralized seams (lignite) are intermingled with barren,

C. F. Chung et al. (eds.), Quantitative Analysis of Mineral and Energy Resources, 375–384.

which means that only one variable-indicator, is available and defined on a geological criterium: the sample takes the value 1 if it belongs to a lignite seam and 0 if it is barren.

Thus the simulation process differs from the previous mentioned technique in the conditioning phase where a gaussian transform of indicator is required.

Inside the simulated geometry of lignite seams it is applied the classical Conditional Simulation for the lignite heat values.

2. SIMULATION OF THE INDICATOR FUNCTION

Consider a stationary random function I(x) defined as

$$I(x) = \begin{cases} 1 - \text{if } I(x) \text{ belongs to a specific ore-type which is defined by a cutoff grade or any geological criterium.} \\ 0 - \text{If } I(x) \text{ belongs to the complementary ore-types.} \end{cases}$$

With the following mean and variogram:

$$E\{I(x)\} = m_I \qquad \frac{1}{2} E\{[I(x+h) - I(x)]^2\} = \gamma_I(h) \tag{1}$$

The objective of simulation is to obtain a spatial realization $i_s(x)$ of the random function I(x) with the same mean and variogram.

The method presented by Journel and Isaaks (1984) follows the fundamental steps of the more general random functions simulation.

. Considering a standard gaussian random function Y(x), the following relation with I(x) can be found:

$$I(x) = \begin{cases} 1 & \text{if } Y(x) > y \\ 0 & \text{otherwise} \end{cases} \tag{2}$$

being y a cutoff value defined on the gaussian random function by the relation:

$$\text{prob}\{I(x) = 1\} = \text{prob}\{Y(x) > y\}$$

which means $E\{I(x)\} = 1 - G(y)$

and consequently $y = G^{-1}(1-m_I)$ (3)

G(y) is the standard gaussian distribution function.

For the same value y other moments can be defined:

variance - $\text{Var}\{I(x)\} = G(y) \cdot [1-G(y)]$

non centered covariance - $K_I(h) = L(y,y;\rho_Y(h))$

variogram - $\gamma_I(h) = 1 - G(y) - L[y,y;1-\gamma_y(h)]$ (4)

with $L(y,y;\ \rho_y(h))$ being the bivariate gaussian distribution function and $\rho_y(h)$ the correlogram of $Y(x)$.

. The experimental indicator data can be considered as a realization of the random function $I(x)$ with a mean m_I an a variogram $\gamma_I(h)$ (1).

Once the cutoff is determined (3), the gaussian variogram $\gamma_y(h)$ is given by the inverse relation of (4)

$$\gamma_y(h) = 1 - L^{-1}(y,y;\ m_I - \gamma_I(h)) \qquad (5)$$

Where $L(y,y;\ \rho_y(h))$ is classical tabulated (Abramowitz and Stegun, 1965) or can be approximated by an hermitian polynominal expansion (Matheron, 1974).

A spatial realization $y_s(x)$ of the gaussian random function $Y(x)$ can thus be simulated using the turning-bands method.

The simulated $y_s(x)$ reproduces the variogram model $\gamma_y(h)$. The corresponding $i_s(x)$ obtained by (2):

$$i_s(x) = \begin{cases} 1 & \text{if } \ y_s(x) > y \\ 0 & \text{otherwise} \end{cases}$$

will have the imposed mean m_I and variogram $\gamma_I(h)$ (1), That is the objective of the simulation.

3. CONDITIONING THE SIMULATION TO INDICATOR DATA

The simulated indicator realization $i_s(x)$ reproduces the experimental mean m_I and variogram $\gamma_I(h)$ but it is not conditioned to experimental data, that is at a data location x_α the simulated $i_s(x_\alpha)$ can be different from the experimental indicator value $i(x_\alpha)$.

In the conditioning process described in Journel, (1974), the conditional simulated realization $y_{cs}(x)$ is defined by:

$$y_{cs}(x) = y^*(x) + \left[\, y(x) - y_s^*(x) \,\right]$$

where $y^*(x)$ and $y_s^*(x)$ are two kriged values.

To calculate $y^*(x)$, the gaussian transformed values $y(x_\alpha)$ of experimental $i(x_\alpha)$ are required.

Note that until conditioning phase no gaussian transform of experimental indicator was required but only the variogram model transform.

4. GAUSSIAN TRANSFORM OF INDICATOR

To transform a variable $z(x)$ with a distribution function $F(z)$ in a gaussian distributed variable $y(x)$, a method based on the bijection between $F(z)$ and $G(y)$ can be used:

if $F(z) = G(y)$ then

$$y = G^{-1}(F(z)) \text{ and } z = F^{-1}(G(y)) \tag{6}$$

This transformation process must account for the spatial variability (characterized by the variogram model) of the transformed variable. When the cumulative histogram of the original variable contains classes with a high number of equal values the bijection is not verified, and consequently, a random correspondence between the original and the gaussian values can bring severe distortions in the variogram model of the transformed values.

To transform an indicator in a gaussian variable - a correspondence between a value 0 (or 1) and a ranking of gaussian values, an approximative method is proposed by Soares (1985).

The idea is to transform each spatial located $i(x_\alpha)$, based on the spatial variability of $i(x)$ and surrounding values of $i(x_\alpha)$.

The solution consists of doing the rank order of the values $i(x_i)$, $(i=1,N)$, based on the estimated spatial probability $px_i=\text{prob}\{i(x_i)=1\}$ that is the probability of existing the value 1 in location x_i:

i) - In each sample location x_i the linear estimator px_i^* is calculated by kriging (ordinary kriging of the value $i(x_i)$ with the surrounding $i(x_j)$, after removing the respective value:

$$px_i^* = [i(x_i)]^* = \sum_j l_j \, i(x_j) \text{ with } j \neq i$$

where l_j are the kriging weights

ii) - The experimental values $i(x_i)$ are rank ordered (separately for each class - 0 and 1) according its p_{xi}^*, and an ordered vector $i(k)$, $(k=1, N)$, is obtained. The corresponding N gaussian values are then calculated by the relation (6):

$$y(k) = G^{-1}(F(i(k))) = G^{-1}(k/(N+1))$$

Note that the kriging values make the ordering (and the gaussian transformation) of each $i(x_i)$ depend on the spatial variability of $i(x)$ and the values of the surrounding samples.

As it is an approximative method, it must be checked by running variograms on the transformed $y(x)$ and comparing with the transformed model obtained by (5). But the important check is the comparison between the variograms of the final conditional simulated values:

$$i_{cs}(x) = \begin{cases} 1 & \text{if } y_{cs}(x) > y \\ 0 & \text{otherwise} \end{cases}$$

and the imposed model (1) $\gamma_I(h)$.

5. CASE STUDY

5.1. The Deposit and Available Data

The Rio Maior deposit located in midle of Portugal, is a lignite/diatomite complex with a sincline attitude. For the purposes of this study it was selected an area A1 (Fig. 1) intersected by 32 vertical drill holes. The heat values are only available in the samples geologically defined as mineralized coal interceptions.

The indicator variable was obtained splitting each drill hole in two meter core samples. For each core, the sample is taken as "1" if it cuts more than 50 cm of coal imposed by technological reasons, and as "0" otherwise.

The resulting 1576 indicator values provide the following statistics:

$$m_I = .15 \qquad \sigma_I^2 = .13 \tag{7}$$

For the 231 heat values the same statistics are:

$$m_h = 1.57 \times 10^3 \text{ Kcal/Kg}$$
$$\sigma_h^2 = 210 \times 10^3 \text{ (Kcal/Kg)}^2 \tag{8}$$

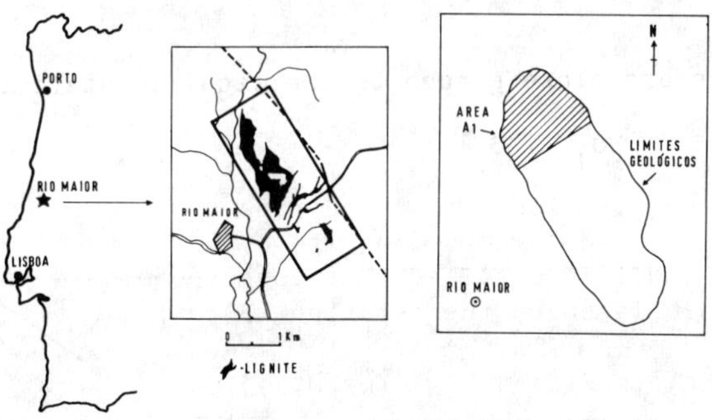

Fig. 1 - Location of RIO MAIOR deposit and selected area A1

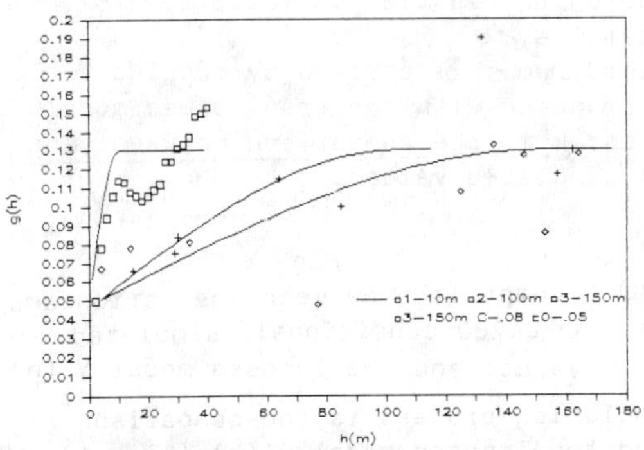

Fig. 2 - Variograms of Indicator - $\gamma_I(h)$.

5.2. Simulation of Indicator

. Fig. 2 shows the experimental variograms of the 1576 indicator values and their fit by an anisotropic spherical structure

. The cutoff y is given by the relation (2):

$$y = G^{-1}(.85) = 1.08$$

. Variograms $\gamma_y(h)$ were calculated by the inverse relation (5), using an hermitian expansion. They were fitted by one anisotropic spherical structure.

. The indicator values were rank ordered and transformed in gaussian values, by the exposed method in 4, and the variograms were calculated.

Fig. 3 gives the model of the transformed variogram $\gamma_y(h)$ and the experimental variograms - $\gamma_x'(h)$, obtained after gaussian transformation of i(v).

. With the model $\gamma_y(h)$, 84280 point values were simulated and conditioned in a 30 x 20 x 2m regular grid.

The resulting $i_{cs}(x)$ present the following mean and variance:

$$m_{I_{cs}} = .144 \qquad \sigma^2_{I_{cc}} = .123$$

Both values are close enough to the required statistics (7):

$$m_I = .15 \qquad \sigma^2_I = .13$$

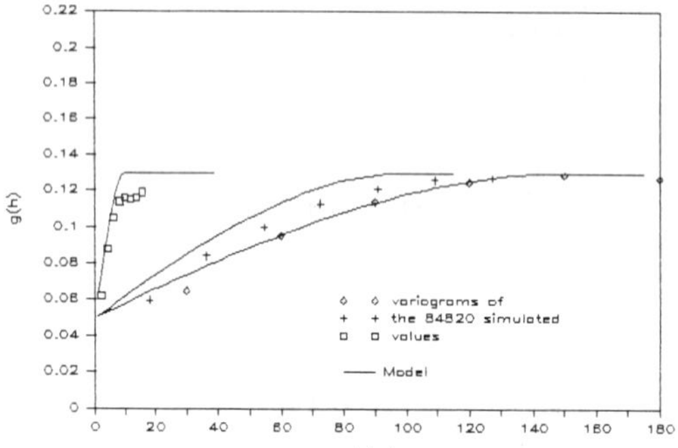

Fig. 3

Comparison between the model $\gamma_y(h)$ and the experimental $\gamma_y'(h)$.

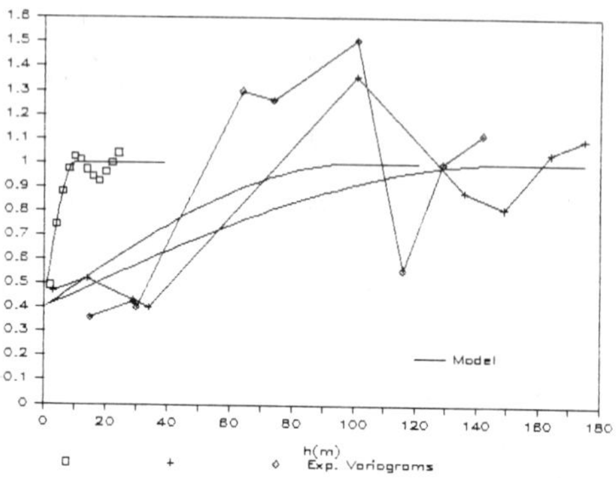

Fig. 4

Comparison between the variograms of 84280 conditional simulated values and the imposed model $\gamma_I(h)$.

the most important check of any simulation process is the comparison between the variograms of $i_{cs}(x)$ and the imposed model $\gamma_I(h)$ (Fig. 4).

It appears that the conditioned values $i_{cs}(x)$ have the same spatial variability as the original data $i(x)$.

Fig. 5 shows two vertical sections where the $i_{cs}(x)$ values (hatched area) reproduces the geometry of the lignite seams. In Fig. 6 it can be seen two perspective views of top and bottom of lignite seams. A contour map of total thickness of lignite is presented in Fig. 7.

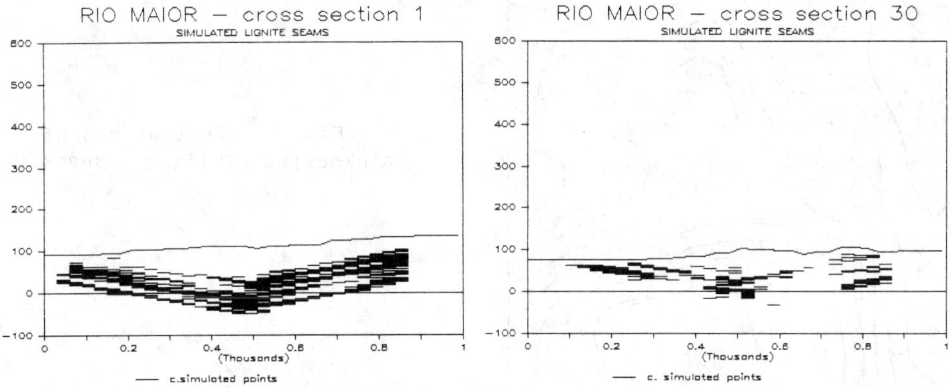

Fig. 5 - Cross sections of simulated lignite seams.

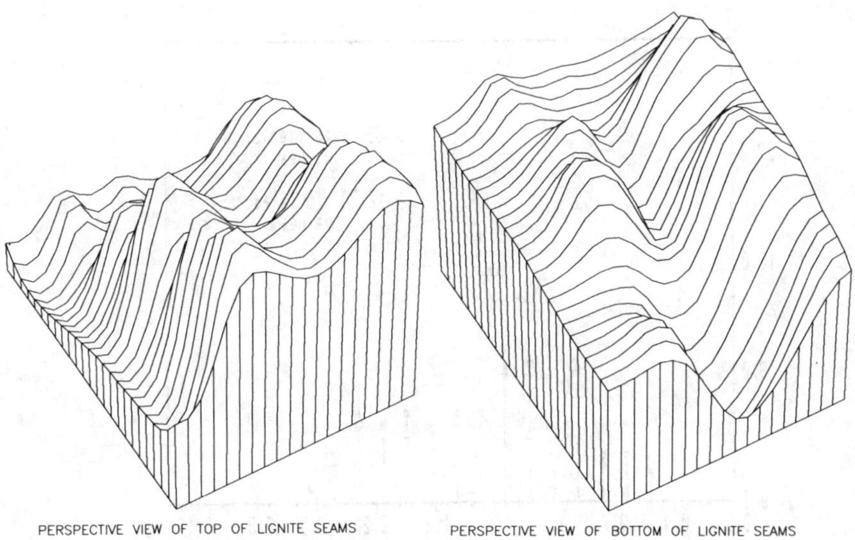

Fig. 6 - Perspective views of top and bottom of lignite seams

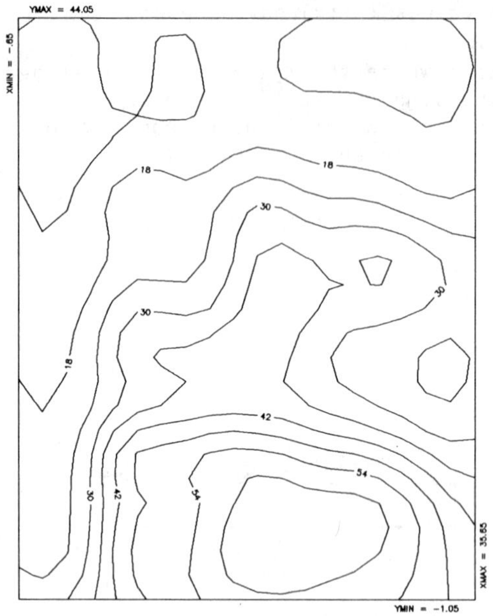

PLOT OF THICKNESSES OF LIGNITE SEAMS

Fig. 7 - Contour map of thicknesses of lignite seams

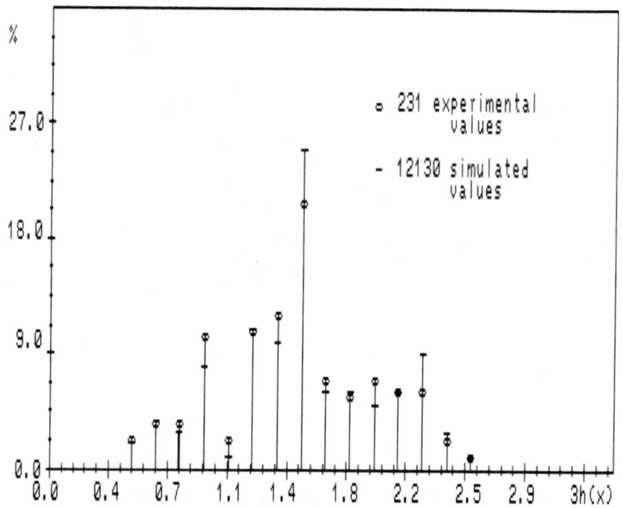

Fig. 8 - Histograms of experimental and simulated values

5.3. Conditional Simulation of the Heat Values - h(x)

Inside the simulated lignite seams the heat values - h(x) were simulated , by the classical Conditional Simulation.

12130 heat values were simulated and conditioned, presenting the final following statistics:

$$m_{hcs} = 1.53 \times 10^3 \text{ Kcal/Kg} \qquad \sigma^2_{hcs} = 184. \times 10^3 \text{ (Kcal/Kg)}^2$$

These values are close enough to the required mean and variance (8):

$$m_h = 1.57 \times 10^3 \text{ Kcal/Kg} \qquad \sigma^2_h = 210. \times 10^3 \text{ (Kcal/Kg)}^2$$

Fig. 8 shows the comparison between the histogram of the 231 experimental data - h(x) and the histogram of the 12130 conditional simulated values $h_{cs}(x)$.

The experimental variograms of the conditional simulated $h_{cs}(x)$ and the imposed model $\gamma_h(h)$ can be seen in Fig. 9.

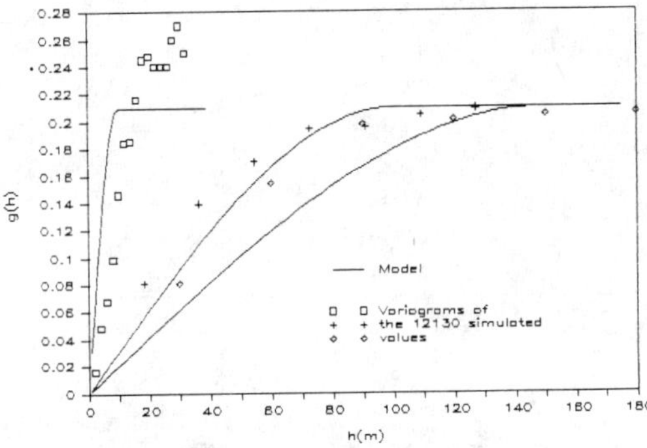

Fig. 9 - Variograms of $h_{cs}(x)$ and the model $\gamma_h(h)$

6. CONCLUSIONS

The method, presented in this paper, of doing the conditioning of indicator data, permits the morphological modelling of such deposits where the rock-types are identified only by geological criteria (that is, grades or another quantity variables are not available).

The simulation and conditioning process, depending on the spatial variability of the indicator data, honors the geological structures of the rock-types.

It is an approximative method that must be checked for each particular case by the variograms of the final values (a very familiar practice in simulation of random functions). For the case study - Rio Maior deposit - very satisfactory results were obtained.

ACKNOWLEDGMENT

The most computational work of this paper was done in LNEC (Laboratório Nacional de Engenharia Civil).

REFERENCES

JOURNEL, A.G., 1974, 'Geostatistics for Conditional Simulation of Ore-bodies' Econ. Geol., V. 69, p. 673-687.

JOURNEL, A.G., ISAAKS, E.H., 1984, 'Conditional Indicator Simulation. Application to a Saskatchewan Uranium Deposit': Math. Geol., Vol. 16, nº 7, p. 685-718.

MATHERON, G., 1974, 'Les Fonctions de Transfert des Petits Panneaux', Les Cahiers du Centre de Morphologie Mathématique de Fontainebleau. N-395, 73 p.

SOARES, A., 1985, 'Simulação Morfológica de Jazigos Minerais: Aplicação a um Jazigo Lignitico', Msc. thesis, I.S.T..

FACTORIAL KRIGING AS A METHOD TO INCLUDE SPATIAL STRUCTURE INTO CLASSIFICATION. A CASE STUDY ON A SULPHIDE OREBODY.

A. JORGE SOUSA
Centro de Valorização de Recursos Minerais
Technical University of Lisbon
Instituto Superior Técnico, Av. Rovisco Pais
1096 LISBOA CODEX

ABSTRACT. It is well known that cluster analysis tend to create groups that shown a poor spatial contiguity. In this paper a method, based on factorial kriging, is proposed, in order to include the regionalized nature of variables into the classification algorithm. The method is applied to a complex sulphide orebody and groups produced compare favorably with classical cluster analysis results.

INTRODUCTION

The increasing importance of polimetallic deposits in the last few years have raised new questions to appraisal and management. Indeed, as commodity prices decreases and shortages of certain mineral products occour simultaneously with overproduction of other ores, it is necessary to diversify exploration and exploitation efforts. On the exploitation point of view, mine planning must be prepared to give flexible answers to quick qualitative changes in the demande of final products. Hence, multivariate statistical techniques must be improved and intensively used, in order to cope with some of the above mentioned problems.

In a polimetallic orebody, the in situ grades for a set of different elements must be, jointly, taken into account by mine planning. So, the issue of ore type definition arises - different combinations of the in situ grades have important consequences on the ore processing response and, therefore, the orebody must be splitted into "homogeneous" zones,to be mined separately and feed, eventually, different processing methods or blending procedures.

The ore type definition based on in situ grades may be achieved by classical cluster analysis (Rogado et al, 1979; Muge et al, 1979) but, except for some particular cases, groups created by this technique turn out to show poor spatial contiguity and, therefore, technological problems may arise in the exploitation stage. On the other hand, it is felt that, using only cluster analysis, an important feature of the mineralization is left out - its spatial structure. To overcome these drawbacks and produce groups with geological and technological meaning, it is compulsory to introduce the regionalized characteristics of variables being

C. F. Chung et al. (eds.), Quantitative Analysis of Mineral and Energy Resources, 385–392.

treated. The novel techique of Factorial Kriging (Matheron , 1982 ; Sandjivy, 1984) allows to account for spatial correlation revealed by variography (Journel and Huijbregts, 1978).

In this paper, a method of handling multivariate data using both Factorial Kriging and Cluster Analysis is proposed and tested on a case study concerning a complex sulphide orebody, which is being studied to produce different concentrates of copper, lead and zinc, in addition to the actual pyrite extraction for sulphur. The planning of the ore dressing plant, the exploitation sequence and eventual blending depend, to a certain extend, on the feed characteristics, and so, it was decided to study carefully classification procedures based on in situ grades , in order to create meaningful ore types.

The proposed method includes four steps: In the first step a Principal Components Analysis is performed in order to obtain an orthogonal basis for classification; the second step consists on the variography of projections on factorial axis, giving rise to nested structures,modelled by spheric schemes (in this case study); in the third step, Factorial Kriging is accomplished and produces the decomposition of data into its spatial components, according to variogram models; finally, the cluster analysis algorithm is applied on factorial kriging components, in order to get groups with spatial meaning. The method compares favorably with classical cluster analysis, on the standpoint of spatial contiguity of samples within groups.

AVAILABLE DATA

The Moinho sulphide orebody is exploited by cut-and-fill methods. From the samples taken for production control and assayed for S, Fe, Cu, Zn, Pb, and As, a set of 499 were selected in a certain level (cf. Fig. 1), in order to illustrate the proposed procedure.

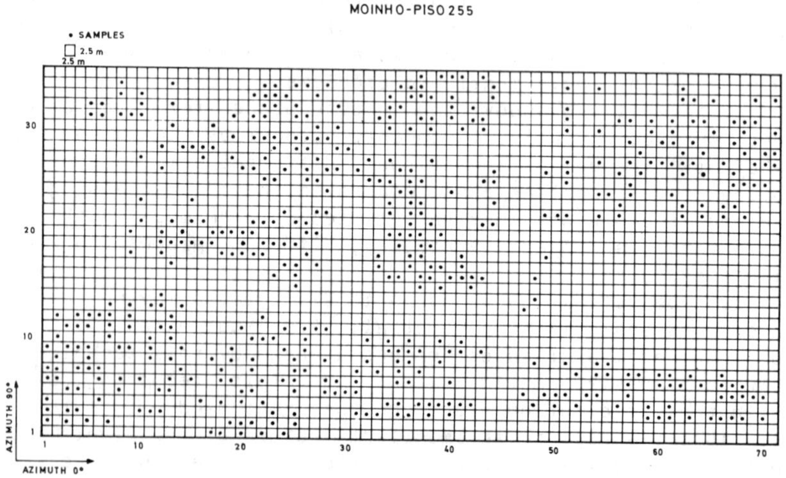

Fig. 1 - Location of samples

Basic statistics of the above mentioned data set are shown in Table I.

	S	Fe	Cu	Zn	Pb	As
Average	46.80	40.59	1.04	3.33	1.47	0.67
Variance	10.68	9.64	0.96	5.46	1.34	0.46

TABLE I - Basic statistics of data

PRINCIPAL COMPONENTS ANALYSIS

The 499 x 6 data matrix was submitted to a Principal Components Analysis (PCA) program, in order to extract orthogonal axis. Projections of variables onto the principal plan are shown in Fig. 2. The main output of this step is a new 499 x 6 matrix containing projections of samples onto factorial axis (projection matrix). It is worth noting that, as no factors was rejected, the information given by the new matrix reproduces, in an orthogonal basis, the original one. So, the PCA transformation does not disturb further comparisons.

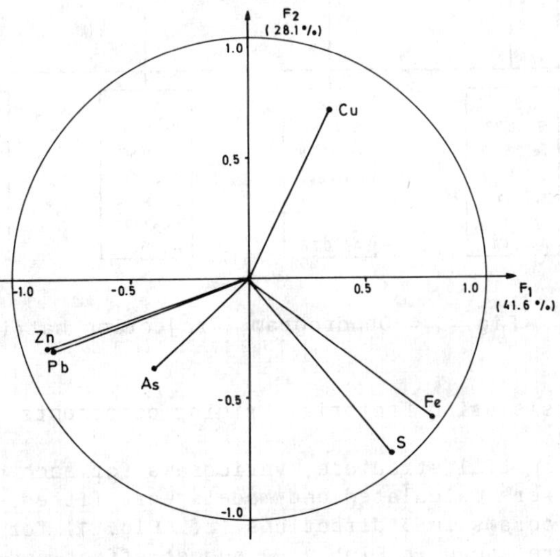

Fig. 2 - Projections of variables onto first and second axis

CLASSIFICATION PROCEDURE

. Cluster analysis on PCA projections

Taking the projection matrix as an input, the UPGMA (unweigthed pair group arithmetic average) cluster algorithm (Sneath and Sokal, 1978)

was applied using the correlation coefficient as similarity criterium. Samples were classified in 4 types, according to a significant cutoff in the dendrogram. In Fig. 3 is sketched the final dendrogram and the average grades for each group.

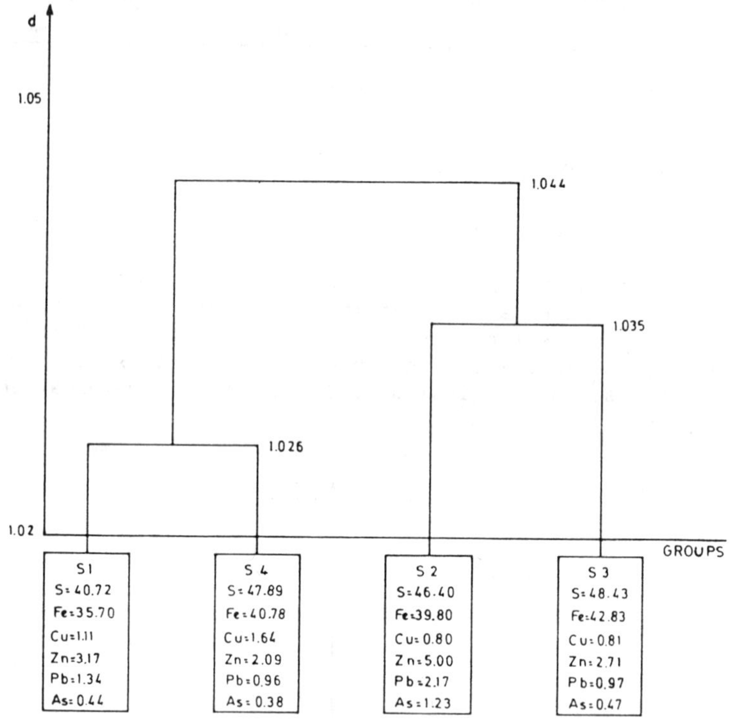

Fig. 3 - Dendrogram (Projection matrix)

. Cluster analysis using Factorial Kriging components

To account for spatial structure, variograms for each column of the projection matrix were calculated and models were fitted. As an example, experimental variograms in 3 directions (cf. Fig. 1) for the projections onto factor 1 are shown in Fig. 4. A nugget effect and two nested anisotropic spherical schemes were fitted (cf. Fig. 4). Results of model fitting for the projections onto the 6 factors are summarized in Table II.

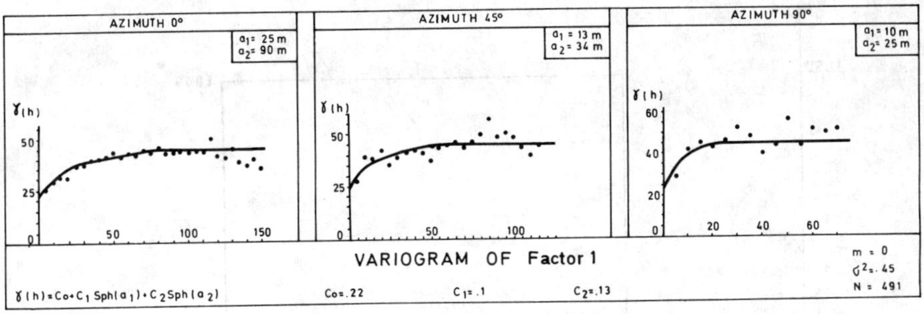

Fig. 4 - Variogram of factor 1

	Nugget effect (%)^2	1st Sill (%)^2	2nd Sill (%)^2	1st range (m)	2nd range (m)	Ratio of anisotropy
Factor 1	0.220	0.100	0.130	25	90	3.60
Factor 2	0.050	0.110	0.120	45	100	2.00
Factor 3	0.040	0.020	0.078	30	120	2.40
Factor 4	0.040	0.020	0.035	40	130	2.17
Factor 5	0.023	0.006	-	100	-	2.86
Factor 6	0.008	0.002	0.002	40	120	2.40

TABLE II - Parameters of variogram models

Each column of the projection matrix is now splitted into 3 components (except for factor 5, which reveals only two structures) - the first component accounts for nugget effect, the second and third are calculated using the correspondent variogram models in the factorial kriging algorithm. The output of this step is a 499 x 17 matrix - factorial kriging components matrix (F_{ijk}). The relationship between F_{ijk} and the projection matrix (F_{ij}) is outlined in Fig. 5.

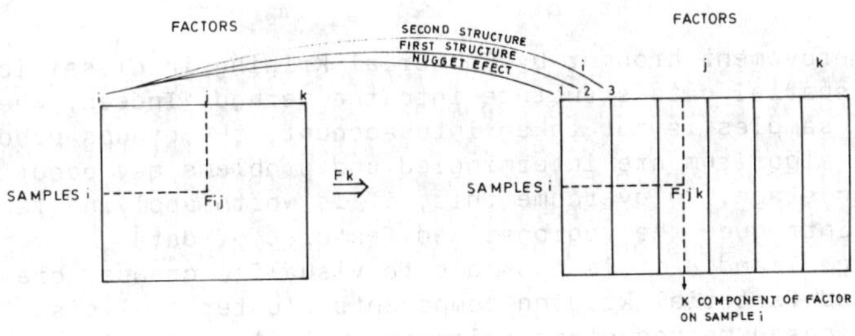

Fig. 5 - Relationship between the projection and the factorial kriging components matrices.

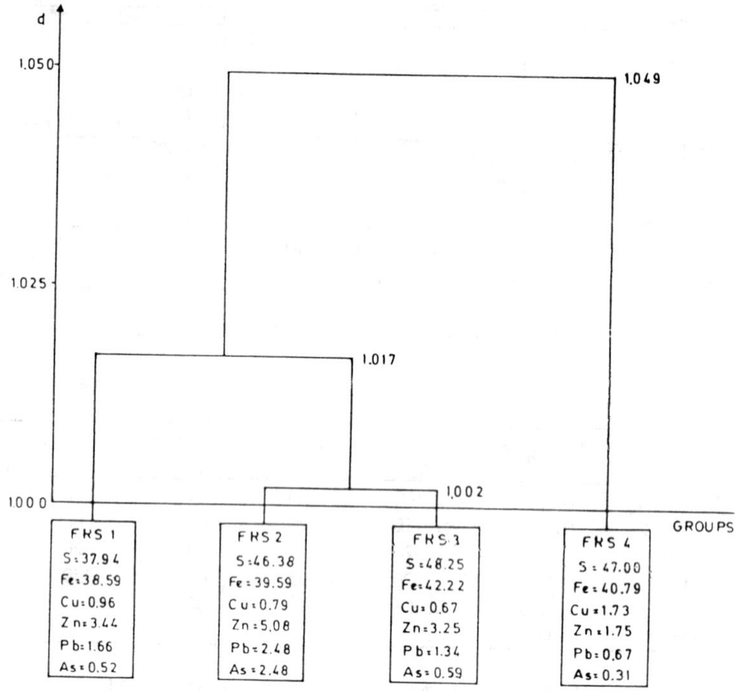

Fig. 6 - Dendrogram (factorial kriging components matrix)

Taking now the factorial kriging components matrix as an input to the same cluster analysis algorithm, new dendrogram was derived and 4 new groups were retained. Results of this procedure are shown in Fig. 6.

CONCLUSIONS

The main improvement brought by Factorial Kriging in classification is to include spatial data structure into the method. Indeed, when spatial location of samples is not taken into account, the groups produced by the cluster algorithm are intermingled and problems may occur in the exploitation stage. To overcome this, it is worth applying factorial kriging to introduce the regionalized features of data.

In Fig. 7 and 8 it is possible to visualize groups created by classical and factorial kriging components cluster analysis. It seems that the proposed method gives raise to spatially more coherent ore types.

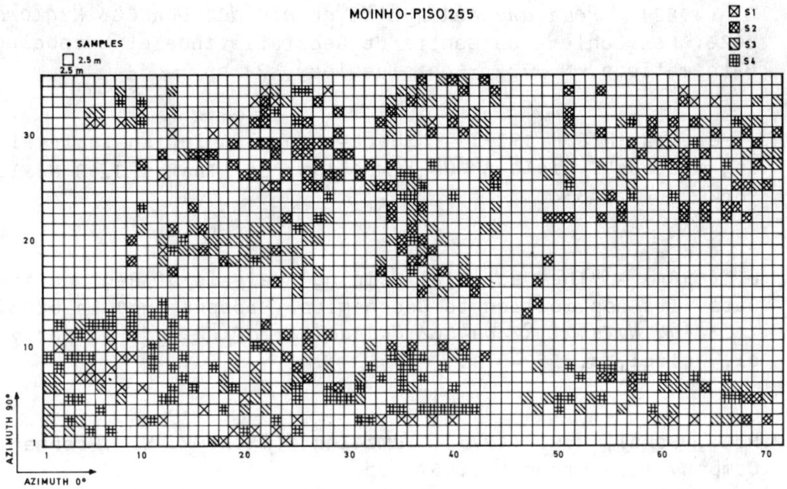

Fig. 7 - Location of groups (projection matrix)

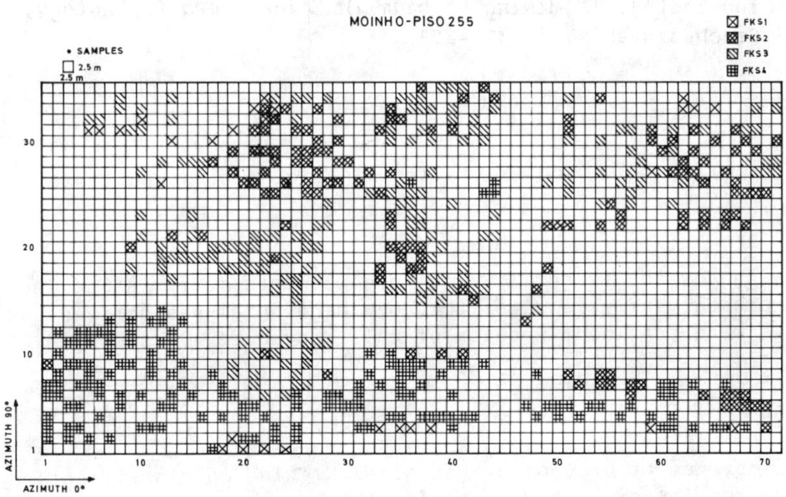

Fig. 8 - Location of groups (factorial kriging components matrix)

REFERENCES

Journel, A., Huijbregts, Ch. J. (1978). *Mining Geostatistics*. Academic Press, London, 600 pp.

Matheron, G. (1982) - *Pour une Analyse Krigeante des Données Regionalisées*. Les Cahiers du Centre de Geostatistique et Morphologie Mathématique, N. 732, Fontainebleau, 21 pp.

Muge, F.H., Pereira, H.G. (1979) - "Short-term planning in sublevel stoping methods". *16th APCOM Symposium*, Ed. Thomas J. O'Neil, New York, p. 323-334.

Sandjivy, L. (1984) - "The Factorial Kriging Analysis of Regionalized Data. Its Application to Geochemical Prospecting" In *Geostatistics for Natural Resources Characterization*, ed. G. Verly et al, Reidel, p. 559-571.

Sneath, P.H.A., Sokal, R.R. (1973). *Numerical Taxonomy*. W.H.Freeman Company, San Francisco, 573 pp.

Rogado, J.Q. et al (a979). "Sampling of Complex Orebodies for Ore Dressing tests". *IX Mining Pribram in Science and Technology*, Czechoslovakia, p. 324-351.

GEOSTATISTICAL TECHNIQUES FOR INTERPRETING MULTIVARIATE SPATIAL INFORMATION

Hans Wackernagel

Centre de Géostatistique
Ecole des Mines de Paris
Fontainebleau, France

ABSTRACT. Several geostatistical techniques for exploring the structure of spatially distributed multivariate data are presented. The techniques are based on a combination of variogram modelling, principal component analysis and cokriging. Some possibilities to map the essential features of the multivariate spatial structure of the data are discussed. An example using geochemical data is given with an interpretation.

1. INTRODUCTION

In the geosciences as well as in other fields, we are concerned with the synthesis and interpretation of multivariate information provided by samples which are characterized by their spatial location. Such samples are not uncorrelated. They are related to one another by physical (geological, biological,...) phenomena. They can describe the same phenomenon in different ways, or, conversely, the same phenomenon can be present to a different extent in several variables.

Thus it seems worthwhile to analyse such multivariate data using a method which takes into account these two aspects—the relation between samples due to their geographical position and the relation between variables due to their partial redundancy.

In geochemical prospecting for example, the final aim is the interpretation of certain anomalous values in the data (the 'anomalies'), which could give indications of the presence of deposits of particular mineral resources nearby. In our opinion, this can be done best if the interpretation of the anomalies is based upon an explanation of the physical phenomena described by the data. In ideal

C. F. Chung et al. (eds.), Quantitative Analysis of Mineral and Energy Resources, 393–409.

situations the parameters of the mathematical model can be related directly to geological facts. In practical applications, it is generally useful to look for a physical interpretation of the parameters although it may not be possible to give a precise explanation at a certain stage of exploration.

Geostatistics is better known as a method for estimating and selecting mineral resources (Journel and Huijbregts, 1978). Matheron (1982) proposed a method of data synthesis and representation that was put into practice by the author (1985). It involves multivariate variogram modelling, principal component analysis and cokriging.

The general method is described first, and then applied to geochemical data from the region of Ancenis (Loire, France) provided by the Bureau de Recherches Géologiques et Minières (BRGM).

2. THE MULTIVARIATE GEOSTATISTICAL MODEL

Consider a set of random functions $\{Z_i(x);\ i = 1, \ldots, N_v\}$ and one realisation of these random functions, the set of regionalized variables $\{z_i(x);\ i = 1, \ldots, N_v\}$.

We possess a set of samples concerning the N_v regionalized variables, which are located at N points,

$$\{z_i(x_\alpha);\ i = 1, \ldots, N_v;\ \alpha = 1, \ldots, N\}$$

where x_α is the vector of the coordinates of a point.

An increment, $z_i(x_\alpha) - z_i(x_\alpha + h)$, is defined as the difference between two values of a variable $z_i(x)$ at two locations x_α and $x_\alpha + h$ separated by a vector h in a physical space.

We can postulate the hypothesis that the mean and the covariance of the increments are finite and independent of location,

$$\begin{matrix} i = 1, \ldots, N_v \\ j = 1, \ldots, N_v \end{matrix} \qquad \begin{cases} E\big(Z_i(x) - Z_i(x+h)\big) = 0 \\ Cov\Big[\big(Z_i(x) - Z_i(x+h)\big)\big(Z_j(x) - Z_j(x+h)\big)\Big] = 2\gamma_{ij}(h), \end{cases}$$

where $\gamma_{ij}(h)$ is the defined as the *variogram*.

This is called the ***intrinsic hypothesis***. It is possible to define even more general hypotheses (Matheron 1973). But for the present purpose of comparing geostatistical with classical statistical results we need only consider the more particular hypothesis of ***second order stationarity***. This states that the means m_i and the covariances σ_{ij} of the variables have finite values,

$$\begin{matrix} i = 1, \ldots, N_v \\ j = 1, \ldots, N_v \end{matrix} \qquad \begin{cases} E\big(Z_i(x)\big) = m_i \\ Cov\big(Z_i(x), Z_j(x+h)\big) = \sigma_{ij}, \end{cases}$$

which are independent of spatial location.

3. EXPERIMENTAL VARIOGRAM

The experimental data can be analysed by considering the locations of samples. Increments are calculated and then classified into vector classes h_k defined by distance and direction classes.

The experimental variogram is defined as the half of the experimental covariance of the increments for a given class h_k,

$$\gamma_{ij}(h_k) = \frac{1}{2N_k}\sum_{\alpha=1}^{N_k}\big(z_i(x_\alpha) - z_i(x_\alpha + h)\big)\big(z_j(x_\alpha) - z_j(x_\alpha + h)\big),$$

with

h belonging to a class of vectors h_k,

N_k being the number of increment pairs in such a class.

Let $\mathbf{A}_k$ be the $N_k \times N_v$ matrix of increments. If all variables have been measured at the same locations (which shall be assumed here) then $\mathbf{\Gamma}_k$, the $N_v \times N_v$ matrix of variograms for a distance class h_k, is positive semi-definite,

$$\mathbf{\Gamma}_k = \frac{1}{2N_k}\mathbf{A}_k^T\mathbf{A}_k,$$

where T denotes matrix transposition.

For every distance class the variogram matrix $\mathbf{\Gamma}_k$ may be compared with the variance-covariance matrix $\mathbf{V}$,

$$\mathbf{V} = \frac{1}{N}\mathbf{Z}^T\mathbf{Z},$$

where $\mathbf{Z}$ is the $N \times N_v$ matrix containing the N samples of the N_v variables, after substraction of the means of the variables.

This comparison makes sense as theoretically under the assumption that the varaibles are uncorrelated for large distances,

$$\gamma_{ij}(h) \rightarrow \sigma_{ij} \quad for \quad h \rightarrow \infty.$$

Further, for every distance class a correlation coefficient $r_{ij}(h_k)$ can be defined and compared to the classical correlation coefficient of statistics,

$$r_{ij}(k) = \frac{\gamma_{ij}(h_k)}{\sqrt{\gamma_{ii}(h_k)\,\gamma_{jj}(h_k)}},$$

with $-1 \leq r_{ij}(h_k) \leq +1$ provided that the variogram matrix $\mathbf{\Gamma}_k$ is positive semi-definite.

4. VARIOGRAM MODEL

The experimental variogram values for a number of vector classes h_k can be modelled by a mathematical function in order to calculate variogram values for any vector h.

The variogram model is a linear combination of $S+1$ coefficients b_{ij}^u with elementary variogram functions $g_u(h)$,

$$\gamma_{ij}(h) = \sum_{u=0}^{S} b_{ij}^u g_u(h).$$

The elementary variogram functions $g_u(h)$ have to be conditionally negative definite (Matheron 1970) and the matrices of coefficients b_{ij}^u, for fixed u, must be positive definite. Then the variance of any linear combination of the data with weights adding up to zero will be positive.

In this paper we shall use only two elementary variogram functions, the *nugget-effect* model and the *spherical* model.

The nugget-effect function is

$$nug(h) = \begin{cases} 0 & if\ |h| = 0, \\ 1 & if\ |h| > 1. \end{cases}$$

It is used to represent a discontinuity at the origin of a variogram, which can be explained physically either by measurement error or by microstructures. The microstructures are spatial structures below the scale defined by the sample spacing.

The spherical variogram function is

$$sph(h,a) = \begin{cases} \dfrac{3|h|}{2a} - \dfrac{|h|^3}{2a^3} & for\ 0 \le |h| \le a, \\ 1 & for\ |h| > a. \end{cases}$$

The parameter a is called the *range* of the spherical model.

In theory it is possible to imagine spheres of equal size distributed in three-dimensional space following a Poisson distribution. Each sphere contains Poisson distributed points x with a value $z(x) = 1$. An experimental variogram calculated for this data has the shape of the spherical function. Its range is the diameter of the spheres. The same result is obtained if the variogram is calculated along a plane or a line set through the space with the spheres.

Each function $g_u(h)$ determines a part of the variogram curve suggested by the experimental variogram values. Serra (1968) has shown that the variogram curve can have a clear geological meaning. In his study of the Lorraine iron ore deposit each $g_u(h)$ could be associated to different geological phenomena at six different spatial scales. The range of each variogram structure indicated the mean diameter of corresponding physical objects generated by some geological process.

The idea of decomposing the variogram curve into several structures leads to an analysis of the relationship between variables at the the spatial scales defined by these structures. This relationship is described by $N_v \times N_v$ matrices $\mathbf{B}_u$ of coefficients b^u_{ij} called ***coregionalization matrices.***

The multivariate model used here considers the phenomenon to be generated by a sum of several multivariate random processes. Experimentally these processes can only be distinguished if their variogram functions $g_u(h)$ have an impact on the shape of the experimental curve. If this is the case it is possible to uncover underlying covariance structures which may be entirely different from that shown by the variance-covariance matrix. This can easily be seen in the frame of a second order stationarity hypothesis in which the variance-covariance matrix turns out to be equal to the sum of the coregionalization matrices as

$$\mathbf{\Gamma}(h) \rightarrow \sum_{u=0}^{S} \mathbf{B}_u = \mathbf{V} \quad \textit{for } h \rightarrow \infty.$$

Correlation coefficients $r_{ij}(u)$ associated with different spatial structures $g_u(h)$ can be calculated,

$$r_{ij}(u) = \frac{b^u_{ij}}{\sqrt{b^u_{ii}\, b^u_{jj}}},$$

where $-1 \leq r_{ij}(u) \leq +1$ if the coregionalization matrix $\mathbf{B}_u$ is positive semi-definite.

The coregionalization matrices can be viewed as variance-covariance matrices of a particular type. So a principal component analysis (in R-mode) can be performed on them.

5. LINEAR MODEL OF COREGIONALIZATION

A set of correlated random functions $\{Z_i(x);\ i = 1, \ldots, N_v\}$ is to be decomposed into a set of uncorrelated random functions $\{Y^u_p(x);\ u = 0, 1, \ldots, S;\ p = 1, \ldots, N_v\}$ such that

$$Z_i(x) = \sum_{u=0}^{S} \sum_{p=1}^{N_f} a^i_{up} Y^u_p(x).$$

This means that we are looking for transformation coefficients a^i_{up} which allow passing from one set of variables to the other.

A coregionalization matrix $\mathbf{B}_u$ may be decomposed by solving the generalized eigenvalue problem

$$\mathbf{B}_u \mathbf{Q}_u = \mathbf{D}_u \mathbf{M}_u \mathbf{Q}_u \qquad \text{with} \quad \mathbf{Q}_u^T \mathbf{M}_u \mathbf{Q}_u = \mathbf{I},$$

where

$\mathbf{Q}_u$ is an $\mathbf{M}_u$-orthonormal matrix of eigenvectors,

$\mathbf{D}_u$ is a diagonal matrix of eigenvalues,

$\mathbf{M}_u$ is a symmetric positive definite matrix representing a metric.

The variance-covariance matrix $\mathbf{V}$ could be used as a metric. But in this paper $\mathbf{M}_u$ shall be an identity matrix which reduces our problem to the classical eigenvalue problem,

$$\mathbf{Q}_u^T \mathbf{B}_u \mathbf{Q}_u = \mathbf{D}_u.$$

The transformation coefficients are contained in matrices $\mathbf{A}_u$ and obtained by setting,

$$\mathbf{A}_u = \mathbf{Q}_u \sqrt{\mathbf{L}_u}.$$

The eigenvalues maximize the variance represented by a matrice $\mathbf{B}_u$ and the eigenvectors can be chosen to span an orthonormal system of axes called *principal axes.* The principal axes associated with the most important eigenvalues explain the essential features of the covariance structure.

The coordinates of the variables on a pair of principal axes lie inside a unit circle centred at the origin because the orthonormal matrix $\mathbf{Q}_u$ satisfies the equation of the unit hypersphere.

6. MAPPING MULTIVARIATE SPATIAL INFORMATION

On the basis of multivariate spatial data it can be interesting to map three quantities,

1) the *regionalized variables* $z_i(x)$,
2) the *spatial components* $z_i^u(x)$, which represent the behavior of the regionalized variables $z_i(x)$ at spatial scales of index u,
3) the *regionalized factors* $y_p^u(x)$, which reflect essential aspects of the multivariate information for given spatial scales.

In each case the estimation of a value at a point x_0 is made using a linear combination of weights λ_α^i wi‹n data at n data points surrounding the point x_0,

$$\sum_{i=1}^{N_v} \sum_{\alpha=1}^{n} \lambda_\alpha^i Z_i(x_\alpha).$$

The problem now is to find in each situation the optimal weights λ_α^i which minimize the estimation variance under the condition that there should be no bias.

7. ESTIMATION OF THE REGIONALIZED VARIABLES

The estimator $\hat{Z}(x)$ of a fixed variable $\hat{Z}_{i_0}(x)$ at a given point x_0 is

$$\hat{Z}_{i_0}(x_0) = \sum_{i=1}^{N_v}\sum_{\alpha=1}^{n} \lambda_\alpha^i Z_i(x_\alpha).$$

The condition for unbiasedness,

$$E\big(Z_{i_0}(x_0) \;-\; \hat{Z}_{i_0}(x_0)\big) \;=\; 0,$$

yields N_v conditions,

$$i = 1,\ldots,N_v \qquad \left\{\sum_{\alpha}^{n} \lambda_\alpha^i \;=\; \delta_{ii_0},\right.$$

where δ_{ii_0} is a Kronecker delta.

The estimation variance is

$$\begin{aligned} Var\big(Z_{i_0}(x_0) \;-\; \hat{Z}_{i_0}(x_0)\big) \;&=\; E\Big(Z_{i_0}(x_0) \;-\; \sum_{i=1}^{N_v}\sum_{\alpha=1}^{n} \lambda_\alpha^i Z_i(x_\alpha)\Big)^2 \\ &=\; C_{i_0 i_0}(0) \;-\; 2\sum_{i=1}^{N_v}\sum_{\alpha=1}^{n} \lambda_\alpha^i C_{ii_0}(x_\alpha, x_0) \\ &\quad+\; \sum_{i=1}^{N_v}\sum_{j=1}^{N_v}\sum_{\alpha=1}^{n}\sum_{\beta=1}^{n} \lambda_\alpha^i \lambda_\beta^j C_{ij}(x_\alpha, x_\beta), \end{aligned}$$

where C_{ij} is the spatial covariance,

$$C_{ij}(h) \;=\; E\big(Z_i(x)Z_j(x+h)\big).$$

The following relation can be established between the variogram and the spatial covariance:

$$\gamma_{ij}(h) \;=\; C_{ij}(0) \;-\; \frac{1}{2}\big(C_{ij}(h) + C_{ij}(-h)\big).$$

Assuming that the spatial covariance is symmetrical in h, it is possible to deduce the spatial covariance from the variogram.

Minimizing the estimation variance subject to the conditions for no bias leads to a system of linear equations called the *cokriging equations*,

$$\begin{matrix} i = 1,\ldots,N_v \\ \alpha = 1,\ldots,n \end{matrix} \qquad \left\{\begin{aligned} \sum_{j=1}^{N_v}\sum_{\beta=1}^{n} \lambda_j^\beta C_{ij}(x_\alpha, x_\beta) + \mu_i \;&=\; C_{ii_0}(x_\alpha, x_0) \\ \sum_{\beta=1}^{n} \lambda_i^\beta \;&=\; \delta ii_0, \end{aligned}\right.$$

where μ_i are Lagrange multipliers.

The minimum estimation variance can be written as

$$\sigma_K^2 = C_{i_0 i_0}(0) - \mu_{i_0} - \sum_{i=1}^{N_v} \sum_{\alpha=1}^{n} \lambda_\alpha^i C_{i i_0}(x_\alpha, x_0).$$

The estimation procedure is repeated at all the nodes of a regular grid. These are then interpolated to produce the isolines of the map of a variable $z_{i_0}(x)$.

8. ESTIMATION OF THE SPATIAL COMPONENTS

A spatial component $z_u^i(x)$ is estimated from the information at n surrounding data points by the estimator,

$$\hat{Z}_{u_0}^{i_0}(x_0) = \sum_{i=1}^{N_v} \sum_{\alpha=1}^{n} \lambda_\alpha^i Z_i(x_\alpha).$$

Starting with a variogram model split into several spatial structures denoted by the index u,

$$\gamma_{ij}(h) = \sum_{u=0}^{S} b_{ij}^u g_u(h) = \sum_{u=0}^{S} \gamma_{ij}^u(h),$$

we assume that the components $\gamma_{ij}^u(h)$ of the variogram model can be related to a decomposition of the random functions $Z_i(x)$ into orthogonal spatial components $Z_u^i(x)$,

$$Z_i(x) = \sum_{u=0}^{S} Z_u^i(x) \quad \textit{with} \quad E\big(Z_u^i(x) Z_v^j(x+h)\big) = 0 \quad \textit{for } u \neq v.$$

The no bias condition is,

$$E\big(Z_{u_0}^{i_0}(x_0) - \hat{Z}_{u_0}^{i_0}(x_0)\big) = 0.$$

Writing m_i for the local means of the variables and $m_{u_0}^{i_0}$ for the local mean of the spatial component to be estimated the condition becomes,

$$m_{i_0}^{u_0} - \sum_{i=1}^{N_v} m_i \sum_{\alpha=1}^{n} \lambda_\alpha^i = 0.$$

As the scale of the spatial component $z_{u_0}^{i_0}$ is arbitrary it is possible to set it through the no bias conditions. It could for example be assumed that the mean is proportional to the coefficients of the variogram model,

$$m_{i_0}^{u_0} = \frac{b_{i_0 i_0}^{u_0}}{\sum_{u=0}^{S} b_{i_0 i_0}^{u_0}} m_{i_0}.$$

Another approach is to estimate the residuals of the spatial components and to add afterwards an arbitrary local mean.

The cokriging equations in this case, for a spatial component with a local mean equal to zero, are,

$$\begin{array}{l} i = 1,\ldots,N_v \\ \alpha = 1,\ldots,n \end{array} \quad \left\{ \begin{array}{rcl} \displaystyle\sum_{j=1}^{N_v}\sum_{\beta=1}^{n} \lambda_j^\beta C_{ij}(x_\alpha, x_\beta) + \mu_i & = & C_{ii_0}^{u_0}(x_\alpha, x_0) \\ \displaystyle\sum_{\beta=1}^{n} \lambda_i^\beta & = & 0. \end{array} \right.$$

Several authors have examined the estimation of spatial components by using only the information from one variable, Sandjivy (1983), Galli & al. (1984), Chilès & Guillen (1984).

9. ESTIMATION OF THE REGIONALIZED FACTORS

The estimator used to map a regionalized factor $y_p^u(x)$ is,

$$\hat{Y}_p^u(x_0) = \sum_{i=1}^{N_v}\sum_{\alpha=1}^{n} \lambda_\alpha^i Z_i(x_\alpha),$$

The variogram model is decomposed into spatial structures and the coefficients b_{ij}^u are decomposed into transformation coefficients a_{up}^i,

$$\gamma_{ij}(h) = \sum_{u=0}^{S} b_{ij}^u g_u(h) = \sum_{u=0}^{S}\sum_{p=1}^{N_v} a_{up}^i a_{up}^j g_u(h).$$

The original variables are represented as a weighted sum of regionalized factors,

$$Z_i(x) = \sum_{u=0}^{S}\sum_{p=1}^{N_v} a_{up}^i Y_p^u(x).$$

It does not seem reasonable to try to give the regionalized factors $y_p^u(x)$ a particular mean via the no bias conditions. The best thing to do is to set this local mean to zero. This way residuals are estimated from an arbitrary mean that will not be specified.

Taking into account the orthogonality of the random functions $Y_p^u(x)$ the estimation variance is,

$$\begin{aligned} Var(Y_{p_0}^{u_0}(x_0) - \hat{Y}_{p_0}^{u_0}(x_0)) = \; & C_{u_0}(0) - 2\sum_{i=1}^{N_v}\sum_{\alpha=1}^{n} \lambda_\alpha^i a_{u_0 p_0}^i C_{u_0}(x_\alpha, x_0) \\ & + \sum_{i=1}^{N_v}\sum_{j=1}^{N_v}\sum_{\alpha=1}^{n}\sum_{\beta=1}^{n} \lambda_\alpha^i \lambda_\beta^j C_{ij}(x_\alpha, x_\beta), \end{aligned}$$

The spatial covariances $C_u(h)$ of the regionalized factors are deduced from the variogram by setting,

$$g_u(h) \;=\; C_u(0) - C_u(h) \;=\; 1 - C_u(h).$$

The optimal weights λ^i_α are obtained by solving the following cokriging equations,

$$\begin{array}{l} i = 1, \ldots, N_v \\ \alpha = 1, \ldots, n \end{array} \quad \left\{ \begin{array}{rcl} \displaystyle\sum_{j=1}^{N_v} \sum_{\beta=1}^{n} \lambda^\beta_j C_{ij}(x_\alpha, x_\beta) + \mu_i & = & a^i_{u_0 p_0} C_{u_0}(x_\alpha, x_0) \\ \displaystyle\sum_{\beta=1}^{n} \lambda^\beta_i & = & 0. \end{array} \right.$$

As the regionalized factors are purely synthetic, the minimum variance is not calculated because it has no physical meaning.

10. GEOCHEMICAL DATA

Over the past ten years the BRGM has undertaken a systematic geochemical survey of France collecting between two and four samples per square kilometer (Wilhelm & Zeegers 1984).

They provided us with a set of 1233 soil samples from the region of Ancenis (Loire). The location of these samples in an area of 29 km by 21 km is shown on Figure 1.

The blank strip at the lower right corner of Figure 1 is due to the Loire river.

The samples have been analysed for their content of fifteen chemical elements: nickel (Ni), chromium (Cr), vanadium (V), iron (Fe), manganese (Mn), cobalt (Co), copper (Cu), lead (Pb), zinc (Zn), barium (Ba), boron (B), phosphorus (P), arsenic (As), yttria (Y), niobe (Nb).

Experimental variograms have been calculated in several directions using distance classes of 1 km width. The variograms in the East-West direction had a sill while they were linear in the North-South direction.

If we suppose that the underlying variograms are isotropic we can assume that the East-West variograms are the underlying variograms (Chilès 1977). So our model assumes the data to be a realization of a second order stationary process affected by a linear drift in the North-South direction.

For brevity we shall show here only three of the 120 variograms obtained in the East-West direction (the entire set in both directions was presented in Wackernagel 1985). Figure 2 shows the simple variograms for nickel and for copper as well as the cross variogram for the two variables.

The fitted variogram model consists of a nugget effect model plus a spherical model with a range of 5 km,

$$\gamma_{ij}(h) \;=\; b^0_{ij}\ nug(h) \;+\; b^1_{ij}\ sph(h, 5km).$$

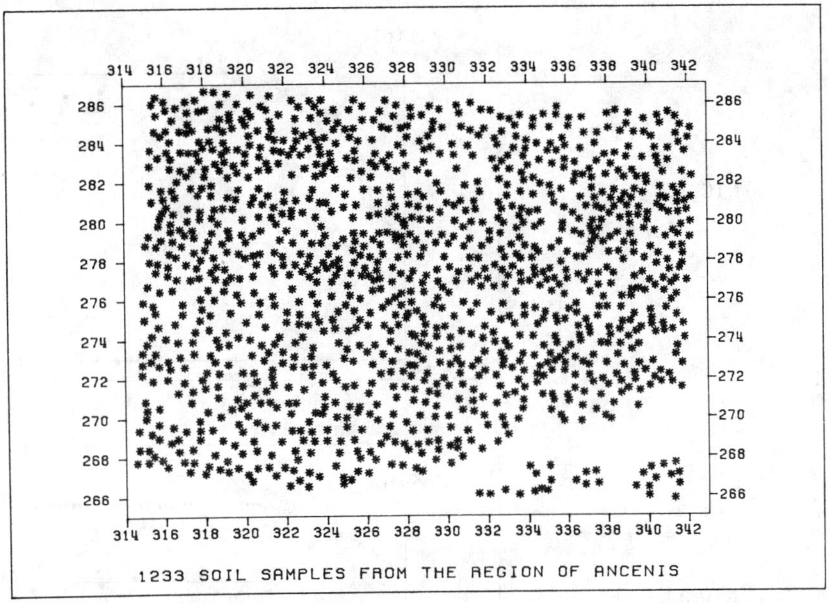

Figure 1. Location of the samples (units are km).

In Figure 2 the experimental variogram values are represented by dots and the variogram model is a continuous curve fitted to these values. The cross variogram of nickel and copper also contains a dotted curve which is obtained after fitting the simple variograms,

$$Hull\left[\gamma_{ij}(h)\right] = \pm \left[\sqrt{b^0_{ii} b^0_{jj}}\; nug(h) + \sqrt{b^1_{ii} b^1_{jj}}\; sph(h, 5km)\right].$$

This curve gives the hull of perfect (positive or negative) correlation between the variables. It allows us to evaluate the goodness of fit of a cross variogram model in terms of the correlation between two variables.

Figure 3 shows for two matrices the position of the variables on the plane defined by the first two principal axes inside the unit circle around the origin. A comparison is made between the graph for the normed variance-covariance matrix and the graph for the normed coregionalization matrix of the spherical model.

The first two principal axes of the variance-covariance matrix extract 56% of the total variance while the corresponding axes of the coregionalization matrix explain 73% of the variance of the data at the spatial scale described by the spherical model.

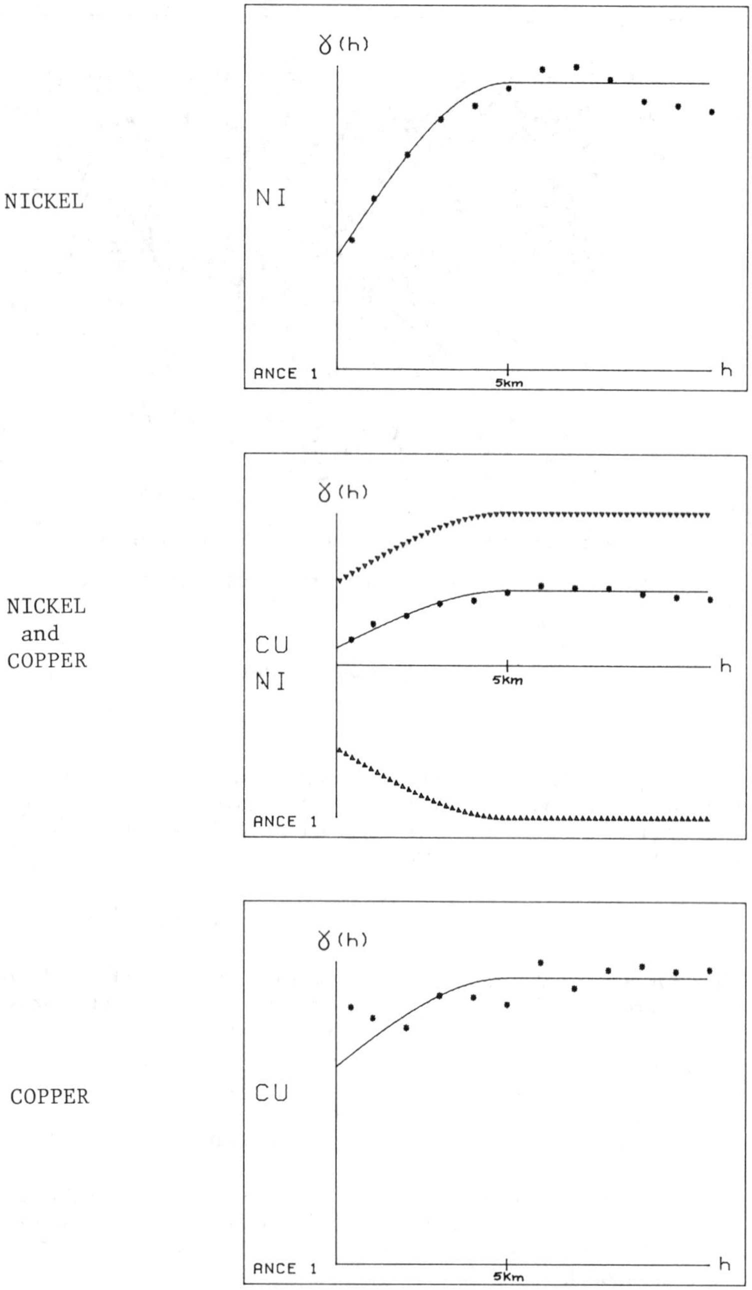

Figure 2. Three variograms.

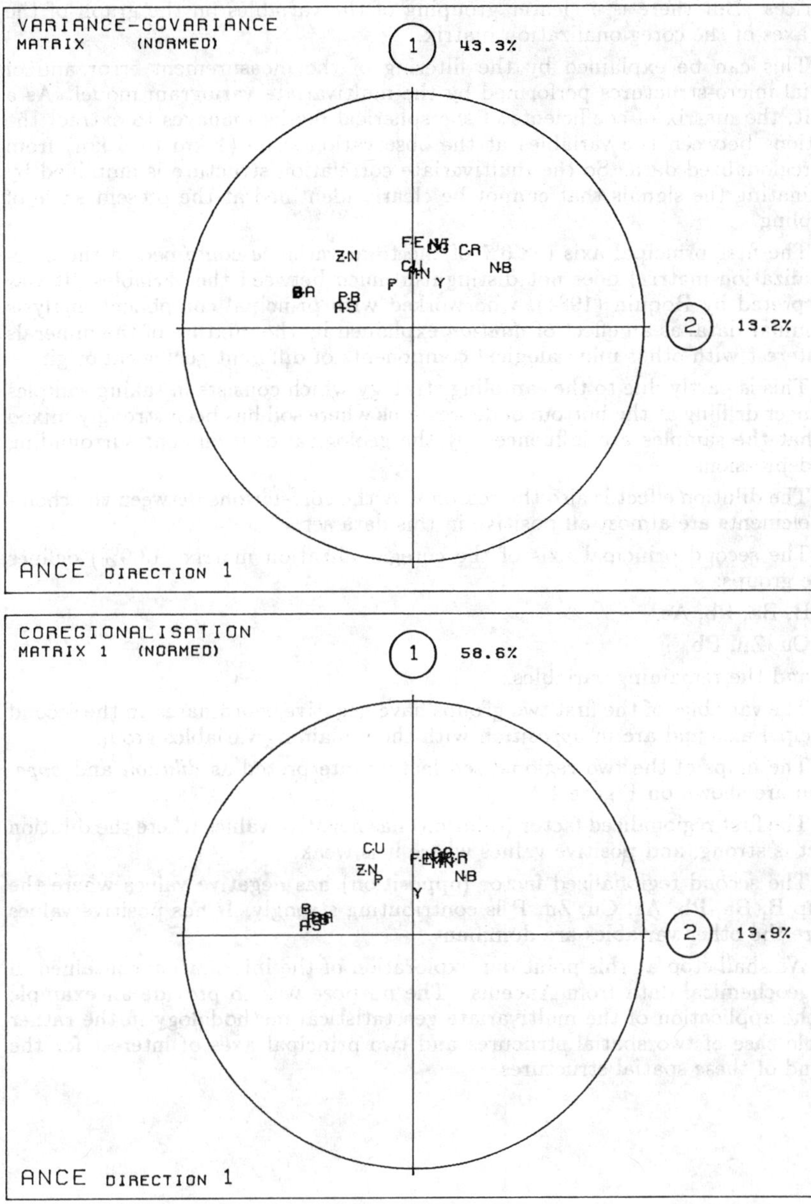

Figure 3. Position of variables on two principal axes.

The position of the variables in the unit circle is very similar for the two matrices. But there is a clearer grouping of the variables on the graph of the two axes of the coregionalization matrix.

This can be explained by the filtering of the measurement error and of spatial micro-structures performed by the multivariate variogram model. As a result, the matrix of coefficients of the spherical model manages to extract the relations between the variables at the observation scale (1 km to 6 km) from the regionalized data. So the multivariate correlation structure is amplified by eliminating the signals that cannot be clearly identified at the present scale of sampling.

The first principal axis (58.6% of the total variance contained in the coregionalization matrix) does not distinguish much between the variables. It was interpreted by Roquin (1984), who worked with principal component analysis on similar data, as an effect of *dilution* explained by the mixture of the minerals of interest with other mineralogical components of different geological origin.

This is partly due to the sampling strategy which consists in taking samples by auger drilling at the bottom of depressions where soil has been strongly mixed so that the samples are influenced by the geological environment surrounding the depressions.

The dilution effect is also the reason why the correlations between the chemical elements are almost all positive in this data set.

The second principal axis of the coregionalization matrix (13.9%) defines three groups:

- B, Ba, Pb, As,
- Cu, Zn, Pb,
- and the remaining variables.

The variables of the first two groups have negative coordinates on the second principal axis and are in *opposition* with the remaining variables group.

The maps of the two regionalized factors interpreted as *dilution* and *opposition* are shown on Figure 4.

The first regionalized factor (dilution) has negative values where the dilution effect is strong, and positive values where it is weak.

The second regionalized factor (opposition) has negative values where the group B, Ba, Pb, As, Cu, Zn, P is contributing strongly. It has positive values where the other variables are dominant.

We shall stop at this point our exploration of the information contained in the geochemical data from Ancenis. The purpose was to provide an example for the application of the multivariate geostatistical methodology in the rather simple case of two spatial strucures and two principal axes of interest for the second of these spatial structures.

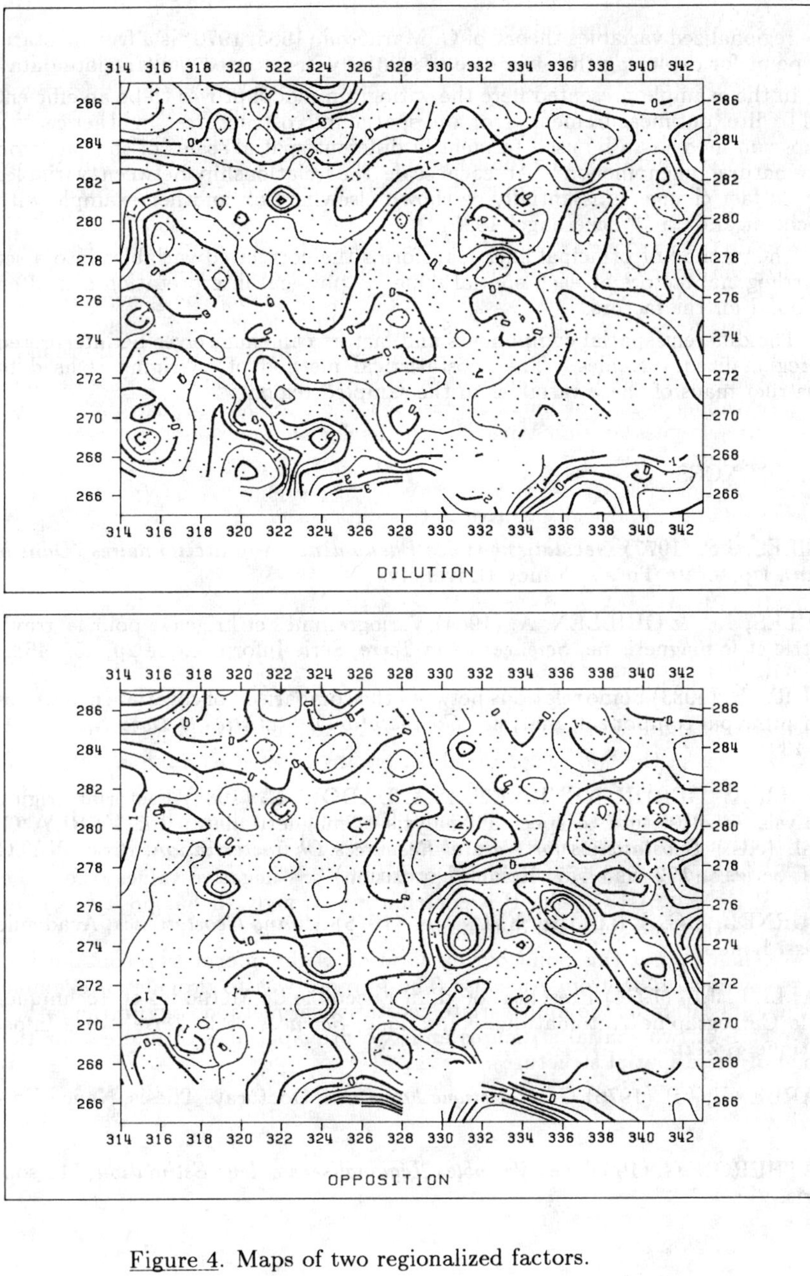

Figure 4. Maps of two regionalized factors.

11. CONCLUSION

The regionalized variables theory of G. Matheron (1965, 1970) is a fruitful starting point for exploring the structure of spatially distributed multivariate data.

In the example presented here the variogram model proves to be an efficient tool to filter out measurement error and spatial microstructures. In other cases a variogram model can be used to identify different spatial scales of the variation of a natural phenomenon. At each scale the relationship between variables may in fact change substantially as shown elsewhere on another example with geochemical data (Wackernagel 1986).

The method of principal axes transforms the correlated variables into a set of orthogonal factors at each spatial scale. A physical interpretation can often be found for the factors.

The different spatial components and factors can themselves be interpreted as regionalized variables. The geostatistical method of cokriging is used to construct maps of these variables in the sampled region.

REFERENCES

CHILES, J.P. (1977) *Géostatistique des Phénomènes Non-Stationnaires (Dans le Plan)*, Doctorate Thesis, Nancy University.

CHILES, J.P. & GUILLEN, A. (1984) Variogrammes et krigeage pour la gravimétrie et le magnétisme, *Sciences de la Terre*, Série Informatique **20**, 455–468.

FLURY, B. (1983) Some relations between the comparison of covariance matrices and principal component analysis. *Computational Statistics & Data Analysis* **1**, 97–109.

GALLI, A., GERDIL-NEUILLET, F. & DADOU, C. (1984) Factorial kriging analysis: a substitute to spectral analysis of magnetic data. In: VERLY, G. & al. (eds.) *Geostatistics for Natural Resources Characterization*, Proc. NATO ASI, Series C **122**, 543–557, Reidel, Dordrecht.

JOURNEL, A.G. & HUIJBREGTS, Ch. (1978) *Mining Geostatistics*, Academic Press, London.

MALLET, J.L. (1974) Présentation d'un Ensemble de Méthodes et Techniques de la Cartographie Automatique Numérique, *Sciences de la Terre*, Série Informatique **4**, 212p.

MARBEAU, J.P. (1976) *Géostatistique Forestière*, Doctorate Thesis, Nancy University.

MATHERON, G. (1965) *Les Variables Régionalisées et leur Estimation*, Masson, Paris.

MATHERON, G. (1970) The Theory of Regionalized Variables and its Applications, *Les Cahiers du Centre de Morphologie* **5**, 211p., Centre de Géostatistique, Fontainebleau.

MATHERON, G. (1973) The intrinsic random functions and their applications, *Advances in Applied Probability* **5**, 439–468.

MATHERON, G. (1982) Pour une analyse krigeante des données régionalisées, Internal Report N-732, Centre de Géostatistique, Fontainebleau, 22p.

ROQUIN, Cl. (1984) Traitement de Données en Prospection Géochimique, *Documents du BRGM* **86**, Orléans, 316p.

SANDJIVY, L. (1983) Analyse krigeante de données géochimiques, *Sciences de la Terre*, Série Informatique **18**, 141–172.

SERRA, J. (1968) Les structures gigognes: morphologie mathématique et interprétation métallogénique, *Mineralium Deposita* **3**, 135–154.

WACKERNAGEL, H. (1985) *L'Inférence d'un Modèle Linéaire en Géostatistique Multivariable*, Doctorate Thesis, Paris School of Mines.

WACKERNAGEL, H. (1986) Description of a computer program for analysing multivariate spatially distributed data, *Computers & Geosciences* **13**, in press.

WILHELM, E. & ZEEGERS, H. (1984) La prospection géochimique au BRGM, *Geologisches Jahrbuch* **A75**, 49–75.

THE SHAPE OF LLOYDMINSTER OIL AND GAS DEPOSIT ATTRIBUTE DATA*,†

G. Andreatta, G. M. Kaufman, R. G. McCrossan, & R. M. Procter
Massachusetts Institute of Technology
Alfred P. Sloan School of Management
50 Memorial Drive
Cambridge, Massachusetts 02139
United States

ABSTRACT. A method for fitting distributional shape proposed by Tukey is applied to a unique set of oil and gas field data: oil in place, volume and area of Lloydminster deposits in Alberta. This data is unique because it is composed of more than 2500 genetically similar deposits in the same geographical area and because a common protocol was used to measure field attributes. The shape of the bulk of the data is approximately lognormal; only the very largest and very smallest observations (one well fields) do not fit the pattern of the remainder of the data. (The large sample size allows study of a fractile range from .00024 to .99976).

1. INTRODUCTION

Relative frequency histograms of some oil and gas deposit attributes are characteristically unimodal and positively skewed with fat right tails. Histograms composed from measurements of deposit area, rock volume, hydrocarbons in place, and recoverable hydrocarbons for deposits in a thoroughly explored or "mature" play are usually so shaped:

*We wish to thank David Hoaglin for providing software for carrying out this study.
†This work was supported by Brookhaven National Laboratory, Associated Universities Inc. Contract No. 533152-S. The work of Giovanni Andreatta was supported in part by the National Research Council of Italy (CNR). Richard Procter is executive director, Petroleum Resources appraisal Secretariate, Geological Survey of Canada. R. G. McCrossan is a consulting research geologist.

C. F. Chung et al. (eds.), Quantitative Analysis of Mineral and Energy Resources, 411–431.

The observation that relative frequency histograms of petroleum deposit sizes and of deposit sizes for minerals other than petroleum are roughly similar in shape whether size is defined as deposit rock volume or as projective area or as volume of mineral in place led researchers to ask these questions. Suppose that deposits in a given petroleum or metallogenic province are classified into descriptively homogenous subpopulations and the sizes of deposits within each subpopulation are regarded as coming from an idealized sampling process that generates values of mutually independent, uncertain quantities possessing a common probability distribution. Is the functional form* for this probability distribution the same from subpopulation to subpopulation for a given mineral? For each possible definition of size? For all minerals? If it is hypothesized that the same functional form characterizes subpopulations of deposit sizes for a given mineral and a particular definition of size, then what specific functional form is suggested by observational data?

The hypothesis most frequently entertained is that mineral deposit size distributions are approximately lognormal; that is, that relative frequency histograms of the logarithms of deposit sizes for a prespecified population of deposits are closely fit by a normal or Gaussian density. The idea that, among an infinite number of possible functional forms, the lognormal distribution is a logical candidate for this role has its root in statistical studies of deposit and mine value, most notable those of Krige; Blondel and Ventura; and Allais.**

Other candidates for fitting field size distributions that have been investigated are the gamma, Pareto, and Zipf's law distributions. Scheunemeyer and Drew (1983) provide an excellent discussion of attempts to test hypotheses about distributional shape. Their discussion sharply etches the difficulties of hypothesis testing to this end: left tail truncation that shifts with economics, non-independence of observations due to sampling proportional to magnitude and measurement error. In contrast to most earlier studies, Scheunemeyer and Drew aim at estimation of the left tail of oil and gas field size distributions. Their rationale is sound: as prices fall, small fields that were economically unattractive at higher prices become viable targets for drilling. In addition, as exploration effort unfolds, smaller fields constitute a larger and larger proportion of remaining undiscovered resources. Their study leads them to conjecture that the shape of the distribution of oil and gas fields may be log-geometric. Davis (1987) [this volume] discusses later work on this problem.

---

*A mathematical formula defining a family of probability distributions.

**Adelman, M. A., Houghton, J., Kaufman G. M., and Zimmerman, M, Energy Resources in an Uncertain Future: Oil, Gas, Coal, and Uranium Supply Forecasting, Ballinger Publishing Co., Cambridge, Massachusetts, 1983.

Adelman et al. (1981) review the history of attempts to confirm or disconfirm the reasonableness of the lognormal distributions as a fit to histograms of oil and gas field size. The issue is vexing. Most past studies are suggestive but inconclusive because, as just stated, the data used is tainted with manifold sources of measurement error, the sampling process may have been mis-specified, and the number of observations employed is at best several hundred. In order to explore distribution shape effectively we might wish for

1. deposit data from a single mature play in which several thousand discoveries have been made, and
2. individual deposit attributes all measured by one group of experts using a uniform measurement method.

A recent study of the Lloydminster play in Alberta and Saskatchewan by McCrossan, Procter, and Ward (1981) provides a publically announced example of play data that meet these criteria: over 2500 Lloydminster pools have been discovered and attributes of each measured according to a well designed protocol by McCallum, Steward and Associates. Measurement variability due to variation across different individuals or groups in interpretation and implementation of measurement protocols is common in most play data but absent here.

Some past studies or petroleum deposit distribution shapes rely on visual interpretation of lognormal probability paper fractile plots to detect departures from lognormality and some hypothesis testing of lognormality against alternatives has also been done. Neither of these approaches provides innterpretive flexibility: "How far away from normality or lognormality is the shape of the Data?" Tukey (1977) proposes a class of distributions, the "g and h" distributions, that can be used to answer this question. The g and h distributions are a two parameter family of shapes, parameterized so as to allow departures from normality or lognormality to be identified with parameters g and h. The parameter g controls skewedness relative to the normal distribution and the parameter h controls tail stretching or squeezing relative to the normal distribution.

Hoaglin and Peters have developed software for exploring distribution shape with the g and h distribution. They provide an excellent discussion of its properties, and the rationale for its use as a descriptor of real data in place of more traditional statistical methods.*

Desirable properties of their procedure for studying distribution shape are that:

- it allows a fit of shape directly to data without having to estimate parameters of a particular functional form;

---

*Hoaglin P. and Peters, S. "g- and h-Distributions", Encyclopedia of Statistical Sciences, Vol. 3, pp. 299-301, N. Johnson and S. Kotz editors, John Wiley & Sons.

- it is resistant: changes in a small fraction of the data should produce small changes in a fit of shape;
- departures from normality can be directly measured.

We begin in section 2 with a brief discussion of the geologic setting and description of the data. Section 3 is a review of properties of the g and h distributions as described by Hoaglin and Peters, along with their suggestions about how to fit them to data. The results of application of their software routines to the Lloydminster data are presented in section 4.

2. THE LLOYDMINSTER DATA

In their 1981 paper McCrossan, Procter, and Ward present a fractile plot of data describing oil in place and recoverable oil in Lloydminister deposits located in Alberta (Figure 2.1).

Notable features of these fractile plots are:

1. The bulk of the data points appear to be well fit by a straight line.
2. Possible deviations from a straight line appear in data points at extreme fractiles: about .9925 and larger and about .005 and smaller.
3. Without further analysis it is not clear that the pattern of the data cleanly supports shape different from a straight line in the extreme tails. However, the visual appearance of the extreme right and left tails forces us to entertain the hypothesis that some other shape for the tails might represent the data better.

These features of the data are unexceptionable. Extreme fractiles of petroleum deposit data and other data often exhibit them. In order to investigate the shape of the data in the extreme tails more carefully, diagnostic devices with more resolving power than fractile plots are needed. Some visual diagnostic insight is given by a move from the fractile plot of Figure 2.1 to a quantile-quantile (Q-Q) plot. Q-Q plots of log oil in place, log area, and log volume (Figures 2.2, 2.3 and 2.4) highlight tail behavior better than the fractile plots in Figure 2.1.

All three Q-Q plots display a fatter than normal extreme right tail and thinner than normal extreme left tail. Traditional explanations of this left tail behavior are:

- Economic truncation: there is a minimum deposit size below which it is not profitable to produce a deposit. Subeconomic deposits are often omitted intentionally and sometimes unintentionally from deposit data files. Minimum deposit size shifts with depth, deposit type, and changes in costs and prices.
- Larger deposits tend to be found early in the evolution of a play (For example, as a sampling shceme, discovery is proportional to a function of "size" and without replacement). Consequently, at any stage in the exploration history of a play prior to discovery of all deposits, the proportion of all very "small" deposits is with high probability smaller than the proportion of all very "large" deposits.

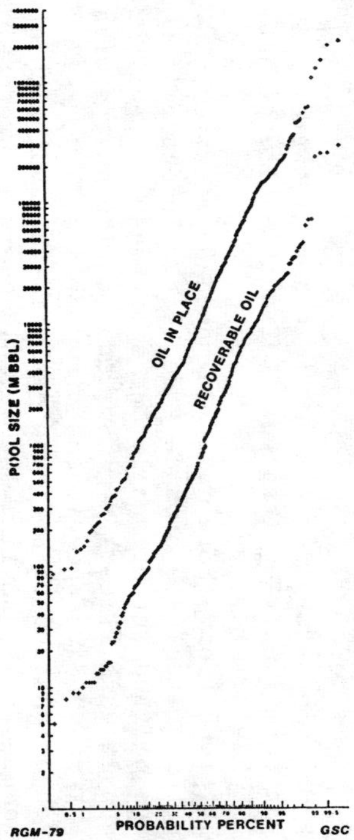

FIGURE 2.1 - LOG PROBABILITY PLOT OF POOL SIZES FOR ALL LLOYDMINSTER POOLS DELINATED BY MORE THAN ONE WELL.

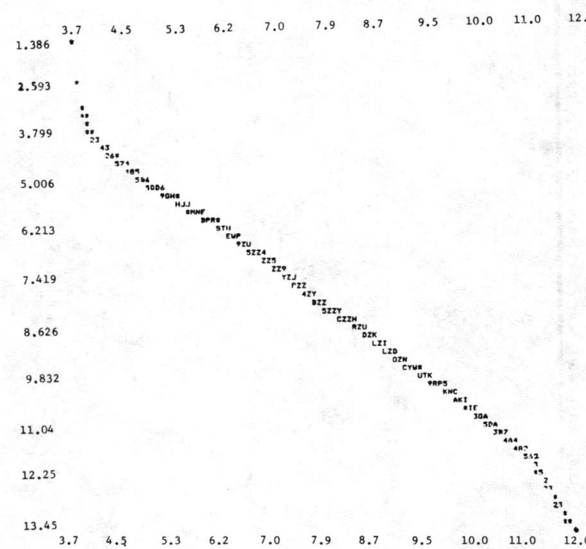

FIGURE 2.2 - Q-Q PLOT OF LOG OIL IN PLACE (VERT. AXIS) VS. NORMAL QUANTILES ($\hat{\mu}$ = 7.857, $\hat{\sigma}$ = 1.746)

FIGURE 2.3 - Q-Q PLOT OF LOG AREA (VERT. AXIS) VS. NORMAL QUANTILES (HORIZ. AXIS) ($\hat{\mu}$ = 6.289, $\hat{\sigma}$ = 1.552).

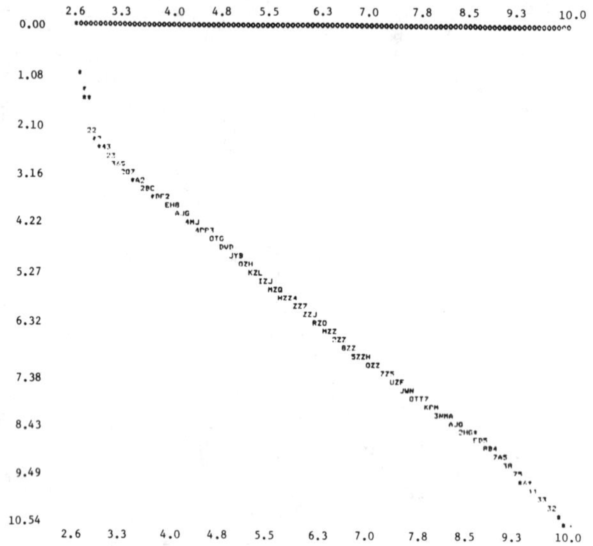

FIGURE 2.4 - Q-Q PLOT OF LOG VOLUME (VERT. AXIS) VS. NORMAL QUANTILES ($\hat{\mu}$ = 15.08, $\hat{\sigma}$ 1.790).

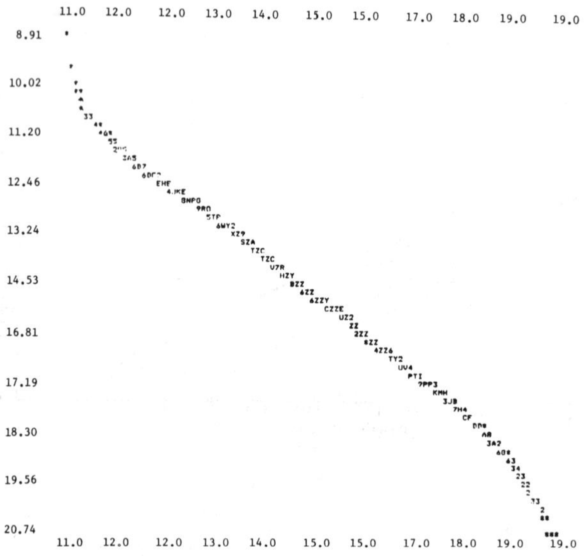

- For some deposit types there is a minimum cutoff volume; i.e. there are no deposits of smaller volume.

In their plot, shown as Figure 2.1, McCrossan et al. omitted one well deposits; i.e. they truncate the data. The fractile plots in Figure 2.1, however, are not adjusted for the effects of this truncation. Economic truncation plays at most a weak role here, as the deposit data we shall examine includes all known deposits, irrespective of whether or not a deposit is profitable to produce. While, it is possible that the second and third explanations in tandem account for left tail behavior, the sharpness with which the left tail visually bends away from normality favors the assertion that there is a geological cutoff volume for deposits of Lloydminster type.

What causal mechanism, if any, underlies stretching of the right tail away from a functional form or shape that fits the bulk of the data? In the context of this data two possible explanations are:

- Very large deposits are genetically different than small and medium sized deposits; i.e. giant oil fields belong in a separate population. Maybe very large (but not giant) fields also possess features that distinguish them. They are outliers.
- Individual accumulations close in distance to one another may, over geological time, elide. When measured at the time of discovery, two or more such accumulations may be recorded as a single deposit. More particularly, where there is superfluous charge and spill-over between anomalies close to one another, these anomalies may be perceived to be a single deposit.

The second explanation suggests that observable data describing deposit volume and area might possibly be generated in a fashion that allows models of spatial "clumping" to be applied (Mack (1954)).

One explanatory model is to conceive of the process generating observable deposit attributes for a petroleum play as unfolding in two stages. At the first stage deposit attributes are conceptualized as values of mutually independent random variables with common probability law. Call these primary attributes. Given a realization of N values of deposit areas and N corresponding individual deposit shapes--primary areas and primary shapes--primary shapes are distributed according to kinematic density within the play area. Some primary deposit shapes may overlap, or clump. Exploration results in observation of deposit shapes and areas of clumps; i.e. of secondary shapes and areas. If deposition adhers to such a model, the shape of primary areal data may differ significantly from that of secondary areal data. Intuitively one expects a fatter right tail for secondary areal data than for primary areal data. This line of reasoning is also applicable to deposit rock volumes, oil in place, and reocverable oil in individual deposits; it will not be developed in detail here.

Before presenting an analysis of the Lloydminster data, we review Tukey's shape fitting procedure as presented by Hoaglin and Peters (1983). Without this review, the discussion of shape fitting in section 4 may appear mysterious to readers.

3. REVIEW OF PROPERTIES OF THE g-AND-h DISTRIBUTIONS*

The distributions of two random variables (rvs) $\tilde{X}_1$ and $\tilde{X}_2$ are defined to have the same <u>shape</u> if $\tilde{X}_1 = a + b\tilde{X}_2$, for some constants a and b, $a\varepsilon(-\infty, \infty)$, $b\varepsilon(0, \infty)$. That is, a shape is defined by specification of a functional form for a distribution and allowable changes in scale and location that keep the functional form invariant. Descriptions of distribution shape are conveniently anchored on a "standard" functional form and the class of distributions with the same shape related to it by scale and location parameter transformations. The most common example is the normal distribution: if $\tilde{Z}$ is a rv with density $e^{-\frac{1}{2}z^2}/\sqrt{2\pi}$, $z\ \varepsilon(-\infty, \infty)$, then $a + b\tilde{Z}$ is also normal (for $b\varepsilon(-\infty, \infty)$, $b \neq 0$, not just $b\varepsilon(0, \infty)$). Here we are concerned with departures from normality (or equivalently lognormality) so the "standard" shape on which we anchor is of more complicated form. Denoting the rv $\tilde{Y}$ as a rv with standard shape, the family of rvs with distributions of the same shape are of the form $A + B\tilde{Y} = \tilde{X}$. Choice of a standard shape determines allowable values for A and B. How A and B are determined is explained in more detail following a definition of the standard form for the g and h distributions.

The g and h distributions are defined relative to a standard (mean zero, variance one) normal distribution N(0,1) in the following way: for $0 < p < 1$ define z_p as the p-th quantile of N(0,1), and $\tilde{Z}$ as a N(0,1) rv. Then

$$P(\tilde{Z} < z_p) = \frac{1}{\sqrt{2\pi}} \int_{-\infty}^{z_p} e^{-\frac{1}{2}x^2}\, dx = p.$$

Consider the family of shapes generated by the quantile function

$$Q_{g,h}(z) = \frac{e^{gz} - 1}{g} \times e^{\frac{1}{2}hz^2}.$$

If z_p is the p-th quantile of N(0,1) then

$$P(Q_{g,h}(\tilde{Z}) < Q_{g,h}(z_p)) = P(\tilde{Z} < z_p) = p$$

and $Q_{g,h}(z_p)$ is the p-th quantile of a g and h distribution. The subfamily of distributions with $g = 0$ are called the "h distributions" and the subfamily with $h = 0$ are called the "g distributions".

---

*This review is based on Hoaglin and Peters, <u>op. cit</u>.

If h = 0, then $Q_{g,0}(z) = (e^{gz} - 1)/g = z[(e^{gz} - 1)/gz]$ and this suggests interpreting the factor multiplying z on the righthand side of these equalities as a skewing operator. If $g > 0$, the right tail is stretched relative to a N(0,1) distribution so that the distribution of $Q_{g,0}(\tilde{Z})$ is positively skewed, and if $g < 0$, it is negatively skewed. At the median of $\tilde{Z}$, $\tilde{Z} = 0$, $Q_{g,0}(0) = 0$.

For any fixed z, as $g \to 0$, $e^{gz} = 1 + gz + 0(g^2)$, so $(e^{gz} - 1)/g \to z$. Thus $Q_{0,h}(z) = ze^{\frac{1}{2}hz^2}$ and the distribution of the rv $Q_{0,h}(\tilde{Z})$ is symmetric about zero. When h = 0, $Q_{0,0}(\tilde{Z}) = \tilde{Z}$, a N(0,1) rv. When $h > 0$, $Q_{0,h}(\tilde{z}) > z$ and the larger z is, the more $Q_{0,h}(z)$ is stretched away from z; i.e. the further a quantile is into the tail, the more it is stretched from its N(0,1) value. Hoaglin and Peters point out that while negative values of h are allowable, then $Q_{0,h}(z)$ is not monotonic for values of $z^2 > - 1/h$.

The relation between the family of all normal distributions and the g and h distributions is straightforward: if $\tilde{X}$ is normal with mean μ and variance σ^2, then

$$\tilde{X} = \mu + \sigma\tilde{Z} = A + B[\frac{e^{g\tilde{Z}} - 1}{g}]e^{\frac{1}{2}h\tilde{Z}^2}$$

The last equality can be true for all values of $\tilde{Z}$ if and only if $A = \mu$, $B = \sigma$, $g = 0$, and $h = 0$.

Of particular interest here is the relation between the family of lognormal distributions and the g and h distributions. If $\tilde{X}$ is lognormal with $\log_e\tilde{X}$ having mean μ and variance σ^2, then $[\log_e\tilde{X}-\mu]/\sigma = \tilde{Z}$ is a standard normal rv and $\tilde{X} = e^{\mu+\sigma\tilde{Z}}$. Hence

$$\tilde{X} = e^{\mu+\sigma\tilde{Z}} = A + B[\frac{e^{g\tilde{Z}} - 1}{g}]e^{\frac{1}{2}h\tilde{Z}^2}.$$

At the median $z_{.5} = 0$, so

$$x_{.5} = e^{\mu} = A.$$

Hence

$$\tilde{X} = e^{\mu+\sigma\tilde{Z}} = e^{\mu} + B[\frac{e^{g\tilde{Z}} - 1}{g}]e^{\frac{1}{2}h\tilde{Z}^2}$$

or equivalently,

$$\frac{e^{\sigma\tilde{Z}} - 1}{B} = \frac{e^{g\tilde{Z}} - 1}{ge^{\mu}} e^{\frac{1}{2}h\tilde{Z}^2}.$$

If $\tilde{Z}$ is a standard normal rv, the last equality can be true if and only if $h = 0$, $g = \sigma$, and $B = \sigma e^{\mu}$.

Thus if the data support a constant value of $g > 0$ and a constant value of $h = 0$, the shape of the data is lognormal-like. If the data are transformed by taking the logarithms of original measurements as the data, and the transformed data support constant values of $g = h = 0$, the shape of the untransformed data is lognormal-like.

3.1 Hoaglin and Peters Fitting Procedure

To find g and h values that fit the data when the sample is large, Hoaglin and Peters recommend use of the median and pairs of quantiles x_p and x_{1-p} with p taking on integer powers of 1/2: $(x_{.25}, x_{.75})$, $(x_{.125}, x_{.875})$, etc. These particular values of p are called "letter values". Pairs of quantile values symmetric about the median lead to simple algebraic formulae for g and h values corresponding to each such ordered pair of quantiles. The g values and h values so computed can be studied in various ways.

A. Fitting the g distribution when h = 0

1) Calculate a value g_p of g that exactly fits the spacing of x_p and x_{1-p} about $x_{.5}$:

$$g_p = -\frac{1}{z_p}\log_e \frac{x_{1-p} - x_{.5}}{x_{.5} - x_p}, \quad p = \frac{1}{4}, \frac{1}{8}, \frac{1}{16}, \ldots$$

with z_p the pth quantile of N(0,1).

2) Examine the behavior of g_p as a function of p. Does g_p vary systematically with respect to p? If not, the median of g_p values is a resistant measure of a constant g. If so, then try a resistant fit of a simple function of z_p^2; e.g. $g_p(z_p) = a + bz_p^2$.

3) Plot the data quantiles x_p against quantiles $Q_{g,0}(z_p)$ of the standard g distribution with g fixed at a value fitted in step 2. If g is in fact a constant function of p, then the slope of this quantile - quantile plot is B, since a constant gimplies that

$$x_p = A + BQ_{g,0}(z_p).$$

This plot provides another check on the reasonableness of assuming that g is a constant: a straight line plot of x_p vs. $Q_{g,0}(z_p)$ is consistent with constant g. Systematic curvature is not.

B. Fitting a Value of g When $h \neq 0$

1) Calculate g_p when $h \neq 0$ in the same way as when $h = 0$. ($e^{\frac{1}{2}hz_p^2}$ cancels out and does not influence g_p as defined in A.1).

C. Fitting a Value of h When $g = 0$

1) When $g = 0$, $x_{1-p} - x_p = -2BQ_{0,h}(z_p)$ so

$$h_p \equiv \log_e\left[\frac{x_{1-p} - x_p}{-2z_p}\right] = \log_e B + \frac{1}{2} hz_p^2.$$

Plot the lefthand side of this equality against the righthand side. A straight line plot supports a constant value of h.

D. Fitting h When $g \neq 0$

1) Fit a value of g as suggested in A, then symmetrize as described next.

2) As $x_{1-p} - x_{.5} = B\left[\frac{e^{gz_p} - 1}{g}\right]e^{\frac{1}{2}hz_p^2}$, given a fitted value of g,

$h_p \equiv \log_e](x_{1-p} - x_{.5})g/(e^{gz_p} - 1)] = \log_e B + \frac{1}{2} hz_p^2$. Plot h_p against z_p^2. A straight line supports a constant value of h.

E. Fitting g and h When One or Both Vary Systematically as Functions of z_p

If plots of g_p and/or h_p against z_p^2 show systematic departure(s) from straight line(s), let g and/or h be functions of z^2 of the form $g(z^2) = g_0 + \phi_g(z^2)$, $h(z^2) = h_0 + \phi_h(z^2)$ with ϕ_g and ϕ_h chosen so as to account for these departures. Then apply B and D.

4. THE LLOYDMINSTER DATA

Is the shape of the data describing oil in place in individual Lloydminster pools lognormal? Is the shape of the data describing the projective surface area of individual Lloydminster pools lognormal? If indeed the answer is "Yes", then we should expect that it makes no difference whether we work with the data either as recorded or as transformed by taking logarithms. The transformed data should lead to constant $g \cong h \cong 0$; the untransformed data to $h \cong 0$, and $g \cong \hat{\sigma}$, $\hat{\sigma}$ a MLE for σ. We examine the shape of both transformed and untransformed data for clues about g and h. The principal displays are:

- A table of spreads $x_{1-p} - x_p$, semi-spreads $x_{1-p} - x_{.5}$ and $x_{.5} - x_p$, z_p and g_p as a function of letter values for p.
- A plot of g_p against z_p^2.
- A plot of h_p against z_p^2.
- A plot of $Q_{g,h}(z_p)$ against x_p with g_p and h_p fixed at median values.

Prior to examining in detail the shape of deposit data describing oil in place (barrels), volume (acre-feet), and area (acres) we offer the following conclusions:

- It is reasonable to assume that the shape of the bulk of the Lloydminster data for each of these attributes is lognormal.
- The very largest and very smallest observations do not fit the pattern of the remainder of the data.
- Oil in place and volume appear closer to lognormal in shape than area.

Denote h and g values for log data by h^T and g^T respectively, and A and B values by A^T and B^T respectively, if the shape of the data is exactly lognormal then $h = h^T = 0$, $g^T = 0$, $\mu = A^T = \log A$ and $\sigma = g = B^T = B/A$. The data leads to the following values:

COMPARISON OF SHAPE PARAMETERS, TRANSFORMED vs. UNTRANSFORMED DATA ASSUMING CONSTANT g AND h

UNTRANSFORMED DATA

	A	B	g	h
Oil in Place (Barrels)	3,594,000	6,617,000	1.812	-.003915
Volume (Acre-feet)	2,646	4,738	1.739	-.005638
Area (Acres)	568	916.6	1.417	-.02252

TRANSFORMED DATA

	A^T	B^T	g^T	h^T
Oil in Place (log barrels)	15.09	1.781	.03103	-.003667
Volume (log Acre-feet)	7.881	1.751	.0242	-.005996
Area (log Acres)	6.342	1.591	-.06643	-.01109

	Oil in Place	Volume	Area
B^T	1.781	1.751	1.591
B/A	1.841	1.791	1.614
g	1.812	1.739	1.417
$\hat{\sigma}$	1.783	1.744	1.550

g_p RANGES

RANGE OF g_p VALUES-UNTRANSFORMED DATA*

	Min	Median	Max
Oil in Place	1.642	1.812	1.825
Volume	1.620	1.739	1.823
Area	1.218	1.417	1.5014

*The min and max values occur at p = .25 and 1-p = .999756 respectively.

RANGE OF g_p VALUES-TRANSFORMED DATA

	Min	Median	Max
Log Oil in Place	-.060478	.031032	.052774
Log Volume	-.042847	.024205	.052111
Log Area	-.12093	-.066434	-.022513

We turn next to a more detailed examination of the data. This is a large sample and it allows examination of extreme tail values. The table shown below displays p-values attached to letter values. Letter value V corresponds to low and high p-values .00024 and .99976 respectively -- very extreme fractiles.

	p-values		
	Low		High
M		.5	
F	.25		.75
E	.125		.875
D	.0625		.9375
C	.03125		.96875
B	.015625		.989375
A	.0078125		.9921875
Z	.00390625		.99609375
Y	.001953125		.999046875
X	.0009765625		.9990234375
W	.0004828125		.9995171875
V	.000244140625		.9997558594

4.1 Oil in Place - Untransformed Data

To illustrate the fitting process, we begin with oil in place data. Exhibit 4.1.1 is a letter value display, beginning with the median value M(p=.5) and ending with the letter value V corresponding to the pair of p-values .000244 and .999756. Columns labelled "low" and "high" show x_p and x_{1-p} respectively. The midpoint of (x_p, x_{1-p}) and the spread $x_{1-p} - x_p$ appear in the next two columns. Semi-spreads $x_{.5} - x_p$, $x_{1-p} - x_{.5}$ are displayed in the first two columns below, followed by the $\log_e$ ratio (Section 3.1, A), z_p, and g_p.

The median g_p value is 1.812, a value close to the MLE 1.782 of σ that obtains when h is assumed to be zero and g a constant (exact log-normal shape). If the g_p value corresponding to p values .000244 and .999756 (letter V) is excluded, the range of g_p values becomes much tighter: 1.752 to 1.826. Clearcut systematic variation in g_p values is not apparent in the plot of g_p versus z_p^2 (Exhibit 4.1.2) until z_p reaches -3.0553 (letter X, p values .000977 and .99902). At letter values W, and V, g_p values appear to decrease systematically with increasing $|-z_p|$. The plot of h_p against z_p^2 is of very similar shape. This signals that the very largest and very smallest observations may not fit the pattern of the remainder of the data.

An examination of the quantile - quantile plot of letter values against quantiles of a g and h distribution with g = 1.812 and h = -.003915 (Exhibit 4.1.3) suggests that the largest two observations are not large enough to force a match of letter values W and V to the straight line generated by the bulk of the data. The extreme right tail is squeezed at p-values .99952 and .99976. The vertical scale of this particular quantile - quantile plot visually de-emphasizes small valued data; thirteen from a total of twenty-three data points are

EXHIBIT 4.1.1 - LETTER-VALUE DISPLAY FOR OIL IN PLACE (10^6 BARRELS)

	DEPTH	LOW	MEDIAN	HIGH	MID	SPREAD
M	1184		3.59			0
F	592.5	.977		12.5	6.76	11.6
E	296.5	.415		29.2	14.8	28.7
D	148.5	.246		55.5	27.9	55.3
C	74.5	.162		98.3	49.2	98.2
B	37.5	.101		164.	82.3	164.
A	19.0	.072		292.	145.	292.
Z	10.0	.054		445.	222.	445.
Y	5.5	.035		627.	313.	627.
X	3.0	.023		928.	464.	928.
W	2.0	.015		967.	484.	967.
V	1.0	.007		1020.	510.	1020.

CALCULATION OF GP

	SEMI-SPREADS		LN RATIO	ZP	GP
F	2.62	8.95	1.23	-0.67	1.83
E	3.18	25.6	2.08	-1.15	1.81
D	3.35	51.9	2.74	-1.53	1.79
C	3.43	94.7	3.32	-1.86	1.78
B	3.49	160.	3.83	-2.15	1.78
A	3.52	288.	4.40	-2.41	1.82
Z	3.54	441.	4.83	-2.65	1.82
Y	3.56	623.	5.17	-2.85	1.81
X	3.57	924.	5.56	-3.06	1.82
W	3.58	963.	5.60	-3.19	1.75
V	3.59	1016.	5.65	-3.44	1.64

concentrated at point D in the upper left hand corner of the graph. From the median to the .999 quantile of the fitted g-and-h distribution the graph of $(x_p, Q_{g,h}(z_p))$ pairs appears to be a straight line.

In order to account for systematic variation in g_p values, we fit a quadratic function of z^2, $g(z) = a + bz^2 + cz^4$ to g_p-values:

$$g(z) = 1.7781 + .0208z^2 - .0023z^4.$$

The resulting Q-Q plot Exhibit 4.1.4 partially straightens out the bend apparent in the Q-Q plot of oil in place with constant g and constant h.

4.2 Oil in Place - Transformed Data

Transforming oil in place data by taking logarithms symmetrizes the data, as can be seen by comparing semi-spreads for the transformed data (Exhibit 4.2.1) and untransformed data (Exhibit 4.1.1).

EXHIBIT 4.1.2

GP (HOR. AXIS) VS. SQUARED NORMAL QUANTILES (VERT AXIS) FOR OIL IN PLACE

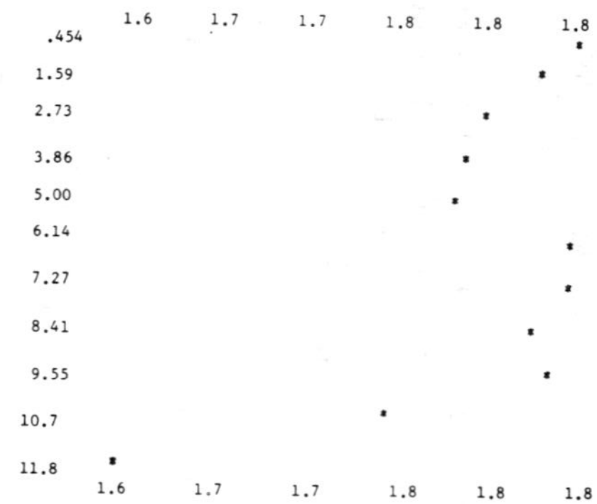

HP VS. SQUARED NORMAL QUANTILES (VERT. AXIS) FOR OIL IN PLACE

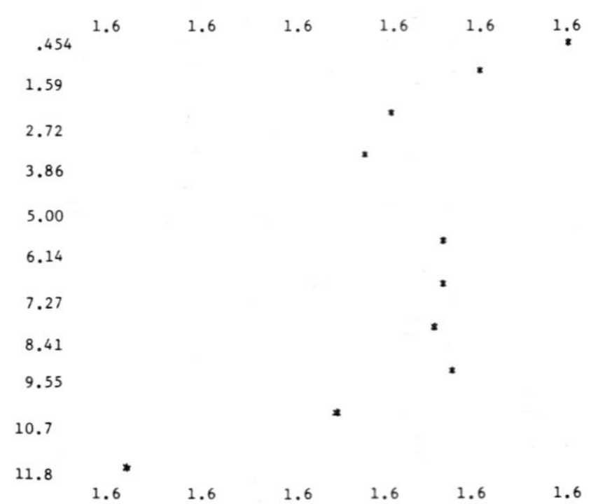

EXHIBIT 4.1.3

Q–Q PLOT OF THE LETTER VALUES (HOR. AXIS) VS. THE QUANTILES OF THE FITTED G–H DISTRIBUTION (VERT. AXIS) FOR OIL IN PLACE (10^8 BOE) WITH $g = 1.812$ AND $h = -.003915$

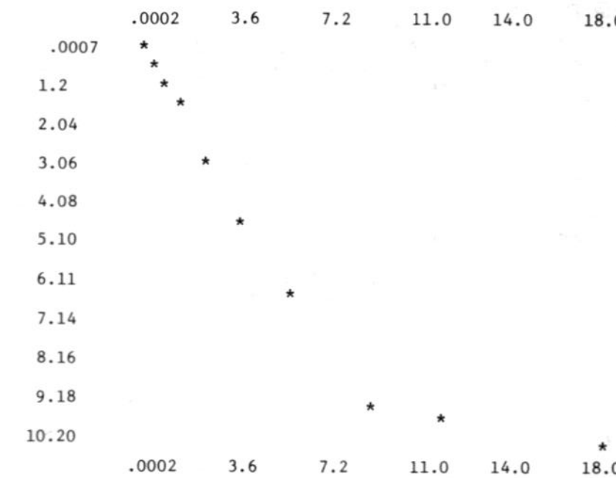

EXHIBIT 4.1.4

Q–Q PLOT OF THE LETTER VALUES (HOR. AXIS) VS. THE QUANTILES OF THE FITTED g and h DISTRIBUTION (VERT. AXIS) FOR OIL IN PLACE (10^6 BOE) WITH $g(z) = 1.7781 + 0.0208z^2 - 0.0023z^4$ AND $h = -.003915$

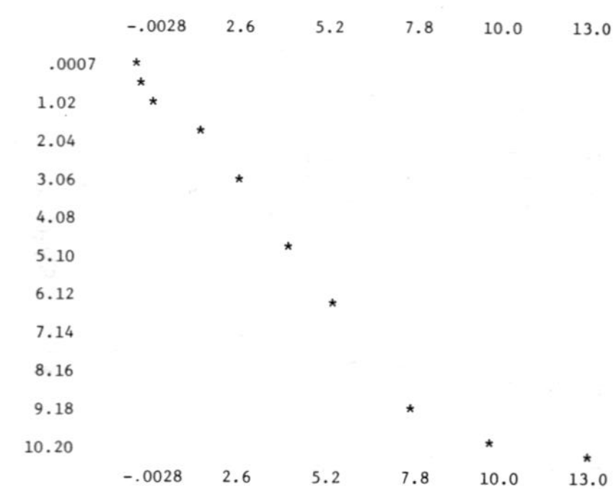

EXHIBIT 4.2.1 - LETTER-VALUE DISPLAY FOR LOG OIL IN PLACE

			MEDIAN	HIGH	MID	SPREAD
M	1184		15.095			
F	592.5	13.79		16.34	15.10	.00000
E	296.5	12.94		17.19	15.07	2.55
D	148.5	12.41		17.83	15.12	5.42
C	74.5	11.99		18.40	15.20	6.41
B	37.5	11.51		18.91	15.222	7.40
A	19.0	11.19		19.49	15.34	8.31
Z	10.0	10.90		19.91	15.41	9.01
Y	5.5	10.45		20.26	15.36	8.80
X	3.0	10.06		20.65	15.35	10.59
W	2.0	9.64		20.67	15.17	11.04
V	1.0	8.91		20.74	14.83	11.83

CALCULATION OF GP

	SEMI-SPREADS		LN RATIO	ZP	GP
F	1.30	1.25	-0.041	-0.67	-.0605
E	2.16	2.09	-0.031	-1.15	-.0267
D	2.68	2.74	0.021	-1.53	-.0135
C	3.10	3.31	0.065	-1.86	-.0350
B	3.58	3.823	0.067	-2.15	-.0310
A	3.91	4.40	0.118	-2.42	-.0487
Z	4.19	4.82	0.140	-2.65	-.0528
Y	4.64	5.16	0.106	-2.85	-.0373
X	5.03	5.553	0.098	-3.06	-.0321
W	5.45	5.560	0.027	-3.19	-.0083
V	6.18	5.658	-0.091	-3.44	-.0261

Recalling that if g_p is a constant function of z_p^2 for untransformed data, the corresponding behavior of g_p as a function of z_p^2 for log data is $g_p = 0$. Exhibit 4.2.2 shows systematic departure of the graph of g_p vs. z_p^2 from a vertical straight line, albeit within a modest range about zero: -.605 to .0528. This range of variation is small enough so that a quantile - quantile plot of x_p against $Q_{g,h}(z_p)$ for constant $g^T = .031032$ and $h^T = -.003667$ (Exhibit 4.2.3) shows only mild departures from a straight line until letter values W and V are reached.

Returning to the graph of g_p vs. z_p^2, the shape of the graph suggests that a function of the form $g(z) = a + bz^2 + cz^4$, $c < 0$ is more appropriate than a constant function $g(z) = a$. A fit is

$$g(z) = -.0694 + .0372z^2 - .0029z^4$$

and the corresponding Q-Q plot of letter values appears in Exhibit 4.2.4.

EXHIBIT 4.2.1

HP VS. SQUARED NORMAL QUANTILES (VERT. AXIS) FOR LOG OIL IN PLACE

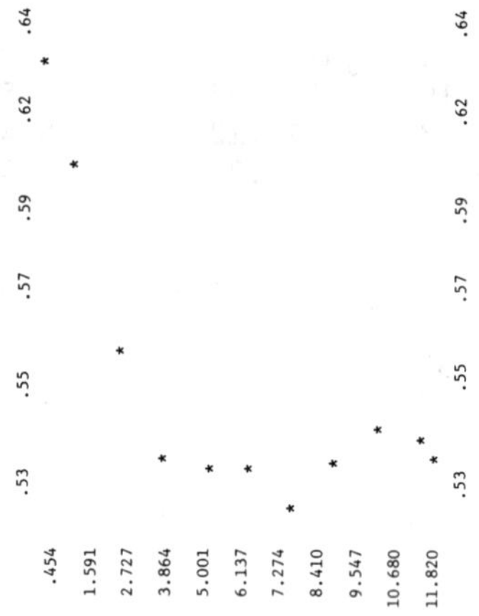

EXHIBIT 4.2.2

GP (HOR. AXIS) VS. SQUARED NORMAL, QUANTILES (VERT. AXIS) FOR LOG OIL IN PLACE

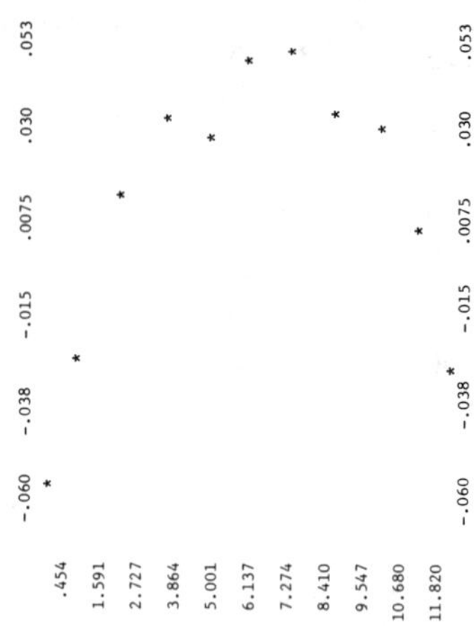

EXHIBIT 4.2.3

Q-Q PLOT OF THE LETTER VALUES (HOR. AXIS) VS. THE QUANTILES OF THE FITTED g-h DISTRIBUTION (VERT. AXIS) FOR LOG OIL IN PLACE WITH $g^T = .031032$ and $h^T = -.003667$

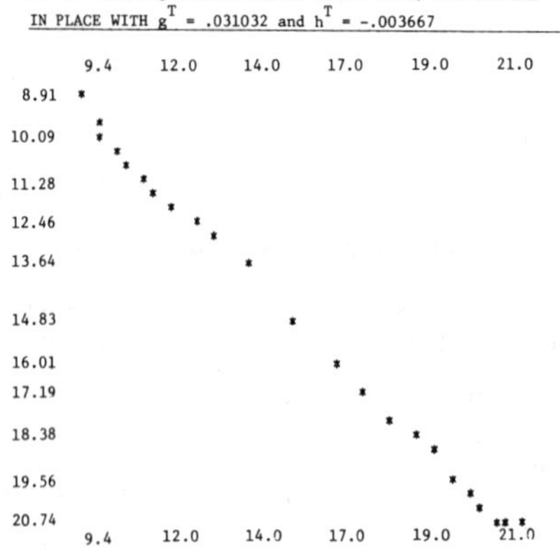

EXHIBIT 4.2.4

Q-Q PLOT OF THE LETTER VALUES (HOR. AXIS) VS. THE QUANTILES OF THE FITTED g-h DISTRIBUTION (VERT. AXIS) FOR LOG OIL IN PLACE WITH $g(z) = -0.0694 + 0.0372z^2 - 0.0029z^4$ AND $h = -.003667$

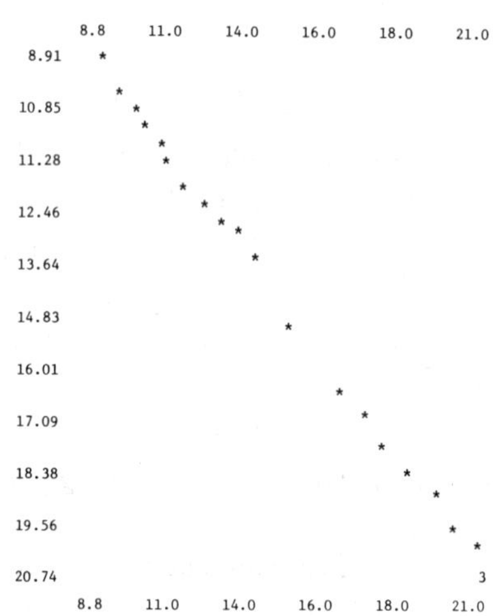

4.3 Comparison of g-h Analysis for Oil in Place With That for Log Oil In Place

Let a supercript T denote shape parameters computed for log data. If both transformed and untransformed data are fit by g and h distributions with constant g values and constant h values, and the data is close to lognormal in shape, we expect that $h \cong h^T \cong 0$, $g^T \cong 0$, a MLE $\hat{\sigma}$ for $\sigma \cong g \cong B^T \cong B/A$. This appears to be the case.

	A	B	g	h
Oil in Place (Barrels)	3,594,000	6,617,000	1.812	-.003915

	A^T	B^T	g^T	h^T
Oil in Place (log Barrels)	15.09	1.781	.03103	-.003667

B^T	1.781
B/A	1.841
g	1.812
$\hat{\sigma}$	1.782

Graphs in Exhibits 4.1.2 and 4.2.2 of h_p versus z_p^2 for oil in place and log oil in place respectively, signal small negative values for h and h^T for values of z_p of 2.0 or less. The graph of log oil in place suggests constant value of h^T for z_p^2 greater than 2.0. The graph of oil in place more strongly visually emphasizes the impact of large z_p^2 values on the behavior of h and suggests a small negative value for z_p^2 above about 9.0. Together these graphs suggest that h and h^T are close to zero for intermediate values of z_p^2 and that both take on small negative value for other z_p^2 values.

Resistant fits of constant h and constant h^T are -.003915 and -.003667, implying a mild tail squeezing of shape relative to (log) normality.

4.4 Area and Volume

A similar analysis of deposit area shows that

- log area is negatively skewed (upper semi-spreads are uniformly smaller than lower semi-spreads and g_p is negative at all letters)

and

- for log area, a resistant fit of the h function yields a slope of $h^T = -.01109$ suggesting that tails are mildly squeezed relative to lognormal shape.

In the interest of brevity the results of g and h distribution fits to log area are summarized in the form of two Q-Q plots: one with $g^T = \hat{\sigma} = .0242$ and $h^T = 0$ (Exhibit 4.4.1) and one with resistant fits of slope $h^T = -.01109$ and $g^T(z) = -.1248 - .0332z^2 - .0028z^4$ (Exhibit 4.4.2). While the latter fit straightens out the graph of the former fit, some slight inflections remain.

The volume of a deposit is defined as deposit area in acres times average net pay in feet so that the shape of deposit volumes is a concatenation of the shapes of area and net feet of pay.

A fit of $h^T = -.005996$ and $g(z) = -.0670 + .0354z^2 - .0028z^4$ (Exhibit 4.4.4) straightens out the left tail, but leaves the (anomalous ?) inflection appearing in the Q-Q plot for log volume (Exhibit 4.4.3). Since $h^T < 0$, the g and h distribution fit to log volume squeezes the tails relative to a lognormal distribution.

EXHIBIT 4.4.1

Q-Q PLOT OF THE LETTER VALUES (HOR. AXIS) VS. THE QUANTILES OF THE FITTED G-H DISTRIBUTION (VERT. AXIS) FOR LOG AREA WITH $g^T = 0.0242$ AND $h^T = 0$.

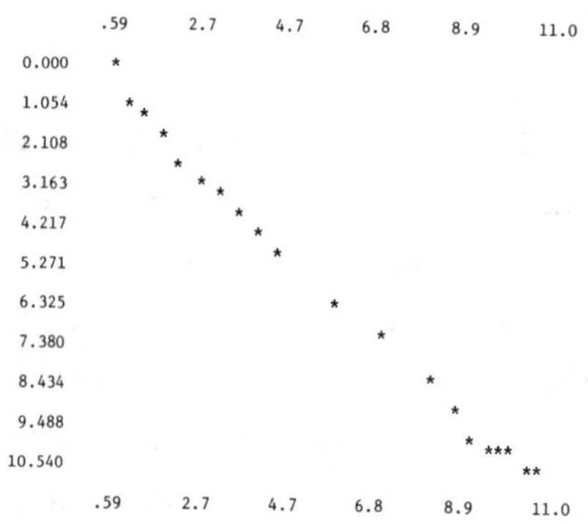

EXHIBIT 4.4.2

Q-Q PLOT OF THE LETTER VALUES (HOR. AXIS) VS. THE QUANTILES OF THE FITTED G-H DISTRIBUTION (VERT. AXIS) FOR LOG AREA WITH $g(z) = -.1248 -.0332z^2 -.0028z^4$ AND $h^T = -.01109$

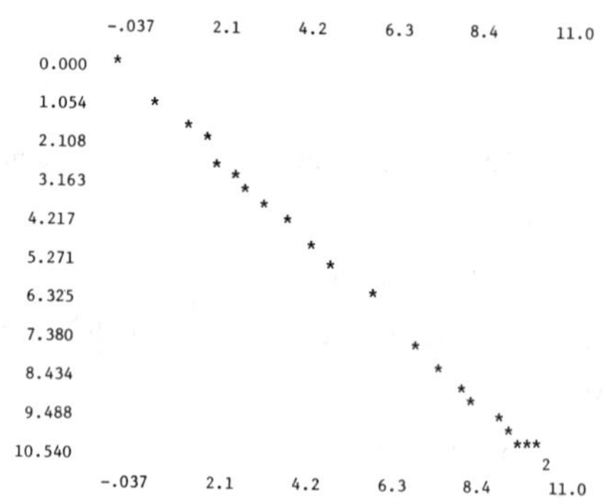

5. CONCLUSIONS

Tukey's shape fitting procedure affords insights into the shape of the empirical distribution of large samples not achievable by standard testing and plotting methods. Application of his method to the Lloydminster data suggests that adoption of the lognormal distribution as an approximation to the empirical distribution of oil in place is an adequate first approximation of shape for applications that do not require great precision in specification of very extreme tail values.

It would be of great interest to conduct a similar analysis of the shape of deposit attributes for other play types and compare them. Unfortunately, the Lloydminster data appears to be unique in both its large sample size and measurement protocol.

EXHIBIT 4.4.3

Q-Q PLOT OF THE LETTER VALUES (HOR. AXIS) VS. THE QUANTILES OF THE FITTED g-h DISTRIBUTION (VERT. AXIS) FOR LOG VOLUME WITH $g(z) = -0.0670 + 0.0354z^2 - 0.0029z^4$ AND $h^T = -.005996$

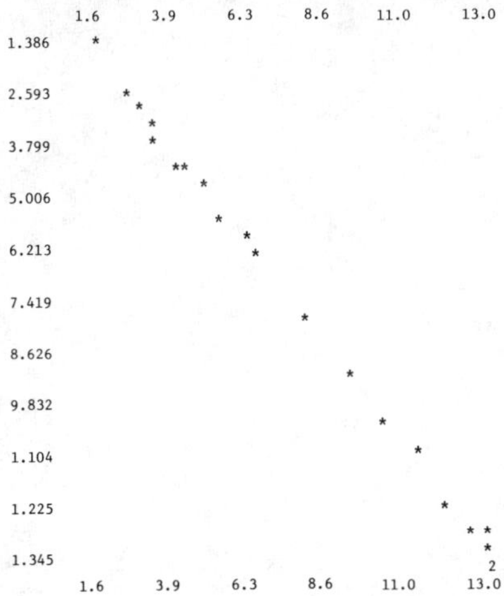

EXHIBIT 4.4.4

Q-Q PLOT OF THE LETTER VALUES (HOR. AXIS) VS. THE QUANTILES OF THE FITTED g-h DISTRIBUTION (VERT. AXIS) FOR LOG VOLUME

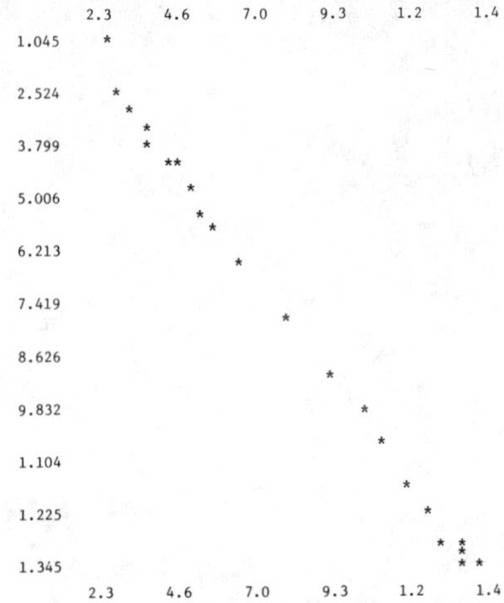

6. REFERENCES

Adelman, M. A., Houghton, J., Kaufman, G. M. and Zimmerman, M. (1983). Energy Resources in an Uncertain Future: Coal, Gas, Oil, and Uranium Supply Forecasting, Ballinger Publishing Co., Cambridge, MA.

Davis, J. C. (1987). Statistical Evaluation of Petroleum Deposits Before Discovery (this volume).

Hoaglin P. and Peters, S. (1983). 'g- and h-Distributions,' Encyclopedia of Statistical Sciences, **3**, pp. 299-301, N. Johnson and S. Kotz ed., John Wiley & Sons, New York, NY.

Mack, C. (1954). "The Expected Number of Clumps When Convex Laminae are Placed at Random and with Random Orientation on a Plane Area,' Proceedings of the Cambridge Philosophical Society, **50**, pp. 581-585.

McCrossan, R. G., Procter, R. M., and Ward, W. J. (1981). 'Estimate of Oil Resources, Lloydminster Area, Alberta,' Proceedings of the First International Conference on the Future of Heavy Crude and Tar Sands.

Schuenemeyer, J. H., and Drew, L. J. (1983). 'A Procedure to Estimate the Parent Population of the Size of Oil and Gas Fields as Revealed by a Study of Economic Truncation,' Mathematical Geology, **15**, No. 1, pp. 145-161.

Tukey, J. W. (1977). 'Modern techniques in data analysis.' NSF-sponsored regional research conference at Southeastern Massachusetts University, North Dartmouth, MA.

CONFIDENCE BANDS FOR THE DISTRIBUTION AND QUANTILE FUNCTIONS FOR TRUNCATED AND RANDOMLY CENSORED DATA.*

Chang-Jo F. Chung
Geological Survey of Canada
601 Booth Street
Ottawa, Ontario K1A 0E8
Canada.

Abstract. For the statistical analysis of many geoscience data, incompleteness of data - either truncated and/or randomly censored, causes much difficulty. The first step toward analysing such data is to estimate the distribution and quantile functions, and to infer about the distribution and quantile functions by constructing the confidence bands for the functions from the truncated and/or censored observations. We not only introduce the methodologies but also apply them to geoscience data as illustration.

1. Introduction

Let X^o be a real-valued random variable (rv) with distribution function $F^o(x) = P\{ X^o \le x \}$. Then the quantile function is defined as

$$(1.1) \qquad Q(y) = \inf\{ x \in R : F^o(x) \ge y \},\ 0 \le y \le 1 .$$

Although the distribution function F^o and the quantile function Q are equivalent from a probabilistic point of view, they represent complementary, in fact opposite, views of the rv from a statistical point of view. In this exposition I am taking an excursion into the both worlds of Q and F^o for truncated and randomly censored data. The world of Q and related statistics are excellently described by Csörgö (1983).

When we have n independent observations of the rv X, the Kolmogorov-Smirnov confidence band for the distribution function is well-known and has been frequently used in pratice. Csörgö and Révész (1984) constructed confidence bands for the quantile function by using a strong approximation theorem (cf. Csörgö and Révész (1978, 1981), and Csörgö (1983)). Csörgö and Horváth confidence band (1986) for the distribution function will be also presented. We will review these confidence bands first.

* Geological Survey of Canada Contribution Number 23787

C. F. Chung et al. (eds.), Quantitative Analysis of Mineral and Energy Resources, 433–457.

We then extend the construction of the confidence bands for the distribution and quantile functions to the situation that the observations are truncated. These extensions are particularly useful for handling geochemical data with observations that are in part below the detection limit.

Next we introduce the random censorship model and generalize the construction of confidence bands for the distribution function (Nair(1984), S. Csörgö and Horváth (1986)) as well as the quantile function (Chung, M. Csörgö and Horváth (1986)). The methodology is then applied to construct the confidence bands for distribution and quantile functions of the lengths of fractures in a granitic rock.

2. Kolmogorov-Smirnov and Csörgö-Horváth confidence bands for the distribution function and Csörgö-Révész confidence band for the quantile function

Consider a set of independent observations $X_1^o, \ldots, X_n^o$ on X^o. In order to describe X^o via estimating and inferring about F^o and/or Q, we produce two different types of pictures of X^o. The empirical distribution function:

$$(2.1) \qquad F_n^o(x) = \begin{cases} 0 & \text{if } X^o_{1:n} > x, \\ k/n & \text{if } X^o_{k:n} \leq x < X^o_{k+1:n} \ (k = 1, \ldots, n-1), \\ 1 & \text{if } X^o_{n:n} \leq x, \end{cases}$$

is a statistic for F^o, where $X^o_{k:n}$ denotes the k-th order statistic of $X_1^o, \ldots, X_n^o$, while the empirical quantile function:

$$(2.2) \qquad Q_n(y) = \inf\{ x : F_n^o(x) \geq y \}, \quad 0 \leq y \leq 1,$$

is a statistic for Q. The two statistics we obtain complement each other, featuring shapes of the distribution from opposite angles. The theories related with the two statistics are interwoven but distinctly different.

The Kolmogorov-Smirnov confidence bands for the distribution function F^o :

$$(2.3) \qquad \lim_{n\to\infty} P\{ F_n^o(x) - c/\sqrt{n} \leq F^o(x) \leq F_n^o(x) + c/\sqrt{n} : -\infty < x < \infty \}$$

$$= \lim_{n\to\infty} P\{ \sqrt{n} \sup_{-\infty<x<\infty} | F_n^o(x) - F^o(x) | \leq c \}$$

$$= P\{ \sup_{0\leq s\leq 1} |B(s)| \leq c \} \text{, and}$$

$$\text{(2.4)} \quad \lim_{n\to\infty} P\{\ F_n^{\,o}(x) - c'/\sqrt{n} \le F^o(x)\ :\ -\infty < x < \infty\ \}$$
$$= \lim_{n\to\infty} P\{\ F^o(x) \le F_n^{\,o}(x) + c'/\sqrt{n}\ :\ -\infty < x < \infty\ \}$$
$$= P\{\ \sup_{0\le s\le 1} B(s) < c'\ \},$$

where { B(s): $0\le s\le 1$ } is a Brownian bridge process (cf. Doob (1949)), are well-known and they have been heavily used in practice.

Recently Csörgö and Horváth (1986) proposed the following bands:

$$\text{(2.5)} \quad \lim_{n\to\infty} P\{\ F_n^{\,o}(x)-\pi_1(c,n) \le F^o(x) \le F_n^{\,o}(x)+\pi_2(c,n)\ :\ -\infty<x<\infty\ \}$$
$$= P\{\ \sup_{0\le s\le 1} |B(s)| \le c\ \},\ \text{and}$$

$$\text{(2.6)} \quad \lim_{n\to\infty} P\{\ F^o(x) \le F_n^{\,o}(x)+\pi_2(c',n)\ :\ -\infty<x<\infty\ \}$$
$$= P\{\ \sup_{0\le s\le 1} B(s) < c'\ \},$$

where

$$\text{(2.7)} \quad \begin{aligned} \pi_1(c,n) &= c(1-c/(\sqrt{n}F_n^{\,o}(x)+c))/\sqrt{n} \\ \pi_2(c,n) &= c(1-c/(\sqrt{n}(1-F_n^{\,o}(x))+c))/\sqrt{n}\ . \end{aligned}$$

Of course Csörgö and Horváth (C.-H.) bands are narrower than K.-S. bands. However the comparison of two bands is beyond the scope of this exposition. By its construction C.-H. bands obtain zero at the left-side tail and one at the right-side tail instead of negative value at the left-side tail and above-one value at the right-side tail in K.-S. bands. This zero and one property is desirable for interpretation of the confidence bands. A simulation study of C.-H. bands for small samples has not been made yet.

Csörgö and Révész (1984) constructed the following confidence band for the quantile function without estimating the density function f^o using a strong approximation theorem on the normed quantile process (cf. Csörgö and Révész (1978, 1981), and Csörgö (1983)):

<u>Theorem A. (Csörgö and Révész (1978)). Assume that the density function $f^o(t) = d/dt\ F^o(t)$ is positive and differentiable on (t_{F^o}, T_{F^o}) with $t_{F^o} = \sup\{\ t : F^o(t)=0\ \}$, $T_{F^o} = \inf\{\ t : F^o(t)=1\ \}$, and for some $r > 0$, if we have</u>

$$\text{(2.8)} \quad \sup_{0\le t\le 1}\ t\ \frac{|f^{o\prime}(Q(t))|}{f^o(Q(t))^2} \le r\ .$$

Then we have,

(2.9) $$\lim_{n\to\infty} P\{ Q_n(y-c/\sqrt{n}) \le Q(y) \le Q_n(y+c/\sqrt{n}) : \delta_n \le y \le 1-\delta_n \}$$
$$= P\{ \sup_{0\le s\le 1} |B(s)| \le c \} \text{ , and}$$

(2.10) $$\lim_{n\to\infty} P\{ Q_n(y - c'/\sqrt{n}) \le Q(y) : \delta_n \le y \le 1 \}$$
$$= \lim_{n\to\infty} P\{ Q(y) \le Q_n(y + c'/\sqrt{n}) : 0 \le y \le 1-\delta_n \}$$
$$= P\{ \sup_{0\le s\le 1} B(s) < c' \} ,$$

where $\sqrt{n}\,\delta_n \to \infty$ as $n \to \infty$.

Remark 2.1. The random variable $\sup(0\le s\le 1)|B(s)|$ has well-known Kolmogorov distribution function and is obtained by:

(2.11) $$P\{ \sup_{0\le s\le 1} |B(s)| \le c \} = 1 - 2 \sum_{k=1}^{\infty} (-1)^{k+1} e^{-2k^2c^2} .$$

Tables 1a - 1d show the critical values of c for the selected probabilities, 0.900, 0.950, 0.975 and 0.990.

Remark 2.2. In order to construct the K.-S. two-sided confidence band for F using (2.3), for a given significance level α, the critical value c_α is chosen from Tables 1a-1d or computed by using the analytic formula in Chung (1986) and the computer program by Chung (1987a). Then the confidence band is constructed by

(2.12) $$\{ [F_n^o(x) - n^{-\frac{1}{2}} c_\alpha , F_n^o(x) + n^{-\frac{1}{2}} c_\alpha] : -\infty<x<\infty \} .$$

Remark 2.3. To construct the C.-H. two-sided confidence band using (2.5), for a given significance level α, c_α is computed, and $\pi_1(c_\alpha,n)$ and $\pi_2(c_\alpha,n)$ are computed. The C.-H. band is obtained by

(2.13) $$\{ [F_n^o(x)-\pi_1(c_\alpha, n) , F_n^o(x)+\pi_2(c_\alpha, n)] : -\infty<x<\infty \} .$$

The band in (2.13) is again narrower than that of (2.12).

Remark 2.4. In order to construct the C.-R. two-sided confidence band for Q using (2.9) for a given significance level α, the critical value c_α is computed as in Remarks 2.2 and 2.3. In addition, a constant δ_n is computed. Then the confidence band is constructed by

(2.14) $$\{ [Q_n(x - n^{-\frac{1}{2}} c_\alpha) , Q_n(x + n^{-\frac{1}{2}} c_\alpha)] : \delta_n<x<1-\delta_n \} .$$

Table 1a. Critical point τ of the distribution of $\sup_{a\leq s\leq b} |B(s)|$ for selected values of a and b, and $P\{ \sup_{a\leq s\leq b} |B(s)| \leq \tau \} = 0.900$.

a	b=0.1	b=0.2	b=0.3	b=0.4	b=0.5	b=0.6	b=0.7	b=0.8	b=0.9	b=1.0
0.00	0.59852	0.81547	0.95967	1.06164	1.13340	1.18126	1.20937	1.22157	1.22383	1.22383
0.05	0.59631	0.81542	0.95967	1.06164	1.13340	1.18127	1.20937	1.22157	1.22383	1.22383
0.10		0.81144	0.95893	1.06144	1.13332	1.18122	1.20934	1.22155	1.22381	1.22381
0.15		0.78909	0.95327	1.05923	1.13222	1.18055	1.20885	1.22112	1.22339	1.22339
0.20			0.93664	1.05198	1.12813	1.17780	1.20670	1.21922	1.22155	1.22155
0.25			0.89946	1.03670	1.11918	1.17152	1.20167	1.21470	1.21714	1.21714
0.30				1.00952	1.10368	1.16057	1.19279	1.20670	1.20934	1.20934
0.35				0.96180	1.07963	1.14389	1.17933	1.19458	1.19753	1.19753
0.40					1.04358	1.12029	1.16057	1.17780	1.18122	1.18122
0.45					0.98707	1.08796	1.13572	1.15585	1.15998	1.15998
0.50						1.04358	1.10368	1.12813	1.13332	1.13332
0.55						0.97874	1.06272	1.09388	1.10070	1.10070
0.60							1.00952	1.05198	1.06144	1.06144
0.65							0.93570	1.00068	1.01461	1.01461
0.70								0.93664	0.95893	0.95893
0.75								0.85148	0.89241	0.89241
0.80									0.81144	0.81144
0.85									0.70742	0.70742
0.90										0.70742
0.95										0.70742

Table 1b. Critical point τ of the distribution of $\sup_{a\leq s\leq b} |B(s)|$ for selected values of a and b, and $P\{ \sup_{a\leq s\leq b} |B(s)| \leq \tau \} = 0.950$.

a	b=0.1	b=0.2	b=0.3	b=0.4	b=0.5	b=0.6	b=0.7	b=0.8	b=0.9	b=1.0
0.00	0.68248	0.92685	1.08682	1.19755	1.27307	1.32108	1.34708	1.35678	1.35809	1.35809
0.05	0.68141	0.92684	1.08682	1.19755	1.27307	1.32108	1.34708	1.35678	1.35809	1.35809
0.10		0.92473	1.08655	1.19749	1.27306	1.32107	1.34707	1.35677	1.35809	1.35809
0.15		0.90770	1.08316	1.19638	1.27255	1.32078	1.34686	1.35659	1.35790	1.35790
0.20			1.07057	1.19154	1.27003	1.31914	1.34558	1.35544	1.35677	1.35677
0.25			1.03841	1.17961	1.26345	1.31464	1.34196	1.35214	1.35353	1.35353
0.30				1.15622	1.25076	1.30583	1.33480	1.34558	1.34707	1.34707
0.35				1.11239	1.22961	1.29139	1.32307	1.33483	1.33649	1.33649
0.40					1.19637	1.26987	1.30583	1.31914	1.32107	1.32107
0.45					1.14240	1.23934	1.28214	1.29778	1.30013	1.30013
0.50						1.19637	1.25076	1.27003	1.27306	1.27306
0.55						1.13249	1.20989	1.23502	1.23911	1.23911
0.60							1.15622	1.19154	1.19749	1.19749
0.65							1.08141	1.13783	1.14712	1.14712
0.70								1.07057	1.08655	1.08655
0.75								0.98154	1.01364	1.01364
0.80									0.92473	0.92473
0.85									0.81138	0.81138
0.90										0.81138
0.95										0.81138

Table 1c. Critical point τ of the distribution of $\sup_{a\leq s\leq b}|B(s)|$ for selected values of a and b, and $P\{\sup_{a\leq s\leq b}|B(s)| \leq \tau\} = 0.975$.

a	b=0.1	b=0.2	b=0.3	b=0.4	b=0.5	b=0.6	b=0.7	b=0.8	b=0.9	b=1.0
0.00	0.75893	1.02821	1.20243	1.32095	1.39966	1.44759	1.47165	1.47943	1.48021	1.48021
0.05	0.75840	1.02821	1.20243	1.32095	1.39966	1.44759	1.47165	1.47943	1.48021	1.48021
0.10		1.02707	1.20233	1.32092	1.39966	1.44759	1.47165	1.47943	1.48021	1.48021
0.15		1.01395	1.20029	1.32037	1.39943	1.44745	1.47156	1.47934	1.48012	1.48012
0.20			1.19067	1.31712	1.39786	1.44646	1.47079	1.47864	1.47943	1.47943
0.25			1.16263	1.30772	1.39297	1.44321	1.46817	1.47621	1.47702	1.47702
0.30				1.28744	1.38248	1.43605	1.46233	1.47079	1.47165	1.47165
0.35				1.24693	1.36376	1.42343	1.45201	1.46121	1.46217	1.46217
0.40					1.33290	1.40369	1.43605	1.44646	1.44759	1.44759
0.45					1.28114	1.37468	1.41333	1.42564	1.42702	1.42702
0.50						1.33290	1.38248	1.39786	1.39966	1.39966
0.55						1.26985	1.34163	1.36211	1.36462	1.36462
0.60							1.28744	1.31712	1.32092	1.32092
0.65							1.21161	1.26106	1.26736	1.26736
0.70								1.19067	1.20233	1.20233
0.75								1.09789	1.12349	1.12349
0.80									1.02707	1.02707
0.85									0.90476	0.90476
0.90										0.90476
0.95										0.90476

Table 1d. Critical point τ of the distribution of $\sup_{a\leq s\leq b}|B(s)|$ for selected values of a and b, and $P\{\sup_{a\leq s\leq b}|B(s)| \leq \tau\} = 0.990$.

a	b=0.1	b=0.2	b=0.3	b=0.4	b=0.5	b=0.6	b=0.7	b=0.8	b=0.9	b=1.0
0.00	0.85120	1.15054	1.34186	1.46960	1.55194	1.59958	1.62137	1.62723	1.62762	1.62762
0.05	0.85098	1.15054	1.34186	1.46960	1.55194	1.59958	1.62137	1.62723	1.62762	1.62762
0.10		1.15002	1.34183	1.46960	1.55194	1.59958	1.62137	1.62723	1.62762	1.62762
0.15		1.14061	1.34079	1.46939	1.55188	1.59955	1.62134	1.62720	1.62759	1.62759
0.20			1.33398	1.46744	1.55103	1.59903	1.62094	1.62683	1.62723	1.62723
0.25			1.31039	1.46054	1.54770	1.59686	1.61920	1.62518	1.62561	1.62561
0.30				1.44360	1.53949	1.59140	1.61469	1.62094	1.62137	1.62137
0.35				1.40686	1.52341	1.58072	1.60590	1.61267	1.61313	1.61313
0.40					1.49527	1.56293	1.59140	1.59903	1.59958	1.59958
0.45					1.44604	1.53568	1.56976	1.57886	1.57956	1.57956
0.50						1.49527	1.53949	1.55103	1.55194	1.55194
0.55						1.43314	1.49854	1.51434	1.51572	1.51572
0.60							1.44360	1.46744	1.46960	1.46960
0.65							1.36646	1.40842	1.41226	1.41226
0.70								1.33398	1.34183	1.34183
0.75								1.23639	1.25577	1.25577
0.80									1.15002	1.15002
0.85									1.01639	1.01639
0.90										1.01639
0.95										1.01639

Remark 2.5. Similarly, the one-sided confidence bands corresponding to (2.12), (2.13) and (2.14) can be obtained using (2.4), (2.6) and (2.10).

The confidence bands (2.3), (2.4), (2.5), (2.6), (2.9) and (2.10) will be extended to the situation of truncated data in the next section.

3. Confidence bands for truncated data.

Suppose that the random sample is doubly truncated at u and v, and instead of $X_1^o, \ldots, X_n^o$, we have $X_1^*, \ldots, X_n^*$ where

(3.1) $$X_i^* = \begin{cases} u, & u > X_i^o, \\ X_i^o, & u \le X_i^o \le v, \\ v, & v < X_i^o, \end{cases}$$

for i = 1, ..., n. Then the empirical distribution $F_n^o(.)$ in (2.1) is no longer defined outside of [u,v]. We would like to construct confidence bands for F^o and Q on [u,v] or to test the hypothesis that

(3.2) $$H_o : F^o(x) = F_o(x), \quad x \in [u, v].$$

Using Donsker's theorem(Donsker (1952)), or the Komlós, Major and Tusnády theorem(Komlós, Major and Tusnády (1975)) which says that the empirical process can be almost surely approximated by a sequence of Brownian bridge processes at a certain convergence rate, we generalize the Kolmogorov-Smirnov confidence bands in (2.3) and (2.4) for truncated data:

(3.3) $$\lim_{n\to\infty} P\{ F_n^o(x) - d/\sqrt{n} \le F^o(x) \le F_n^o(x) + d/\sqrt{n} : u < x < v \}$$
$$= P\{ \sup_{F^o(u)\le s\le F^o(v)} |B(s)| \le d \}, \text{ and}$$

(3.4) $$\lim_{n\to\infty} P\{ F_n^o(x) - d'/\sqrt{n} \le F^o(x) : u < x < v \}$$
$$= \lim_{n\to\infty} P\{ F^o(x) \le F_n^o(x) + d'/\sqrt{n} : u < x < v \}$$
$$= P\{ \sup_{F^o(u)\le s\le F^o(v)} B(s) < d' \} .$$

Similarly, from (2.5) and (2.6), we also obtain

$$(3.5)\qquad \lim_{n\to\infty} P\{ F_n^o(x)-\pi_1(d,n) \le F^o(x) \le F_n^o(x)+\pi_2(d,n) : u<x<v \}$$
$$= P\{ \sup_{F^o(u)\le s\le F^o(v)} |B(s)| \le d \} , \text{ and}$$

$$(3.6)\qquad \lim_{n\to\infty} P\{ F^o(x) \le F_n^o(x)+\pi_2(d',n) : u<x<v \}$$
$$= P\{ \sup_{F^o(u)\le s\le F^o(v)} B(s) < d' \} ,$$

where $\pi_1(d,n)$ and $\pi_2(d',n)$ are defined in (2.7).

Using the same strong approximation theorem on quantile process as in Theorem A (cf. Csörgö and Révész (1978, 1981), and Csörgö (1983)), the confidence bands for quantile function of truncated data are constructed. Here we only need the condition that the density function is positive and differentiable in Theorem A to be satisfied instead of all the conditions in Theorem A including (2.8) required for the confidence bands for the quantile function in (2.9) and (2.10). Now we construct the confidence bands as follows:

$$(3.7)\qquad \lim_{n\to\infty} P\{ Q_n(y - d/\sqrt{n}) \le Q(y) \le Q_n(y + d/\sqrt{n}) : F^o(u)<y<F^o(v) \}$$
$$= P\{ \sup_{F^o(u)\le s\le F^o(v)} |B(s)| \le d \} , \text{ and}$$

$$(3.8)\qquad \lim_{n\to\infty} P\{ Q(y) \le Q_n(y + d'/\sqrt{n}) : F^o(u)<y<F^o(v) \}$$
$$= \lim_{n\to\infty} P\{ Q_n(y - d'/\sqrt{n}) \le Q(y) : F^o(u)<y<F^o(v) \}$$
$$= P\{ \sup_{F^o(u)\le s\le F^o(v)} B(s) < d' \} .$$

However, contrary to c and c' in the non-truncated case in (2.3), (2.4), (2.5), (2.6), (2.9) and (2.10), d and d' in (3.3), (3.4), (3.5), (3.6), (3.7) and (3.8) depend upon the unknown distribution function F^o, and thus we cannot immediately use (3.3) - (3.8) to construct confidence bands for F^o and Q in practice.

Nevertheless (3.3), (3.4), (3.5) and (3.6) can be used for testing the null hypothesis in (3.2) on [u,v]. Namely, we first define $\tau := \tau(\alpha,F(u),F(v))$ and $\tau' := \tau'(\alpha,F(u),F(v))$ for any $\alpha \in (0,1)$ and any distribution function F by

$$(3.9) \qquad P\{ \sup_{F(u)\le s\le F(v)} |B(s)| \le \tau \} = 1 - \alpha \text{, and}$$

$$(3.10) \qquad P\{ \sup_{F(u)\le s\le F(v)} B(s) \le \tau' \} = 1 - \alpha \ .$$

Then, for a given level of significance α, we compute $\tau(\alpha,F_o(u),F_o(v))$ and $\tau'(\alpha,F_o(u),F_o(v))$ from (3.9) and (3.10), and reject the hypothesis in (3.2) if, for two-sided K.-S. statistics,

$$\sqrt{n} \sup_{u\le x\le v} |F_n{}^o(x)-F_o(x)| \ge \tau(\alpha,F_o(u),F_o(v)),$$

and, for one-sided K.-S. statistics,

$$\sqrt{n} \sup_{u\le x\le v} (F_n{}^o(x)-F_o(x)) \ge \tau'(\alpha,F_o(u),F_o(v)).$$

Unlike the K.-S. tests in (2.3) and (2.4) for the non-truncated case, the critical values τ and τ' depend on the null hypothesis. Of course, we can also use the C.-H. test in (3.5) and (3.6).

For singly truncated data, Barr and Davidson (1973), and Koziol and Byar (1975) proposed to use the special cases of (3.3) and (3.4) when $u = -\infty$ (i.e., $F^o(u) = 0$) for testing the hypothesis in (3.2). Barr and Davidson (1973) suggested to use $\tau(\alpha,0,F_o(v))$ and $\tau'(\alpha,0,F_o(v))$ as the critical values, while Koziol and Byar (1975) suggested to use $\tau(\alpha,0,w)$ and $\tau'(\alpha,0,w)$ where w is slightly greater than both $F_o(v)$ and $F_n{}^o(v)$.

Chung (1987b) justifies the use of $\tau(\alpha,F_n{}^o(u),F_n{}^o(v))$ for $\tau(\alpha,F^o(u),F^o(v))$ in (3.3) - (3.8) via proving that asymptotically the substitution would not make any difference. This will be the topic of the next section.

We conclude this section by remarking that a detailed discussion on the formulae for the probability statements in (3.9) and (3.10) can be found in Chung (1986). Also, Chung (1987a) developed a computer program package called WIENER PACK written in FORTRAN 77 for these formulae. As in Remark 2.1 Tables 1a-1d include the critical values d for the selected probabilities, 0.900, 0.950, 0.975 and 0.990.

4. Modified K.-S. and C.-H. confidence bands for the distribution function and modified C.-R. confidence band for the quantile function for truncated data.

Let us define:

$$(4.1) \qquad \varphi_1(a,b,p) := \sup\{ c : P\{ B(s)\le c : a\le s\le b \} \le p \} ,$$

$$(4.2) \qquad \varphi_2(a,b,p) := \sup\{ c : P\{|B(s)|\le c : a\le s\le b \} \le p \} ,$$

where $a \in [0,1]$, $b \in [a,1]$ and $p \in (0,1)$. Then by using the continuity of φ_1 and φ_2 (c.f. Chung, Csörgö and Horváth (1986)) and the strong law of large numbers of F_n^o, Chung (1987b) constructed the following confidence bands for the distribution function which can be used in practice:

Theorem B. Let $F_n^o(.)$ be the empirical distribution function of a continuous distribution function $F^o(.)$ on $[u,v]$ where $u > t_{F^o}$ and $v < T_{F^o}$ and $u < v$. Then, using φ_1 and φ_2 as defined in (4.1) and (4.2), we have, for any $\alpha \in (0,1)$,

$$\lim_{n\to\infty} P\{ \sqrt{n} \sup_{u\le x\le v} (F_n^o(x)-F^o(x)) \le \varphi_1(F_n^o(u),F_n^o(v),1-\alpha) \} = 1 - \alpha , \text{ and} \tag{4.3}$$

$$\lim_{n\to\infty} P\{ \sqrt{n} \sup_{u\le x\le v} |F_n^o(x)-F^o(x)| \le \varphi_2(F_n^o(u),F_n^o(v),1-\alpha) \} = 1 - \alpha . \tag{4.4}$$

Also

$$\lim_{n\to\infty} P\{ (F_n^o(x)-F^o(x)) \le \pi_2(\varphi_1(F_n^o(u),F_n^o(v),1-\alpha),n) : u\le x\le v \} = 1 - \alpha , \text{ and} \tag{4.5}$$

$$\lim_{n\to\infty} P\{ F_n^o(x)-\pi_1(\varphi_2(F_n^o(u),F_n^o(v),1-\alpha),n) \le F^o(x) \le F_n^o(x)+\pi_2(\varphi_2(F_n^o(u),F_n^o(v),1-\alpha),n) : u<x<v \} = 1 - \alpha , \tag{4.6}$$

where $\pi_1(.,n)$ and $\pi_2(.,n)$ are defined in (2.7).

For the confidence bands for the quantile function for the truncated data from (3.9) and (3.10), since $Q_n(.)$ does not have any significance outside $[F_n^o(u), F_n^o(v)]$, we can have at most the following theorem:

Theorem C. Let $Q_n(.)$ be the empirical quantile function of a quantile function $Q(.)$ on $[u',v']$ where $u' > F^o(u)$ and $v' < F^o(v)$ and $u' < v'$. Then, using φ_1 and φ_2 as defined in (4.1) and (4.2), we have, for any $\alpha \in (0,1)$,

$$(4.7) \quad \lim_{n\to\infty} P\{ Q_n(y - \varphi_1(u', v',1-\alpha)/\sqrt{n}) \le Q(y) : u'<y<v' \}$$
$$= \lim_{n\to\infty} P\{ Q_n(y + \varphi_1(u', v',1-\alpha)/\sqrt{n}) \ge Q(y) : u'<y<v' \}$$
$$= 1 - \alpha \text{ , and}$$

$$(4.8) \quad \lim_{n\to\infty} P\{ Q_n(y - \varphi_2(u', v',1-\alpha)/\sqrt{n}) \le Q(y) \le Q_n(y + \varphi_2(u', v',1-\alpha)/\sqrt{n}) : u'<y<v' \}$$
$$= 1 - \alpha .$$

Remark 4.1. In order to construct a two-sided confidence band for F^o on [u,v], where $F_n{}^o$ are well-defined, for a given significance level α, $\varphi_2(F_n{}^o(u),F_n{}^o(v),1-\alpha)$ is either chosen from Tables 1a-1d or computed by using the formula discussed by Chung (1986) and the computer program by Chung (1987a). Then the two-sided confidence band is constructed by

$$(4.9) \quad \{ [F_n{}^o(x) - n^{-\frac{1}{2}} \varphi_2(F_n{}^o(u),F_n{}^o(v),1-\alpha), \quad F_n{}^o(x) + n^{-\frac{1}{2}} \varphi_2(F_n{}^o(u),F_n{}^o(v),1-\alpha)] : x \in [u,v] \} .$$

Of course, when $u \le t_{F^o}$ and $v \ge T_{F^o}$, then the confidence band in (4.11) coincides with the K.-S. confidence band in (2.3).

Remark 4.2. To construct the modified two-sided C.-H. confidence band, for a given significance level α, $\varphi_2(F_n(u),F_n(v),1-\alpha)$ is computed and the C.-H. band is obtained by

$$(4.10) \quad \{ [F_n(x)-\pi_1(\varphi_2(F_n(u),F_n(v),1-\alpha),n) , \quad F_n(x)+\pi_2(\varphi_2(F_n(u),F_n(v),1-\alpha),n)] : u<x<v \} .$$

Remark 4.3. In order to construct a two-sided confidence band for Q on [u',v'] as in Theorem C, for a given significance level α, $\varphi_2(F_n{}^o(u),F_n{}^o(v),1-\alpha)$ is computed. Then the two-sided confidence band is constructed by

$$(4.11) \quad \{ [Q_n(x - n^{-\frac{1}{2}} \varphi_2(u',v',1-\alpha)) , \quad Q_n(x + n^{-\frac{1}{2}} \varphi_2(u',v',1-\alpha))] : u'<x<,v' \} .$$

5. Random censorship model from the right.

Let $X_1{}^o, \ldots , X_n{}^o$ be independent identically distributed(i.i.d.) random variables (rv) with continuous distribution function $F^o(x) = P\{ X_1{}^o \le x \}$. Let $Y_1, \ldots , Y_n$ be i.i.d. rv with distribution function H. Suppose that the two sequences $\{X_i{}^o\}$ and $\{Y_i\}$ are independent. In the random censorship model from the right, the $X_i{}^o$ may be censored on the right by the Y_i, so that one observes only the pairs

(X_i,δ_i), $i=1,2,\ldots,n$, where X_i is the minimum of X_i^o and Y_i, and δ_i is the indicator function of the event $X_i^o \le Y_i$, i.e.,

$$(5.1)\qquad X_i = X_i^o \wedge Y_i \quad \text{and} \quad \delta_i = \begin{cases} 1 & \text{if } X_i^o \le Y_i, \\ 0 & \text{if } X_i^o > Y_i, \end{cases}$$

$i=1,2,\ldots,n$. Thus the $\{X_i\}$ are i.i.d. rv with distribution function F given by

$$(5.2)\qquad 1 - F(t) = (1-F^o(t))(1-H(t)), \quad -\infty < t < \infty$$

and the subdistribution function of the uncensored observations is given by

$$(5.3)\qquad \tilde{F}(t) := P\{\, X_i<t \text{ and } \delta_i=1 \,\} = \int_{-\infty}^{t} (1-H(s))\, dF^o(s).$$

Then, by (5.2), $T_F = T_F{}^o \wedge T_H$ and $t_F = t_F{}^o \vee t_H$ (T and t functions are defined in Theorem A).

To estimate F^o in the above model, the product-limit(PL) estimator (cf. Kaplan and Meier (1958)), $F_n{}^o$ has been widely used and defined by

$$(5.4)\qquad 1 - F_n{}^o(t) := \begin{cases} \prod_{\{\, 1\le i\le n: X_i<t \,\}} \left[\dfrac{n-R_i}{n-R_i+1} \right]^{\delta_i} & \text{if } t \le X_{n:n} \\ 0 & \text{if } t > X_{n:n} \end{cases}$$

where $X_{n:n}$ = Max (X_1, ... , X_n) and R_i is the rank of $(X_i,1-\delta_i)$ in the lexicographic ordering of $\{\, (X_i,1-\delta_i) \,\}$, $i=1,2, \ldots , n$.

6. Product-limit(PL) process and confidence bands for distribution function.

$F_n{}^o$ has been generally accepted as a substitute for the empirical distribution function in the random censorship model from the right. The confidence bands for the product limit estimator $1-F_n{}^o(t)$ in (5.4) had been extensively discussed by Efron(1967), Hall and Wellner (1980), Nair(1984) and S. Csörgö and Horváth(1986) among many others. While Nair(1984) has made a comparative study among the bands and also several simulation studies, S. Csörgö and Horváth(1986) have proposed the bands in Theorem D and discussed that their bands include most existing bands which can be regarded as special cases of their bands.

To construct the confidence bands, the PL-process, an analogue of the empirical process, is defined as

$$(6.1) \qquad \beta_n(t) := \sqrt{n}\,(\,F_n^o(t) - F^o(t)\,).$$

Weak convergence of β_n to a Gaussian process over a finite interval was obtained by Breslow and Crowley (1974). Burke, S. Csörgö and Horváth (1981, 1986) have established strong approximations of β_n in terms of appropriate Gaussian processes in a general setting over right closed infinit intervals, expanding to infinity on the right too, and using this theorem, S. Csörgö and Horváth(1986) have obtained the following theorem to construct the confidence bands for the distribution function F^o:

Theorem D. Let $p < T_{F^o}$. Then we have

$$(6.2) \quad \lim_{n\to\infty} P\{\, F_n^o(t)-a_n(t,\beta_1,\beta_2) \le F^o(t) \le F_n^o(t)+a_n(t,\beta_1,\beta_2) \;:\; -\infty<t\le p \,\}$$
$$= P\{\, |W(s)| \le \beta_1 s+\beta_2 \;:\; 0\le s\le 1 \,\},$$

where $\beta_1 > 0$, $\beta_1+\beta_2 \ge 0$, $W(.)$ is a Wiener process, and

$$(6.3) \qquad a_n(t,\beta_1,\beta_2) := \frac{1-F_n^o(t)}{\sqrt{n}} \left[\, \beta_1 \frac{d_n(t)}{\sqrt{d_n(p)}} + \beta_2 \sqrt{d_n(p)} \,\right],$$

$$(6.4) \qquad d_n(t) := n \sum_{\{1\le i\le n:X_i<t\}} (n-R_i)^{-1}(n-R_i-1)^{-1}\delta_i \,.$$

We note that $n^{-1}(1-F_n^o(t))d_n(t)$ is a consistent estimate of the variance of $1-F_n^o(t)$ in (5.4) (c.f. Kaplan and Meier (1958), Breslow and Crowley (1974) and Nair (1984)). For the left truncated and right censored data we have the following theorem.

Theorem E. Let $a > t_{F^o}$ and $p < T_{F^o}$. Then we have

$$(6.5) \quad \lim_{n\to\infty} P\{\, F_n^o(t)-a_n(t,\beta_1,\beta_2) \le F^o(t) \le F_n^o(t)+a_n(t,\beta_1,\beta_2) \;:\; a\le t\le p \,\}$$
$$= P\{\, |W(s)| \le \beta_1 s+\beta_2 \;:\; d(a)/d(p) \le s \le 1 \,\},$$

where $\beta_1 > 0$, $\beta_1+\beta_2 \ge 0$, $W(.)$ is a Wiener process, and $a_n(t,\beta_1,\beta_2)$ is defined in (6.3).

Remark 6.1. Contrary to (6.2) of Theorem D, the limiting distribution in Theorem E depends on the unknown $d(a)/d(p)$. Consequently, (6.5) cannot be used directly for constructing confidence bands for F^o as in (3.3) - (3.6). To overcome this difficulty, we follow arguments identical to those discussed in Section 4 in terms of Wiener process instead of Brownian bridge process. Let us define

$$(6.6)\quad \begin{aligned} w_1(\delta,a,b,p) &:= \sup\{\ \tau \in (0,\infty) : P\{\ |W(t)|\le\tau+\delta t : a\le t\le b\ \} \le p\ \}, \\ w_2(\tau,a,b,p) &:= \sup\{\ \delta \in (0,\infty) : P\{\ |W(t)|\le\tau+\delta t : a\le t\le b\ \} \le p\ \}. \end{aligned}$$

Now on estimating $d(a)/d(p)$ by $d_n(a)/d_n(p)$, we have

$$(6.7)\qquad d_n(a)/d_n(p) \xrightarrow{\text{a.s.}} d(a)/d(p) \text{ as } n \to \infty.$$

Since Chung, M. Csörgö and Horváth(1986) showed that w_1 and w_2 are monotone strictly continuous functions in δ and τ, respectively, for given δ, a and α, we have

$$(6.8)\qquad w_1(\delta,d_n(a)/d_n(p),1,1-\alpha) \xrightarrow{\text{a.s.}} w_1(\delta,d(a)/d(p),1,1-\alpha),$$

and for given τ, a and α, we have

$$(6.9)\qquad w_2(\tau,d_n(a)/d_n(p),1,1-\alpha) \xrightarrow{\text{a.s.}} w_2(\tau,d(a)/d(p),1,1-\alpha).$$

Hence we have the following theorem:

Theorem F. Let $a > t_{F^o}$ and $p < T_{F^o}$. Then we have, for given α and β_2

$$(6.10)\quad \lim_{n\to\infty} P\{\ F_n^o(t)-a_n(t,\beta_{1,n},\beta_2)\le F^o(t)\le F_n^o(t)+a_n(t,\beta_{1,n},\beta_2) : a\le t\le p\ \} = 1 - \alpha,$$

and for given α and β_1,

$$(6.11)\quad \lim_{n\to\infty} P\{\ F_n^o(t)-a_n(t,\beta_1,\beta_{2,n})\le F^o(t)\le F_n^o(t)+a_n(t,\beta_1,\beta_{2,n}) : a\le t\le p\ \} = 1 - \alpha,$$

where $\beta_1 > 0$, $\beta_2 > 0$, $a_n(t,.,.)$ is defined in (6.3) and,

$$(6.12)\quad \begin{aligned} \beta_{1,n} &:= w_2(\beta_2,d_n(a)/d_n(p),1,1-\alpha), \text{ and} \\ \beta_{2,n} &:= w_1(\beta_1,d_n(a)/d_n(p),1,1-\alpha). \end{aligned}$$

7. PL-quantile process and confidence bands for the quantile function.

Under the model of random censorship from the right, a parallel problem to estimating F^o is that of estimating the quantile function Q of F^o where Q is defined in (1.1).

The counterpart of the PL estimator F_n^o of F^o for Q is the PL-quantile function Q_n defined by

(7.1) $$Q_n(y) := \inf\{t : F_n^o(t) \ge y\}, \quad 0 < y < 1.$$

An analogue of the normed quantile process defined in Csörgö and Révész (1978) for the complete observation, is the PL-normed quantile process ρ_n which is defined as

(7.2) $$\rho_n(y) := \sqrt{n}\, f^o(Q(y))\,(Q(y) - Q_n(y)), \quad 0 < y < 1,$$

where f^o is the derivative of F^o and Q and Q_n are as in (1.1) and (7.1), repectively. For this process, Aly, M. Csörgö and Horváth (1985) have proved a strong approximation theorem and using the theorem we have:

Theorem G. (Chung, M. Csörgö and Horváth (1986)). Assuming that the density function is positive and differentiable on (t_{F^o}, T_{F^o}) as in Theorem A, for $Q(a) > t_{F^o}$, $a < p$ and $Q(p) < T_{F^o}$,

(7.3) $$\lim_{n\to\infty} P\{\, Q_n(t-b_n(t,\tau_1,\tau_2) \le Q(t) \le Q_n(t+b_n(t,\tau_1,\tau_2)) : a\le t\le p \,\}$$
$$= P\{\, |W(s)| \le \tau_1 s+\tau_2 : \frac{d(Q(a))}{d(Q(p))} \le s \le 1 \,\},$$

where $\beta_1 > 0$, $\beta_1+\beta_2 \ge 0$, W(.) is a Wiener process, and

(7.4) $$b_n(t,\beta_1,\beta_2) := \frac{1-t}{\sqrt{n}} \left[\beta_1 \frac{d_n(Q_n(t))}{\sqrt{d_n(Q_n(p))}} + \beta_2\sqrt{d_n(Q_n(p))} \right].$$

In addition, for some $r > 0$ and $(p <)\ p^* \in [0,1]$, if we have

(7.5) $$\sup_{0\le t\le p^*} t\, \frac{|f^{o\prime}(Q(t))|}{f^o(Q(t))^2} \le r .$$

Then, we have

(7.6) $$\lim_{n\to\infty} P\{\, Q_n(t-b_n(t,\beta_1,\beta_2))\le Q(t)\le Q_n(t+b_n(t,\beta_1,\beta_2)) : \varepsilon(n)\le t\le p \,\}$$
$$= P\{\, |W(s)| \le \beta_1 s+\beta_2 : 0\le s\le 1 \,\},$$

where $\sqrt{n}\,\varepsilon(n) \to \infty$ as $n \to \infty$.

Theorem H. Assuming that the density function is positive and differentiable on (t_{F^o}, T_{F^o}) as in Theorem A, for $Q(a) > t_{F^o}$, $a < p$ and $Q(p) < T_{F^o}$, for a given level of significance α,

$$\lim_{n\to\infty} P\{\ Q_n(t-b_n(t,\beta_1,\beta_{2,n}) \le Q(t) \le Q_n(t+b_n(t,\beta_1,\beta_{2,n}) \ :\ a\le t\le p\ \} = 1-\alpha, \tag{7.7}$$

$$\lim_{n\to\infty} P\{\ Q_n(t-b_n(t,\beta_{1,n},\beta_2) \le Q(t) \le Q_n(t+b_n(t,\beta_{1,n},\beta_2) \ :\ a\le t\le p\ \} = 1-\alpha, \tag{7.8}$$

where

$$\beta_{1,n} := w_2(\beta_2, d_n(Q_n(a))/d_n(Q_n(p)), 1, 1-\alpha), \text{ and}$$

$$\beta_{2,n} := w_1(\beta_1, d_n(Q_n(a))/d_n(Q_n(p)), 1, 1-\alpha).$$

Remark 7.1. We note here that the condition (7.5) of Theorem G regarding the tail behavior of density function f^o, is no longer required for Theorem H.

Remark 7.2. When the data are truncated on left-side of distribution, (7.8) is still applicable where a would be the known truncation point.

Remark 7.3. Remark 6.1 is also applied to Theorem H. The only difference is that the limiting distribution in (7.8) depends on the unknown $d(Q(a))/d(Q(p))$ instead of $d(a)/d(p)$ in (6.10). By the exactly identical arguments in Remark 6.1, we can prove Theorem H.

We conclude this section by remarking again that analytic formulae for $P\{\ |W(s)| \le \beta_1 s+\beta_2\ :\ a\le s\le b\ \}$ for any $a \ge 0$ and $b > a$ are discussed in detail by Chung (1986) and a computer program written in FORTRAN 77 has been developed by Chung (1987a).

8. Applications.

The data to be considered here for the confidence bands for truncated data discussed in Section 4, come from GSC Open file 868 (Goodfellow, 1981), which covers a geochemical reconnaissance survey of the Nahanni area, Yukon, Canada (Map - NTS 1051). There are 211 samples of stream waters in the north-eastern quarter of NTS 1051 and all 211 samples were analyzed in order to observe the amount of Mn(manganese) in stream waters.

When the amount of Mn in the samples is below the detection limit of 10 ppm(part per million), then it has been recorded as " < 10ppm ". Among 211 samples, only 27 samples had Mn above the detection limit of 10ppm. The other 187 samples contained less than 10ppm of Mn. The 27 values above 10ppm are:

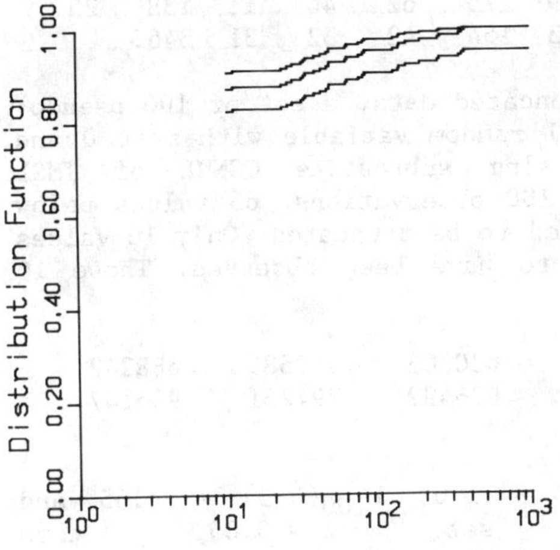

Figure 1. Truncated empirical distribution function and the modified C.-H. 95% confidence band in (4.10) for the distribution function of Mn in stream water, Nahanni area, Yukon, Canada.

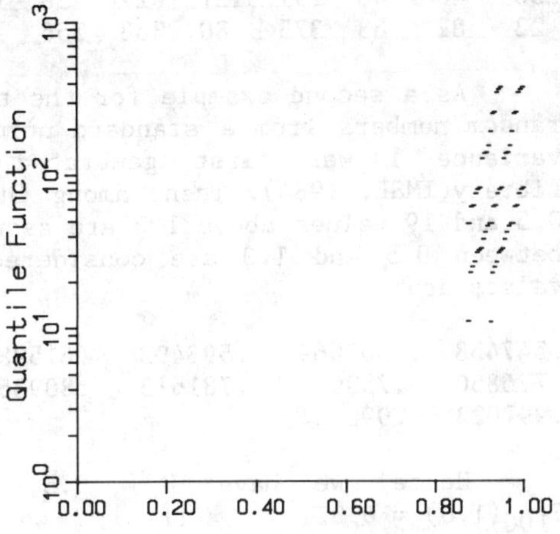

Figure 2. Truncated empirical quantile function and the modified C.-R. 95% confidence band in (4.11) for the quantile function of Mn in stream water, Nahanni area, Yukon, Canada.

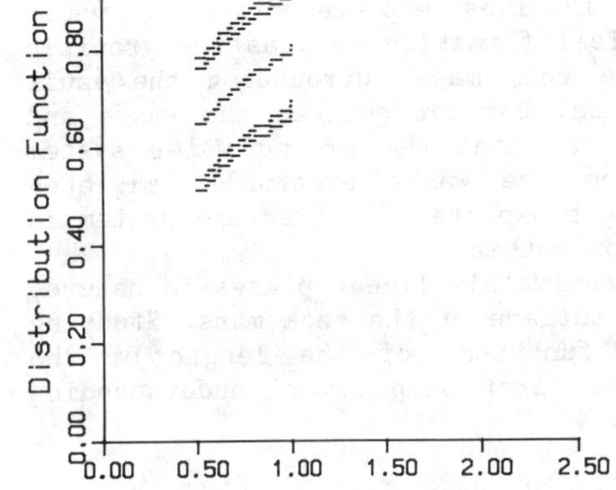

To apply (4.6) in Theorem B, we have $u = 10$, $v = \infty$, and then $F_{211}(10) = 187/211 = 0.8863$ and $\varphi_2(0.8863,1,0.95) = 0.724$ for $\alpha = 0.05$. The latter 0.724 is to be compared with $c = \varphi_2(0,1,0.95) = 1.358$ of the K.-S. band in (2.12). Figure 1 illustrates the truncated empirical distribution function on $[10,\infty)$ defined in (2.1) and the modified confidence band (4.10). Figure 2 illustrates the truncated empirical quantile function and the modified C.-R. confidence band for the quantile function in (4.11).

Figure 3. Truncated empirical distribution function of 100 simulated values from the standard normal distribution N(0,1) which are truncated at both sides of 0.5 and 1.0, with the 95% K.-S. confidence band in (2.12) and the modified K.-S. 95% confidence band in (4.9).

250	48	48	191	127	27	38	139	75	62	46	11	138	25
23	82	63	375	80	363	256	33	154	42	32	31	346.	

As a second example for the truncated data, a set of 100 pseudo-random numbers from a standard normal random variable with mean 0 and variance 1 was first generated using subroutine GGNML of IMSL Library(IMSL, 1984). Then, among the 100 observations, 65 values below 0.5 and 19 values above 1.0 are assumed to be truncated. Only 16 values between 0.5 and 1.0 are considered to have been observed. These 16 values are:

.547458	.562069	.593493	.615984	.620805	.669589	.688242
.726850	.753925	.781613	.809352	.886492	.894751	.946147
.997023	.998144					

Hence, we have $u = 0.5$, $v = 1.0$, $F_{100}(0.5) = 0.65$ and $F_{100}(1.0) = 0.81$. If we set $\alpha = 0.05$, then $\varphi_2(F_{100}(0.5),F_{100}(1.0),0.95) = \varphi_2(0.65,0.81,0.95) = 1.140$, which is to be compared to $\varphi_2(F(0.5),F(1.0),0.95) = \varphi_2(0.6915,0.8413,0.95) = 1.093$. This comparison implies that, in this example, the modified band in (4.6) using 1.140 is wider than the theoretical band in (3.3) using 1.093.

Using the empirical value 1.140 and (4.9), we now construct a 95% confidence band for F on [0.5, 1.0] which is shown in Figure 3 together with the empirical distribution and the K.-S. confidence band in (2.12).

For the applications of the methodology developed in Section 6 and 7, we construct the confidence bands for the distribution function and the quantile function of the length of fractures in the granitic plutons near Lac du Bonnet, Manitoba, Canada for the nuclear waste disposal program in Canada which involves emplacement of a vault containing waste in a stable geological formation such as the granitic plutons in the Canadian Shield. The rock mass surrounding the vault containing the waste acts as a natural barrier between the waste and the biosphere. It is assumed, however, that the ground water system contaminated with radionuclides from the waste eventually migrates through the rock mass and reaches the biosphere. The fracture system in the host rock forms the main migration pathway.

Although the fractures are approximately linear planes in nature, we only observe linear lines on the surface of the rock mass. Studying the distribution and the quantile functions of the length of the fracture lines on the surface is the first step toward understanding the fracture system in the host rock.

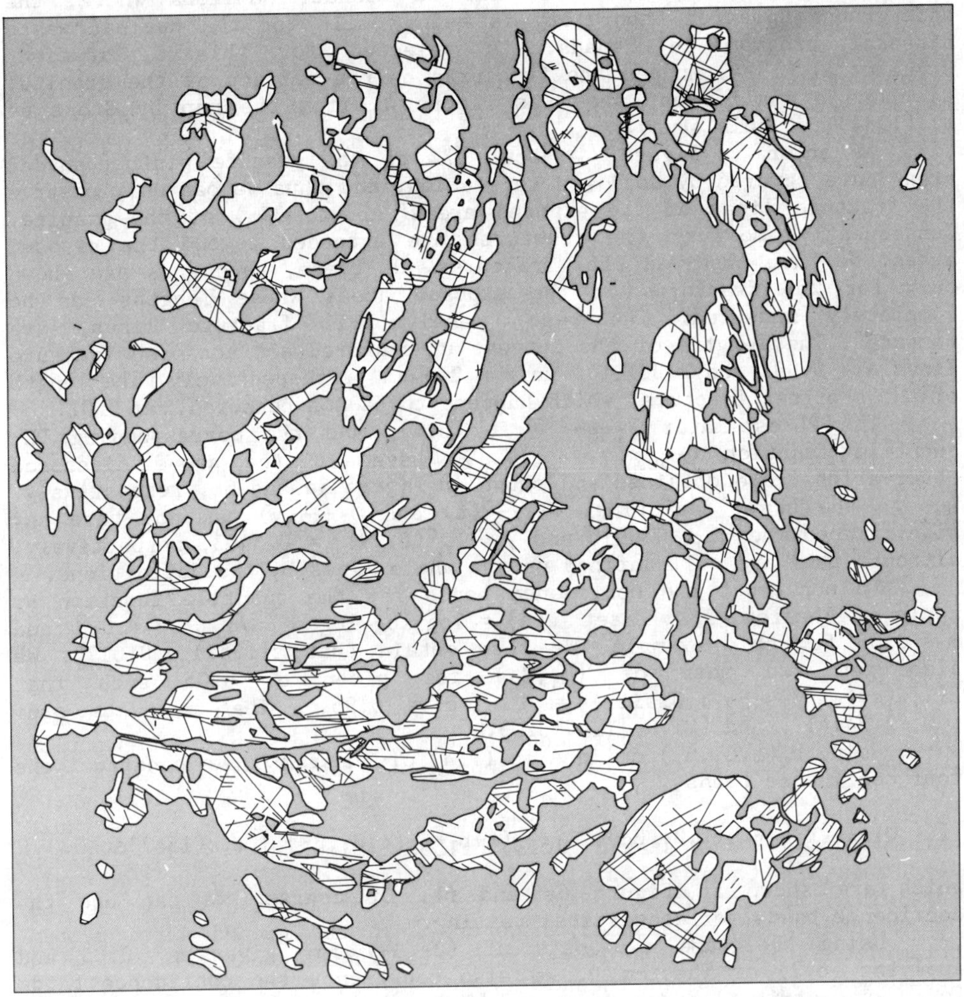

Figure 4. Fracture patterns in a part of Stone et.al.(1984) map of a granitic pluton near Lac du Bonnet, Manitoba, Canada. There are 1567 fractures. Of these, both ends are shown only for 256 fractures in the exposed areas whose lengths can be completely measured. The rest, namely 1311 fractures are assumed to be randomly censored.

The granitic pluton near Lac du Bonnet, Manitoba where, the Underground Research Laboratory is being built for the nuclear waste disposal program in Canada, is selected for this application. Stone et al.(1984) mapped the fractures on the surface of the granite. We chose a small area, shown as Figure 4, from the map by Stone et al.(1984) for this study.

As shown in Figure 4, about 67% of the granite pluton(shaded area) are covered by soil and vegetation, and thus we can only observe the fracture lines on the exposed areas(about 33%) of the granite. Consequently, most of the fractures are extended beyond the exposed areas. We have observed 1567 fractures. Of these, both ends are shown only for 256 fractures in the exposed areas whose lengths can be completely measured. The rest, namely 1311 fracture lines, are censored. The lengths of the longest un-censored and censored fracture lines are 0.991 and 2.361 (1 unit = 2.54 m), respectively. The length of the shortest fracture, which happens to be non-censored, is 0.01.

The PL-estimator F_{1567}^{o} (cf. (5.4)) and the corresponding PL-quantile function Q_{1567} (cf. (7.1)) based on the above discussed observations, are obtained and shown in Figure 5a and 5b, respectively. Due to the heavy censoring, $F_{1567}^{o}(t)$ and $Q_{1567}(y)$ can not have any significance when $t > 0.9910$ and $y > F_n^{o}(0.991) = 0.3403$, respectively, although they are well defined on the whole lines by the definitions.

To construct the confidence bands for the quantile function by applying Theorem G, we set $\varepsilon(n) = \log\log\log n / \sqrt{n}$, $p = 0.3$ and $\alpha = 0.05$ for this example. We then obtain that $\varepsilon(1567) = 0.0175$. We also get two sets of (β_1, β_2) for given $\alpha = 0.05$ such that $P\{ |W(s)| \le \beta_1 s + \beta_2 : 0 \le s \le 1 \} = 1 - \alpha = 0.95$. The two sets are: (0, 2.24141) and (1.0, 1.45132).

Using $b_n(t, \beta_1, \beta_2)$ defined in (7.4) of Theorem G, we construct the four confidence bands,

$$\{ [Q_{1567}(t - b_{1567}(t,\beta_1,\beta_2)),\ Q_{1567}(t + b_{1567}(t,\beta_1,\beta_2))] : \varepsilon(1567) \le t \le 0.3 \}$$

which are shown in Figure 6a and 6b. Of course, we can use the confidence bands for statistical testing.

Using the same two sets of (β_1, β_2) for given $\alpha = 0.05$ and applying (6.2) of Theorem D, we also construnct the confidence bands for the distribution function on [0.0, 0.9] and they are shown in Figure 7a and 7b.

For the examples as applications of (6.11) in Theorem F and (7.7) in Theorem H, we wish to construct the 95% confidence bands for the distribution on [0.5, 0.9] and quantile functions on a fixed interval [0.1, 0.25].

To construct the confidence bands for the quantile function on [0.1, 0.25] for $\alpha = 0.05$, we first obtain that $d_{1567}(Q_{1567}(0.1)) = 0.1333$ and $d_{1567}(Q_{1567}(0.25)) = 0.6777$. To apply (7.7) we also set $\beta_1 = 1$, and then we compute $\beta_{2,1567} = w_1(\beta_1, 0.133/0.678, 1.0, 0.95)$ by using (6.6) (cf. Chung(1986)). The value for $(\beta_1, \beta_{2,1567})$ is (1.0, 1.4509). Applying the value, we obtain

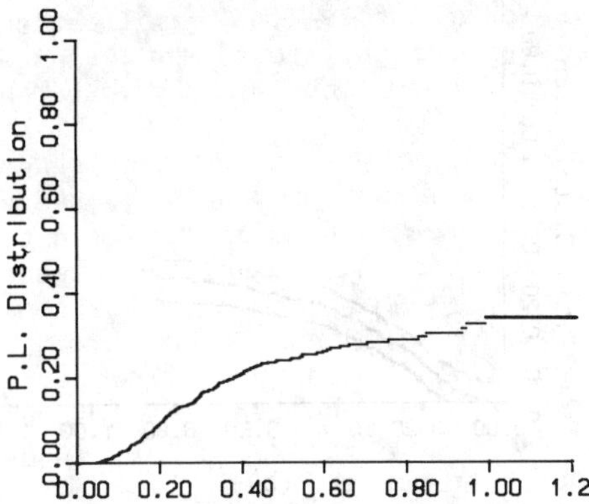

(a)

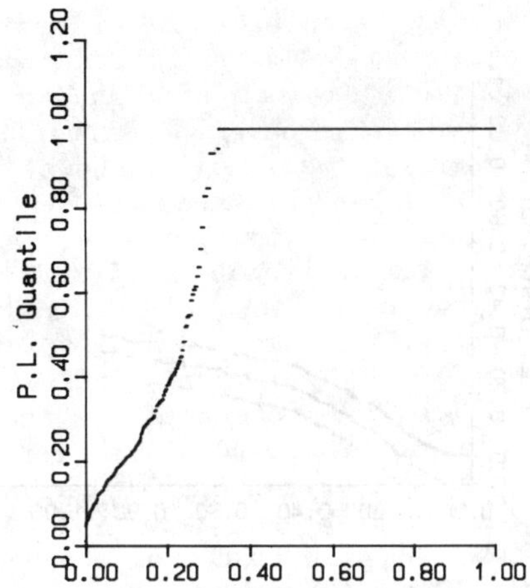

(b)

Figure 5. (a) PL-estimator $F_{1567}{}^{o}$ defined in (5.4) of the distribution function F^{o} of the length of fractures shown in Figure 4. (b) PL-quantile function Q_{1567} defined in (7.1) of the quatile function Q of the length of fractures shown in Figure 4.

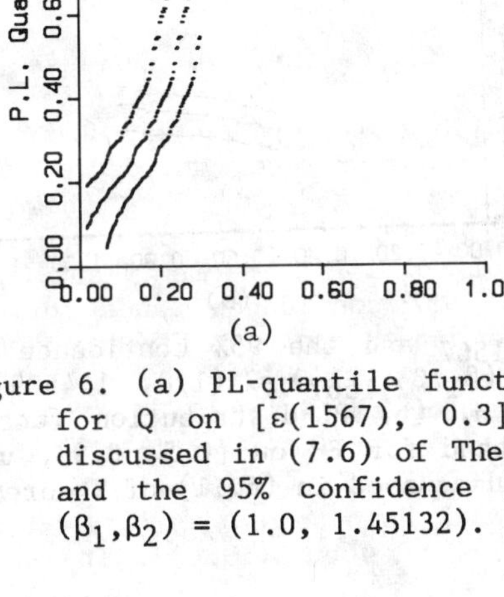

(a)

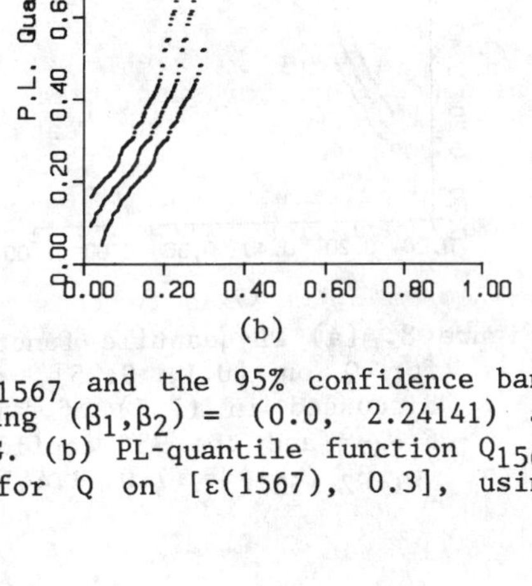

(b)

Figure 6. (a) PL-quantile function Q_{1567} and the 95% confidence band for Q on $[\varepsilon(1567),\ 0.3]$, using $(\beta_1,\beta_2) = (0.0,\ 2.24141)$ as discussed in (7.6) of Theorem G. (b) PL-quantile function Q_{1567} and the 95% confidence band for Q on $[\varepsilon(1567),\ 0.3]$, using $(\beta_1,\beta_2) = (1.0,\ 1.45132)$.

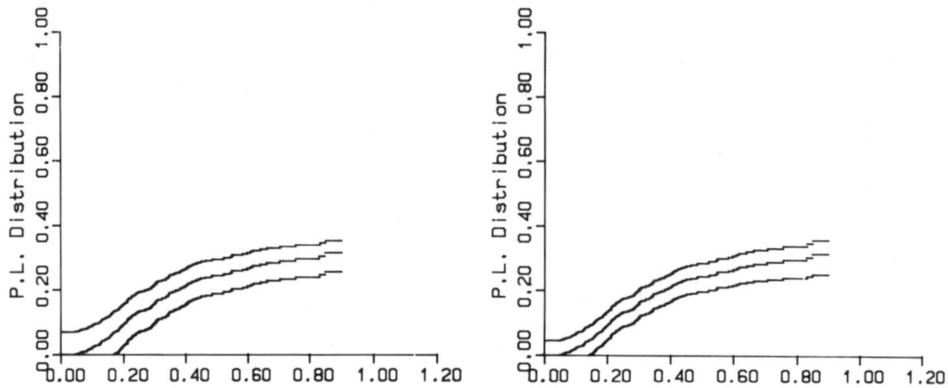

(a) (b)

Figure 7. (a) PL-distribution function $F_{1567}{}^{o}$ and the 95% confidence band for F^{o} on [0.0, 0.9], using (β_1,β_2) = (0.0, 2.24141) as discussed in (6.2) of Theorem D. (b) PL-distribution function $F_{1567}{}^{o}$ and the 95% confidence band for F^{o} on [0.0, 0.9], using (β_1,β_2) = (1.0, 1.45132).

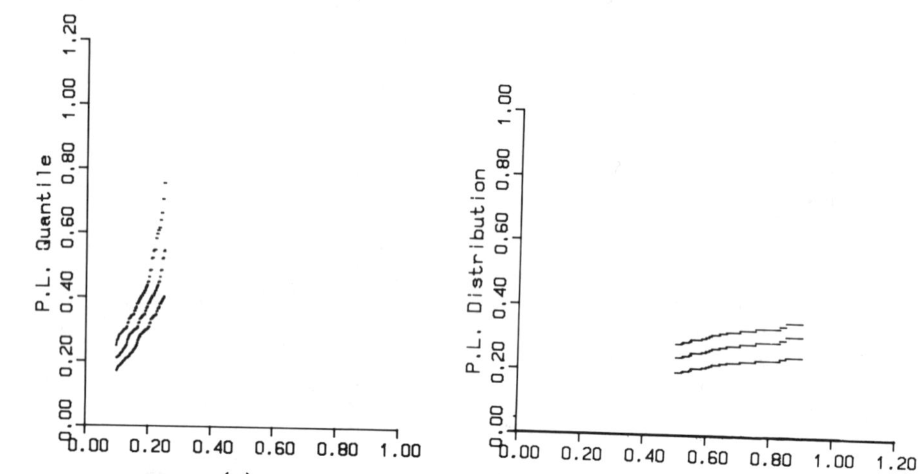

(a) (b)

Figure 8. (a) PL-quantile function Q_{1567} and the 95% confidence band for Q on [0.1, 0.25], using $(\beta_1,\beta_{2,1567})$ = (1.0, 1.4509) as discussed in (7.7) of Theorem H. (b) PL-distribution function $F_{1567}{}^{o}$ and the 95% confidence band for F^{o} on [0.5, 0.9], using $(\beta_1,\beta_{2,1567})$ = (1.0, 1.4415) as discussed in (6.11) of Theorem F.

the confidence band for $Q(t)$ on [0.1, 0.25] which is shown in Figure 8a. It should be noted that the two critical values for Wiener processes for the two intervals (0, 0.3] and [0.1, 0.25] are very similar, since both $P\{ |W(t)| \le \beta_1 t+\beta_2 : 0 \le t \le 1 \}$ and $P\{|W(t)| \le \beta_1 t+\beta_2 : 0.2 \le t \le 1 \}$ are approximately 0.95 for those (β_1, β_2). Consequently, the widths of the confidence bands in Figure 8a is nearly similar to the corresponding confidence bands in Figure 6b.

We now construct the confidence bands for the distribution function on [0.5, 0.9] for $\alpha = 0.05$ as in (6.11), we obtain that $d_{1567}(0.5) = 0.61$ and $d_{1567}(0.9) = 1.70$. To apply (6.11) we again set $\beta_1 = 1$, and we compute $\beta_{2,1567} = w_1(\beta_1, 0.61/1.70, 1.0, 0.95)$ by using (6.6) (cf. Chung(1986)). The value for (β_1, $\beta_{2,1567}$) is (1.0, 1.4415). Applying the value, we obtain two confidence bands for $F^o(t)$ on [0.5, 0.9] and they are shown in Figure 8b. We note that the width of the confidence band in Figure 8b is only slightly narrower than that of Figure 7b, although the interval [0.5, 0.9] for the band in Figure 8b is less than the half of the interval [0.0, 0.9] for the band in Figure 7b. The reason is again that the limiting distribution Wiener process has little random fluctuation near zero compared with near one.

9. Concluding remarks.

FORTRAN 77 computer programs on IBM PC and HP Plotter (or their compatible computers and plotters) were written to construct and plot the confidence bands for the distribution and quantile functions for truncated data and randomly censored data and the source code is published in Computer and Geosciences (Chung (1987c, 1987d) in Special Issue for this NATO ASI).

10. References.

Aly, E.-E., Csörgö, M. and Horváth, L. (1985) Strong Approximations of the quantile process of the product-limit estimator. Journal Multivariate Analysis, V. 16, N. 2, p.185-210.

Anderson, T.W. (1960). A modification of the sequential probability ratio test to reduce the sample size. Annals of Mathematical Statistics, V. 31, p.165-197.

Barr, D.R. and Davidson, T. (1973). A Kolomogorov-Smirnov test for censored samples. Technometrics V. 15, N. 4, p.737-757.

Breslow, N. and Crowley, J. (1974). A large sample study of the life table and product limit estimates under random censorship. Annals of Statistics, V. 2, p.437-453.

Burke, M.D., Csörgö, S. and Horváth, L. (1981). Strong approximations of some biometric estimates under random censorship. Z. Wahrsch. Verw. Gebiete 56, p.87-112.

Burke, M.D., Csörgö, S. and Horváth, L. (1986). An improved approximation rate for the product-limit process. Technical Report No.79, Carleton Univ., p.26-35.

Chung, C.F. (1986). Formulae for probabilities associated with Wiener and Brownian bridge processes. Technical Report No.79, Carleton Univ..

Chung, C.F. (1987a). WIENER PACK - A subroutine package for computing probabilities associated with Wiener and Brownian bridge processes. Geological Survey of Canada Paper 87-xx, in Press.

Chung, C.F. (1987b). An extenstion of the Kolmogorov-Smirnov confidence band for the distribution function on truncated data. To appear.

Chung, C.F. (1987c). FORTRAN 77 program for constructing and plotting confidence bands for the distribution and quantile functions for truncated data. Computer and Geosciences, Special Issue for NATO ASI on "Statistical Treatment for Estimation of Mineral and Energy Resources".

Chung, C.F. (1987d). FORTRAN 77 program for constructing and plotting confidence bands for the distribution and quantile functions for randomly censored data. Computer and Geosciences, Special Issue for NATO ASI on "Statistical Treatment for Estimation of Mineral and Energy Resources".

Chung, C.F., Csörgö, M. and Horváth, L. (1986). Confidence bands for quantile function under random censorship. To appear.

Csörgö, M. (1983). Quantile Processes with Statistical Applications. SIAM CBMS-NSF No. 42, 156p.

Csörgö, M. and Révész, P. (1978). Strong approximations of the quantile process. Annals of Statistics, V. 4, p.882-894.

Csörgö, M. and Révész, P. (1981). Strong Approximations in Probability and Statistics. Academic Press, 284p.

Csörgö, M. and Révész, P. (1984). Two approaches to constructing simultaneous confidence bounds for quantiles. Probability and Mathematical Statistics (Warsaw/Wroclaw) 4, p.221-236.

Csörgö, S. and Horváth L. (1985). Confidence bands from censored samples. Technical Report No. 44, Carleton University, p.39-76.

Donsker, M. (1952). Justification and extension of Doob's hueristic approach to the Kolmogorov-Smirnov theorems. Annals of Mathematical Statistics, V.23, p.277-281.

Doob, J.L. (1949). A heuristic approach to the Kolmogorov-Smirnov theorems. Annals of Mathematical Statistics, V. 20, p.393-403.

Efron, B. (1967). The two-sample problem with censored data. In Proceedings of Fifth Berkeley Symposium in Mathematical Statistics and Probability, V. 4, p.831-853.

Gillespie, M.J. and Fisher, L. (1979). Confidence bands for the Kaplan-Meier survival curve estimate. Annals of Statistics, V. 7, N. 4, p.920-924.

Goodfellow, W.D. (1981). Regional stream sediment and water geochemistry reconnaissance data, Yukon and Northwest Territories; Geolocal Survey of Canada Open file 868, NTS 1051.

Hall, W.J. and Wellner J.A. (1980). Confidence bands for a survival curve from censored data. Biometrika V. 67, N.1, p.133-143.

IMSL (1984). IMSL Library - User's Manual V.1-4.

Kaplan, E.L. and Meier, P. (1958). Nonparametric estimation from incomplete observations. Journal of American Statistical Association, V.53, p.457-481.

Komlós, J., Major, P. and Tusnády, G. (1975). Weak convergence and embedding. In Limit Theorems of Probability Theory (P. Révész ed.), p.149-165.

Koziol, J. A. and Byar, D.P. (1975). Percentage points of the asymptotic distributions of one and two sample K-S statistics for truncated or censored data. Technometrics V. 17, p.507-510.

Nair, V.J. (1984). Confidence bands for survival functions with censored data: A comparative study. Technometrics V. 26, N.3, p.265-275

Stone, D., Kamineni, D.C. and Brown, A. (1984). Geology and fracture characteristics of the Underground Research Laboratory lease near Lac du Bonnet, Manitoba. T.R. 243, Atomic Energy of Canada Ltd Research Co..

GEOSTATISTICAL CRUSTAL ABUNDANCE RESOURCE MODELS

DeVerle P. Harris
Department of Mining and Geological Engineering
The University of Arizona
Tucson, Arizona 85721 USA

ABSTRACT. Geostatistical crustal abundance models of metal resources rest upon the description of metal concentration by a probability relationship. Initially, probability relations were univariate, primarily lognormal and binomial. More recently, probability relations have been bivariate (deposit tonnage and grade) and multivariate (deposit tonnage, grade, depth, and intradeposit grade variance). Recently, developments have been at two very different levels: one is the use of more specific measures of crustal abundance in simple univariate models to estimate mineral endowment at the local or regional level; the other is the use of multivariate crustal abundance models in larger systems to estimate the potential supply of the metal for specified economic circumstances.

CRUSTAL ABUNDANCE MODELING IN GENERAL

Basic Perspective

Crustal abundance refers to the concentration, usually stated in parts per million (ppm) or parts per billion (ppb), of an element or compound in the earth's crustal material, i.e., in the outer 6 - 50 km of rock of the earth. In crustal abundance modeling of the endowment of a region in an element, e.g., copper, or chemical compound, e.g., U_3O_8 or hydrocarbons, *all crustal material is potential ore, and all ore is crustal material, although differing significantly by level of element concentration.*

Motivations for Crustal Abundance Modeling

The appeal of crustal abundance as a foundation for mineral resources estimation appears to be due to the following features, some of which may be relevant only in a limited sense:

- Mineral resources are linked to a very basic and fundamental physical feature of the earth's crust -- the concentration of the element (compound) of interest.

C. F. Chung et al. (eds.), Quantitative Analysis of Mineral and Energy Resources, 459–488.

Crustal abundance methods of appraisal are quantitative and objective, as opposed to the traditional subjective evaluations by geologists.

Crustal abundance transcends the troublesome details of geologic information, thereby avoiding the need for explicit analysis by geologists.

Because of aggregation beyond geological particulars, crustal abundance models permit (allow) modes (types) of deposit occurrence not currently recognized in the region.

Kinds of Crustal Abundance Models

The task of estimating mineral resources and potential supply from crustal abundance is complex and belies the simplicity of the crustal abundance concept. Complexity arises from (1) the use of limited and diverse information on known deposits to identify and estimate a mathematical model that bridges from crustal abundance, the foundation, to the stock of mineral and ore deposits that comprise the resource base, and (2) the consideration of economics in the transformation of unknown endowment[1] to reserves and potential supply.

Three main lines of inquiry to crustal abundance modeling have emerged:

- the relationship across elements of crustal concentration to ore reserves (McKelvey, 1960; Erickson, 1973; Sekine, 1963; Ovchinnikov, 1971) or to ore reserves plus cumulative production (DeNapoli, 1976; Garrett, 1978)

- quantity-quality (tonnage-grade) relationship for a single element or compound (Singer and DeYoung, 1980; Harris, 1984a)

- the probability (geostatistical) modeling of the "parent" distribution of element (compound) concentration (Brinck, 1967; Agterberg and Divi, 1978; Deffeyes and MacGregor, 1980; Harris 1984a, 1984b; Harris and Chavez, 1981, 1984; Garrett, 1986).

The subject matter of this paper is the last of the above, which is referred to generally as *geostatistical crustal abundance (CA) resource models*.

---

[1] Endowment refers to the metal present in occurrences that meet minimum threshold conditions referred to as endowment cutoff criteria, e.g., minimum size of the accumulation and minimum concentration. A quantity of a metal resource is the metal present in those occurrences that constitute endowment and can be produced economically for the specified prices and costs. Potential supply is that part of resources that is also discoverable for the stated economic circumstances.

GEOSTATISTICAL CA RESOURCE MODELS

Basic Structure

The geostatistical approach to crustal abundance modeling of mineral endowment rests upon the use of a mathematical law, usually the lognormal probability density function, of the distribution of element concentration. The idea is clearly grasped by imagining the following situation. The concentration of an element has been determined for each of a large number of rock samples collected in the region of interest; these sample results are summarized by a histogram which shows the quantity of crustal material having grades within each of the specified grade intervals. Suppose that the amount of material represented by each bar of the histogram were divided by the total material in all samples, giving a relative frequency for each bar, and that a mathematical function, f(q), were fitted to these relative frequencies such that $\int_{-\infty}^{\infty} f(q)\,dq = 1.0$. Then by a geostatistical approach to crustal abundance modeling, the mineral endowment, m, of a region would be a function of cutoff grade (concentration) and the amount of crustal material, W, in the region:

$$m_q = (\bar{q}_{q'}) \cdot W \cdot \int_{q'}^{q\,max} f(q)\,dq, \qquad (1)$$

where

$$\bar{q}_{q'} = \int_{q'}^{q\,max} q \cdot f(q)\,dq \Big/ \int_{q'}^{\bar{q}\,max} f(q)\,dq \qquad (2)$$

$\bar{q}_{q'}$ = average grade of that crustal material having grades of q' or higher.

Forms of f(q) that have been used include the lognormal (Agterberg and Divi, 1978; Harris, 1984a,b; Brinck, 1967) and the binomial (Brinck, 1972; DeWijs, 1951, 1953; Garrett, 1986; Harris and Chavez, 1981,1984). Although the mathematical calculations using a CA resource model and estimation of parameters may be complex, these models are simple in principle, as indicated by the foregoing mathematical expressions.

A Brief Review of Some Important Early Developments and Ideas

Brinck's Initial Model - Lognormal. Brinck (1967), drawing upon the classic papers of Ahrens (1953,1954) on the lognormal distribution of element concentration, McKelvey's (1960) examination of reserves and crustal abundances, and the variance-volume relations of DeWijs (1951, 1953) and Matheron (1971), was the first to base an estimate of mineral resources upon a probability model of element concentration. His approach, in its most straightforward application, was demonstrated on Oslo, Norway, for which he had metal concentrations in samples obtained from a geochemical survey. The parameters of the normal distribution of the logarithms of element concentrations were estimated directly from moments computed on the geochemical samples. Brinck then modified this distribution, using the Matheron-DeWijs variance-volume relation, to one that described the distribution of concentration (grade) in "typical" deposit-size accumulations. The quantity of mineralized rock (R) comprising resources of the element within the region was estimated by multiplying the fraction of all deposits of the typical size that have grades above a specified cutoff grade, q' and W, the weight of the crustal material to the specified depth:

$$R = W \cdot \int_{\ln q'}^{\infty} f(\ln q;\mu,\sigma^2)\, d\ln q, \tag{3}$$

where $f(\ln q;\mu,\sigma^2)$ is the lognormal probability density function (pdf) of grade in "typical" deposit-size accumulations

μ,σ^2 are the parameters of the lognormal pdf

μ, as defined by Brinck, is ln(A), where A is crustal concentration.

Brinck described a second approach (here referred to as "stock relations") to estimate the parameters of f when no geochemical data are available. This approach requires (1) a measure of cumulative production plus reserves ($w_{q'}$) and (2) the identification of the cutoff grade (q') of the known inventory (known deposits). By making the assumption that all deposits having grades above the current mining grade (q') have been discovered, the ratio of $w_{q'}/W$ can be interpreted to lie on the distribution function, F( ), of crustal grades:

$$\frac{w_{q'}}{W} = \int_{\ln q'}^{\infty} f(\ln q;\mu,\sigma^2)\, d\ln q = 1 - F\left(\frac{\ln q' - \mu}{\sigma}\right), \tag{4}$$

where μ is assumed known (or reasonably well estimated) and equal to the logarithm of crustal concentration: $\mu = \ln A$

$$F\left(\frac{\ln q' - \mu}{\sigma}\right) = P(\ln Q < \ln q')$$

Since μ is assumed known, σ is calculated directly:

$$\sigma = \frac{\ln q' - \mu}{z'} = \frac{\ln q' - \ln A}{z'}$$

Thus, information on the known stock, given the assumptions, provides estimates of the parameters for the lognormal CA model of endowment described by equation (3). Quantity of mineralized rock (R) for cutoff grades other than q' is easily computed by evaluating equation (3).

Agterberg and Divi's Modification. The theme of this early work by Brinck was employed by Agterberg and Divi (1978) for estimating the distributions of crustal material by grades of copper, lead and zinc in the Canadian Appalachian Region -- see also Agterberg (1980). Unlike Brinck, however, Agterberg and Divi interpreted the available measures of crustal concentration, A, to be the *mean* concentration of a lognormal distribution having parameters μ and σ^2:

$$\text{Given} \quad \ln Q \sim N(\mu, \sigma^2)$$

$$\text{Then,} \quad A = e^{\mu + \sigma^2/2} .$$

Therefore, $\mu = \ln A - \sigma^2/2$, whereas Brinck equated μ to $\ln A$.

Employing $w_{q'}$ separately for each of eight cutoff grades, Agterberg and Divi (1978) made eight estimates of σ, one for each grade:

$$w_{q'}/W = P(\ln Q \geq \ln q') \rightarrow Z$$

and

$$\sigma = z - [z^2 + 2A - 2\ln(q')]^{\frac{1}{2}}, \ z > \sigma$$

$$\sigma = z + [z^2 + 2A - 2\ln(q')]^{\frac{1}{2}}, \ z \leq \sigma$$

Finding estimated σ's to be very similar for the four highest cutoff grades but significantly smaller for the lower ones, the average of the estimates for the four high cutoff grades was taken as the estimate of σ, reasoning that estimates for the lower grades are affected (biased) by incomplete exploration.

An Asymptotic $\hat{\sigma}$. Harris and Chavez (1984), using the 1978 pre-production inventory of \$50.00 U_3O_8 reserves plus production reported by the U.S. Department of Energy (DOE), extended the approach demonstrated by Agterberg and Divi (1978) of estimating σ for each of several cutoff grades to the use of these multiple estimates to estimate an asymptotic value of $\hat{\sigma}$, i.e., the value to which $\hat{\sigma}$ tends as cutoff grade increases. They reasoned that exploration is not really complete for any grade; consequently, even an average of $\hat{\sigma}$'s for high grades is negatively biased. Furthermore, if exploration were progressively less complete for successively lower cutoff grades (q'), $w_{q'}$ also would be progressively less than it should be as q' is decreased. Such an effect in $w_{q'}$ would produce estimates of σ that would be successively smaller as q' is decreased. Because of thorough data acquisition and reporting by DOE, data were available on the magnitude of the pre-production inventory for many cutoff grades, permitting the testing of this hypothesis.

Figure 1 shows the estimated standard deviations associated with each cutoff grade. This figure suggests that as the value of the cutoff grade increases, the associated estimate of σ approaches asymptotically the true value of σ. A mathematical curve that exhibits a pattern similar to that of the estimates of σ is the modified exponential: $\hat{\sigma} = \sigma^* + ab^{q'}$, $a < 0$, $0 < b < 1$. Thus, as q' approaches infinity, $\hat{\sigma}$ approaches σ^*. Figure 1 shows the modified exponential fitted to the standard deviation estimates. The asymptotic value of the equation of this fitted function is 1.437, a value slightly larger than 1.3561, which was obtained by averaging the standard deviations for the high cutoff grades.

Estimates of U_3O_8 endowment using models based upon an averaging of selected estimates of σ and an asymptotic $\hat{\sigma}$ are plotted in Figure 2, as also is the DOE inventory of U_3O_8. Figure 2 shows that the endowment model based upon the asymptotic $\hat{\sigma}$ has the desirable property that for all cutoff grades, estimated U_3O_8 exceeds the known inventory. Inasmuch as exploration for uranium in the U.S. is far from exhaustive, this is a desirable result. Estimates of U_3O_8 endowment by the model based upon the average of estimates of σ for the high cutoff grades do not possess this property: endowment estimates for cutoff grades in excess of 0.175 are less than the known inventory, which is a disturbing result. By this criteria of evaluation, the model based upon the asymptotic $\hat{\sigma}$ is superior to that based upon the average of selected estimates of σ.

Binomial Models. Application of a binomial law for describing mineralized material was first proposed and demonstrated by DeWijs (1951,1953) on the Pulacayo lead vein in Peru. In this model, the grade, q, of a mining block of ore of size $T/2^\beta$ was described by a binomial expression:

$$q_k = \bar{Q}(1 + d)^{\beta-k} (1 - d)^k ,$$

$$k = 0, 1, \ldots, \beta$$

where $\bar{Q}$ = average grade of the deposit
d, β are parameters estimated from the mineralized material
T = total tonnage of the deposit.

The number of mining blocks having grade of q_k is given by C_k^β, and $\sum_{k=0}^{\beta} C_k^\beta$ is the total number of mining blocks in the deposit. It follows that the total tonnage, T, is $\left(\frac{T}{2^\beta}\right) \sum_{k=0}^{\beta} C_k^\beta$ and that $\bar{Q} = \sum_{k=0}^{\beta} q_k C_k^\beta \Big/ \sum_{k=0}^{\beta} C_k^\beta$, or equivalently

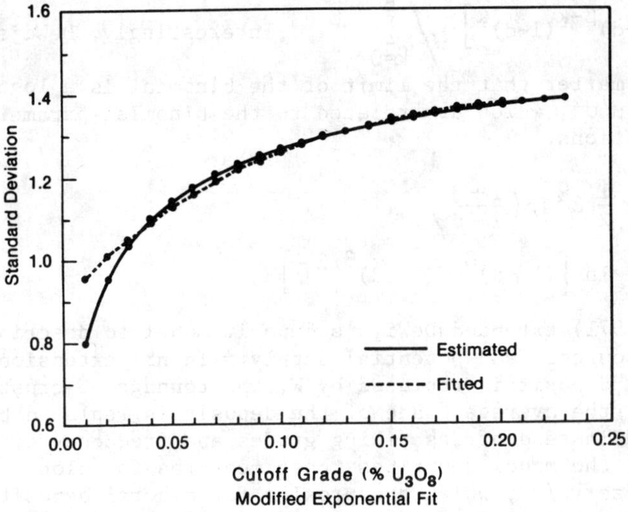

Figure 1. Modified exponential curve fitted to estimated standard deviations of U.S. uranium inventory grade categories (Source: Harris and Chavez, 1981).

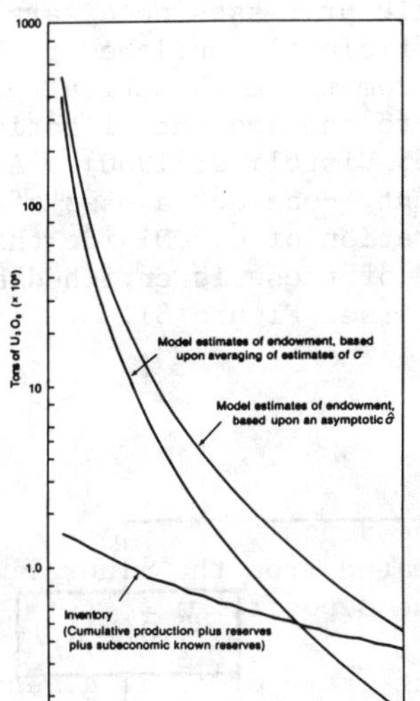

Figure 2. Comparison of estimates by lognormal models of uranium endowment inventory (Source: Harris, 1984a).

$\bar{Q} = \sum_{k=0}^{\beta} C_k^{\beta} \left[(1+d)^{\beta-k}(1-d)^k\right]\bar{Q} \bigg/ \sum_{k=0}^{\beta} C_k^{\beta}$. Interestingly, DeWijs showed as an incidental matter that the limit of the binomial is a lognormal, having parameters of μ, σ^2, which are related to the binomial parameters by the following equations:

$$\sigma = \frac{1}{2}\beta^{\frac{1}{2}} \ln\left(\frac{1+d}{1-d}\right)$$

$$\mu = \ln\left[(1+d)^{\beta/2}(1-d)^{\beta/2}\bar{Q}\right]$$

Brinck (1971) extended DeWijs's deposit model to describe mineral endowment, resources, and potential supply. In his extension, the tonnage of the deposit is replaced by W, the tonnage of crustal material in the region; the average grade of the deposit is replaced by crustal abundance; and instead of describing grades and frequency of grades of mining blocks, the model describes these measures for blocks of crustal material of size $W/2^{\beta}$, which are proxies for mineral deposits. Of course, such an extension is accompanied by an appropriate procedure for the estimation of parameters β and d, $\bar{Q}$ being taken as an existing measurement of crustal abundance. In practice these are estimated from geochemical samples (Brinck, 1967; Chavez, 1983) or from the inventory of mined material plus ore reserves (Harris, 1984a,b).

Use of the binomial in the generation of mineral endowment has a special appeal because it can be viewed as a mathematical analogue of the sequential operation of the geologic processes necessary to create accumulations of material that are sufficiently enriched to be economically exploited. Quite apart from the mathematical perspective, geologists attribute grade enrichment to the sequential action of earth processes (Singer, Cox, and Drew, 1975; Vistelius, 1960). As a mathematical analogue of such enrichment, consider a mass of crust having weight W and average concentration of $\bar{Q}$. Divide this mass into two blocks, each of size W/2; one of these is enriched by factor d_1 by depleting the other by factor d_1 (see Figure 3).

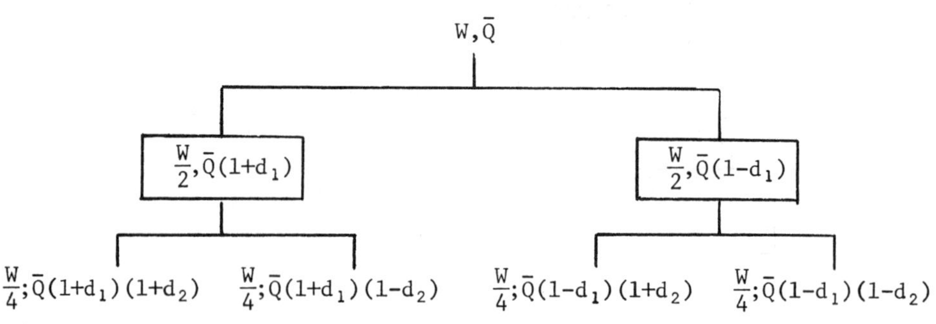

Figure 3. Schematic Representation of Grade Depletion and Enrichment.

Suppose that each of these two blocks is subjected to a second enrichment process, which results in each of these two blocks being similarly divided into two blocks and enriched and depleted by factor d_2 of the second process, giving four blocks, each of size $W/4$ and each having a different grade, ranging from the high grade of $\overline{Q}(1+d_1)(1+d_2)$ to the low grade of $\overline{Q}(1-d_1)(1-d_2)$. Continuing this process of dividing β times results in a population of deposits (blocks) each of the size $W/2^{\beta}$ but each having a different grade.

In the absence of information on the d_i, $i = 1, 2, \ldots, \beta$, the simplifying assumption is made that $d_1 = d_2 = \ldots = d_{\beta}$. Letting d represent an overall average degree of depletion and enrichment permits the description of this analogue model by the foregoing binomial expressions. This is the approach demonstrated by Brinck (1971).

An Important Change in Model Structure and Parameter Estimation -- Drew

Drew (1977), striving to improve the usefulness of crustal abundance models in describing uranium resources and in the examination of the potential to meet future uranium demands, made the following extensions in crustal abundance modeling:

Replaced the univariate lognormal model of concentration (grade) by a *bivariate* model of deposit size and grade.

Obviated the use of $w_{q'}$, the weight of the inventory, and W, the weight of the crust in estimating unknown parameters in the bivariate model by adopting a more conventional statistical approach in which statistical data on deposit size represent a biased and truncated sample.

Identified (1) the need for a biased bivariate distribution to describe discoverable deposits and (2) the relationship between unbiased and biased distributions.

Identified a relationship of cost to deposit size and grade; this relationship served to truncate the population of discoverable deposits, excluding those having sizes and grades so low that production cost is higher than price.

Man's experience is that a mineral occurrence in nature exhibits and is defined by *both* size and grade, and that both of these show great variation in nature. Consequently, a population of deposits described by a joint (bivariate) distribution of t and q should in some regards be a vast improvement over univariate models which express only element concentration.

Replacing the use of $w_{q'}$ for parameter estimation by a statistical procedure requires considerably more structure in the crustal abundance model. Much of this structure reflects the introduction of exploration performance and mining cost. Since statistical data on deposit size and grade exist primarily because of economic activities, introducing economics is unavoidable; it is ever present in exploration experience, and its influence is present in basic deposit data. Thus, as in research in general, efforts to resolve one problem or issue identifies or highlights a new one. Drew's extension to a bivariate model is no exception to the rule. In this case, the new problem (issue) is correlation between deposit size and grade. Does it exist? If so, how should it be modeled and estimated?

The assumption made by Drew (1977) of statistical independence seems reasonable as a first approximation, for it is supported by statistical studies of some kinds of deposits -- see Singer et al. (1975) and Harris (1984a). However, independence of tonnage and grade in a crustal abundance model, which by definition represents all modes of occurrence of an element or compound, may be far less innocuous and less justifiable than it seems on the surface. Simple logic and the persuasion of experience cast doubt on the existence of independence over the *entire range of element concentration*. Some of these issues are reconsidered in a later section that deals with a 4-dimensional CA model, one that is estimated by statistical analysis, avoiding the use of stocks, an idea that was first demonstrated by Drew.

RECENT DEVELOPMENTS

Overview

Developments over recent years have taken place at two very different levels. One of these, which is represented by the works of Garrett (1986) and Ruzicka (1977), is towards greater specificity of crustal abundance measures with the objective of employing them in one or more ways to make estimates of mineral resources at the local or regional levels. The other level retains the national or global feature that has characterized many CA models, but seeks to increase the dimensions of the models to include the description of the deposits by those physical features that affect production cost, exploration cost, and discovery efficiency. This latter level embeds the CA model, both for parameter estimation and for later inference, within a system that depicts the economic relations that over time have interacted with mineral endowment to produce known discoveries. In this way the biases that economics imparts to basic data on known deposits can be diminished.

Local Models for Resource Appraisal - Garrett (1986)

Garrett (1986) reports on investigations by the Geological Survey of Canada to generate local models, as contrasted to the usually large scale, sometime global, CA models, for the appraisal of mineral resources. Everything else being equal, there has been less interest in the application of CA models on small areas. One reason for this is that a large region, e.g., continent, is likely to be a fairly representative sample of the earth's crust and may, therefore, contain geologic features of many different geological environments and metallogenic provinces. Consequently, available measures of crustal abundance may be used with reasonable assurance that they are acceptable as estimates of the crustal abundance of the region to be appraised.

Use of a local CA model for resource appraisal requires information that is more specific to the area to be appraised. Consider, for example, the measures of crustal abundance ($\bar{Q}$) and the mineralizability coefficient of Brinck's model (d): it is possible for a small area that $\bar{Q}$ and d can be estimated directly from geochemical samples. Or, if such samples are not available on the area, $\bar{Q}$ may be estimated from detailed information on lithologies of the area and the geochemical values that characterize them in other regions exhibiting the same geologic environment. By weighting these abundances by the volumes of the various lithologies, a weighted average value for $\bar{Q}$ may be calculated that is a good estimate of crustal abundance for the area of interest.

When the area is known to be part of an already recognized metallogenic province, information on the most frequent or typical mineral occurrences of a training area may be useful in estimating other parameters of the binomial CA model. Suppose, for example, that the training area and the appraisal areas are parts of a metallogenic province for uranium deposits in which the principal ore mineral is pitchblende. The uranium content of pure pitchblende is about 840,000 ppb. Suppose also that the crustal abundance of uranium in the area has been estimated to be 2,500 ppb. Following Garrett (1986), the 840,000 ppb could be selected as a barrier (maximum) grade. From the binomial equation, it can be seen that the maximum grade occurs when the value of k is zero:

$$q = \bar{Q}(1+d)^{\beta-k}(1-d)^{k}$$

$$q_{max} = \bar{Q}(1+d)^{\beta-0}(1-d)^{0} = \bar{Q}(1+d)^{\beta}$$

Using the 840,000 for q_{max} and 2,500 for $\bar{Q}$ we can write the following:

$$\ln\left(\frac{840{,}000}{2{,}500}\right) = \beta\ln(1+d)$$

$$5.8171 = \beta\ln(1+d)$$

Consideration of the barrier grade and crustal abundance reduces parameter estimation to satisfying this equation. Even so, without additional information there still are many acceptable pairs of d and β. This set can be reduced through the use of information on the analogous training area. One kind of such information is deposit size, for a feature of the binomial law is that t, deposit size, is determined by size of region, W, and the binomial order, β:

$$t = \frac{W}{2^{\beta}}$$

Suppose, for example, that most of the deposits fall within the size range of 4×10^{5} to 4×10^{6} tonnes and that the size of W on the training region is 10^{15} tonnes. Using this information and the above relation, a restricted set of possible values of β and d can be generated:

β	d
28	0.2309
29	0.2221
30	0.2140
31	0.2064
32	0.1994

Other, very useful information is the quantity of cumulative production plus known reserves and resources and the associated cutoff grade for the training region.

Extrapolation from a training area to an appraisal area for which the crustal abundance varies significantly from that of the training area presents a problem, for to preserve the barrier grade, either d or β must be changed. Garrett (1986) found that it is better to preserve the value of β estimated on the training region and allow d to vary as necessary to preserve the barrier grade.

Another use of a local binomial model is to estimate the magnitude of resources on a well-explored area at grades lower than cutoff grades of the present or past. Garrett (1986) provides an example of such an application for the Klondike alluvial gold fields. Through the analysis of information from various sources, the area was described as follows:

$W = 1.517 \times 10^{12}$ tonnes
$\bar{Q} = 5$ ppb
$q' = 0.5$ ppm $= 500$ ppb
maximum fineness = 880
$m_{500} = T_{500} \cdot \bar{q}_{500} = 311$ tonnes

Using the above information, Garrett (1986) estimated β to be 131 and d to be 0.156. This model describes 3.45×10^9 tonnes of material having grades in the interval of 100 ppb to 500 ppb, with an average grade of 250 ppb, giving 851 tonnes of gold.

The Binomial as Part of a Uranium Deposit-Generating Model

Deposit Morphology and Economics. The generalization of DeWijs's binomial model of mining blocks of a deposit to deposits of the earth's crust runs into difficulty when the objective is a credible modeling of the economics of exploration and mine development, because the population of deposits (crustal blocks) for a given β are all of the same size and exhibit no grade structure within them; i.e., all block material is of one grade, the grade of the block. This feature is of no consequence in the modeling of mining blocks of a deposit, for man controls the size of mining blocks and often does use a uniform size. However, prices for metals justify mining of only those accumulations within a deposit that have been sufficiently enriched. *The earth processes that cause this enrichment do not produce deposits of equal size, nor do they generally produce deposits of uniform grade.* Thus, if such properties are important in analysis, as they are in the estimation of economic resources and potential supply, the binomial model alone is not sufficient as a deposit generating model.

Extension of the Binomial Model to Deposit Generation. In spite of some difficulties presented by the binomial model, it was retained by Harris and Chavez (1981,1984) as a part of their uranium endowment model. This decision reflects the desire to preserve the generality of a crustal abundance model and the facility of the binomial in subdividing the crust of a large region, e.g., U.S., into smaller crustal blocks. Figure 4 shows schematically how the binomial was used to (1) partition the continental U.S. into basin-sized crustal blocks, and (2) to partition each basin-sized block into a population of smaller blocks referred to as geologic deposits. Exploration targets and ore deposits were generated by subdividing the geologic deposit into mining blocks and then clustering the blocks to produce exploration targets.

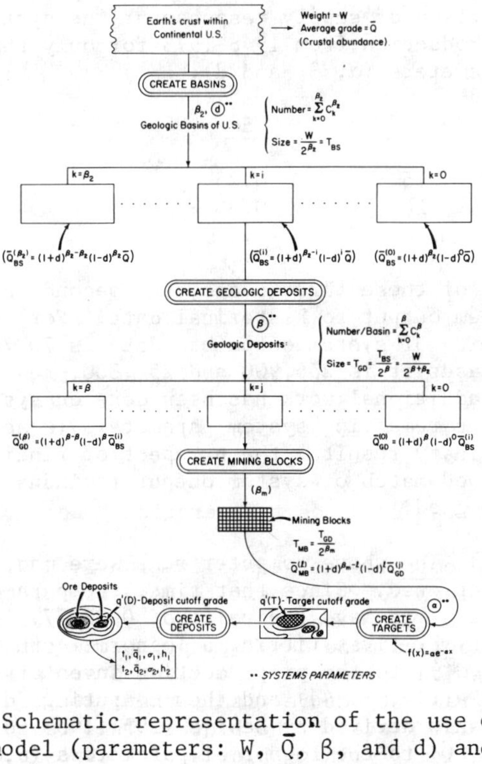

Figure 4. Schematic representation of the use of a binomial crustal abundance model (parameters: W, $\bar{Q}$, β, and d) and a clustering function (parameter: α) to create exploration targets and ore deposits that are functions of cutoff grade (Source: Harris and Chavez, 1984).

Each target was further seeded with subtargets, which served as ore deposits.

Characteristics of the deposits that were thus created obviously reflect (1) the size of the geologic deposit, (2) the concentration of uranium within the geologic deposit, and (3) the clustering parameter, α. Given d and $\bar{Q}$, both the size and concentration of the geologic deposit are controlled by β.

Parameter Estimation by System Search. Before this deposit generating model can be employed, parameters $\bar{Q}$, d, β, and α must be estimated. In the study by Harris and Chavez (1981), the crustal abundance of 3 ppm was taken as $\bar{Q}$ leaving d, β and α undefined. The approach taken by Harris and Chavez (1981) was to construct the potential supply system so that it imitates reasonably well the economics and efficiencies of exploration, development, and production and then use the system to search for those values of d, α and β that yield the best match of system output with historical features of the domestic uranium industry. Figure 5 illustrates schematically this approach.

A Preliminary Demonstration. A useful and definitive demonstration is not possible at this time because the system is not complete. All that can be offered is a comparison of supply response of the system to the time series of industry production from 1948-1978 for only three cases of the unknown system parameters (α, β, and d):

α	β	d
0.02	15	0.25
0.01	16	0.30
0.05	14	0.21

As indicated by Figure 6, of these three cases, the second one provides the closest match of system output to historical data. For this case, the measure of nonconformity of system output to data is 70,760, while for the first and third cases it is 256,900 and 252,300, respectively. Due to lack of funds, no additional work has been done on system design and on the estimation of the remaining system parameters (α and β). On the basis of these preliminary results, the prospect of finding values for α and β that give a good match of system output to industry performance appears quite good.

Independent Estimate of d. The above computer runs were made some time ago, before funding was depleted. Since that time, the parameter d has been independently estimated to have the value of 0.22337. Parameter d was estimated indirectly by first fitting a lognormal crustal abundance model [$\ln q \backsim N(\mu,\sigma^2)$] to the preproduction inventory of uranium reported by DOE (Harris, 1984a) and then computing d from the estimated σ by a relationship derived by DeWijs (1951,1953) of the binomial parameters (β, d, $\bar{Q}$) to the lognormal parameters (μ,σ^2):

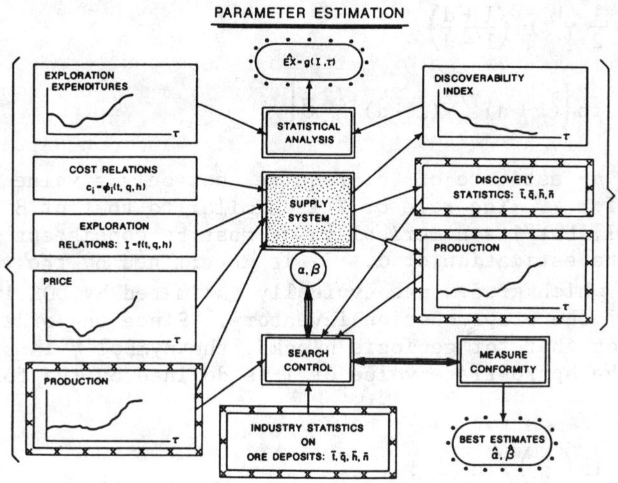

Figure 5. Stylized use of supply system and time series data to estimate parameters (α,β) of the system - asterisks denote those items that either control the search or are products of the search.

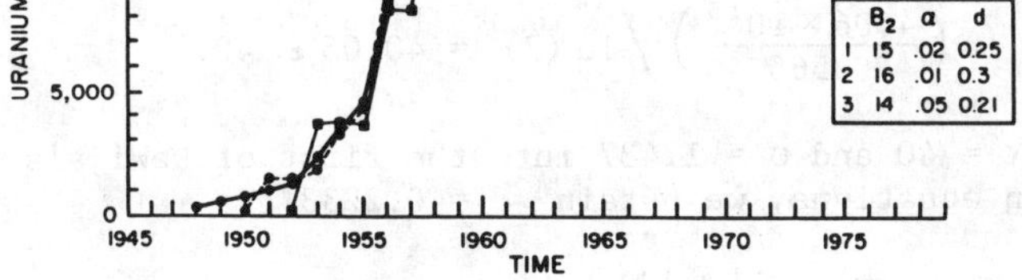

Figure 6. Actual and estimated U.S. uranium production, 1948-1978 (Source: Chavez, 1983).

$$\sigma = \frac{1}{2}\,\gamma^{\frac{1}{2}}\,\ln\left(\frac{1+d}{1-d}\right)$$

$$\mu = \ln\left[(1+d)^{\gamma/2}(1-d)^{\gamma/2}\,\bar{Q}\right]$$

Clearly, an estimate of d requires not only a value for σ, but also a value for γ. The role of γ is similar to that of β in the binomial model, but its numerical value must be consistent with the data employed in the estimation of α. γ is determined by the volume of material upon which grades are typically estimated by DOE in the description of the preproduction inventory. Since the order of subdivision is not that for geologic blocks, the symbol γ is used to avoid confusion. The appropriate value of γ is defined by the following relationship:

$$\gamma = \ln\left(\frac{W}{S}\right)\Big/\ln(2),$$

where W is the weight of the environment (crust of the continental U.S. to a depth of 600 feet);
S is the weight of the sample.

The average area of influence ascribed to a gamma log or grade determination ranges from 2500 to 10,000 square feet, with more frequency attached to the high end of this range (Maurice Hansen, US DOE, July 1983, personal communication); consequently, 8100 ft^2 was selected as a representative value. Taking an average thickness of deposit to be 6 feet and converting this volume to weight of typical uranium bearing rock gives S = 3565 m.t. While a few uranium deposits in the San Juan Basin occur to depths of up to 3000 feet, the bulk of preproduction inventory derives from deposits within the first 600 feet of depth. Using 600 feet as the appropriate depth, surface area to be 7.82×10^6 m^2, and the average density of rock to be 2.5 grams/cm^2 gives a value of 4.06×10^{15} m.t. for W. These values for W and S produce a value of approximately 40 for γ:

$$\gamma = \ln\left(\frac{4.06 \times 10^{15}}{3565}\right)\Big/\ln(2) = 40.05 \approx 40.$$

Substituting $\gamma = 40$ and $\sigma = 1.437$ into the first of DeWijs's transformation equations, we obtain $d = 0.22337$:

$$\sigma = 0.5(\gamma)^{\frac{1}{2}}\,\ln\left(\frac{1+d}{1-d}\right)$$

$$1.437 = (0.5)(40)^{\frac{1}{2}}\,\ln\left(\frac{1+d}{1-d}\right)$$

$$d \approx 0.22337$$

A Four-Dimensional Comprehensive Crustal Abundance Model for U.S. Uranium

Motivation. The comprehensive model here described was motivated by two general considerations: (1) the desire to avoid the use of stocks to estimate model parameters and (2) the need to describe mineral occurrences by those intrinsic features that affect exploration and production.

The importance of depth (h) as a cost determinant and intradeposit grade variance (v) in mining economics strongly indicates that a four-dimensional (q,t,h,v) pdf is preferred to one that is bivariate (t,q). Of course, if concerns about correlation are accommodated, this expanded model must be generalized to a *dependent four-dimensional* model. Finally, accounting for the translation[1] effect requires two transformation functions, one for ore grade and one for ore tonnage, with communication between them.

The finding by Harris and Chavez (1981) of a negative bias in the estimate of σ invites further critical examination of the ratio $w_{q'}/W$ (ratio of cumulative production plus reserves to total crustal material of the region) as a probability from the crustal abundance model, a procedure demonstrated by Brinck (1967). The suggestion was made in a previous section that the negative bias results from considering $w_{q'}$ to be complete, i.e., all deposits having grades of at least q' have been found. It is noteworthy that $w_{q'}$ is only the numerator of a ratio and if incomplete data on $w_{q'}$ can lead to bias in an estimate of σ, so also can an inappropriate specification of W, which is a function of area (A), depth (h), and weight factor (f): $W = A \times h/f$. The solution to these problems that is presented here is to avoid them entirely by following the example of Drew (1977) and employing a statistical approach, rather than reference stocks, to parameter estimation.

Theory. Consider mineral occurrence to be four-dimensional: t, quantity of mineralized rock; q, average grade (%) of t; v, the variance of grade within t (intradeposit grade variance); and h, depth (ft) within the earth's crust to the occurrence. Then, the foundation for a comprehensive crustal abundance model is the joint pdf, w( ), for t, q, h and v:

$$w(\ln t, \ln q, \ln v, h)$$

Following Drew (1977), the preferential discovery by exploration performance of the largest and richest (highest q) deposits is represented by purposefully biasing w( ) to form a pdf for discoverable deposits -- see Charles River Associates, Inc. (1978).[2] However, in

[1]This refers to the effect of profit maximization on cutoff grade, average ore grade, and ore tonnage; specifically, the orebody that is mined is of lower tonnage and higher average grade than the original deposit.

[2]The author originally developed some of these ideas while a consultant to CRA on a research contract for the National Science Foundation.

this case, rather than discovery being proportional to $t \cdot q$, the proportionality is to $q^{0.48} \cdot t^{0.80}$. Of course, unlike Drew's model, this one contains depth to deposit, h, for h is a strong determinant of discoverability. This depth effect is represented by a weighting function which weights probabilities for occurrence inversely to depth: $0.00175e^{-0.00175h}$. The parameter (0.00175) was derived from the exponent of h in the discoverability function:

$q^{0.48}t^{0.8}h^{-0.64}$ = Index of discoverability.

The pdf for discoverable deposits, $w^*(\ )$, is a biased form of $w(\ )$, as indicated by the following proportionality relationship:

$$w^*(\ln t, \ln q, \ln v, h) \propto (q^{0.48}t^{0.8}) \cdot w(\ln t, \ln q, \ln v, h) \cdot (0.00175e^{-0.00175h})$$

Or, since h is taken to be independent of lnt and lnq, we can write the following:

$$w^*(\ln t, \ln q, \ln v, h) \propto (q^{0.48}t^{0.8}) \cdot f(\ln t, \ln q, \ln v) \cdot \alpha(h) \cdot (0.00175e^{-0.00175h})$$

Specifically,

$$w^*(\ln t, \ln q, \ln v, h) = f^*(\ln t, \ln q) \cdot y_t(\ln v) \cdot \alpha^*(h),$$

where

$$f^*(\ln t, \ln q) = \left[\frac{1}{(2\pi)\sigma_t \cdot \sigma_q\sqrt{1-\rho^2}}\right] \cdot \exp\left\{\left[-\frac{1}{2(1-\rho^2)}\right] \cdot \left[\frac{\left[\ln t - (\mu_t + \beta_1\sigma_t^2)\right]^2}{\sigma_t^2} - \frac{2\rho\left[\ln t - (\mu_t + \beta_1\sigma_t^2)\right] \cdot \left[\ln q - \mu_q + \beta_2\sigma_q^2\right]}{\sigma_t \cdot \sigma_q} + \frac{\left[\ln q - (\mu_q + \beta_2\sigma_q^2)\right]^2}{\sigma_q^2}\right]\right\}$$

$$y_t(\ln v) = \left[\frac{1}{\sigma_v\sqrt{2\pi(1-\rho^2)}}\right] \cdot \exp\left\{\left[-\frac{1}{2\sigma_v^2(1-\rho^2)}\right] \cdot \left[\ln v - \mu_v - \rho \cdot \sigma_v\left(\frac{\ln t - \mu_t}{\sigma_t}\right)\right]^2\right\}$$

$$\alpha^*(h) \propto (h) \cdot (0.00175e^{-0.00175h})$$

where $\alpha(h) = 1/\text{DEPTH}$, $0 < h \le \text{DEPTH}$, DEPTH is maximum depth to deposit.

Specifically,

$$\alpha^*(h) = \alpha(h) \cdot (0.00175e^{-0.00175h}) \Big/ \int_0^{\text{DEPTH}} \alpha(h) \cdot$$

$$\cdot (0.00175e^{-0.00175h})\, dh$$

Thus,

$$\alpha^*(h) = (0.00175e^{-0.00175h}) \Big/ (1.0 - e^{-0.00175\,\text{DEPTH}}),$$

$$0 < h \le \text{DEPTH}$$
$$= 0, \text{ otherwise.}$$

The pdf $f^*(\ )$ is related to $f(\ )$ as follows:

$$f^*(\ln t, \ln q) \propto (q^{\beta_1}) \cdot (t^{\beta_2}) \cdot f(\ln t, \ln q)$$

Since $f(\ )$ is a dependent bivariate normal pdf, $f^*(\ )$ is $f(\ )$ with a shift in its location parameters, i.e., μ_t and μ_q. Specifically, the parameters of $f^*(\ )$ and $f(\ )$ are related as follows:

$$\mu_q^* = \mu_q + \beta_1 \sigma_q^2$$

$$\mu_t^* = \mu_t + \beta_2 \sigma_q^2$$

$$\sigma_q^* = \sigma_q$$

$$\sigma_t^* = \sigma_t$$

As noted previously, $\beta_1 = 0.48$ and $\beta_2 = 0.80$.

Define two transformation functions that describe ore tonnage ($\tilde{t}$) and grade ($\tilde{q}$) as functions of t, q, v, h and price (p):

$$\ln\tilde{q} = \phi_q(\ln t, \ln q, \ln v, h, p)$$
$$\ln\tilde{t} = \gamma(\ln t, \ln q, \ln v, \ln\tilde{q}, h, p)$$

By substitution, lnq can be purged from $\Upsilon(\ )$, defining a new function $\phi_t(\ )$:

$$\ln\tilde{t} = \phi_t(\ln t, \ln q, \ln v, h, p)$$

Given $\phi_t(\ )$ and $\phi_q(\ )$, lnt and lnq can be described as inverse functions $\phi_t^{-1}(\ )$ and $\phi_q^{-1}(\ )$:

$$\ln t = \phi_t^{-1}(\ln\tilde{t}, \ln q, \ln v, h, p)$$

$$\ln q = \phi_q^{-1}(\ln t, \ln\tilde{q}, \ln v, h, p)$$

Employing the inverse functions of $\phi(\ )$ and transformation of variables, $\psi(\ln\tilde{t}, \ln\tilde{q}, h; p)$, the joint pdf for ore tonnage, ore grade, and ore depth, given p, is derived from the basic model for discoverable deposits:

$$\psi(\ln\tilde{t}, \ln\tilde{q}, h; p) = \int_Q \int_T \int_V w^* \Big[\phi_t^{-1}(\ln\tilde{t}, \ln q, \ln v, h, p),$$

$$\phi_q^{-1}(\ln t, \ln\tilde{q}, \ln v, h, p),$$

$$\ln v, h\Big] \, |J| \, dv \, d\ln t \, d\ln q$$

Now, define function $\lambda(\ln\tilde{t}, \ln\tilde{q}, h)$ which describes total unit cost per lb U_3O_8 (c) as a function of ore tonnage and grade and depth to deposit:

$$c = \lambda(\ln\tilde{t}, \ln\tilde{q}, h)$$

Then, combining $\psi(\ )$ and the inverse functions of $\lambda(\ )$, given c = p, four model expectations can be described.

$$E[\tilde{Q}|p] = \int_{\tilde{Q}} \int_{\lambda_t^{-1}(p, \ln\tilde{q}, h)}^{\infty} \int_{0}^{\lambda_h^{-1}(\ln\tilde{q}, \ln\tilde{t}, p)} \tilde{q} \cdot$$

$$\cdot \psi(\ln\tilde{t}, \ln\tilde{q}, h; p) \, dh \, d\ln\tilde{t} \, d\ln\tilde{q} / K$$

$$E[\tilde{T}|p] = \int_{\tilde{T}} \int_{\lambda_q^{-1}(\ln\tilde{t},p,h)}^{\infty} \int_{0}^{\lambda_h^{-1}(\ln\tilde{q},\ln\tilde{t},p)} \tilde{t} \cdot \psi(\ln\tilde{t},\ln\tilde{q},h;p)\, dh\, d\ln\tilde{q}\, d\ln\tilde{t} / K$$

$$E\left[(\tilde{Q} - E[\tilde{Q}|p])^2 | p\right] = \int_{\tilde{Q}} \int_{\lambda_t^{-1}(p,\ln\tilde{q},h)}^{\infty} \int_{0}^{\lambda_h^{-1}(\ln\tilde{q},\ln\tilde{t},p)} \tilde{q}^2 \cdot \psi(\ln\tilde{t},\ln\tilde{q},h;p)\, dh\, d\ln\tilde{t}\, d\ln\tilde{q} / K - \{E[\tilde{Q}|p]\}^2$$

$$E\left[(\tilde{T} - E[\tilde{T}|p])^2 | p\right] = \int_{\tilde{T}} \int_{\lambda_q^{-1}(\ln\tilde{t},p,h)}^{\infty} \int_{0}^{\lambda_h^{-1}(\ln\tilde{q},\ln\tilde{t},p)} \tilde{t}^2 \cdot \psi(\ln\tilde{t},\ln\tilde{q},h;p)\, dh\, d\ln\tilde{q}\, d\ln\tilde{t} / K - \{E[\tilde{T}|p]\}^2$$

where

$$K = \int_{\tilde{Q}} \int_{\lambda_t^{-1}(p,\ln\tilde{q},h)}^{\infty} \int_{0}^{\lambda_h^{-1}(\ln\tilde{q},\ln\tilde{t},p)} \psi(\ln\tilde{t},\ln\tilde{q},h;p)\, dh\, d\ln\tilde{t}\, d\ln\tilde{q}$$

Approach to Parameter Estimation - Computer Search. Suppose that from actual data on ore deposits, statistics ($\bar{\tilde{q}}$, $\bar{\tilde{t}}$, $s^2_{\tilde{q}}$, $s^2_{\tilde{t}}$) can be computed that are comparable in concept to the expectations. Then, define an error measure, Δ.

$$\Delta = \left[1 - \frac{E[\tilde{Q}|p]}{\bar{\tilde{q}}}\right]^2 + \left[1 - \frac{E[\tilde{T}|p]}{\bar{\tilde{t}}}\right]^2 + \left[1 - \frac{E[(\tilde{Q} - E[\tilde{Q}|p])^2|p]}{s^2_{\tilde{q}}}\right]^2 + \left[1 - \frac{E[(\tilde{T} - E[\tilde{T}|p])^2|p]}{s^2_{\tilde{t}}}\right]^2$$

Of course, each of these expectations changes if the value of any one of the four[1] parameters changes; consequently, it is useful conceptually to consider the measure of error and the expectations to be functions of the parameters.

$$\Delta = E(\sigma_q,\mu_t,\sigma_t,\rho) = \left[1 - \frac{\varepsilon_q(\sigma_q,\mu_t,\sigma_t,\rho)}{\bar{\tilde{q}}}\right]^2 + \left[1 - \frac{\varepsilon_t(\sigma_q,\mu_t,\sigma_t,\rho)}{\bar{\tilde{t}}}\right]^2 + \left[1 - \frac{\varepsilon_{s_q^2}(\sigma_q,\mu_t,\sigma_t,\rho)}{s_{\tilde{q}}^2}\right]^2 + \left[1 - \frac{\varepsilon_{s_t^2}(\sigma_q,\mu_t,\sigma_t,\rho)}{s_{\tilde{t}}^2}\right]^2$$

With this functional description as a reference, the approach of this study is to use a generalized computer search routine to find the values of σ_q, μ_t, σ_t, and ρ that minimize function E( ).

Overview of Requirements of the Model. A version of the crustal abundance model was programmed to support the estimation of the unknown parameters σ_q, μ_t, σ_t and ρ. Required inputs include the following:

A	crustal abundance
$y_t(\ln v)$	pdf for intradeposit grade variance, conditional upon deposit size
$\alpha(h)$	pdf for depth to deposit (h)
$q^{\beta_1} \cdot t^{\beta_2}$	exploration bias coefficients (β_1,β_2)
ke^{-bh}	influence of depth to deposit on discoverability
$\phi_t(\ln t,\ln q,\ln v,h,p)$, $\phi_q(\ln t,\ln q,\ln v,h,p)$	ore transformation functions
$\lambda(\ln\tilde{t},\ln\tilde{q},h)$	cost function
p	price
$\bar{\tilde{t}}$, $\bar{\tilde{q}}$, $s_{\tilde{t}}^2$, $s_{\tilde{q}}^2$	deposit statistics

[1]Although f(lnt,lnq), a dependent bivariate pdf, requires five parameters, only four need estimation, for two of the five are related through crustal abundance (A): $\ln A = \mu_q + \sigma_q^2/2$.

Of these, only the following were readily available from previous studies or routinely computable:

$A = 0.0003\%\ U_3O_8$	$\bar{\tilde{t}} = 656{,}908$ s.t.
$\beta_1 = 0.46$	$s^2_{\tilde{t}} = 3503 \times 10^9$
$\beta_2 = 0.80$	$\bar{\tilde{q}} = 0.184\%\ U_3O_8$
$k = 0.0019065$	$s^2_{\tilde{q}} = 0.0161$
$b = 0.00175$	$p = \$50.00/\text{lb}\ U_3O_8$

β_1, β_2, k, and b were obtained from a previous study on uranium exploration, which produced a discoverability relation (Harris, 1984b). Statistics $\bar{\tilde{t}}$, $s^2_{\tilde{t}}$, $\bar{\tilde{q}}$, $s^2_{\tilde{q}}$ were calculated from data.

Lacking data or model for $\alpha(h)$, it was *assumed* to be rectangular over the interval of 0 to 2000 ft; therefore,

$$\alpha(h) = 1/2000, \quad 0 \leq h \leq 2000$$

$$= 0, \text{ otherwise.}$$

The remaining inputs were not available; consequently, accommodating the model's input requirements required considerable supplementary analysis and indirect relations were needed to obtain usable estimates. The model estimated and employed for intradeposit variance, v, was the following (see Harris, 1984b):

$$y_t(\ln v) = \left(\frac{1}{2.486}\right) \cdot \exp\left\{(-1.845)\left[\ln v - (-1.65) - 0.445\left(\frac{\ln t - \mu_t}{\sigma_t}\right)\right]^2\right\},$$

$$-\infty \leq \ln v \leq \infty$$

$$-\infty \leq \ln t \leq \infty$$

The nonavailability of $\phi_t(\ )$ and $\phi_q(\ )$, ore transformation functions, and of data for their estimation forced the design of a different approach to representing translation and truncation due to economics of profit maximization. Specifically, a computer subroutine for mine optimization was designed as an integral part of the computerized crustal abundance system. This subroutine takes as input the deposit characteristics (t, q, h and v) and determines that cutoff grade for which the net present value of profits is a maximum. The ore tonnage ($\tilde{t}$) and grade ($\tilde{q}$) associated with maximum net present value are included in the model's inventory, given that maximum net present value is positive. Those deposits having a negative maximum net present value are considered to belong to the truncated part of the deposit population, i.e., that part which either is not known or not reported by industry.

The foundation for determining net present value consists of price (p), which is an input value, and a relation that describes c, production cost per lb U_3O_8, as a function of ore tonnage ($\tilde{t}$), grade ($\tilde{q}$), and depth (h).

Estimates of Parameters. The crustal abundance system was linked to a general search program which searched the four-dimensional parameter space for those values which, when employed by the system, produced expectations that approximated the statistics for actual ore deposits. This search yielded the following parameters for f(lnt,lnq):

$$\mu_q = -9.1491 \qquad \mu_t = 11.7065 \qquad \rho = -0.25$$

$$\sigma_q^2 = 2.0747 \qquad \sigma_t^2 = 0.60477$$

$$\text{Given } A = 0.0003 = e^{\mu_q + \sigma_q^2/2}$$

Associated with these parameters are the expectations of the model for ore deposit characteristics, which compare quite well with the actual statistics from ore deposit data, as shown in Table I.

TABLE I. Comparison of Model Expectations with Deposit Statistics

	Statistics of Known Deposits	Expectations of the Model
$\bar{q}$	0.184%	0.192%
s_q	0.1269	0.1256
$\bar{t}$	656,908	715,388
s_t	1.8716×10^6	1.153×10^6

Source: Harris 1984b

It is noteworthy that the parameters of the grade distribution produced by this model and analysis are nearly identical to those estimated by Harris and Chavez (1981) using the ratio of stocks ($w_{q'}/W$) and employing an asymptotic variance estimate:

$$\ln Q \sim N(-9.1442, 2.065)$$

Of course, additional insight about the sizes of mineral occurrences and the relationship of size to grade has been gained and incorporated in the model. The impact of this gain is fully appreciated only when the crustal abundance models are used to support economic analysis and estimation of mineral resources and potential supply. Such analysis now is feasible, while a simple model that does not include deposit size and depth could not be similarly used without making strong assumptions or without auxiliary relations.

Comparison. An important, but limited, comparison is the model's U_3O_8 endowment and the DOE inventory of cumulative production plus reserves. Since DOE's inventory is basically a description by grade of potential supply for a forward cost of $50/lb U_3O_8, it is not strictly comparable to U_3O_8 endowment. Figure 7 shows these two stocks and their divergence as cutoff grade is lowered. Of course, the proposition of this study is that when size, grade, depth, cost and price are allowed for, this endowment is compatible with DOE's stock. It is satisfying that the contradiction found in an earlier model, based upon stock analysis only, of model endowment being less than the known stock for the higher grades, is not present in this model.

Table II shows the parameters of the various CA models that have been developed for U.S. uranium. As duly noted earlier, the structures of these models differ, diminishing the usefulness of simple comparisons of parameters.

TABLE II. Parameters of Crustal Abundance Models of U.S. Uranium Endowment

	Harris and Chavez			
	Stock	Asym. $\hat{\sigma}^2$	Drew	This Study
$\bar{q}$	-9.0312	-9.1442	-10.0917*	-9.1491
σ_q^2	1.839	2.065	3.96	2.075
$\bar{t}$	--	--	7.517	11.7065
σ_t^2	--	--	1.362	0.605
ρ	--	--	--	-0.25

*In Drew's analysis grade is a proportion; consequently, the value reported (-15.10) has been adjusted to % equivalent. Additionally, Drew used 0.0002, instead of 0.0003, for crustal concentration. The value of -10.0917 is a value adjusted for both effects.

Source: Harris, 1984b

Perhaps a more interesting exercise is to compare estimates of endowment by the different models. To that end, suppose that each bivariate model is represented by its marginal distribution of grade, i.e., it describes U_3O_8 endowment in terms of concentration only.

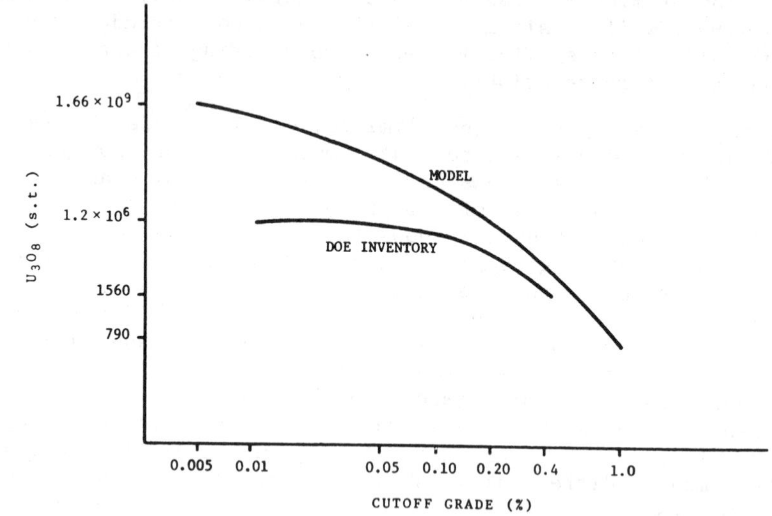

Figure 7. U_3O_8 endowment (model) and U.S. DOE inventory (Source: Harris, 1984b.

TABLE III. Comparison of Model Estimates of U_3O_8 Endowment

Cutoff Grades (%)	Drew U_3O_8 ($\times 10^6$)	Drew Ave. Grade	Harris and Chavez $\bar{\sigma}$ U_3O_8 ($\times 10^6$)	Harris and Chavez $\bar{\sigma}$ Ave. Grade	Harris and Chavez Asym. $\hat{\sigma}$ U_3O_8 ($\times 10^6$)	Harris and Chavez Asym. $\hat{\sigma}$ Ave. Grade	This Study U_3O_8 ($\times 10^6$)	This Study Ave. Grade
0.20	57	0.40	0.37*	0.33	0.9	0.27	1.1	0.33
0.10	152	0.17	2.6	0.18	5.4	0.14	2.9	0.19

	U.S. DOE Inventory U_3O_8 ($\times 10^6$ s.t.)	U.S. DOE Inventory Ave. Grade (% U_3O_8)
0.20	0.43*	0.36
0.10	0.74	0.21

Source: Harris, 1984b.

Employing this procedure, endowment was estimated by each of four crustal abundance models. Table III shows these estimates as quantity and average grade of endowment for cutoff grades of 0.10 and 0.20% U_3O_8.

The estimates in Table III take on more meaning when they are considered in light of the chronology of the methodologies by which they were estimated. In this regard, note that the quantity of endowment (0.37×10^6 s.t.) estimated by Harris and Chavez using an average of estimated σ for high grades is less than the DOE inventory, but average grades match quite closely those of the inventory. Because of this anomaly, Harris and Chavez next employed an asymptotic value of estimated σ; quantities of endowment by this model are greater than the inventory quantities for both cutoff grades, which is a desired result, but now the average grades do not match nearly so well those of the inventory.

Drew's bivariate model, which was the first attempt to model both concentration and deposit size and to estimate parameters by a statistical approach, produces endowments that are very large when compared to the DOE inventory and the other model estimates, but average grades compare reasonably well with those of the inventory.

The univariate form of the model of this study produces quantities of endowment more similar to those of the asymptotic-$\hat{\sigma}$ model than those of Drew's model. Average grades of endowment compare more closely to those of the inventory than do estimates of the other three models. The close match of expectations of the full model with statistics of ore properties, the close agreement of endowment average grades for specified cutoff grades and the fact that quantity of endowment exceeds considerably the known inventory for all cutoff grades gives this crustal abundance model considerable appeal.

The results reported from this study should be considered preliminary and subject to revision as better data are assembled. These results undoubtedly reflect compromises made because of the state of information available at the time of the initial investigation. For example, the representation of mining costs by depth and size was based upon a simplified cost analysis reported in 1974 for deposits of the San Juan Basin of New Mexico. Clearly, there are some differences in costs among regions and deposit types, e.g., the roll-type sandstone deposits of Wyoming and Texas in contrast to the tabular sandstone deposits of the San Juan Basin.

Everything else being equal, credibility of a crustal abundance model is diminished by representing costs of all modes of occurrences by costs of a single deposit type. Effort at present is being made to model costs more comprehensively.

One of the motivations for crustal abundance modeling is the estimation of resources and potential supply as a function of price and other economic variables. Only when this has been done will the implication of the model be clearly revealed. Such work already is underway, but results are not yet available.

REFERENCES

Agterberg, F. P., 1980, Lognormal models for several metals in selected areas in Canada, In: Guillemin, C. and Lagny, P. (eds.), Colloque C-1, Ressources minérales - mineral resources; 26th CGI, Bur. Recherches Géol. et Minières Mém. 106, Orleans, France, pp. 83-90.

Agterberg, F. P. and Divi, S. R., 1978, A statistical model for the distribution of copper, lead, and zinc in the Canadian Appalachian region; Econ. Geol., v. 73, no. 2, pp. 230-245.

Ahrens, L. H., 1953, A fundamental law of geochemistry; Nature, v. 172, no. 4390, p. 1148.

Ahrens, L. H., 1954, The lognormal distribution of the elements; Geochim. Cosmochim. Acta, v. 5, no. 2, pp. 49-73, v. 6, no. 2/3, pp. 121-131.

Brinck, J. W., 1967, Note on the distribution and predictability of mineral resources; Eur. 3461.e, European Atomic Energy Community - Euratom, Brussels, 25 p.

Brinck, J. W., 1971, MIMIC, the prediction of mineral resources and long-term price trends in the non-ferrous metal mining industry is no longer Utopian; Eurospectra, v. X, no. 2, pp. 46-56.

Brinck, J. W., 1972, The prediction of mineral resources and long-term price trends in the non-ferrous metal mining industry, In: Section 4 - mineral deposits, twenty-fourth session International Geological Congress Montreal, 24th International Geological Congress, Ottawa, pp. 3-15.

Charles River Associates, Inc., 1978, The economics and geology of mineral supply: an integrated framework for long-run policy analysis; CRA Rept. no. 327, Boston, Massachusetts, 468 p.

Chavez-Martinez, M. L., 1983, A potential supply system for uranium based upon a crustal abundance model; Unpubl. Ph.D. dissertation, Univ. of Arizona, Tucson, Arizona, 491 p.

Deffeyes, K. S. and MacGregor, I. D., 1980, World uranium resources; Sci. American, v. 242, no. 1, pp. 66-76.

DeNapoli, F., 1976, Mineral resource adequacy; Unpubl. Masters thesis, The Pennsylvania State Univ., University Park, Pennsylvania, 217 p.

DeWijs, H. J., 1951, Statistics of ore distribution, Part I. Frequency distribution of assay values; Geologie en Mijnbouw, n.s. 13 (v. 30), no. 11, pp. 365-375.

DeWijs, H. J., 1953, Statistics of ore distribution, Part II. Theory of binomial distribution applied to sampling and engineering problems; Geologie en Mijnbouw, n.s. 15 (v. 32), no. 1, pp. 12-24.

Drew, M. W., 1977, U.S. uranium deposits - a geostatistical model; Resources Policy, v. 3, no. 1, pp. 60-71.

Erickson, R. L., 1973, Crustal abundance of elements, and mineral reserves and resources, In: Brobst, D. A. and Pratt, W. P. (eds.), United States mineral resources, U.S.G.S. Professional Paper 820; U.S. Geological Survey, Washington, D.C., pp. 21-25.

Garrett, R. G., 1978, An abundance model resource appraisal for some Canadian commodities; Math. Geol., v. 10, no. 5, pp. 481-494.

Garrett, R. G., 1986, Geochemical abundance models: an update, 1975 to 1984, In: Prospects for Mineral Resource Assessments on Public Lands: Proc. of Leesburg Workshop, U.S.G.S. Circular 980, pp. 207-220.

Harris, D. P., 1984a, Mineral resources appraisal - mineral endowment, resources, and potential supply: concepts, methods, and cases; Oxford Univ. Press, New York, 448 p.

Harris, D. P., 1984b, Crustal abundance modeling of mineral resources: recent investigations and preliminary results; Paper presented at 27th International Geological Congress (Colloquium C.20 Mathematical Geology and Geological Information, C.20.1.3 Geostatittics), Moscow, USSR, Aug. 4-14, 1984.

Harris, D. P. and Chavez, M. L., 1981, Crustal abundance and a potential supply system, Part II, Systems and economics for the estimation of uranium potential supply; U.S. Department of Energy, Res. Rept. prepared under subcontract 78-238-E, Grand Junction Office, Colorado, pp. 385-506.

Harris, D. P. and Chavez, M. L., 1984, Modelling dynamic supply of uranium -- an experiment in the integration of economics, geology, and engineering; Proc. 18th International Symposium on Application of Computers and Mathematics in the Mineral Industries, Institution of Mining and Metallurgy, London, pp. 817-892.

Matheron, G., 1971, The theory of regionalized variables and its applications; Les Cahiers du Centre de Morphologie Mathématique, No. 5, Fontainebleau, France, 211 p.

McKelvey, V. E., 1960, Relation of reserves of the elements to their crustal abundance; Am. Jour. Sci., Bradley volume 258-A, pp. 234-241.

Ovchinnikov, L. N., 1971, Estimate of world resources of metals in terrestrial deposits; Dokl. Akad. Nauk. S.S.S.R., v. 196(1-6), pp. 200-203.

Ruzicka, V., 1977, National favourability studies, Canada; Unpubl. International Uranium Resource Evaluation Project (IUREP) Rept. for Nuclear Energy Agency/International Atomic Energy Agency (NEA/IAEA), 28 p.

Sekine, Y., 1963, On the concept of concentration of ore-forming elements and the relationship of their frequency in the earth's crust; Int. Geol. Rev., v. 5, no. 5, pp. 505-515.

Singer, D. A., Cox, D. P., and Drew, L. J., 1975, Grade and tonnage relationships among copper deposits, U.S.G.S. Professional Paper 907-A; U.S. Geological Survey, Washington, D.C., pp. A1 -- A11.

Singer, D. A. and DeYoung, J. H., Jr., 1980, What can grade-tonnage relations really tell us? In: Guillemin, C. and Lagny, P. (eds.), Colloque C-1, Ressources minérales - mineral resources; 26th CGI, Bur. Recherches Géol. et Minières Mém. 106, Orleans, France, pp. 91-101.

Vistelius, A. B., 1960, The skew frequency distributions and the fundamental law of geochemistry; Jour. Geol., v. 68, pp. 1-22.

THE USAGE OF ARTIFICIAL INTELLIGENCE IN REMOTE SENSING: A REVIEW OF APPLICATIONS AND CURRENT RESEARCH

Andrea G. Fabbri, Ko B. Fung, and Susan M. Yatabe*
Canada Centre for Remote Sensing, Ottawa, Canada K1A 0Y7
*Intera Technologies Limited, Ottawa, Canada K1S 5H4

ABSTRACT. This paper reviews current research on the usage of artificial intelligence in remote sensing and related sub-fields. Various knowledge-based systems have been built for processing images in the fields of robotics, natural language interface, computer vision and geographic information systems. Problems being considered by the remote sensing community at present are: taxonomies of terms of LANDSAT MSS image analysis, knowledge bases for contingency analysis, and the identification of rules for specific applications. Among the applications, the following have been developed: (1) land cover analysis (including forest clear-cut monitoring and vegetation change detection); (2) map-assisted photointerpretation; (3) image segmentation; (4) hyperspectral image analysis; (5) cartographic feature extraction; (6) analysis of geological structures; and (7) structural analysis of complex aerial photographs. Considerations are made on the advantages in the new approaches and of the difficulties in the application of artificial intelligence.

1. INTRODUCTION

Artificial intelligence (AI) is the field of research which develops computer programs to solve problems in ways that appear to emulate human intelligence. While general purpose programs for AI have proven too difficult to develop, it was found useful, at times, to store separately information that could be used to make a computer program behave intelligently. This was done only in extremely well-defined and narrow applications by programming expert systems. Organized information about a problem domain is called a knowledge base. "Experience" is stored separately in knowledge-based systems, or expert systems, from the knowledge about organized problem-solving, which is called an inference engine. It is mainly due to the programming of expert systems that the practical usefulness of AI was successfully demonstrated. Many expert systems have been developed for scientific and industrial applications, including medicine, military science, engineering, geology, space

C. F. Chung et al. (eds.), Quantitative Analysis of Mineral and Energy Resources, 489–512.

technology, pattern recognition, mathematics, and physics. Numerous examples are described in a recent guide to expert systems provided by Waterman (1986).

The human approach to problem-solving uses abstract symbols to refer to certain concepts and to the relationships between them in terms of type, time or space. AI programs manipulate characters representing names or symbols by symbolic reasoning. It is not a simple task to transform such relationships into symbols. Generally, in applications such as remote sensing and image processing, it is not completely known in advance what an image represents or contains. Few computer vision or geographic information system studies have investigated the problem of defining spatial relationships or computing spatial transforms in order to study the information content of an image.

A comprehensive study on the development of image analysis of remotely sensed data was made by Nagao and Matsuyama (1980). Their synoptic view is shown in Figure 1. Region-based analysis has developed from statistical classification and has led to using the concept of "object", spatial properties of regions, and multispectral properties. On the other hand, target detection has emphasized the importance of contextual and semantic constructs among objects in their recognition. AI methods have been employed for representing knowledge and for developing control structures for organizing such knowledge in the performance of complex analyses of airborne remotely sensed images.

Of particular interest to remote sensing are intelligent information systems based on workstations that handle both symbolic and spatial data through graphic interaction with raster and vector data structures. The problems currently being investigated with AI techniques by the remote sensing community at the present are: taxonomies of terms of LANDSAT MSS image analysis, knowledge bases for contingency analysis and the identification of rules for specific applications. These applications include land cover analysis, geological structure analysis, map-assisted photointerpretation, image segmentation and cartographic feature extraction. This paper reviews current research on the usage of expert systems for remote sensing and the related sub-fields of computer vision and geographic information systems. Some of the difficulties in applying artificial intelligence methods of remote sensing are described.

2. APPLICATIONS OF AI TO REMOTE SENSING

2.1. General Background

During the past few years, a new attention has been paid to the application of artificial intelligence (AI) in the remote sensing field. Several papers were published introducing the basic concepts of AI and expert systems to the remote sensing and earth science community. Tables 1 and 2 (after Wilkinson and Fisher, 1984) summarize the main characteristics required for an intelligent knowledge-based system and some ideal expert system application domains in remote sensing.

Figure 1. Development of analyses of methods in remote sensing (after Nagao and Matsuyama, 1980, p. 9, Figure 1.1).

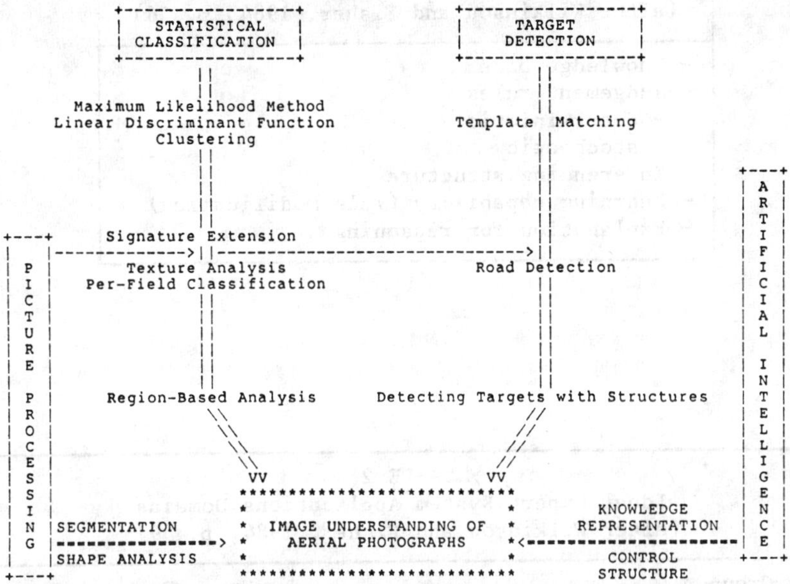

Intelligent data management systems, intelligent spatial data processing and automated systems to perform time-consuming tasks were reviewed by Campbell and Roelofs (1984). At present, no data management is available for storing and managing the large amount of spatial data needed in remote sensing. Few existing systems can handle such data in a very limited manner and are overly complex to use by non-experts. The same complexity is typical of intelligent data ingest (IDI), i.e., the automation of radiometric and geometric correction and registration, and data set evaluation, feature identification and classification. Spatial pattern understanding is an extremely difficult task whose automation requires the development of goal-directed processes able to identify the goal and properly utilize the processing resources to achieve it. Pattern directed inference systems (PDIS) model human cognition with some ability to learn. To function, they require initial conditions, parameters, dynamic interactive responses with a human operator, digitized spatial data in addition to the outcomes of earlier computational sessions.

Lambird et al. (1984) paid particular attention to the extraction of cartographic features from remotely sensed images. They review problems of fusion of information from different sensors by non-interaction and group-interaction approaches in consensus theory, distributed problem-solving in remote sensing and control mechanisms relevant to distributed architectures based on novel algorithms for

TABLE 1 Principal Characteristics of Expert Systems (after Wilkinson and Fisher, 1984, p. 56)

- Knowledge base
- Judgement rules
  - deterministic
  - stochastic
- Inferencing structure
- Learning capability (rule modification)
- Explanation for reasoning

TABLE 2
Ideal Expert System Applications Domains
(after Wilkinson and Fisher, 1984, p. 59)

Group A	Group B
Meteorological nowcasting and forecasting	Geological surveys
Disaster control	Biomass mapping
Hydrogeological monitoring and flood control	Land use inventories
Crop state monitoring	Soil surveys
Sea state monitoring	Intelligent computer-aided remote sensing education

A: Applications requiring multisource data assimilation for quasi real-time monitoring and control of dynamic events.

B: Applications requiring extensive knowledge base and archive of expert skills.

parallel search of AND/OR graphs. This approach is generic and does not provide sufficient insight on how to apply AI.

Engle (1985) described an unsupervised image classification system developed at NASA's Ames Research Center, CERBERUS. The system, which includes fifty-five rules, is able to distinguish among five of the seven Level-I land-cover types: water, agriculture, range, bare ground and forest. The operator manually enters pixel information about the spectral responses in MSS channels 5 and 7 (plus ancillary data on elevation, slope, etc.). CHESHIRE, an enhanced version of CERBERUS, will operate on a XEROX 1108 LISP workstation, using a VAX-11/780 to store LANDSAT MSS images and to perform computations and analysis. Further refinements envisage an expert system able to examine an image as configurations of pixels instead of on a pixel-by-pixel basis.

Tinney et al. (1983) reviewed some space applications of AI that relate to remote sensing, database query, intelligent on-board processing and automated image analysis. Methods of pattern recognition are particularly well-suited for implementation in expert systems. Examples of automated image analysis systems outside the realm of earth resources are ACRONYM (Standford University), HAWKEYE (Standford Research Institute), and ARGOS (Carnegie-Mellon University).

Swain (1985) described a computer reasoning scenario in the context of mapping forest resources in mountainous terrains. Low-level reasoning is analyzed for classification of multi-attribute and multiple types of data (stacked vector approach, data source-specific prior probabilities, a generalization of the latter from an arbitrary number of independent data sources and contextual classification). Here again, the difficulties in such approaches are apparent in the generic description of the problems and applications.

From the above-mentioned articles, it seems clear that only a few original ideas have been put forward on the application of AI methodology in remote sensing: mostly, they are analyses of some of the remote sensing applications that might be amenable to building expert systems. More interesting are a few examples of approaches to very simple-looking problems and of some clearer analyses for partitioning a remote sensing application into simpler and more solvable (at first) sub-problems. The narrowing of the field of application appears to be a solution in the construction of an expert system.

2.2. Applications

The conceptual design of an expert system for unsupervised classification in land cover assignment was analyzed by Erickson and Likens (1984) at NASA. While in the system PROSPECTOR (Duda et al., 1978), each pixel is considered on a point-by-point basis as a separate problem, the size of a full LANDSAT image makes the method impractical. Alternatively, a given spectral class can be treated as a group of pixels by using contingency tables to represent the relations between other ancillary data sets and the spectral classes (e.g., spectral class versus elevation range). A first step would determine which ancillary data sets are available and generate a table of pixel counts or of percentages of

coincidence. A second step would use the contingency table and the contents of a knowledge base to determine appropriate land-cover assignments. A third step could be the analysis of the table for indication of "spectral confusion" leading to a reclassification of the confused spectral classes. The authors describe a LANDSAT MSS image analysis expert system that uses a Xerox 1108 workstation and the object-oriented language LOOPS. Their contribution is also of interest for establishing a taxonomy of remotely-sensed imagery for land-use cover classification. The taxonomy uses terms and relations applicable to LANDSAT and associated collateral data sets of both discrete and continous values. This is, perhaps, the only paper that provides a thorough base for building an expert system for land-cover identification.
This is, perhaps, the only paper that provides a thorough base for building and expert system for land-cover identification.

The problem of classifying surface materials in multispectral imagery by a knowledge-based approach was treated by Carlotto et al. (1984) for LANDSAT Thematic Mapper (TM) imagery. Multispectral signature analysis and histogram analysis are the two relative measures that describe the materials. A scene is partitioned into three "superclasses", water, vegetation, and soil-like materials, by relatively simple rules of recognition of the type:

$$[A \ \& \ B] \dashrightarrow [C],$$

i.e., "if A and B are true, then C is true". A water threshold extends the darkest modes of the histogram of band 4 of the TM. Additional rules were developed to identify more specific land-cover classes: plowed fields, concrete-silt, and crops. The following are some of the rules in simple terms:

(1) If: (band-4-relative-intensity is dark)
then: (assert water);

(2) If: (band-4 $\geqslant$ band-3) and (band-4 $\geqslant$ band-5)
then: (assert vegetation);

(3) If: (band-4 $\leqslant$ band-5)
then: (assert soil-like);

(4) If: (band-4-relative-intensity is bright)
then: (assert crop).

The authors used topographic maps and aerial photos to infer ground truth for assessing the accuracy of the classification summarized in Figure 2. This contribution represents the beginning of an application inits simplest logical terms and uses a form of pattern recognition procedure.

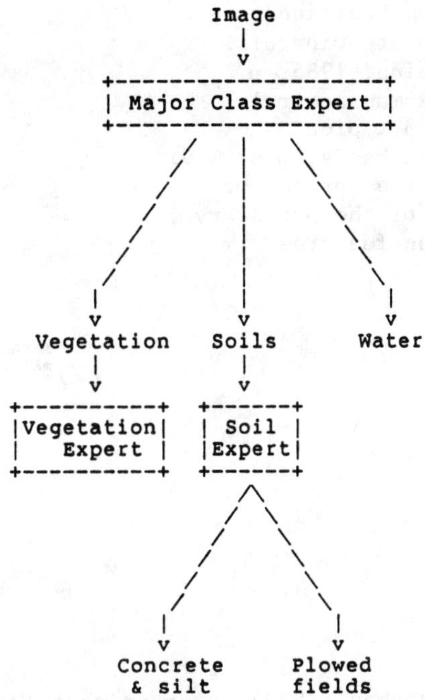

Figure 2. Expert system modules organized into a hierarchy for classification of surface materials in multi-spectral imagery (after Carlotto et al., 1984, p. 52).

Another simple-looking approach is an expert system for the analysis of remotely-sensed hyperspectral image-data developed by Chiou (1985) at NASA's Ames Research Centre. Its purpose is to assist a geologist in the identification of earth surface mineral properties directly fom the airborne and spaceborne imaging-spectrometer data. Typical spectral curves consist of 32 spectral points and are used to analyse how an expert geologist would perform a mineral analysis by studying the spectral curves and to attempt to find the significance of the reflectance and absorption characteristics. A knowledge engineer has codified and tested ten steps in the decision-making process of an expert geologist for spectral mineral#identification.

The expert system can extract automatically the absorption features from a given spectral curve (and allow the user to question the reasoning!). An example in Figure 3, illustrates how to apply the selection criteria to the radical group of carbonate minerals. A decision tree for search domain classes (the minerals) for 3 typical absorption bands IV, III and I (the informative bands in this instance) indicates with 0 or 1 the absence or presence of the necessary absorption features. The selection algorithms are written on the basis of decision trees like the one in Figure 3. The elements in the "unsorted group", $|G|$, are separated by other criteria such as the wavelength differences for band I. Each search step is essentially a binary operation. An interesting feature of the approach is that the users can edit their data bases for fine tuning. Thus, a learning capability is built in.

Figure 3. Decision tree for carbonate minerals (after Chiou, 1985, p. 2088). Roman numerals indicate 3 typical absorption bands; 0 or 1 indicate the absence or presence of the necessary absorption features.

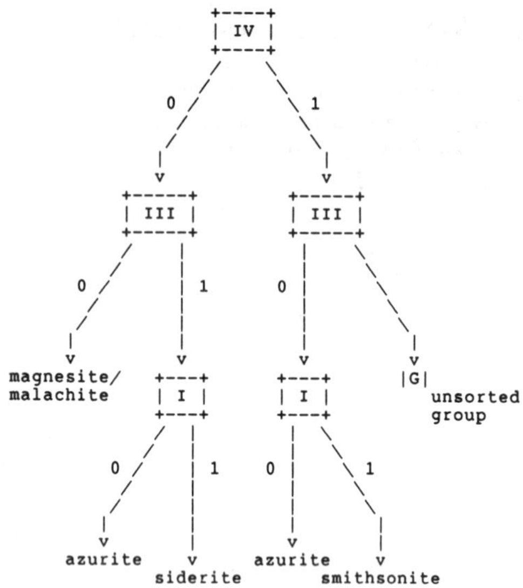

Mooneyhan (1983) reported that NASA conducted a review of the role that AI might play in the implementation of a large multisensor, multi-purpose remote sensing system. Two specific areas for considering AI are: "Spacecraft Autonomy" and "Image Data Analysis". That author visited three institutions to describe the status of the following expert systems for image interpretation: (i) MAPS, Map-Assisted Photo-Interpretation System (Dave McKeown, Carnegie-Mellon University); (ii) VISIONS (A.R. Hanson and E.M. Riseman, University of Massachusetts), and (iii) ACRONYM (Tom Binford, Stanford University). MAPS is a prototype interactive aid for photo-interpretation; VISIONS and ACRONYM will be described later in the section on computer vision. None of the expert systems were designed to deal with multichannel, multispectral digital image data. In addition the same author analysed the potential of expert systems to monitor changes in land-surface cover and land use. The components of an knowledge-based image-interpretation system can be classified as follows: (1) image interpreter; (2) change interpreter; (3) cause interpreter; and, (4) global data base. An example of a hypothetical area in the tropics where deforestation has occured was used.

Goldberg et al. (1983) proposed an expert system based on production rules for the interpretation of multitemporal sequences of LANDSAT images. The system, tested on an forested site in Newfoundland, identified change over time due to fire and insect damage. It applied pattern recognition techniques to classify each pixel and expert knowledge to update the classification.

The following contributions applied pattern recognition to the extraction of spatially-characterized features in remotely sensed images. Such approaches seem to be particularly promising for the construction of expert systems.

Nguyen et al. (1984) designed algorithms to assist the geologist the identification of linear structures and their directional information from LANDSAT data. After the automatic extraction of linear features, their selection was based on shape parameters (length, linearity, and direction) and rose diagrams were produced to represent the principal directions of the features. Contours were detected that had a strong local gradient in four steps: (1) detection of edges; (2) elimination of points where the gradient was not a local maximum in it) direction; (3) search of the nearest neighbour at each point taking into account the gradient direction; and, (4) construction of the contours from the image containing the neighbouring pixels. Linear features could be detected by: (a) selection of operators (Sobel, texture edge, etc.); (b) selection of neighbours (to link edge points) and, (c) selection of shape parameters. Contours were described in their feature space. Feature spaces could be segmented to identify linear structures or structures in a certain direction. The contour image could be "edited" ad used to mask the angle image used in the generation of the rose diagrams.

The same authors proposed an interactive approach to introduce the geologist's decision by providing an "image editor" in which the users can select, modify or add contour limits or patterns according to their own criteria. Figure 4 shows the approach as a lineament recognition algorithm. The difficult task here is the formalization of identification criteria which, in general, are not entirely objective. The algorithm is to propose "linear structure candidates" to be selected by the geologists. Pattern recognition approaches of this kind are the basis for the application of artificial intelligence. A thorough analysis of how a geologist makes decisions while interacting with the recognition program has good potential for the construction of an exper system for lineament detection.

Merchant (1984) proposed a strategy for spatial/spectral classification of LANDSAT TM which uses a "spatial logic" emulating important aspects of visual image interpretation. Following spectral classification, a definition of fields is derived for areas of similar land composition. A supervised iterative contextual classification is applied to assign final land use/land cover labels and to arrive at a generalized map using spatial post-processing techniques. That author used an interesting concept of "geographic window" that includes a fiel and all its neighbouring fields. The effective "field of view" of the geographic window was not constrained, but varied with the field size being considered. Such processing strategy could be enhanced by greated formal incorporation of the logic and techniques of artificial intelligence.

Figure 4. A lineament recognition algorithm (after Nguyen et al., 1984, p. 976).

```
                        +----------------+
+-----------------------| Original Image |
|                       +----------------+
|                               |
|                               *----------<------ Gradient
|                               |                  Operator
|                               v
|               +-------------------------------+
|               |                               |
|               v                               v
|      +----------------+              +-----------------+
|      | Gradient Image |              | Direction Image |
|      +----------------+              +-----------------+
|               |                               |
|               v                               v
|               +---------------*---------------+
|                               |
|                               v
|                               *----------<------ Contour
|                               |                  Detection
|                               v
|                       +---------------+
|                       | Contour Image |
|                       +---------------+
|                               |
|                               v                  Geometrical
|                               *---------<------- Feature
|                               |                  Extraction
|                               v
|                          +----------------+
|                        +----------------+ |
|                      +----------------+ |-+  / Programmable
|                      | Feature Images |-+   /  Internal
|                      +----------------+     |  Criteria
|                               |             v
|                               *---------<---*--- Lineament
|                               |                  Recogntion
|                               v
|                 +---------------------------+
|                 | Lineament Candidate Image |
|                 +---------------------------+
|                               |             /  Internal Complex
+-------------->----------------*             |  Criteria
                                |             v
                                *-----*---<---*--- Lineament
                                |     |
                                |     +---<-- External
                                v             Criteria
```

An expert system has been designed by Goldberg et al. (1985), and Goodenough et al. (1987), at the Canada Centre for Remote Sensing, CCRS, to estimate forest depletion by logging using LANDSAT imagery. The forestry geographic information system, consisting of digital map data, is subsequently updated. The expert system is composed of a number of hierarchically-arranged experts. It is written in PROLOG and represent high-level processing; the lowest level experts communicate with FORTRAN image processing tasks through a task interface.

Also CCRS, work has been initiated on the automatic sequencing of a image processing algorithms for integrating digital data, including remotely sensed imagery and ancillary data for resources assessment.

Previous studies on applying digital image processing to geological data, including geophysics, geochemistry, lithology and deposit informa tion have been carried out by Fabbri (1984) and Kasvand (1983). This new approach will represent data by sets of binary images, instead of square cells, as was described by Agterberg et al. (1981). The new approach will exploit the concurrent capabilities of spatial data processors, as proposed by Kasvand and Fabbri (1985) and will involve the use of inference networks for spatially distributed data. Examples of inference networks for spatially-distributed data were programmed in the expert system PROSPECTOR (Duda et al. 1978; Duda, 1980) which had graphical interface to an image processing system. Such an approach to spatial characterization can be extended to digital image data by sequentially computing spatial transforms for binary images (Fabbri, 1984). The experiments attempt to simulate the map-making procedures from experts in mineral commodities and in reconnaissance mapping for mineral exploration.

2.3 Conclusions

In the remote sensing community, several efforts have been made to evaluate the usefulness of AI and to find ideal application domains to remote sensing. Research is being performed in the following fields: (a) analysis of tentative scenarios for AI applications; (b) construction of taxonomic networks for land use and land cover classification; (c) contingency analysis and spectral confusion analysis; (d) classification of surface material in multispectral imagery; (e) analysis of hyperspectral imagery; (f) monitoring of changes in forest cover; (g) pattern recognition procedures for automatic lineament extraction; and, (h) the development of spatial logic.

From the description of these research endeavours, it becomes evident that while much work is being done in several governmental and academic institutions, the usage of AI in remote sensing is at an initial phase of development in which the real difficulties and advantages are only beginning to surface.

3. AI IN COMPUTER VISION

3.1 Introduction

This section examines some of the research work done in the field of "computer vision" that, according to the authors of this paper, has a potential impact on the analysis of remotely sensed data because the two fields are closely related in terms of image processing. In the study of computer vision (Shapiro, 1985), computers identify the objects in an image and define their spatial relationships. Processing the input image data into regions and lines is termed low-level processing. The higher levels of processing include evaluating the "goodness" of regions and lines, setting strategies of next processing steps, splitting and merging regions and lines and finally, at the top level, deciding when to stop, generating the output result which a symbolic image properly labeled.

Most of the published papers in computer vision that deal with natural scenes can be grouped into two approaches in handling recognition of objects (Kim et al., 1984). They are the "data-driven" and the "model-driven" approaches. These two approaches are summarized in Figures 5 and 6. The data-driven approach is commonly found in the low-level processing modules; while the model-driven approach is used for higher level reasoning. The model-driven approach has its shortcomings: when the number of models is not small, the control and sequencing of the expert system becomes complex and generally the processing time is greatly affected.

Figure 5. Data-driven hypothesis generation.

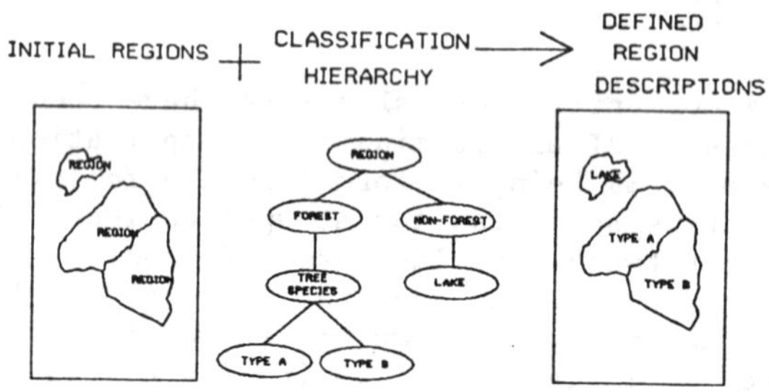

3.2 Expert Systems

Expert systems are one aspect of artificial intelligence employed in recent experiments in computer vision. One of the early expert systems for the interpretation of complex aerial photographs of urban areas was implemented by Nagao and Matsuyama (1980). It attempted to use sophisticated AI techniques and has been implemented in FORTRAN. A mechanism of focus of attention, a blackboard model and error recovery by backtracking were employed in situations where image processing and AI were combined. Figure 7 shows a block diagram of the system. Since this is a significant early system, its structure is described briefly in the following paragraphs.

A common data base, termed a blackboard, communicates with several processing modules. Following edge preserving smoothing, the image is segmented into elementary regions, using the multispectral properties of each pixel. Such regions are the basic units to be subjected to higher level analysis. The blackboard stores the properties of the regions. Next, characteristic regions are extracted from the segmented picture,

Figure 6. Model-driven hypothesis generation.

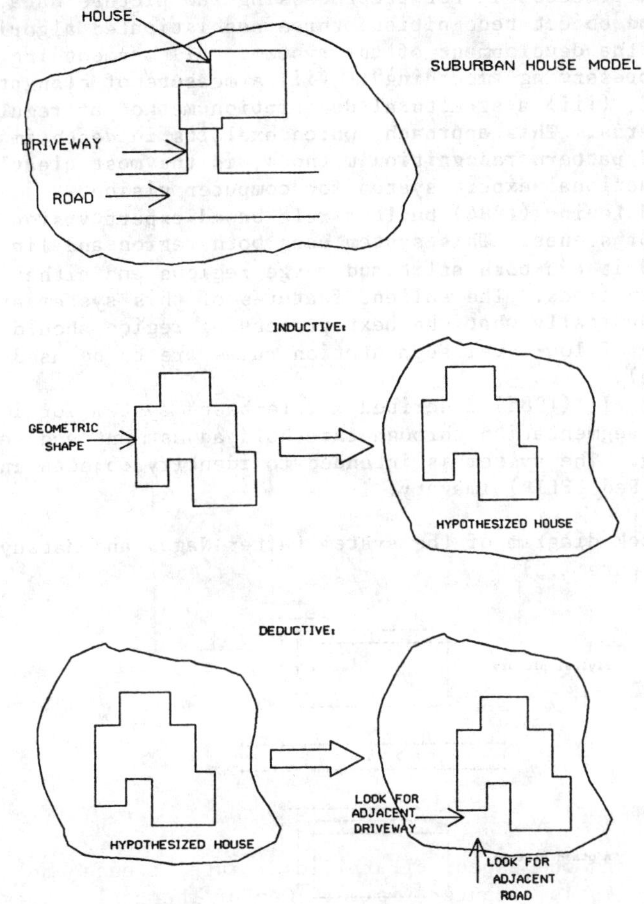

using general knowledge about the photograph (direction of sun, size of pixels, spectral bands, etc.). Then, the object-detection subsystems use the knowledge about specific objects (roads, fields, houses, cars, etc.) to locate them in the scene. Each of these subsystems focuses its attention on particular local areas and combines several characteristic regions.

The stored knowledge in a subsystem is consulted for the existence of specific objects. It is at this step that intrinsic properties and contexts of objects are analyzed and used. Properties of objects and of regions are all stored in the blackboard and all subsystems communicate indirectly within it. The expert system manages the information in the blackboard and controls the information flow. Conflicts are resolved among subsystems and segmentation errors are corrected by backtracking to lower level processes. For preprocessing raw picture data, feature extraction, and object recognition, three sophisticated algorithms were programmed in the development of the system: (i) a smoothing method termed "edge-preserving smoothing"; (ii) a measure of elongatedness of a region; and, (iii) a structural description method of regularly arranged patterns. This approach, which exploits in depth image processing and pattern recognition methods, is the most clearly documented and functional expert system for computer vision.

Nazif and Levine (1984) built a rule-based expert vision system to analyse outdoor scenes. This system uses both region and line segmentation rules and it can both split and merge regions and either add, delete, or join lines. The salient features of this system are that it can decide dynamically what the next processing region should be and then what type of low-level segmentation rules are to be used (be it region or line).

Venale et al. (1985) described a rule-based system for improving edge-oriented segmentation through threshold adjustment and segmented region merging. The system is intended to identify objects in Forward-Looking Infra-Red (FLIR) imagery.

Figure 7. Block diagram of the system (after Nagao and Matsuyama), 1980 p. 17, Figure 2.2).

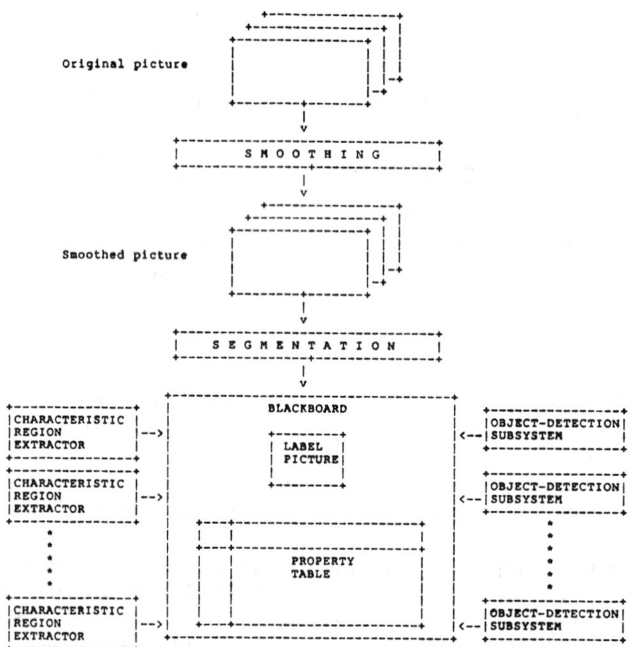

The VISIONS system, "Visual Integration by Semantic Interpretation Of Natural Scenes" (Hanson and Riseman, 1978), was initiated in 1974 at the University of Massachusetts. It uses both high-level interpretation processes and low-level segmentation processes to interpret multi-spectral images of outdoor scenes. The knowledge of the system is stored as multi-level semantic network structures. VISIONS is written in LISP on a VAX-11/780 computer; it uses FORTRAN for image manipulation. A subsystem, SYMBOLS (Strategy Guided Model Building Operating on Multileveled Structures) is designed for several levels of strategy.

The ACRONYM system, as reported by Shapiro (1985) is a model-driven expert system which uses symbolic reasoning in analyzing aerial scenes. During its iterative interpretation process, it makes predictions, generates constraints, evaluates the current results, verifies, corrects and reiterates. ACRONYM is written in LISP and operates on a symbolic LISP machine.

The SPAM system (McKeown, 1984) is a rule-based expert system which interprets monochromatic imagery of airport scenes. It interacts with an image/map database which can relate map features to it through map-to-image projection models. The image processing tools that can be manipulated by the expert system are a region growing segmentation program, a road/road-like feature follower, a linear feature extraction and 3-D junction analysis program and an interactive segmentation system.

Gilmore (1984) described a model-driven image understanding system for mobile vision system work which can be implemented by a parallel architecture. The system consists of a number of independent image processing and pattern recognition tasks for extracting independent, complementary context information, decomposing the dynamic scene into high-level constituents. The extracted data are stored in symbolic form and in a relational data structure so that the rule-based expert system can use them. The system attempts to make a total scene description and perform object classification.

Tucker (1984) has constructed an expert system for cell identification based on a quadtree data structure (Samet et al., 1984; see also Figure 9). Knowledge about a cell was constructed as a hierarchical tree with each node and leaf representing a certain level of knowledge. The quadtree approach saves storage space, despite the difficulty of searching differently-sized objects.

3.3 Conclusions

The field of computer vision has been developing rapidly in the recent years along with artificial intelligence. There are research projects on identifying objects using aerial photographs, multiple frames of photographs (such as movie frames), and on using techniques to derive 3-D models from photographs taken at ground level and at an angle to the object. It is the image vision aspect which is related to remote sensing, in which objects and fields of interest are recognized using photographs acquired from airborne and satellite platforms. Many recent published papers on this subject have successfully demonstrated the concepts on a limited scale. The model-driven systems can recognize a few stored objects at a time; and, furthermore, the size of the image they can process is in the neighbourhood of a few hundred pixels by a few hundred lines. This limitation on the volume of data is shared with the data-driven approaches. In applying image vision techniques to remote sensing for monitoring natural resources, we need to deal with far larger volumes in far less computation time. That means, in physical terms that we need computational parallelism in software and hardware for the future expert systems. Gilmore's (1984) system is a step in the right direction.

What is needed is an expert system that can learn from its own experience or that of other expert systems. Currently, there has been no published work on such a system in image vision. Although limited planning or strategy formation has been reported (Nazif and Levine, 1984), an expert system which can plan for its own computational resources, optimizing its strategy with efficiency and effectiveness, has not been reported. Most of the image vision systems reported did not mention any facility for the user or the system engineer to trace the reasoning used by the system. Expert systems should allow the domain expert to transfer his knowledge to an expert system without the action of an knowledge engineer. Along the same line, explanation facilities should be provided to establish the credibility of the system and also as an aid for debugging. Finally, to make the expert system effective and efficient, it is necessary to have proper representation formats for the knowledge and data and to have an efficient data base system to store and retrieve the needed information. Above all, to use expert vision systems successfully in remote sensing and to monitor natural resources, we need timely output, reliable results, an increase in the number of objects to be recognized, better knowledge base development and better expert-system development-tools.

4. AI AND GEOGRAPHIC INFORMATION SYSTEMS

4.1. Introduction

A geographic information system (GIS) is a computerized archive that incorporates large amounts of spatial data derived from various sources. From it, one should store, edit, retrieve, and manipulate the data so that they are transformed into useful information. Available systems cover the following range of information:

- engineering mapping;
- property/land parcel information;
- thematic (land use, forestry, and soil) and statistical mapping systems;
- bibliographic systems (data catalogued in document form);
- geographic base file systems (such as street networks);
- image processing (of geographical data);
- topographic analysis of slope, contour, and drainage patterns.

As reviewed by Dangermond (1983), the spatial data being handled in a GIS consist of points, lines and areas. Each unit of information exists within a coordinate system that enables it to be positioned within earth space. A GIS system, in general, consists of data entry, data manipulation (including graphical manipulation), and preparation of output. Additionally, a GIS system must often deal with non-spatial attributes, such as population sizes, flow rates, or timber volumes.

The topic of GIS's and remote sensing has been covered extensively in a review by Marble and Peuquet (1983). It must be stressed that remotely-sensed data do not represent a primary data source. An exception to this is the Image-Based Information System, IBIS, developed at the Jet Propulsion Laboratory, JPL, in California (Zobrist et al., 1978). It is based on an image analysis system and also possesses powerful GIS capabilities. Nearly all of the present operational GIS's make use of maps (some of which have been derived from aerial photography, or less commonly, from other remote-sensing imagery) as primary input of spatial data.

The scale of investigation available for spaceborne and many airborne sensors is very broad and regional, whereas many of the current GIS's are concerned with smaller areas and greater levels of detail. Too little attention is being focused on the use of ancillary information which can substantially improve classification accuracy.

4.2. AI and GIS's

Peuquet (1984) analyzed the spectral characteristics of geographical data. Geographic boundaries tend to be irregular; digital data are often incomplete, imprecise and error prone; spatial relationships tend to to fuzzy and application-specific. Advances are needed in three areas on GIS: (i) to develop data models that represent geographical phenomena more efficiently and completely; (ii) to develop procedures for searching complex geographic data bases flexibility and efficiently; and, (iii) to develop methods for dealing with imprecision.

Figure 8 shows the architecture of the latest version of the Knowledge-Based Geographic Information System (KBGIS), described by Peuquet (1984) and by Smith and Pazner (1984). The dualism of functions that relate to spatial objects and to the distribution of those objects in space is evident in the illustration. Because of this duality, storage of information allows queries to have both object-based and spatially-based components. Initially, queries have been restricted to the use of remote-sensing data for natural-resource problems: topography, rainfall, land use, and vegetation for a small test area. The KBGIS, illustrated in Figure 8, consists of two major components: (a) a spatial data base for organizing properties of spatial objects by locations; and (b) a spatial knowledge-base for organizing locations of spatial objects by properties. The spatial data-base component consists of a set of quadtree data structures and a set of procedures to operate such structures (boolean operators and search procedures). The knowledge base component of the KBGIS is programmed in LISP; however, a learning procedure is programmed in PASCAL, and the data-base component is programmed in C, within a UNIX environment.

Symbolic analysis of spatial data was used by Palmer (1984) to describe point, linear, and areal topographic features in a watershed data base. These included peaks, pits, flows, valleys, and ridges. According to the author, both symbolic and numerical analyses can represent complex spatial relationships in maps. This article is a good introduction to dissecting map data for building an expert system to manipulate geographical objects.

Samet el al. (1984) describes in detail the applications of hierarchical structures such as quadtrees to geographic information systems. Figure 9 shows an example of binary quadtree encoding for a simple region. A quadtree memory management system for large quadtrees (30,000 to 40,000 leaves or leaf nodes) for images of up to 4096 X 4096 pixels was developed by those authors.

Mark and Lauzon (1984) proposed a new file structure for representing key-based quadtrees, termed two-dimensional run-encoding (2DRE). Many tasks typical of image processing applications useful for GIS must still be evaluated for feasibility within this quadtree approach.

Freeman and Ahn (1984) described the development of AUTONAP, an expert system for automatically placing name labels on maps, based on

Figure 8. Latest development of KBGIS-II (Theodore Albert, U.S. Geological Survey, personal communication).

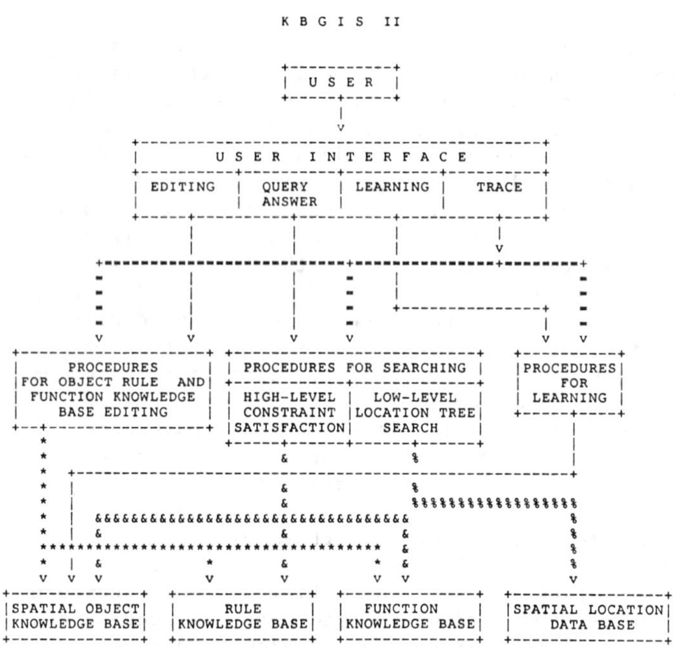

well-established cartographic conventions for high map quality. Point, line and area features are considered and 25 name placement rules have been developed. The cartographic level is not yet comparable to that of a good human cartographer.

Figure 9. A region, its binary array, its maximal block and the corresponding quatree (after Samet et al. 1984, p. 409, Figure 9).

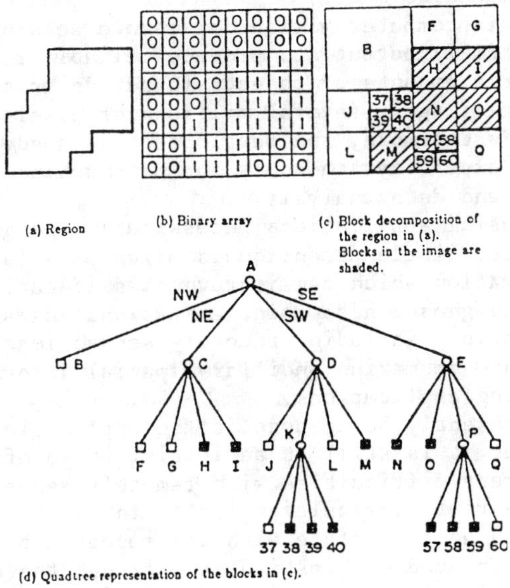

4.3. Concluding Remarks

As it can be seen in the material reviewed, much research will be needed not only on the use of AI in geographic information systems, but also on the practical interfacing of remotely sensed data and GIS data. This can be understood by considering the applications of AI to remote sensing that have been described earlier. The most successful and practical expert systems designed for GIS-related applications examine a very small subset of the problem at hand, such as the systems described by Freeman and Ahn (1984) and Palmer (1984).

The usage of ancillary geographical data in remote sensing is still minor, even if most applications require the control of a topographic ground truth. However, current trends in geoscience data processing indicate that within the next two decade, national data banks on geographical, remote sensing, renewable and non-renewable resources will be accessible for guidance in monitoring land use and estimating resource potential. To this end, new programming techniques will contribute to making geographic information systems essential instruments for study and routine work in remote sensing.

5. CONCLUSIONS

From this review of AI applications to remote sensing, including the related fields of computer vision and geographic information systems, some general remarks can be made.

Expert systems in computer vision for remote sensing require the ability to provide timely output, to produce reliable results, to recognize and analyze large amounts of objects (hundreds or thousands), and to process very large images (several millions of pixels for several different bands). Furthermore, the development is needed of better knowledge bases and expert system tools for man-machine interaction, program development and data analysis.

Most remote sensing applications necessitate a geographical control. However, too little attention is being paid to the usage of ancillary map information which can improve classification accuracy. Current trends in the geosciences point at national data banks of geographical information, including remotely sensed imagery. AI technology will contribute in making ancillary spatial information accessible to remote sensing applications.

Much work is presently being done in the application of AI methods to remote sensing but it is still at an initial phase of development because of the inherent difficulties with remotely sensed data and their interpretation. The real difficulties for AI in remote sensing and its advantages are beginning to surface with the research being done in either governmental or academic institutions around the world.

The likely solution to the problem of efficiently using expert systems for digital image processing of remotely sensed data will be achieved by integrating special-purpose processors for AI programming with parallel or pipelined processors that are tailored for spatial data manipulation. Such solutions are still far from sight because of the logical difficulties of representing problems in terms of pattern recognition processes and symbolic reasoning: AI is presently used only for the analysis of specific and narrow problems and applications.

While many AI technologies were developed since the 70's and many expert systems were programmed, the establishment of robust and predictable procedures for machine reasoning and learning in general problems is and will be a formidable challenge for many years to come. It is by no means clear if image processing tasks (leading to object recognition or to automated learning procedures) are best developed by constructing expert systems or by designing query programs - that analyze images interactively and generate automatic procedures by storing and evaluating the results of such man-machine interactions. From the isolation of primitive image processing steps, it may be possible to develop more general approaches to automated manipulation of digital images in remote sensing, as was done in the field of pattern recognition. The complexity of using remote sensing data and the stringent need to monitor natural resources should represent sufficient reasons to encourage research efforts to exploit any form of automatic processing and interpretation.

6. REFERENCES

Agterberg, F.P., Chung, C.F., Divi, S.R., Eade, K.E. and Fabbri, A.G., 1981, 'Preliminary Geomathematical Analysis of Geological, Mineral Occurrence and Geophysical Data, Southern District of Keewatin, Northwest Territories', Geol. Surv. Canada, Open File Report **718**, 31 p.

Campbell, W.J. and Roelofs, L.H., 1984, 'Artificial Intelligence Applications Concepts for the Remote Sensing and Earth Science Community', Proc. IEEE 1984 IX Pecora Symposium on Spatial Information Technologies for Remote Sensing Today and Tomorrow, p. 232-242.

Carlotto, M.J., Tom, V.T., Baim, P.W. and Upton, R.A. 1984, 'Knowledge-Based Multispectral Image Classification', SPIE, v. **504**, Applications of Digital Image Processing VII, p. 47-53.

Chiou, W.C., 1985, 'NASA Image-Based Geological Expert System Development Project for Hyperspectral Image Analysis', Applied Optics, v. **25**, n. 14, p. 2085-2091.

Dangermond, J., 1983, 'Software Components Commonly Used in Geographic Information Systems', 1983 Int. Geoscience and Remote Sensing Symposium Digest, San Francisco, California, Aug. 31 - Sept. 2, 1983, p. FP1 3.1-3.5.

Duda, R.O., 1980, 'The Prospector System for Mineral Exploration', SRI International, Final Report, April, 1980, 120 p.

Duda, R.O., Hart, P.E., Barrett, B., Gashnig, J.G., Konolige, K. Reboh, R. and Slocum, J. 1978, 'Development of the Prospector Consultation System for Mineral Exploration', SRI International, Final Report, covering October 1, 1976 to Sept. 30, 1978, 193 p.

Engle, S.W., 1985, 'The Development of an MSS Satellite Imagery Classification Expert System', DIGEST, v. **2**, Proc. 1985 Intl. Geoscience and Remote Sensing Symp. (IGARSS '85), Oct. 7-9, 1985, Amherst, Mass., p. 693-696.

Erickson, W.K. and Likens, W.C. 1984, 'An Application of Expert Systems Technology to Remotely Sensed Image Analysis', Proc. IEEE 1984 IX Pecora Symposium on Spatial Information Technologies for Remote Sensing Today and Tomorrow, p. 258-277.

Fabbri, A.G., 1984, 'Image Processing of Geological Data', New York, Van Nostrand-Reinhold, 244 p.

Freeman, H. and Ahn, H., 1984, 'AUTONAP - An Expert System for Automatic Map Name Placement', Proc. Intl. Symp. on Spatial Data Handling, Zurich, Aug. 20-24, 1984, p. 544-569.

Gilmore, J.F., 1984, 'Artificial Intelligence in Automatic Target Recognizers: Technology and Timelines', SPIE, vol. **504**, Applications of Digital Image Processing VII, 1984, p. 33-39.

Goldberg, M., Karam, G. and Alvo, M., 1983, 'A Production Rule-Based Expert System for Interpreting Multi-Temporal LANDSAT Imagery', Proc. IEEE Computer Society Conference on Computer Vision and Pattern Recognition, Washington, D.C., p. 77-82.

Goldberg, M., Goodenough, D.G., Alvo, M. and Karam, G., 1985, 'A Hierarchical Expert System for Updating Forestry Maps with LANDSAT Data', IEEE Proc., v. **73**, n. 6, June, 1985, p. 1054-1063.

Goodenough, D.G., Goldberg, M., Plunkett, G. and Zelek, J., 1987, 'An Expert System for Remote Sensing', IEEE Trans. on Geoscience and Remote Sensing, v. **GE-25**, n. 3, p. 349-359.

Hanson, A.R. and Riseman, E.M., 1978, 'Segmentation of Natural Scenes', Computer Vision Systems, edited by A.R. Hanson, E.M. Riseman, Academic Press, 1978, p. 129-163.

Kasvand, T., 1983, 'Computer Vision for the Geologist', Mathematical Geology, v. **15**, p. 3-23.

Kasvand, T. and Fabbri, A.G., 1985, 'Considerations on Pyramidal Pipelines for Spatial Analysis of Geoscience Data', in, H. Freeman and G.G. Pieroni, eds., Computer Architectures for Spatially Distributed Data, NATO ASI Series, Vol. **F18**, Berlin, Springer-Verlag, p. 295-311.

Kim, J.H., Payton, D.W. and Olin, K.E., 1984, 'An Expert System for Object Recognition in Natural Scenes', Proc. IEEE Computer Society, The First Conference on Artificial Intelligence Applications, Denver, Dec. 5-7, 1984, p. 170-175.

Lambird, B.A., Levine, D. and Kanal, L.N., 1984, 'Distributed Architecture and Parallel Non-Directional Search for Knowledge-Based Cartographic Feature Extraction Systems', Int. J. Man-Machine Studies. v. **20**, p. 107-120.

Marble, D.F. and Peuquet, D.J. eds., 1983, 'Geographic Information Systems and Remote Sensing', in Coldwell, R.M., ed., Manual of Remote Sensing, Second Edition, V. **I**, Theory, Instruments and

Techniques, Falls Church, Virginia, American Society of Photogrammetry, The Sheridan Press, p. 923-958.

Mark, D.M. and Lauzon, J.P., 1984, 'Linear Quadtrees for Geographic Information Systems', Proc. Intl. Symp. on Spatial Data Handling, Zurich, Aug. 20-24, 1984, p. 412-430.

McKeown, D.M., Jr., 1984, 'Knowledge-Based Aerial Photo Interpretation', Photogrammetria, Special Issue, Pattern Recognition in Photogrammetry, 1, vol. **39**, no. 3, Sept., 1984, p. 91-123.

Merchant, J.W., 1984, 'Using Spatial Logic in Classification of LANDSAT TM Data', Proc. IEEE 1984 IX Pecora Symposium on Spatial Information Technologies for Remote Sensing Today and Tomorrow, p. 378-385.

Mooneyhan, D.W., 1983, 'The Potential of Expert Systems for Remote Sensing Application', 1983 Int. Geoscience and Remote Sensing Symposium Digest, San Francisco, California, Aug. 31 - Sept. 2, 1983, p. FP1 4.1-4.5.

Nagao, M. and Matsuyama, T., 1980, 'A Structural Analysis of Complex Aerial Photographs', New York, Plenum.

Nazif, A.M. and Levine, M.D., 1984, 'Low-Level Image Segmentation: An Expert System', Pattern Analysis and Machine Intelligence, Vol. **PAMI-6**, No. 5, Sept., 1984, p. 555-577.

Nguyen, P.T., Asfar, L. and Simon, S. 1984, 'An Application of Pattern Recognition for Structural Analysis in Geology', Proc. of 7th Intl. Conference on Pattern Recognition, Montreal, Canada, July 30 0 August 2, 1984, P. 975-977.

Palmer, B., 1984, 'Symbolic Feature Analysis and Expert Systems', Proc. Intl. Symp. on Spatial Data Handling, Zurich, Aug. 20-24, 1984, p. 465-478.

Peuquet, D.J., 1984, 'Data Structure for Knowledge-Based Geographic Information System', Proc. Intl. Symp. on Spatial Data Handling, Zurich, Aug. 20-24, 1984, p. 372-391.

Samet, H.A., Rosenfeld, A., Shaffer, C.A. and Webber, R.E., 1984, 'Use of Hierarchical Data Structures in Geographical Information Systems', Proc. Intl. Symp. on Spatial Data Handling, Zurich, Aug. 20-24, 1984, p. 392-430.

Shapiro, L.G., 1984, 'The Role of AI in Computer Vision', Proceedings of the Second Conference on Artificial Intelligence Applications, IEEE Computer Society, Miami Beach, Florida, Dec. 11-13, 1985, p. 76-81.

Smith, T.R. and Pazner, M.I., 1984, 'Knowledge-Based Control of Search and Learning in a Large-Scale GIS', Proc. Intl. Symp. on Spatial Data Handling, Zurich, Aug. 20-24, 1984, p. 498-519.

Swain, P.H., 1985, 'Approaches to Computer Reasoning in Remote Sensing and a Survey of Geographical Information Processing', in, DIGEST, v. 2, 1985, Intl. Geoscience and Remote Sensing Symp. (IGARSS '85), Oct. 7-9, 1985, Amherst, Mass., p. 687-692.

Tinney, L.R., Sailer, C. and Estes, J.E., 1983, 'Applications of Artificial Intelligence to Remote Sensing', Proc. 7th Intl. Symp. of Remote Sensing of Environment, Ann Arbor, Michigan, May 9-13, 1983, p. 255-269.

Tucker, L.W., 1984, 'Control Strategy for an Expert Vision System Using Quadtree Refinement', Proceedings of the IEEE Workshop on Computer Vision: Representation and Control, Annapolis, MD., USA, April 30 - May 2, 1984, p. 214-218.

Venable, S.F., Richter, D.G. and Wiedeman, M., 1985, 'A Rule-Based System for Improving on Image Segmentation', Proceedings of the 2nd Conference on Artificial Intelligence Applications, Dec. 11-13, 1985, p. 94-99.

Waterman, D.A., 1986, 'A Guide to Expert Systems', Reading, Addison-Wesley, p. 419.

Wilkinson, G.G. and Fisher, P.F., 1984, 'The Impact of Expert Systems in Future Operational Remote Sensing', in Satellite Remote Sensing - Review and Preview, Proc. 10th Anniversary Intl. Conf., Reading, U.K., Sept. 1984, The Remote Sensing Society, p. 53-60.

Zobrist, A.L., Bryant, N.A., Friedman, S.Z. and Angelici, G.L., 1978, 'Image-Based Information System (IBIS) System Guide', JPL Publication 900-909 (Available from COSMIC).

GEOVALUATOR, AN EXPERT SYSTEM FOR RESOURCE APPRAISAL: A DEMONSTRATION PROTOTYPE FOR KAOLIN IN GEORGIA, U.S.A.

George S. Koch, Jr., and
Demetrios Papacharalampos
University of Georgia
Department of Geology
Athens, GA 30602
U.S.A.

ABSTRACT. GEOVALUATOR is an expert system to help geologists and other specialists provide an estimate of the mineral endowment of large areas for mineral exploration and governmental planning. When completed, GEOVALUATOR will estimate the mineral endowment of the United States by combining estimates from the 621 two-degree quadrangles of the National Topographic Map Series. Developed from the well-known expert system PROSPECTOR (of SRI International), the new system will help geologists use their geological knowledge and available data to evaluate mineral endowment. GEOVALUATOR will prompt geologists to use <u>fully</u> their knowledge of geological principles, processes, and facts, together with available data, published and unpublished. The current version of GEOVALUATOR is a demonstration prototype to evaluate the kaolin resources in part of the kaolin belt of Georgia, U.S.A. The demonstration prototype is written in the muLISP language for an IBM personal computer. Working at the keyboard, the user defines the sizes, categories, and numbers of deposits that may be present for a quadrangle. After the user has evaluated the complete set of quadrangles, GEOVALUATOR provides a summary table and a map to display graphically the estimated kaolin endowment of the trial area.

1. INTRODUCTION

1.1. Purpose

Estimates of mineral endowment are needed for National planning and for many other purposes. Some desirable attributes of endowment estimates are these:

* Timely -- completed promptly and updated regularly.
* Congruent with the development of a maintained data base.
* Capable of capturing and preserving current information that is available while mines are operating, geological surveys are in progress, etc.

C. F. Chung et al. (eds.), Quantitative Analysis of Mineral and Energy Resources, 513–527.

Our expert system GEOVALUATOR has these attributes. When completed, it will estimate the mineral endowment of the United States by combining estimates from each of the 621 two-degree quadrangles of the National Topographic Map Series.

This paper reports on our demonstration prototype expert system to evaluate the kaolin resource in part of the kaolin belt in the state of Georgia, U.S.A. (Expert systems terminology is defined in the Glossary). We choose this commodity because kaolin is economically important, Georgia is the world's largest producer, and the geological situation is relatively simple.

GEOVALUATOR has similarities to PROSPECTOR, an expert system developed by the Stanford Research Institute (SRI), with funding from the National Science Foundation of the U.S.A. and the U.S. Geological Survey, to help exploration geologists search for ore deposits. The major difference is that GEOVALUATOR has a data base in addition to PROSPECTOR's rule base.

1.2. Previous work

From early times, people have been concerned with the quality and quantity of mineral resources. Yet it was not until computers capable of processing the large data bases characteristic of geological resources became available that progress was made beyond cataloging resources.

Harris (1984) clearly summarizes the present state of knowledge in his comprehensive book, "Mineral resources appraisal." He writes (1984, p. v) that

> Paramount among considerations of content and style of this book is the identification and description of a conceptual framework that defines the field and enriches understanding of the highly varied contributions, most of which are applications of one or more quantitative methods. By virtue of the compound nature of mineral and energy resources, this framework must bring together elements from economics, geology, and technology. Furthermore, because of uncertainty about the existence and characteristics of undiscovered mineral deposits, appraisal of mineral resources involves issues of statistical estimation.

Harris recognizes many models within his conceptual framework. Some of these are:

1. Models of economic activities.
2. Quantity-quality models.
3. Deterministic geological models.
4. Geostatistical models.
5. A multivariate model for wealth.
6. Occurence models.
7. A crustal abundance geostatistical model.

8. Univariate lognormal crustal abundance models.
9. A bivariate lognormal deposit model.
10. Geological inference and probability estimation models.

PROSPECTOR is an example of the last category in this list, representing "formalized geoscience which supports active analysis of data by the geologist" (Harris, 1984, p. xiv). In the language of computing and information sciences, it is an "expert system," developed through concepts of "artificial intelligence."

Waterman (1986, p. 3-60) provides a clear explanation of expert systems, using PROSPECTOR as an example. He points out that, in the 1960's, artificial-intelligence scientists tried without success to develop general purpose programs. Then, in the 1970's, these scientists made a "conceptual breakthrough" by recognizing this principle: "To make a program intelligent, provide it with lots of high-quality, specific knowledge about some problem area," (Waterman, 1966, p. 4). The resulting programs were named "expert systems," developed through "knowledge engineering," involving collaborative work among "knowledge engineers," who design and build the expert systems, and "domain experts," who have subject matter expertise.

Features of expert systems are these (Waterman, 1986, p. 6-8):

(1) Knowledge is accumulated and codified; it is explicit and accessible.

(2) The knowledge represents "the best knowledge of the top experts in the field, leading to problem solutions that are imaginative, accurate, and efficient."

(3) An expert system provides an institutional memory, of the knowledge supplied by individuals who are no longer working in the organization.

(4) A training facility is provided.

While PROSPECTOR was the first expert system in geology, others have been developed since; Waterman (p. 45, 262-264, 312-314) provides descriptions and references.

Richard Duda, John Gaschnig, and Peter Hart, three of SRI's principal contributors to PROSPECTOR, describe it in several publications, cited under their names in the list of references. The geological situation is throughly explained in an article by Hart and others (1978), which includes a discussion by a geologist, M.T. Einaudi. Another paper by Campbell and others (1982) explains PROSPECTOR's role in the discovery of a molybdenum deposit in the state of Washington, U.S.A.

Duda, Gaschnig, and Hart describe PROSPECTOR's design in an anthology chapter (Duda and others, 1979). They write (p. 154) that:

> The Prospector system is intended to emulate the reasoning process of an experienced exploration geologist in assessing a given prospect site or region for its likelihood of containing an ore deposit of the type represented by the model he or she designed. The performance of Prospector depends on the number of models it contains, the types of deposits modeled, and the quality and completeness of each model.

Continuing with an explanation of methodology, Duda and others (1979, p. 155-156) write that "for use in Prospector an ore deposit model must be encoded as a so-called inference network, a network of connections or relations between field evidence and important geological hypotheses." They express these connections as probabilities; therefore "the 'connections` in the inference network determine how a change in the probability of one assertion will affect those of other assertions."

PROSPECTOR was developed as a "cooperative enterprise involving an exploration geologist, who [was] an authority on the type of deposit being modeled, and a computer scientist who [understood] the operation of the Prospector system" (Duda and others, 1979, p. 161).

Summarizing their work, Duda and others (1979, p. 166) write that:

> Besides the running program, there appear to be several other benefits to this type of expert system approach. The model design process challenges the model designer to articulate, organize, and quantify his expertise. Without exception, the economic geologists who have designed Prospector models have reported that the experience aided and sharpened their own thinking on the subject matter of the model.

1.3. Acknowledgements

R.B. McCammon has helped us develop GEOVALUATOR by demonstrating and discussing PROSPECTOR, and by providing us with a copy of a version of PROSPECTOR that he wrote for an IBM-PC computer using muLISP. S.M. Pickering, Jr., Director of the Lands and Exploration Division, Georgia Kaolin Company, discussed kaolin geology with us. Neither of these men is responsible in any way for errors that we may have made.

2. GEOVALUATOR: A DEMONSTRATION PROTOTYPE

2.1. The geological situation

In this section, we review the geology of Georgia's kaolin deposits, the mining history, and the size, shape, and locations of the mineable deposits. Most of the information has been published; a little comes from conversations with knowledgeable people. (Buie and others (1979) and Patterson and Murray (1984) provide full reference lists; Smith (1983) gives a clear, popular account.) Some of our inferences will be wrong; one purpose of GEOEVALUATOR is to evaluate inferences and correct the wrong ones. Table I, for the entire area, and table II, for individual 7-1/2-minute topographic quadrangles, summarize the numerical data.

In 1985, Georgia produced 5.8 million tonnes of kaolin, valued at about half a billion dollars, according to the U.S. Bureau of Mines (table I). This kaolin accounted for 82 percent of the U.S. tonnage and 27 percent of the world tonnage. About 80 percent of the kaolin was used in the paper industry.

Georgia kaolin is produced from a belt that is from 8 to 50 kilometers wide, located seaward from the Fall Line, which marks the waterfalls between the crystalline rocks of the Piedmont to the northwest, and the sedimentary rocks of the Coastal Plain to the southeast (figure 1). Nearly all the Piedmont rocks are schists, gneisses, and granites of Precambrian and Paleozoic age. Most geologists believe that Georgia kaolin formed by weathering of rocks in the Piedmont; the resulting clays were then transported and deposited in lakes, streams, and tidal swamps at the edge of the shoreline during Late Cretaceous and Early Eocene time. Some of the kaolin is mined at or near the surface, but much of that presently mined is covered by up to 50 meters of younger rocks, mainly the Clinchfield Sand, Tivola Limestone, and Barnwell Formation, all of Late Eocene age.

An average deposit is from 1.6 to 4 hectares in area and from 2 to 3 meters thick; the largest are from 160 to 200 hectares in area and from 10 to 13 meters thick. While distributions of deposit sizes have not been published, those of areas are approximated by distributions of areas for which mining permits have been issued. (Georgia law requires that firms mining surface deposits obtain permits that provide for reclamation of the mined land.) Figure 2 shows that the distribution of permitted areas is more or less lognormal. Aggregate mined areas are large, accounting for about 3.1 percent of the kaolin-bearing area in Wilkinson County and about 2.7 percent in Washington County.

Kaolin has been mined in Georgia since Colonial times; production since 1935 is summarized in figure 3 and table III, which show that production has remained more or less constant for the past five years. Table III also shows that, assuming an average thickness of 3 meters and a specific gravity of 1.5, the total area mined for kaolin is about 26 km sq, which is equal to about 18 percent of the area of a 7-1/2-minute quadrangles. The continued success of the industry through difficult economic times in recent years is said to come from supplying a product at a low cost to the paper industry with high quality control. The total yearly production comes from about 300 hectares, assuming that the area mined is equal to the area restored. This is equal to nearly 2 percent of a standard 7-1/2 minute topographic quadrangle at this latitude. Mining has advanced northeastward, and downdip; in both directions, the cover is generally greater, so that stripping ratios are becoming larger. Also, markedly lower grade deposits are being mined today than even a few years ago.

The two largest producing companies are Englehard Minerals and Chemical Co, and Georgia Kaolin Co. (a subsidiary of Combustion Engineering, Inc.); each can produce some 2 million tonnes per year. In

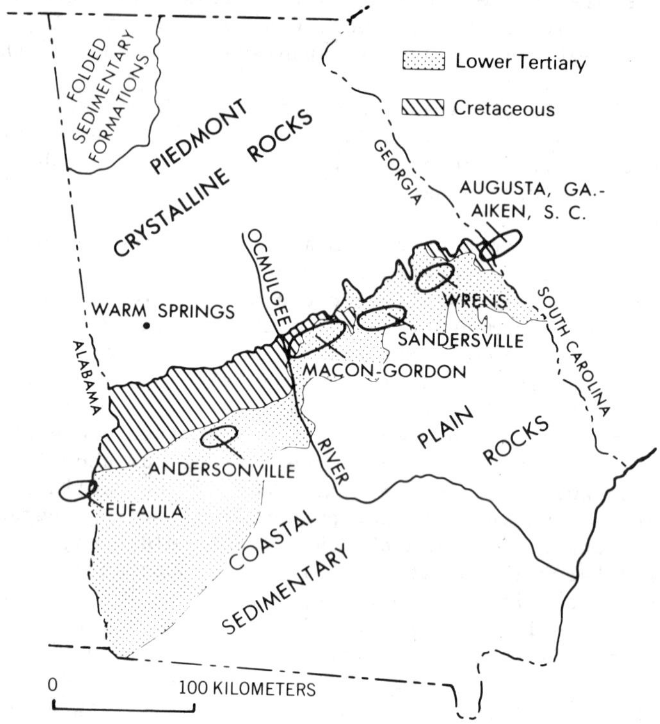

Figure 1. Kaolin deposits in Georgia (from Patterson and Murray, 1984)

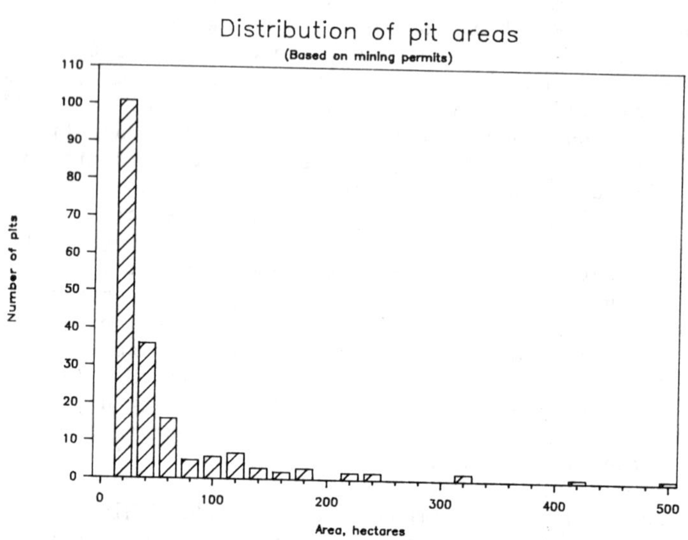

Figure 2. Distribution of pit areas

Table I. Basic data for Georgia kaolin deposits

Item	Value
Production statistics for 1985	
Millions of tonnes	5.8
Dollar value	554
Percent of U.S. tonnage	82
Percent of world tonnage	27
Percent of U.S. value	89
Summary production and resource data	
Total production (1935-1985), millions of tonnes	123
Total estimated resources, billions of tonnes	6-9
Size of the kaolin belt	
Length of the entire belt, kilometers	300
Length of the more productive northeastern part	150
Minimum width, kilometers	8
Maximum width, kilometers	50
Size of average deposits	
Thickness, meters	2-3
Area, hectares	1.6-4.0
Tonnes, minimum	76800
Tonnes, maximum, millions	0.72
Size of largest deposits	
Thickness, meters	10-13
Area, hectares	160-200
Tonnes, minimum, millions	24
Tonnes, maximum, millions	39
Aggregated areal data, hectares	
Area mined per year (1)	300
Wilkinson County estimates	
Permitted area	2865
Percentage of Coastal Plain beds (2)	3.1
Percentage of exposed kaolin-bearing beds (2)	5.5
Washington County estimates	
Permitted area	2403
Percentage of Coastal Plain beds (2)	2.7
Percentage of exposed kaolin-bearing beds (2)	13.6
Area mined by the largest companies, 1982-83 (3)	
Englehard	175
Georgia Kaolin	130
Ore reserve data for Englehard	
Tonnage, millions of tonnes (4)	140
Volume, millions of cubic meters (5)	93
Area, square kilometers (6)	47

Table I. (Continued). Basic data for Georgia kaolin deposits

Item	Value
Tonnage-volume factors	
Million tonnes per sq km, 2-m-thick bed	3
Million tonnes per sq km, 3-m-thick bed	4.5
Exploration and development data	
Exploration factors	
Grid spacing, feet	400
Grid spacing, meters	120
Number of holes per sq km	81
Cost of a single hole, $	300-600
Cost of drilling 1 sq km, $ thousands	$25-50
Probability of finding a mineable deposit, %	20
Development factors	
Grid spacing, feet	100
Grid spacing, meters	30
Mining factors	
Grid spacing, feet	50
Grid spacing, meters	15
Definitions of reserves and resources (7)	
Reserves	
Maximum overburden thickness, meters	60
Maximum percentage of sand impurities	20
Maximum ratio of overburden to clay	10:1
Sub-economic resources	
Maximum percentage of sand impurities	40
Undiscovered resources	
Maximum overburden thickness, meters	75

Footnotes

(1) For 1982-1983, assuming that the area restored was equal to the area mined.
(2) Estimated from the Geologic map of Georgia.
(3) Assuming that the area restored was equal to the area mined.
(4) According to 10-K reports to the U.S. Securities and Exchange Commission.
(5) Assuming that the density of kaolin is 1.5.
(6) Assuming a thickness of 2 meters.
(7) From Buie and others (1979), p. 30.

Table II. Data for the 28 quadrangles in the trial area. (Explanation in text.)

Quad. No.	Quadrangle Name	Mining areas, hect.			Number of areas active	Geology	
		Total	Active	Mined 82-84		Coastal Plain,%	Exposed Kaolin Bearing,%
306	Dearing	20	12	4	1	93	75
307	Harlem	0	0	0		98	15
331	Jewell	0	0	0		15	1
332	Beall Springs	0	0	0		30	0
333	Bastonville	0	0	0		85	3
334	Bowdens Pond	389	334	44	9	95	35
335	Avondale	0	0	0		100	22
357	Lake Sinclair East	13	10	3	1	24	0
358	Devereux	0	0	0		15	0
359	Warthen NW	0	0	0		25	0
360	Warthen NE	0	0	0		90	0
361	Mitchell	0	0	0		90	20
362	Gibson	38	36	0	1	97	8
363	Wrens	0	0	0		100	2
364	Matthews	0	0	0		100	6
386	Milledgeville	0	0	0		15	8
387	Friendship	107	80	8	4	92	20
388	Deepstep	430	263	10	12	100	50
389	Warthen	31	11	0	1	100	5
390	Downs	15	15	0	1	100	8
391	Grange	0	0	0		100	15
392	Louisville	0	0	0		100	15
418	Gumm Pond	1567	1175	166	14	100	20
419	Tabernacle	386	251	37	8	100	40
420	Sandersville	0	0	0		100	0
421	Riddleville	0	0	0		100	0
449	Toomsboro	448	424	18	2	100	20
450	Oconee	11	10	1	1	100	10

Table III. Georgia kaolin production, 1935-1985.

Year	Kaolin		Estimated
	Quantity	Value	area, sq.km.
1935	308	2350	0.07
1940	517	4835	0.11
1945	637	6685	0.14
1950	1107	17340	0.25
1951	1201	18670	0.27
1952	1205	18802	0.27
1953	1217	19660	0.27
1954	1184	20526	0.26
1955	1356	23376	0.30
1956	1510	26605	0.34
1957	1505	28220	0.33
1958	1616	30774	0.36
1959	1850	35684	0.41
1960	2009	39599	0.45
1961	1948	39557	0.43
1962	2164	46823	0.48
1963	2259	50294	0.50
1964	2344	54520	0.52
1965	2468	57411	0.55
1966	2908	67156	0.65
1967	2730	69327	0.61
1968	2871	79448	0.64
1969	3335	87815	0.74
1970	3401	99580	0.76
1971	3340	------	0.74
1972	3598	120496	0.80
1973	4091	144726	0.91
1974	4320	183611	0.96
1975	3644	177612	0.81
1976	4468	251000	0.99
1977	4521	261864	1.00
1978	5019	325266	1.12
1979	5497	406186	1.22
1980	5725	463700	1.27
1981	5739	519497	1.28
1982	4779	445389	1.06
1983	5340	523407	1.19
1984	5904	562797	1.31
1985	5824	554100	1.29

Quantity in thousands of tonnes.
Value in thousands of dollars.
Area assuming an average bed thickness of three meters.
Data from U.S. Bureau of Mine's Minerals Yearbook

1982 and 1983, Englehard restored 175 hectares of land, and Georgia Kaolin 130 hectares; we may assume that this represents roughly the mined areas. Engehard reserves are about 140 million tonnes, across an area of 47 hectares (assuming a thickness of 2 meters).

Table II gives data for the trial area, consisting of twenty-eight 7-1/2-minute topographic quadrangles in the kaolin belt of (figure 5). The quadrangle numbers were assigned by the Georgia Geologic Survey. The total mining areas are the sum of the areas permitted for mining by the state of Georgia since the land rehabilitation program was established at the beginning of 1969, providing relative estimates of the amount of mining in the quadrangles. The active areas are those for which rehabilitation was incomplete in 1984; the difference between the total and active areas represents the area mined between 1969 and 1984. The areas mined between 1982 and 1984 are those for which rehabiliation was completed in a two-year period between 1982 and 1984, corresponding roughly to the area mined during these years, assuming that rehabilitation kept pace with mining. The number of active areas indicate the number of deposits for which permits are still outstanding, approximately equal to the number of deposits being mined. The two right-hand columns provide percentage estimates from the 1:500,000-scale Geologic Map of Georgia (Georgia Geologic Survey, 1976). The percentage of area in Coastal Plain is the maximum area in which kaolin deposits may occur; the percentage of exposed kaolin-bearing beds corresponds to a unit designated "Lower Tertiary-Cretaceous Undifferentiated."

2.2. The demonstration-prototype expert system

Our demonstration prototype for GEOVALUATOR is based on a flowchart (figure 4). Working at a keyboard, the user selects a commodity (for the demonstration prototype limited to the single commodity, kaolin). Then, quadrangle by quadrangle, GEOVALUATOR searches its rule base and list of facts for pertinent information to display on the computer screen. The user next edits, adds, or deletes facts based on his or her knowledge. Finally, the program outputs the results in the form of maps and reports.

In the version that we demonstrated at the NATO Advanced Study Institute at IL CIOCCO, the user followed these steps:

1. For a specified quadrangle, define the sizes and categories of deposits that may be present.

2. Specify the expected number of deposits of each type.

3. Based on a printed report for the quadrangle, validate the evaluation, or redo the analysis by returning the step (1).

4. Select additional quadrangles until the total number have been evaluated.

GEOVALUATOR then concluded the analysis by printing a table and a map to display graphically the kaolin endowment of the trial area.

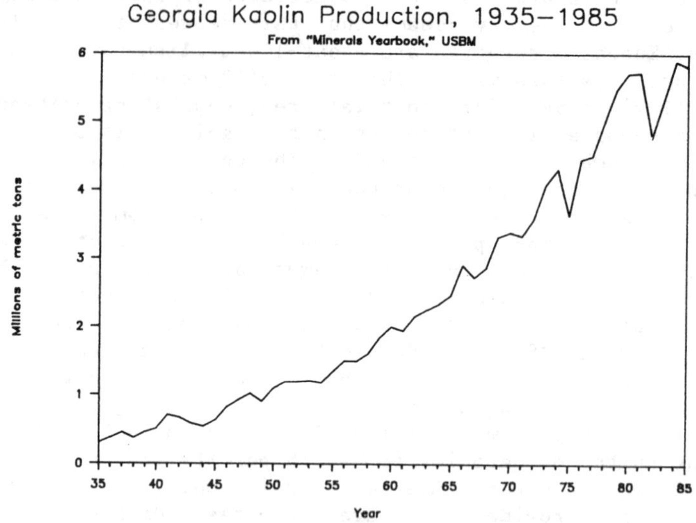

Figure 3. Georgia kaolin production

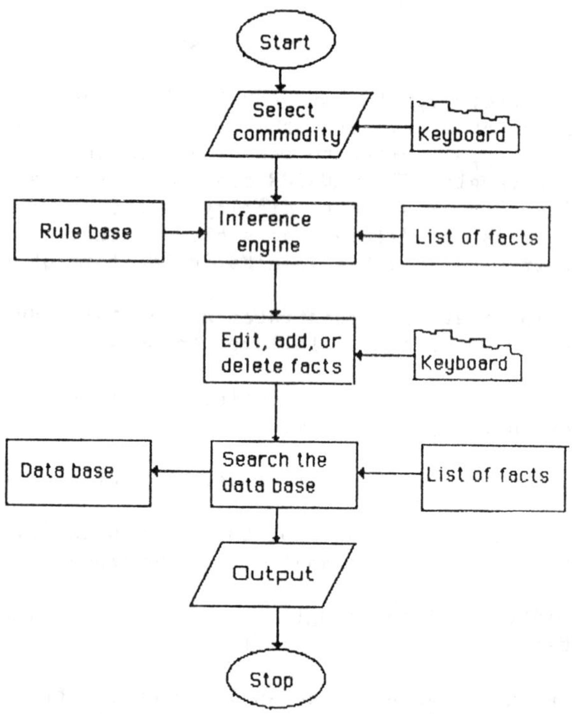

Figure 4. GEOVALUATOR flowchart

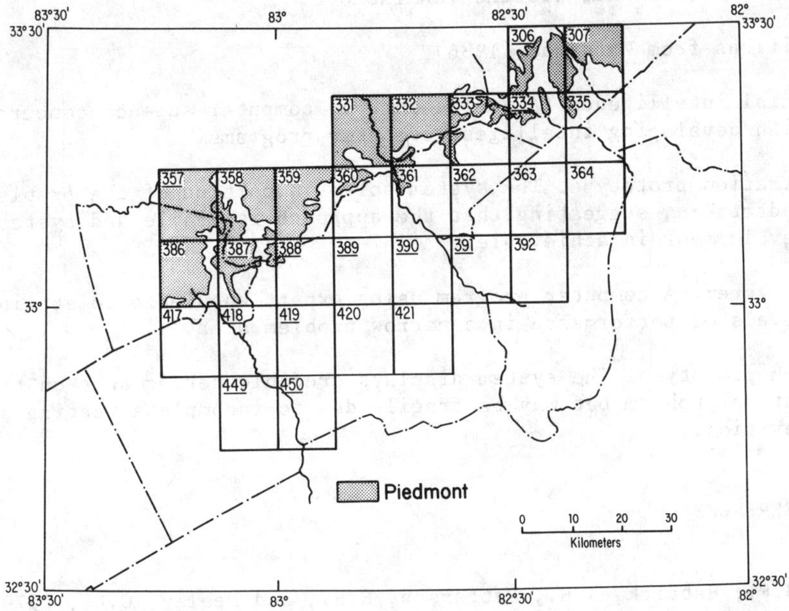

Figure 5. Kaolin data for the test area

2.3. Testing the system

We intend to have GEOVALUATOR used by a group of kaolin geologists, who will meet to learn the purpose of GEOVALUATOR, its operation, and finally to make evaluations.

The results will be analyzed to determine the effectiveness of the GEOVALUATOR demonstration prototype, and to point the way toward the next step of developing a research prototype. The analysis will also indicate the consistency of the geologists, as indicated by their evaluations.

3. CONCLUSIONS

We draw these conclusions:

(1) The demonstration prototype of GEOVALUATOR shows that development of a research prototype is justified.
(2) Development of the data base (tables 1 and 2) took more time than any other part of the work.
(3) Completion of GEOVALUATOR for resource evaluation of the United States is a practicable goal.

4. GLOSSARY OF EXPERT SYSTEMS TERMINOLOGY

(Definitions from Waterman (1986)

Artificial intelligence (AI). The part of computer science concerned with developing intelligent computer programs.

Demonstration prototype. The system solves a portion of the problem undertaken, suggesting that the approach is viable and system development is achievable.

Expert system. A computer program using expert knowledge to attain high levels of performance in a narrow problem area.

Research prototype. The system displays credible performance on the entire problem but may be fragile due to incomplete testing and revision.

5. REFERENCES

Buie, B.F., Hetrick, J.H., Patterson, S.H., and Neeley, C.L., 1979, Geology and industrial mineral resources of the Macon-Gordon kaolin district, Georgia: U.S. Geological Survey, Open-file report 79-526, 36 p.

Campbell, A.N., Hollister, V.F., Duda, R.O., and Hart, P.E., 1982, 'Recognition of a hidden mineral deposit by an artificial intelligence program,' Science, Vol. 217, p. 927-929.

Duda, R., Hart, P.E., Nilsson, N.J., Reboh, R., Slocum, J., and Sutherland, G., 1977, Development of a computer-based consultant for mineral exploration, SRI Report, Stanford Research Institute, 333 Ravenswood Avenue, Menlo Park, Calif.

Duda, R., Hart, P.E., Nilsson, N.J., Barrett, P., Gaschnig, J.G., Konolige, K., reboh, R., and Slocum, J., 1978, Development of the PROSPECTOR consultation system for mineral exploration, SRI Report, Stanford Research Institute, 333 Ravenswood Avenue, Menlo Park, Calif.

Duda, R., Gaschnig, J., and Hart, P., 1979, 'Model design in the PROSPECTOR consultant system for mineral exploration,' Expert systems in the Microelectronic age, Michie, D., ed., Edinburgh, Edinburgh University Press, p. 153-167.

Duda, R.O. and Reboh, R., 1984, 'AI and decision making: the PROSPECTOR experience,' Artificial Intelligence Applications for Business, Reitman, W., ed., Norwood, N.J.: Ablex.

Gaschnig, J., 1981, 'PROSPECTOR: an expert system for mineral exploration,' Machine Intelligence, Infotech State of the Art Report 9, no. 3.

Gaschnig, J., Reboh, R., and Reiter, J., 1981, Development of a knowledge-based expert system for water resource problems, Report SRI 1619, AI Center, SRI International, Menlo Park, Calif.

Gaschnig, J.,1982, 'PROSPECTOR: an expert system for mineral exploration,' Introductory Readings in Expert Systems, Michie, D., ed., Gordon and Breach, Science Publishers.

Georgia Geologic Survey, 1976, Geologic map of Georgia, 1:500,000, Atlanta, Georgia, Department of Natural Resources, Georgia Geologic Survey.

Hart, P.E., Duda, R.O., and Einaudi, M.T., 1978, 'PROSPECTOR--A computer-based consultation system for mineral exploration,' Math. Geology, Vol. 10, no. 5, p. 589-610.

Harris, D.P., 1984, Mineral resources appraisal Oxford, England, Clarendon Press, 445 p.

Patterson, S.H., and Murray, H.H., 1984, Kaolin, refractory clay, ball clay, and halloysite in North America, Hawaii, and the Caribbean region, U.S. Geological Survey, Prof. Paper 1306, 56 p.

Pickering, S. M., Jr., 1986, personal communication.

Smith, W.H., 1983, Kaolin deposits of central Georgia, Sandersville, Georgia, Whitehall Press, 91 p.

Waterman, D.A., 1986, A guide to expert systems, Reading, Mass., Addison Wesley, 419 p.

GEMS: A MICROCOMPUTER-BASED EXPERT SYSTEM FOR DIGITAL IMAGE DATA

D. Papacharalampos
University of Georgia
Department of Geology
Athens, GA 30602
U.S.A.

ABSTRACT. Digital images recorded by satellite-based sensor systems provide an inexpensive and widely available source of data for the compilation of preliminary geologic maps. An expert system which uses reasoning techniques and incorporates comprehensive knowledge in its classification rules can extract information about rock-types from digital images. Today's microcomputers are powerful enough to address digital image analysis problems and are inexpensive enough to be standard equipment. The Geological Exploration and Mapping System (GEMS) is a microcomputer-based expert system that analyzes multispectral digital images typically obtained from satellite sensors. The menu structure of GEMS is designed to allow fast, error-free selection of program functions. In addition to that, an on-line help capability, graphical representation of objects, instructions, and messages make the system simple to use. GEMS is divided into two major segments: the user/computer interface which handles the communications between the user and the machine, and the applications segment which includes a digital-image-analysis program. This program contains three modules: one to produce a descriptive report, another to generate a gray tone image of the data, and a third to classify rock-types and plot maps. GEMS can help geologists planning exploration, by developing a preliminary lithologic map.

1. INTRODUCTION

In the past decade, a tremendous effort has been made to collect data about the earth's surface and resources. One of the well known programs is the Landsat project of NASA which was started in 1972. Over the years, the satellites involved in this program evolved to a degree which can record images from space with adequate resolution for mapping purposes. Until recently, digital images from satellites were treated and interpreted as any other aerial images. The main perceivable differences were that a traditional image was stored on film and was handled with optical equipment, while the digital one was stored on magnetic tape or disk and was handled with electronic equipment. Devices

C. F. Chung et al. (eds.), Quantitative Analysis of Mineral and Energy Resources, 529–542.

to convert from one form to the other are widely available. Video digitizers convert images on film to digital ones, and film-writing devices transfer digital images to film. The typical processing of a digital image involves some photo-interpretation performed visually by an operator and some numerical processing performed by a computer. In the past, little attention has been paid in the automation of this process for two main reasons: the computers were not powerful enough to handle the enormous amount of computations, and it was difficult to incorporate in an automated process the operator's knowledge and experience.

The Geological Exploration and Mapping System (GEMS) addresses the automation problem by incorporating an expert system for digital image processing. Expert knowledge, incorporated in the rules, reduces the number of computations to a minimum, making the system suitable for today's typical microcomputers.

Designed for a microcomputer, GEMS is an expert system that analyzes raster data, typically captured by satellite-based sensors. GEMS is a structured, modular system with an open frame architecture. The system is divided into two major segments: the user interface which handles the communications between the user and the computer, and the application segment which performs the data analysis. The application segment contains a collection of packages to address a variety of problems. The architecture of GEMS ensures that new packages can be easily appended in the system. The expansion capabilities are limited only by hardware considerations such as memory and disk space.

2. HARDWARE/SOFTWARE CONSIDERATIONS

When a computer is needed to solve a problem, one of the questions to be answered is the selection of the machine to implement the algorithm. The common options are mainframe-, mini-, and micro-computers. The distinction is based on factors such as speed, addressable memory, number of peripheral devices, and cost, with the microcomputers being the least expensive. Until recently there were clear cutting points in performance between mainframes, minis and micros, but technology in the recent years has dramatically narrowed the differences to a point that a typical microcomputer is ten times slower than a typical mainframe but only one-hundredth its cost. For a typical research situation in an academic environment, a mainframe serves about 100 to 200 users simultaneously, with perceivable speed from 10 to 20 times slower than a microcomputer which serves only one user.

GEMS runs on a microcomputer from the IBM-PC family or compatibles that operates under the DOS 2.0 or later operating system. A hard disk, although not required for a small demonstration area, is necessary for any realistic area. GEMS, being a complex system, was developed using a number of programming languages. For the majority of the system, the Pascal language was used, because its programs are easily read, structured, and self-documented. Additionally, long programs can be divided into smaller manageable units, which can be developed and debugged separately. Libraries of routines written in other languages such as Assembly can be used to expand the language. The choice was based also on the fact that a good Pascal compiler was available for the PC-DOS environment which was able to address all the available memory, take advantage of the 8087 math co-processor and use the Virtual Device Interface (VDI) routines provided from IBM for the development of graphics oriented systems. In addition to Pascal, Assembly language was used mainly to give access to some specific functions of the operating system such as the date function. The expert modules of GEMS were modeled in LISP which is considered as the best language for developing rule-based system.

3. USER/COMPUTER INTERFACE

The user communicates with GEMS using the keyboard, an optional pointing device, and a segmented display. Because the display is static, meaning that it does not scoll up or down during operation, the user always has a familiar view of the system. Furthermore, the display is divided into segments or windows, each one with a dedicated use. The "Function key" window shows graphically the on-line available options which can be accessed at any time. One of the most useful options is a help facility which provides specific explanations making the operation of the system smoother and eliminating the need for a guide-to-operation-type manual. A "System messages" window informs the user about the status of the system with meaningful statements. An "Instructions" window gives complete information about all the possible actions at any time. In the "Interaction" window, the user-computer interaction takes place; menus are displayed, and the user makes selections. Assistance and special instructions, if required, are displayed in this window. In addition to these, "Dataset" and "Date" windows display the current dataset and date. An attempt was made to represent input items with icons and to give special instructions in graphical form.

4. DIGITAL IMAGES

A digital image is a collection of cells or picture elements (pixels) organized into rows and columns. Each pixel corresponds to a particular area on the ground and has a value proportional to some property, typically the reflection of electromagnetic radiation. The value of each pixel usually ranges from 0 (black) to 255 (white). Numbers in this range can be encoded in 8 bits, using the binary system, a convenient quantity for processing with computers. Furthermore, more data can be packed using only 8 bits (typically integer numbers require 16 bits of storage while real numbers require either 32 or 64 bits depending on the precision). This saving is very important, given the considerable number of pixels in a typical digital image. For example, the pixel size on the ground of the Landsat 5 Thematic Mapper images is 30 X 30 meters, resulting in more than 1000 pixels per square kilometer. Each pixel can be addressed by its row and column coordinates.

Digital images can be captured by a variety of technologies, the most common being optical scanners, radar, and video digitizers. An interesting digital "image" can be produced by sampling the area and measuring some property such as elevation, soil type, geology, metal concentration in sediments, and so forth. Those synthesized images can be combined with real ones recorded over specific ranges of the electromagnetic spectrum and produce a multi-image data set which is usually called an Information System.

5. CLASSIFICATION

Classification procedures are used to convert the row information from digital images into meaningful classes. Although a computer is always employed because of the nature of the digital images, its role varies widely from being only an output device to convert the numbers into perceivable quantities such as spots of light, to itself classifying the image.

Under this perspective two main approaches to classification become evident: classification by humans aided by computers, and classification by computers aided by humans.

5.1. Classification by humans

5.1.1. Looking over the numbers: This is the most primitive classification. The analyst clusters together pixels with similar values and draws boundary lines when two adjacent pixels differ significantly. This approach is impractical, because the human eye does not perform well in looking at arrays of numbers. This method is the basic concept behind any attempt to have the computer classify the image.

5.1.2. Using black and white images: A better presentation of a digital image is by displaying the image on the screen (or printing) as tones of gray. In this way a black and white image is generated, with dark areas corresponding to pixels with low values and bright areas to pixels with high values. Standard photo-interpretation techniques can be used to classify the image.

5.1.3. Using color images: By assigning one layer from an information system to the color blue, another to green, and a third to red, a color image can be generated by displaying (or printing) the three superimposed layers. The image can be of true color, if the three layers correspond to the blue, green, and red ranges of the electromagnetic spectrum, or false color if one or more layers corresponds to an invisible range such as the infrared, or even to a non-electomagnetic measurement, such as elevation. Again photo-interpretaion techniques can be used for classification.

The advantages of using black and white or color images derived from digital images are few. The main one is the ability to make odd combinations of layers that can produce interesting false color images.

5.1.4. Using image enhancement techniques: In the previous methods, the role of the computer is to read the pixel values from the information system and send them to an output device such as a display or a printer. Pixels in digital images are typically recorded as 8-bit numbers having a potential range of 256 (from 0 to 255). Recorded features rarely span the whole range. Instead, they cover a restricted range. Enhancement techniques typically try to expand the range of the recorded features using some function (either linear or non-linear) in order to amplify the variability in the image. Filtering techniques are also employed, especially to reduce (and if possible to remove) any noise present in the image, as well as to emphasize certain features, particularly the edges. The results are crisp, colorful images with extremely high contrast. Classification can be performed using photo-interpretation techniques, although some adjustments must be made to take into account the artificially produced colors.

5.2. Classification by computers

Using a computer to classify an image is similar to looking at the pixels and grouping them visually. The key difference is that the operation is fast and precise because computers are designed to work with numbers. Each pixel is a multivariate observation with its variables corresponding to layers of the information system. When the values of each variable are plotted against the layer number, they give a profile or signature for that particular pixel. Almost all computer-based classification techniques try to match the pixel's profile with a list of known or derived profiles. If the profiles match, then the pixel is classified as belonging to that particular class. Depending on the method used to match, computer-based classification can be divided into statistical and rule-based.

5.3. Statistical methods

Statistical methods employ a classifier which is applied to each pixel. The criterion commonly used is a Bayes decision rule which selects as the best solution the one that maximizes the gain under certain conditions (Haralick and Fu, 1983). With this approach two major classification methods can be seen, a supervised or an unsupervised method.

5.3.1. Supervised method: The operator must have some information, usually called ground truth, about the area which is being analyzed. The operator then defines the known areas, and their statistics are calculated. If the areas have significantly different statistics, the operator can devise a classification. Usually, the signature of each pixel is compared with the signatures of the known areas using a distance formula. A pixel is assumed to belong to the area with the smallest distance.

Supervised methods have the benefit of being fast, accurate, and are already implemented in many computer programs. However, there are some problems and limitations associated with such methods: a trained analyst must be present to orchestrate the entire operation; ground truth is required but may not be available; it is difficult to incorporate comprehensive knowledge in a distance function; and there is a need for expensive specialized hardware such as high resolution color displays (Papacharalampos and Koch, 1986).

5.3.2. Unsupervised methods: The machine itself tries to classify the image using some clustering procedure. It usually assigns one pixel to class 1 and searches the image looking for pixels with similar signatures, classifying them as class 1. Then it assigns another pixel which was unclassified to class 2 and repeats the procedure until all pixels have been classified. The system cannot give meaningful names to classes, assigning this responsibility to the user. The benefits from this method are that ground truth is not required, although later the user somehow must match the computer generated classes with real ones, the system can be operated by untrained users, and no special hardware is required since the user-computer interaction is minimized. The disadvantages are that this approach is less accurate, and it is difficult to incorporate comprehensive knowledge in the classifier which again is a distance formula.

5.4. Rule-based methods

A promising alternative appeared with the introduction of artificial intelligence approaches to classification problems in the form of expert systems (Appendix A). An expert system uses hierarchical rules instead of statistical classifiers. The typical structure of a rule is:

```
IF ( (condition 1 is TRUE)
     (condition 2 is TRUE)
            .
            .
     (condition n is TRUE))
THEN ((perform action 1)
      (perform action 2)
            .
            .
      (perform action k))
```

In a typical rule-based system that interprets aerial imagery, a number of phases can be incorporated (McKeown, D.M.,Jr., Harvey, W.A., and McDermott, J, 1984):

* Build phase where individual pixels are classified as belonging in one of the predefined classes.
* Local evaluation phase where adjacent pixels that belong in the same class are aggregated.
* Consistency phase where the accuracy of the previous step is checked and a consistency score is assigned.
* Functional areas phase where segments derived from the local evaluation phase are combined to create more general classes.

* Global evaluation phase where the classification results are further generalized.

The benefits from using a rule-based approach are many:

* A compact solution is given to the problem because each pixel is tested for membership in a subset of classes, while in a traditional system all classes are used.
* Educated guesses can be included in the classification model, permitting a solution even with incomplete data.
* The knowledge that is embedded in the rules and related to parameters such as the time of the year the image was recorded, allows rule-based systems to have some clues about the appearance of certain classes such as vegetation in the image.
* The classification procedure can be developed and tested in steps, starting with the recognition of general categories and in a later time extending to smaller classes.
* The consistency of the model can be tested and the rules can be modified easily to accomodate changes.
* Classes can be described by rules in relative terms, so absolute measures are not necessary. For example, because a basaltic outcrop usually appears darker than a granitic one, a simple rule that differentiates the two outcrops needs to consider only the difference in brightness rather than the absolute brightnesses of both.
* Rules that extract new rules can be employed in the model, so the system can learn from experience.
* Hierarchically structured rule-based systems resemble the organization of natural features and processes.

The major problem with this method is that applying rules is a slow process. New, faster machines and especially machines using parallel processing will contribute to the success of this method.

6. DIGITAL IMAGE PROCESSING ON GEMS

The current version of GEMS includes a package for image processing. This package contains three modules: one to describe the image, another to print a gray-tone image on a graphics printer, and a third to perform classification.

6.1. Describe

This module generates a descriptive report on the printer. The report includes acquisition data such as the sensor and the platform used to capture the image, the date and time as well as the band number. Summary statistics include minimum, maximum, range, number of pixels, mean, standard deviation, and mode. The location summary includes the coordinates of the center of the image in latitude and longitude, its location with the larger image (if the image under analysis is a sub-image), its number of lines and pixels per line, as well as its row and path number in a global coordinate system. In addition, a graph shows the location of the image (if it is a sub-image) with respect to the entire scene. Finally a detailed histogram is plotted with 256 classes, illustrating the relative frequency of the pixel values in the scene. The class with the highest frequency is scaled to 100 percent and the remaining classes are scaled accordingly. The prime use of the descriptive report is for an archive, although the statistics and the histogram can be used as indicators for certain features.

6.2. Picture

This module prints a gray-tone image of the data on the printer. The image, which is intended only for archival purposes, is printed pixel by pixel using 5 gray levels, each one spanning about 51 values. The gray patterns are formed using a 2 X 2 matrix of dots and have a graduated density. In other words, the pattern for the second interval is twice as light as the first, the third three times, and so on. The resolution of an average graphics printer is 480 dots in the horizontal and 792 dots in the vertical direction for a typical page. Allowing some space for titles and legends the vertical dimension is reduced to about 640 dots. This resolution suffices for images as large as 240 X 320 pixels. Smaller images are centered, while larger ones are sampled every other pixel or so, to give a final image within the limits but with smaller scale. Below the image, a gray level bar is plotted with the cutting points as well as the approximated scale of the image. Due to the difference in the aspect ratio of most printers, the image is shorter in the vertical direction resulting in slightly different horizontal and vertical scales. Nevertheless, the printed image can be used for quick identification of archived images.

6.3. Classify

This is the main module of the image analysis package. The user can ask the system to classify specific rock types or let the system try to classify everything. The operation is divided into five distinct stages:

6.3.1. Background stage. In this stage, all the known information about the area, from an information file created at the time the image was imported to the system, is displayed. The user is asked to provide more information. Any supplied information is saved for future sessions. The range of the available information can vary with the lowest bound being no information at all.

6.3.2. Planning stage. The background information which was gathered in the previous stage, combined with the basic statistics and histograms of all the available bands, is used as input to the planning stage. This is an expert system which, using knowledge-based techniques, decides the number of the distributions present in the histograms, as well as the possible rock types. The inference engine uses a combination of forward and backward chaining depending on the objective. If a specific rock type is desired, then the engine uses backward chaining logic to prove the hypothesis, that with the available data, this particular rock type is possible. If it is true, the inference engine suggests the best procedure for classification. If all features are to be searched, then the system uses forward chaining logic in an attempt to generate more facts from the available data until it reaches some conclusions regarding the possibility of the existence of particular features. The result from this stage is a planning file that contains the strategy to be used by the next stage.

6.3.3. Classification stage. In this stage, each pixel is classified as belonging to one of the classes predetermined by the planning stage. Because a digital image contains many pixels, this stage is the most time consuming and is the bottleneck for most systems. The strategy developed in the previous stage tries to solve this problem. Rather than testing each pixel for membership in any one of the rock types that is included in the systems database, it is tested only against the probable types to be expected in such an area as determined by the expert system in the planning stage. In this way, the search space is drastically reduced. Moreover, different approaches are used for different types. For example, when trying to classify granites it is more efficient to classify whole blocks rather than individual pixels. Certainly, granitic outcrops in nature extend more than one pixel in area. A second pass through the image looking at the edges, can define the outline of the outcrop within a pixel unit. Similar approaches can be applied to the classification of other features. The result from the classification stage is a set of layers holding the outlines of the classified features.

6.3.4. Generalization stage. This stage combines the different layers from the classification into a single composite image. Any overlaps or gaps between classes are averaged, and any non-rock type is replaced with the most probable rock type. For example, if a body of water is found in the middle of a sandstone outcrop, then the area covered by water is assumed to be sandstone. The results from this stage are a composite image and a key that describes the various classes.

6.3.5. Output stage. The composite image is read and is plotted either as a color map or as a map with symbols depending on the capabilities of the output device. Finally a complete report describing the whole task is printed.

7. PERFORMANCE AND FUTURE TRENDS

Preliminary timing tests on GEMS show that the generation of gray-tone images is printer bound, meaning that the image is generated as fast as the printer can print it. The descriptive report, incorporating a highly optimized algorithm, is disk bound, meaning that the slowest part of the operation, is reading the data from the disk; a hard disk substantially improves the performance of this module. All tests were performed using an IBM PC/XT computer.

Digital image analysis poses specific demands on computer architecture. First, the machine must have a large memory to hold the images being analyzed; second, it must be fast in order to have a reasonable processing time. The solution for the first problem is easy because modern microprocessors can address large amounts of memory, and memory devices are dense and inexpensive. The second problem is far more difficult. Today's computers are not fast enough to analyze digital images with sizes required for reasonable use. Future machines will be faster, however, the speed required for real world applications is far higher than the theoretical limits of the machines with today's architecture (serial machines). Besides, as the machines become faster, the users naturally ask for more functions, adding more work to the system. The apparent solution is the development of machines with other (non serial) architectures.

A promising alternative is to use more than one processor to handle the task. (In a typical serial machine the problem is described by series of simple operations performed one after the other by a single processor.) The additional processors can be classified into four major categories (Krajewski, 1985):

Pipeline processors simultaneously execute different parts of different instructions in an assembly-line fashion. One typical example of pipelining of modern microprocessors is the look-ahead or prefetch capability.

Co-processors are special processors, like floating point or text processors, which help speed up execution because of their specially optimized architecture.

Array processors process the same instruction on an array of data and give a result on every clock cycle as opposed to the multiple cycles required by general processors.

Parallel processors are general processors which work independently on the same problem, handling either different parts of the data (data partition) or different parts of the algorithm (algorithm partition). They can be further differentiated into coarse, medium, and fine grain processors. In coarse grain a few high speed processors usually belonging to the supercomputer class are incorporated. In medium grain a number of general purpose microprocessors is used. In fine grain hundreds or even thousands of simple processors are employed.

Parallel processing, the approach adopted for the fifth generation of machines, promises to solve most of the speed problems. Theoretically, it resembles the way the human brain operates, using an extreme number of slow neurons acting as parallel processors. A problem with parallel machines is that all the existing algorithms have been developed for serial machines and need to be converted or preferably redesigned to take full advantage of the capabilities of parallel processing.

8. SUMMARY AND CONCLUSIONS

The expert system GEMS can assist geologists planning exploration by identifying rock-types present in a digital image. Comprehensive knowledge, incorporated in the rule base of the system, makes the identification process accurate. Moreover, GEMS uses logic rather than brute force approaches to achieve a compact and elegant solution to the classification problem.

8. ACKNOWLEDGMENT

This work was supported in part by a grant from the Anaconda Minerals Co.

9. REFERENCES

Charniak, E., and McDermott, D., 1985, Introduction to Artificial Intelligence, Addison-Wesley, Reading, MA.

Duda, R.O., and Gaschnig, J.G., 1985, 'Knowledge-Based Expert Systems Come of Age,' Applications in Artificial Intelligence, Andriole, S.J., ed., Petrocelli Books, Inc., Princeton, NJ, pp. 45-85.

Haralick, R.M., and Fu, K.S., 1983, 'Pattern Recognition and Classification,' Manual of Remote Sensing, Simonett, D.S., ed., second Edition, Vol. 1, ASP, Falls Church, VA, pp.793-805.

Harmon, P., and King, D., 1985, Expert Systems, John Wiley & Son, New York.

Krajewski, R., 1985, 'Multiprocessing: An Overview,' BYTE, Vol. 10, No. 5, pp. 171-181.

McKeown, D.M.,Jr., Harvey, W.A., and McDermott, J, 1984, 'Rule Based Interpretation of Aerial Imagery,' Proceedings: IEEE Workshop on Principles of Knowledge-Based Systems, IEEE Computer Society, Denver, CO., pp. 145-157.

Papacharalampos, D., and Koch, G. S.,Jr., 1986, 'Automating Rock-Type Classification from Satellite Data: Developing an Expert System,' 19th International Symposium on Applications of Computers and Operation Research in the Mineral Industry, Ramani, R.V., ed., Society of Mining Engineers, Littleton, CO, pp. 809-815.

Weiss, S.M., and Kulikowski, C.A., 1984, A Practical Guide to Designing Expert Systems, Addison-Wesley, Reading, MA.

APPENDIX A. EXPERT SYSTEMS

A.1. An overview

The earlier computers were intended to perform tasks related to numerical computations and reasoning. For some time the implementation of reasoning was not very successful for two main reasons. The power of the early machines was inadequate to handle complex tasks such as reasoning. In addition, the architecture of almost every computer was designed to maximize only the performance of numerical computations until recently.

Nevertheless, researchers were active in a field they called Artificial Intelligence (AI), with the objective of building an intelligent machine, one that could mimic some of the activities traditionally associated with humans. The fundamental assumption was that the processes which take place in the human brain can be viewed as some kind of computation (Charniak and McDermott, 1985).

The AI field is divided into three broad areas:

* Natural language processing with the goal to develop a machine that can read, speak, and understand everyday language.
* Robotics with the goal to develop a machine capable of performing complex mechanical tasks guided by information collected by its sensors.
* Human-reasoning-emulation aiming to develop a system which can mimic human behavior in solving problems by employing symbolic knowledge.

A successful implementation of the third area is a knowledge-based system which utilizes a set of rules, stored in its rule base, to solve complex problems for which traditional algorithms either do not exist or are impractical.

An expert system can be defined as a knowledge-based system with the knowledge acquired from humans who are experts in their field. The distinction between a knowledge-based and an expert system is not clear, and some researchers prefer to use the term "knowledge-based system" for all rule-based systems including the expert ones.

The motivations for building expert systems are that expert knowledge is formalized, diverse sources of knowledge can be integrated, and rare and expensive expertise becomes widely available (Weiss and Kulikowski, 1984).

Expert systems differ in concept and structure from conventional programs, where an algorithm, which includes all the required knowledge, is applied to the input data to draw some predetermined conclusions. A conventional program, for example, which calculates the average of some numbers will always give an answer which is indeed the average. In contrast, an expert system separates the knowledge required to solve a particular problem and keeps it in its rule base. The program itself is a rule interpreter which, incorporating some standard methodology, applies the rules and draws conclusions. One obvious benefit is that one can use the same program to solve a variety of problems, simply by changing its rule base. An expert system can be thought of as a general-problem solving tool.

Table I compares the conventional programming used in traditional programs with the symbolic programming used in expert systems. Generally, a conventional program has a starting point, a clear flow of control predetermined by the algorithm, which before reaching the ending point assigns a value in an already predetermined objective. In contrast, the structure of an expert system is far more complex; neither the value nor the objective itself is known before the program runs.

Duda and Gaschnig (1985) summarize the advantages of using an expert system instead of a conventional program:

* The rule base can be developed using an incremental process where new rules can be added at any time and old rules can be deleted or changed.
* The same general system can solve a variety of problems by changing its rule base.
* The knowledge which is encoded in the rules can be used in different ways with different rule interpreters.
* The rules can be used to explain system actions.
* The system can check the consistency of its own rules as well as derive new ones and add them in its rule base.
* The system can mimic experts by using rules that are not conclusive but rather suggestive, with a certainty level attached to them.

TABLE I. Differences in conventional and symbolic programming (after Harmon and King, 1985)

Conventional Programming	Symbolic Programming
Algorithms	Heuristics
Numerically addressed data base	Symbolically structured knowledge base in a global working memory
Oriented toward numerical processing	Oriented toward symbolic processing
Sequential, batch processing	Highly interactive processing
Mid-run explanation impossible	Mid-run explanation easy

A.2. Expert system components

An expert system has three major components: the user/computer interface, the rule base, and the inference engine.

A.2.1. User/computer interface: This is the part of the system a user sees. It handles the communications between humans and machine. The implementation of the interface plays a significant role in the success of any system. The key concept is to be user friendly. For instance, the system can incorporate the same sequence of steps for similar tasks, on-line help, on-line utilities, input from menus, error checking, clear messages, graphical representation, segmented displays, and instructions on the screen (Papacharalampos and Koch, 1986).

A.2.2. Knowledge base: This includes the rules that represent the knowledge which is known to the system and a list of facts about the problem. There are many ways to encode knowledge, each one having positive and negative points (Harmon and King, 1985):

* Semantic networks
* Object-Attribute-Value triplets
* Rules
* Frames
* Logical expressions

A.2.3. Inference engine: This is the program that applies the rules to the input data, draws conclusions, and solves internal conflicts. Among other things, the inference engine must decide where to start and what path to follow if more than one alternative exists. The two most popular methodologies used in inference engines are backward and forward chaining. Furthermore, both can be differentiated in either depth-first or breadth-first search (Harmon and King, 1985).

Using backward chaining, the system selects a goal and tries to find rules that prove it. Additional facts required to prove the goal become sub-goals. The system continues recursively until either the goal is proved or no applicable rules remain, leading the system to ask the user for advice.

Using forward chaining, the system tries each rule on the input data. If the rule applies, then new facts are generated and placed in the input data. The system continues repetitively until a conclusion is reached or no applicable rule remains.

Backward chaining is prefered when the goal is known and the user wants to prove it. Forward chaining is prefered when the goal is not known initially, and the user needs advice about the existing facts.

On the Understanding of Geological Cross-Sections by Expert Systems

Amir Toister
Department of Geophysics and Planetary Sciences
Tel Aviv University
Ramat Aviv
Tel Aviv 69978
ISRAEL

ABSTRACT. Some aspects important for the implementation of Expert Systems capable of understanding geological cross-sections are reviewed. Terminology which should be used for the translation of geometric relationships between the various elements to temporal expressions is discussed. The use of Expert System Shells for overcoming the knowledge acquisition bottleneck is advocated. A preliminary model for the determination of time relationships between elements in geological cross-sections is presented.

1. Introduction

The understanding of geological history depicted in the rock record is the first and most crucial step in any geological analysis, such as the identification of basin types, location of oil and gas traps, etc. The lithologic cross section can be described as a sequence of elements, each element defined by its lithic characteristics.

The contacts between the various elements have an important role for the understanding of the history of the section. The contact can be conformable, depicting a change of an environment that has left its imprint by a diastem, the result of a relatively short period of non-deposition. In other cases, an unconformity between two sequences reveals a period of uplift and erosion. Faults and intrusions can part lithostratigraphic units into separate rock bodies, making identification of the sequence of deposition much more difficult to the untrained eye. Each element in the cross-section, can have complex geometric relationships with the surrounding rock bodies. These relations provide evidence for the reconstruction of the sequence of processes which shaped the geological pattern at the locality portrayed by the cross-section.

As the understanding process is characterized by the use of expert knowledge and expertise, it is plausible to suggest that the interpretation of geological cross-sections should be implemented as

C. F. Chung et al. (eds.), Quantitative Analysis of Mineral and Energy Resources, 543–551.

a Knowledge Based System. There have been several attempts to implement knowledge based scene understanding systems (Ballard & Brown 1982, McKeown 1984). In geology, scene understanding can be applied to the understanding of maps (Fabbri & Kasvand 1987), or of cross-sections (Simmons 1983).

There are two main "types" of cross sections in which the tools used for arriving at the interpretation dictate slightly different approaches and different terminology: Seismic Stratigraphy and Lithological cross-sections. Seismic stratigraphy has become a very important source for cross-sections in the last few years, mainly in the oil industry. The analysis of seismic attributes,(see e.g., Justice et al. 1985), is the basis for the creation of an Expert System for the interpretation of seismic cross sections. The geometric relationships between the seismic horizons, such as: lapout, truncation or concordance (Vail et al. 1977), could use finger printing identification techniques as described by Isenor & Zaky (1986).

We shall discuss the understanding of lithological cross-sections compiled mainly from field studies, in which the eye of the geologists was the most important tool for acquiring the data. The eye serves as a preprocessor to the interpretation process by improving the quality of the image, by extracting the main features and segmenting the image into meaningful parts that can be modeled.

The evaluation of a section in searching for a potential oil field takes into consideration the complete network of events and processes. There is no published Expert System that we know of, that solves this type of problem. We shall describe the first steps in creating such a system. First we shall review the tool employed for the formulation of the knowledge, in order to explain the terms used in the Knowledge Structuring Process. It is an Expert System Shell called The 'Deciding Factor' (programmed by Allan Campbell and Todd Glover, copyrights reserved by Software Publishing Corporation). Then the preliminary model for handling binary relationships between elements will be discussed.

2. TERMINOLOGY

In field work, the basis for what is called "traditional stratigraphy", is the mappable rock unit, distinguished and delimited on the basis of lithic characteristics and stratigraphic position (NACSN 1983). A mathematical formalization of stratigraphic terminology, using set theory, has been made at the Hungarian Geological Survey (Dienes & Mann 1977). As indeed claimed by the authors, for most geologists the stratigraphic definitions expressed in mathematical terms are unusual and difficult to understand. An alternative formalization that does not require special mathematical training and is closer to the procedures undertaken by geologists is needed. This formalization has to be easily used by man and computer alike. We shall follow the terminology used by Carimati et al.

(1982) to describe the relationships between sedimentary strata. The terminology proposed in the mathematical formalization assumes all reference lines to be infinite and horizontal and therefore is not appropriate for use with "real life" sedimentary examples.

The most significant principle determining the understanding of relations between strata is the Principle of Superposition (Weller 1960). The principle states that overlying beds are younger that the underlying beds. Superposition brings forward the most important relation between two elements in the cross section : 'overlying' and its antonymic relation 'overlain'. These relations can be extended to have more information about the surrounding elements by adding the prefixes 'totally- (e.g. overlying)' and 'partially-'.

The simplistic model in which 'overlying' and 'overlain' relations are the only relations presented in the section, and in which the layers extend evenly beyond the examined field of study is called the "Layer Cake" model. In more realistic models the situation is much more complicated. Several lithological facies can be found to be formed side by side as part of the general setting of the environments of deposition. The expression of this phenomena in the field is a heterotopic geometrical relation between elements, known in geological terms as interfingering. The 'heterotopic' relation is actually the combination of the two relations defined above. This relation can be extended as well, in the same manner. The third basic relation is the 'including' and 'included' relationships. The extension to 'totally-' and 'partially-' has more meaning than the extensions of the other basic relations demonstrated in the model below.

A large part of the 'understanding' of a geological cross-sections is the ability to transfer the geometric relations into the time domain to define the time relationships. These relations are easier to define as they can be described on an linear axis (e.g. Vilain 1982). The terms 'younger' and 'older' describe successive events with no break inbetween. If an hiatus or any depositional events exist between the two elements, the terms 'much older' and 'much younger' should be used. Other terms include 'during', 'contemporaneous', etc.

3. THE EXPERT SYSTEM

The bottleneck in the process of creating an Expert System is at the stage of Knowledge Acquisition (Buchanan et al. 1983). In the early days of Expert Systems, the expert had to converse with the programmer responsible for the Expert System in order to have his knowledge incorporated into a Knowledge Base. The knowledge had to be formulated in such a way that the program could handle it and create the knowledge structure. Nowadays, experts can converse with specialized knowledge editing programs, which maintain the knowledge bases' structure and help the expert in the knowledge acquisition process.

The Expert System software was usually created for the specific application in which it was first used. These systems led to the creation of Expert System Shells. These shells contain the inference engine written using the techniques developed in the dedicated expert system, and different levels of sophistication of Input and Output. The shell is usually limited to solve one classification problem, which is one of the more frequent tasks required of an expert. It can help in extracting knowledge from the expert, to formulate it in a computer manageable form. Using the shell can clarify rational paths and means of documenting the reasoning that leads to the decisions can be provided. Flaws of reasoning become more explicit and expertise can be transmitted in a clear and self-explanatory manner. The system can test the validity of assertions including single ideas, forcing the expert to define the knowledge in a clear way.

One Expert System Shell, running on a personal computer, is The 'Deciding Factor'. The main advantage of this shell over many others is that in addition to the usual interface that handles the conversation with the user of the system during a consultation session, it has an editor for extracting knowledge.

The 'EDITOR' helps entering, organizing and storing knowledge in files called the 'decision models'. The main idea of the model is refined by adding 'key ideas' to support it. Each of these ideas is supported by more hypotheses (ideas that receive support from other hypotheses or from facts) or by stand-alone facts. The 'EDITOR' leads the expert in the knowledge structuring process by requesting the proper type of knowledge at each stage. Default values are provided whenever a value can be given and the expert prefers not to provide it till later, while tuning the model, inorder not to interfere with the stream of thought. Several logical relationships between the ideas of the model can be specified. The 'ALL' logic relationship, dictates that all the supporting ideas contribute to the evaluation of the main idea according to their predefined weight and importance.The 'MOST' relationship will identify a trend in the supporting ideas allowing the majority of the ideas to rule. The 'BEST' logical relationship is used when only the strongest supported factor in a group should rule on the hypothesis. The antonymic logical relationship to 'BEST' is 'WORST', which is used when the least strongly factor should rule the hypothesis. Flexibility and ease of use can be added by using range tests on answers to improve focusing on the model in hand (Campbell & Fitzgerrel 1985).

The 'Deciding Factor's user interface, the 'CONSULTANT', allows changing previous answers given during the consultation. This facility allows the user and the expert to experiment with various answers and test alternative models.

4. THE PRELIMINARY MODEL FOR THE DETERMINATION OF TIME RELATIONSHIPS BETWEEN ELEMENTS IN GEOLOGICAL CROSS SECTIONS

The first task of an 'Expert System for the interpretation of geological cross sections is to define the sequence of events and processes that created the cross section. In many cases, the shape of the elements and the geometrical relationships between the rock-bodies provide enough information to tell an experienced geologist of the events that brought forward the structure of the studied section. In other cases, more information about the lithic material of the unit has to be provided in order to understand the correct relationships between two formational units.

The transformation from the geometrical relationships to the temporal relationships is by no means trivial and requires expertise and experience. Occasionally, all the gathered data and knowledge are not enough to make a reliable judgement concerning the history of the cross-section and more than one scenario is possible.

The outline of the topmost levels of the preliminary model, is represented using one of the 'Deciding Factor's output routines (Appendix A). The model solves only half of the temporal relationships as they are symmetric. The main idea or judgment of the model, is the establishment of the relationships between two elements in the cross-section. If the judgment will turn out to be unsupported by the facts, the elements A and B, the order been arbitrarily chosen, should be interchanged. The supporting ideas, the 'key ideas', are the temporal relationships defined for the system. Since only one of them can support the main idea for a given pair, the logical relationship related to these ideas is the 'BEST' relationship.

Each of these 'key ideas' is supported by one or more levels of ideas. These ideas describe the geometric relations between the elements. In each of the geometric situations, all facts contribute to the justification of the hypothesis. A negative answer (anything from 'DON'T KNOW' to 'NO'), regarding the existence of any one of the facts, will completely refute the hypothesis. In some cases the geometric terminology defined above, needs support from lithic data, and in other cases more knowledge about the nature of contacts (faults and unconformities) is required. This knowledge, together with knowledge that will translate terms like 'above' into coordinate specifications, will be added as a refinement of the Knowledge Base, by using the 'Deciding Factor's editor.

SUMMARY

The understanding of geological cross-sections is the basis for many geological analyses. It is repeatedly performed by experts and therefore an attempt to implement an Expert System for solving the problem could be of benefit to many. Expert System shells can serve as an important aid in the knowledge acquisition process.

ACKNOWLEDGEMENTS

The first model was tried out in a short course on Knowledge Based Systems given by Dr. Allan Campbell during the conference's siesta time. His help in guidance and providing the software is greatly appreciated.

REFERENCES

Ballard D.H. and C.M. Brown, 1982, Computer Vision: Prentice-Hall Inc., N.J. 523 p.

Buchanan, B.G., D. Barstow, R. Bechtel, J. Bennet, W. Clancey, C. Kulikowski, T. Mitchell and D.A. Waterman, 1983, 'Constructing an Expert system'. in: Hayes-Roth F., D.A. Waterman and D.B. Lenat, 1983. Building Expert Systems: Addison Wesly, Reading, Mass. 444 p.

Campbell,A., and S. Fitzgerrell, 1985, The Deciding Factor: User's Manual ,Software Publishing Corporation, Mountain View, California. 51 p.

Carimati, R., A. Marini and R.G. Potenza, 1982, 'The Mathematical Formalization of the Geological Relations Identifying the Basic Structure of a Geological Data Bank'. in: Cubbitt, J.M. and R.A. Reyment (eds.), Quantitative Stratigraphic Correlation: John Wiley and Sons.pp:13 - 18.

Dienes, I., and C.J. Mann, 1977, 'Mathematical Formalization of Stratigraphic Terminology'. Mathematical Geology 9: 587 - 603.

Fabbri A.G. and T. Kasvand, 1987, 'Automatic Integration of Mineral Resource Data By Image Processing and Artificial Intelligence': This volume.

Isenor, D.K., and S.G. Zaky, 1986, 'Fingerprint Identification Using Graph Matching'. Pattern Recognition 19 :113 - 122.

Justice, J.H., D.J. Hawkins and G. Wong, 1986, ' Multidimensional Attribute Analysis and Pattern Recognition for Seismic Interpretation'. Pattern Recognition 18:391 - 405.

McKeown, D.M. Jr., 1984, 'Knowledge Based Aerial Photo Interpretation': Photogrammetria 39: 91 - 123.

North American Commission on Stratigraphic Nomenclature (NACSN) 1983, 'North American Stratigraphic Code'. The American Association of Petroleum Geologists Bulletin 67: 841 - 875.

Simmons, R.G., 1983, 'Representing and Reasoning about Change in Geological Interpretation'. MIT AI-TR-749 (Master's Thesis) 130p.

Vail, P.R., R.M. Mitchum Jr., R.G. Todd, J.M. Widmeir, S. Thompson III, J.B. Sangree, J.N. Bubb, and W.G. Hatelid, 1977, 'Seismic Stratigraphy and Global Changes of Sea level, in: Seismic Stratigraphy - Applications to Hydrocarbon Exploration: The American Association of Petroleum Geologists Memoir 26: 49 - 212.

Vilain, M.B., 1982, 'A System for Reasoning About Time', Proceedings of AAAI-82, August 1982, Pittsburgh, PA. pp. 197 - 201.

Weller, J.M., 1960, Stratigraphic Principles and Practice: Harper Brothers, N.Y. 725 p.

Appendix A

A PRELIMINARY MODEL FOR THE DETERMINATION OF TIME RELATIONSHIPS BETWEEN ELEMENTS IN GEOLOGICAL CROSS SECTIONS

```
the temporal relation between element a and b is established
I
I
I-----a and b belong to the same formational unit
I     but are separated by preceding topography, an intrusion,
I     or a fault
I     I
I     I
I     I-----a has the same  lithic characteristics as b
I     I
I     I
I     I-----a is in the same sequence setting as b
I           I
I           I
I           I-----a and b are overlain by the same unit
I           I
I           I
I           I-----a and b overly the same unit
I
I
I
I
I-----a was formed during part of the time of b
I     I
I     I
I     I-----a is fully heterotopic to a part of b
I     I
I     I
I     I-----a is totally included in b
I
I
I-----a is  fully comtemp. to b
I     I
I     I
I     I-----a has the same stratigraphic setting as b
I     I
I     I
I     I-----a is fully heterotopic to b
```

```
I
I-----a is younger and contemp. to b
I     I
I     I
I     I-----a is overlying and is heterotopic to an
I     I     eqivalent of b
I     I
I     I
I     I-----a is overlying and is heterotopic to b
I
I
I-----a is younger than b
I     I
I     I
I     I-----a is overlying b
I
I
I-----a is much younger than b
      I
      I
      I-----a is overlying a unit which is younger or
      I     much younger than b
      I
      I
      I-----a is partly included in b
      I
      I
      I-----a is on the footblock of a reverse fault
      I     while b is higher
      I
      I
      I-----a is on the down thrown block of a normal
      I     fault while b is higher
      I
      I
      I-----a intrudes through b
      I
      I
      I-----a is younger than an unconformity which is
            younger than b
```

EXAMPLES OF SPATIAL ANALYSIS AND MANAGEMENT IN THE GEOGRAPHICAL AND CONFORMAL PROJECTION DOMAINS

Giovanni Bortoluzzi and Marco Ligi
Istituto di Geologia Marina - Consiglio Nazionale Ricerche
Via Zamboni, 65
40127 Bologna (ITALY)
L56A@ICINECA.EARN.BITNET

ABSTRACT. This contribution describes the technical experience of Istituto di Geologia Marina in cartographic mapping and spatial data display and analysis. The importance of using conformal cartography in geology and geophysics is stressed in terms of readability, transportability, and standardization of data. Some problems are discussed that arise when attempting to link good standard cartography with spatial processing, i.e. the procedures of datum change and the accuracy of capturing, storing and converting geographical coordinates. Interactive computer programs were developed that: (1) perform coordinate transformation and accurate datum shifts, (2) capture geographical coordinates from conformal projections, (3) digitize and manage data from continuous analog profiles, and (4) produce good-quality composite maps in conformal projections from data sets in geographical coordinates, while performing spatial analysis by interfacing graphics systems (e.g. SURFACE II). The results from the programs indicate their usefulness in geological and geophysical applications.

1. INTRODUCTION

In geology and geophysics most of the interpretative work is carried out on maps by analyzing spatial relationships. In particular, the use of map analysis added to sophisticated mathematical tools increases the amount of information revealed by simple data presentation. This is true for automatic contouring, trend analysis, and kriging.

In economic geology, it is important to represent the territory as accurately as possible, and this can only be done by using good quality cartography. When this representation also conforms to some standard, data readability and transportability are improved. Moreover, since there is no mapping functions that apply well to the totality of the earth surface, one should be able to switch to the cartographic representation that gives the best plotting of data or that is required by the application.

A satisfactory link between spatial processing and good cartographic standards, e.g. National Topographic Systems, can only be achieved

C. F. Chung et al. (eds.), Quantitative Analysis of Mineral and Energy Resources, 553–564.

by using the geographical (ellipsoidal) coordinates, because they are absolute and can be converted to rectangular cartesian coordinates. In turn, cartesian coordinates can be input to graphics systems for processing and plotting by machine. The superposition of a geographical grid provides the cartographic standard.

However, the use of geographical coordinates poses some problems, particularly in terms of data homogenization, because of the great number of reference systems. Even if there is a tendency towards a general, planetary geographical reference system, there will remain a huge amount of cartographic information linked to national systems. This problem arises when capturing coordinates from maps or when data are to be merged from different sources, while building cartographic data bases. These operations consume time and money, and any attempt should be made to preserve the maximum amount of flexibility.

We present some experience of the Istituto di Geologia Marina (IGM) in cartographic mapping and data presentation and analysis. After a short discussion of the problems faced with when trying to link standard cartography and spatial processing, we will describe interactive computer programs that: (1) perform coordinate transformation and accurate datum change shifts; (2) capture data in geographical coordinates from maps in the most widely used conformal projections; (3) create composite maps from different data sets of geographical coordinates, while performing spatial analysis by interfacing to graphics systems such as SURFACE II; and (4) digitize and manage data from continuous analog profiles.

These procedures are easy to use even by unexperienced operators, and can be transported on most common micro, mini and maxi computers.

2. METHODS

2.1 Cartographic representations, problems and data types

The earth's surface can be represented by mapping functions that establish a unique correspondence between points on a plane and on an ellipsoid that approximates the earth's shape. The mathemathical and geometrical properties of the mapping functions are well known (Richardus and Adler, 1972). An important class of projections is the one of the conformal projections, including the Direct Mercator, UTM or Gauss-Kruger, Lambert conformal and Polar Stereographic. The term conformal indicates that the maps maintain angles and geometrical shapes, and are widely used as cartographic standards (Fig.1). Accurate direct and inverse geographic-to-projection coordinate conversion procedures are available (Snyder, 1982; Bortoluzzi and Ligi, 1986; Evendon, 1986).

The geometry of the selected ellipsoid (semimajor axis and amount of flattening) and the adoption of an initial point characterize a cartographic datum (Snyder, 1982). Because of the different values of the used ellipsoid or different coordinates of the initial point, these networks may not precisely overlay. As an example, there are differences of 5-6 seconds of latitude between the Italian National System and the European Datum Network (Fig.2). Larger differences may be encountered when using other ellipsoids. The extremely accurate conversion from one datum to another may be a very complex procedure involving the knowledge

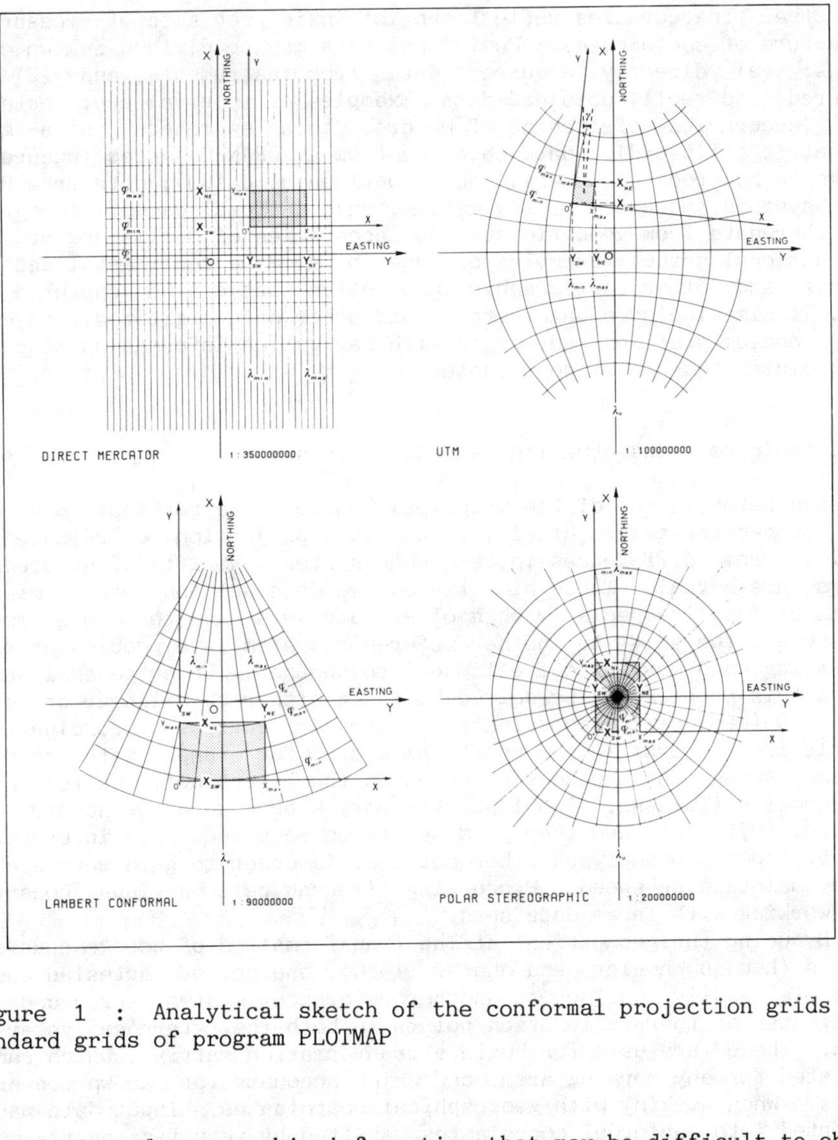

Figure 1 : Analytical sketch of the conformal projection grids and standard grids of program PLOTMAP

of geodetic and topographic information that may be difficult to obtain. However, acceptable precision over large areas, such as an entire country, may be achieved with simpler methods, if assumptions are made for the acceptable errors.

When working with cartographic media, an acceptable treshold or lower error limit could is the graphic precision at the working scale; that is, the thinnest line that is drawn on a map. This normally is 0.2 mm, representing a lower error limit of 2 meters at the scale 1:10000. This scale is possibly the smallest useable scale for geological and geophysical purposes, so it is unrealistic to try to achieve accuracies better than this figure at least on a national scale.

Other inaccuracies reflect the intrinsic precision of measurement or capture of coordinates. Positional data may be divided in two cathegories: (a) directly acquired data from instruments, and (2) map-captured, indirectly obtained data. Examples of type (a) are: satellite data (accuracies of tenths of meters, to a few meters on a global geocentric ellipsoid); data obtained from a LORAN-C system (accuracy of tenths to hundreds of meters, on a continental datum); data from Range-Range systems (accuracy of few meters, on a national datum); topographic and other data from geodetic surveys (accuracies of decimeters or less, on a national datum). Examples of type (b) are the continental and state oulines and other cartographic information that can be captured from maps. It also includes bathymetric and other data digitized from continuous analog profiles and merged with navigation information to provide each point with spatial coordinates.

2.2 Coordinate transformations and datum change

As said before, one of the principal sources of error that may occur when processing geographical (or conformal projection) coordinate data results from differences in reference systems. Shifts of hundreds of meters are normal. These bias the entire data set and cause misalignments of maps, especially when other data sets have to be overimposed. In general, the shift to another reference system is a problem of rototranslating and scaling one ellipsoid to another. Under general assumptions this problem may be solved by using either curvilinear or rectangular 3-D (geocentric) coordinates. Moreover, cartesian coordinates are useable if one works in the conformal plane, for example with the U.T.M. system. Among the methods proposed in the literature, we focused on Molodenski's (Tomassi, 1985) and Vincenty's procedures (Vincenty, 1965; Bomford, 1971), because they promised to be easy and rapid in execution. However, we also analyzed other methods, in order to gain more accuracy while maintaining speed. Processing time can be a serious constraint when working with large data sets.

By using the geometrical affine transformation of homogeneous coordinates (Loudon,Wheeler, and Andrew, 1980), one set of cartesian coordinates is easily shifted to another reference system, provided that coordinates of four calibration points in both the reference systems are known. These are used for build a tranformation matrix, which can be tabulated for any working area, and which accounts for known conversion errors. When working with geographical coordinates, input data may be converted into conformal coordinates, shifted by very fast matrix multiplication, and finally converted back to geographical coordinates in the new reference system.

Ligi and Bortoluzzi (1986,a) presented a method that gains a better accuracy, while maintaining speed. By tabulating rototranslation parameters for any area, change of datum (or of reference system) is easily performed. The four methods were tested by shifting points in the Italian National System (with origin point M.Mario,near Rome) to ED50 (with origin point Postdan, near Bonn). Ellipsoid parameters for the two systems are the same. The data set consists of the coordinates in two

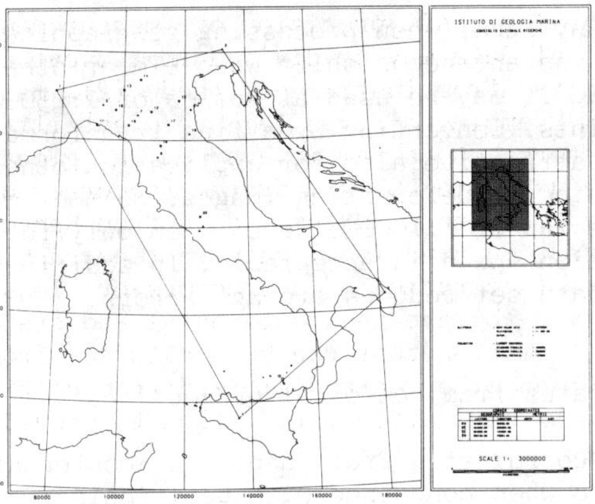

Figure 2: First order geodetic points used for testing datum change procedure, triangles represent calibration points, asterisks represent test points. Points 1 to 4 were used when calibrating with four points only.

systems of 20 first order geodetic points distributed all over the country (courtesy of Istituto Geografico Militare Italiano) (Fig.2). Table 1 summarizes the result. When compared to the graphic precision, error statistics may indicate the larger useable map scale, which is the scale where the error is lower than the graphic precision. Method (a) was determined to be fast and easy to perform, because the procedure requires only the input of shifts between the ellipsoids and differences between semimajor axes and amount of flattening. However, a maximum error was determined that limits the use of the procedure to maps at scales of 1:100000 and smaller. Method (b), is the fastest one and has an accuracy appropriate for use with maps at 1:50000.

	DATUM CHANGE $\varphi_{old},\lambda_{old} \dashrightarrow \varphi_{new},\lambda_{new}$	# of points Calib.	Test	ERRORS (mt) Max	Mean	Var.	Ex.time (msec)	Useable scale
(a)	tabulated geometrical differences between ellipsoids	0	20	16.70	10.40	3.00	1.5	1:100000
(b)	computation by least square and tabulation of geometr. diff. between ellipsoids	10	20	5.90	2.10	1.70	1.0	1:50000
(c)	computation and tabulation of affine transformation matrix in the conformal plane (e.g. UTM)	4	20	3.30	1.40	1.10	12.0	1:25000
(d)	Computation by least square and tabulation of roto-translation parameters in geocentric coordinates	10	20	2.00	0.90	0.40	3.0	1:10000
		4	20	1.80	0.95	0.40	3.0	

(a) Molodenski, as reported by Tomassi (1985); (b) Vincenty, as reported by Bomford (1971); (c) Affine transformation, Loudon, Wheeler and Andrew (1980); (d) present work. Execution times refer to a DEC micro VAX II installation.

Table 1 : Summary data of error analysis of four datum change procedures discussed in text. Calibration and test points are shown in Fig.1.

Method (c), although slower than any other when processing geographical coordinates, achieves a virtually instantaneous shift when UTM or other conformal coordinates are converted. It may be used at scales of 1:25000 and needs only four calibration points. Converting by affine transformation of homogeneous coordinates is attractive also for application other than datum shifts, as when stretching or deformating images. Method (d) is sufficiently accurate at a scale map of 1:10000, even when only four calibration points are used. Execution speed is acceptable. In addition, the use of a larger control point data set reduces average errors.

2.3 Capture of geographical coordinates from conformal projections

The digitization of geographical coordinates from maps in conformal projections is of great importance when concerned with positioning or cartographic information not stored in machine-readable media.

In theory, this problem can be solved by shifting the digitizer coordinates to the reference system of the map, if one assume that the geographical map frame is right-handed cartesian. However, this assumption is correct only for large-scale maps. As it can be seen in Fig. 1, the deformation of the plane transforms of the geographical grid does not permit one-to-one correspondence between the two reference systems. Conversely, a one-to-one correspondence exists between the conformal projection coordinates, that by definition are rectangular cartesian, and the geographical ones. Thus, by shifting the digitizer's coordinates to the conformal plane, one obtains the geographical coordinates by inverting the mapping function of the map projection.

Bortoluzzi and Ligi (1986) presented a computer program that acquires geographical coordinates from the most used conformal projections. The use of the affine transformation of homogeneous coordinates allowed the creation of a flexible program, that potentially can recover part of the map deformations. The independence of the procedure from scale and projection was also tested, showing that the main conntraints are the map conditions and operator's selection of calibration points. Error analysis indicated that, in general, the converted coordinates approach those expected at twice the graphic precision.

The procedure is extensively used for digitizing of continental boundaries, bathymetric observation, sampling stations, and navigation lines, and is suitable for any geophysical or geological variable or feature drawn on a map. The original version was updated to conform to the software environment of Ligi and Bortoluzzi (1986,a,b).

2.4 Digitization of continuous analog profiles

The digitization of continuous analog profiles (bathymetry, magnetometry, reflection seismic lines) is not only a mean for capturing and processing previous data, but possibly is the only way to capture information that pertain to certain interpretations of the raw data.As in the treatment of seismic profiles, different geophysicists may provide diffe rent interpretation, each of which have their own spatial relationships.

Even in the situation of direct digital acquisition of data, the digitization of analog profiles may be necessary to capture features that are not revealed by the adopted sampling rate. The problem is again a matter of geometric transformations between different reference systems (Fig.3), and can be solved easily if the assumptions are made that : (a) the speed of the vehicle was constant between two contiguous control points, and (b) the deformations induced by scale length when working with geographical coordinates are lower than graphic precision.

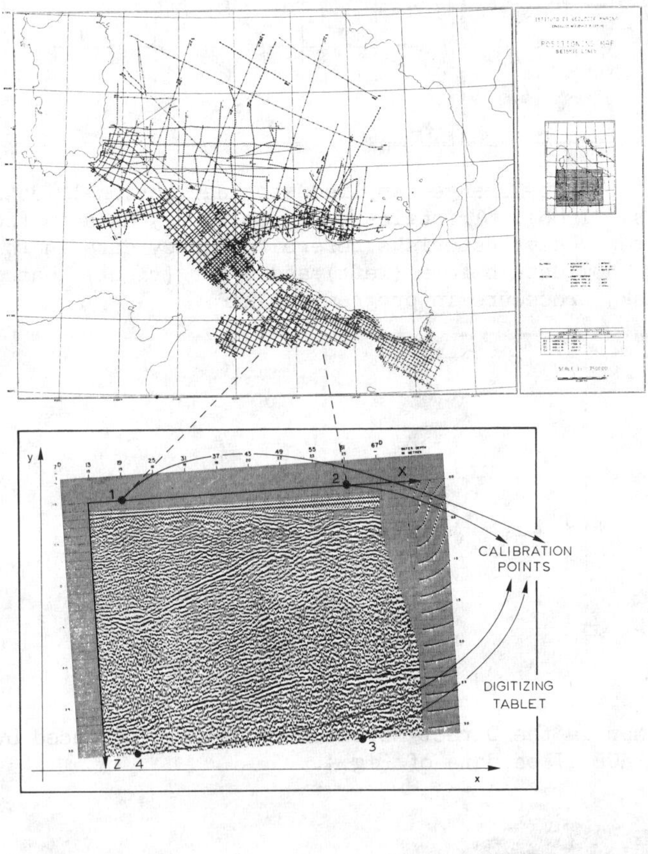

Figure 3 : Imaginative sketch of digitization of continuous analog profiles. The digitizer reference system (x,y), the profile reference system (X,Z) and the navigation reference system (Latitude, Longitude) are shown.

The procedure we developed (Bortoluzzi, Ligi, and Prosperi, in preparation) emphasized man-machine interaction, and provide the operator a high flexibility in the organization of his task. The acquisition phase is separated from the recontruction phase, while digitized information are stored and managed by a specialized DBMS. A large number of horizons can be acquired per single profile. Final coordinates of the reconstructed profiles may be geographical or UTM, corresponding to the coordinate system of the control points.

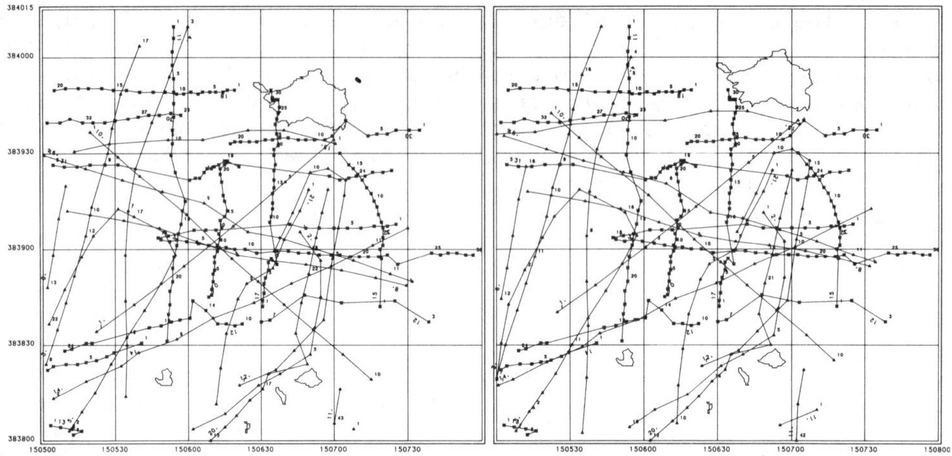

Figure 4: Example of datum change applied to navigational data (Gabbianelli and others, 1986). Triangles refer to survey data in the Italian Topographic System, heavy asterisks refere to survey data in the ED 50 system. The maps show data before (left) and after (right) datum change with the Molodenskj procedure in program PLOTMAP.

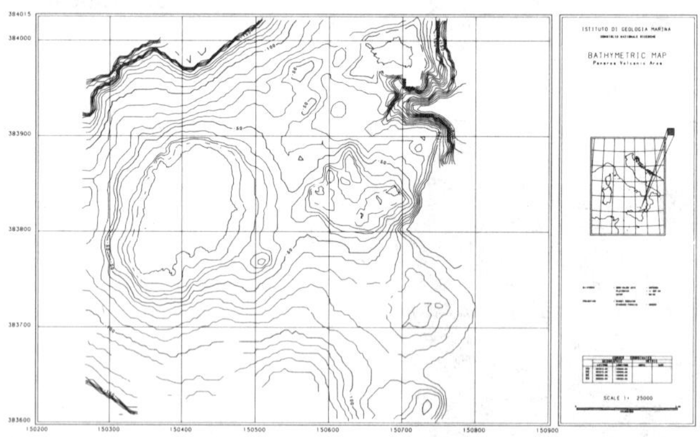

Figure 5: Bathymetric map in the Direct Mercator Projection produced by program PLOTMAP and SURFACE II on data of Fig.4.

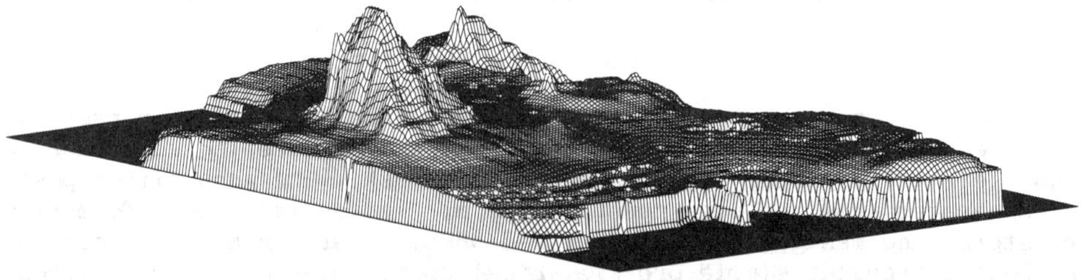

Figure 6 : Perspective diagram produced by SURFACE II during processing of data of Fig. 4.

2.5 Plot and spatial analysis of data sets in geographical coordinates

The problem of linking standard cartography with spatial processing, can be solved by: (1) creating a geographical grid under the operator control, that is suitable also for pertinent cartographic information; and (2) converting input data coordinates to projection coordinates, and transferring them to either on-line or external spatial analysis package.

Ligi and Bortoluzzi (1986,b) developed a procedure that performs good quality standard cartographic mapping, with interactive selection of projection and drawing options, and also performs spatial processing by interactively interfacing with graphics systems. Options include geographical grid creation, inclusion of coast and bathymetry outlines, posting of points and lines, and the creation of small-scale reference maps with continental outlines. The small reference map can be drawn on a separate reference box and automatically reference the selected frame. Each user response is written on a logfile, that can be edited and submitted as a job batch for either reproducing output or performing a different analysis. Published software, such as Watson (1982), is suitable to produce some spatial processing when inserted in the program.

Because of its flexibility and powerful spatial analysis capabilities, we interfaced the SURFACE II graphics system (Sampson, 1978). Other graphics systems, like GPCP (Calcomp, 1971) were interfaced as well. We process SURFACE II main commands, such as those controlling the input, boundaries, blanking, gridding and contouring. The projection-converted data sets and command file are available on disk upon completion of the analysis. Changes of parameters, or addition of commands can be made with minor effort. In addition to give allowance to the user to input blanking areas to the system, an automatic blanking procedure is available, which creates the minimum convex hull that contains samples.

3. EXAMPLES OF APPLICATIONS AND CONCLUDING REMARKS

Gabbianelli and others (1986) studied the morphology of the marine area surrounding the island of Panarea. Two detailed bathymetric surveys were carried out by range/range positioning under two different datums. The 1984 survey was linked to the the Italian National Topographic System, whereas the 1985 study was refered to ED50. The maps in fig. 4 give a comparison between the data plotted with and without datum shift. The entire, homogeneous data set was input to the program PLOTMAP, that produced a bathymetric map in the Direct Mercator Projection (fig.5). In addition to contour the data, SURFACE II intermediate files were used to produce perspective block-diagrams (fig.6).
Bortoluzzi, Frascari and Ravaioli (1986) collected 45 surface samples in front of the Tiber River mouth. Grain-size analyses were made and a Q-mode factor analysis was performed. Fig. 7 shows automatic contouring of Factor 4 (silty clay) in the UTM projection produced by SURFACE II.

Bedetti (1986) and Grassi (1986) produced the morphostructural map of the Thyrrenian bathial plain shown in Fig.8, from horizons digitized on 30KJ sparker seismic profiles.

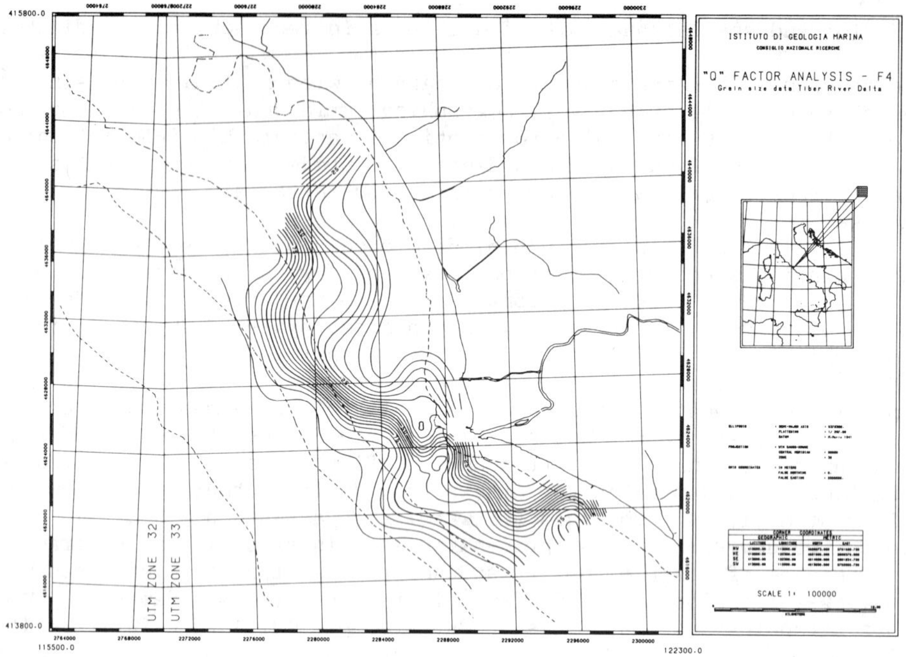

Figure 7: Automatic contouring by SURFACE II of factor 4 of factor analysis of grain-size data (Bortoluzzi, Frascari and Ravaioli, 1986). Note the zonechange of UTM graticule produced by PLOTMAP.

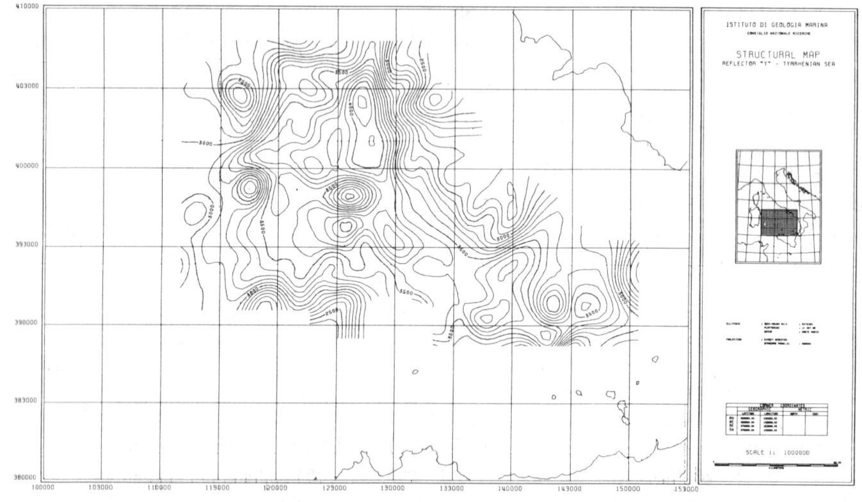

Figure 8: Morphostructural map in the Thyrrenian bathyal plain.Seismic horizons digitized by 30Kj sparker profiles and processed by SURFACE II to produce transformation of scale (time/thickness) and contouring.

By interfacing SURFACE II we intended primarily to make its use easier for the operators. Our experience shows that this goal was achieved, and that "raw" maps with basic properties can be made with a minimum knowledge of both the computer Operating System or of SURFACE II structure. However, a good knowledge of both spatial analysis and the SURFACE II system is required to insure data reliability and optimum performance. As a matter of fact, all the maps shown in the examples were produced after extensive fine tuning of the "raw" maps.

The examples illustrate that some of the problems related to mapping and spatial analysis in the conformal projection domains are solved. In a similar way, a higher integration level could include data base management and statistical processing. As a matter of fact, once the information is stored and can be retrieved easily and efficiently, and different tasks can be performed on the data by simple procedures, productivity increases of some orders of magnitude. DBMS such as GRASP of the USGS (Bowen and Botbol, 1975) may be good candidate to serve as the core of a data processing system. A direct interface to the computer operating system could allow both data retrieval and storage and processing by other software components. However, particular problems of data storage, compression and fast retrieval, may require the production of specialized DBMS's, as in the instance of geographical coordinates obtained by digitizing continental boundaries and bathymetry from large-scale standard cartography, or when acquiring data from continuous analog profiles.

ACKNOWLEDGEMENTS. This work represents contribution N. 596 of the Istituto di Geologia Marina (IGM) of the Italian National Council for Research (CNR). Dr. M. Ligi was partially supported by a fellowship of the CNR's Comitato per le scienze Geologiche e Minerarie. The support and encouragement of Dr. F. Frascari and Dr. N. Zitellini of IGM are acknowledged. Ing. Arca of the Istituto Geografico Militare Italiano kindly provided data test information.

REFERENCES

Bomford, G. (1971) Geodesy, Clarendon Press, Oxford, p.593.

Bedetti A. (1986) 'Carta morfostrutturale dell'orrizzonte X ed isopache del pliocene inferiore-medio P.P. della piana batiale del M.Tirreno', Tesi di laurea, Ist.Geol.Paleont., Universita', Bologna.

Bortoluzzi, G., Frascari, F., and Ravaioli M. (1986) 'Ricerche sedimentologiche in aree marine adriatiche e tirreniche finalizzate alla comprensione dei fenomeni di inquinamento marino', Mem. Soc. Geol. It., Vol. 27, 499-525.

Bortoluzzi, G., and Ligi, M. (1986) 'DIGMAP: a computer program for accurate acquisition by digitizer of geographical coordinates from conformal projections', Computers & Geosciences, Vol.12, no.2, 175-197.

Bortoluzzi G., Ligi M., and Prosperi M. (1986) 'DIGPROF: a computer program for the acquisition and management from continuous analog profiles', manuscript in preparation.

Bowen R.W. and Botbol, J.M. (1975) 'The Geological Retrieval and Synopsis Program (GRASP)', U.S. Geological Survey, Prof. Paper, no.966, 87pp.

Evendon G.I. (1986) 'UNIX cartographic projections', U.S.Geol.Survey, technical report in press., Woods Hole, Massachusetts.

Gabbianelli G., Gillot, P.J., Lanzafame, G., Ligi M., Postpitschl D. and Rossi P.L. (1986) 'Controllo strutturale nell'evoluzione vulcanica di Panarea (Isole Eolie)', submitted to Marine Geology.

Grassi K. (1986) 'Carta morfostrutturale dell'orrizzonte X ed isopache del pliocene medio P.P-attuale. della piana batiale del M.Tirreno', Tesi di laurea, Ist.Geol.Paleont., Universita', Bologna.

Ligi, M., and Bortoluzzi, G. (1986,a) 'DATUM: a computer program for accurate and fast datum shift of geographical coordinates between different cartographic systems', manuscript submitted to Computers & Geosciences.

Ligi, M., and Bortoluzzi, G. (1986,b) 'PLOTMAP: geological and geophysical applications of good standard quality cartographic software', manuscript submitted to Computers & Geosciences.

Loudon , T.V., Wheeler, J.F. and Andrew, K.P. (1980) 'Affine transformation for digitized spatial data in geology', Computers & Geosciences, Vol.6, no.4, 397-412.

Richardus , P. and Adler, R.K. (1972) Map projections for geodesists, cartographers and geographers, North-Holland, Amsterdam, p.73-101.

Sampson, R. (1978) 'SURFACE II Graphics system', Kansas Geological Survey Series on Spatial Analysis, No.1-A, update 7, Univ. Kansas, Lawrence, Kansas, 240p.

Snyder, J.P. (1982) 'Map projections used by the U.S.Geological Survey', U.S.Geol.Survey Bulletin, no.1532, 310p.

Tomassi, G. (1985) 'Radioposizionamento in mare', AGIP, Interim Report, S.Donato Milanese, Italy, 85p.

Watson D.F. (1982) 'ACORD: Automatic contouring od raw data', Computers & Geosciences, Vol.8, No.1, 97-101.

Vincenty, T. (1965) 'Transformation of coordinates between geodetic systems', Survey Review, XVIII(137), 128-133.

A CLASSIFICATION PROCEDURE FOR EXPLOITABLE RESERVES

R. Bruno
Università degli Studi di Roma "La Sapienza"
Dipartimento di I.C.M.M.P.M.
Via Eudossiana, 18
00184 - Roma

M. Guarascio
Università degli Studi di Trieste
Istituto di Arte Mineraria
Piazzale Europa, 1
34127 - Trieste

D. Lipari
Società Italiana Miniere S.p.A.
Miniera di Monte Agruxiau
09016 - Iglesias (Ca)

ABSTRACT. This paper proposes a classification procedure for minable reserves for use by mining companies. After reviewing resource/reserves concepts and definitions, the inter-relationship between reserves and mining project is examined with a view to establishing a classification based on: a) considerations of economic feasibility, and b) level of confidence of the design itself. The described approach relies on the probabilistic evaluation of Net Present Value (NPV) and is presented as an operational procedure for recognizing and characterizing the random technical and economic magnitudes of mine designs with particular reference to the reserves.

1. INTRODUCTION

Before attempting to classify ore reserves it is necessary first to define what is understood by this term, especially with regard to the distinction between "resources" and "reserves" in the context of a mining project.

In fact, the word "reserves" is generally taken to mean those resources that can be technically and economically exploited. According to the classical reference system where: Degree of geological assurance/Level of economic feasibility, the reserves are the resources that have been defined as being "demonstrated and economically feasible" (Fig. 1).

This definition, which is essentially common sense, is not sufficiently precise for the evaluation needs of a mining company, however. For example, even

C. F. Chung et al. (eds.), Quantitative Analysis of Mineral and Energy Resources, 565–580.

though a detailed analysis may have confirmed, with a high level of confidence, the presence of high and therefore economic grades, an orebody may be located in an area subject to political, social or military constraints that effectively preclude its exploitation: it is clear that a company cannot consider such an orebody to be a reserve. On the other hand, an orebody with low metal grades may not be economically feasible, yet its exploitation may be justifiable in the context of a strategy for the verticalization of a company's production or for national self-sufficiency in a given industrial sector.

The dynamic nature of any classification must also be emphasized since it is perfectly possible to obtain entirely different classifications for the same project simply because of variations over time of the base parameters.

The above considerations highlight the vagueness of the generic term, reserves, but they cannot indicate how the problem can be solved once and for all because other elements (political, social, etc.) are involved that are quite beyond the scope of a normal quantitative evaluation. It is, therefore, necessary to go deeper into the concept and definition of reserves, and this task is less difficult where the pertinent realm of validity and interest can be clearly specified. In our case this is the mining company.

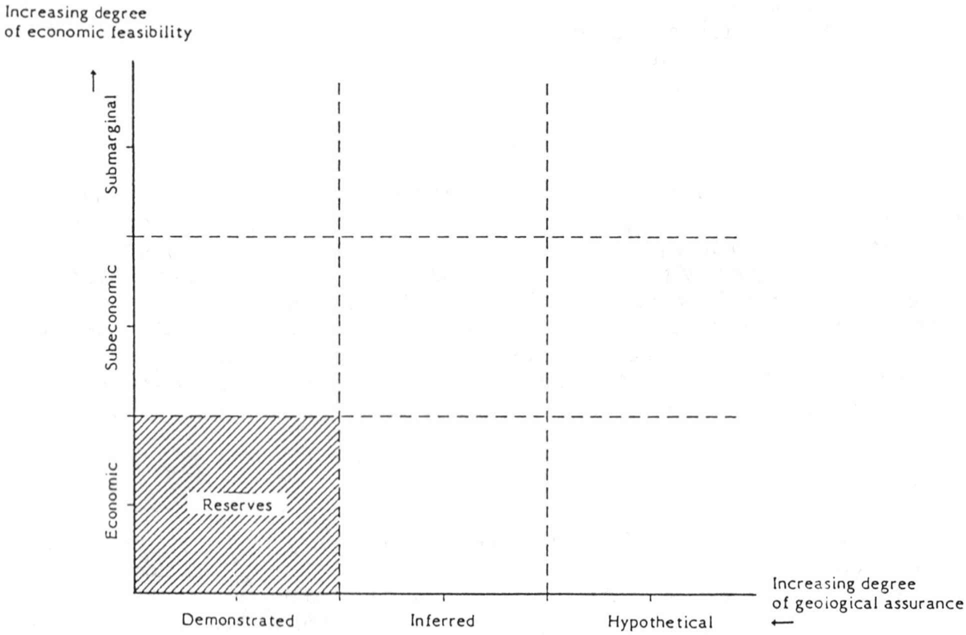

Fig. 1 - **Resources classification according to degree of geological assurance and economic feasibility**

2. THE RESERVES/PROJECT RELATIONSHIP AND CLASSIFICATION

As already mentioned, a resource is normally classified on the basis of two characteristics:

a) measurements of the intrinsic physical magnitude(s) peculiar to each orebody, i.e. grades and mineralized volumes;
b) the degree of geological assurance, i.e. the level of knowledge of these magnitudes or of the orebodies themselves.

For a resource to be considered a reserve, it is necessary to examine, in addition, the technical and economic feasibility of its exploitation, and hence to define a scale of magnitudes that can measure that exploitability. In reality, the elements for evaluating the true exploitability of an orebody can only be provided by a well prepared, positive feasibility study. This implicit dependence on the existence of feasible designs suggests that a classification of reserves is actually a classification of the alternative designs that can be prepared on the basis of the available knowledge of the orebody, namely the resources.

By way of example, consider the case of an orebody for which two or more designs have been prepared on the basis of different technologies or philosophies of production. The differences in the foreseeable economic results will allow us to distinguish between them, with one of the designs appearing clearly more promising than the other(s). It is even possible that one design may indicate a positive outcome and another a negative one: although both necessarily refer to the same physical entity, i.e. the orebody, and dispose of the same us information for all the parameters normally considered (degree of geological assurance, quantities of useful minerals, etc.), the first one "transforms" the resource into a reserve while the second one does not.

This could be the case of the difference between an orebody that is exploited selectively, involving low investments but also low production levels, and the same orebody exploited with highly productive but poorly selective technologies. In this case, the differences between production features (lower outputs/high grades; higher outputs/low grades) will be evident, but the differences in terms of economic results cannot always be evaluated in absolute terms nor guessed at a priori because they are dependant - with often complex links - on the characteristics both of the orebody and the design itself.

Because some measure of exploitability is needed - though it must always be considered in connection with the economic feasibility of the project - it is necessary to take one of the typical economic and financial magnitudes used for project evaluation. An example is the Net Present Value (NPV) of the Discounted Cash Flow (DCF), which is obtained by applying the company's internal discounting rate to the tentative cash flow drawn up on the basis of envisaged project implementation.

Money is an excellent unit of measurement as it allows us to compare totally different realities: e.g. exploitation of an orebody containing Pb, Zn and Cu deposited in seams with the exploitation of a massive deposit of pyrite.

$$NPV = \sum_{i=1,\ N_a} F_i \times (1 + r)^{-i}$$

where:
- NPV = Net Present Value of the Discounted Cash Flow
- r = Discounting rate
- N_a = Foreseen project lifetime
- F_i = Cash flows by years i

It should be noted that the cash flow, generally presented in terms of actual flows of money (returns, costs and investments), is heavily conditioned by certain features of the orebody, like grades and mineralized tonnages, that are the fundamental parameters for the first magnitude, NPV, adopted for reserves classification.

It has been seen that the second magnitude adopted for classifying resources is the degree of geological assurance of the resource itself, i.e. the level of confidence with which the first magnitude, e.g. tonnages, has been estimated. The same is true for the classification of reserves/projects, where a second magnitude is needed to define the level of confidence of the first one.

Having decided to take NPV as the basis for measuring the economic viability of a reserve/project, the necessary next step is to consider the random nature of NPV and hence of the magnitudes and parameters used to calculate it. In fact, the values assumed for tonnages and grades are not actual figures but estimates, evaluations or forecasts; the same is true of the values taken for metal prices or costs, although these are parameters of a completely different nature. Thus, the levels of confidence can be measured by calculating the variances.

Basically, the generic flow F_i is a function of random parameters, and a random variable will of course be any index, like the NPV, that is calculated from the flows. Thus, the correct meaning of the NPV associated with a project will be that of an expected value; consequently, where the confidence level of a project is high, the variance of its NPV will be small and vice versa.

The proposed classification is therefore expressed in terms of NPV/RSD where RSD = Relative Standard Deviation; thus, RSD = SQRT(variance) /NPV, or the ratio between the standard deviation and the expected value of NPV for each individual project (Fig. 2).

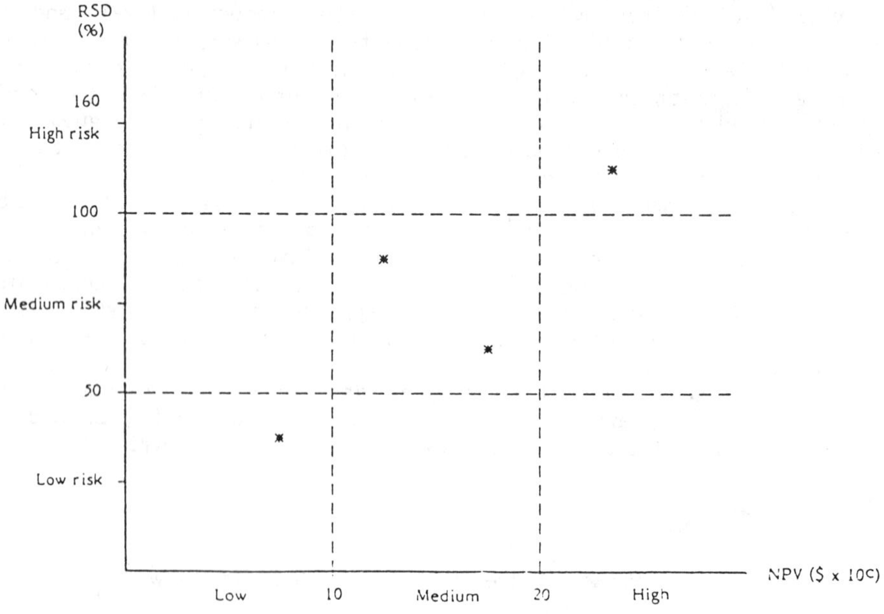

Fig. 2 - Reserves (projects) classification by expected economic result (NPV) and corresponding level of confidence (RDS)

At this point, it is possible to qualify each and every project on the basis of the expected economic result and level of risk (high, medium or low) associated with it.

The combined analysis with output, i.e. the amount of material that can be produced during total project lifetime or year by year, may prove to be fundamental and discriminating. Fig. 3 shows how the NPV/RSD of a project "A" is plotted against the size of the reserves themselves. It is thus possible to determine:

a) the specific profitability (NPV/T) of the project (high, medium, low);
b) the size of the reserves in absolute terms (T) or in terms of productivity (low, medium, high);
c) the sensitivity of the project to variations in the classifications, measured by plotting the intervals of the standard deviations of NPV against the size of exploitable reserves.

For the cases where it is important to evaluate a project in terms of the impact of investment, the data on reserves are replaced by data on foreseeable investments Fig. 4 shows how a project "B" is analysed interms of investment.

Similarly, it may be essential to evaluate the results of a project in terms of the quantity of useful material contained in the end-product. In fact, this parameter, which could be called the supply capability, is of significant interest for the industrial or market operators immediately downstream from the mine (smelters, etc.). For example, two projects, "A" and "B" (Fig. 5) showing comparable levels of profitability may be differentiated on the basis of differences in their capabilities in this respect. In the context of the mining/metallurgical process as a whole, the considerations based on supply capability could well justify the selection of a project with a lower profitability rating than that of the others.

In summary, the proposed approach, presented schematically, is as follows:

a) classification of the reserves/projects in terms of their economic feasibility and reliability or level of confidence of estimates;
b) use of the NPV to measure economic feasibility, and its variance to measure level of confidence;
c) highlighting of the fundamental parameter(s) or magnitude(s) that complete the fundamentally important reference framework for the characterization of reserves/projects.

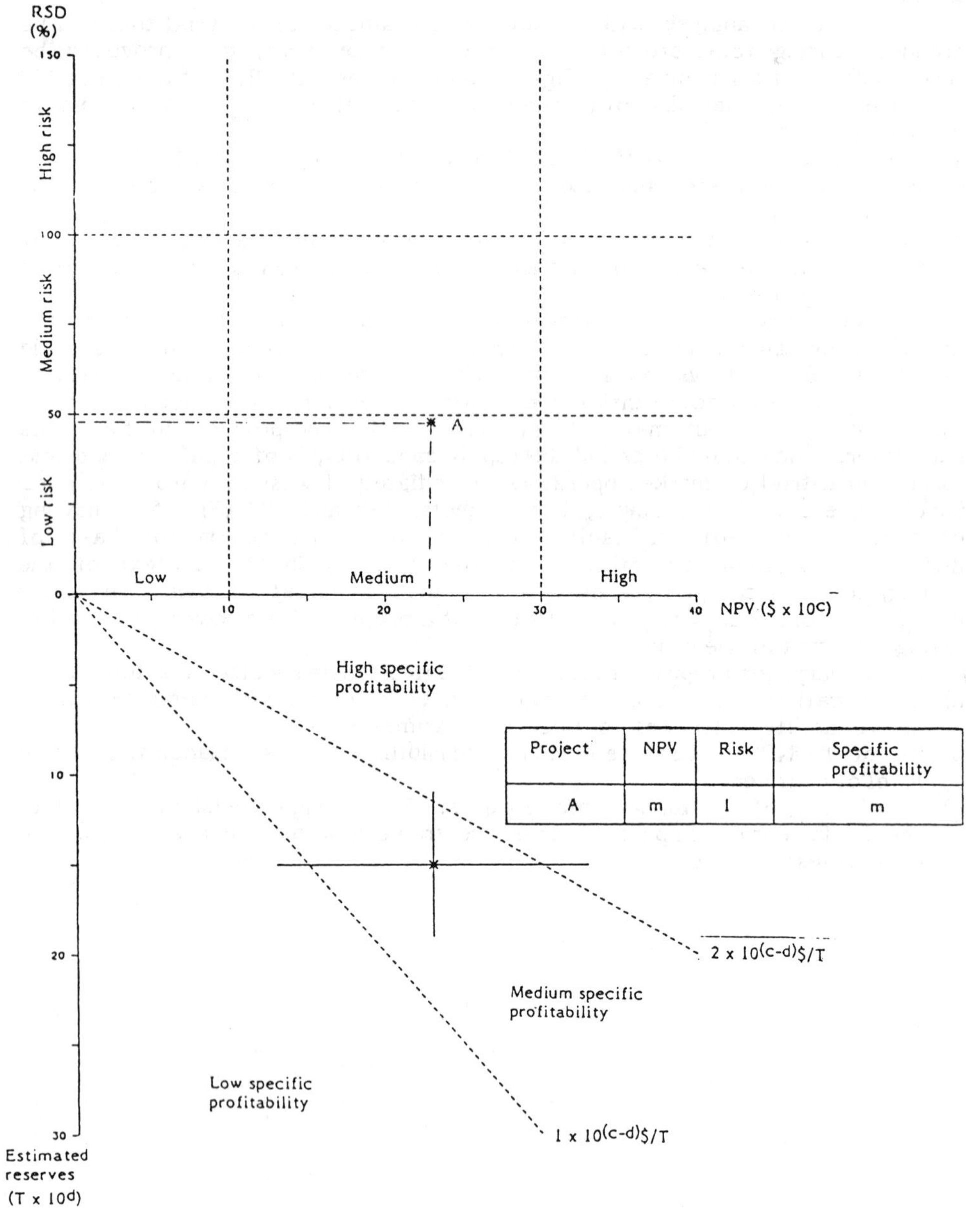

Project	NPV	Risk	Specific profitability
A	m	l	m

Fig. 3 - Analysis of projects classification based on reserves

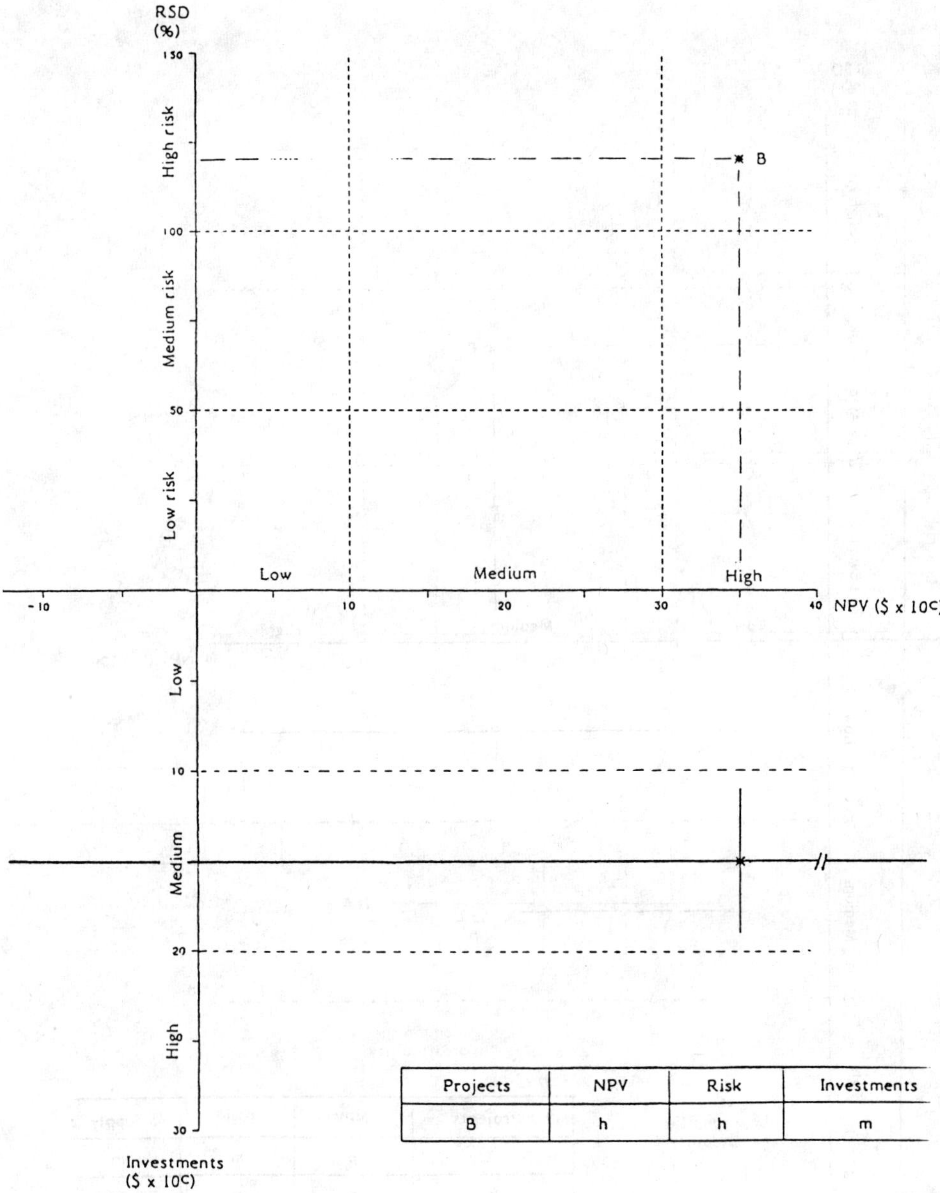

Projects	NPV	Risk	Investments
B	h	h	m

Fig. 4 - Analysis of projects classification based on investments

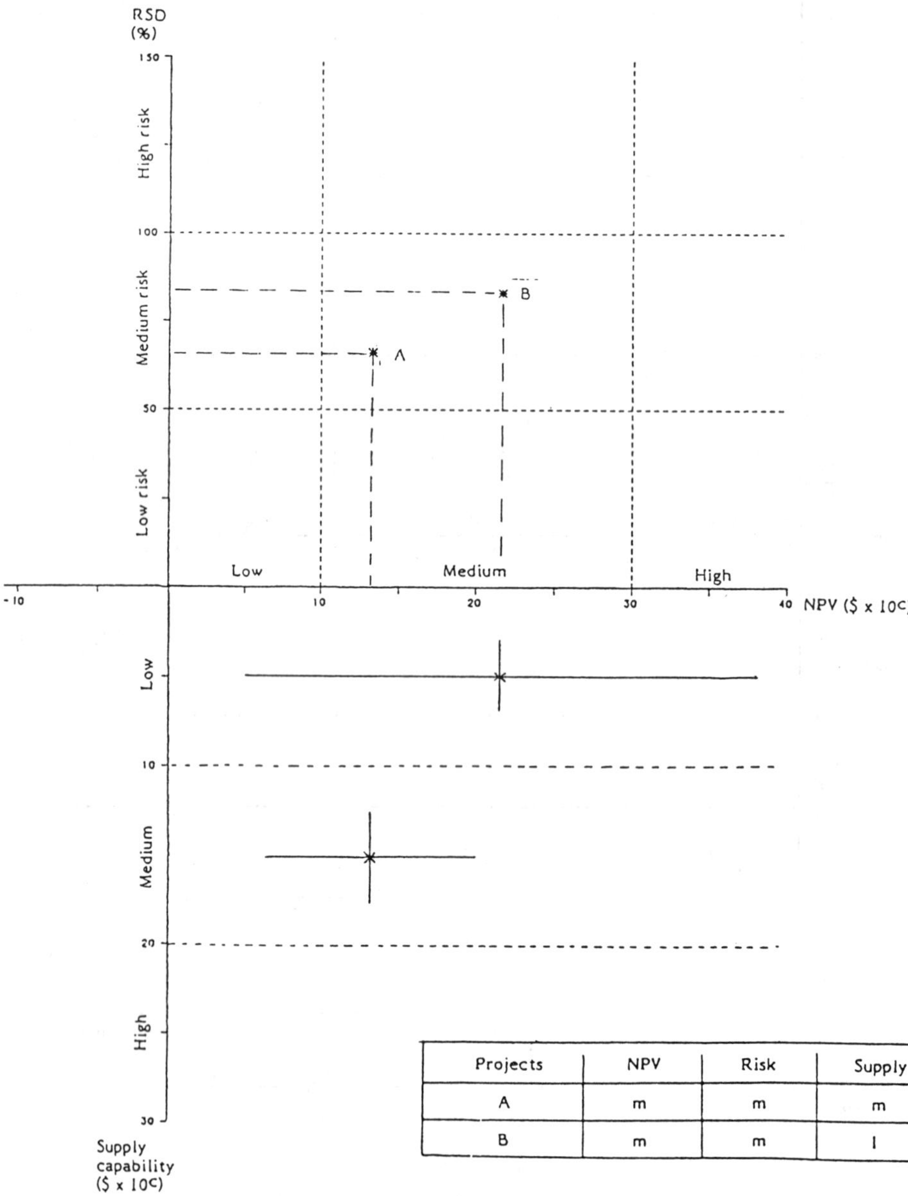

Projects	NPV	Risk	Supply
A	m	m	m
B	m	m	l

Fig. 5 - Analysis of projects classification based on supply costs

3. COMPUTING PROCEDURE

To make the described classification system operational, the calculation of NPV is expressed as a function of project magnitudes (tonnages, grades, etc.), that must be characterized both as project values and as probability distributions.

The flow F_i that appears in the calculation of NPV (Fig. 6) can be simplified as follows:

$$F_i = R_i - C_i - I_i$$

where R_i, C_i and I_i represent the foreseeable returns, costs and investments during year i of the project.

Whereas the investment figures of a project's initial cash flow can be assumed relatively constant, since investments can be considered to be independent variables, models can be built to take account of the dependence of the flows of costs and returns on a number of elementary magnitudes of the orebody and/or the project. Costs, for example, may easily be associated with the production level P_i of the project during year i, by calculating the classical ratio of fixed costs/variable costs (Fig. 7).

For the returns, the ratios may have to be somewhat more complex but account must always be taken of the fundamental features of the orebody (tonnages, grades), of the planned mining and beneficiation processes (productivities, recoveries, yields), and of market considerations (metal prices, exchange rates).

$$R_i = R(P_i, t_i, rec(t_i), V, \dots)$$

where
R_i = Returns during the year i
R = Returns function
P_i = Production during the year i
t_i = Grade of feed ores during year i
$rec(t_i)$ Recovery of useful material, expressed as a function of t_i
V = Value of concentrate

Each one of the independent magnitudes defined for the project can be considered as a random variable with a related probability distribution, and the project value for it must correspond to a characteristic parameter of the distribution: mean/expected value, mode/most probable value, median/50% mark, etc. Once these distributions are determined, the NPV distribution can be calculated using, for example, classical risk analysis by the Montecarlo Simulation method; the expected value and the variance of NPV are thus calculated.

The problem now consists in characterizing the probability distributions of the random variables of the project that has been designed to exploit the resource. At this point, it is necessary to distinguish between three different types of variables:

a) there are the internal variables of the project, which can be estimated and recognized with a desired precision (that clearly has a cost): e.g. the grades/tonnages curve can be calculated to any desired precision and the related sampling programme planned to the correspondingly required detail;
b) there are the internal variables of the project that are linked to future

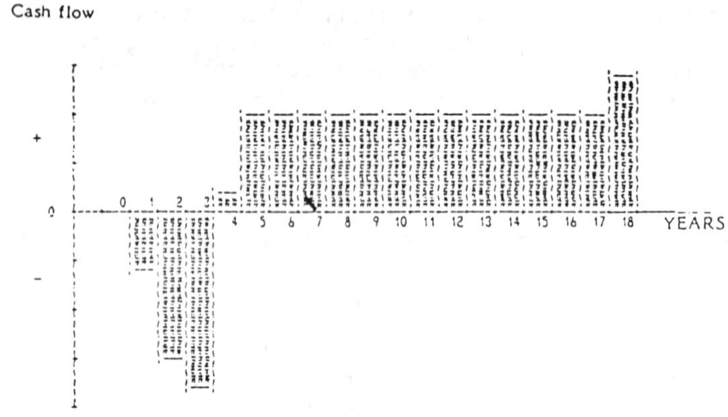

Fig. 6 - Cash flow of a project

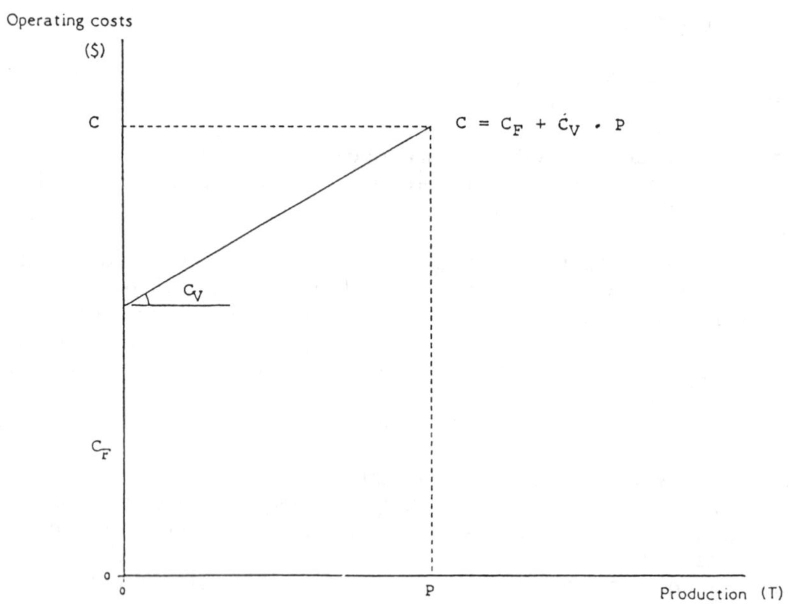

Fig. 7 - Graph of operating costs in fixed and variable terms events, the probability distributions of which are only partially dependent on present-day actions: production levels, recoveries, etc., and therefore can only be estimated on the basis of comparisons with similar processes, pilot tests, analyses of historical series, etc.;

c) finally, there are the external variables, which can only be forecast because they are linked with future facts and situations and more or less independant of present-day actions: metal prices, exchange rates, costs of labour.

For the first type of variables, characterization of the probability distribution can be performed by applying, to existing sampling data, geostatistical estimating techniques that provide the foreseeable values for grades and the related estimating variances for the different volumes identified by the project for each year of its lifetime. For a stationary punctual kriging run, for example:

$$t_i = \sum_{j=1} pj \times tj$$

where: t_i = Estimated grade of feed ore during year i
j = Index of the given generic
pj = Kriging weights
tj = Known grades data
nd = Number of available data elements

and $$VS \sum_{j=1,nd} pj \times \bar{G}(j,V) + m - \bar{G}(V,V)$$

where VS = Estimating variance for grades
$\bar{G}$ = Average variogram
V = Volumes mined
m = Lagrange's parameter

It is thus possible to formulate hypotheses concerning the type of distribution that is associated with this estimate; this is done on the basis of the statistical distribution observed in sampling data and the known impact of the estimating and "regularizing" operations on the grades distribution.

With regard to the magnitudes of type b), it is basically not reasonable to expect to arrive at an objective definition of their distributions. Rather, the proposed approach is heuristic, consisting in exploiting the designer's capacity to synthesise all the information that is available from hetergeneous sources (test results, historical series, similar cases, forecasts) processing them by means of a PERT-type approach:

a) assumption (arbitrary) of a beta-type probability distribution;
b) subjective identification, by the designer, of the three characteristic values for each variable: minimum (Min), maximum (Max) and most likely (Mode);
c) definition of the parameters of the distribution, mean and variance (Var) by applying the following arbitrary equation, whose theoretical significance is well known:

$$\text{Var} = (\text{Max} - \text{Min})/6$$
$$\text{Mean} = (2\ \text{Mode} + (\text{Min} + \text{Max})/2)\ /3$$

With regard to the magnitudes of type c), though it is possible to proceed as described for magnitudes of type b), it must be noted that the design engineer is rarely the most qualified person to make such forecasts. There is a fairly wide range of possible solutions. By way of example, models developed by expert sources (operators in the sector, specialized companies, government agencies or international agencies, etc.), some of which are already available on Expert Systems, may be applied. Alternatively, in-house models may be constructed (on the basis, for example, of trend analyses), preferably in collaboration with specialized company staff.

Whichever solution is adopted, this is not a simple task because it is necessary to take account of the effects of inflation and corresponding rates, of the evolution of exchange rates, of structural changes in the market, etc.

The example is provided below of an analysis of Pb and Zn prices, performed by applying geostatistical interpretations to equivalent "instruments" (correlogram/variogram) normally applied for time series analysis.

Visualization of the historical series immediately brings to light the non-stationary nature of metal prices for there is clear evidence of a structural trend since the 70s (Fig. 8). The annual variation of these prices was therefore analyzed: the average value corresponds to the average yearly trend and the time series appears to be quite stationary, in the geostatistical sense of course, even for this case of not deflated prices (Fig. 9). Variographic analysis was then performed to verify the degree of temporal correlation between the variations, i.e. the level of ressemblance between the yearly values at distances of 1, 2, 3,... years. The results are illustrated in Fig. 9, where the variograph shows a "nugget" effect, well known to geostatisticians. This result means that each annual increase is independent in respect of the preceding values, especially the variations observed the preceding year: in other words, if a small increase occurs in any given year, it will have no effects on the probability of an increase or decrease or the size of the variation the next year. Basically, knowledge of the variations that occurred during a given year does not provide useful information for forecasting those of the following year.

Although provided only as an exemplifying case, other significant magnitudes not being taken into account, the above result is of considerable help for applying the so-called Montecarlo simulation techniques to the cash flow of a given project because the flow characteristics are normally simulated year by year.

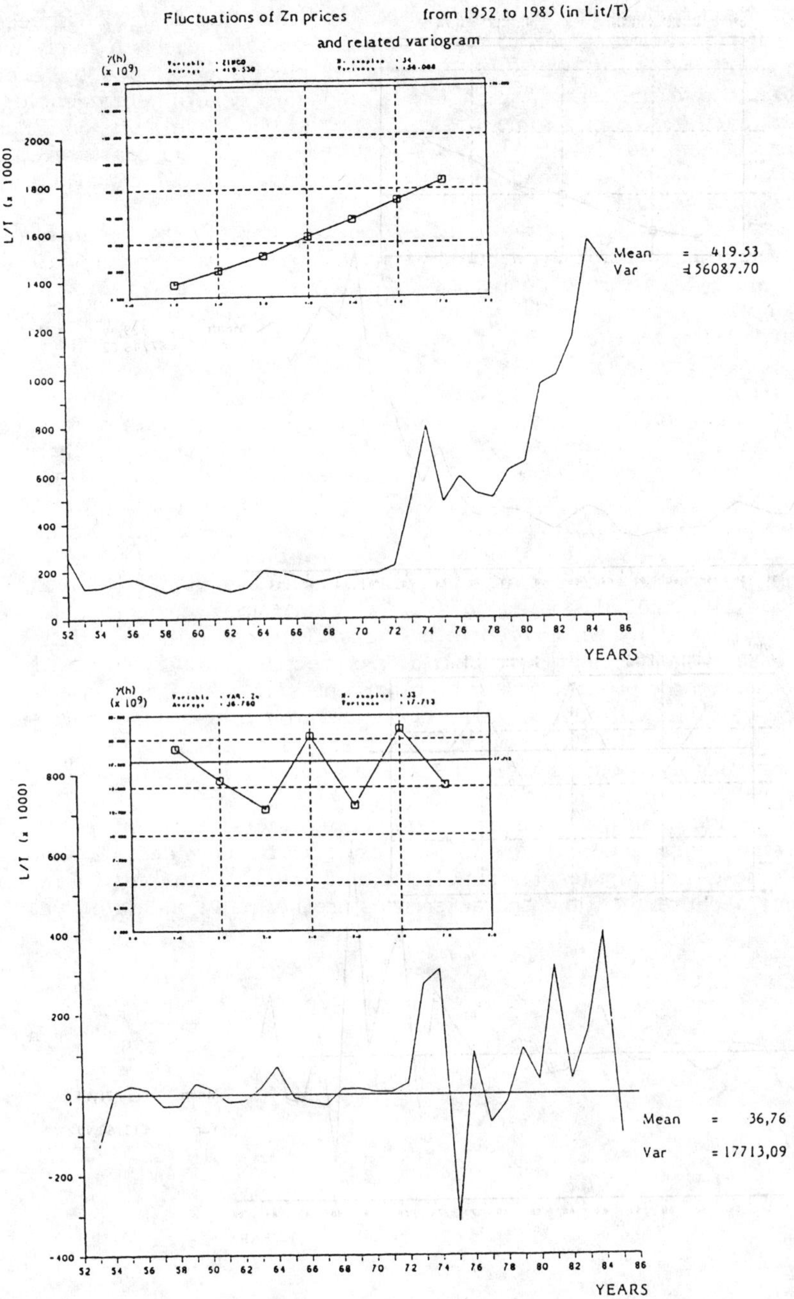

Fig. 8 - **Fluctuations of Zn prices from 1952 to 1985 (in Lit/T) and related variogram**

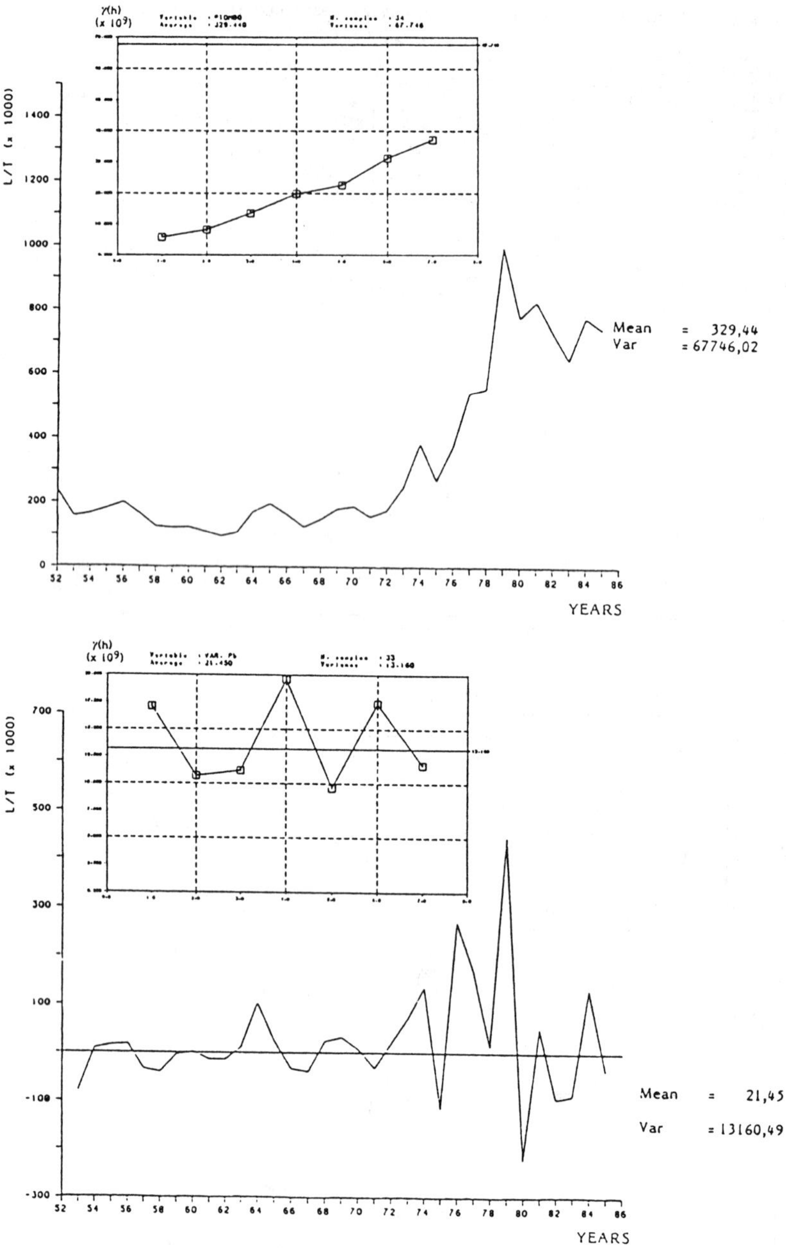

Fig. 9 - Fluctuations of Pb prices from 1952 to 1985 (in Lit/T), and related variogram

4. CONCLUSIONS

For a mining company, the classification of available reserves is, in practice linked to the classification of its projects for the exploitation of those reserves.

The evaluation of mining projects is based on considerations of economic feasibility and level of confidence of the project itself, and therefore on the technical and economic magnitudes that characterize it.

The described approach bases such classifications on the NPV interpreted as a random variable, and on the Relative Standard Deviation of the feasible NPV. To make this procedure operational, it is necessary to recognize and characterize the random technical and economic magnitudes of the project with special reference to the parameters of the reserves.

REFERENCES

1. Diehl, P. and David, M., 'Classification of ore reserves/resources based on geostatistical methods', , CIM Bulletin, Feb. 1982, p. 127

2. United Nations Economic and Social Council, The International Classification of Mineral Resources, E/C.7/104, Ankara, March 1979.

3. U.S. Bureau of Mines and U.S. Geological Survey, Resource/Reserve Classification System, U.S. Geological Survey, Circular C831, Washington, D.C., June 1980.

4. Energy, Mines and Resources Canada, Departmental Terminology and Definitions of Reserves and Resources, Interim Doc., Ottawa, Jan. 1975.

5. Australian Inst. Min. Metall. and the Australian Min. Ind. Council, Report by the Joint Committee on Ore Reserves, Victoria, 1972, Unpubl.

6. Sabourin, R., Définitions des réserves (Reserves evaluation), Energy, Mines and Resources Canada, Energy Research Program, Rep. ERP/MRL, Ottawa, 1978.

7. Institution of Mining and Metallurgy, Standards and Definitions, London, 1902.

8. Cummins, A.B. and Given, I.A. (Ed.), SME - Mining Engineering Handbook, New York, 1973.

9. David, M., Geostatistical ore reserve estimation, Elsevier, 1977.

10. Journel, A.G., and Huijbregts, Ch. J., Mining Geostatistics, Academic Press, 1978.

11. Taylor, H.K., 'General background theory of cut-off grades', Inst. Min. Metall. Trans., Sec. A., vol. **81**, 1972.

12. Matheron, G., Les variables régionalisées et leur estimation (Regionalized variables and their estimation), Masson, Paris, 1965.

13. Mao, J.C.T., 'The internal rate of return as a ranking criterion', Engineering Economist, 1966, 11, No. 4

14. Teichroew, D., Robicheck, A.A. and Montalbano, M., 'Mathematical analysis of rates of return under certainty', Management Science, 11, No. 3, 1965.

DRAINAGE AND DIVIDE NETWORKS DERIVED FROM HIGH-FIDELITY DIGITAL TERRAIN MODELS

Alberto Carrara
National Research Council
c/o Dept. Civil Engineering, University of Florence
Florence, Italy

ABSTRACT. Through a new contour-specific interpolation algorithm, high fidelity, grid-based, digital terrain models (DTMs) are generated from contour lines digitized by an electronic raster scanner. Starting from DTM data, drainage lines and related divides are automatically identified by means of a set of procedures which also calculates the fundamental geomorphologic and hydrologic parameters of the basin, such as stream-line order, magnitude and length, sub-basin area, perimeter and shape. These techniques find useful application in many fields, ranging from theoretical geomorphology/hydrology to the evaluation of environmental resources and constraints.

1. INTRODUCTION

The potential of digital terrain models in contributing to the solution of a wide spectrum of geomorphological, hydrological and cartographic problems has long been known (Smart, 1970; Chorley, 1972; Evans, 1972). However, only in the late seventies has computer technology (hardware/software) made it possible to acquire, process and display elevation data efficiently and cost-effectively, through the application of sophisticated electronic raster scanners, interactive graphic workstations and plotters (Boyle, 1980; Leberl & Olson, 1982).

Nowadays, in many countries governmental and private cartographic agencies employ contour lines as a data source for the production of grid DTMs. In Italy, for example, the Military Cartographic Institute (IGMI) is undertaking a major project aimed at digitizing contour and drainage lines from 1:25,000 scale topographic maps - which cover the whole country - by using electronic raster scanner technology (Ammannati & Grassi, 1982).

A crucial step in the procedure concerns the method by which DTM elevations are interpolated; many different interpolation algorithms have been proposed or implemented (cf. Clarke et al., 1982); however, few or none of them appear to provide true high-fidelity data.

Among the different possible applications of DTMs (cf. Doyle, 1978), those related to hydrology and geomorphology play an important role. In particular, automatic identification of stream-lines and divides allows the construction of distributed hydrologic simulation mo-

C. F. Chung et al. (eds.), Quantitative Analysis of Mineral and Energy Resources, 581–597.

dels (Beven & Kirkby, 1979). Likewise, the partitioning of watersheds into sub-basins and main slope-units greatly facilitates those investigations aimed at assessing regional slope-instability hazard by means of geostatistical models (Carrara, 1983, 1984).

A few investigators have long attempted to extract drainage networks and basin/sub-basin morphologic parameters from digital terrain models (Puecker & Douglas, 1975); however, only recently has the problem received increased attention (O'Collaghan & Mark, 1984; Jenson, 1985; Palacios-Velez & Cuevas-Renaud, 1986; Band, 1986); these last two works being of particular interest. Yet the complete achievement of this task seems to require further efforts and experimental tests.

Within the framework of two different nation-wide projects - the first aimed at assessing the potential of hydro-power resources of some Italian regions; the second oriented towards the control and the prevention of natural geologic-hydrologic catastrophes - an investigation on the development and application of DTMs has been initiated. The philosophy behind this study is illustrated elsewhere (Carla' et al., 1984, 1987; Carrara et al., 1987); hence, it will suffice here to point out that first the quality of the data obtained by digitizing, with an electronic raster scanner, contour lines and drainage networks of topographic maps, was tested (Carla' et al., 1984, 1987). Then, a new procedure for generating high fidelity, grid-based, digital terrain models, starting from digitized contour lines, was set up. Using DTM data, an attempt was made to identify automatically stream lines and related divides and determine the main geomorphologic and hydrologic parameters of a watershed (Fig. 1).

In the present paper, the procedures for both producing grid-based DTMs and digitally generating drainage-divide networks along with basin/sub-basin parameters, are briefly described. The results obtained from the application of these techniques to a sample area, nearly 100 sq. km in size, are shown.

2. GENERATION OF A DIGITAL TERRAIN MODEL

2. 1. Method

As previously mentioned, for each topographic sheet at 1:25.000 scale (named "Tavoletta"), a digital contour data file is produced by IGMI, whose structure is specifically designed to meet the needs of this cartographic agency (Ammannati & Grassi, 1982).

In order to reduce both data volume storage and computer central memory requirements, the IGMI data organization was first transformed into a structure more feasible for handling and processing data by means of a 16 bit minicomputer (such as the HP 2117/F, interfaced with a VDS image-processor system, used for this phase of the work). Hence, contour data were partitioned and reorganized into subdomains (named "meshes") of 1x1 km (4x4 cm on the map), each corresponding to one grid unit of the UTM network; UTM coordinates were then converted into x/y Cartesian coordinates, referenced to the origin of each mesh and expressed as multiples of 4/255 cm (cf. Bonfatti et al., 1982).

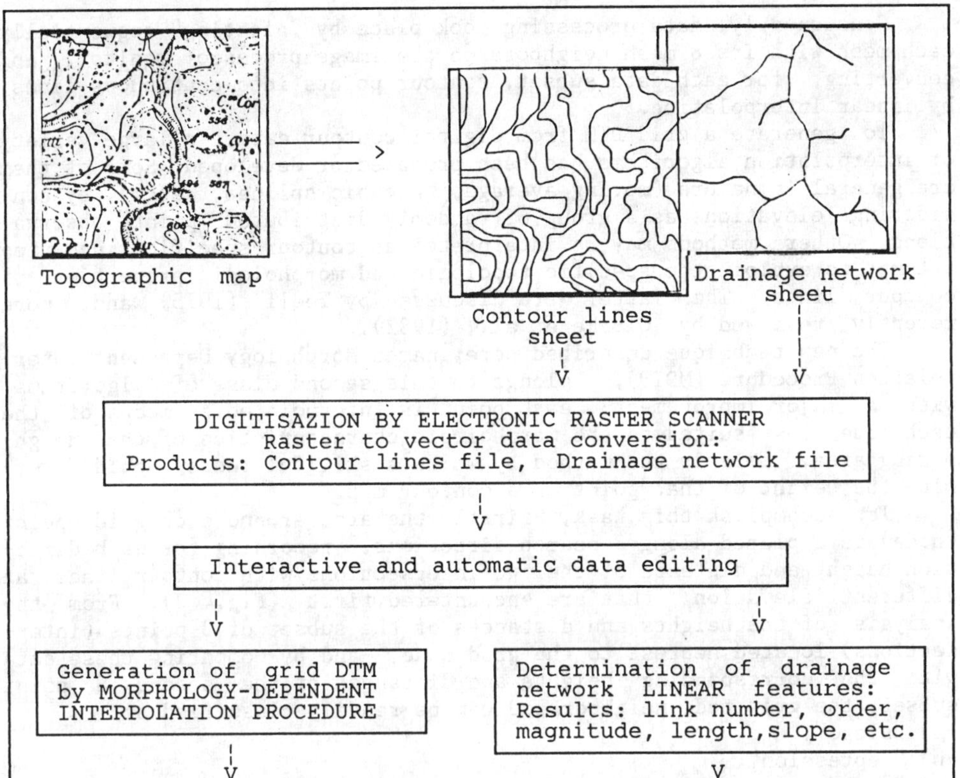

Fig. 1. Schematic representation of the different phases of the investigation.

Consequently, data processing took place by "plotting" sequentially each mesh with its 8 mesh neighbors on the image-processor monitor, and converting, for each data subset, contour points into continuous lines, by linear interpolation.

To generate a grid DTM from digital contour data, a large variety of interpolation algorithms has been proposed or developed. Some of them are general in nature (moving average, bicubic splines, etc.), by considering elevations as a group of randomly distributed point observations. Other methods may be interpreted as contour-specific since they attempt to exploit the specific topologic and morphologic properties of contour lines. The latter were discussed by Yoeli (1975) and, more recently, reviewed by Clarke et al., (1982).

The new technique described here, named Morphology-Dependent Interpolation Procedure (MDIP), belongs to this second class of algorithms, with a major improvement: each point is interpolated by means of the technique most suited to the morphologic characteristics of the neighboring area. This is the method by which a skillful reader would determine the height of that point on a contour map.

To accomplish this task, firstly the area around each grid point (node) is explored along 8 search directions, recording for each direction height and distance of the two intersections with contour lines, at different elevation, that are encountered first (Fig. 2). From the analysis of the heights and distances of the subset of 8 points (intersections) located nearest to the grid node, and by comparing these data with the corresponding heights and distances of the second 8 point subset, the grid node neighborhood can be readily classified as:

-a) "slope";
-b) "depression";
-c) "peak".

The first term refers to any topographic configuration, which in general reflects a slope unit, where at least two points of the first subset have different elevations, whose difference is necessarily equal to the contour interval of the map. The second term applies when the grid node is surrounded by the 8 points of the first subset having the same elevation, which is lower than the heights of the second subset. This outcome indicates either a true geomorphological depression (such as a sinkhole) or, more frequently, inadequate information provided by the search along the 8 directions; the latter is likely to occur when the grid point falls along the course of a narrow, meandering valley with shallow longitudinal profile. Being equivocal, this situation requires further inspection, to be carried out in a subsequent phase of the procedure. The third term applies where the 8 heights of the first subset are equal and greater than those of the second subset of points; that is, the grid point lies on a hill top.

Therefore, the great variety of morphological configurations found in nature are reduced to the above listed basic classes: each one processed by a different method, as follows:

-a) Interpolation of a point on a "slope": in this case, which is the most frequent, the space surrounding the grid node is divided into 8 sectors (a, b, c,.., in Fig. 2) by 4 axes coinciding with the previously mentioned search directions. Each sector is progressively explored up to the maximum value of the search radius (in Fig. 2, equal to 1/2 mesh size), recording height and distance of the two points which are nearest to the grid node and have elevation equal to the upper (Emax) or lower (Emin) values of the contour interval within which the node falls. Consequently, comparing sequentially each pair of opposite sectors, it is readily possible to identify the pair of sectors where the distance between Emax and Emin is smallest: such points define the direction of steepest slope. Then, elevation of the grid node is simply calculated by linear interpolation within these two data points (see example A in Fig. 2).

Since each sector is fully explored during this phase of the procedure, it is also readily possible to distinguish grid nodes actually falling into a "true" depression from those only apparently isolated (see example B in Fig. 2). The latter are interpolated using the general algorithm for "slopes".

-b) Interpolation of a point within a "depression": grid points appertaining to this class may reflect two different situations: the first corresponds to natural (karst morphology) or man-made (pits) true closed depression; the second relates to the case in which the contour line at Emin height lies beyond the search radius of the grid node (see example D in Fig. 2). In both cases the following interpolation technique is used.

First the ratio (Ro) of the mean distance of the first 8 point subset to the mean distance of the second subset is calculated. For Ro smaller than a threshold value T (at present set equal to 0.3), the elevation of the grid node is calculated as follows. Each slope, measured along the 8 search directions between the two points appertaining to the first and second subsets, is linearly extrapolated to the grid node. Then, the resulting eight heights are averaged and weighted by a factor that is dependent on the maximum and minimum distances of the node from the nearest contour line. Otherwise, for Ro greater than T, the grid node is calculated by linear extrapolation of the lowest gradient measured along the 8 directions. In the latter case, a further test is performed: the grid node elevation obtained in this way is compared with that calculated by extrapolating the slope from the search direction whose first point is nearest to the grid node. Of the two elevations the higher is selected.

Behind this "adaptive" procedure is the attempt to minimize the probability of occurrence of artificial "pits" along a river course. As later discussed, these features may cause serious errors for some algorithms that use DTM data for hydrologic applications.

-c) Interpolation of a point on a "peak": grid nodes located on a hill top are simply calculated by linearly extrapolating the gradients of the neighboring area, and averaging with a weighting factor the resulting heights; namely, the first step of the previous procedure is applied. This simplified approach appears satisfactory because the occurrence of small elevation errors in such zones does not significantly hamper the use of DTM data for most hydrologic and geomorphologic applications.

2. 2. Experimental results

The procedure outlined has been developed and implemented using the contour data from an area, located in Central Italy (Spoleto, Umbria), that includes the upper course of the Maroggia river. The area displays geological and geomorphological features common to a large portion of the hilly/mountainous zones of Italy. Rock types range from Mesozoic carbonates to Pliocene-Quaternary sediments. The former exhibit ragged topography, high relief and karst morphology; the latter show gently sloping hills or flat-lying fluvial plains. Therefore, the results obtained should be readily applicable to a much larger region.

As discussed elsewhere (Carla' et al., 1984, 1987), a DTM with 50m point spacing was found to be the best trade-off between costs (memory storage) and benefits (spatial accuracy).

To perform a quantitative evaluation of the DTM data, for each mesh the area lying between each pair of adjacent contour lines was computed, on the image-processor system, by means of a specifically-designed algorithm, of the "region growing" type (Pratt, 1978). Then, these surface

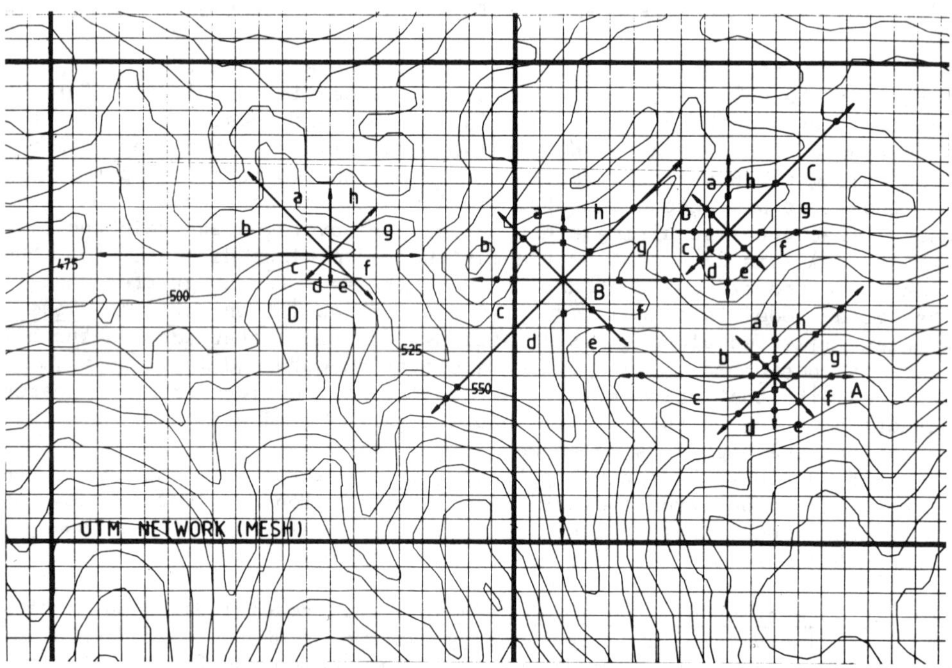

Fig. 2. Portion of topographic map of the study area. Examples of DTM grid nodes falling on/within: a "slope" (A), a "depression" identified as "false" (B), a "peak" (C), and a "depression" not recognized as false (D).

values were compared with those obtained by grouping DTM elevations into classes whose limits correspond to the map contour interval. The difference between these two sets of values provides a direct measure of the fidelity of the DTM generated by the procedure outlined. The results of this comparison, listed in Tab. 1, refer to two meshes located in the western portion of the map, which is morphologically the most complex of the study area.

In order to better assess MDIP results, the same surface data were determined from a second set of DTM elevations (Tab. 1) which was computed by means of a previous procedure, named "weighted mean" (Carla' et al., 1984). This procedure was derived modifying the "weighted average" method proposed by Yoeli (1975). The major refinement of Yoeli's method consists in forcing point sampling to be homogeneous within the

Tab. 1. Examples of quantitative evaluation of the quality of DTM data obtained by means of the MDIP and "weighted mean" procedures for two meshes of 1 sq. km each.

Contour interval	Reference Area	Computed DTM Area (MDIP)		Computed DTM area (WEIGHTED MEAN)	
m	Km %	Km %	Dif.%	Km %	Dif.%
550-575	.430	.500	-.070	0.000	.430
575-600	2.396	2.250	.146	1.500	.896
600-625	3.966	4.250	-.284	4.500	-.534
625-650	13.033	13.000	.033	6.250	6.783
650-675	15.530	15.750	-.220	21.250	-5.720
675-700	18.246	17.250	.996	17.000	1.246
700-725	17.888	18.250	-.362	19.250	-1.362
725-750	12.212	12.500	-.288	12.500	-.288
750-775	11.754	11.500	.254	12.750	-.996
775-800	4.427	4.750	-.323	4.750	-.323
800-825	.119	0.000	.119	.250	-.131
		RMSE = 0.375%		RMSE = 2.773%	
450-475	.674	.250	.424	0.000	.674
475-500	7.112	7.500	-.388	3.500	3.612
500-525	15.002	15.250	-.248	14.750	.252
525-550	23.318	23.500	.182	20.750	2.568
550-575	21.089	20.750	.339	25.250	-4.161
575-600	14.584	15.250	-.666	15.750	-1.166
600-625	6.940	6.500	.440	7.750	-.810
625-650	6.795	6.750	.045	7.250	-.455
650-675	3.572	3.250	.322	4.000	-.428
675-700	.912	1.000	-.088	1.000	-.088
		RMSE = 0.359%		RMSE = 1.997%	

neighborhood of the grid node (Carla' et al., 1984, 1987; cf. also Harbaugh et al., 1977).

Comparison between these two sets of data clearly shows the better quality of the DTM generated by the MDIP procedure, as witnessed by a root mean square error (RMSE) smaller than 0.5% (for each single mesh). Conversely, RMSEs for the "weighted mean" algorithm are at least 5 times greater. When this type of comparison is extended to an area progressively larger up to 16 meshes, both sets of RMSEs decrease exponentially to much smaller and slightly closer values: 0.082% and 0.507%, respectively.

However, an assessment of the local accuracy of the two methods is better achieved inspecting the relative error percentage for each area difference within a single mesh : for the MDIP data this error is less than 10%, while for the "weighted mean" data it can exceed 50% (see for example area of contour interval 625-650, Tab. 1).

As illustrated elsewhere (Carla' et al., 1987), the simultaneous display on the image-processor monitor of contour lines and DTM elevations produced either by the MDIP or by the "weighted mean" methods, confirms the outcomes of the statistical analysis: every contour configuration is much better reflected by the MDIP data than by the "weighted mean" elevation data. The former indeed are not affected by those "smoothing effects" which generally lead to a loss of morphological information.

3. DIGITALLY DERIVED DRAINAGE AND DIVIDE NETWORKS

The first application of the DTM data consisted in an attempt to identify automatically basin drainage lines and divides, and to calculate basin/sub-basin geomorphologic parameters.

In order to accomplish this task, a set of procedures (named BACINI-1) was set up; while some of them are derived from the literature, most constitute an original approach to the problem. Since the method is still currently under development and implementation, and experimental tests are not fully completed, only the philosophy behind the overall procedure is illustrated here, while a detailed discussion of the different techniques and algorithms will be provided elsewhere (Carrara et al., 1987).

3. 1. Method

The method is based on a simulation model of "overland flow" concentration generated by an hypothetical spatially homogeneous rainfall event, throughout a completely impermeable area. Under such simple circumstances, the problem, which in nature clearly depends on a very large set of complex factors, reduces to assessing the control exerted by ground morphology on flow lines.

A similar simulation model was recently suggested by Moniod (1983) for a very different purpose: the estimation of lag-time and specific peak discharge.

In order to analyze flow lines and terrain morphology relationship, the area to be investigated is divided into square cells (grid-units or land-unit parcels). In the center of each grid-unit, elevation is determined by means of the previously described interpolation algorithm; hence the resulting elevation matrix (DTM) constitutes the input information of the whole procedure.

-a) Data pre-processing: before starting to treat elevation data, a quality test is performed to identify those points ("pits") whose height is lower than that of their eight-connected neighbors. In nature such a circumstance occurs in some landscapes (karst areas, recently glaciated zones, etc.); however, "pits" most frequently reflect either failure of the interpolation algorithm used, or inadequacy of the grid-size selected, when too large for the morphological features of the area investigated (cf. Band, 1986). Whatever the cause of these closed depressions, they can significantly hamper any hydrological data processing; in particular, the present method requires that no "pit" occur in the DTM matrix. Because standard techniques for numerically smoothing the terrain model invariably lead to a loss of information (O'Callaghan & Mark, 1984), a simple local-filtering procedure was designed.

Using a moving 3x3 point spatial window, the elevation of the central point is compared with the heights of the eight-connected neighbors; when one of them is lower, the spatial window is moved to the next point, and the test repeated. Otherwise, the point is classified as "pit" and corrected by an algorithm which, exploring the neighborhood of the "pit", both "adjusts" its height by increasing it to a value nearly equal to that of the lowest adjacent point, and marks the point at the lowest elevation among the sixteen-connected neighbors. By centering the spatial window on the flagged point, the process is repeated until a point at an elevation lower than that of the "pit" is not found. Then, heights of all previously flagged points are replaced by values obtained by linear interpolation between the "adjusted" elevation of the "pit" and the elevation of the last point found.

The procedure outlined is controlled by a set of parameters (maximum search path, maximum difference in heights, etc.) whose values are assigned by the user, depending on the morphology of the study area. If very conservative (low) parameter values are assigned, a few "pits" may be left uncorrected. In this case, a second algorithm is used, which explores progressively larger neighborhoods of the "pit", searching at each step for a point with height lower than the "adjusted" elevation of the "pit"; when the point is found, the shortest path between this and the "pit" is identified and marked; then the elevations of the points flagged are modified as above.

This two-step procedure proved to be capable of readily resolving the very few closed depressions found in the elevation matrix of the study area, since the quality of DTM data is high and its grid-spacing fairly adequate to the topography of the zone. In order to test the procedure under much more severe conditions, a second pre-existing DTM was used. This data set was produced in 1978 by means of semi-automatic techniques, within the framework of a previous project (Carrara et al., 1978; Bottero et al., 1985). The DTM, having a 100x100m spacing, covers a large basin, located in Calabria (Southern Italy), which is characterized both by extremely rugged topography (basin head) and by flat-lying large river terraces. The area is currently re-investigated for

hydrological purposes, applying the techniques outlined in this paper.
As a result of this test, it was found that all of the several hundred "pits" were adequately corrected by the procedure.

-b) Data processing: starting from the upper left corner of the map (matrix), the elevation of the central cell of a moving 3x3 point spatial window is compared with the heights of the eight-connected neighbors. Then, the central cell is both flagged with a code equal one, and marked with the coordinates of the neighboring cell in the direction of steepest slope, that is, where water would be drained off from the land-unit parcel. Moving and centering the window on the neighboring flagged cell, the process is sequentially repeated throughout the elevation matrix up to the point where an edge of the matrix is encountered. This occurrence is classified and stored as a "relative" outlet.

The next point of the upper left corner of the matrix becomes the center of the spatial window and the process is repeated up to the point where either an outlet is reached or a cell already flagged and marked is encountered. In the latter case, the spatial window moves following the path already defined by the previously marked cells, simply increasing the code of each cell by one unit.

By repeating the process for all the grid-units of the elevation matrix, two data sets are obtained, each one stored as a separate matrix (map); the first (named pointer matrix) indicates the direction along which water from each land-unit parcel would flow toward one of its eight-connected neighboring land-units; the second (named counter matrix) provides for each grid-unit the total number of cells that "spill" into it, that is, the "upstream" contributing area drained by each land-unit parcel.

The above simple (but computationally costly) data sets constitute the information necessary to identify drainage lines and divides.

Visual inspection of the counter matrix already reveals the main features of each basin/sub-basin occurring in the study area: drainage lines, divide zones, and slope/sub-basin areas can be approximately identified by cells with high, low and intermediate counter values, respectively. However, converting this qualitative information into a fully connected and topologically ordered network of stream lines and divides requires a further step, conceptually simple but algorithmically complex.

Before performing this task, it is necessary to predefine a set of threshold parameters, the most important of which being the minimum first-order sub-basin size and the minimum first-order channel length (outer link).

Depending on the value of these parameters, the resulting drainage-divide networks will be characterized by a different degree of generalization, in a similar manner as happens when drainage lines of an area are displayed on topographic sheets of different scales.

Having assigned to the above parameters "appropriate" values, the procedure starts choosing one of the matrix edge outlets and recursively "climbs" up the stream network that drains into that outlet. This operation is repeated for all the outlets found.

Using the data of the pointer and of the counter matrices, a 3x3 spatial window is centered on the first outlet, and, among the neighboring cells pointing to window center, the one with a counter value greater than the predefined threshold is selected: this cell is then classified as appertaining to a link. The step is sequentially repeated, every time moving the spatial window upstream. When two cells of the neighborhood have both counters greater than the threshold value - that is, when a junction is reached - the link at lower contributing area (at lower counter) is first climbed; then, accordingly, all its tributaries up to a point where a cell surrounded by neighbors having counters smaller than the threshold is encountered (that is, when a source is reached). Then, the spatial window is moved back to the first node found, and the second link is processed in the same way up to its outer links and sources.

It may happen that a four-link junction occurs; namely, that 3 cells neighboring the one under process have counters greater than the selected threshold value. This occurrence was found to be very uncommon for the data of the study area; however, as pointed out elsewhere (Jenson, 1985; Band, 1986), four to five-link junctions may occur frequently when using a low-fidelity or a coarse DTM. Therefore, a technique was defined to avoid this outcome, which seriously hampers subsequent data treatment. Depending on the local morphological setting around the four-link junction, either the upstream link with lowest contributing area is deleted or its end point (node) is displaced one grid-unit upstream or downstream.

At the end of this operation for each outlet, a fully connected drainage network is obtained, whose links, nodes and sources are progressively numbered for facilitating subsequent analyses.

The next phase of the work consisted in identifying basin/sub-basin surfaces. This is achieved by using the pointer and the drainage lines matrices. Starting from the upper left corner of the pointer map (matrix), the path defined by the pointers of the grid-units is sequentially followed and stored up until a link cell is encountered; then all cells appertaining to the path are marked with the code number of the link. The operation is repeated starting with the next cell of the matrix upper left corner; in this case, and in all the subsequent ones, the path stops when either a link cell is found or a previously defined path is met. In both cases, cells are marked as belonging to the sub-basin area. In this way, all the cells are sequentially scanned and correctly assigned to their respective links. Hence, the resulting map exhibits the following features:

- -a) a fully connected drainage network rooted at and progressively numbered from each outlet;
- -b) a fully connected divide network;
- -c) the sub-basin area appertaining to each drainage link.

It is worth noting that divides are defined as boundary between four-connected cells belonging to different basins/sub-basins (Fig. 4). The decision to use grid-unit edges to define the divide network may be somewhat questionable, because it generates a dissymmetry with drainage lines that are identified by cells (Fig. 4). However, it enables the spatial resolution imposed by the grid-based model (which is generally low or relatively low) to be better exploited.

-c) Data post-processing: having acquired the above groups of data, it is readily possible to produce a large spectrum of morphologic parameters: some correspond to those routinely acquired by traditional methods (link length, number, order, etc.), others can be considered unique (sub-basin area, aspect, mean slope, etc.) because their acquisition by manual techniques would require a tremendous amount of time and work.

In order to generate the parameters outlined, a set of procedures was designed, most of which are already implemented, while others are still under development. At present, for an area of any shape, starting from the matrices of elevations, pointers, and drainage-divide networks, the following parameters can be readily calculated (for terminology cf. Jarvis, 1984):

-a) outlet number
-b) link number
-c) number of the two tributaries of the link
-d) coordinates of the link up/down-stream nodes (junctions)
-e) link value of topologic vector
-f) link order (Strahler)
-g) link magnitude
-h) number of cells constituting the link
-i) link length
-j) link mean slope
-k) link sub-basin area
-l) total basin area drained by the downstream node of the link
-m) area of right/left sides of link sub-basin (main slopes)
-n) perimeter of right/left sides of link sub-basin
-o) mean aspect and slope of right/left sides of link sub-basin

The above parameters are stored in tabular form, ready to be further processed for generating more refined, problem-oriented, morphologic or hydrologic parameters. For example, since the topologic vector of the link network is already available, streams can be easily ordered following criteria selected by the user.

In order to meet the specific display requirements of the data, high-quality, raster-based, graphic programs were developed (Fig. 3 and 4) or obtained modifying software produced during a previous investigation (Carrara et al., 1978: Reali, 1980).

All the programs described, written in standard FORTRAN, were developed and implemented on a Honeywell DPS7 (SIRIS8 oper. sys.); most of them have recently been transferred on a MicroVaxII (VMS oper. sys.).

3. 2. Experimental results

The method for automatically extracting basin/sub-basin linear and areal characteristics was implemented with and tested on the data from the previously outlined Maroggia basin area. Since for this area digitized drainage lines were available (see Fig. 1), a direct comparison between digitally-generated and "blue-lines" derived drainage networks was possible.

Before comparing these two sets of data, "blue-lines" were first

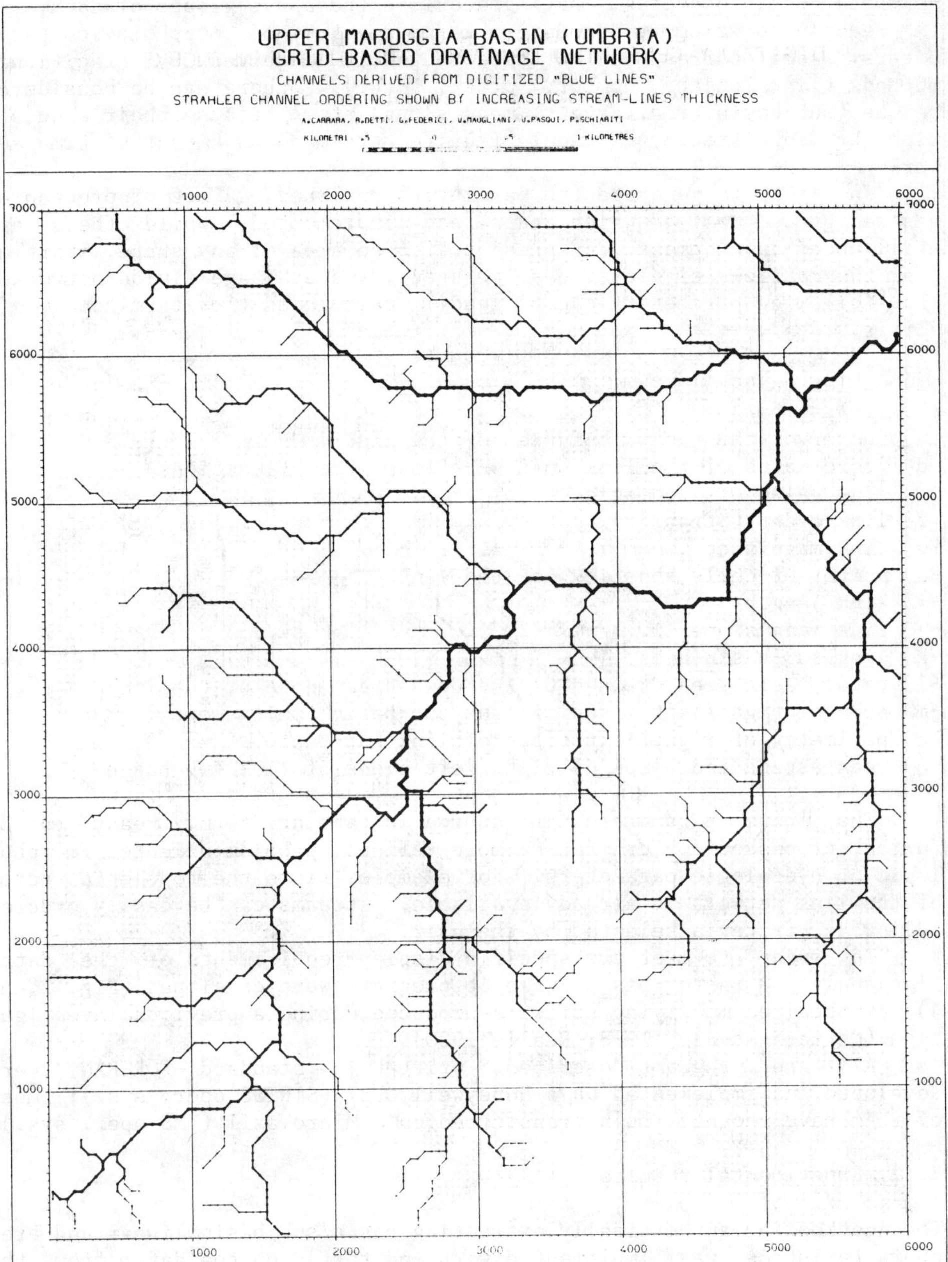

Fig. 3. Grid-based drainage network derived from digitized "blue-lines" of the topographic map of a portion of the study area.

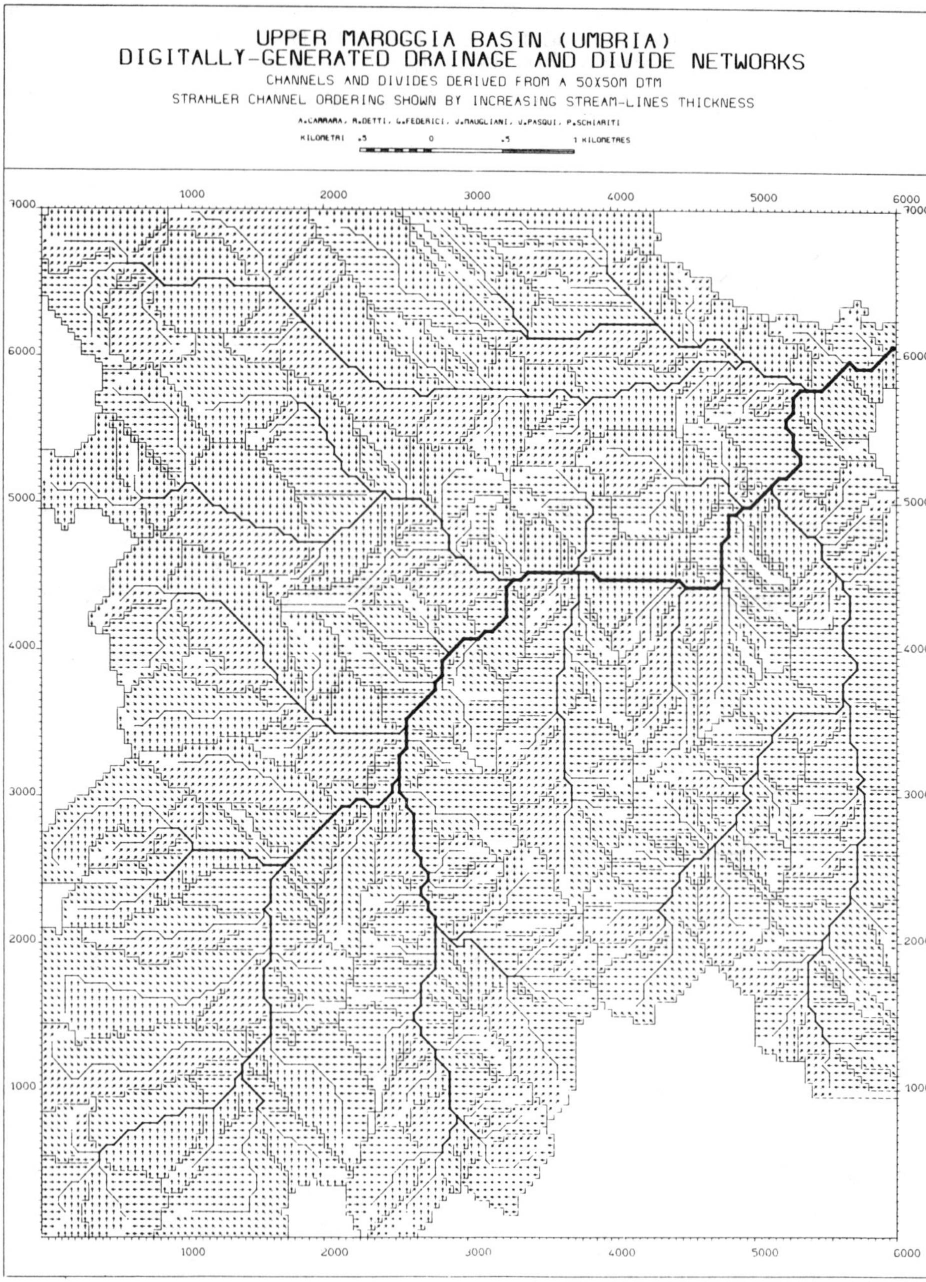

Fig. 4. Digitally-generated drainage and divide networks derived from a high-fidelity DTM. Arrows indicate mean aspect of right and left sides (main slopes) of each subbasin. Area as in Fig. 3.

processed by various programs aimed at both correcting errors generated during the digitizing phase, and calculating the main geomorphic parameters (stream order, magnitude, length, etc.; cf. Carla' et al., 1984). In addition, continuous drainage-lines data were converted into raster form with a 50x50m grid size - that is, equal to that selected for implementing and testing the BACINI-1 program.

In Fig. 3 and 4, for the upper portion of the Maroggia basin, the grid-based blue-lines network, and the digitally-extracted drainage and divides networks are displayed, respectively. The latter map was produced setting the minimum first-order sub-basin size and the minimum outer-link length parameters equal to 10 and 5 grid-units, respectively. In both maps stream-lines are ordered after the Strahler method.

Visual comparison of these two maps reveals an overall good similarity between the two drainage networks. The major difference between them refers to a higher degree of smoothing (generalization) of the second as compared with the first. This feature is, in part, inherent to the logic of the method selected; however, it is also related to the rather wide contour interval of the map, which does not provide adequate information on the local topography of gently sloping zones.

Comparison of two main geomorphic parameters clearly confirms the similarity of the two networks: being Strahler maximum order equal (fifth) and magnitude very similar (130 vs 135).

A systematic assessment of the fidelity of divides, drawn in the map of Fig. 4, has been carried out; careful visual inspection of the topographic map indicates that the digitally identified divides well reflect ridge-crests contour configurations.

4. CONCLUDING REMARKS

On the basis of the preliminary results, derived from the application of these techniques to the sample area of the "Tavoletta" Baiano di Spoleto (Upper Maroggia basin), the following concluding remarks can be put forth.

-a) The new procedure set up for interpolating a grid DTM from digitized contour lines proved to be capable of producing high quality data, which faithfully reflect the most common topographical configurations found in nature.

-b) More statistical work is required to assess quantitatively, in different morphological conditions, the role exerted by both contour-lines spatial resolution and sampling density on the quality of DTMs; in particular, when they are to be applied for identifying automatically drainage-divide networks.

-c) Drainage-divide networks produced by the method illustrated reflect to a great extent the corresponding features shown on the 1:25,000 scale topographic map.

-d) Owing to the logic of the main algorithms, the method is likely to work properly only when starting from high-fidelity DTMs derived from detailed contour maps. Yet correct identification of stream courses flowing through wide flood plains may prove difficult, and therefore represents an intrinsic restriction on this technique.

Where digitized stream-lines are available - as for part of the topographic coverage of Italy - this limitation can be circumvented by integrating specific information provided by DTM data and "blue-lines" data. The latter permit much more accurate spatial detection of stream-courses flowing over flat-lying zones, while the former allow both divides to be automatically identified, and sub-basin parameters to be readily calculated throughout the watershed.

The next phase of this investigation will attempt to develop a technique able to incorporate and use the two types of data efficiently and accurately.

Acknowledgments

The present paper outlines the preliminary results obtained from a multidisciplinary investigation currently being carried out, at the Department of Civil Engineering of the University of Florence and at the CNR-IROE of Florence, by R. Carla' (CNR-IROE), A. Carrara (CNR c/o Dept. of Civil Engineering), R. Detti (TECNOWARE, Florence), G. Federici (Dept. of Civil Engineering), V. Pasqui (University Computer Center), V. Maugliani and P. Schiariti (formerly students at the University of Florence). The work is supported by grants from the "Progetto Finalizzato Energetica II" of the National Research Council (CNR) and from the "Gruppo Nazionale per la Difesa dalle Catastrofi Idrogeologiche".

References

Ammannati F. & Grassi S., 1982. "L'automazione cartografica presso l'Istituto Geografico Militare". Boll. Geod. Sci. Affini, 41, n. 3, p. 215-258.

Band L.E., 1986. "Topographic partition of watersheds with digital elevation models". Water Resour. Res., 22, p. 15-24.

Beven K.J. & Kirkby M.J., 1979. "A physically based, variable contributing area model of basin hydrology". Hydrol. Sci. Bull., 24, p. 43-69.

Bonfatti F., Cavazza L. & Schianchi G.P., 1982. "Trattamento automatico di dati geografici". Riun. Annuale Assoc. Elettrotec. It., p. 1-10.

Bottero M., Braulin N. & Carrara A., 1985. "Ambiente e tecnologia: risorse e vincoli nel bacino del Petrace, Calabria". Quad. Dip. Progr. Proge. Prod. Edil., 2, Politecnico Milano.

Boyle A.R., 1980. "Development in equipment and techniques". in: D.R.F. Taylor (ed.), The computer in contemporary cartography, p. 39-57, Wiley, New York.

Carla' R., Carrara A. & Federici G., 1987. "Generazione di modelli digitali del terreno ad alta precisione". Quad. Dip. Ing. Civile, Univ. Firenze, 0/87, 30 p., Firenze.

Carla' R., Carrara A., Federici G. & Maugliani V., 1984. "Modelli digitali del terreno nell'indagine geomorfologica e idrologica". Atti Conv. Aicographics, Milano, 12-15 nov., p. 221-241.

Carrara A., 1983. "Multivariate models for landslide hazard evaluation". Mathematical Geol., 15, p. 403-426.

Carrara A., 1984. "Thematic cartography and water resources". First Adv. Course on Water Resour. Manag., Univ. Foreign. Perugia.

Carrara A., Catalano E., Sorriso-Valvo M., Reali C. & Osso I., 1978. "Digital terrain analysis for land evaluation". Geol. Appl. Idrogeol., 13, p. 69-127.

Carrara A., Detti R., Federici G., Pasqui V. & Schiariti P., 1987. "Automated derivation of watershed geomorphologic and hydrologic parameters". (in preparation).

Chorley R.J.(ed.), 1972. Spatial analysis in geomorphology. Methuen, London.

Clarke A.L., Gruen A. & Loon J.C., 1982. "The application of contour data for generating high fidelity grid digital elevation models". Proceed. AUTOCARTO-V, p. 213-221, A.S.P. & A.C.S.M., Washington.

Doyle F.J., 1978. "Digital terrain models: an overview". Photogramm. Eng. Remote Sensing, 44, p. 1481-1485.

Evans I.S., 1972. "General geomorphometry, derivatives of altitude and descriptive statistics". in: Chorley R.J. (ed.), Spatial analysis in geomorphology, p. 17-90, Methuen, London.

Harbaugh J.W., Doveton J.H. & Davis J.C., 1977. Probability methods in oil exploration. Wiley, New York.

Jarvis, R.S., 1984. "Topology of tree-like networks". in: Gaile G.L. & Willott C.J. (eds.), Spatial statistics and models, p. 271-292, Reidel Publ., Boston.

Jenson S.K., 1985. "Automated derivation of hydrologic basin characteristics from digital elevation models". Proceed. AUTOCARTO-VII, p. 301-310, A.S.P. & A.S.S.M., Washington.

Leberl F.W. & Olson D., 1982. "Raster scanning for operational digitizing of graphical data". Photogramm. Eng. Remote Sensing, 48, p. 615-627.

Moniod F., 1983. "Deux parametres pour caracteriser le reseau hydrographique". Cha. ORSTROM. Ser. Hydro., 20, p. 191-203.

O'Collaghan J.F. & Mark D.M., 1984. "The extraction of drainage networks from digital elevation data". Comput. Vision Graph. Image Proc., 28, p. 323-344.

Pratt W.K., 1978. Digital image processing. Wiley, New York.

Puecker T.K. & Douglas D.H., 1975. "Detection of surface specific points by local parallel processing of discrete terrain elevation data". Comput. Graph. Image Proc., 4, p. 375-387.

Palacios-Velez O.L. & Cuevas-Renaud B., 1986. "Automated river-course, ridge and basin delineation from digital elevation data". Jour. Hydro., 86, p. 299-314.

Reali C., 1980. "MONLOCA: un programma grafico per la produzione di carte tematiche degli aspetti fisici del territorio". CNR-IRPI, Cosenza.

Smart J.S., 1970. "Use of topologic information in processing data for channel networks". Water Resour. Res., 6, p. 932-936.

Yoeli P., 1975. "Compilation of data for computer-assisted relief cartography". In: Davis J.C & McCullagh M.J. (eds.), Display & analysis of spatial data, p. 352-367, Wiley, New York.

AN ALGEBRAIC MODEL FOR DATA STRUCTURE AND MANIPULATION AS A DESIGN BASIS FOR INTEGRATED GEOSCIENCE PROCESSING SYSTEMS

M. T. Holroyd
Dataplotting Services
(BP Resources Canada)
225 Duncanmill Rd.
Don Mills
Ont. M3B 3K9
CANADA

ABSTRACT: As the more easily detectable deposits become exhausted, minerals exploration must move into areas where deposits are more difficult to find. In such areas, several different types of geoscience survey data are often employed in combination to provide an optimal geological interpretation. Exchange and integration of data continues to be hampered by incompatibility between data from different sources. To overcome this, it is necessary that data structures be independent from the software that accesses them. This has been achieved to some degree in data base management systems (DBMS). No single one of the principal DBMS data models, however, is capable of describing the wide variety of geoscience survey data structures that are used. A new data model, based on vector algebra, has been developed which is capable of describing a wide variety of geoscience data structures. Furthermore, unlike the existing models, the algebraic model can also be used to provide a formal definition of the means whereby data independence can be achieved.

1. INTRODUCTION

As the more easily detectable deposits of petroleum and mineral resources become exhausted, exploration must move into increasing more cryptic areas. The cost of drilling test boreholes increases generally with time, and specifically with remoteness and hostility of the site. Geophysical and geochemical exploration techniques can increase the probability of success in finding deposits, and can hence help to minimise cost increases.

In geologically cryptic areas, a single exploration method is often inadequate and increasing attention is being paid to the integration of several methods within an exploration area.

Li et. al., [1] made metalogenic predictions in new exploration areas by extension from integrated studies of an area of known metaliferous mineralisation in China (the Dachang ore field in the

C. F. Chung et al. (eds.), Quantitative Analysis of Mineral and Energy Resources, 599–615.

Danchi mineralisation belt). A broad suite of different data types was employed, including aeromagnetic total-field and second vertical derivative, pole-reduced gravity, geochemical analyses of Sn, Pb, Zn and Cu, and geological data obtained by digitisation of 10 stratigraphic-lithologic units. Image analysis techniques were used to define "training areas" over known mineral deposits. Then areas with similar statistical properties were sought in unexplored regions.

Eliason et. al., [2] applied similar methods for petroleum exploration studies, also using image processing techniques, to data over the Cement oil field in Oklahoma U.S.A. Integration of at least seismic, gravity and magnetic data is becoming increasingly common in petroleum exploration (Sexton et. al. [3]; Awad [4]).

1.1. The problem of data independence.

Different types of geoscience survey are generally conducted by different groups of specialists. i.e. a geochemical team will usually conduct geochemical surveys in several different exploration regions, whilst a separate group will conduct the aerogeophysical surveys over the same regions, etc. These different groups can belong to different and independent organisations and can even be from different countries.

In order to facilitate integrated interpretation studies, it is important to ensure not merely that data can be exchanged between different groups, but more importantly, that the data can easily be read and correctly interpreted by the recipient's computer software. i.e. with minimum expenditure of programming effort to adapt to a new source of data and minimum risk of mistakes in scale, datum and position, etc.

This problem is no longer one of physical compatibility between digital data storage media. The problem now concerns the logical content and structural organisation of the data. It is a common problem which exists in many areas other than geoscience. It is the problem of providing "data independence". That is, independence of the logical and physical structure of data from the applications software which accesses and processes the data.

Jeffery and Gill [5] make the point clearly when they state that:

> "Independence of data from the processing software is essential for any synthesis of scientific data because it is impossible to define all possible combinations of data format and structure at the time of writing the software".

The problem can be illustrated by contrasting the ideal case with the common reality. Ideally, any data from any source would be loaded to the computer by a universally applicable loading program. Then the application program would simply "ask" the data access interface for specific types of data from specific geographical locations. The system would then supply the requested data, or at worst, would inform the program that no such data was available.

In the above case, the applications program requires no "knowledge" whatsoever of how the data set is structured, or what else it contains other than the data of interest - data independence has been achieved.

In common practice, however, when one group recieves a magnetic tape of data from another group, they also recieve documentation defining the data structure and content of the tape. They must then write a program tailored to this structure to read the data. In this case the data and applications program are clearly not independent.

1.2 Data independence in data base management systems

The closest approach to true data independence in currently available software systems is exhibited by Data Base Management Systems (DBMS). Current implementations of the DBMS generally fall into one of three categories on the basis of the underlying structural model. The categories are "Hierarchical", "Network" and "Relational". The mathematical bases of such structures are described in detail by Berztiss [6]. Hierarchical systems are based upon the "tree" structure, and network systems on the network or plex structure. Relational systems are based upon Relational calculus. A comprehensive description and comparison of the properties of the three types is to be found in Date [7].

Commercially avaiable DBMS's, however, serve only as a demonstration that solutions exist to certain aspects of the general problem. They do not provide a solution to the problems of integration of geoscience data from diferent sources. They offer data independence within the confines of a single data base or across a network of data bases all of the same type and all managed by the same software system.

1.3 The complexities of geoscience data sets

Geoscience data sets exhibit a wide variety of different structures and have significantly different characteristics from the commercial and administrative data sets that the proprietary DBMS's were originally designed for. Examination and analysis of a large number of different geoscience data sets clearly reveals that no single one of the standard DBMS models is singularly appropriate for geoscience data. Figure 1 shows schematically the structures of four typical geoscience data sets which illustrate this point. These are:-

1) Aerogeophysical data in an archival form. It contains all flight path (track point coordinate) data for a sequence of navigational fixes along a flight line, followed by the geophysical measurements at stations along the track in between the navigation points. Each segment of line data is preceded by a "line header" containing information common to all track and geodata records in the segment that follows. Lines are grouped by map. Each map data set is preceded by a "map header" containing information common to all lines that follow.

2) Aeromagnetic leveling data set. Data used to "level" the survey. It consists of a list of all traverse (ordinary flight line) numbers, followed by a list of all control line numbers, followed by data from the intersections of all traverses with

all control lines.

3) Geochemical analysis data in report form as a set of records each containing a sample number followed by the concentrations of various elements within this sample.

4) Geochemical analysis results as ouput from an automated atomic absorption spectrometer. All samples are analysed for a single element in one batch run. The first record contains a list of the sample numbers of the samples analysed. The second record contains the measured Cu concentration for all samples, etc. This structure is a transformation of data set no. 3 above.

Data set 1 has a hierarchical structure (a "tree" or "one-to-many" structure.) Data set 2 is essentially a network ("many-to-many" structure.) Data set 3 has a structure the same as that of a Relation in a Relatonal DBMS (tabular.) Data set 4 does not fit easily into any of the three standard models. Hence the search for means to provide data independence to geoscience data sets can not begin with any one of the existing DBMS data structure models and must begin at a more fundamental level.

2. THE ALGEBRAIC DATA MODEL

Holroyd [8] analysed digital compilation and cartographic systems for five geoscience survey disciplines, namely aeromagnetics, airborne gamma ray spectrometry, gravimetry, regional geochemistry and drift sedimentology. Comparitive examination of the system processes and the various data sets employed, as they were described by workers in the respective fields, initially revealed little in common upon which an integrated system could be based.

When these data sets were broken down to their simplest components, however, and examined as purely logical objects isolated from the jargon of their parent disciplines, it became possible to construct a new type of data model which could faithfully describe all of the features of all of the many different geoscience data types studied. As these data types are widely distributed across the spectrum of geoscience disciplines, and the model is at the most fundamental, discipline-independent level, it is considered that the model is generally applicable to geoscience data and not merely specific to the particular data sets studied.

The model is known as the "Algebraic model" as it represents data sets by algebraic expressions and data manipulation processes by algebraic manipulation. A description of the model and its properties follows.

2.1 Terminology

The terminology employed in works on data structure is complex. Different writers use different words to describe the same thing in different contexts (e.g. "record" = "segment" = "tuple") and sometimes use the same word for different things.

Sutterlin et. al. [9] illustrate this variety of terminology by providing the commonly used alternatives for many of the terms employed. For example, some works refer to some types of data items as "attributes" and others as "entities" and yet others as "identities". This is not only confusing, it is superfluous and restrictive. Qualitative differentation between data items at such an early stage tends to channel the ways in which they are thought about at later stages. This in turn tends to prejudice the methods designed to manipulate data. The terminology to be used herein is simple and is defined in the following paragraphs.

2.1.1 The data set and entities. We will define the data set as a digitally encoded description of a set of entities. The entities remain strictly in the real world. Data sets do not contain entities. Geoscience entities are generally "samples" - material as in geological or geochemical samples, or immaterial as with aerogeophysical measurements.

2.1.2 Properties and attributes. We will state that entities possess properties which can be measured or observed and that the measurement or observation is an "attribute" of the sample. Any and all things that can be recorded which are pertinent to the sample are simply attributes.

2.1.3 Attribute components. An attribute, therefore, is a specific value of a geoscience or related property. As such, an attribute must consist of two components, the "attribute type" (the property being measured of observed) and the "attribute value" (the result of the measurement or observation). When the attribute is digitally encoded, the attribute value becomes a "data item value" the attribute type a "data item type".

Mass and colour are two examples of properties of physical objects. "50 kilograms" is a possible value for the attribute type mass, "blue" is a possible value for the attribute type "colour".

2.1.4 Aggregates of attributes. A group of attributes all related to single entity, when put together, constitute an "aggregate". When digitally encoded, the aggregate is a simply a data record.

2.1.5 The cardinal form. A data set is in "cardinal form" if all of the entities described by the data set belong to the same class (i.e. all are described by the same set of attribute types), and there is a unique one-to-one correspondence between entities and aggregates. Such a data set is a simple table. Each row is an aggregate describing one and only one entity. Each column contains a single type of attribute. This is similar to a normalised relation in a relational data base. The difference is that there is no need for a unique key.

2.1.6 Surveying and data aquisition. Surveying consists of sample (entity) collection. Data aquisition is observing or measuring and recording the attributes of the entities. The data set is the collection of all the aggregates of attributes recorded.

2.1.7 Summary of terminology

i) ENTITIES possess PROPERTIES

ii) SPECIFIC VALUES of properties are measured or observed.

iii) the specific value of a particular property is an ATTRIBUTE of the entity.

iv) An attribute has two components its TYPE (the particular property) and its VALUE (the observation or measurement as quantified).

v) A group of attributes of an individual entity constitute an AGGREGATE which defines and describes the entity.

vi) the digitally encoded form of an attribute is a DATA ITEM. Attribute type becomes DATA ITEM TYPE. Attribute value becomes DATA ITEM VALUE.

vii) the digitally encoded form of the attributes of a set of entities is the DATA SET.

viii) A data set is in CARDINAL FORM if all the entities it describes belong to the same class, and there is a strict one-to-one relationship between aggregates and entities.

2.2 Logical relationships as arithmetical operators.

Logical relationships between data components are expressed as arithmetical operators. This confers the algebraic properties on the data set model:-

i) The logical relationship between attributes of the same entity is equivalent to the simple multiplication operator.

ii) The logical relationship between aggregates describing members of the same entity set is equivalent to the addition operator.

iii) The logical relationship between aggregates describing members of different entity sets is the null operator.

2.3 Notational conventions for the data model

An attribute is represented by a letter for the type and an integer for the specific value. e.g. -

$$A7 \qquad (1)$$

A letter followed by a lower case alphabetic subscript represents the general case. e.g., if "A" represented the property of colour, then "some or any particular colour" is represented by:-

$$Ai \qquad (2)$$

An aggregate of attributes of a single entity is represented by a string of attribute symbols separated by commas to indicate the multiplicative relationship. e.g description of a particular entity by specific values of three types of attributes is represented as: -

$$A7,B5,C2 \qquad (3)$$

A data set in cardinal form is represented by a sequence of aggregates separated by plus signs to indicate the additive relationship. e.g a data set describing three entities of the same class is represented as:-.

$$A7,B5,C6 + A2,B1,C1 + A1,B1,C9 \qquad (4)$$

A data set can be characterised by type simply by the attribute types it contains, regardless of any specific attribute values. Parentheses are used to indicate replication. e.g. the data set in (4) above can be summarised as:-

$$(A,B,C) \qquad (5)$$

3. EXAMPLES OF ALGBRAIC MODELING OF GEOSCIENCE DATA SETS

3.1 Algebraic representation of cardinal form data

A simple aerogeophysical data set will be used as an example. This data set would be created by digitisation of the flight path maps of a survey. The entities are navigational fixes along flight lines on maps. The X/Y coordinates of these points have been omitted from the example for the sake of brevity. The attributes of each entity are therefore as follows:-

M map number
L line number
F fiducial (serial) number.

The model expression representing this data set is :-

$$M1,L1,F1 + M1,L1,F2 + \ldots M1,L2,Fi + \ldots M2,L1,Fj + \ldots Mk,Lk,Lk \ldots \text{etc.} \qquad (6)$$

The first aggregate (M1,L1,F1) describes the first entity - the first fix on the flight path - by the map and line upon which it lies and by its fiducial number. The second aggregate (M1,L1,F2) describes the second flight path point, etc, etc.

Note that, according to the notational convention, the second fix is on the same map and line as the first, but has a different fiducial number. Successive aggregates describe fixes on different lines, then on different maps.

The elipsis (...) in the expression indicates that a sequence of aggregates, unspecified in number, has been omitted for brevity. The lower case subscripts ("i", "j" and "k") indicate the general-case of the aggregate - "some/any specific value".

All of the information available for description of each entity resides in one aggregate and this aggregate contains no information descriptive of any other entity. Hence by definition this is the cardinal form of the data set. The summary representation of (6), as in (5), is:-

$$(M,L,F) \tag{7}$$

3.2 Algebraic modeling of data manipulation processes

The above representation of the cardinal form of the data set was derived simply by observing an actual data set then describing it symbolically according to the notational convention.

Noting that the same map number is common to many successive aggregates, and that the same line number is common to many successive aggregates within one map, and that the relationship between attributes is multiplicative, then these common values can be factored out of (6) to produce the expression:-

$$\begin{aligned} &M1(\ L1(\ F1 + F2...Fi) + L2(Fj...)\ ...\ + Lk(..)) \\ &+ M2(\ L1(..)...)\ ...\ M1 \end{aligned} \tag{8}$$

- which, in summary notation is:-

$$(M(L(F))) \tag{9}$$

Expression (9) is, in fact, a hierarchical structure. This same structure is developed within data compilation programs by the first process applied to the initial digitised track data in order to economise on storage and improve retrieval efficiency.

Here, the structure was developed solely by formal algebraic manipulation of the data set expression. This briefly demonstrates the simulation capabilities if the model.

3.3 The geochemical data sets

The capability to derive the hierarchical (factorised) form of the aeromagnetic flight-path data set from the flat-file original (cardinal form) solely by algebraic manipulation of the model has been demonstrated above. Other types of data exist, however, whose structure can not properly be described or formally developed solely by this type of manipulation. Further properties of the model must be explored to accomplish this.

3.3.1 The atomic absorption spectrometer data. The cardinal form of the data set requires that each entity be described by a single record. Data set no. 4 in figure 1 describes an entity set consisting of a suite of geochemical samples. The first record is a list of the sample numbers. Each of of the records which follow describes the abundance of a single element for all entities (=samples.) The data set, therefore, is not in cardinal form. Nor is it in the factorised form of expression (9) as the separate aggregates do not represent common factors removed to a higher order of placement.

As each record describes all entities, then according to the definition, each record is in fact a "sub data set" - a data set containing sub-aggregates each consisting of but a single attribute.

If we represent the sample number attribute as "A", copper concentration as "B", lead as "C" and zinc as "D", then according to the model notation, this "data set" looks like:-

$$(A)\ (B)\ (C)\ (D) \qquad (10)$$

i.e. A list of all sample numbers, followed by Cu concentrations repeated for all samples, followed by Pb concentrations, etc.

Note, however, that these are attributes of but a single entity set, hence the null operator between the sub-aggregate groups in expression (10) violates the definitions and this expression is therefore incorrect. Note also, though, that the individual members of each group are in one-to-one correspondence. The N'th member in each group is a attribute of the N'th entity. Hence, the N'th member in each group has a multiplicative relationship with the N'th member in each other group. This, in fact, is simply a vector inner product or "dot" product relationship. The correct form of expression (10) is therefore:-

$$(A).(B).(C).(D) \qquad (11)$$

This expression correctly and completely describes the structure of data set no. 4.

3.3.2 The geochemical report data. Expression (11) does more than describe the data set. It also provides the means to model the manipulation process that would be applied to produce the eventually required form as shown in data set no. 3. i.e. if the dot pruduct is multiplied out we get:-

$$(A,B,C,D) \qquad (12)$$

- which is the model expression for data set no. 3, the cardinal form.

3.4 The aeromagnetic levelling data set.

Data set no. 2 in figure 1 is the aeromagnetic levelling network. The entity set in this case is a set of intersections between traverses and control lines. The cardinal form of this data would be:-

$$(C,T,I) \tag{13}$$

- where C is a control line number, T is a traverse number and I describes the intersection of the traverse and control line. All possible traverse-control line pairs are covered by the data set. Hence, each individual specific value of C will be repeated for every value of T. This means that each value of C is a common factor in many aggregates, hence the data set can be restructured in a factorised form as -

$$(C (T,I)) \tag{14}$$

i.e. each unique value of C occurs once only in the data set, followed by its intersections with all traverses. Then the next unique control line number followed by all its intersections, etc.

The same logic applies to traverses also, however, and an equally valid factorisation would be -

$$(T (C,I)) \tag{15}$$

The form that the actual data set takes, however, is apparently -

$$(C) (T) (I) \tag{16}$$

i.e. All unique values of C stated once only, followed by all unique values of T, followed by a value of I for all possible combinations of values of C and T (all possible intersections.) This is not a cardinal form, nor is it a factorisation in the form of expressions (14) and (15), nor is it the same structure as that of expression (10) as there is no necessary one-to-one correspondance between (C) and (T) or between (T) and (I).

The numbers of aggregates in (T) and (C) (the numbers of traverses and control lines respectively) are independent. The number of aggregates in (I), however, is not an independent variable. It is the product of the number of traverses and the number of control lines (e.g. 10 traverses and 4 control lines makes 40 possible intersections.)

As the entity is the "intersection", then the traverse and control line numbers of a given intersection are attributes of the same entity and hence are related by the multiplication operator. The cardinal form contains all possible traverse / control line pairs. The multiplicative combination of all possible pairs of members of two independent groups is achieved by taking the star product (vector outer product) of the groups. Hence the expression -

$$(C) * (T) \tag{17}$$

- when the star product is taken, would produce a sub-aggregate set

$$(C,T) \tag{18}$$

- containing all possible pairs of the individual members of (C) and (T). This sub-aggregate set is now in serial, one-to-one correspondence with the members of (I) in (16) and is therefore related to it by the dot product operator. Hence a model expression can be written which correctly and completely describes the aeromagnetic intersection data set structure. i.e:-

$$(C) * (T) . (I) \tag{19}$$

Expanding this expression according to the rules of star and dot product multiplication and expressing the result in model notation, produces -

$$(C,T,I) \tag{20}$$

- which is the cardinal form. Hence, the aeromagnetic levelling data set structure can be derived from the cardinal form solely by formal algebraic manipulation, and vice-versa.

3.5 The aerogeophysical archival data set.

The aerogeophysical archival data set structure (data set no. 1 in figure 1.) can be summarised by the model expression -

$$(M (L (T) (G))) \tag{21}$$

It is a factorised form similar to expression (8), grouped hierarchically first by Map (M), then by flight line (L), then with the lowest level sub-aggregate groups (T) and (G). The first of these, (T), describes the set of navigational fixes along the flight line. The second group (G) describes the in-flight geophysical measurements at points in space along the flight line. The coordinates of the measurement points were derived from the navigational fixes but the measurement points differ in number and location from the navigation points.

Hence, there is no one-to-one correspondence between (T) and (G). Furthermore, taking all possible pairs of navigational and measurement points is meaningless. Therefore, neither the dot nor the star product operator can be used to express formally the relationship between these two groups. What, therefore, is the relationship ?

To determine this, we observe that M and L are higher order factors of both (T) and (G), and therefore we can expand expression (21) by multiplying out the higher order factors to produce -

$$(M,L,F) \; (M,L,G) \tag{22}$$

This result is exactly what we should expect. It consists of two separate, independent, cardinal forms. The aeromagnetic compilation process process does indeed begin with two independently acquired data sets - the track recovery data and the in-flight data. These two data sets describe two different entity sets - navigational fixes on a map, and in-flight measurements. Hence, two separate cardinal forms are unavoidable.

Although we have two separate cardinal forms, there are common factors (M and L) between the two forms. The legitimacy of grouping common factors from different cardinal forms to produce expression (21), is demonstrated by the fact that the independent components retain their independence in (21) and that the original independent cardinal forms are recovered on expansion.

The legitimacy can be formalised by stating that two sub-aggregate groups at the same level, which can not be related by any other operator are related by the "null" operator. The operations which change the levels of sub-aggregate groups (factorisation and expansion) can be applied across the null operator but it will not, itself, be affected by these operations. The null operator is indicated simply by the absence of any other operator between two adjacent lowest level aggregate groups.

Hence, the sought for relationship between the lowest level sub-aggregate groups of expression (21) is that of the null operator - an indication that the sub-aggregates were originally derived from separate, independent cardinal forms.

4. PHYSICAL AND LOGICAL DATA DEFINITION

If we are to have efficient interchange and integration of different data types, we require clear and complete definition of the logical and physical contents of the data set. "Logical definition" means defining the logical significance of each individual data item, and the logical relationships between items or groups of items. This type of definition is sometimes refered to as a "data dictionary". "Physical definition" means defining such things as the location of items within the data set and the digital code employed.

A truly transportable data set would be essentially "self-defined". That is, the definitions would not be a sheaf of typescript accompanying a magnetic tape which had somehow to be translated into a program to read and interpret the data, instead, the definitions would reside with the data in computer readable form. A pre-existing, standardised "interpreter" program would first read the data definition, then would adapt its input processes to match the logical and physical structure of the data which followed.

Some self-defined data set systems actually exist today but are generally inflexible and restricted to a narrow range of variability in data content and structure.

4.1 Logical definition by the algebraic model

The model, as established to this point, already provides a basic logical definition of the data set. i.e. the logical content of the data is defined by the attribute types and the logical relationships between attributes and aggregates are defined by the arithmetical operators and parentheses.

Logical structures of other types of model such as trees and networks can be drawn as a diagram showing the identity of the data items in the nodes. e.g. a tree diagram could be drawn for the track data set in expression (9) with M (Map) as the "parent" branching down to many L (Lines) as first generation "children", then each L in turn branching down to many F (track point) children.

Tree and network diagrams, whilst depicting logical structure, provide no information whatsoever about the physical storage of the data and hence how individual data items might be accessed. The algebraic model, however, can provide valuable information on the physical storage of the data.

4.2 Physical definition capabilities.

4.2.1 Sequential access structure. Many geoscience data sets possess two particular characteristics:-

1) they are large and hence demand efficient access methods; but

2) sequential access satifies a majority of the processing needs.

Accordingly, the optimum storage medium is digital magnetic tape. Magnetic tape is a linear storage medium. Regardless of the logical relationships between data items, the physical manifestation of data item values is as groups of bytes stored in linear sequence along the strip of magnetic tape. There is no way to access the Nth item without first passing physically over the preceeding N - 1 items in sequence. Hence the term "sequential access".

The algebraic model, as is clearly demonstrated by the non-summarised form shown in expression (6), is also strictly linear - a linear sequence of characters representing attributes and their relationships to each other.

The relationship between attributes of the same entity is multiplicative. Hence it is commutative. i.e. in terms of content and relationships;

$$(A,B) = (B,A) \tag{23}$$

Therefore, we can make the order of attributes within an aggregate the same as the physical order of the corresponding data items in a data record without violating any of the rules of the model. Hence the model can serve as a physical description of the data as well as a logical model.

5.0 FORMAL DERIVATION OF THE DATA DEFINITION STRUCTURE

The logical and physical data definitions discussed in the preceding sections are essential features of DBMS implementations. They reside in "data dictionaries" or in "schema". In DBMS's based on the three principal models, however, they are all ad-hoc features created independently of the model concerned. i.e. the models themselves provide, at most, a graphical description of logical relationships between data items.

With the algebraic model, however, physical and logical data definitions can be derived by formal manipulation and the result can serve as the design for the actual implementation.

In the model notation, the attribute descriptor is a type-value pair (e.g. "A1".) The attribute type provides the logical definition of the value with which it is associated. In most real data sets, however, the data item type is not explicitly specified within the data set. Only data item values are present. Type-value association is made via names of variables within the program to which data item values are assigned by input.

This is data-dependence. i.e. the logical types of data items in the data set must match the corresponding types of variables defined within the program or either the data or the program must be changed to suit.

Data could be allowed to assume a variety of forms and be independent of software if the data possessed a structure exactly the same as that of the model. i.e. each digital data item value would be preceded with a digitally encoded type definition. For example, a set of analysis results would be recorded digitally as:-

Cr-ppb, 103, Cu-ppm, 22, Ni-ppm, 36, Pb-ppm, 125... (24)

This would work in practice but has the distinct disadvantage that it would greatly increase the size of the data set as compared to the common form where only values are recorded.

The ideal case would be to have the attribute type separated from its value for the sake of economy but the connection somehow retained within the data for the sake of efficiency. This can be done within the model by further expansion of its algebraic properties.

5.1 The attribute type/value relationship.

So far, no arithmetical relationship has be defined between attribute type and attribute value. A relationship can be defined, however, which permits separation of type and value and permits formal derivation of a data definition. The structure thus derived matches the ideal case described above.

We will state that the attribute type and value are an "ordered pair" (e.g. complex numbers and two dimensional Cartesian coordinates are ordered pairs) whose relationship is multiplicative and associative but not commutative.

In the model, whole attributes of the same entity have an ordinary multiplicative relationship are therefore both commutative and associative e.g:-

A1, B5 = B5, A1 (25)

(A(B,C)) --> (A,B,C) --> (B(C,A)) (26)

Expression (25) shows commutativity and means simply that "colour =

blue and weight = 3.0 grams" is equivalent to "weight = 3.0 grams and colour = blue". Expression (26) shows associativity in the manipulation processes.

If a normal commutative multiplicative relationship were permitted between attribute type and value, then this would allow that:-

$$A,2,B,5 = A,B,5,2, = 2,B,5,A \ldots \qquad (27)$$

- which could produce results such as "colour = 3.5 grams, weight = blue". or worse still "weight = colour, 3.5 = blue". i.e. the integrity of the information is lost. Hence commutativity can not be a property of attribute type/value pairs and they must therefore be regarded as an ordered pair for which this property is forbidden.

5.2 Separation of attribute type and value

Let us investigate the effects the type/value property by manipulation of the cardinal form of the "Map, Line, Track" data set of expression (7). In non-summarised form this is:-

$$(M1,L1,F1 + M1,L1,F2\ldots + M1,L1,Fi\ldots + M1,L2,Fj\ldots) \qquad (28)$$

In this expression, the attribute types M, L and F are identical in every aggregate - only attribute values can exhibit differences from one aggregate to the next. Assuming a multiplicative and associative but not commutative relationship between type and value allows the development of the following:-

$$(M,L,F)\ (1,1,1 + 1,1,2\ldots + 1,1,i\ldots + 1,3,j\ldots) \qquad (29)$$

The same process can be applied to the hierarchicaly factorised form

$$(M1(L1(F1 + F2\ldots +Fi\ldots) + L2(Fj\ldots)\ldots \qquad (30)$$

- to produce:-

$$(M(L(F)))\ (1(1(1 + 2\ldots + i) + 2(j\ldots)\ldots)\ldots) \qquad (31)$$

In both cases, the logical definition of the data set, the attributes types, has "floated" to the head of the data set and been separated from the attribute values. The attribute type and value still remain attached by logical association but repetitive values of attribute type have been eliminated. This is the sought for ideal case.

6.0 CONCLUSION

The algebraic model provides a formal means for describing the structures and contents of geoscience data sets. It does so for every member of a very wide variety of different data set types. No single one of the principal DBMS data models, hierarchical, network or relational is capable of this. Furthermore, valid data structures exist which can be described by the algebraic model but which are not describable by any of the DBMS models.

The algebraic model is dynamic, unlike the DBMS models. It provides a means to describe the manipulation processes by which data structures are changed.

The major theoretical advantage of the model, however, in the context of data exchange and integration, is that it formally defines the means whereby a data set can be made internally self-defined and no longer have to rely ad-hoc programming to permit correct interpretation.

The practical advantages of the model are that it represents data as a linear sequence - exactly the form in which real data is stored on a magnetic tape or disc - and that it possesses inherent properties which simplify the implementation tasks.

In any implementation a standard would, of course, have to be established for the digital syntax and semantics of the data defininition groups. This, however, is feasible. The alternative - standardisation by pre-definition of all possible combinations of data format and structure - is, as stated by Jeffery and Gill (op. Cit.), impossible.

REFERENCES

[1] Li Gong, Zhi Chihang, 'On the application of image analytical technique of comprehensive geoscientific data for prognostics of tin polymetalic mineral deposits', Proceedings: Beijing International Symposium on Remote Sensing. BISRS '86 , Beijing, PRC, October 1986.

[2] Eliason, P. T., Donovan T. J., Chavez P. S., 'Integration of geologic, geochemical, and geophysical data of the Cement oil field, Oklahoma, using spatial array processing', Geophyics, Vol. 48 No. 10, October 1986.

[3] Sexton J. L., Braile L. W., Hinze W. J., Campbell M. J., 'Seismic reflection profiling studies of a burried Precambrian rift beneath the Wabash valley fault zone', Geophysics, Vol. 51 No. 3, March 1986.

[4] Awad M. G., 'A geophysical study on the Abu Gharadig basin, Egypt' Geophysics, Vol. 50 No. 1, January 1985.

[5] Jeffery K. G., Gill E. M., 'The geological computer', Computers and Geoscience, Vol. 2, pp 347-349, 1976.

[6] Berztiss, A.T., Data Structures: Theory and Practice, Academic Press, New York 1975.

[7] Date, C.J., An Introduction to Data Base Systems, Addison-Wesley Publishing Co., Reading Mass. 1976.

[8] Holroyd, M.T. A System for Automated Compilation and Cartography of Earth Science Data With Special Provision for Aerogeophysical Data, Phd Thesis, University Of Ottawa, Canada, 1984.

[9] Sutterlin P. G., Jeffery K. G., Gill E. M., "Filematch: a format for the interchange of computer based files of structured data", Computers and Geosciences, Vol. 3 pp 429-441, 1977.

1) AEROGEOPHYSICAL ARCHIVAL

MAP HEADER

LINE HEADER

TRACK DATA RECORD

.

.

GEO DATA RECORD

.

.

NEXT LINE

.

.

NEXT MAP

.

.

<END OF DATA>

2) AEROMAGNETIC LEVELLING

ALL TRAVERSE NUMBERS

ALL CONTROL LINE NUMBERS

INTERSECTION TR #1 - CL #1

INTERSECTION TR #1 - CL #2

.

.

INTERSECTION TR #2 - CL #1

INTERSECTION TR #2 - CL #1

.

.

<END OF DATA>

3) FINAL GEOCHEMICAL

SAMPLE No. 1, Cu ppm, Pb ppm...
SAMPLE No. 2, Cu ppm, Pb ppm...
SAMPLE No. 3, Cu ppm, Pb ppm...

.

.

.

<END OF DATA>

4) A/A SPECTROMETER

Sample no. S# 1, Sample no. S# 2
Cu ppm S# 1, Cu ppm S# 2...
Pb ppm S# 1, Pb ppm S# 2...
Zn ppm S# 1, Zn ppm S# 2...

.

.

<END OF DATA>

Figure 1: Four Typical Geoscience Data Sets.

CASE STUDY ON APPLICATION OF QUALITATIVE DATA ANALYSIS TECHNIQUES TO AN URANIUM MINERALIZATION

H.G.PEREIRA
Centro de Valorização de Recursos Minerais
Technical University of Lisbon
Instituto Superior Técnico, Av. Rovisco Pais
1096 LISBOA CODEX

ABSTRACT. Several data analysis techniques were applied to qualitative attributes in order to visualize their connections and calculate a discriminant function to classify samples into two groups. Recoding numerical values into intervals, exploratory data analysis techniques produce different views of the relationship pattern among variables and discriminant analysis quantify the capability of each variable to differenciate two groups of samples.

THE PROBLEM

During the late stages of an exploration program for Uranium, a set of 50 samples was collected in the Central-Northern region of Portugal (Pereira et al., 1984). On each sample, the following in situ attributes were observed:

- lithology
- redox condition
- metamorphism
- presence of a fault
- expressed mineralization
- presence of quartz

Also, the U_3O_8 grade and depth were measured. Furthermore, a leaching test on standard operational conditions was conducted for each sample, and the recovery was calculated.

Using this data set as an input, a battery of data analysis techniques was applied, in order to get a proper insight into the relationships among variables, and identify targets for further investigation.

The main particularity of this data set is the heterogeneity of variables - it is believed that numerical and nominal data must be, jointly, taken into account by the processing methodology and, therefore, a problem of statistical coherence arises. Hence, numerical values (depth, U_3O_8 grade and recovery) were recoded into intervals and qualitative data analysis techniques were applied to the data set as a whole, taking

C. F. Chung et al. (eds.), Quantitative Analysis of Mineral and Energy Resources, 617–624.

globally the available information.

Moreover, it is worth noting that there is a "special" measure - recovery - which should be treated as an "external variable" to be related to in situ attributes, and account for economic favorability.

METHODOLOGY

Bearing in mind the previously mentioned particularities of the problem, a method of combining several data analysis techniques was devised, in order to process this type of data.

The method includes two main steps: an EXPLORATORY DATA ANALYSIS to describe variables and visualize their connections and a DISCRIMINANT ANALYSIS to derive an objective criterium for classification of samples into one of two groups, related to "high" and "low" recovery.

Concerning the techniques of the first step (Exploratory Data Analysis), it was decided to use Q-Analysis (Atkin, 1974, Griffiths, 1983) in order to produce a connectivy pattern among variables, where the most common links are visualized; also, Characteristic and Correspondence Analysis were tried in this step, to obtain graphic diagrams depicting the projection of variables onto the principal factorial axes. Characteristic Analysis (Botbol, 1971) was applied in its simplest formulation, calculating and incidence-coincidence matrix (Geoffroy et al, 1972) and finding its eigenvalues/eigenvectors.

Correspondence Analysis (Benzécri, 1973; Greenacre, 1984) was used to check stability of results to the recoding procedure. Indeed, Correspondence Analysis is based on a special similarity matrix, denoted "Burt Matrix", which can be seen as the analogue of the incidence-coincidence table, in the case where all states of variables are considered.

In the second step, the objective is to derive a discriminant criterium for qualitative data. As the Classical Fisher linear function does not apply, the correspondence analysis technique was modified in order to produce only one factorial axis, where all categories of "explicative" variables are projected, according to their capability to differenciate two groups of samples. The discriminant power of this axis was assessed through the proportion of misclassified samples in the original training set. Also, a discriminant weight was calculated for each variable and an equation was derived to assign any new sample (defined by a vector of "explicative" attributes) to one of the groups, without conducting further leaching tests.

RECODING DATA

As the data set contains nominal and numeric data, it is necessary to transform variables such as depth, U_3O_8 grade and recovery into intervals, in order to assure statistical homogeneity. The interval limits were selected by a trial and error procedure, controling results through geological and metallurgical interpretation.

After several attempts, it was decided to summarize data in a

50 x 9 boolean matrix, each column of which representing the state of a dichotomous variable (presence/absence of a certain attribute or assignment to a certain interval). The recoded variables are shown in Table 1.

TABLE 1 - Recoded variables

VARIABLE	BOOLEAN CODE 1	BOOLEAN CODE 0
LITHOLOGY	SHALE	OTHER
REDOX CONDITION	REDUCED	OXIDIZED
METAMORPHISM	ABSENCE	PRESENCE
FAULT	PRESENCE	ABSENCE
EXPRESSED MINERALIZATION	PRESENCE	ABSENCE
QUARTZ	PRESENCE	ABSENCE
U_3O_8 GRADE	> .2%	< .2%
DEPTH	> 36.7m	< 36.7m
RECOVERY	> 85%	< 85%

RESULTS

Exploratory Data Analysis

The first step of the methodology was applied to the data set, recoded according to Table 1. Denoting T the matrix of recoded data, the incidence coincidence matrix C was calculated by C = T'T (where T' is the transposed of T). C. contains the frequences of matching for all pairs of positive categories and can be depicted in a valuated graph, the vertices of which are variables and the arcs represent connections. The length of each arc is simply given by the absolute frequence of matches. Using a thereshold of 10 links, the graph obtained is sketched in Fig. 1.

It is worth noting, in Fig. 1, the central position occupied by the vertex "recovery > 85%", which links to all pertinent variables (presence of quartz and fault were disregarded applying the threshold).

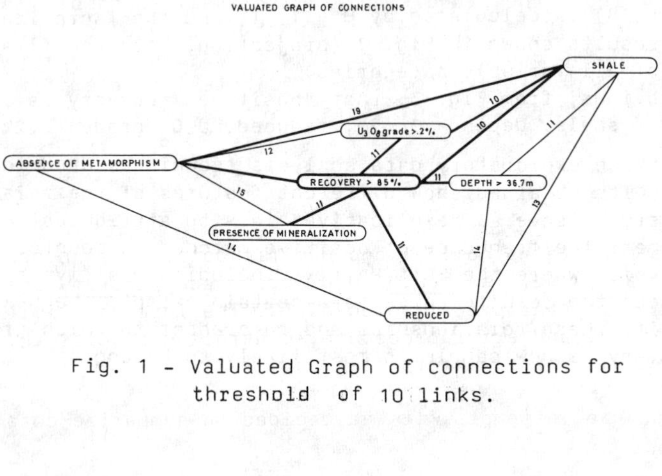

Fig. 1 - Valuated Graph of connections for threshold of 10 links.

If the eigenvalues and eigenvectors of matrix C are calculated, it is possible to project each variable onto the principal axes (eigenvectors for greater eigenvalues). These projections are shown if Fig. 2, using the first and second factors, which account for 70% of the data set variability.

Fig. 2 depicts, in a different way, the same pattern of relationships among variables. In particular, recovery is related with the pertinent "explicative" variables - shale, absence of methamorphism, depth, reducted, U_3O_8 grade, and presence of mineralization.

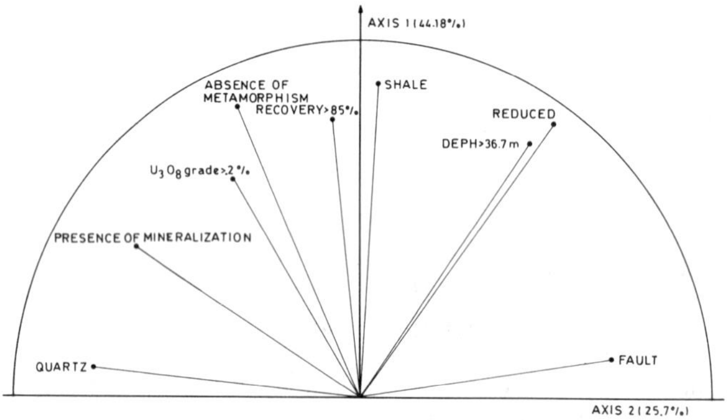

Fig. 2 - Projection of variables ontu 1st and 2nd factors using incidence-coincidence matrix.

The correspondence analysis technique takes as input the Burt Matrix calculated from a new table of logical codes (T_1). In T_1, each row is a sample and there are two columns for each variable - Values 1 denote the presence of the category and 0 denote the absence (Table 1). The Burt Matrix B (18 x 18) is calculated by $B = T_1' T_1$ and the factorization of B produces the results shown in Fig. 3 (projection, onto the first and second axis, of all variable categories).

It is obvious, from Fig. 3, that "positive" recovery is linked to the presence of shale, depth > 36.7m, reduced, U_3O_8 grade > .2%.

In the three exploratory data analysis techniques used in this case study, the results obtained show different features of the relationship linking recovery to several "explicative" in situ attributes. From Q-analysis (where the frequence of positive matches is counted) to correspondence analysis (where the Burt Matrix, including positive and negative matches is used) the results agree to a certain extent. The next step aims to quantify these relationships and to predict to which group (low and high recovery) a new sample is most likely to belong.

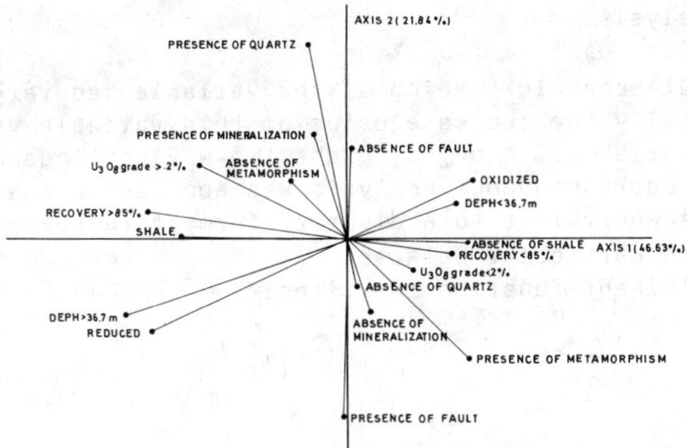

Fig. 3 - Output of correspondence analysis applied to Burt Matrix.

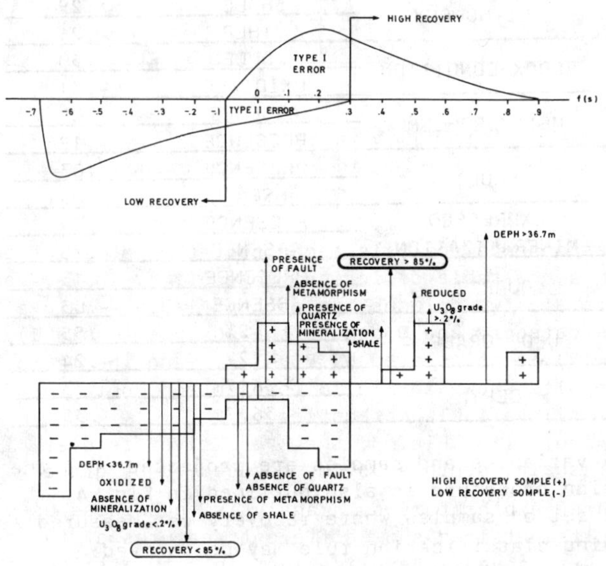

Fig. 4 - Projection of samples and variables onto the discriminant axis

Discriminant Analysis

Given the special economic meaning of the variable recovery, it was decided to cross classify the two categories of this variable versus the states of explicative variables. A set of eight two-way contingency tables were constructed and correspondence analysis was applied to that input matrix. This approach is equivalent to a discrete form of regression, adapted to the nature of the particular data set.

The discriminant function is writen:

$$f(s) = \frac{1}{\sqrt{\lambda}\ p}\ SW \qquad [1]$$

where λ is the eigenvalue assigned to the discriminant axis (λ = .09122)
p is the number of variables (8).
S is the boolean vector of the sample.
W is the vector of weights for each category of variable, given by their projection onto the discriminant axis (TABLE 2).

TABLE 2 - Weights of variable categories in the discriminant function

VARIABLE	CATEGORY	WEIGHT
LITHOLOGY	SHALE	.29
	OTHER	-.21
REDOX CONDITION	REDUCED	.50
	OXIDIZED	-.31
METAMORPHISM	ABSENCE	.09
	PRESENCE	-.19
FAULT	PRESENCE	.03
	ABSENCE	-.06
EXPRESSED MINERALIZATION	PRESENCE	.39
	ABSENCE	-.28
QUARTZ	PRESENCE	.13
	ABSENCE	-.03
U_3O_8 GRADE	> .2%	.55
	< .2%	-.24
DEPT	> 36.7m	.73
	< 36.7m	-.35

In Fig. 4 variables and samples are projected onto the discriminant axis. The decision procedure is also outlined in Fig. 4.

Using the set of samples where recovery was measured as a training set, the following classification rule may be derived:

IF $f(s) > .3$ THEN THE SAMPLE BELONGS TO HIGH RECOVERY GROUP
IF $f(s) < -.1$ THEN THE SAMPLE BELONGS TO LOW RECOVERY GROUP

For the region between -.1 and .3, a study of errors was performed according to the experimental results available in the trainning set, and the conclusion was drawn that the point f(s) = -.1 leads to a minimum sum of misclassification type I and II errors (33%, as calculated in TABLE 3).

TABLE 3 - Assessment of discriminant power by the proportion of misclassified samples.

DISCRIMINANT LIMIT	TYPE I ERROR (%)	TYPE II ERROR (%)	SUM OF ERRORS (%)
-.1	0	33	33
0	6	30	36
.1	29	27	56
.2	47	15	62
.3	59	0	59

So, assuming a risk of 33% of misclassification, a simple and practical decision rule may be derived:

Any sample, taken at a depth greater than 36.7m, has 2 chances in 3 to provide a "high recovery" product if one of the following conditions is met:

Reduced	⇒ f(s) > .22
U_3O_8 grade > .2%	⇒ f(s) > .2
Presence of expressed mineralization	⇒ f(s) > .07
Shale	⇒ f(s) > -.09

CONCLUSIONS

In order to take into account, *jointly*, qualitative and quantitative information in a data analysis problem, numeric values must be recoded into ordinal intervals. After the recoding procedure, correspondence analysis satisfies its basic theoretical assumptions and can be correctly applied to describe the relationship pattern among variables and to discriminate groups of samples.

In this case study, it is shown how to articulate a set of qualitative data analysis techniques in order to give a proper insight into a decision problem in the late stages of an exploration program for Uranium.

REFERENCES

ATKIN, R. (1974) - 'An approach to structure in architectural and urban design'. Environment and planning B, 1974, vol. 1, pp. 51-67.

BENZECRI, J.P. (1973) - L'analyse des Données, Dunod, Paris, vol I and II

BOTBOL, J. (1971) - 'An application of characteristic analysis to mineral exploration', CIM Special vol 12, pp. 92-99.

GEOFFROY, J. and WIGNALL, T. (1972) - 'A statistical study of geological characteristics of porphyry-copper-molybdenum deposits in the Cordilleran Belt - Application to the rating of porphyry prospects'. Ec. Geol. vol. 67, 1972, pp. 656-668.

GREENACRE, M. (1984) - Theory and applications of correspondence analysis, Academic Press.

GRIFFITHS, J. (1983) - 'Geological Similarity by Q-Analysis', Math. Geol. vol 15, nº 1, pp 85-108.

PEREIRA, H.G., SOUSA, A.J., FIUZA, A., COSTA, L. (1984) - 'Improving of interpretation of uranium leaching tests by inclusion of geological information', in 18th APCOM, IMM, pp 441-447.

A SHELL FOR MICROCOMPUTER EXPLORATION SOFTWARE USED IN DEVELOPING COUNTRIES

S. Rehder
Bundesanstalt für Geowissenschaften und Rohstoffe
Postfach 510153
3000 Hannover 51
Deutschland

ABSTRACT. This paper describes some design concepts for exploration software regarding the special situations of German geological joint projects in developing countries. It will show that the hardware is (must be) of less importance than a powerful operating system, an easy to use but high leveled programming language and a well designed software toolbox. Starting from actual problems with existing hardware and software examples and their solutions, using the operating system EUMEL and the programming language ELAN, are discussed.

1. Introduction

1.1. Global Goals

Following the goals of the Federal Government of Germany, research and technology are of growing importance in international coorperation with developing countries. "It is not only a must in terms of solidarity, of provision for the future on the basis of partnership and of ecology awareness but also an important prerequisite for the peaceful coexistence of peoples to help further develop our scientific and technological potential, apply it to the problems of the countries and thus contribute to a favourable economic development, to more humane living and working conditions and to the establishment of scientific and technical capacities (Bundesminister für Forschung und Technologie, 1986).

Within this framework, the German Geological Survey (BGR) carries out exploration projects in developing countries. The principal aim, beside the substantial geological work, is the training of counterparts in order to qualify their institution to increasing use of science and technology independently for their own development. These aims have to be related to all components of the joint project like, geological techniques, lab equipment and also the methods for data evaluation.

Evidently, the transfer of high technology like computerized data processing as an essential part of exploration methods, causes specific problems, and it is not advisable to set up big computer centers with mainframes in the geological surveys of developing countries:

Generally, the infrastructure is not sufficiently developed

C. F. Chung et al. (eds.), Quantitative Analysis of Mineral and Energy Resources, 625–631.

(concerning hardware and software availability and service, dustfree buildings and stable power supply as well as trained operators) and the commercial software has been found to be not always suitable for this kind of application.

1.2. Principal Application

The first task of a computer system will be the support of the actual geological joint project. This requires "closed" programs for standard data evaluation and wordprocessing with short training time where the user interface causes no lingual or other technical difficulties. The user can concentrate on "what to do" rather than on "how to do it". At this level, the user is automatically guided through the program system by a dialog in his colloquial language. He will learn to solve his basic geological data problems very fast because he does not need to know anything about the internal structure, such as operating system or programming language.

Secondly, the program system has to provide an "open" shell for the modification and extension of closed programs, mainly by personel of the developing country, because this part of the technology transfer must not be a black box and no system can offer everything.

In the field of advanced software technology the program system has to include the following basic concepts:

A module concept which allows the creation of standard tools like procedures, functions, operations, and compiled packets.

Strong support of verbalisation and top down programming in order to facilitate the analysis of existing modules and algorithms.

High orthogonality of all software tools, so that the modification of one tool does not effect any other tool at the same level.

High error detection rate during compilation, in order to learn how to write good and accurate programs.

From the pedagogical point of view this "opening" takes place at different levels. The easiest levels are the configuration of a program where the program itself supplies all tools and the definition of own simple expressions like a selection query or an arithmetic expression in a database system. At the next level the user will write his own programs using the programming language and the existing software tools which can be seen as an expansion of the programming language with defined interfaces (= parameters for procedures and procedures for packets).

This software design technique allows a rapid prototyping. One aspect of the great acceptance of modern software like spreadsheets and expert systems is the fact that the main structures (for reasoning and dialog components in artificial intelligence programs) exist already in the program shell as a software tool. The same concept can be applied to programs for geological problems. The advantage is not only the clear structure of the programs but also the possibility of very fast development of new programs.

The analysis of this special software development problem will show, that the success depends on the right choice of the operation system and the programming language. Consequently, this paper describes also parts of the operating system EUMEL (for extendable multi user microprocessor ELAN system) (Hahn 1985) and the programming language ELAN (Klingen, Liedtke 1983).

The EUMEL operating system is a German software development resulting from a cooperation between the University of Bielefeld and the GMD (= Gesellschaft für Mathematik und Datenverarbeitung) in 1977.

The ELAN programming language was designed in 1974 by C.H.A. Koster and his colleagues at the Technical University of Berlin. ELAN is a member of the ALGOL 68 family. Regardless of the fact that ELAN has therefore certain facts in common with languages such as PASCAL and MODULA, it shows some advanced concepts (Klöckner 1985).

2. REDUCTION OF EXECUTION TIME

2.1. Fast Processors

The most fascinating and important feature for many users is the speed of a computer and they refer to the central processing unit. The reasons are firstly the advertisements and demonstrations of hard- and software companies and secondly the historical development during the last few years. But from the practical point of view one should consider two other facts:

There is no difference in speed whether a wordprocessor is using a very fast processor or just a fast enough processor. Here, like in many other single user applications the execution time is limited by the user and by the file handling system.

In high speed applications (like in most of computer graphics) the fast development of new processors causes automatically the problem of software portability, especially if software production takes a long time. The solution will be discussed later.

2.2. Code Sharing, File Organisation and Security

The memory of computers is always too small to store all information. Operating systems usually solve this problem by partitioning the memory into the machine code and the information on disks or tapes, both with different organisation structure and characteristics. The access to machine memory is very fast but limited by the address space, the access to peripherial memory is very slow and the information has to be converted before it can be used by the CPU. On the other hand, the information on disks and tapes is very secure and the capacity is much bigger.

One way to solve this problem is the concept of "demand paging". It is used in EUMEL as well as in some advanced operating systems for mainframes.

Addresses in EUMEL0 code do not refer to the main memory, but to a so called virtual address tree. This virtual memory shows in general the same behaviour as the main memory, it is however much larger.

In a virtual memory system the main memory is used for locating the working set. The unused remainder is located on a secondary non-volatile memory, the so-called "backing memory". From there, anything required is loaded into the main memory. All dataspaces, whether they contain programs, data or files, admit demand paging. The system will serve all pages automatically in the real memory which are actually required at the moment (after Klöckner 1985). Usually the backing memory is a Winchester disk.

Even though this is a highly sophisticated implementation, the user only becomes aware of the advantages:

a) The size of program code can be very large (current version: maximum of 252 dataspaces of 1 MByte each; next generation: maximum of 16000 dataspaces of 1 Gigabyte each (Starcke 1986)) and does not depend on the actual processor.

b) A "fixpoint rerun logic" will automatically save the complete system code every 15 minutes (or any other time) on the backing memory by copying only those (not empty) pages which have been changed during the last 15 minutes. If the power supply breaks down, -and that happens very often in developing countries-, the system will continue at exactly the fixpoint. In the same manner a user can leave his task at any point and continue whenever he wants, regardless whether the system had been switched off or not.

c) Different tasks use the same physical code when they attach the same dataspace. As an example: if a second user wants to copy the dataspace of a first user, this procedure takes nearly no time even if the dataspace is very large because at this stage, the system does not make any physical copy. It merely notes, that there is a second user. As soon as one user will be writing into "his" dataspace, the system will automatically write it into the right page. This technique is known as "lazy copy".

All this background knowledge is an explanation why the data handling is so fast. The user (and also the programmer) will only notice that it is very fast and secure.

2.3. Multitasking

Multitasking is the third way to reduce computer time. A time sharing system enables the user to run different tasks simultaneously and it allows different terminals and users at the same time. Many applications on a microcomputer like printing, plotting, wordprocessing or keyboard input do not require the total capacity of the processor.

Let`s assume an every day situation in an exploration project in a developing country: A geologist has to write a letter with a wordprocessor, calculate and draw some contour maps and his colleague wants to add new data to an existing file. If they execute these jobs sequentially one by one, they would waste a lot of time: with the wordprocessor program the geologist is very busy but the processor is not fully used. The contour maps require the parameters from the geologist first ,then the processor is very busy to calculate the new interpolated values, and then the plotter is very busy while the processor serves slowly command by command. The geologist has to wait until he can start with the next contour map. After he has finished, his colleague can start with his data typing. The whole procedure runs much faster, when the system allows a multitasking.

3. REDUCTION OF TRAINING TIME

3.1. Uniformity

The technology transfer does not only consist of a nice looking and powerful software system but also of the counterpart training. One feature to reduce the training time is the uniformity of the user interface. The design of a user-friendly dialog (such as status texts, information texts, windows, masks, and dialogs) as well as the functions for using of these components should be the same in all programs. The most elegant way is the use of software tools which create these components automatically.

Two particularities are of special importance:

1) The user dialog in his colloquial language, with only a minimum set of technical terms. These terms must be the same in every application program.

2) Help functions must be available at every step of the program in order to explain the dialog and the used method in more detail. Two different types of notebook will help the user to write his own remarks regarding his program or his project. The first type of notes will be available, whenever this program is executed again. It contains additional private explanations for the program. The second type of notebook will assist to write a report of what has been done in this project. The user can copy parts of his dialogs, results and own remarks into this file with a powerful and easy to use wordprocessor.

3.2. Transparency of Programs and Reduction of Programming Time

The following example of the sorting procedure "Quicksort" will show some aspects of the programming language ELAN:

```
PROC quicksort (INT CONST first, last):
   IF first < last
    THEN choose pivot element and take it out of the list;
         sort rough small elements to the left and great to the right;
         insert pivot element into location between the zones;
         quicksort (first, location - 1);
         quicksort (location + 1, last
   FI.

choose pivot element and take it out of the list:
   change in list (first, (first + last) DIV 2);
   INT CONST pivot := list (first).

*
*
*
ENDPROC quicksort;
```

Firstly the algorithm becomes very clear by the strong verbalisation. Initially of course, the compiler does not know, what "choose a pivot element..." means, but this is explained at a later step in the program. These "refinements" are more than just comments, because they are active parts within the program structure. Furthermore it enables the program development from top to bottom.

Secondly, the example shows the advantages of recursive algorithms, wherein the procedure "quicksort" calls itself.

The next very useful tool after "refinement" and "procedure" is the "packet" (= module in other programming languages). A "packet" is the (compiled) closed collection of data and algorithms.

A packet "interpolation routines" can define the procedures "rolling mean", "splines", "polynom", "kriging",....,"3-dim map", "symbol map", "iso lines",... Whenever one of these procedures is required, the programmer or a managertask can insert the packet and make use of the procedure like a "normal" procedure.

In opposite to most other operating systems and programming languages, there is no difference between programming language, job control language and implementation language. The programmer only needs to learn ELAN.

4. REDUCTION OF TRANSFERENCE TIME

4.1. Portability to New Computers

As mentioned earlier, software development, especially for micro computers, is faced with the problem to transfer the software to new processors and hardware configurations. This problem is getting even worse if the software development is carried out by a small group and if it has to be applied for a long time considering the high costs of developing special software.

Because it is the objective to provide developing countries with software the programs have to run on computers which are available in those countries.

EUMEL solved this problem by a very high degree of abstraction from the characteristics of the various processors. The concept is called "virtual machine" since the programmer is in a way provided, via software, with a processor of an extremely powerful instruction set. This will however also entail a certain loss of efficiency which is, by far, not so considerable as intuitively assumed (Klöckner 1985).

All layers, down to the virtual machine, are completely identical regardless of the used processor. Newly introduced computers only require an adaptation in very low layers. The effort depends on whether adaptation to the processor had been done already. In any, case adaptation is much easier with EUMEL than with UNIX because UNIX application software uses many hardware specifics.

4.2. Peripherials

The second problem of transference refers to peripherials like printer, plotter, digitizer and probably graphic terminals. Usually these devices offer different options (like resolution, fonts, etc.) and command languages.

Here again the programmer has to design the software in a device independent way. The main program will create a command file using standardized commands like GKS, and a driver will interpret these commands in order to execute them on the peripherials. This technique guaranties a strict distinction between the problem orientated construction or manipulation on one hand and the presentation on the other hand. In a multitasking operating system like EUMEL, the peripherials are connected to a spooler.

Every new program in such an software environment is written for the virtual peripherials. Every new peripherial keeps the program unchanged but may require a new driver.

5. CONCLUSIONS

The transfer of new technologies like advanced gelogical methods to developing countries includes adequate data processing methods. To be able to meet their special demands high leveled software tools are necessary. That signifies a very careful and difficult development of the basic system but an easy to use and exhaustive tool for the geologists in developing countries.

REFERENCES

Bundesminister für Forschung und Technologie, 1986, New Technologies for the Third World: Bonn 2, Heinemannstr. 2, 182 p.

Hahn, R., 1985, EUMEL user manual version 1.7, Hochschulrechenzentrum Universität Bielefeld, edited by GMD Schloß Birlinghoven D-5205 Sankt Augustin 1, 500 p.

Klingen, L. H., Liedtke, J., 1983, Programmieren mit ELAN, Teubner Stuttgart, 207 p.

Klöckner, K., 1985, The EUMEL operating system: software developing on microcomputers, Selbstverlag GMD Schloß Birlinghoven D-5205 Sankt Augustin 1, 9 p.

Starke, K.W., 1986, EUMEL aktuell, Selbstverlag GMD Schloß Birlinghoven D-5205 Sankt Augustin 1, 91 p.

OPTIMAL EXPLORATION STRATEGIES: THE SYNTHESIS OF RESOURCE ASSESSMENTS AND OPERATIONS RESEARCH

Michael J. Shulman
Exploration Decision Software and Consulting
Dernburgstrasse 15
D-1000 Berlin 19
Federal Republic of Germany

ABSTRACT. Resource appraisals or assessments provide a quantitative and rational foundation upon which subsequent decisions may be based. The appraisals are, however, generally relegated to or accorded a passive role in decision making. Eschewing this quantitative facet of resource appraisals diminishes the options open to the decision maker in formulating and comparing different exploration strategies or policies and hence his effectiveness. To illustrate the synthesis of resource appraisals and Operations Research to determine an optimal exploration strategy, a multiobjective decision problem is formulated to aid in allocating exploration effort in the search for Kuroko type Zn-Ag-Pb deposits in Greece. The decision model consists of two parts: 1. The optimal exploration budget was calculated for each area by maximizing the expected posterior expected monetary value. 2. Based on the area budget, exploration effort is divided between standard geologic field procedures and geophysical studies by simultaneously maximizing the probability of detection and the expected gain in information. In part 2 there are two decision variables. The multiobjective problem was solved by assuming that there was no articulation of preference, thus allowing a global criterion to be used.

1. INTRODUCTION

Resource appraisals and assessments[1] provide a quantitative and rational foundation upon which subsequent decisions may be based. The appraisals are, however, generally relegated to or accorded a passive role in decision making. That is, the quantitative aspects of an appraisal are used qualitatively. Prospective areas are, for example, classified or ranked according to their favorability; quantitative differences in the favorability among areas are neglected and are not used to explicitly formulate policy or strategy. It is, in particular, this quantitative

1. The terms 'appraisal' and 'assessment' are used interchangeably throughout; however, their connotations differ, appraisal implying expert judgement whereas assessment implies a critical appraisal for the purpose of understanding or interpreting, or as a guide in taking action (see Webster's Seventh New Collegiate Dictionary,1966).

C. F. Chung et al. (eds.), Quantitative Analysis of Mineral and Energy Resources, 633–650.

facet which makes such appraisals so interesting and valuable. Eschewing it drastically diminishes the options open to the decision maker in formulating and comparing different strategies or policies and hence diminishes his effectiveness.

It is seldom emphasized that an action or consequence not only is associated with and the logical outcome of a resource appraisal, but that they furnished the incentive for the appraisal. (This incentive is evident in the choice of appraisal goal, e.g. number of deposits, net present value, etc.) In the Earth Sciences, two types of decisions - policy and exploration - each leading to specific actions or consequences can be distinguished.

In policy decisions the resource appraisal is used to help formulate mineral policy; to help select concession areas; to help in concession negotiations, etc. Some of the problems and controversies inherent in linking policy making and resource appraisals have been discussed by Harris(1984).

In an ongoing exploration program or during the generating stage of exploration, resource appraisals aid the decision maker in classifying prospects and in determining whether the prospects warrant further study. Three alternatives or actions are usually available to the decision maker - go, no go or put-on-hold. Provided that exploration should proceed, the decision maker must allocate his exploration effort optimally among the favorable areas.

Decisions are, however, not based solely on resource appraisals; indeed, the importance assigned to the general appraisal or to particular areas may be subordinate to other criteria, be it operational, economic or political. The resource appraisal is thus inextricably intertwined with other criteria. This leads to a multiobjective decision (MOD) problem. Mathematically, such problems have been treated within the framework of Operations Research.

The goal of this paper is to show how the quantitative results obtained from a resource appraisal can be used in conjunction with Operations Research methods to scientifically study the options, risks and rewards of alternative strategies or policies open to the decision maker. Emphasis has been placed primarily on its role in formulating optimal exploration strategies. Its extension to policy decisions is straightforward.

2. CASE STUDY: EXPLORATION FOR Zn-Ag-Pb DEPOSITS IN GREECE

In 1981 a Zn-Ag-Pb deposit was discovered in the vicinity of Molai, Southeast Peloponnese, Greece. Subsequent studies tentatively classified the deposit as being of Kuroko type. A general description of the geology and mineralization can be found in Galanopoulos(1985) and Angelopoulos et al.(1985).

The polymetallic mineralization occurs within the Triassic volcanosedimentary series of the Tiros Beds. The Tiros Beds are divided into an Upper Paleozoic clastic series and the Triassic volcanosedimentary series. The latter are composed of metavolcanites, metapelites, metasandstones and intercalcated lenses of carbonates. The Tiros Beds igneous rocks consist of extrusive rocks (andesite lavas), pyroclastic rocks and tuffites.

The ore is composed of the following minerals: sphalerite, pyrite and galena. In addition, tetrahedrite, tenantite, hematite, chalcopyrite, pyrrotite, freibergite, arsenopyrite, acanthite, argentite, bornite and native copper are present; their percentages decrease accordingly. The gangue minerals are barite and quartz.

The southeastern Peloponnese, situated within the External Metamorphic Belt of the Hellenides Orogeny, is comprised of the following litho-tectonic units: 1. Plattenkalk Unit, 2. Phyllite-Quartzite Unit, 3. Tiros Beds, and 4. Carbonates of the Tripolitza Zone. The Plattenkalk forms the basement of the various zones and nappes in Peloponnese and Crete. It is overlain by the Phyllite-Quartzite Unit which is, in turn, overlain by the Tripolitza nappe. The Tiros Beds which are considered to be the underlying layer of the calcareous Tripolitza series is assumed to have been overthrusted on the Phyllite-Quartzite Unit. The Tiros Beds have, however, been assigned to the Phyllite-Quarzite Unit by many authors because of its tectonic position, lithology and age.

The discovery and delineation of the Molai deposit was the result of the systematic application of the following methods

- geological mapping
- follow-up and detailed soil geochemistry
- geophysical prospecting
- core drilling

2.1. Resource Appraisal

Constantinides et al.(1986) presented a pragmatic ore discovery model for the Molai type deposits. The model is subjective and is based on sets of discovery criteria, each set corresponding to a specific information level.

Three levels of information were defined consisting of 9, 12 and 20 criteria. Furthermore, a weighting factor was assigned to each criterion reflecting its importance as an ore indicator. The probability of ore occurrence, <u>the target location probability</u>, was then defined for each level of information, i.e.,

$$f = \sum_{i=1}^{n} w_i P(c_i) \qquad (1)$$

where $P(c_i)$ is the probability that criteria i occurs and w_i is the weighting factor associated with criteria i. The criteria and their respective weighting factors are shown in Table I.

The ore discovery model was programmed using Monte Carlo methods. Qualitative as well as quantitative data could therefore be used. A probability index was derived from each simulation curve by specifying the importance of over- or underestimating the index (Bayesian Statistical Decision Theory), by calculating the average entropy of the simulation curve or by taking the mean value of the curve. The mean probability of ore occurrence for information levels 2 and 3 are shown in Fig. 1.

In general, a regional analysis is characterized by a patchwork of information levels - some areas have been more intensely studied than others. It is imperative that the probabilities derived from different information levels be placed on a common fundament. Using only the lowest level of information, the common denominator, neglects relevant data.

The following method based on the highest level of information alleviates this problem: The expected monetary value is calculated for the highest level of information. A linear relationship is assumed between the expected monetary value and the probability of ore occurrence. For each lower level of information three doublets (EMV,p) must be furnished by the explorationists. The doublets correspond to the estimated EMV given probability p and the uncertainty inherent in the

TABLE I (after Constantinides et al., 1986)

CRITERIA	WEIGHTING FACTOR
INFORMATION LEVEL 1	
1. Pyroclastics more than extrusives	0.10
2. Fine grained tuffs (ash tuffs)	0.15
3. High % of breccias vs. altered lava flows	0.10
4. Stratiform mineralization	0.30
5. Presence of red (Fe-) tuffs	0.05
6. Transition between Tripolis and Tyros beds	0.05
7. Thickness of Tripolis limestones	0.05
8. Presence of altered tuffs (propylitized)	0.15
9. Regional scale geophysical anomalies	0.05
INFORMATION LEVEL 2	
1. Pyroclastics more than extrusives	0.05
2. Fine grained tuffs (ash tuffs)	0.05
3. High % breccias vs. altered lava flows	0.05
4. Stratiform mineralization	0.20
5. Presence of red (Fe-) tuffs	0.03
6. Transition between Tripolis and Tyros beds	0.03
7. Thickness of Tripolis limestones	0.03
8. Alteration zones with Qu-Ser-Chl-Alunite	0.20
9. Follow-up geophysical anomalies	0.15
10.Other tuffs	0.05
11.Relation of Pb-Zn sulf. to Cu-Fe-Mn and barites	0.10
12.Presence of structural systems NW-SE to N-S	0.05
INFORMATION LEVEL 3	
1. Pyroclastics more than extrusives	0.04
2. Fine grained tuffs (ash tuffs)	0.08
3. High % of breccias vs. altered lava flows	0.05
4. Stratiform mineralization	0.15
5. Presence of red (Fe-) tuffs	0.01
6. Transition between Tripolis and Tyros beds	0.01
7. Thickness of Tripolis limestones	0.01
8. Alteration zones with Ser-Qu	0.10
9. Pyritization	0.08
10.Detailed geophysical anomalies	0.10
11.Other tuffs	0.01
12.Relation of Pb-Zn sulf. to Cu-Fe-Mn and barites	0.04
13.Presence of structural systems NW-SE to N-S	0.04
14.Structural systems NE-SW	0.01
15.Folding with b axes oriented NW-SE	0.04
16.Clastic series	0.01
17.Cu-Fe mineralization	0.04
18.Contact between alt. lavas and alt. cryst. tuffs	0.04
19.Dispersion halos of Hg	0.10
20.Detailed geophysical anomalies of Pb-Zn	0.04

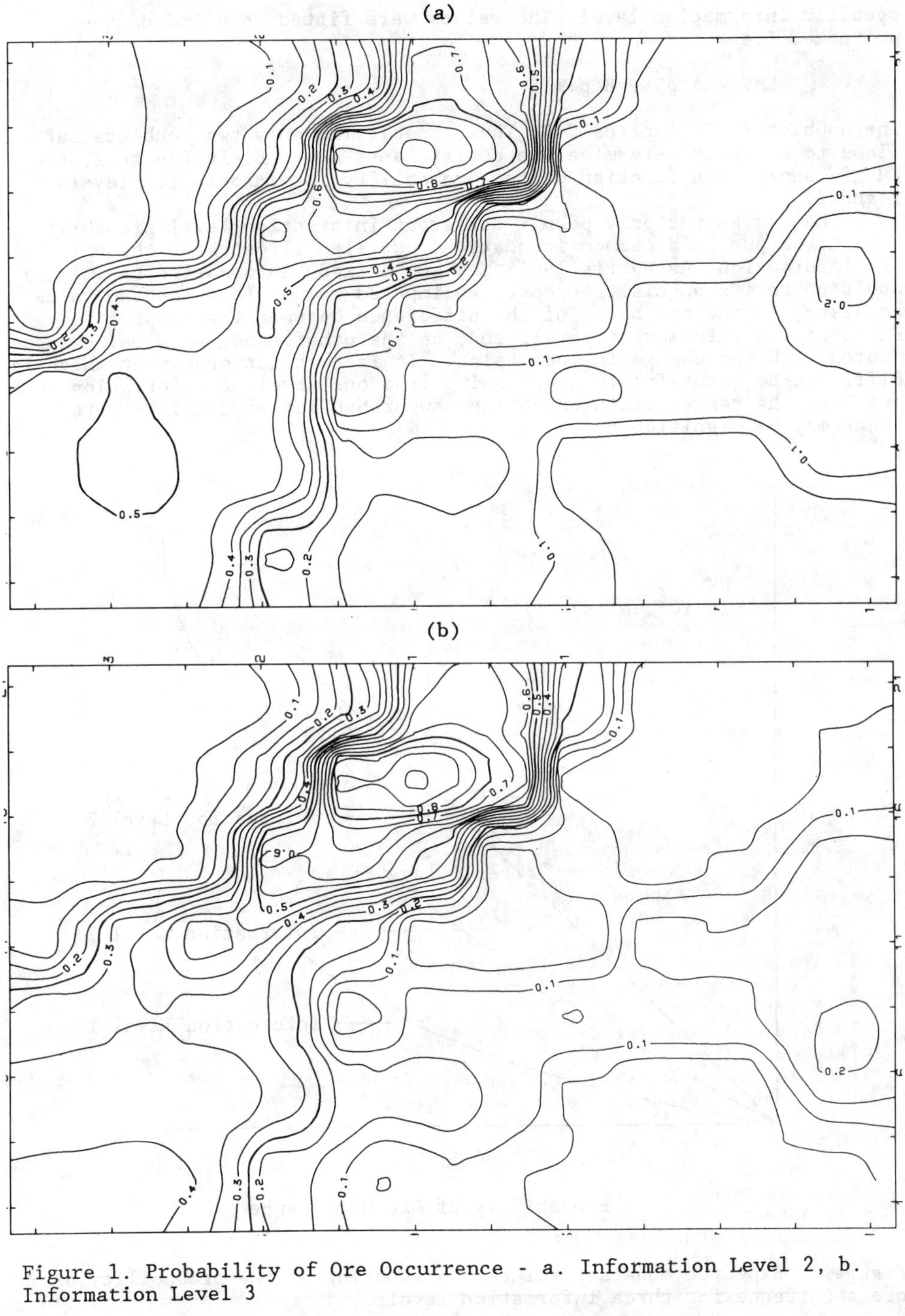

Figure 1. Probability of Ore Occurrence - a. Information Level 2, b. Information Level 3

specific information level. The values were fitted to a 2nd degree polynome, i.e.

$$EMV = A\ p^2 + B\ p + C \qquad (2)$$

The doublet (0,0) implies C=0. Thus, specifying only two doublets suffices to uniquely determine the coefficients A and B. In Figure 2, the EMV is shown as a function of the probability for information levels 1, 2 and 3.

The curves not only permit different information level probabilities to be put on a common foundation, but also illustrates the value of the information. As Harris(1984) has emphasized, information has, in addition to its acquisition cost, an imputed value. This imputed value consists, on the one hand, of the difference between the respective EMV's of the information levels and, on the other hand, of a value associated with the change in uncertainty. It is this latter aspect which is difficult to quantify; in progressing from one level of information to another, the range within which the new probabilities are likely to occur may be significant.

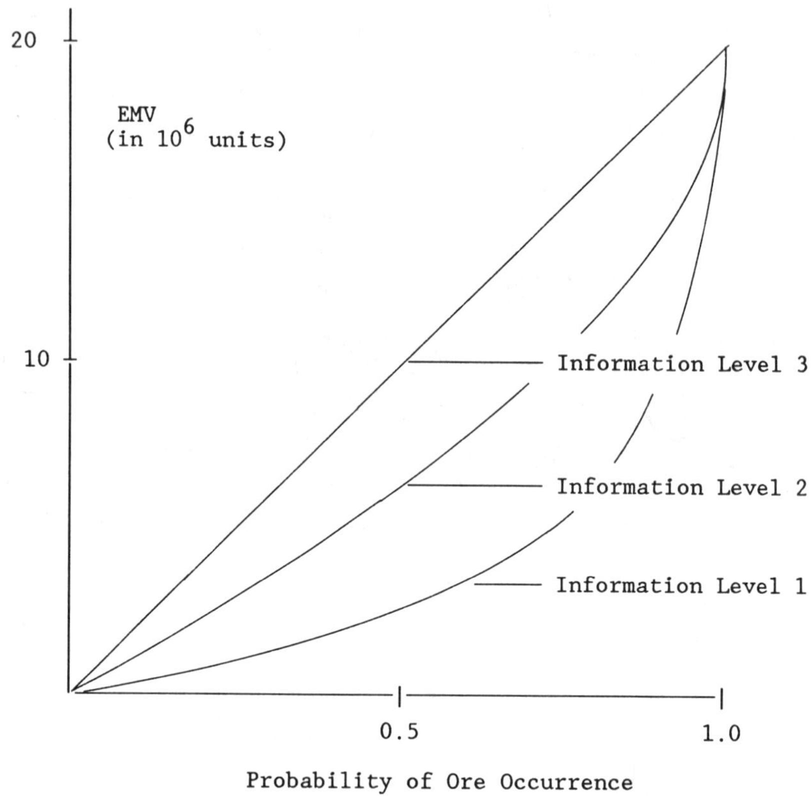

Figure 2. Expected monetary value as a function of the probability of ore occurrence for three information levels.

For information levels 2 and 3, the Molai area was divided into 68 areas. A minimum acceptable value of 20 million units was used. Fig. 3a shows the level 2 probabilities converted into EMV's based on the following doublets which were deemed, by the explorationists, to be commensurate with the risk and cost involved: (0,0),(5.5,0.5) and (20,1.). The contours were drawn at 1 million unit intervals. Using utilities instead of probabilities (see Fig. 1a) suppresses those areas with low probabilities, thus intensifying the general behavior of the EMVs and of specific trends (here N-S and NW-SE). Since level 3 is the highest information level and increases linearly with the ore occurrence probabilities, a contour map of its EMV does not differ from a map displaying the ore probabilities. In Fig. 3b, the cells were randomly assigned to either level 2 or 3 to simulate the patchwork occurrence of information levels. The resulting map of the EMV's again maintains the general behavior of the level 3 probabilities (Fig. 1b) while simultaneously suppressing the least favorable areas.

In TABLE II the EMVs and the mean probabilities are shown for information level 2 and 3 as well as for the random simulation of the information level. It is evident that cells possessing high probabilities are relatively unaffected when extrapolated to the next highest information level (e.g. 29 and 39). The exact sequence may vary slightly. It is only when the probabilities decrease to approximately 0.6 that extrapolating to the highest information level significantly transposes the cell order. For example, cell 18's probability doesn't change with the information level, its EMV however increases by 50%. This difference may be the imputed value of the additional information. By assigning it to level 2 the risk inherent in this level of information interacts with the EMVs of other cells whose probability may be lower (for the specific ore model) but whose EMV is higher such as cells 10 and 11.

TABLE II

INFORMATION LEVEL 2			INFORMATION LEVEL 3			RANDOMLY ASSIGNED INFORMATION LEVELS		
AREA	PROB.	EMV	AREA	PROB.	EMV	AREA	LEVEL	EMV
29	.92	17.22	29	.92	18.40	27	3	17.53
27	.89	16.00	27	.88	17.57	29	2	17.34
40	.83	14.01	40	.76	15.21	40	3	15.35
28	.73	11.12	39	.68	13.63	28	3	13.08
39	.65	9.00	28	.65	13.07	39	2	9.19
30	.62	8.31	30	.65	12.94	30	2	8.45
18	.61	8.02	18	.61	12.23	10	3	8.30
9	.53	6.10	19	.56	11.17	11	3	8.27
32	.52	6.03	32	.51	10.19	18	2	7.98
19	.52	5.94	11	.42	8.39	17	3	7.97

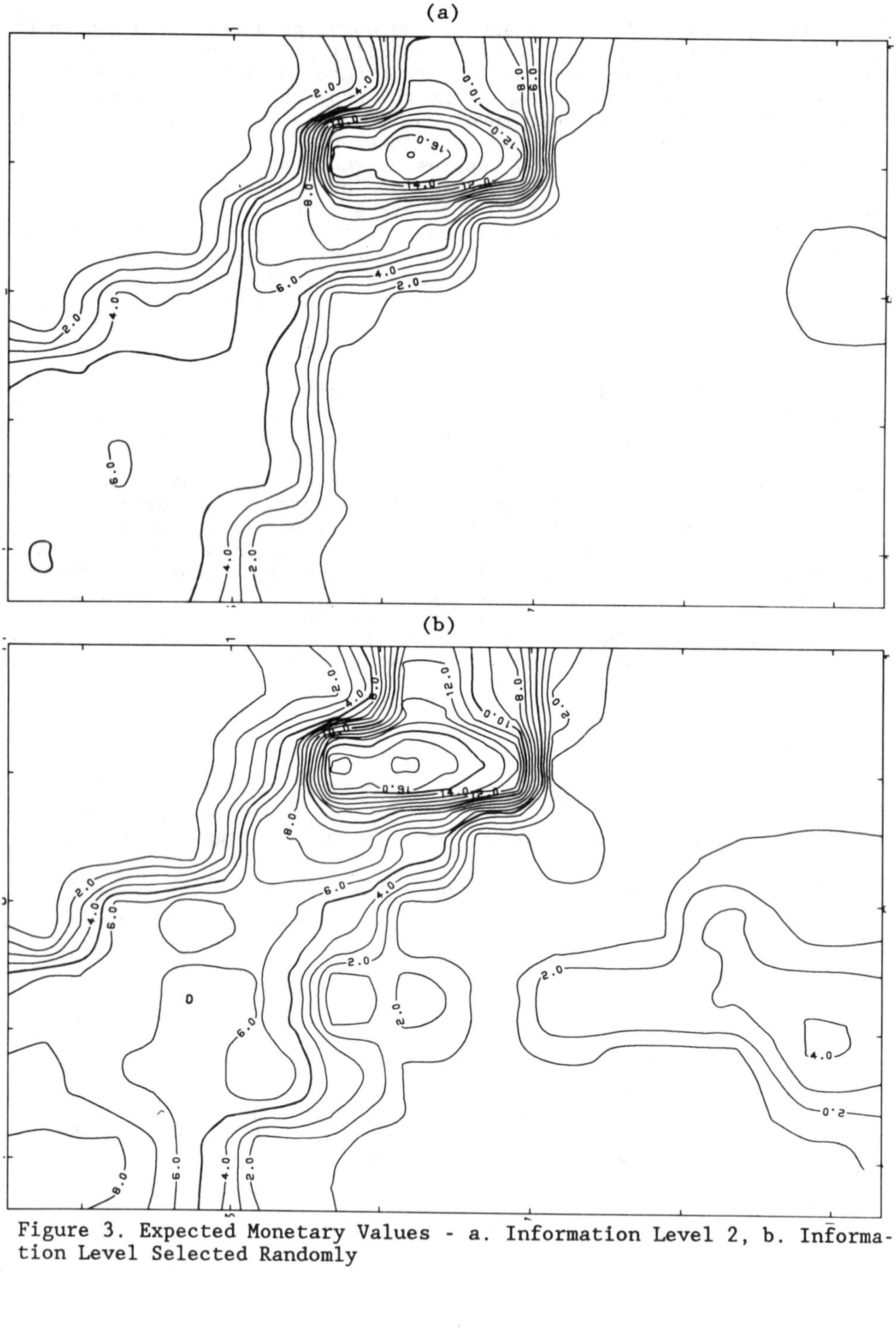

Figure 3. Expected Monetary Values - a. Information Level 2, b. Information Level Selected Randomly

2.2. A Multiobjective Decision Model to Allocate Exploration Effort

In order to reduce overall complexity and computational requirements, which increase exponentially with the number of objectives, the allocation problem was divided into two parts:

a. Exploration effort is optimally allocated among areas based on one objective function and one decision variable.

b. The optimal exploration effort calculated for each area is then optimally divided among exploration methods. Two objective functions and two decision variables were used.

2.2.1. Maximizing the Expected Posterior EMV: The Optimal Area Budget. The optimal exploration budget for each area is calculated by maximizing the expected posterior EMV. The expected posterior probability of ore occurrence is a function of the intensity of exploration and of the area covered. Intensity of exploration is used to denote the amount of effort expended in a given area, e.g. average time spent per sq. km, line spacing for geophysical surveys, etc.

The expected posterior probability can be written as

$$P(T_e|A) = \frac{P(T_e)P(A|T_e)}{P(T_e)P(A|T_e)+P(T_1)P(A|T_1)+...+P(T_n)P(A|T_n)} \tag{3}$$

In Eq. 3 $P(T_e)$ is the probability of an economic deposit occurring; $P(A|T)$ is the probability that the set A of exploration criteria defined by the explorationists are discerned provided that object T is present; and $T_1, T_2, .., T_n$ are false objects, i.e., objects, possibly present, which would produce similar geological and geophysical signatures as the specific target sought. Prior to exploration, the P(T)'s are known and constant, whereas the P(A|T)'s are obviously an increasing function of the intensity of exploration and of the area covered. Multiplying Eq. 3 by the minimum acceptable value of the target and subtracting exploration expenditures yields the expected posterior EMV as a function of effort.

The probability of an economic deposit being present is

$$P(T_eT_{min}) = P(T_{min})P(T_e|T_{min}) = P(T_e) \tag{4}$$

where $P(T_{min})$ is the probability of mineralization, i.e., the probability furnished by the resource appraisal. The joint probability, $P(T_eT_{min})$, of both mineralization and an economic deposit occurring is simply the probability of an economic deposit being found; this empirical value is usually known as is the probability of occurrence conditional on the presence of mineralization. The two values differ, with the latter being larger. Knowing that mineralization is present increases the probability of discovering an economic deposit relative to having previously not been cognizant of the presence of mineralization.

The probability of recognizing positive indicators provided that either a true or false target is present (however not both) is dependent on the geometry and orientation of the geological body, on the total area under consideration, on the physical properties/conditions of the geological body, i.e. on its geological and geophysical contrast with the surrounding rocks, and, finally, on the amount of effort expended.

Consider the total area A to be divided into n hypothetical squares, each approximately the size of the target a. If exploration is conducted in one of the squares, the probability of registering positive results would be

$$P(A|T) = \alpha(g,m,i)\ (a/A) = \alpha(g,m,i)\ (1/n) \tag{5}$$

where α is a contrast and recognizability parameter. It is a function of the geology, exploration methods and intensity of exploration. The cost is, in turn, a function of the intensity of exploration and of the area covered. If the geometry and physical properties of the rocks were known precisely, the intensity of effort required to obtain positive indicators would also be known. Consequently, α would be a constant and the probability of detecting the target would solely be a function of the area covered.

If the selection of the hypothetical squares is independent of detection and the a 's do not differ among squares, we get

$$P(A|T) = \alpha(g,m,i) \sum_{k=1}^{K} 1/n \qquad \text{for } 0 < K \leq n \tag{6}$$

Although K, the number of squares to be explored, is also a decision variable, it was not incorporated into the formulation, but was explicitly provided by the explorationists based on the magnitude of the calculated EMVs.

The parameter α as a function of exploration intensity may be derived by utilizing the law of diminishing returns. That is, a point is reached where any change in the probability of detecting the target is no longer commensurate with the change in exploration intensity and therefore with the change in costs. The change in probability is therefore dependent on the amount of effort already expended. This implies that the first derivative of the probability with respect to intensity must be positive to ensure that the function increases from left to right. Furthermore, the second derivative must be negative since the probability curve is concave downwards (the law of diminishing returns).

It was assumed that the change in probability is inversely proportional to the effort, i.e.

$$\frac{dP}{dE} = \frac{1}{\alpha \times E + K} \tag{7}$$

where α is a constant (although a function of the geology and exploration methods) and K is the minimum effort which must be expended before a change in the probability can be realistically expected. It is obvious that the change in probability is not solely dependent on the effort expended, but on the geological contrast α and on the effort; otherwise the change in probability would not differ from target to target.

Integrating Eq. 7, we get

$$P = (1/\alpha)\ln(c(\alpha \times E + K)) \tag{8}$$

Since P=0 when E=0, cK must equal 1 and hence c=1/K. In addition, specifying the maximum effort, E_{max}, necessary to guarantee recognition of target indicators, P=1, determines α,

$$K \exp(\alpha) - \alpha \times E_{max} = K \tag{9}$$

Eq. 9 was solved for α by taking the first derivative, determining the minimum and iteratively increasing α until the above conditions were fulfilled. Due to the exponential increase beyond the minimum, α did not differ significantly from α_{min}. The trivial solution, α =0, was not considered. (The effort E_{max} need not guarantee detection; specifying any pair, E_{max} and P, uniquely determines α and implies that irregardless of the effort expended there is always a finite probability of overlooking the target.)

Defining the intensity of exploration I as

$$I = E/K, \tag{10}$$

we get

$$P(A|T) = (1/\alpha) \ln(\alpha I + 1) \tag{11}$$

E_{max} differs for each type of target assumed present, e.g. economic or noneconomic mineralization.

The optimization problem can be written as

$$\text{Max } F(I) = \sum_{j=1}^{J} P_j(T_e|A)(V - g_j(I))$$
$$\text{subject to: } I > 0;\ G(I) = \sum_{j=1}^{J} g_j(I) = C \tag{12}$$

where J is the total number of areas, V is the minimum acceptable value for an economic deposit, C is the exploration budget which is to be allocated among the areas and g_j is a function relating exploration intensity to costs in area j. Since F(I) is the sum of functions and since the results in one area do not influence other areas, it is a separable optimization problem.

In the stage of exploration under consideration, the positive indicators A are not simply a set of criteria whose absence or presence is sought, but a set of criteria possessing certain values which are contingent on economic and technological conditions. For example, it does not suffice to determine whether mineralization is present; more important is that the mineralization fulfill specific conditions such as minimum depth, dip of ore body, etc. The presence of ore criteria are a necessary but not sufficient prerequisite for an economic deposit. The minimum cost or intensity of exploration K corresponds to confirming or denying the presence/absence of criteria. It is only after this minimum effort has been expended that a change in the posterior probabilities occurs.

For the Molai area, only two targets - economic and noneconomic mineralization - can be expected to be genetically related to the criteria. Other feasible false targets would have been weeded out during previous exploration stages. Due to the complex nature of the geology and ore body, no unequivocal rejection or confirmation of a target can be obtained. (Confirmation/rejection can, even after drilling, be indecisive.) Thus, noneconomic mineralization will possess a finite and possibly significant probability of being economic.

The EMV's calculated for information levels 2 and 3 (Fig. 3b) were reconverted into probabilities by dividing by 20 million units. It was assumed that the probability of an economic deposit given mineralization is 0.02, that the minimum time which must be spent in an area to determine the presence/absence of criteria is 10,000 units and that by expending 100,000 units (E_{max}), there is only a probability of 0.8 and 0.4 of the criteria being fulfilled given economic and noneconomic mineralization, respectively. This resulted in α's of 4.8862 and 11.9862, respectively. The exploration budget was 500,000 units. Only the twelve most favorable areas were considered.

TABLE III shows the optimal allocations as well as the initial and posterior probabilities and EMVs. It is obvious that the

TABLE III

CELL	INTENSITY (E+K)/K	COST $\times 10^4$	PROBABILITIES INITIAL	PROBABILITIES POSTERIOR	EMV INITIAL $\times 10^5$	EMV POSTERIOR $\times 10^5$
27	5.54	5.5	0.0175	0.033	3.51	6.61
29	5.51	5.5	0.0173	0.033	3.47	6.54
40	5.14	5.1	0.0154	0.029	3.07	5.78
28	4.71	4.7	0.0131	0.025	2.62	4.90
39	3.90	3.9	0.0092	0.017	1.84	3.39
30	3.73	3.7	0.0084	0.016	1.69	3.11
10	3.70	3.7	0.0083	0.015	1.66	3.05
11	3.69	3.7	0.0083	0.015	1.65	3.04
18	3.62	3.7	0.0080	0.015	1.60	2.93
17	3.62	3.7	0.0080	0.015	1.60	2.93
7	3.44	3.4	0.0072	0.013	1.45	2.65
8	3.39	3.4	0.0070	0.013	1.40	2.55

magnitude of the allocations are dependent on the initial probabilities and that the allocations are not proportionally assigned to the areas according to their probability.

2.2.2. Maximizing the Probability of Detection. Maximizing the expected posterior EMV determined the optimal amount of exploration effort which should be allocated to each area. Subsequently, this effort must be further allocated among the different exploration methods. Although, in many cases, it may be difficult to specifically divide exploration into methods, e.g. geophysical, geochemical or geological, one exploration team performing all studies, in other cases, exploration methods and hence exploration teams proceed simultaneously and independently of one another. For the former, the exploration budget suffices; for the latter, however, the exploration effort must be optimally divided among the methods to assure that certain goals or objectives are fulfilled.

Various exploration objectives can be formulated, each leading to an optimization problem whose solution furnishes an optimal division of exploration effort. Consequently, conflicting allocations arise given different objectives or the objectives themselves may conflict. If more than one objective is deemed necessary or worthwhile, a multiobjective decision problem results. A trade-off among the conflicting allocations or objectives is inevitable and the literature on multiobjective decision theory is generally concerned with formulating methods or criteria to reconcile conflicting objectives and results (see, for example, Hwang and Masud, 1979).

In exploration two objectives which, although themselves not conflicting, may furnish conflicting allocations are maximizing the probability of detection (see Shulman,1984a,1984b,1985b,1986) and maximizing the expected gain in information (Shulman,1985a).

The probability of detecting a target can be written as

$$D(z) = pb(z) \tag{13}$$

where b(z) is the detection function, i.e., the conditional probability of detecting the target provided the target is present and effort z is expended; p is the target location probability.

Assume that a negative exponential function adequately describes the detection process, then the probability of detection becomes

$$D(z_1,z_2) = p(1-\exp(-\beta_1 z_1)) + p(1-\exp(-\beta_2 z_2)) \tag{14}$$

where z_1 has been introduced to denote geological/geochemical effort, z_2 denotes geophysical effort and β_1 and β_2 are constants with dimension per exploration unit. The constants β correspond to the unit effectiveness of the method multiplied by the area covered per exploration unit divided by the entire area.

It should be noted that the detection probabilities for the two methods are additive, thereby neglecting the interaction or relationship between the results obtained by the methods. Since the methods are employed independently of one another, Eq. 14 is a reasonable approximation.

Maximizing Eq. 14 subject to a linear budget constraint, $z_1+z_2=w$, can be accomplished analytically by converting the constrained multivariate optimization problem into an unconstrained univariate problem, i.e.,

$$D(z_1) = p(1-\exp(-\beta_1 z_1)) + p(1-\exp(-\beta_2(w-z_1))) \tag{15}$$

Taking the first derivative and solving for z_1, we get

$$\exp(-(\beta_1+\beta_2)z_1) = \beta_2 \exp(-\beta_2 w)/\beta_1 \tag{16}$$

The above equation defines the optimal allocation for a given β_1 and β_2. It should be noted that the optimal allocation is dependent on the effectiveness of the exploration methods, not on the probability of mineralization occurring. If $\beta_1 = \beta_2$, the effort is equally divided between z_1 and z_2 (an intuitive result). If β_2/β_1 approaches 0, the right hand side of Eq. 16 also approaches 0. The effectiveness of exploration method 1 is much greater than that of exploration method 2 which, consequently, can be neglected. If the right-hand side is less than 1, negative effort is obtained for one of the methods. This is dependent on the magnitude of the β's. (The side condition assumed that the geological and geophysical effort was expressed in units such that their costs do not differ significantly from one another.)

Finally, the optimization problem implicitly contains not only positivity constraints, $z_1,z_2>0$, but is subject to more stringent conditions, i.e. $a<z_1<b$, $c<z_2<d$. This further limits the set of feasible solutions.

2.2.3. Maximizing the Expected Gain in Information. The second objective, maximizing the expected gain in information, requires different input data. Based on information theory, the expected gain in information can be derived, i.e.,

$$c \sum_{i,k}^{2} p_i p_{ik} \log(q_{ik}/p_i) \tag{17}$$

with

c = constant

i = 1 economic mineralization; 2 noneconomic mineralization

k = 1 exploration signature associated with economic mineralization; 2 exploration signature associated with noneconomic mineralization

The probability of an economic/noneconomic mineralization is p_i; p_{ik} is the probability that a target of kind i produces a signature of type k. For example, p_{21} is the probability of noneconomic mineralization producing an exploration signature associated with economic mineralization. If signature of type k is, in fact, registered, q_{ik} is the probability that it was produced by a target of kind i. Thus, q_{12} is the probability that an exploration signature associated with noneconomic mineralization was caused by economic mineralization. The q_{ik}'s are derived from Bayes Theorem

$$q_{ik} = \frac{p_i p_{ik}}{\sum_{i=1}^{2} p_i p_{ik}} \tag{18}$$

In Eqs. 17 and 18, the following concepts must be distinguished and precisely defined: targets can be classified as true and false. A true target corresponds to the ore characterization model sought. It produces a distinctive geologic, geochemical and geophysical signature. A false target, on the other hand, refers to geological conditions differing from the true target which, however, also produce the same geologic, geochemical and geophysical signature as the true target. Thus, both true and false targets possess a probability that it will cause the sought after exploration signature, the p_{ik}'s. This probability is a function of the exploration effort expended. (We are not concerned with the relationship between occurrence of mineralization and the target sought, but between the probability of registering a specific signature or set of exploration criteria fulfilling certain values and the kind of target.) The probability that the signature k is produced by target i, i.e. p_{ik}, therefore differs from the probability that signature k was caused by target i, i.e. q_{ik}. Signature k could have been caused by the true target or by any of the false targets present in an area.

(It should be emphasized that whereas p_{ik} is the probability that k is produced by i, the ore characterization model and the mineralization model, b(z) is the probability of detecting the target and only the target given effort z; $P(A|T_i)$ is the probability of positive indicators being detected by the combined efficacy of both methods provided target T_i is present.)

In Eq. 18 only p_{ik} is a function of the exploration effort. The objective function is

$$I(z_1,z_2) = c_1 \sum_{i,k}^{2} p_i p_{ik}(z_1)\log(q_{ik}(z_1)/p_i) + c_2 \sum_{i,k}^{2} p_i p_{ik}(z_2)\log(q_{ik}(z_2)/p_i) \qquad (19)$$

For a geophysical survey, p_{ik} may, for example, be taken as the probability of intersecting the target and is, therefore, a function of the line spacing. Assuming that the target is randomly oriented, the probability of intersecting the target by parallel search is given by

$$P(I) = \begin{cases} 2/\pi\ (L/S) & \text{for } S>L \\ 2/\pi\ (L/S)(1-\sqrt{(1-(S^2/L^2))}+2/\pi \cos^{-1}(S/L) & \text{for } S<L \end{cases} \qquad (20)$$

where S is the line spacing and L is the length of the target. Its complementary probability, P(I'), is simply 1-P(I). However, Eq. 20 also holds for a noneconomic mineralization provided that the length of the target axes are equal. Consequently, the expected gain is a constant if P(I) is taken as p_{ik}.

Intersecting the target is a necessary but not sufficient condition; other geophysical parameters, criteria, etc. differentiate between economic and noneconomic mineralization. Such criteria are based on specific and distinct ore characterization models. Thus, p_{i1} becomes

$$p_{11}(z_2) = P(C|M_1)P(I)$$
$$p_{21}(z_2) = P(C|M_2)P(I) \qquad (21)$$

where $P(C|M_i)$ is the probability that the given criteria is produced by the characterization model i. Eq. 21 is equivalent to Eq. 5, the recognizability and contrast parameter now being a constant and a/(A-ka) being approximated by P(I).

Indications of an economic/noneconomic target being ascertained as a function of geological/geochemical effort can also be derived analogous to Eq. 5, i.e.,

$$p_{ik}(z_1) = 1/\zeta_i \ln(\zeta_i I + 1) \qquad (22)$$

However, instead of considering the change in the probability of detection as a function of effort, we are concerned here with the change in the probability of the occurrence of indicators as a function of effort.

2.2.4. Multiobjective Decision Making: The Global Criterion. To reconcile the different allocations resulting from the two objectives and to determine an optimal solution for the multiobjective decision problem, an additional criterion is necessary. If both objectives are equally important, i.e., no preference between the two objectives can be articulated, a global criterion suffices.

The method of global criterion consists of minimizing the sum of the relative deviation of the objective functions from the feasible ideal points, that is, from the solution derived if each objective function is optimized separately, to a power p. Mathematically, the global criterion optimization problem can be written as

$$\min \sum_{i=1}^{k} [(f_i(x^*) - f_i(x))/f_i(x^*)]^p$$

$$\text{s.t. } g_i(x) < 0 \qquad (23)$$

where the x^*'s are the ideal feasible points.

The method of global criterion requires that each objective be solved separately, thus determining $f_i(x^*)$, followed by solution of Eq. 27. It should be noted that the choice of a specific p may yield results for $f_i(x)$ which are unrealistic, and consequently unacceptable. Therefore, the sensitivity of the optimal solution to p should be studied and, if necessary, a different criterion applied.

2.2.5. Optimal Allocation of Geological/Geochemical and Geophysical Effort. For the Molai case study the following input parameters were used: Each area encompasses 2.64 sq. km. The geological/geochemical and geophysical decision variables were expressed as exploration intensity and as the distance in meters between survey lines, respectively. The length of the target, i.e. economic mineralization, is assumed to be 145 m. The entire area is covered, intensity of exploration implying mapping scale, sampling density, etc. The cost of geological/geochemical exploration is 10,000 units/intensity, that of a ground geophysical survey 1.548 units/meter.

The values of the input parameters for the exploration efficiency models are, in general, derived from both objective, e.g. the numerical simulation of the interaction between the sensor and the target, and subjective , e.g. the experience and knowledge of the explorationists, criteria. A geophysical survey is deemed to be more effective, at this exploration stage, than geological/geochemical methods in detecting the target.

Objective 1, maximizing the probability of detection, requires estimates of β_i. For geological/geochemical effort β_1 was taken to be 1.78×10^{-5} per unit, for a ground geophysical survey β_2 was 3.33×10^{-5} per unit. If a one to one relationship is assumed between time and exploration unit, the mean time ($= 1/\beta$) to detect the target is 56,000 units and 30,000 units, respectively.

Objective 2, maximizing the expected gain in information, requires estimates of ζ_1, ζ_2 , $P(C|M_1)$ and $P(C|M_2)$. The geological/geochemical input parameters, ζ_1, ζ_2 , were taken to be 7.8892 and 14.8992 respectively, giving a maximum p_{ik} of 0.556 and 0.3368. (A maximum exploration intensity of 10 was assumed.) The geophysical parameters $P(C|M_1)$ and $P(C|M_2)$ were taken as 0.6 and 0.3, respectively. (It should again be emphasized that the above parameters reflect the effectiveness of the individual methods taken separately; the higher values $\alpha_1 = 0.8$ and $\alpha_2 = 0.4$ used in maximizing the expected posterior EMV reflect the increased effectiveness of the simultaneous use and interaction between the two methods.) A least square global criterion, i.e. $p = 2$, was used.

The optimal allocation of exploration effort is shown in TABLE IV for search theory, information theory and for the multiobjective decision problem. The difference in allocations between search and information theory is obvious; more effort would be devoted to geological/geochemical exploration if search theory is used than had information theory been used. Furthermore, the optimal allocation of geophysical effort intersects the target with a high probability, the distance between survey lines being significantly less than the length of the target given information theory and ranging from less to more than the length of the target for search theory. The multiobjective global criterion effects a trade-off between the results derived from the individual optimizations. The allocations therefore fall between the ideal feasible points, however show a greater affinity to information theory.

TABLE IV

AREA	PROBABILITY	SEARCH THEORY		INFORMATION THEORY		MULTIOBJECTIVE	
		z_1	z_2	z_1	z_2	z_1	z_2
27	.0175	2.38	129.48	1.58	103.28	1.89	111.8
29	.0173	2.36	129.91	1.57	103.62	1.87	112.1
40	.0154	1.99	135.45	1.35	107.85	1.62	116.1
28	.0131	1.83	142.53	1.10	113.42	1.33	121.1
39	.0092	1.31	158.05	0.65	125.93	0.81	132.1
30	.0084	1.20	161.75	0.56	128.87	0.69	134.6
10	.0083	1.18	162.42	0.54	129.36	0.67	135.0
11	.0083	1.18	162.64	0.54	129.56	0.67	135.2
18	.0080	1.13	164.24	0.49	130.76	0.62	136.2
17	.0080	1.13	164.24	0.49	130.76	0.62	136.2
7	.0072	1.10	168.47	0.38	133.72	0.50	138.8
8	.0070	0.98	169.68	0.35	134.46	0.46	139.5

3. SUMMARY

Combining resource assessments with operations research methods enlarges the scope of numerical applications open to the exploration geologist or to the policy analyst in objectively studying and enhancing the efficiency of decision making. The emphasis and advances which resource assessments have undergone in the last twenty years coupled with the concomitant advances in operations research furnish a foundation upon which options, alternatives, etc. can be discussed. The methods do not obviate the geologist nor usurp his responsibilities, but contribute to a rational evaluation of procedures.

Among the manifold advantages of synthesizing resource estimates and operation research methods to determine an optimal strategy the following cannot be overestimated: 1. It provides a scientific and logical basis for decisions, enabling various exploration models, ore characterization models, assessments, etc. to be compared and analyzed. 2. It furnishes a training tool for new explorationists. 3. It can be used for hindsight studies to investigate the optimal sequence or combination of exploration methods, the optimal exploration budget, etc. 4. It inculcates a quantitative approach to exploration decision making and forces the explorationist to substantiate and defend his intuition.

4. ACKNOWLEDGEMENTS

This work was conducted while the author was employed as research geologist at the Institut fuer Geologie, Freie Universitaet Berlin, Working Group Mathematical Geology.

5. REFERENCES

Angelopous, C., Constantinides, D., and Papaspirou, D., 1985,'Molai Ore Deposit, Lakonia,' internal report, Institute for Geology and Mineral Exploration, Athens, p. I-V.

Constantinides, D., Faltsetas, H., Papastavrou, St. and Shulman, M.J., 1986, Zn-Ag-Pb Deposits, Greece,' Nineteenth International Symposium on the Application of Computers in the Mineral Industries , Society of Mining Engineers, p.687-694.

Galanopoulos, V.,1985,'The Tyros Beds and Phyllite Unit and the Related Massive Sulphide Mineralization (SE Peloponnese and Kreta),' Internal Report, Institute for Geology and Mineral Exploration, Athens, p. 1-11.

Harris, D.P.,1984, 'Mineral Resorce Appraisal and Policy: Controversies, Issues and the Future,' Resources Policy , June, p. 81-100.

Hwang, C.-L. and Masud, A.S.,'Multiple Objective Decision Making - Methods and Applications,' Lecture Notes in Economics and Mathematical Systems , vol. 164, Springer, Berlin, 351 pp.

Shulman, M.J.,1984a,'Initial Optimal Allocation of Effort in Mineral Exploration,' Transactions SME/AIME , vol. 274, p. 1904-1909.

Shulman, M.J.,1984b,'Exploration Strategy for Phosphates in the Sinai Peninsula, Egypt,' Eighteenth International Symposium on the Application of Computers and Mathematics in the Mineral Industries ,Institution of Mining and Metallurgy, London, pp. 79-87.

Shulman, M.J.,1985a,'Allocation of Airborne Reconnaissance Effort: Maximization of Gain in Information,' Society of Mining Engineers, Littleton, Co., preprint 85-305.

Shulman, M.J.,1985b,'Reconnaissance Exploration, Search Theory, and Stopping Criteria,' Transactions SME/AIME , vol. 278, pp. 1801-1806.

Shulman, M.J.,1986, 'Assessment and Exploration Strategy for Iron in the Nubian Sandstone Nubiann Sandstone Series, Southern Egypt,' Nineteenth International Symposium on the Application of Computers in the Mineral Industries , Society of Mining Engineers, p.695-705.

Webster's Seventh Collegiate Dictionary,1966, G. & C. Merriam Co.,Springfield, 1220 pp.

Advantages and Limitations of Discovery Process Modeling: The Case of the Northern West Siberian Gas Plays

John D. Grace
Basin Evaluation
Exploration Research
ARCO Oil and Gas Company
2300 W. Plano Pkwy
Plano, TX 75075

ABSTRACT. Discovery process modeling represents one of the most advanced tools available for hydrocarbon resource assessment of plays in semi-mature to mature areas. Information on the underlying size/frequency distribution of hydrocarbon pools and on the efficiency of exploratory effort are combined to estimate the number and sizes of undiscovered fields. This methodology, along with complimentary statistical validation tests, has been most fully developed in the computer software written at the Geological Survey of Canada (GSC).

Direct incorporation of geologic judgement is viewed as a key advantage to the GSC methodology. The problems of play definition and of accurately reflecting the incorporated subjectivity in the analysis are of critical importance in its application by those engaged in hydrocarbon exploration. Several ad hoc procedures are developed to handle those problems. These issues are examined in the case of assessing the undiscovered potential of the world's richest gas province: the northern West Siberian Basin.

1. Introduction

Discovery process modeling represents one of the leading analytic tools available for the assessment of the quantity and characteristics of undiscovered hydrocarbon resources in semimature to mature areas. Since the early 1980s, the basic model has been adapted and extended at the Geological Survey of Canada (GSC) for the purpose of evaluating petroleum potential for the Canadian government (Lee and Wang, 1983a, 1983b, 1985). Within the last year, the GSC model has become commercially available and has been acquired by North American oil companies. An exploration company and a government agency will necessarily approach evaluation of even the same area with different purposes and require a different suite of output parameters.

The case study described here, the gas bearing northern third of the West Siberian Basin, illustrates how some of the basic parameters of

C. F. Chung et al. (eds.), Quantitative Analysis of Mineral and Energy Resources, 651–668.

exploration interest are derived and some of the difficulties encountered. Two particular problems are specifically addressed:

1) Deriving a play definition that is sensible from the points of view of geology and the exploration history of the region.

2) Accurately reflecting in the final analysis the sources, nature, and consequences of uncertainties inherent in the subjective input to the study.

2. Geologic and Exploration Settings

The gas fields of the northern West Siberian Basin represent the largest single concentration of those resources in the world. As of the early 1980s, proved and probable gas reserves of this area (Soviet categories $A+B+C_1$) amounted to at least 22 Tcm -- approximately a third of the world's reserves. Another 10.3 Tcm are classified as inferred (C_2) reserves. At least 66 fields have been discovered in the northern basin, including most of the ten largest gas fields in the world.[1]

The study area is the northern 920,000 km^2 of the West Siberian Basin (WSB) (Figure 1). The WSB is a composite basin, 3.4 million km^2 in area, with 10 million km^3 of Mesozoic and Cenozoic sedimentary fill overlying a partially folded and metamorphosed Paleozoic section on top of crystalline Proterozoic basement. The basin is bounded on the west by the Ural mountains, on the east by the highlands of the Siberian platform, on the south by the highlands of the northern Kazakhistan Plate and is open to the north, extending under the Kara Sea. The regional dip is to the north, increasing in the study area where the maximum thickness of the Mesozoic-Cenozoic section reaches approximately 12 km.

The discovered fields in the northern basin generally have reservoirs in two productive sections: the Neocomian, at an average depth of 2,800 m and the Cenomanian at 1,100 m. The fields have almost all been discovered on major anticlines which follow three structural trends: 1) north-south through the central part of the northern basin; 2) northwest-southeast along the Nurmin Anticlinorium across the Yamal Peninsula and 3) northeast-southwest in the area of the mouth of the Yenisey River in the northeast corner of the basin (Figure 2).

In the Early Cretaceous, uplift was localized unevenly across the structures in these trends, leading to the formation of multiple, unconnected reservoirs in the Neocomian section of most fields. These reservoirs are sealed by individual interbedded shales. Growth in Cenomanian time tended to be even across the major structures, creating sin-

---

[1]The regional and gas geology of the northern basin and data on field sizes is based on Grace and Hart, 1986.

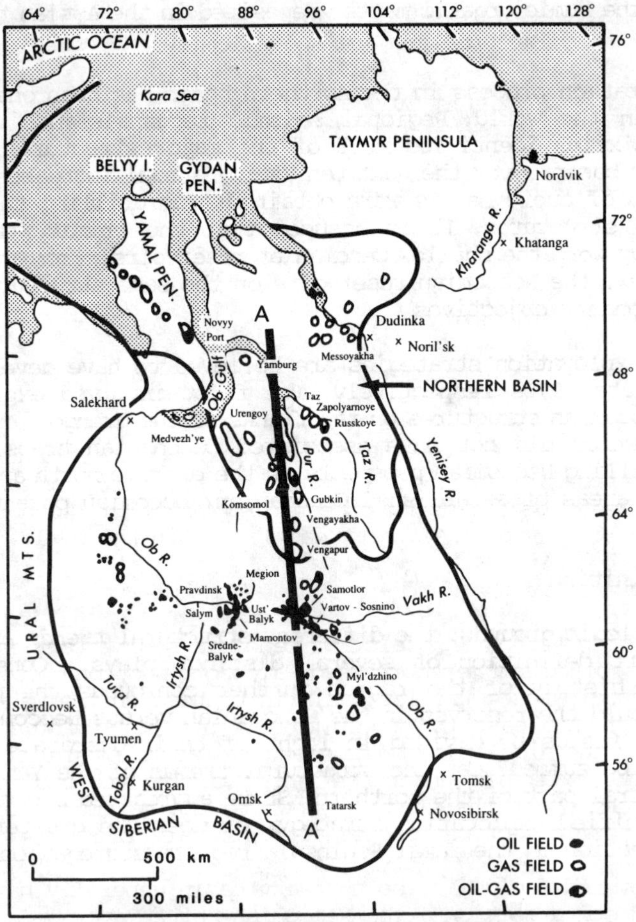

Figure 1. Location of the West Siberian Basin. (Grace and Hart, 1986, reprinted with permission).

gle massive traps in the Cenomanian section. All Cenomanian reservoirs are sealed by a single regional shale of Turonian age.

Though the region is overwhelmingly gas prone, the hydrocarbons of the Cenomanian and Neocomian suites can be sharply distinguished. The deeper, Neocomian reservoirs contain 88% methane on average, have high condensate content and often have thin oil rims. The source of these hydrocarbons was thermal cracking of the sapropelic organic matter in the shales of the Middle Jurassic to Neocomian section, particularly the Upper Jurassic Bazhenov Suite. The Cenomanian reservoirs contain approximately two-thirds of the region's gas and are 98% methane on average, with no liquids present. Some fraction of the Cenomanian gas was also sourced from the Jurassic, but the remainder is biogenic gas, produced from the humic organic matter deposited in the Aptian to Cenomanian section.

The exploration process in the northern basin has been one of the most efficient in the world. Regional seismic reconnaissance completed in the early sixties identified most of the major structural features in the northern basin. Over the next ten years, annual increases in reserves of trillions of cubic meters were obtained from drilling the backlog of these major structures. It is assumed that the primary targets under this strategy were the shallow Cenomanian reservoirs. From an exploration point of view, the Neocomian reservoirs on the same structures were bonuses or secondary objectives.

Two more exploration strategies are believed to have developed in the early and late 1970s respectively. The first involved exploration for separate Neocomian structures on the flanks of the Urengoy and Medvezh'ye Megaswells which did not have associated Cenomanian traps. The second involved drilling Neocomian prospects in the extreme north and southeast, outside the areas of established Cenomanian/Neocomian potential.

3. Play Definition

Based of geologic grounds, the different structural trends and reservoir units support definition of several distinct plays. Considering the exploration history of the region further confounds the problem. For example: Should the reserves in the Cenomanian versus Neocomian sections of the same fields be divided in light of their desparate sources, or should they be summed? Can the structural trends of the Yamal Peninsula and the central part of the northern WSB be aggregated because they were part of a unified exploration strategy, or does the greater uplift and shallower section on the Yamal Peninsula require segregation on geologic grounds?

Out of the larger set that might have been defined, three plays were analyzed to examine sensitivity to different play definitions. The fields included in each play are given in Table I and the areas included are shown in Figures 3-5 and the discovery histories are plotted in Figures 6-8.

1) The Cenomanian Gas Play. This play was defined on a strict geologic basis. Only the gas reservoired in the Cenomanian section was included, and the fields were limited to those in the principle north-south structural trend of the central part of the northern basin.

2) The Northern Basin Gas Play. This play included all of the fields in the northern WSB, aggregating Neocomian and Cenomanian reservoirs. The basis for this definition was that explorationists approaching the basin first drilled the structures that were best basin-wide. It also recognizes that Neocomian pays were bonuses to drilling Cenomanian targets

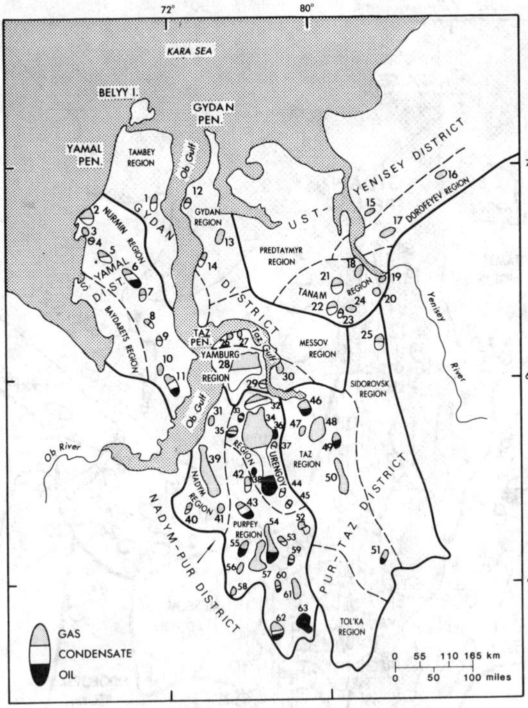

Figure 2. Field location map, with boundaries of Petroliferous Districts and Regions. (Grace and Hart, 1986, reprinted with permission). Key to numbered fields:

1 - South Tambey, 2 - Kharasavey, 3 - Kruzenshtern, 4 - South Kruzenshtern, 5 - Bovanenko, 6 - Neyto, 7 - Arktik, 8 - Nurmin, 9 - Sredneyamal, 10 - Maloyamal, 11 - Novyy Port, 12 - Utro, 13 - Gydan, 14 - Geofizika, 15 - Deryabin, 16 - Dzhagot, 17 - Ozero, 18 - Kazantsev, 19 - Nizhnekheta, 20 - Zima, 21 - Pelyatka, 22 - North Solenin, 23 - South Solenin, 24 - Messoyakha, 25 - Suzun, 26 - Semakov, 27 - Antipayuta, 28 - Yamburg, 29 - Nakhodka, 30 - Yurkharov, 31 - Kharvuta, 32 - North Urengoy, 33. Yanyakha, 34. Urengoy, 35 - Pestsov, 36 - Samburg, 37 - South Samburg, 38 - Noyabr' 39 - Medvezh'ye, 40 - Nadym, 41 - Pangody, 42 - Yubiley, 43 - Yamsovey, 44 - East Urengoy, 45 - Pyrey, 46 - Taz, 47 - West Zapolyarnoye, 48 - Zapolyarnoye, 49 - Russkoye, 50 - South Russkoye, 51 - Ust-Chasel'ka, 52 - East Tarkosalye, 53 - West Tarkosalye, 54 - Gubkin, 55 - North Komosomol, 56 - Verkhnepur, 57 - Komsomol, 58 - Muravlenkov, 59 - Tarasov, 60 - Vengayakha, 61 - Yety-Pur, 62 - Vengapur, 63 - Yarayner.

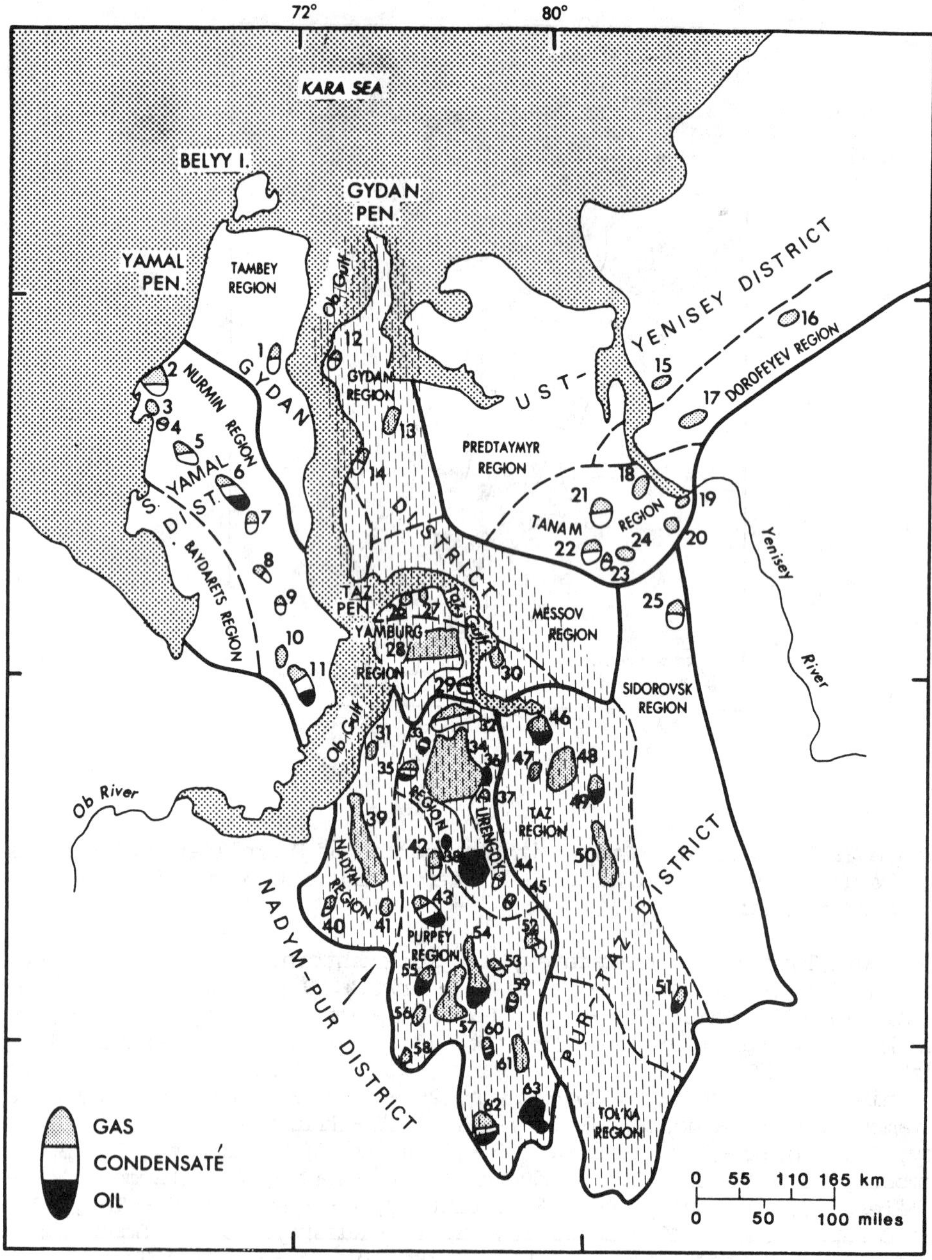

Figure 3. Boundaries of the Cenomanian Gas Play (Hatched)

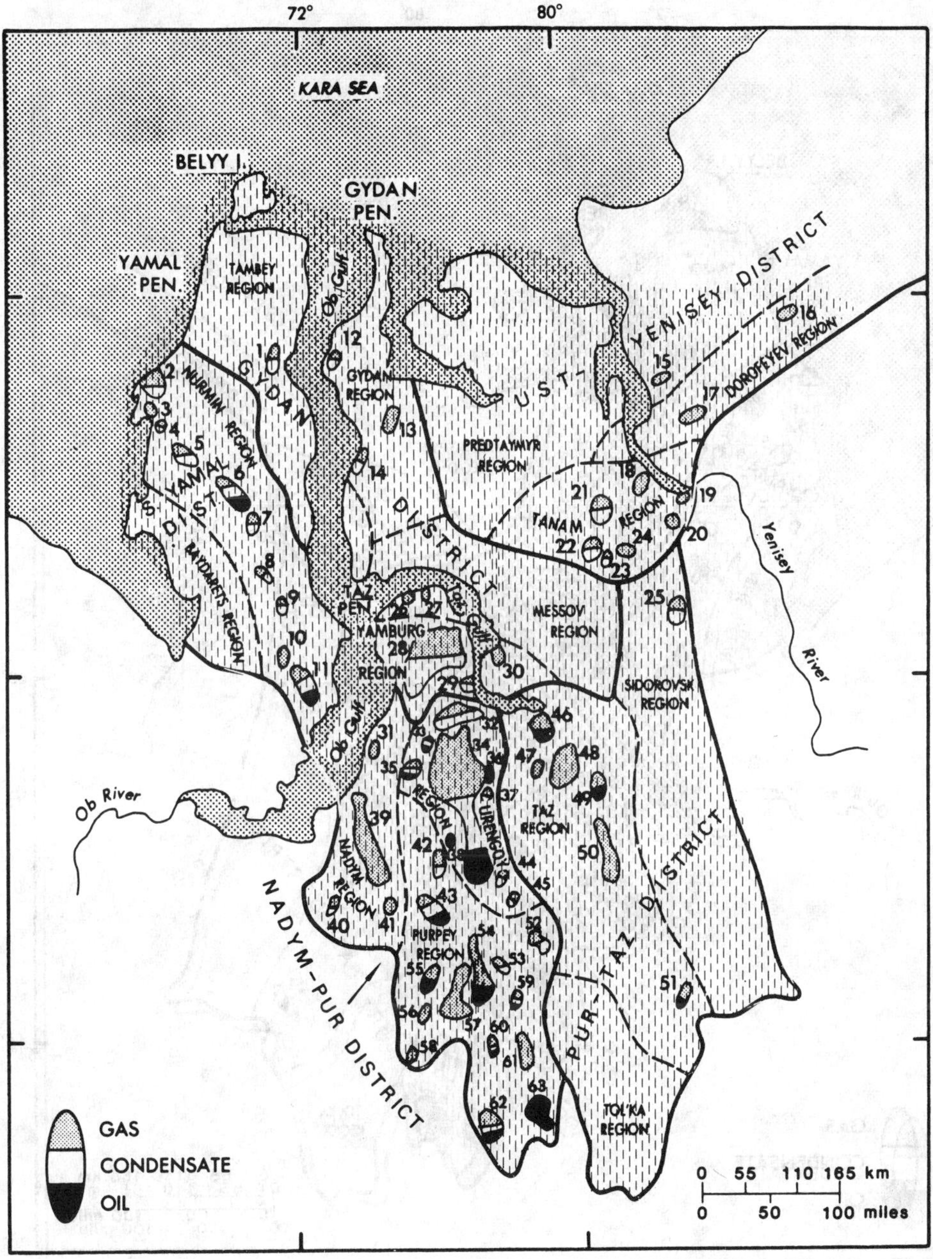

Figure 4. Boundaries of the Northern Basin Gas Play (Hatched)

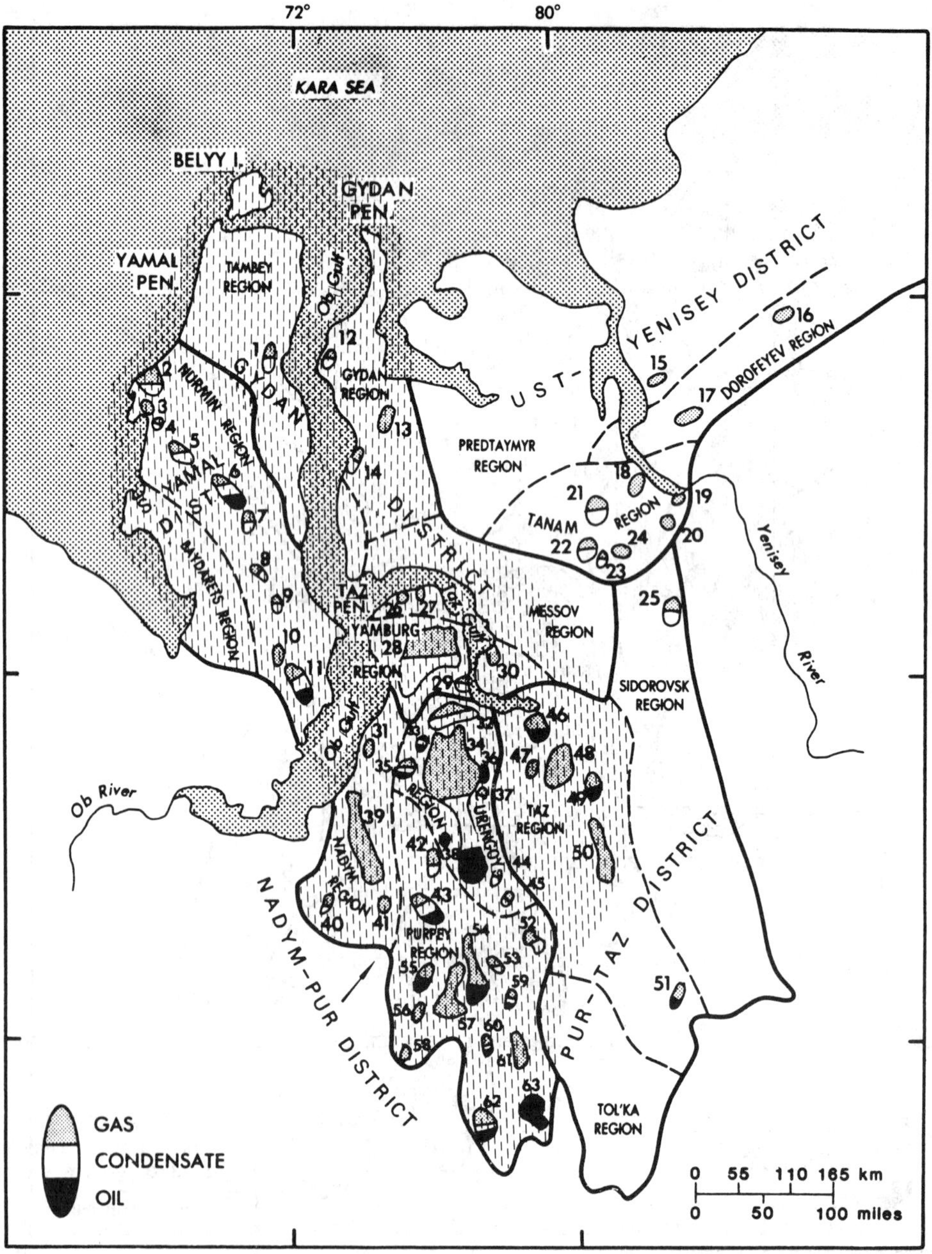

Figure 5. Boundaries of the Central and Yamal Gas Play (Hatched)

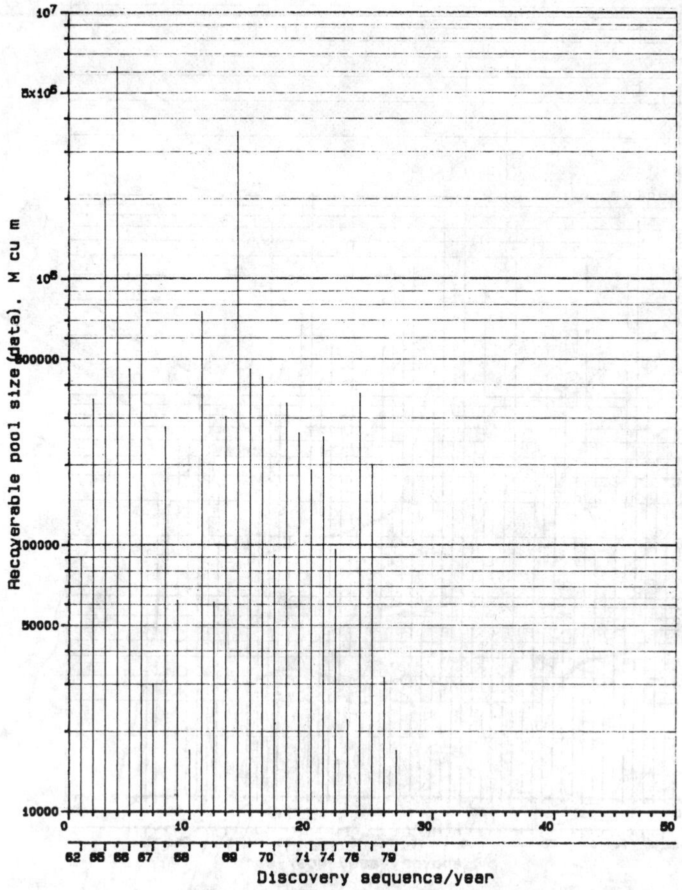

Figure 6. Discovery History of the Cenomanian Gas Play

and did not in themselves represent a separate set of exploration prospects.

3) <u>The Central and Yamal Gas Play</u>. This is a refinement on the Northern Basin Gas play. The eleven fields of in the northeastern corner of the basin (the Ust'-Yenisey District) were eliminated because of fundamentally different geology. Two more sets of fields were also removed. First is the five Neocomian fields on subordinate structures of the flanks of the Urengoy Megaswell and one on the Medvezh'ye Megaswell. Second is the Neocomian fields of the northern Gydan and Yamal Peninsulas and the extreme southeastern part of the Pur-Taz District. The rationales for these decisions were that the presence of a Cenomanian reservoir in the included fields was an important distinguishing geologic characteristic and that the eliminated discoveries resulted from exploration con-

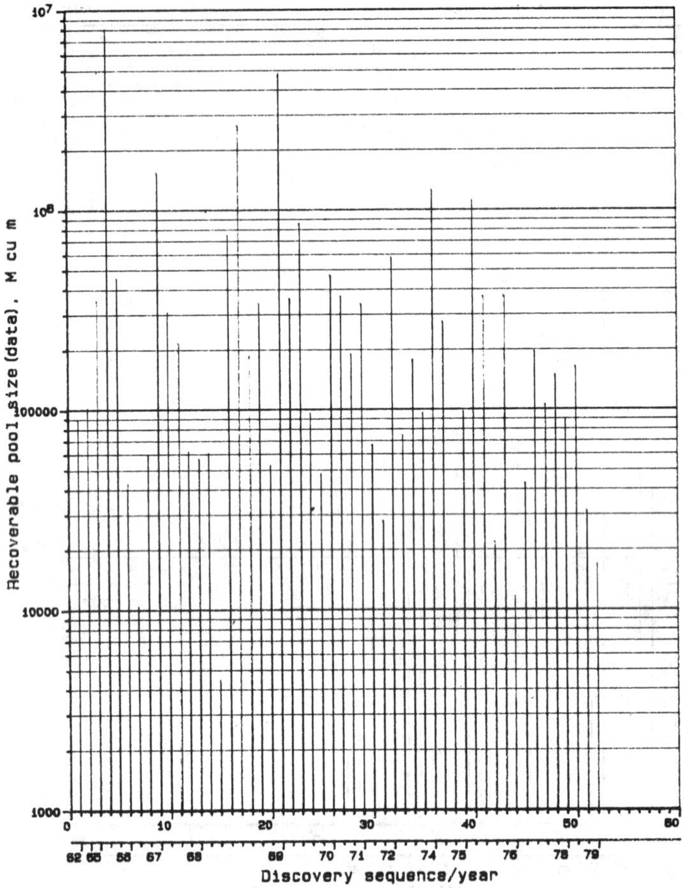

Figure 7. Discovery History of the Northern Basin Gas Play

cepts that were distinctly different from the original strategy of drilling giant Cenomanian structures.

The problems of play definition in this case are representative of those found in the analysis of any developing area. Exploration concepts change in response to additional information and exhaustion of prospects developed in the initial strategy for the region yields to secondary objectives on the intensive margin (e.g., such as the Neocomian discoveries around Urengoy and Medvezh'ye) and the extensive margin (e.g., on the northern Gydan and Yamal Peninsulas).

4. Analysis of the Plays

The GSC system (called Petroleum Resources Information Management and Evaluation System or PRIMES) is comprised of several modules used in

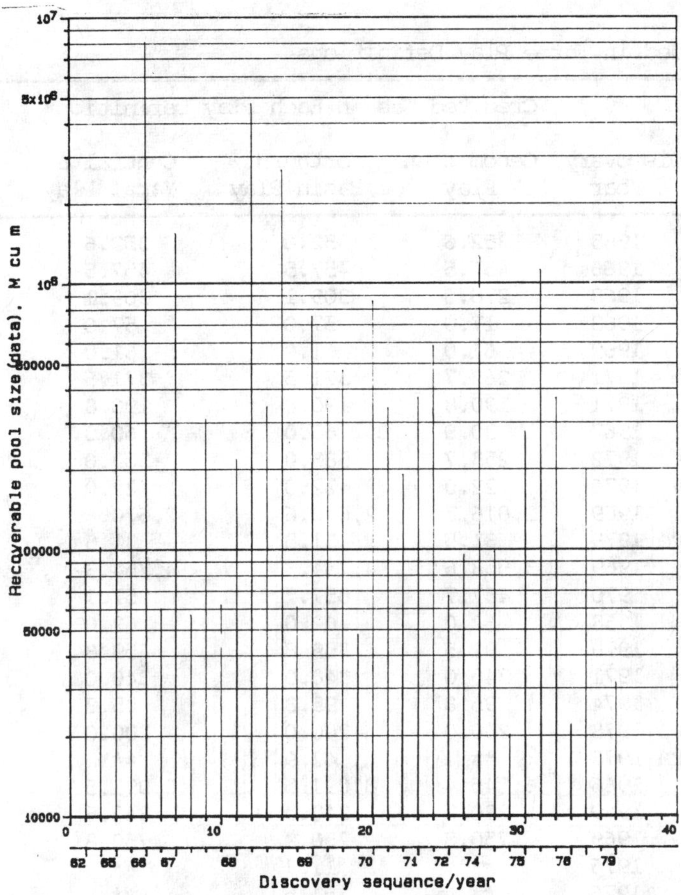

Figure 8. Discovery History of the Central and Yamal Gas Play

resource assessment for mature to frontier regions (Lee and Wang 1983a, 1983b, 1985). The analysis conducted here employed the discovery process model, which has been extended beyond the application given by Kaufman, et al (1975) to include provision for subjective choice by the analyst between competing scenarios. The inputs at the first stage of the discovery process analysis are the date of discovery and size of each field in the play. Search spaces are also nominated on the maximum total number of pools in the play (N) and the coefficient beta, a size-bias parameter. A reasonable search range for beta can be easily determined from inspection of the discovery history plot. A priori expectations on the total number of pools in the play is not, however, straightforward. The following types of conclusions must be reached in order to bound N:

Table I
Fields Included in Three Play Definitions

Field	Discovery Year	Credited Gas in Each Play Definition* Cenomanian Play	Northern Basin Play	Central & Yamal Play
Gubkin	1965	352.6	352.6	352.6
Komsomol	1966	457.5	457.5	457.5
Vyngapur	1968	276.6	309.1	309.1
Vyngayakha	1968	17.0	57.0	57.0
N. Komsomol	1969	61.0	61.0	61.0
E. Tarkosalye	1971	262.7	371.5	371.5
Yety-Pur	1971	190.8	190.8	190.8
Tarasov	1967	30.0	60.0	60.0
W. Tarkosalye	1972	253.7	583.0	583.0
Verkhnepurpey	1976	22.0	22.0	22.0
Zapolyarnoye	1969	2,016.7	2,670.6	2,670.6
W Zapolyarnoye	1979	31.5	31.5	31.5
Yamburg	1969	3,558.6	4,754.3	4,754.3
N. Urengoy	1970	457.7	857.7	857.7
Russkoye	1968	62.0	62.0	62.0
Yurkharov	1970	91.5	359.8	359.8
Semakov	1971	340.0	340.0	340.0
Nakhodka	1974	95.8	95.8	95.8
Antipayuta	1978	200.0	200.0	200.0
Ust'-Chasel'ka	1978	**	42.9	**
Urengoy	1966	6,245.4	8,091.5	8,091.5
Yubiley	1969	320.8	342.5	342.5
S. Russkoye	1969	750.8	750.8	750.8
Pyrey	1976	**	11.7	**
Yan'yakha	1976	**	354.5	**
Medvezh'ye	1967	1,243.0	1,548.0	1,548.0
S. Samburg	1978	**	106.3	**
E. Urengoy	1978	**	89.5	**
Yamsovey	1970	428.5	471.9	471.9
Nadym	1972	**	28.0	**
Pestsov	1974	**	178.0	**
Kharvuta	1976	370.2	370.2	370.2
Pangody	1979	16.9	16.9	16.9
Novoport	1964	**	102.0	102.0
Arktik	1968	**	218.0	218.0
Nurmin	1970	**	48.0	48.0
Sredneyamal	1974	**	67.0	67.0
S. Tambey	1974	**	74.7	**
Maloyamal	1975	**	20.0	20.0
Neito	1975	**	98.0	98.0
Geofizika	1975	**	277.5	**
Gydan	1978	**	150.0	**
Utro	1979	**	165.0	**

Table I (cont.)
Fields Included in Three Play Definitions

Field	Discovery Year	Credited Gas in Each Play Definition* Cenomanian Play	Northern Basin Play	Central & Yamal Play
Kharasavey	1976	**	1,266.3	1,266.3
Bovanenko	1976	**	1,120.4	1,120.4
Taz	1962	90.1	90.1	90.1
Messoyakha	1969	**	10.5	**
S. Solenin	1970	**	52.4	**
N. Solenin	1969	**	96.3	**
Pelyatka	1969	**	185.7	**
Kazantsev	1969	**	4.5	**
Zima	1966	**	43.0	**

* Proved, probable and inferred reserves measured in Bcm.
** Not included in the play.

1) What has been the exploration efficiency in the play to date? How many more undrilled prospects of suitable size exist in the play area?

2) From a materials-balance point of view, given the total resources of the discovered pools, how much of the original hydrocarbon charge remains "unaccounted"?

3) Given more mature analogs within or outside the basin, how many pools can be expected in the play?

The output from the first stage of analysis is search ranges on the mean (mu) and variance (sigma squared) of the underlying lognormal distribution of pool sizes, and a new, narrower search range on values of N. The second stage of analysis involves the generation of the lognormal populations created by the combinations of mu, sig. sq. and N provided from the first stage. The estimated mean sizes of the pools in each of the distributions is matched to the actual discovery record.

In addition to inputs on mu, sig. sq. and N, it is possible to restrict the matching process conditional on the hypothesis that the first n pools in the distribution have already been discovered. The choice between the generated scenarios is statistically guided by the deviation of estimated field size from the observed, a log likelihood ratio and other measures, but ultimately requires a subjective assessment by the geologist that certain scenarios are closer than others to the true state of nature.

The final output of the discovery process analysis is a set of very precise point estimates (with forecast intervals) of the sizes of remaining undiscovered pools in the play (Figure 9).[2] The precision of the individual size estimates is of relatively small consequence to a government agency, because it is the aggregates that are used in policy analysis. For an exploration company, undiscovered pools represent specific drilling targets for which millions of dollars may be spent. It is vital, therefore, that the individual pool size estimates be reported in a fashion which reflects the uncertainty surrounding subjective decisions in the analysis and discourages a false sense of precision, which is naturally engendered by the precision of point estimates.

Subjective input by the geologist enters the discovery process analysis at four critical junctures:

1) In defining of what pools are included within the play.

2) In nominating a geologically reasonable search range for N, the total number of pools in the play.

3) In restricting the scenarios to those in which the first n pools of the distribution are assumed to be discovered.

4) In choosing the scenario which is believed to best represent the actual state of nature in the play.

This involvement of geologic judgement is one of the greatest strengths of the GSC approach to statistical appraisal of undiscovered resources. It leads not only to geologically sensible results, but facilitates acceptance of the methodology by explorationists and managers for whom prospect development is intuitive as well as scientifically objective. It is essential to reconcile the uncertainty of geologic interpretations with the analytically rigorous component of the system. At a minimum, the estimated sizes of undiscovered fields cannot be reported with more precision than we have in estimates of the sizes of discovered fields. Consequently, it is necessary to "fuzz" the estimated size/frequency of undiscovered pools if they are to be credible in practice.

Consistent with industry practice in evaluating the sizes of individual prospects, an effective way to convey the uncertainty about the state of nature in the play is to give ranges judged to represent the low (90% probability of occurrence), median and high (10%) scenarios. The choices represent not just ranges on the total quantity of hydrocarbons in the play, but consideration of the entire scenario: including both the total number of pools and where the gaps (undiscovered pools) fall in the distribution.

---

[2]In Figure 9, the boxes represent the members of the lognormal distribution chosen as the state of nature in the play. The height of the boxes represents the 25%/75% forecast interval around the estimated mean field size for each rank. The boxes filled with dots represent the discovered fields, those remaining empty represent the undiscovered fields.

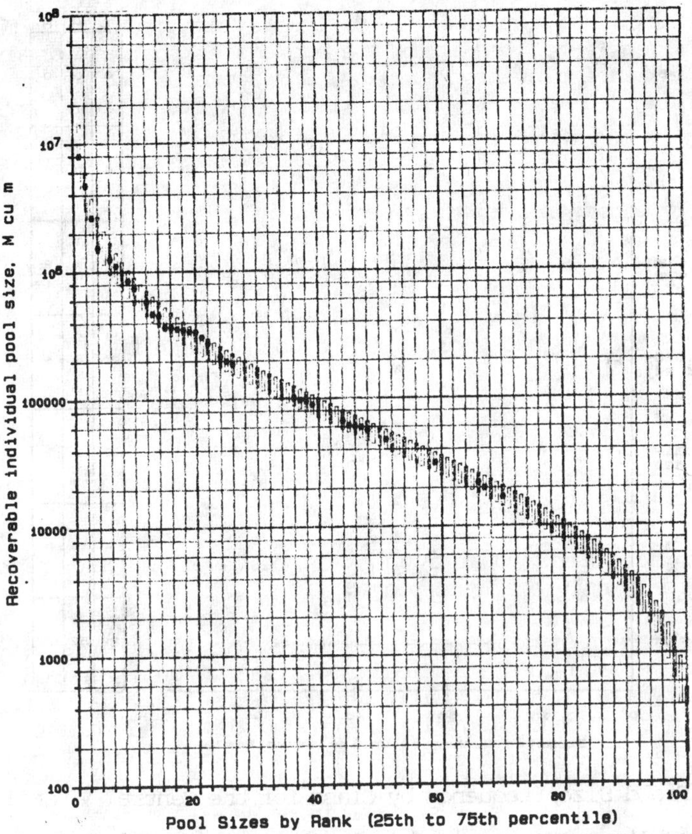

Figure 9. Rank/Size Plot for the Central and Yamal Gas Play

A further step is to report the estimated sizes of undiscovered pools by class, and not individually. Taking the scenario reported in Figure 9 as the best estimate for the Central and Yamal play, the 5th, 8th and 11th fields (with resources of 1.546, 0.925 and 0.635 Tcm respectively) remain undiscovered. When bracketed by low and high scenarios, and reported by class, the analysis reads: no fields expected above 2.5 Tcm, 1 to 2 from 1.25 to 2.5 Tcm and 2 to 4 in the .625-1.25 Tcm class (Figure 12). This preserves the essential information on the size/frequency distribution of remaining targets, the total number of undiscovered pools and total hydrocarbon richness of the play.

5. Results on the Three Plays

The discovery process analysis was independently conducted on all three plays defined for the northern WSB. For each, three scenarios were chosen corresponding to subjective estimates of the high, median and low cas-

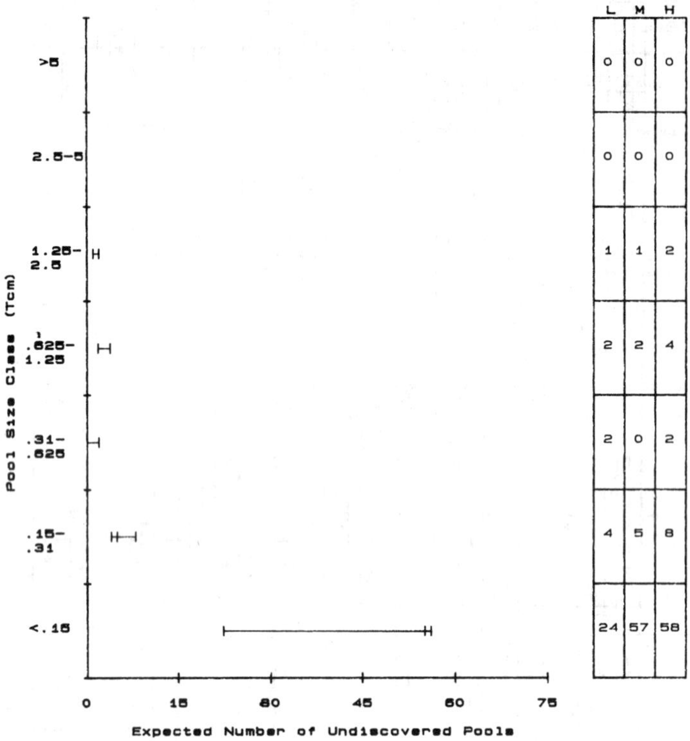

Figure 10. Field Size/Frequency by Class for the Central & Yamal Gas Play

es, these results are summarized in Table 2 and graphically for the Central and Yamal play in Figure 10. All three analyses are close with respect to the ranges on the total richness of the plays and on the sizes of remaining targets. Consistent with a highly efficient exploration process in an environment of clear structural traps, only a very few of the largest pools (>1 Tcm, occupying structures averaging >2,000 km^2) remain fourteen years after the onset of exploration. The distribution of undiscovered fields is dominated by the smallest size class (less that 150 Bcm, which is small only by West Siberian standards!). These fields are not only harder to find (ceteris paribus), but would have been low priority exploration targets as long as a backlog of major structures remained to be drilled.

Judging only from the gross volume of sedimentary rock included in each play, it was expected that the estimates on the richness of the plays and total number of pools would run from the Cenomanian play on the low end (one horizon, one structural trend), to the Central and Yamal play (two horizons, two trends), to the Northern Basin play (two horizons, three trends) on the high end. That was the result for the range of estimates of the volume undiscovered recoverable hydrocarbons (Table II). It is not the case for the estimated number of undiscovered pools.

Table II
Summary Parameters in Assessments

Play /Probability	Undiscovered Fields 90%	50%	10%	Undiscovered Volumes (Tcm) 90%	50%	10%
Cenomanian	63	73	93	5.1	11.5	18.1
N. Basin	63	93	103	9.4	13.5	22.0
Cent. & Yamal	34	64	74	8.3	13.8	21.1

The Northern Basin play had, as expected, the highest number, but the range for the Cenomanian play exceeded the Central and Yamal.

Comparison of the median estimates of remaining volumes of gas gives a more consistent picture. The Central and Yamal play estimate is 20% larger than that for the Cenomanian play, which agrees with the ratio for discovered volumes between the two plays. The median estimate for the Northern Basin play is, however, slightly smaller than for Central and Yamal. By aggregating the fields of the Ust'-Yenisey District into the Northern Basin play, rather than analyzing them as a separate play (which would extend further to the northeast along the Yenisey-Khatanga Trough), probably underestimated the potential of that region. That decision and the small contribution of the Ust'-Yensiey District to the volume of discovered hydrocarbons best explains the apparent inconsistency in the estimates of undiscovered volumes in the two plays. With respect to the difference in the estimated number of remaining pools in the Cenomanian versus the Central and Yamal plays, there is no clear reason. This is not a particularly disturbing result because the plays were analyzed <u>independently</u> and not as a system wherein the smaller, included volume of rock was used as a base for estimating the larger unit.

A similar problem arises in examination of the low, median and high estimates of the number of pools in a specific size category (Figure 10). In the Central and Yamal play, for fields in the 310 to 625 Bcm range, the estimates are low=2, median=0, high=2. Because each of the scenarios was chosen on all three criteria (number of pools, total volume and field ranks), rather than in order to make each criteria consistent, it is not surprising that all details do not exactly fall in line.

Among the three plays, from a geologic and exploration standpoint, the Central and Yamal play is the most sensible and has been chosen as the best estimate of the undiscovered gas potential for the northern WSB. Though their contributions would be small, to be complete, separate analyses would be required for the Ust'-Yenisey District, as well as the two new Neocomian plays on the intensive and extensive margins of exploration.

6. Conclusions

The results of the discovery process analysis consist of three indicators which are critical to exploration planning: the number of remaining exploration targets, their size/frequency distribution and the total richness of the play measured in volumes of undiscovered hydrocarbons. Modifications by the GSC which incorporate geologic input in the analytic process are viewed as a major strength to the system, increasing the reliability of the estimates as well as the likelihood that they would actually be relied upon.

Play definition, which is at the heart of many resource assessment techniques, remains a problem, but the results of this analysis tended to clarify, rather than obfuscate the plays. Greater statistical guidance on what pools belong and do not belong in a play would be of great value (e.g., cluster analysis, multiple discriminant analysis), but progress along that line should not be used as a substitute for geologic judgement.

For use by explorationists, there is a real problem of false precision in any system that can, in evaluating mathematical functions, produce estimates of the sizes of undiscovered pools with better resolution than we have for estimating of the sizes of well understood, discovered pools. The procedures used here (running high, low and median scenarios, reporting pool sizes by class) are clearly ad hoc and produce some apparent inconsistencies. It is felt, however, that those deficiencies are sufficiently compensated by the increased credibility and usefulness of the results.

8. References

Grace, J.D. and G.F. Hart, 1986. 'The Gas Giants of the Northern West Siberian Basin,' Bulletin of the American Association of Petroleum Geologists, vol. 70, no. 7, pp. 830-852.

Kaufman, G.M., Y. Balcer and D. Kryut, 1975. 'A Probabilistic Model of Oil and Gas Discovery,' in J.D. Huan (ed) Methods of Estimating the Volumes of Undiscovered Oil and Gas Reserves. Studies in Geology, vol 1. Tulsa: American Association of Petroleum Geologists.

Lee, P.J. and P.C.C. Wang, 1983a. 'Probabilistic Formulation of a Method for the Evaluation of Petroleum Resources,' Journal of Mathematical Geology, vol. 15, no. 1, pp. 163-181.

____________________, 1983b. 'Conditional Analysis for Petroleum Resource Evaluations,' Journal of Mathematical Geology, vol. 15, no. 2, pp. 353-365.

____________________, 1985. 'Prediction of Oil or Gas Pool Sizes When Discovery Record is Available.' Journal of Mathematical Geology, vol.- 17, no. 2, pp. 95-113.

Some Aspects of Multivariate Analysis

Donald E. Myers
Department of Mathematics
University of Arizona
Tucson, AZ 85721 U.S.A.

ABSTRACT. Classical multivariate methods for analyzing data tableaux do not explicitly incorporate spatial correlation. In contrast geostatistical techniques such as kriging, dual kriging and cokriging are primarily estimation methods. The multivariate formations of dual kriging and disjunctive kriging, dual cokriging and co-disjunctive kriging, provide methods which bridge the gap between the two approaches. Basic properties of both are presented.

INTRODUCTION

The importance of multivariate techniques has been recognized for a number of years. In particular, Cluster Analysis, Principal Components and Discriminant Analysis have had widespread use. However, these tools which had their origin in other disciplines other than the geosciences do not fully respect some characteristics of multivariate data sets arising out of problems in the geosciences, mining, environmental monitoring, hydrology and remote sensing. In particular the models underlying these do not specifically incorporate spatial correlation. The dichotomy between these techniques and estimation techniques such as kriging can be seen in the context of two perspectives of a data tableau. Consider

$$\begin{matrix} Z_1(x_1), & \ldots, & Z_m(x_1) \\ \cdot & & \cdot \\ \cdot & & \cdot \\ \cdot & & \cdot \\ Z_1(x_n), & \ldots, & Z_m(x_n) \end{matrix}$$

C. F. Chung et al. (eds.), Quantitative Analysis of Mineral and Energy Resources, 669–687.

In general multivariate techniques interpret the tableau as a (random?) sample of size n from a vector valued random variable whereas geostatistics a la Matheron would model the tableau as a non-random sample from one realization of a vector-valued random function where the spatial correlation is of principal interest. The above multivariate techniques do not require position coordinates and do not reflect the support of the sample. In general they are not estimation techniques but rather seek to identify underlying phenomena or discern patterns. The various forms of kriging have been developed and used for estimation almost exclusively. There have been a few instances where a multivariate technique such as principal components has been applied in conjunction with kriging, for example Borgman and Frahme (1976), Myers and Carr (1984), Myers (1984), Davis and Greenes (1983). In each of these the objective was to simplify or to avoid cokriging. Additional interesting applications are given in Bopp and Biggs (1981) or Hopke et al (1976) neither of whom used geostatistics but simply contoured the derived factors to identify patterns. Although pattern recognition is a tool much desired it is as yet not fully developed.

The results presented here incorporate some aspects of classical multivariate methods as well as incorporating spatial correlation and are intended to stimulate further thought.

I. POSSIBLE GENERALIZATION

A. Consider a data tableau with an infinite number of rows, i.e. one for each point in a domain of interest, i.e., each column is a function, not just a finite number of points and which is written simply as

$$Z_1, \ldots, Z_m$$

For a finite number of points, i.e. a finite number of lines, a uniform distribution is assumed, that is each line is equally weighted. For the infinite case a probability distribution must be assumed for each column. The analogy to choosing a vector which carries the largest amount of variance is to choose a random variable such that the sum of the squares of the projection is maximized. In the finite version this is given as the inner product, for random variables it is the conditional expectation. Let the conditional expectation $E(Z_i/U)$ be written in operator form $D_U Z_i$ which easily extends to the vector $Z = [Z_1, \ldots, Z_m]$. Writing the inner product of two random variables V, W as $\langle V, W\rangle = E(VW)$, the problem is to maximize

$$\mathrm{Tr}\ \langle D_U Z, D_U Z\rangle \tag{1}$$

which is a positive quadratic form. However, unlike the finite dimensional case where the bivariate densities are implied by the data, in this case the operator D_U is characterized by bivariate densities that are unknown. One possible resolution of this problem is to assume that each pair Z_i, U is bivariate Gaussian and then it is sufficient to know the correlation for each pair. This is essentially the disjunctive kriging approach but recast in multivariate form, and will be discussed in Section IV.

B. Because the above formulation does not lead to a numerical solution without the use of strong distributional assumptions it is useful to re-consider the problem from another perspective. Principal components produces a representation of the data in the following form

$$Z = FA \tag{2}$$

where $F^T F$ is diagonal and the columns of F are ordered by percent of variance explained. One of the objectives of this transformation is to remove inter-variable correlation however this method only incorporates the inter-variable correlation at the same location. It would be more appropriate to remove spatial inter-variable correlation.

This suggests the following: if

$$\bar{Z}(x) = [Z_1(x), \ldots, Z_m(x)] \tag{3}$$

is a vector of second order stationary functions, find

$\bar{Y}(x) = [Y_1(x), \ldots, Y_p(x)]$ and C such that

$$\bar{Z}(x) = \bar{Y}(x)C \tag{4}$$

and the components of $\bar{Y}(x)$ are uncorrelated. Written in terms of the variograms and cross-variograms this becomes

$$\bar{\gamma}_Z(h) = C^T \bar{\gamma}_Y(h)C \tag{5}$$

and $\bar{\gamma}_Y(h)$ is to be a diagonal matrix. Here $\bar{\gamma}_Z(h)$, $\bar{\gamma}_Y(h)$ denote

$$\frac{1}{2} E([\bar{Z}(x + h) - \bar{Z}(x)]^T[\bar{Z}(x + h) - \bar{Z}(x)]) \tag{6}$$

and

$$\frac{1}{2} E([\bar{Y}(x + h) - \bar{Y}(x))]^T[\bar{Y}(x + h) - \bar{Y}(x)]) \tag{7}$$

$\bar{\gamma}_Z(h)$ corresponds to a correlation matrix. The representation

$$C^T\bar{\gamma}_Y(h)C \tag{8}$$

is analogous to the diagonalization of the correlation matrix as in Principal Components Analysis.

If all of the variograms of the components of $\bar{Z}(x)$ have sills then the total inertia of $\bar{Z}(x)$ is the sum of the sills, i.e.

$$\mathrm{Tr}\ \bar{\gamma}_Z(\infty) = \Sigma(\Sigma\ C_{jk}^2)\gamma_Y(\infty)_k \tag{9}$$

The columns of C can be normalized by requiring

$$\sum_{j=1}^{m} C_{jk}^2 = 1 \tag{10}$$

then the component of $\bar{Y}(x)$ with the largest percent of variance is the component of $\bar{Y}(x)$ whose variogram has the largest sill.

When the objective is pattern recognition as opposed to co-kriging, then the sample variogram matrix may substitute for the true variogram matrix. After the sample variogram matrix is computed for various lags and diagonalized to find the eigenvalues, the sills for each variogram are estimated by the eigenvalues. When C is non-singular the equation

$$\bar{Z}(x) = \bar{Y}(x)C \tag{11}$$

can be solved as

$$\bar{Y}(x) = C^{-1}\,\bar{Z}(x) \tag{12}$$

and hence at least at data locations the values of $\bar{Y}(x)$ are determined. This would permit separate kriging of the components of $\bar{Y}(x)$ with subsequent reconstruction of kriged values of $\bar{Z}(x)$. This decomposition would also allow simulation of $\bar{Z}(x)$ by simulation of the components of $\bar{Y}(x)$.

The model formulation given by (11) is similar to that used by Wackernagel (1985) but the method of achieving it is different. In Wackernagel the procedure is to model the variogram (or covariances) of the components of $\bar{Z}(x)$ by nested models. It is assumed that the components of these nested models represent the components of $\bar{Y}(x)$ then the cross-variograms are modeled as linear combinations of the variogram models. In particular it is often assumed that the principal difference in the variogram models is reflected in the ranges. The approach proposed above allows a more general model for $\bar{\gamma}_Z(h)$. Not all such models can be diagonalized whereas the construction in Wackernagel ensures a diagonalizable model and hence is more restrictive in particular it assumes all variograms and cross variograms are proportional to the same model.

II. DUALITY - AN ALTERNATIVE

Before introducing a multivariate version it is useful to recall the formulation of the univariate kriging estimation in dual form. Not only is this form analogous to the use of splines for interpolation but it also provides a method for removing or reducing the spatial correlation.

Recall the equations determining the ordinary Kriging

estimator. Let $x_1, \ldots, x_n$ be data locations with data $Z(x_1), \ldots, Z(x_n)$. Then the kriging equations are written as

$$Z^*(x_0) = \sum_{i=1}^{n} \lambda_i \, Z(x_i) \tag{13}$$

where

$$\begin{aligned} &\sum_{i=1}^{n} \lambda_i \, \gamma(x_i - x_j) + \mu = \gamma(x_0 - x_j) \\ &\Sigma \, \lambda_i = 1 \end{aligned} \tag{14}$$

In matrix form we have

$$\begin{aligned} &K\lambda + F\mu = K_0 \\ &F^T \lambda = 1 \end{aligned} \tag{15}$$

which we can write as

$$\begin{bmatrix} K & F \\ F^T & 0 \end{bmatrix} \begin{bmatrix} \bar{\lambda} \\ \mu \end{bmatrix} = \begin{bmatrix} K_0 \\ 1 \end{bmatrix} \tag{16}$$

where

$$K = \begin{bmatrix} \gamma_{11} & \cdots & \gamma_{1n} \\ \gamma_{n1} & \cdots & \gamma_{nn} \end{bmatrix}, \quad \gamma_{ij} = \gamma(x_i - x_j)$$

$$F = \begin{bmatrix} 1 \\ \vdots \\ 1 \end{bmatrix}, \quad \bar{\lambda} = \begin{bmatrix} \lambda_1 \\ \vdots \\ \lambda_n \end{bmatrix}, \quad K_0 = \begin{bmatrix} \gamma_{01} \\ \vdots \\ \gamma_{0n} \end{bmatrix} \tag{17}$$

$$Z^*(x_0) = [Z(x_1), \ldots, Z(x_n), 0] \begin{vmatrix} \bar{\lambda} \\ \mu \end{vmatrix} \tag{18}$$

$$= [K_0^T \; 1] \begin{bmatrix} K & F \\ F^T & 0 \end{bmatrix}^{-1} \begin{bmatrix} \hat{Z}^T \\ 0 \end{bmatrix} \tag{19}$$

Let

$$\begin{bmatrix} K & F \\ F^T & 0 \end{bmatrix} \begin{bmatrix} \hat{Y} \\ b \end{bmatrix} = \begin{bmatrix} \hat{Z}^T \\ 0 \end{bmatrix} \tag{20}$$

If unique neighborhood kriging is used then $[\hat{Y}\ b]^T$ need only be computed once since all the information about the point to be estimated is contained in $[K_0^T\ 1]$. $\hat{Y}$ is a data vector "dual" to $\hat{Z}$. First introduced by Matheron, Dubrule (1983) showed that this formulation allowed for the use of a unique neighborhood in cross-validation and more recently Dubrule and Kostov (1986) have shown how it facilitates the incorporation of inequality constraints. Royer and Vieira (1984), Galli, Murillo and Thomann (1984) have suggested interpretations of $\hat{Y}$.

$\hat{Y}$, b have some properties which have perhaps not been noted. For example

$$\hat{Y}F = 0 \quad \text{and} \quad E(\hat{Y}) = [0,\ 0,\cdot\cdot 0,\ 0]$$

So that $\hat{Y}$ is centered in two different senses.

If $E(Z(x)) = m$ then $E(\hat{Z}) = mF^T$ and $E(b) = m$ hence b is an unbiased estimator of m. By writing

$$\begin{aligned} U &= K^{-1}K, \quad V = K^{-1}F \\ W &= K^{-1}\ \hat{Z}^T \end{aligned} \tag{21}$$

then we obtain

$$\begin{aligned} \mathrm{Var}(b) &= \sigma^2 - (F^T W)^{-1} \\ &= \sigma^2 - (\hat{Z}\ V)^{-1} \end{aligned} \tag{22}$$

In a crude sense $\hat{Z}$ is replaced by new vector in which each entry is a linear combination but the components of $\hat{Y}$ are less correlated. If we compute

$$E[\hat{Z}^T Z] = (\sigma^2 + m^2)FF^T - K \tag{23}$$

then

$$E[\hat{Y}^T\hat{Y}] = V(F^TV)^{-1}V^T - K^{-1} \tag{24}$$

which is analogous to a correlation matrix.

III. CO-DUALITY

Although dual kriging is not quite a multivariate technique it suggests an extension of Cokriging that can properly be called multivariate.

Recall from Myers (1982, 1983, 1984) the general form of cokriging. As above $x_1, \ldots, x_n$ denote sample locations, $\bar{Z}(x_1), \ldots, \bar{Z}(x_n)$ are the data vectors where $\bar{Z}(x) = [Z_1(x), \ldots, Z_m(x)]$. Then the cokriging estimator is given by

$$\bar{Z}^*(x_0) = \sum_{i=1}^{n} \bar{Z}(x_i)\Gamma_i \tag{25}$$

where

$$\begin{aligned} \sum_{j=1}^{n} \bar{\gamma}(x_i - x_j)\Gamma_j + \bar{\mu} &= \bar{\gamma}(x_0 - x_i) \\ \sum_{j=1}^{n} \Gamma_j &= I \end{aligned} \tag{26}$$

and

$$\bar{\gamma}(h) = \frac{1}{2}([\bar{Z}(x + h) - \bar{Z}(x)]^T[\bar{Z}(x + h) - \bar{Z}(x)]) \tag{27}$$

Unlike the kriging estimator $Z^*(x_0)$, $\bar{Z}^*(x_0)$ is not a scalar and hence is not equal to it's transpose but since

$$[[\bar{Z}^*(x_0)]^T]^T = \bar{Z}^*(x_0) \tag{28}$$

we shall work with $[\bar{Z}^*(x_0)]^T$ using notation analogous to that above for dual kriging

$$K = \begin{bmatrix} \bar{\gamma}(x_1-x_1)\cdots\bar{\gamma}(x_1-x_n) \\ \vdots \qquad\qquad \vdots \\ \bar{\gamma}(x_n-x_1)\cdots\bar{\gamma}(x_n-x_n) \end{bmatrix} \tag{29}$$

$$F = \begin{bmatrix} I \\ \vdots \\ I \end{bmatrix}, \quad \text{each } I \text{ an } m \times m \text{ identity matrix}$$

$$\bar{\Gamma} = \begin{bmatrix} \Gamma_1 \\ \vdots \\ \Gamma_n \end{bmatrix}, \quad K_0 = \begin{bmatrix} \bar{\gamma}(x_0-x_1) \\ \vdots \\ \bar{\gamma}(x_0-x_n) \end{bmatrix} \tag{30}$$

then the cokriging system becomes

$$\begin{bmatrix} K & F \\ F^T & 0 \end{bmatrix} \begin{bmatrix} \bar{\Gamma} \\ \bar{\mu} \end{bmatrix} = \begin{bmatrix} K_0 \\ I \end{bmatrix} \tag{31}$$

We have then

$$[Z^*(x_0)]^T = [K_0^T \ I] \begin{bmatrix} K & F \\ F^T & 0 \end{bmatrix}^{-1} \begin{bmatrix} \hat{\bar{Z}}^T \\ 0 \end{bmatrix} \tag{32}$$

where

$$\hat{\bar{Z}} = [\bar{Z}(x_1), \ldots, \bar{Z}(x_n)] \tag{33}$$

Let

$$\hat{\bar{Y}} = [\bar{Y}(x_1), \ldots, \bar{Y}(x_n)]$$

with

$$\begin{bmatrix} K & F \\ F^T & 0 \end{bmatrix} \cdot \begin{bmatrix} \hat{\bar{Y}}^T \\ B \end{bmatrix} = \begin{bmatrix} \hat{\bar{Z}}^T \\ 0 \end{bmatrix} \tag{34}$$

Then $\hat{\bar{Y}}$ is the "dual" to $\hat{\bar{Z}}$ in that it is a vector obtained from $\hat{\bar{Z}}$ by incorporating the spatial and intervariace correlation quantified by the variogram matrix. We see immediately that

$$\hat{\bar{Y}}^T = [I - K^{-1}F(F^TK^{-1}F)F^T]K^{-1}\hat{\bar{Z}}^T \tag{35}$$

and

$$B = (F^TK^{-1}F)^{-1}F^TK^{-1}\hat{\bar{Z}}^T \tag{36}$$

or alternatively

$$\hat{\bar{Z}} = K[I - K^{-1}F(F^TK^{-1})F^T]^{-1}\hat{\bar{Y}} \tag{37}$$

The original data tableau is then given as a linear combination of the entries in the new data tableau. Moreover, B is an estimator of $E[\bar{Z}(x)]^T$ when $\bar{Z}(x)$ is second order stationary and

$$\text{Trace } E[B - E(\hat{\bar{Z}}^T)][B - E(\bar{Z}^T)]^T = Tr[C(0) - K^{-1}F(F^TK^{-1}F)^{-1}] \tag{38}$$

is the sum of the variances of the estimation errors.

IV BIVARIATE GAUSSIAN MODEL

The first model proposed to generalize principal components would require minimizing the quadratic form given by equation (1) which is determined by the conditional expectation operator. In turn this would require knowledge of (at least) bivariate densities. Matheron (1976) extended the kriging estimator from a linear to a non-linear form by utilizing a bivariate Gaussian density function. Specifically the disjunctive kriging estimator is written in the form

$$Z^*_{DK}(x_0) = \sum_{j=1}^{n} f_j(Z(x_j)] \tag{39}$$

The f_j's are obtained by Hermite polynomial expansions which in turn are the eigenfunctions of the conditional expectation operator. By utilizing the general cokriging formulation given in Eq. (26), the DK estimator can be fully generalized.

A. Preliminaries

Let L^2_{ij} denote the space of functions f, such that

$f(Z_i(x_j))$ is square integrable, i.e. $E[f(Z_i(x_j))]^2 < \infty$. Then form the sum $L^2 = \sum_{i=1}^{m} \sum_{j=1}^{n} L_{ij}^2$, which is not a direct sum since the constant functions are in each L_{ij}^2. If $\| \ \|_{ij}$ is the norm in L_{ij}^2 then $[\sum_{i=1}^{m} \sum_{j=1}^{n} \| \ \|_{ij}^2]^{1/2}$ is a norm on L^2. The non-linear estimator for $Z_j^*(x_0)$ is then the projection of $Z_j(x_0)$ onto the linear subspace of L^2 generated by the elements of the L_{ik}^2, that is,

$$Z_j^*(x_0) = \sum_{k=1}^{n} \sum_{i=1}^{m} f_{ij}^k(Z_i(x_k)) \tag{40}$$

where $f_{ij}^k \in L_{ik}^2$.

B. The Projections

Since $Z_j^*(x_0)$ is a projection, the errors $Z_j(x_0) - Z_j^*(x_0)$ are orthogonal to all the $f_i^k(Z_i(x_k))$ and hence we have

$$E[Z_j(x_0) f_{ij}^k(Z_i(x_k))] = E[Z_j^*(x_0) f_{ij}^k(Z_i(x_k))] \tag{41}$$

for all i, j, k. Since $Z_j^*(x_0)$ is a linear combination, each of these expectations only requires a bivariate density. However the functions f_{ij}^k are in general unknown and the problem is insoluable in this generality.

C. Transformations

As with single variable disjunctive kriging we assume that each random function may be transformed to one with a Gaussian distribution. More specifically we assume that for each Z_j there is a random function Y_j which is Gaussian and a function ϕ_j such that $Z_j(x) = \phi_j(Y_j(x))$ for each j, x. Moreover, we assume that for each i, j and each pair x, y

$$Y_i(x), \quad Y_j(y)$$

are bivariate Gaussian. If $i = j$ then this is the usual assumption for disjunctive kriging.

D. Hermitian Expansions

If $Y(x)$ is a Gaussian random function and f is a function such that $E[f(Y(x))]^2 < \infty$ then f may be represented by an expansion in terms of Hermite polynomials. In particular if f_{ij}^k is a function in L_{ik}^2 then

$$f_{ij}^k (Y(x)) = \sum_{p=0}^{\infty} f_{ij}^{kp} H_p(Y(x)) \tag{42}$$

where

$$f_{ij}^{kp} = E[f_{ij}^k(Y(x))H_p(Y(x))] \tag{43}$$

This provides a method for representing the unknown functions f_{ij}^k. Moreover the functions ϕ_j can be represented as

$$\phi_j(Y(x)) = \sum_{p=0}^{\infty} c_j^p H_p(Y(x)) \tag{44}$$

E. Bivariate Densities

Under the assumptions that each pair $Y_i(x)$, $Y_j(y)$ is bivariate Gaussian, the joint densities can be written in the form

$$g_{ij,xy}(u,v) = \sum_{p=0}^{\infty} (\rho_{ij}^{xy})^p H_p(u)H_p(v)g(u)g(v) \tag{45}$$

where $g(x)$, $y(y)$ are standard Gaussian densities, ρ_{ij}^{xy} is the correlation coefficient of $Y_i(x)$, $Y_j(y)$. As with univariate disjunctive kriging we could also consider more general bivariate densities such as the Hermitean models.

F. The Orthogonality Conditions

Combining Eq. (41), (42) we will have

$$E[Z_j(x_0)|Z_\ell(x_r)] = \sum_{i=1}^{m} \sum_{k=1}^{n} [f_{ij}^k(Z_i(x_k))|Z_\ell(x_r)] \tag{46}$$

for all j, ℓ, r

Using the transformations $Z_j(x) = \phi_j(Y_j(x))$ we obtain

$$E[\phi_j(Y_j(x_0))|Y_\ell(x_r)] = \sum_{i=1}^{m} \sum_{k=1}^{n} E[f_{ij}^k(\phi_i(Y_i(x_k)))|Y_\ell(x_r)] \tag{47}$$

Since both the ϕ_j's and the f_{ij}^k's are unknown we replace $f_{ij}^k(\phi_j(Y_j(x_k)))$ by simply $f_{ij}^k(Y_j(x_k))$ where of course the f_{ij}^k's are not the original ones and then Equation (40) becomes

$$Z_j^*(x_0) = \sum_{k=1}^{n} \sum_{i=1}^{m} f_{ij}^k(Y_i(x_k)) \tag{48}$$

or

$$Z_j^*(x_0) = \sum_{k=1}^{n} \sum_{i=1}^{m} \sum_{p=0}^{\infty} f_{ij}^{kp} H_p(Y_i(x_k)) \tag{49}$$

If we write $\bar{Z}^*(x_0) = [Z_1^*(x_0), \ldots, Z_m^*(x_0)]$ and

$$\bar{H}_p(x_k) = \begin{bmatrix} H_p(Y_1(x_k)) \\ \vdots \\ H_p(Y_m(x_k)) \end{bmatrix}^T \tag{50}$$

$$F_k^p = \begin{bmatrix} f_{11}^{kp} & \cdots & f_{1m}^{kp} \\ \vdots & & \\ f_{m1}^{kp} & \cdots & f_{mm}^{kp} \end{bmatrix} \tag{51}$$

then we have

$$\bar{Z}^*(x_0) = \sum_{p=0}^{\infty} \sum_{k=1}^{n} H_p(x_k) F_k^p \tag{52}$$

where

$$\bar{Z}(x_0) = [\phi_1(Y_1(x_0)), \ldots, \phi_m(Y_m(x_0))]$$

$$= \sum_{p=0}^{\infty} \bar{H}_p(x_0) \begin{bmatrix} C_1^p & & 0 \\ & \ddots & \\ 0 & & C_m^p \end{bmatrix} \tag{53}$$

Now if we let F_k^p be written as

$$\begin{bmatrix} b_{11}^{kp} & \cdots & b_{1m}^{kp} \\ \vdots & & \\ b_{m1}^{kp} & \cdots & b_{mm}^{kp} \end{bmatrix} \begin{bmatrix} C_1^p & & 0 \\ & \ddots & \\ 0 & & C_m^p \end{bmatrix} = B_k^p \, C^p \tag{54}$$

then

$$\bar{Z}(x_o) - \bar{Z}^*(x_0)$$

$$= \sum_{p=0}^{\infty} [\bar{H}_p(x_0) - \sum_{k=1}^{n} \bar{H}_p(x_k) B_k^p] \bar{C}^p \tag{55}$$

and the estimation variance in L^2 is given by

$$\text{Trace } E[\bar{Z}(x_0) - \bar{Z}^*(x_0)]^T [\bar{Z}(x_0) - \bar{Z}^*(x_0)]$$

$$= \sum_{p=0}^{\infty} \text{Trace } (\bar{C}^p)^T \, W_p \, C^p \tag{56}$$

$$W_p = E(\bar{H}_p(x_0))^T (\bar{H}_p(x_0))$$

$$- \sum_{k=1}^{n} (B_k^p)^T E[(\bar{H}_p(x_k))^T \bar{H}_p(x_0)]$$

$$- \sum_{k=1}^{n} E[(\bar{H}_p(x_0))^T (\bar{H}_p(x_k))] B_k^p$$

$$+ \sum_{\ell=1}^{n} \sum_{k=1}^{n} (B_\ell^p)^T E[(\bar{H}_p(x_\ell))^T \hat{H}_p(x_k)] B_k^p \tag{57}$$

Since the C^p's are determined only by the transformations ϕ_i, to minimize the estimation variance it is sufficient to minimize, for each p separately,

$$\mathrm{Tr}(C^p)^T W_p \bar{C}^p \tag{58}$$

Moreover, for each p, it is sufficient to consider

$$C^p = \begin{bmatrix} 1 & & 0 \\ & \ddots & \\ 0 & & 1 \end{bmatrix}$$

But minimizing

$$\mathrm{Tr}\ W_p$$

is exactly the problem of cokriging $\hat{H}_p(x_0)$ using the data $\bar{H}_p(x_1), \ldots, \bar{H}_p(x_n)$. Hence the matrices $B_1^p, \ldots, B_n^p$ are the solutions of the (simple) cokriging systems

$$\sum_{k=1}^{n} G_{jk}^p B_k^p = G_{0j}^p \tag{59}$$

$$G_{jk}^p = E[(\bar{H}_p(x_j))^T \bar{H}_p(x_k)] \tag{60}$$

as formulated in Myers (1982)

For the bivariate Gaussian case the entries in G_{jk}^p are

$$(\rho_{jk}^{i\ell})^p$$

where $\rho_{jk}^{i\ell}$ = correlation of $Y_i(x_j), Y_\ell(x_k)$

G. <u>The Undersampled Case</u>

If $\bar{Z}$ is under-sampled, i.e. if for i and some k, $Z_i(x_k)$ is missing then in equation (10) for all j, $f_{ij}^k \equiv 0$ and hence for all p, $f_{ij}^{kp} = 0$. In the context of equation (27) this is exactly the under-sampled formulation of the cokriging problem as given in Myers (1984) and in Carr, Myers and Glass (1985) hence the algorithm given there still suffices.

V. DATA or MODEL DRIVEN

Principal Components is essentially a data driven technique, that is, no parameters need be separately fitted nor are there assumptions such as normality or stationarity. Some of the interpretations of these components and their eigenvalues follow only from the geometric formulation without the use of external models. Cokriging, dual kriging and cokriging in contrast require fitting variograms and cross-variograms, these fitting techniques are not on completely solid ground statistically and also require implicit modelling of the phenomena by random functions and the varograms must satisfy certain conditions. The extension from linear to non-linear techniques require very specific statistical models and assumptions these additional conditions are often untestable. It would appear, that multi-variate techniques that truly incorporate spatial correlation are either not possible or at least not as yet developed.

References

Armstrong, M., (1984) 'New Types of Disjunctive Kriging', Part I, N-872 Fontainebleau

Armstrong, M. and Matheron, G., (1985) 'Isofactorial models for granulo-densimetric data', N-944 Fontainebleau

Armstrong, M. and Matheron, G., (1985) 'New types of Disjunctive Kriging', Part II, N-943 Fontainebleau

Bopp, F. III, and Biggs, R. Metals in Estuarine Sediments. Factor Analysis and its Environmental Significance, Science v 214, p 441-443

Borgman L. and Frahme, R. B., (1976), 'Multivariate Properties of Bentonik in Northeastern Wyoming', in Advanced Geostatistics in the Mining Industry, Guarascio, M. et al (eds) D. Reidel Publishing, Dordrecht

Carr, J., Myers, D. E. and Glass, C., (1985) 'Co-Kriging: A Computer Program', Computers and Geosciences, **11**, No 2, p 111.

Carr, J. and Myers, D. E., 'COSIM: (1985) A FORTRAN IV Program for Co-Conditional Simulation', Computers and Geosciences, **11**, No. 6 p 675.

Davis, B. and Greenes, K., (1983), 'Estimation using spatially distributed multi-variate Data; An example with coal quality data', Math. Geology, **15**, p. 287

Dubrule, O., (1983), 'Two methods with different objectives: splines and kriging', Math. Geology, **15**, p 245

Dubrule, O. and Kostov, C., (1986) 'An interpolation method taking into account inequality constraints I. Methodology', Math. Geology, **18**, No. 1, p 33

Galli, A., Murillo, E. and Thomann, J., (1984), 'Dual Kriging. Its properties and its uses in direct contouring', in Geostatistics for Natural Resource Characterization, G. Verley et al (eds.), D. Reidel Publishing, Dordrecht

Gupta, A. K., (1983) 'On the expansion of the bi-variate gamma distribution', Tech. Report, Bowling Green St. Univ.

Hopke, Ph., Ruppert, D.F., elute, P. R., Metzger, W.J. and Crowley, D.J., (176) 'Geochemical Profile of Chautauqua Lake Sediments,' J. Radioanalytical Chem. 29, p. 73.

Jackson, M. and Marechal, A., (1979) 'Recoverable reserves estimator by disjunctive kriging; a case study', in 16th Application of Computers and Operations Research in the Mineral Industry (O'Neil, T. J. ed.) Society of Mining Eng. of American Institute of Mining and Mettallurgy, N.Y. 240-249

Kim, Y. C., Myers, D. E. and Knudsen, H. P., (1977) 'Advanced geostatistics in ore reserve estimation and mine planning', U. S. DOE, Grand Junction 237 p

Lantuejoul, Ch. and Rivoirard, J., (1984) 'Une methode de determination d'anamorphose' N-916, Fontainebleau

Marechal, A., (1974) 'Generalities sur les fonctions de transfert', N-384 Fontainebleau

Marechal, A., (1975) 'Analyse numerique des anamorphoses gaussiennes', N-418 Fontainebleau

Marechal, A., (1976) 'The practice of transfer functions: Numerical methods and their applications'. M. Guarascio et al (eds) Advanced Geostatistics in the Mining Industry 253-206, D. Reidel Pub. Co., Dordrecht-Holland

Matheron, G., (1973) 'Le Krigeage Disjonctif', N-360/N-432 Fontainebleau

Matheron, G., (1974) 'Les fonctions de transfert des petits panneaux' N-395, Fontainebleau

Matheron, G., (1975) 'Modeles isofactoriels discrets et modele de Walsh', N-449, Fontainebleau

Matheron, G., (1975) 'Modeles isofactoriels pour l'effet zero' N-659, Fontainebleau

Matheron, G., (1976) 'A simple substitute for conditional expectation: The disjunctive kriging', in M. Guarascio et al (eds) Advanced Geostatistics in the Mining Industry, D. Reidel Pub. Co., Dordrecht-Holland

Matheron, G., (1976) 'Forcasting block grade distributions: The transfer functions' M. Guarascio et al (eds) Advanced Geostatistics in the Mining Industry, D. Reidel Pub. Co., Dordrecht-Holland 237-251

Matheron, G., (1977) 'Peut-on imposer des conditions d'universalite au krigeage disjonctif', N-539, Fontainebleau

Matheron, G., (1981) 'Quatre familles discrete', N.-703, Fontainebleau

Matheron, G., (1982) 'Pour une analyse krigeante des donnees reginalisees,' N-732, Fontainebleau

Myers, D. E., (1982), 'Co-Kriging: The Matrix Form', Math. Geology **14**, No 3 p. 249

Myers, D. E., (1983), 'Estimation of Linear Combinations and Co-Kriging', Math Geology, **15** No. 5 p 633

Myers, D. E., (1984), 'Co-Kriging: New Developments', in Geostatistics for Natural Resource Characterization, G. Verley et al (eds), D. Reidel Publishing Co, Dordrecht

Myers, D. E. and Carr, J., (1984), 'Co-kriging and Principal Components, Bentonite Data re-visited', Sciences de la Terre, **21**, p. 65

Myers, D. E., (1985) 'Co-kriging methods and Alternatives', in the Role of Data in Scientific Progress, P. Glaeser, (d), Elsevier Pub.

Rendu, Jean-Michel, (1980) 'Disjunctive Krigings: Comparison of Theory with actual results', Math. Geology, **12**, No. 4, p. 305

Rivoirard, J., (1985) 'Convergence des developpements en polynome d'Hermite', N-955, Fontainebleau

Royer, J. J. and Vieira, P.C., (1984) 'Dual Formalism of Kriging' in Geostatistics for Natural Resource Characterization, G. Verly et al (eds.), D. Reidel Publishing, Dordrecht

Wackernagel, H., (1985) "The inference of the linear model of the Coregionalization in the case of a geochemical data set,' Sciences del la Terre, No. 24, p. 81

REPORT OF THE WORKSHOP ON CURRENT PROBLEMS AND FUTURE DEVELOPMENTS IN MULTIVARIATE ANALYSIS

J.C. Griffiths (Leader), F.P. Agterberg (Rapporteur),
L. Carbognin, A. Carrara, DeVerle P. Harris, M. Ozcelik,
R. Potenza, U. Rossi, W. Skala, C. Stanley, Zhao Pengda

1. INTRODUCTION

Multivariate analysis is a scientific method in which the combined effect of many variables is studied. In statistics, the principal techniques of multivariate analysis are (multiple) regression, discriminant analysis, factor analysis and canonical correlation. In the earth sciences, special consideration must be given to the spatial settings of the rocks in a study area as well as to the dimension of geologic time. Multivariate mineral resource estimation should be performed according to the following steps. First the objective of the study is defined. This leads to the definition of a population from which one or more samples will be drawn. The objects in a sample have sizes, shapes, spatial arrangements and other attributes such as chemical composition allowing the definition of specific variables. This, in turn, leads to measurements. Each observation bears the values of many variables. The frequency distributions of the variables will be analysed in order to create new information as well as heuristics for solving the problems defined at the beginning of the study.

2. ADVANTAGES AND DRAWBACKS

The advantage of the multivariate approach is that it is objective; it allows for controlled experiments; and the results are reproducible. On the other hand, the success of multivariate analysis in mineral resource estimation, so far, has been limited because of the diversity of geologic information, restricted access, and difficulties in defining the populations and samples: we believe that the construction and selection of variables to be used in multivariate analysis requires additional research. The definition of relevant variables which reflect well-established geological processes is likely to improve the results that can be obtained. Publications regarding the application of multivariate analysis for mineral exploration should state as clearly as possible on which geological

C. F. Chung et al. (eds.), Quantitative Analysis of Mineral and Energy Resources, 689–693.

model the variables are founded. Accordingly, each variable selected for the analysis should be given a short explanation concerning the role it plays within the framework of the problem at hand.

It is necessary to distinguish between the use of multivariate analysis for resource appraisal and for exploration decision-making. Initially, these two objectives require a similar methodology. Later, the lines of approach differ. The purpose of resource appraisal is to obtain estimated numbers of mineral deposits of different types, their sizes, grades, and other characteristics. Exploration, on the other hand, consists of different stages requiring different types of multivariate analysis which is often focussed on relatively few specific targets. In the earth sciences, multivariate analysis is also used for other purposes including, for example, environmental problems such as prediction of landslides.

During the past 20 years, many applications of multivariate analysis have been made and considerable experience has been gained. Figure 1 illustrates the three types of input required for performing a regional mineral resource appraisal by multivariate statistical methods. Dependent variables expressing the characteristic features of the mineral deposits in a region can be correlated with explanatory variables that have been systematically quantified for the geologic framework. In general, the finding of orebodies requires major expenditures for local exploration consisting of detailed surveying and drilling. Areas that are well explored can be selected as control (or training) areas for extrapolation toward relatively less explored target (or study) areas. The explanatory variables are based on data for the geologic framework that have been gathered systematically according to the same methods in both control and target areas. A basic assumption is that the geologic environments of the mineral deposits (which may contain orebodies) are similar in control and target areas (geologic analogy). A drawback of past applications of the procedures illustrated in Figure 1 has been that they were time-consuming. This disadvantage is gradually being diminished because computer technology is moving at a rapid pace. Large regional data bases containing patterns for many types of geoscience data are becoming more readily available. Automated methods of relating these patterns to one another through metallogenic concepts (deposit models) should be further developed. There is the need for new research in mathematical statistics, e.g., for considering the spatial cross-correlations of the patterns.

3. FUTURE DEVELOPMENTS

In the future, multivariate statistical methods will become more useful, when large databanks and good data processing facilities for digitizing and image analysis are widely available. In addition to the improved choice of variables which will capture geologic-process models in a more meaningful way, a capacity for coping with missing data will also be necessary. Many previous applications of multivariate analysis (cf. Figure 1) have relied heavily on existing information without collection of new data in accordance with the

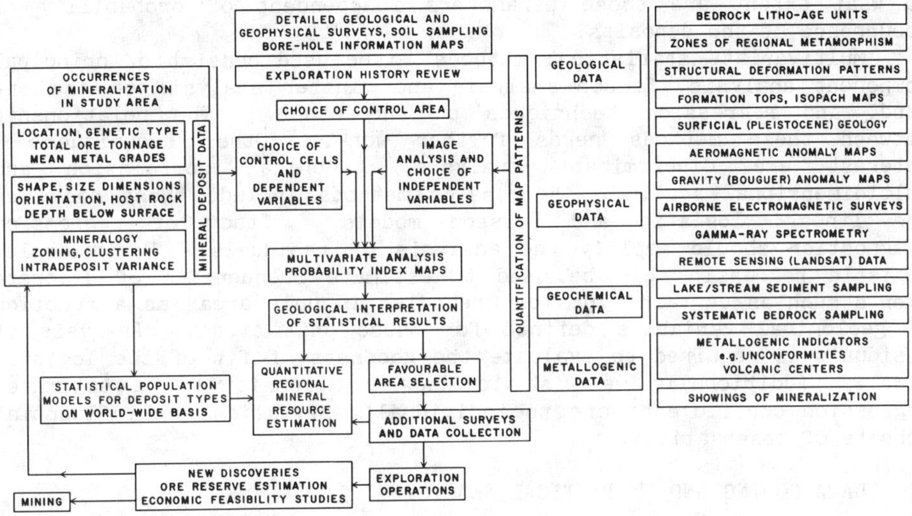

Figure 1. Schematic representation of the three types of input required for multivariate analysis in mineral resource estimation. (Source – F.P. Agterberg: Computers as an aid in mineral-resource evaluation; in D.F. Merriam, editor; "Computer Applications in the Earth Sciences", Plenum, New York, 1981, pp. 43-61).

specifications of a population defined on the basis of the objective at the beginning of the study. We believe that considerable improvements can be gained by incorporating field work and the sampling of rocks as integral parts of a multivariate statistical analysis.

The following topics require further development: (1) Definition of mineral deposit types and the characteristics of their environments; (2) Definition of control and target areas which involves the modelling of intensity of mineral exploration through time; (3) Establishment of regional geoscientific data bases to be used for creating explanatory variables in regression analysis; (4) Choice of cells or domains to be used for coding; (5) Choice of multivariate methods, including preprocessing techniques; (6) Incorporation of spatial statistics into multivariate analysis; (7) Use of three-dimensional aspects including geometry of folds and faults as well as depths of mineral deposits; (8) Use of concepts of artificial intelligence and expert systems in interactive graphic computer program packages for multivariate analysis; and (9) Representation of final results in map form with expected values accompanied by estimates of uncertainty.

Compilation and study of frequency distributions of parameters of mineral deposits should be continued; this includes work on bivariate size-grade distributions. An important question to investigate is: to what extent are these parameters independent of probability of occurrence of the deposits?

Multivariate statistical methods to be used consist of principal component analysis, factor analysis and cluster analysis on the one hand, and regression techniques on the other. The relationship between these methods needs further work. Other techniques of interest are discriminant analysis, canonical correlation and multidimensional scaling. Regression methods include those based on the linear, logistic and Poisson models. Study of regression diagnostics should amplify the analysis of residuals. For example, logistic regression can be used to estimate frequencies of discrete events such as certain types of orebodies in unit areas as a function of geological variables defined for these unit areas. Analysis of residuals is required to evaluate the goodness of fit of the logistic model. Additionally we wish to know how sensitive the estimated regression coefficients are to individual observations or influential subsets of observations.

4. DATA CODING AND STATISTICAL ANALYSIS

The quantification of geological, geochemical, geophysical and LANDSAT data should be performed on the basis of metallogenic criteria. The recent increase in use of automatically recorded remote sensing data is resulting in the development of new technology for map processing. The pixels for different types of variables can be kept in registration with one another using one of several possible cartographic projection systems. Digitizing and image analysis make use of the vector as well as the raster. Techniques to stack digitized map patterns on top of each other using logical OR, AND, NOT data structures operators and IF expression can be employed more extensively. The choice of the sizes and shapes of the cells, or natural domains for coding both dependent and explanatory variables should be carefully made. For example, in landslide prediction such domains are determined by the topographic relief, and in geochemical exploration using stream sediment data by the shapes of the drainage basins.

Methods of preprocessing coded data is generally useful prior to multivariate analysis; this includes separation of regional and local variation and use of "second derivative" maps in order to eliminate broad regional trends. Spatial analysis should be combined with multivariate analysis; for example, the frequency distributions of variables coded for cells depend on the sizes and shapes of these cells. Proximity analysis provides a spatial extension to principal component analysis and correspondence analysis. Because mineral deposits can often be regarded as points on a map, stochastic point-process models can be applied. In construction of the final maps which may, for example, represent probabilities or expected values and their associated uncertainties, the dependency among variables from adjoining areas should be appropriately expressed.

Multivariate analysis in mineral resource estimation presents a challenge not only to geoscientists but also to mathematical statisticians. The actual procedures involved are highly sophisticated and extremely demanding in that they require very large data sets. Not just 'any old data' will do; the measurements (even if zero/one type) must eventually conform to a comprehensive model. For example, appropriate transformations must be made on the basis of known frequency distributions. Such data hardly exist at present in geological investigations. If space- and time-dependent variables are included, we are enlarging the problem to four or more dimensions, each dimension demanding extensive new data. We rarely have the good fortune to be sampling a single homogeneous population, so that the populations must be separated out first. A considerable amount of exploratory data analysis should precede the actual sampling and model application. Final results should be subjected to confirmatory data analysis. In general, very careful experimental designing is needed prior to commencing an investigation involving multivariate analysis for mineral resource evaluation. Increasingly, this will mean the systematic measurement of newly defined variables in the field.

OIL AND GAS RESOURCE MODELING AND FORECASTING

G.M. Kaufman (Leader), R.A. Crovelli (Rapporteur), S. Chow, J.D. Grace, R. Sinding-Larsen, B.H. Sollie, P.C.C. Wang

1. INTRODUCTION

The fundamental objective of oil and gas resource assessment is to estimate the amount of hydrocarbons present in an area. Although an assessment can be conducted at several geologic levels (e.g., prospect, play, basin, geographic region), we judge the petroleum play to be the primary unit of analysis for resource assessment. Estimation of the size of individual deposits, their productive potential, and their spatial distribution are important elements of the oil and gas assessment task. Reporting of uncertainties associated with estimates is essential.

The group first considered the current state of the art of hydrocarbon resource assessment as a springboard for developing a research agenda armed at development of new assessment methods likely to produce a high return to research effort. To a lesser extent, practical problems that arise in implementation of current methods were considered.

The state of the art of hydrocarbon assessment can best be summarized by a taxonomy of models and methods currently in use. Although these models and methods may be classified along a number of dimensions (e.g., data intensity, data objectivity), a simple dichotomy was found to be descriptively useful.

Resources are turned into reserves by the interaction of two fundamental processes: (1) the natural process of hydrocarbon generation and accumulation, and (2) the human process of exploration for and subsequent discovery of accumulated hydrocarbons. Genetic deposit models are examples of models of natural processes. Rate of effort models and discovery process models are example of models of human processes. A model classification based on the natural and human processes is portrayed in Figure 1.

In addition to forming a framework for classifying present models, this simple dichotomy highlights key challenges for future research, among which are a strengthening and formalization of linkages between deposit formation and deposit discovery. More dimensions can be easily added to this basic scheme; objective vs.

C. F. Chung et al. (eds.), Quantitative Analysis of Mineral and Energy Resources, 695–700.

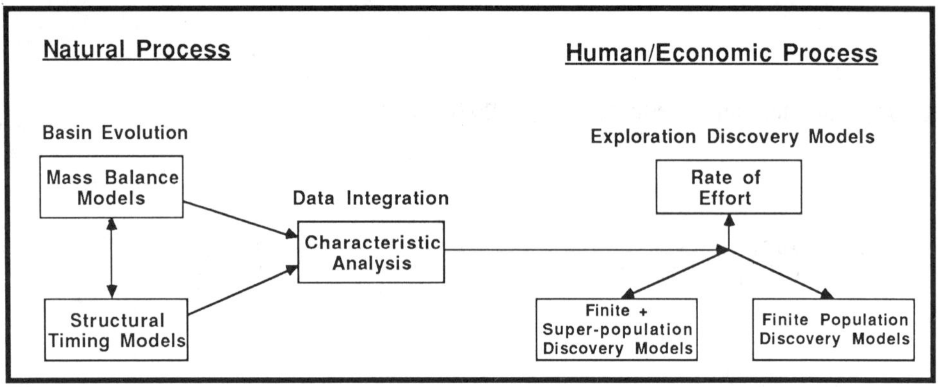

Figure 1: Model Classification

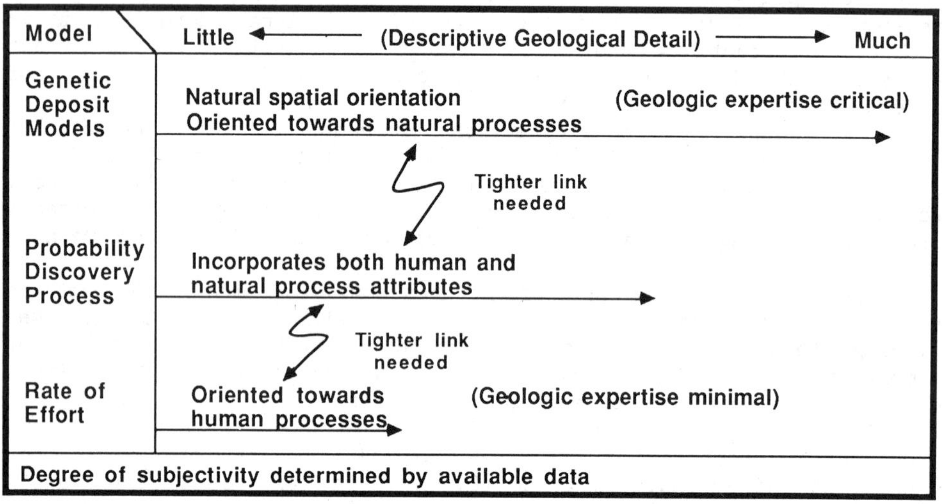

Figure 2: Model Characteristics

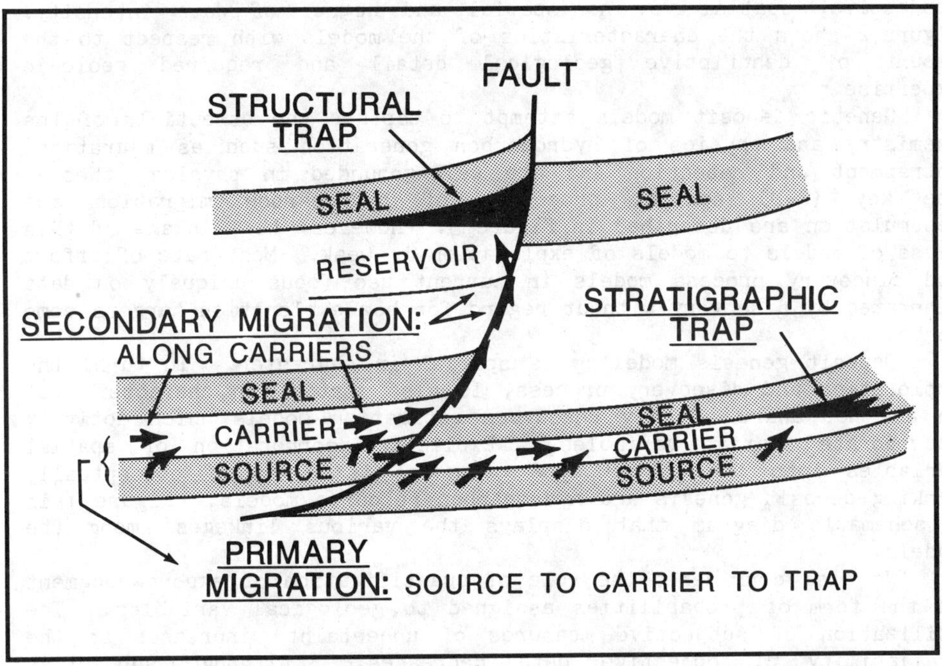

Figure 3: Key (2-D) Spatial Processes of Hydrocarbon Migration and Accumulation

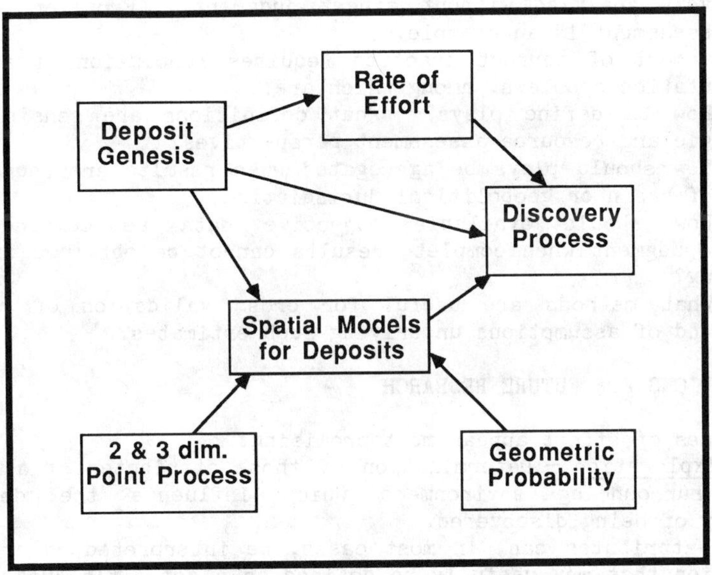

Figure 4: Model Linkages

subjective, spatial vs. non-spatial and degree of data intensity. Figure 2 shows the characteristics of the models with respect to the amount of descriptive geological detail and required geologic expertise.

Genetic deposit models attempt to capture the essentials of the chemistry and physics of hydrocarbon generation such as migration, entrapment and retention and are well-grounded in physical theory. The key (2-D) spatial processes of hydrocarbon migration and accumulation are described in Figure 3. However, the linkage of this class of models to models of exploration is weak. Most rate of effort and discovery process models in current use focus uniquely on data generated by discovery without regard for how exploration targets come to be.

Deposit genesis modeling is spatial in character. In turn, the exploration and discovery process, is also spatial in character, but we are not aware of many resources estimation models which actively incorporate spatial variables. Explicit incorporation of spatial variables into the latter type of model is a way of conceptually linking deposit genesis and hydrocarbon resource models. Figure 4 is a schematic diagram that displays the various linkages among the models.

We also considered the role of quantification of expert judgment in the form of probabilities assigned to geological variables. The utilization of subjective measures of uncertainty increases as the availability of objective data decreases. Although subjective assessment may be employed as a validation tool where objective data are present, it is presently used most often when objective well and deposit data are scarce and/or information is too diffuse to be quantitatively analyzed without expert judgment. Frontier province resource assessment is an example.

Improvement of current practice requires resolution of a number of implementation problems, among which are:

(1) How to define plays? What definitions are sensible from both geologic and resource assessment perspectives?

(2) How should plays be aggregated when results are required at the level of basin or geopolitical jurisdiction?

(3) How should available objective data be combined with subjective judgment when complete results cannot be obtained from the former along?

(4) What methods are useful for cross validation of resource estimates and of assumptions underlying such estimates?

2. DIRECTIONS FOR FUTURE RESEARCH

Three avenues of effort appear most promising:

1) <u>Exploration</u> – Determination of those attributes of a deposit and its surrounding environment which influence the deposit's probability of being discovered.

These attributes can, in most cases, be interpreted as arguments of a function that may usefully be defined as a deposit's "magnitude".

To date, most models have taken deposit size (area, rock volume, B.O.E.) as a measure of magnitude, but the relationship between probability of discovery and magnitude is clearly more complex than such simple definitions of magnitude imply. That complexity must be represented in the next generation of models. Identification of attributes that determine magnitude is a first step toward estimating the functional relation between magnitude and probability of discovery. A second step is to determine how definition of magnitude should change with changes in geologic environment, technology, economics and resources remaining to be discovered.

2) Deposit Genesis – What macro-geologic controls form individual deposit attributes?

A clear understanding of this process would contribute to a more precise definition of magnitude in particular geologic settings and would aid in understanding how magnitude influences the probability of deposit discovery. Among topics which might be studied to this end are: relationships between field size distributions and basin types; how trapping style affects seismic detectability; how geochemical information may be used to predict hydrocarbon phase and aggregate quantities of hydrocarbon's generated; how to employ hydrodynamic, paleothermal and structural history data to infer preferred orientation and clustering of accumulations.

This type of research might proceed by additional work typing standard frameworks for basin classification and statistical analyses. It would contribute to our understanding of links between individual attributes that determine deposit magnitude and models of hydrocarbon generation and accumulation.

3) Spatiality – Given the central role of spatial configuration in both deposit genesis and exploration, our understanding of it must be strengthened and formalized. Several approaches are possible. These include, but are not limited to, exploitation of models employed in other disciplines, e.g., biological migration patterns.

Application of probabilisitc spatial models to data describing observed anisotropy in hydrocarbon accumulation distance from source rocks and the clustering of exploratory effort that follows would possibly yield insights. The role of time in a geological scale hydrocarbon as it influences generation and accumulation needs to be better understood. So does the role of time as an influence on exploration success.

3. CONCLUSIONS

The workshop resulted in the following recommendations:

(1) The principal focus of research in oil and gas resource assessment should be at the play level

(2) Play taxonomys need refinement:
   (a) Distinguish between common and specific attributes;
   (b) Devote particular attention to deposit and play attributes that determine discoverability.

(3) Research on formulation of basin evolution and deposit genesis concepts in a manner compatible with mathematical models of exploration and discovery is needed.

(4) More intensive use of spatial modeling and spatial statistics should be undertaken.

(5) Work should continue on development or properties of currently available probabilistic resource models along two lines:

(a) Understanding of properties of finite and finite plus super-population models,

and

(b) creation and use of tests of model structure and model predictive accuracy.

In general, linkages between geological concepts and resource models need to be strengthened.

We also recommend a future workshop:

(1) Designed to do research aimed at implementing the above recommendations;

(2) Composed of people from a wide variety of earth sciences and mathematical disciplines;

(3) Scheduled to meet several times for two week periods.

WORKSHOP ON SPATIAL STATISTICS AND IMAGE PROCESSING

B.D. Ripley (Leader), A.R. Solow (Rapporteur), C.F. Chung, J.C. Davis, H.K. Omre, P. Switzer, A. Grassia, D.E. Myers, J.J. Royer, H. Wackernagel, F. Wurzer, A. Yalniz

1. INTRODUCTION

Problems in the earth sciences are often spatial in nature, and many involve images either on a macroscopic scale (e.g., satellite imagery) or on a microscopic scale (e.g., electron micrographs). Many of the methods applied to such problems are either explicitly statistical (e.g., kriging) or can be interpreted in terms of spatial statistics. There is no doubt about the importance of the topic today or in the foreseeable future.

Successful applications of spatial statistics are usually driven by specific problems. These problems occur on a wide range of scales and with very different objectives in different industries. Thus, we were only able to touch on the most widely known and widely used ideas within the field.

2. SPATIAL SURFACES

Examples of problems involving spatial surfaces include mapping reservoir thickness and porosity or ore grade. Objectives can be as simple as estimating volume-in-place or total ore content, but many uses call for an estimated map of the variable of interest: for example, as input to a numerical simulation of an enhanced oil recovery scheme or to determine which blocks of a copper mine are economically recoverable.

Considerable discussion was devoted to spatial interpolation by kriging. Under certain conditions, linear kriging provides the minimum variance linear unbiased estimator of the value of a variable at a point in space (or a linear functional of pointwise variables such as block averages). Some concern was raised about the distinction between theoretical optimality, which requires that the variogram and the form of the drift (if any) be known, and practical situations, in which neither the variogram nor the form of the drift is known. It was pointed out that any notion of strict optimality may be of little practical interest. The question of the sensitivity or robustness of kriging to mis-specification in the variogram or the

C. F. Chung et al. (eds.), Quantitative Analysis of Mineral and Energy Resources, 701–706.

drift function was felt to be important, and further research is needed on this subject. Concern was also raised about the use of kriging to estimate an entire map, instead of pointwise values. Mapping criteria may differ from that associated with pointwise estimation. In particular, it may be important to reproduce spatial variability or to identify peaks in mapping.

Because of its importance in kriging, emphasis was placed on improving methods for variogram estimation. The variogram cloud (Chauvet, 1982) was felt to be a promising tool. The importance of statistical inference in variogram estimation, as described by Switzer (1984) and Mardia and Marshall (1984), was also brought up. The use of formal Bayesian methods to incorporate prior information in kriging was felt to have great potential use. Three other areas needing further study were the proper interpretation of the estimation variance, the problems posed by preferential sampling, and methods for dealing with the support effect (particularly in the presence of highly skewed distributions).

Simulation of spatial surfaces can be extremely useful for assessing variability and also for estimation under asymmetric loss functions. As simulation has come to be used more widely, a number of interesting issues have arisen. These include the problem of assessing the success of a given simulation algorithm and the problem of determining the equivalence or non-equivalence of different algorithms. Ripley reports some success at simulating large systems directly, by treating the problem as one of simulating a large multivariate normal vector. He cautions, however, that care should be taken in the numerical aspects of this and other methods. One advantage of the direct method is that it works equally well in one, two, or three dimensions.

3. SPATIAL AUTOCORRELATION AND PROXIMITY ANALYSIS

Searching for spatial autocorrelation in its own right has little purpose, but knowledge of correlations can be of great value in assessing variability and when regressing on explanatory variables. The key feature of spatial autocorrelation methods is that they use non-Euclidean distances to represent proximity between locations. This gives the possibility of geological input, so that (for example) locations separated by a fault are not regarded as adjacent. The exact definition of proximity may be somewhat tendentious, but any sensible solution is likely to be much more satisfactory than ignoring spatial autocorrelation altogether. Spatial autocorrelation methods have been used quite successfully in economic geography (Cliff and Ord, 1973) and medical geography (Cook and Pocock, 1983), and deserve more attention in the geosciences.

4. POINT PATTERNS

For mineral resource evaluation in a large region, the traditional first step is to delineate "favorable areas" by a group of geologists, based on their subjective geological judgement about the region.

The second step is to predict the number of undiscovered deposits in the delineated areas. It is common to assume that the number of deposits follows a Poisson distribution and to estimate the rate parameter based on data from well-exposed areas. In that case, the predicted number of undiscovered deposits is found by multiplying the rate by the size of the delineated areas. However, if the deposits of a given type are regarded as points in a well-explored area, the point patterns observed are not at all random. Instead, the patterns have distinct "footprints." If we could model this characterization of the point patterns, then we could certainly improve the prediction of the number and locations of undiscovered deposits. For example, massive sulfide deposits may occur in a ring pattern around volcanic centers. Other types of deposits may occur in a linear pattern along faults. A few attempts have been made by geologists to correlate the point pattern of deposits on a large scale with other geologic features, such as faults. Further development of these approaches is timely.

One interesting aspect of point patterns of deposits is that the pattern is incompletely observed, the probability of discovery being related to the intensity of search locally. Thus, we actually observe a selectively thinned pattern and are interested in the original pattern. We know of no treatment of this potentially important problem.

5. MOSAICS

Some of the most promising applications of spatial statistics are in mosaic problems. Mosaics are collections of "blobs" which may join, such as shale bodies in sandstone, as well as grains in sedimentary rock. A good stochastic model of pore geometry up to a scale of, say, one meter would be of great value in the understanding of the operation of well-logging instrumetns. Although this may be over-ambitious, there is some hope for better models than those used currently.

On a larger scale, there has been some interest in the effect of impermeable inclusions on fluid flow through oil reservoirs. In Prudhoe Bay, Alaska, there are thought to be thin, discontinuous shale layers laid down as estuarine mud flats. Conversely, in some North Sea fields thin reservoir sands occur in an impermeable shale volume. The problem in both cases is to model the distribution of two (or more) facies with dissimilar flow properties. One way to do so is to consider facies as consisting of discs whose centers are distributed according to some stochastic model (e.g., a spatial Poisson process). As always, care is needed not to use overly simple models.

Questions related to this model include the size, shape, and orientation of the dispersed facies, and the point pattern of the centers. Limited information is available at the boreholes which suggests that the vertical distribution is non-stationary. Once a stochastic model is established, it can be used to simulate realizations of the reservoir, and these realisations should reproduce the heterogeneity of reservoir properties. It is important that these simulations be conditioned on the observations at the available wells to ensure consistency with the reservoir under study. These simulations may then be used to study flow properties numerically.

Finally, it is worth mentioning that it may be possible to use percolation theory to make some additional analytical progress.

6. IMAGE PROCESSING AND ANALYSIS

Statistial methods for the treatment of images are necessarily spatial, and many of the same problems are encountered and similar solutions may be applied. The work of Switzer has been instrumental in incorporating spatial concepts into the analysis of satellite imagery. Image problems occur in resource assessment at both microscopic and global (remote sensing) scales. Whereas remote sensing has had success in the earth sciences, statistical ideas have played a relatively minor role so far. Almost all statistical techniques in remote sensing have been aimed at image enchancement to aid interpretation by a human interpreter.

6.1 Evaluation

The absolute value of gray-level images may not be as important as jumps in intensity and intensity gradients. Procedures to optimize the estimation of intensity tend to depress gradients and smooth discontinuities. New criteria are needed based on gradient estimation and boundary detection. Given the criteria, an image estimation procedure can be evaluated using model considerations and cross-validatory procedures. Local error estimates are desirable, possibly even at sub-pixel scale. The whole problem of evaluation of image-processing algorithms appears ripe for statistical treatment.

6.2 Algorithms

Bayesian algorithms appear to be a powerful way to incorporate prior information about image structure. Since images can contain very different scenes, local parameter estimation seems desirable.

6.3 Data on different grids

It is increasingly common to have data on different grids from different remote sensing scanners on the same scene. These data differ both in their sampling rate and in the areas averaged to give individual pixel values. Some thought is needed to devise methodologies for integrating these different types of information.

6.4 Image time series

Successive images of the same scene may be available over periods of months or years. Registration is important, and may be data-driven. Correcting for changes such as illumination angle is also important. At each pixel, there is then a time series. When good registration is possible, consideration of changes in image intensity may be revealing and may form the basis of a more formal statistical analysis. Seasonal changes can be particularly important in identifying vegetation type, which can in turn help interpret the underlying geology.

6.5 Uses of remote sensing

Passive remote sensing, as available from civilian satellites, produces superficial images and seems of limited use in oil and mineral exploration. Active sensors may be of greater value.

6.6 Microscopic images

It is easy (with the proper equipment) to take many electron micrographs of rock cross-sections. Statistical techniques are needed to extract average characteristics of such images, and to characterize their spatial variation. Ripley reports some success at extracting pore and grain shape and spacing using the techniques of mathematical morphology (Serra, this volume).

7. Geochemical sampling

Spatially sampled geochemical data have some unique features which require special consideration. One common form of sampling is to collect samples of stream water or sediments. These reflect the geochemistry not just at the sampling location but of the entire upstream basin. Soil samples also have a large support in this sense. Thus, geochemical samples reflect the composition within a certain neighborhood of the sampling location. The size and shape of this neighborhood depend on the sampling technique and on the underlying geologic and geochemical processes.

Geochemists have concentrated on the multivariate aspects of their data. It should be possible to incorporate spatial information by using non-Euclidean distances. This has already been attempted by Royer under the name of proximity analysis. We believe that further work in this direction will yield valuable benefits in the analysis of geochemical data.

8. REFERENCES

Chauvet, P., 1982: The variogram cloud, 17th APCOM Symposium Colorado School of Mines, Golden, Colorado.

Cliff, A.D. and J.K. Ord, 1973: Spatial Autocorrelation, Pion, London.

Cook, D.G. and S.J. Pocock, 1983: Multiple regression in geographical mortality studies with allowance for sptailly correlated errors, Biometrics, 39 p.

Mardia, K.V. adn R.J. Marshall, 1984: Maximum likelihood estimation of models for residual covariance in spatial regression, Biometrika, 71 p.

Switzer, P., 1984: Inference for spatial autocorrelation functions, in Geostatistics for Natural Resoruces Characterization, G. Verly, et al., editors, Reidel, Dordrecht, Holland.

WORKSHOP ON MINERAL AND ENERGY RESOURCE EXPERT SYSTEM DEVELOPMENT

G.S. Koch, Jr. (Leader), A.N. Campbell (Rapporteur), A.G. Fabbri, W. Lanyon, H.G. Pereira, M.J. Shulman, K.S. Sinding-Larsen, P.R. Stokke, A. Toister, W.L. Watney

1. SUMMARY AND RECOMMENDATIONS

Expert Systems are computer programs which store expertise and judgemental knowledge for the purpose of guiding users through evaluations or diagnoses. Expert System Software tools are becoming readily available, so a wide range of projects are now possible at reasonable cost. This report reviews the needs of the industry and recommends the following six research topics which are most likely to result in practical applications and useful technology.

1. Evaluating Petroleum Basins
2. Assessing Mineral, Energy and Environmental Resources
3. Processing Digital Images and Making Maps
4. Publishing Knowledge
5. Developing Sensor Advisors

The theme which ties these subject areas together is that Expert Systems have promise for storing expert knowledge, putting it into usable form, and distributing it. In contrast to this theme, projects for replacing experts with computer programs are of little practical value, at least within the mineral resource and energy fields.

Expert System research is in its infancy, yet it has been effectively driven forward by demands for specific services from resource industry and government clients. This report proposes topics that are likely to yield significant results in the near term, given proper funding. In order to get useful results from Expert System Research and Development projects, we recommend clearly defined projects in each of the reviewed subject areas. Research should focus on the development of specific products which have high value to the clients. For example, Hydrothermal Energy well operators frequently encounter difficulties that require the experience and judgement of an expert. Some times the expert is away. By the time the expert arrives, a full scale emergency may be in progress. However, this kind of emergency might be avoided by the right kind of expert system program. A hydrothermal well expert system program could continually read sensor outputs. Early indications of potential trouble would be

C. F. Chung et al. (eds.), Quantitative Analysis of Mineral and Energy Resources, 707–713.

promptly identified. The program could give advice and, as a bonus, new expertise could be captured and retained at the well, even as operators and experts move on.

2. INTRODUCTION

Expert systems are computer programs which store information for dealing with facts, inferences and judgements. They carry on consultations, produce maps, give advice, answer questions and explain reasoning. The purpose of this report is to review the utility of expert systems for the mineral and energy resource industries and propose research and development projects likey to yield valuable results in the near future.

3. EVALUATING PETROLEUM BASINS

Decision making in petroleum basin analysis provides a particularly fruitful area for exploiting expert system technology. The most receptive potential clients are petroleum exploration companies. Petroleum information is continually updated. Similarly, Expert Systems for decision makers will also be updated as new expertise is gained. The Systems may become a dynamic and robust repository for corporate knowledge. At the present time, some knowledge is lost when people move, and some knowledge is under utilized because of bureaucratic inefficiencies in decision-making. By reducing these losses, Expert System products have a potentially high value for the petroleum industry.

To be useful in the petroleum exploration industry, an expert system project must address certain criteria. The system must be easily updated because information will be added or modified at frequent intervals. The system must be amenable to orderly management. The system must eventually comprise a management information system of a scale and nature required by petroleum exploration firms. This would include a separate geological expert system package, including criteria for assessing the presence of constraints for accumulation of petroleum, namely, the characteristics of petroleum source rocks, the maturation level of organic matter, the timing and paths of petroleum migration, the size, geometries, and expected frequecies of reservoir rocks, and the trapping mechanisms. Systems should address at least the following topics:

1. Evaluating Tectonic Settings
2. Evaluating Basin Geology
3. Considering Politics
4. Assessing Environmental Constraints
5. Reviewing Paleogeoclimates and Environments.

We consider that there are potential Petroleum Industry clients for this kind of system, and that the technology currently exists for creating and delivering the product. The principal issues are to insure proper design and efficient management of the system. Once implemented, we speculate that clients other than petroleum exploration companies will emerge. These include: banks and financing firms, governmental agencies (for example a Department of Energy), research organizations and consulting firms.

4. ASSESSING MINERAL, ENERGY AND ENVIRONMENTAL RESOURCES

Expert System applications are particularly well suited to Resource and Environmental Evaluation. The data bases needed to make decisions are already in place and the public, government and exploration industry need clear, rational and unbiased explanations of decisions that a good expert system gives. Exploration and resource analyses attempt to estimate or appraise the probability of occurrence of one, more than one, or many commodities in a given area. Environmental evaluations, in contrast, are made to determine favorable sites for toxic waste disposal, landfills, commercial development and other non-consumptive uses. In both areas, the objectives of government agencies on one hand and industrial developers on the other might be quite different. However, in either case, the clear definition and justification of decisions is essential for effective management.

For resource evaluation and planning processes, the formulation of a rational, definable mineral policy is the focus of much activity. Many methods exist for quantitative analysis of resoruces. Economic modelling, environmental modelling, statistical estimation, and environmental impact statements are all examples. The integration of these studies into an easily explained and understood policy is difficult and cumbersome. As a result, wise policies are difficult to formulate and even more difficult to implement. Expert System technology may clarify the rationale of policies. With Expert Systems, these can be tested, improved and updated as problems are found. These techniques can humanize bureaucracies by removing inefficiencies and mistakes which come from the current difficulty in formulating, presenting and understanding resource management policy.

We recommend that government agencies focus their research funding on projects which aid policy formulation, and help explain and justify management decisions. This focus will be particularly cost-effective because the basic quantitative information needed to make good policies already exists in agency data bases and area studies.

5. PROCESSING DIGITAL IMAGES AND MAKING MAPS

5.1 Processing Images

The first task in using expert sytem technology for image processing is to identify the "client" which may be such diverse agencies as the Ministry of Forests in British Columbia, Agriculture Canada or consultants in remote sensing. In designing image processing software, it is advisable to assume that the clients will need expert advice in the choice and application of software routines. The second task is the identification of the type of images to be studied (satellite imagery, airborne data, digitized maps, microscopic images, etc.) that contain different information both in terms of detail and quality of data. The processing techiques may be similar but the meaning of the processing steps may be different.

Image processing software should be designed for users who have no knowledge of how to interact with images and with the processing algorithms. Image processing transforms images into new, more informative images, as well as lists and tables of properties (global and local) from which decisions can be taken on further processing steps. The main issue is guiding inexperienced users. Expert System methods promise the development of "friendly" user interfaces. These front ends would respond to "high-level" questions in non-image-processing terms. Parsed requests from the user would correctly formulate image processing leading to the desired results.

Data integration of ancillary maps on resources and digital images (satellite or airborne) is a complex and error prone task. Expert System programming methods might be very instrumental in the automatic correction of digitizing errors (from scanning or from the interaction with a graphic digitizer), geometric and radiometric corrections, and in image registration.

5.2 Making Maps

The understanding of the object relationships, implicit in geological cross section (see Toister in this Volume, and other work under Randall Davis at MIT) will require expert system programming methods for practical implementation. Mapping, for example, in engineering geology, is a subject that requires Expert System techniques (see Varnes, 1974, USGS, Prof Paper 837, 48 p.). Expert systems are being developed for the following applications in computer vision, remote sensing and geographic information systems.

1. Interpretation of aerial photographs of city and airport scenes;
2. Forest monitoring (including change detection, clear cuts and the spreading of diseases);
3. Hyperspectral classification in remotely sensed imagery:
4. Image segmentation;
5. Intelligent front-ends for data integration (different sensors, ancillary data from maps and images, digitization etc.);
6. Analysis of data from digital terrain models (detection of catchment basins, identification of valleys, ridges, saddles, and the structuring of relationships between geographical objects.)

In summary, it is now technically feasible to exploit PROSPECTOR'S "drilling site" mode of interaction for automatic map making. Government agencies can now build routines for evaluating maps and spatial information. These might be used by government agencies, petroleum exploration companies, mining companies, and consulting firms. Automatic map making is the natural application of artificial intelligence and spatial data analysis.

6. PUBLISHING KNOWLEDGE

The publishing of books, technical journals, magazines, and computer software provides an important foundation of modern civilization. We speculate that eventually, the publishing of expert knowledge and judgement will add considerably to the foundation, although the exact form and methods of this kind of publication is unknown.

6.1 Users of Published Knowledge

We propose that the place to begin planning the process of publishing expert knowledge is to identify the potential customers, and to try to determine what would best suit their needs. Clients for published knowledge might include:

1. Students and teachers
2. Organization Decision Makers
3. Scientists and Researchers
4. Anyone who requires Expert Opinions and Advice

6.2 Expected Features of Published Expert Systems

Expert system techology is so new that notions of publishing expertise are speculative. We expect that good expert systems will supplement the factual knowledge in textbooks and maps with experience-based inference and judgement. For example, students and teachers of ore deposit geology might benefit from the PROSPECTOR deposit models, because finding out how seasoned experts evaluate mineral properties is an important supplement to the descritpive information in economic geology texts. Similarly, most managers require understandable, defensible standards, for selecting garbage and waste disposal sites. Some workers in the earth sciences require advice on choosing appropriate statistical techiques. The list of potential applications is quite long. But in each knowledge publication, the descriptive and factual knowledge in most technical books might be supplemented in a very beneficial way by expert systems which guide readers through the judgemental aspects of the topics in the book.

6.3 Expected Advantages

Expert systems contain judgements, diagnoses, inferences, and supporting facts. Good expert systems show the relationships between all of these to pass on understanding at a level not found in print. We believe that expert systems can add new dimension to texts by conveying methods of judgement, and making it easy to view knowledge from different persepctives.

Two possible scenarios for distributing expert knowledge might be considered. (1) Publishing knowledge, in much the same way that technical books are currently published, and (2) Building centralized data bases from which expert knowledge may be accessed by computer.

We propose that an important advantage of publishing expert systems as opposed to maintaining large knowledge data bases will be to facilitate frequent and regular uses, which will lead to the improvement and updating of expert knowledge. The social process will be much the same as the process of publishing technical papers and books to record and update facts and theories. This contrasts with the frequent disuse, mismanagement and collapse of centralized information data bases.

6.4 Anticipated Problems

Expert systems are a new medium for storing and transmitting knowledge. We expect the delicate and erasable nature of electronic media will create distribution problems. Expert Systems require computer to deliver their knowledge. Therefore, the market is restricted to people who have access to the right kind of computers.

7 DEVELOPING SENSOR ADVISORS

The newer machines in the mineral and energy resource industries are controlled by electronic sensors. These measure temperature, pressure, voltage, frequency, weight, velocity, and other values. Computers interpret the sensor values to regulate machines and processes. The same computers can warn operators when sensor values go beyond acceptable limits.

In modern systems, numerous sensors quickly deliver large quantities of information. When things go awry, most operators have trouble interpreting the copious readings. For example, seismic cruise ships tow large arrays of geophones. The geophone readings are recorded for later processing by large computers. Often, evidence of tangled or damaged geophones shows up only at this late date. The cost of these late error detections is high, both in time and money. A geophone sensor advisor, on board the cruise ship might identify problems, as they occur, and suggest possible remedies.

7.1 Clients For Sensor Advisors

Clients for sensor advisor systems might include:

1. Geophysical companies
2. Gas, Oil and Hydrothermal well operators
3. Oil well monitoring personnel
4. Suppliers of technology which requires quick decisions and reactions

7.2 Expected Advantages

Sensor advisors might be expected to give the quality of advice that an experienced operator gives on his better days. This is advantageous because advice will be available when the experienced operator is at home sleeping or away on vacation. For example, at a hydrothermal production well, an expert system would be set up to continuously monitor all the electronic sensors. When the first signs of trouble emerge, the proposed sensor advisor might comment about the developments and offer advice. In contrast, at the present time, experienced hydrothermal well operators may get late night calls when emergencies are already highly developed. In fact, in some emergencies, the experts have not been available at all.

Sensor advisors, by their very nature will be in constant use, and therefore will serve as a growing repository for expertise through the operating life of the equipment they monitor. It is feasible that during the normal course of running the well, operators might formulate new pieces of knowledge as they are discovered. And these, after review and testing might be added to operating knowledge. This contrasts favorably with the present situation where good operators tend to be promoted and move on.

WORKSHOP ON CURRENT PROBLEMS AND FURTHER DEVELOPMENTS IN THE USE OF MICROCOMPUTERS.

D.F. Merriam (Leader), P.G. Sutterlin (Rapporteur), U. Artan, G. Bortoluzzi, M.T. Holroyd, F. Muge, D. Papacharalampos, S. Rehder, T. Testa

1. HARDWARE

A "fourth generation" microcomputer, in the context of this Workshop, is regarded as a physically small machine, with the Central Processing Unit (CPU) and other specialized logic circuits contained on single Very Large Scale Integration (VLSI) chips. For most applications, the "system", which includes the microprocessor, will be assumed to be configured with a minimum of 640 kb of Random Access Memory (RAM), a fixed or "hard" disk of 10-20 mb storage capacity, a flexible (floppy) removable diskette of (usually) at least 360 kb of storage capacity, and a printer-plotter (dot-matrix). Ideally, the microprocessor should have the ability to handle both 8-bit and 16-bit data and instructions. This system configuration now costs less than $5000 (US), and costs are being reduced rapidly. Peripherals useful to a geologist include a plotter, a digitizing tablet, a color monitor, and communication devices, the addition of which may bring the cost of the system to about $10,000 (US). Machines with full 32-bit capabilities, although more costly, are becoming available.

The "fifth" generation microcomputers will likely incorporate the ability to do vector processing, parallel or "pipeline" processing, and array processing. Geological applications which will take special advantage of these additional capabilities include:

- finite-element solutions
- simulation
- graphics (image analysis)
- seismic processing
- partitioning of programs and data sets
- organizational problems (expert systems)

Fifth generation microcomputers will be smaller, faster, and less expensive.

"Sixth" generation" microcomputers will consist of networked CPU's and associative memories within the same "box". Optical disk storage (up to one gigabyte), videodisks, and specialized peripherals will enable the geologist to do large-scale simulations and

C. F. Chung et al. (eds.), Quantitative Analysis of Mineral and Energy Resources, 715–720.

three-dimensional problems. This is important because of the "time" dimension which is an integral aspect of all geological investigations.

The developmental history of mainframe computers has resulted in the evolution of both "generalized" machines and more "specialized" computers. This history is likely to be paralleled in the development of microcomputers, with the result being generalized microcomputers with "built-in" specialized functions either as hardware or firmware.

There is a need for:

- faster raster-plotting devices, both elecrostatic and laser driven
- more sophisticated optical readers and digitizers
- standardized peripherals interfaces
- holography capabilities
- graphics standards

Present problems include:

- hardware compatibility
- lack of standards
- lack of adherence to de facto standards

2. GEOLOGICAL UTILIZATION OF MICROCOMPUTERS

The geological uses of microcomputers, in order of perceived frequency of application, are:

- word processing
- data analysis
- data management, including
  - data acquisition
  - process control
  - data communication
- program development
- data display
- data entry and verification.

In general, geological users no longer write the code for sophisticated programs, but act as an interface. The costs of hardware is continuing to decrease, but it must be realized that the cost of software development will continue to rise. Users will have to face the fact that software is going to be expensive in the future.

3. SOFTWARE

3.1 Word-Processing Software

Among all the various word-processing packages available, and despite some serious shortcomings, WORDSTAR is seen as the most popular, and probably is used by more geoscientists than any other package. Nevertheless there is a need for improvement of word-processing software in the following areas:

- graphics capabilities within the software
- the ability to specify different fonts
- the ability to generate mathematical symbols

- the inclusion of dynamic spelling checks and thesauri
- simplification of the command language
- the ability to be included as subroutines
- the inclusion of a task protocol
- the generation of an audit trace.

3.2 Data Analysis Software

Data analysis software for microcomputers, examples of which are LUPAC (Geomath, Denver, CO.) and STATPAC (U.S.G.S.) must be "portable" - that is implementable on a wide range of machines - and it must operate efficiently. Most microcomputer programs satisfy these conditions, although some improvement is desirable. The user interface must be simplified and made more "friendly"; error detection routines must be improved. In many situations, the output of data analysis programs is virtually uninterpretable except by the programmer. Output must be annotated in clearer and more easily understood form. There should be periodic messages to the user that, despite indications to the contrary, the program is executing or, for that matter, has ceased execution.

A partial list of new techniques in geological data analysis for which microcomputer programs undoubtedly will be developed are:

- artificial intelligence
- expert systems
- fractals
- fuzzy analysis
- advanced image analysis
- geostatistiques
- geotomography
- analog morphocorrelation
- map comparison
- quadtree analysis
- coencorrelation curves
- point/line analysis
- characteristic analysis
- logistic/poisson regression
- factorial kriging analysis
- proximity analysis
- integral geometry
- 3-D interpolation
- simulation
- robust and rank statistics.

3.3 Operating Systems Software

The most used operating systems for microcomputers are CP/M and MS-DOS. The most severe limitations are the limits in the amount of RAM which can be addressed. This will soon be rectified.

There are a number of multiuser multitasking operating systems for use with microcomputers. Examples are UNIX and its variants, and EUMEL, a German-developed system which uses a concept analogous to some mainframe Virtual Memory (VM) systems, and which can address up to a gigabyte of RAM. Although these systems are not in widespread use, in future they will become more popular.

3.4 Firmware

It is apparent that, for ease of use, more firmware will be incorporated into microcomputers. This will include graphic packages, compilers, interpreters, input-output drivers, and libraries of various types.

3.5 Graphics Software

Although screen and plotter graphics are becoming an important aspect of microcomputer applications, graphic packages tend to be nonstandard. It is recommneded that graphic programs for geological applications make use of the GKS (Graphics Kernel System) graphic standard.

3.6 Network Software

There are two levels of networking software available. ETHERNET is an example of a Local Area Network (LAN) for use within a single organization. BITNET and EARN are networks which permit communication among many organizations. There are 350 organizations in North America and Europe which are connected on this network, and geological users are to be encouraged to take advantage of this efficient service for electronic mail and file transfer applications.

3.7 Programming Language Software

On the basis of the published literature, 80% of all geologically oriented software has been written in FORTRAN; the microcomputer standard is Microsoft FORTRAN, which actually complies with ANSI FORTRAN 1977. The diagnostics in general are good, but the major drawback is that FORTRAN does not permit recursive functions and data-type definitions. In comparison to some other languages, it is more difficult to design structured and modular programs in FORTRAN.

Pascal is increasing in popularity because of its ability to produce clear, well-structured, and modular code. Recursion and memory pointers make Pascal capable of handling complex programming problems. Diagnostics for MICROSOFT and Borland's Turbo Pascal are good.

It is irrelevant which programming language is used as long as it is a well-designed language. The use of BASIC should be limited. LISP and PROLOG are recommended for the development of expert systems. MODULA 2 and ELAN are used both for program development and for documentation. APL and SIMSCRIPT are used by specialists. It is important for the geologist to define precisely what is to be done, and then decide how this should be done via one of the available languages.

3.8 Communication Software

There are many good communication software packages available which enable a microcomputer to emulate a terminal and require no additional hardware other than a MODEM. One which is suitable is KERMIT, which is available free-of-charge from Columbia University. KERMIT is able to transfer both binary and text files on all mainframes, minis and micros on the market.

3.9 Database Management Software

Bibliographic database software generally is developed and maintained by the database vendor (e.g. Lockheed Systems "DIALOG"). The databases are stored centrally and accessed through a network such as INFONET. In Europe, the European Space Agency (ESA) has developed and provided a common query language (ESA-QUEST) for all databases available through ESA-ESRIN. This permits the user to gain access to the most important databases such as NASA, NTIS, ENVIROLINE, PASCAL, etc., using a single query command language. In contrast, retrieval languages of the various software packages elsewhere are not standard and therefore a user must learn several query languages to access all available databases. The services of someone specially trained in bibliographic database retrieval techniques is the optimal way to make use of these systems.

Many microcomputer-based systems to manage nonbibliographic data are available. The most advanced are "relational" database-management systems such as dBASE III. There are however a number of shortcomings. The internal system languages of these DBMS's are different from other programming languages. There usually is insufficient and ambiguous documentation. Accessibility to the system on the part of the uninitiated user may be complex. Query languages among the various packages is not standard. Future development should be concentrated on more user-oriented front ends for these systems. File size is no longer a significant consideration, and partitioning databases using devices such as a "Bernoulii Box" is practical usig a microcomputer.

The question of capture and storage of "historical" data, some of which is qualitative, has not been resolved. Manual digitization of these data is expensive and the costs are seldom justifiable. Few organizations can afford optical scanners which could handle the wide range of applications needed.

4. PROGRAM SOURCES

A wide range of geologically oriented computer programs is available through the Kansas Geological Survey, the Computer Oriented Geological Society (COGS), and the U.S. Geological Survey to name a few, as well as in publications such as Computers & Geosciences and Geobyte. However, access to the availability of such programs would be enhanced if an international catalog was available in which programs were described in standard form. Commercially available programs could be included for a fee to assist in defraying costs. Copyright questions also should be investigated. Exchange under the auspices of an international organization might circumvent some of the patent or copyright concerns.

5. EDUCATION

Computer-Assisted Instruction (CAI) is an excellent teaching tool and its use should be encouraged especially with the wide availability of micros. CAI can be used at all educational levels and for simple exercises or interation with complex realistic learning models. A good background in mathematics, statistics, and computer science is necessary.

INDEX